T0180456

Birkhäuser

Contemporary Mathematicians

Gian-Carlo Rota†
Joseph P.S. Kung

Editors

For further volumes:
http://www.springer.com/series/4817

Peter Duren • Lawrence Zalcman
Editors

Menahem Max Schiffer: Selected Papers Volume 2

Birkhäuser

Editors
Peter Duren
Department of Mathematics
University of Michigan
Ann Arbor, Michigan, USA

Lawrence Zalcman
Department of Mathematics
Bar-Ilan University
Ramat-Gan, Israel

ISBN 978-1-4939-3955-8 ISBN 978-1-4614-7949-9 (eBook)
DOI 10.1007/978-1-4614-7949-9
Springer New York Heidelberg Dordrecht London

Mathematics Subject Classification (2010): 30-XX, 31-XX, 35-XX, 49-XX, 76-XX, 20-XX, 01-XX

Printed on acid-free paper

Springer is part of Springer Science+Business Media (www.birkhauser-science.com)

Max Schiffer in his Stanford office, ca.1976

Contents

Publications of M. M. Schiffer

1. Ein neuer Beweis des Endlichkeitssatzes für Orthogonalinvarianten. *Math. Z.* 38 (1934), 315–322.
2. Sur un principe nouveau pour l'évaluation des fonctions holomorphes. *Bull. Soc. Math. France* 64 (1936), 231–240.
3. Un calcul de variation pour une famille de fonctions univalentes. *C. R. Acad. Sci. Paris* 205 (1937), 709–711.
4. Sur un problème d'extrémum de la représentation conforme. *Bull. Soc. Math. France* 66 (1938), 48–55.
5. A method of variation within the family of simple functions. *Proc. London Math. Soc.* (2) 44 (1938), 432–449.
6. On the coefficients of simple functions. *Proc. London Math. Soc.* (2) 44 (1938), 450–452.
7. Sur les domaines minima dans la théorie des transformations pseudoconformes. *C. R. Acad. Sci. Paris* 207 (1938), 112–115.
8. Sur un théorème de la représentation conforme. *C. R. Acad. Sci. Paris* 207 (1938), 520–522.
9. (with S. Bergmann) Familles bornées de fonctions de deux variables complexes dans des domaines avec une surface remarquable. *C. R. Acad. Sci. Paris* 207 (1938), 711–713.
10. Sur la variation de la fonction de Green de domaines plans quelconques. *C. R. Acad. Sci. Paris* 209 (1939), 980–982.
11. Sur la variation du diamètre transfini. *Bull. Soc. Math. France* 68 (1940), 158–176.
12. On the subadditivity of the transfinite diameter. *Proc. Cambridge Philos. Soc.* 37 (1941), 373–383.
13. Variation of the Green function and theory of the p-valued functions. *Amer. J. Math.* 65 (1943), 341–360.
14. The span of multiply connected domains. *Duke Math. J.* 10 (1943), 209–216.
15. (with S. Bergman) Bounded functions of two complex variables. *Amer. J. Math.* 66 (1944), 161–169.
16. Sur l'équation différentielle de M. Löwner. *C. R. Acad. Sci. Paris* 221 (1945), 369–371.
17. Hadamard's formula and variation of domain-functions. *Amer. J. Math.* 68 (1946), 417–448.
18. On the modulus of doubly-connected domains. *Quart. J. Math. Oxford Ser.* 17 (1946), 197–213.
19. The kernel function of an orthonormal system. *Duke Math. J.* 13 (1946), 529–540.
20. (with S. Bergman) A representation of Green's and Neumann's functions in the theory of partial differential equations of second order. *Duke Math. J.* 14 (1947), 609–638.
21. (with S. Bergman) On Green's and Neumann's functions in the theory of partial differential equations. *Bull. Amer. Math. Soc.* 53 (1947), 1141–1151.
22. An application of orthonormal functions in the theory of conformal mapping. *Amer. J. Math.* 70 (1948), 147–156.
23. (with S. Bergman) Kernel functions in the theory of partial differential equations of elliptic type. *Duke Math. J.* 15 (1948), 535–566.

24. (with R. von Mises) On Bergman's integration method in two-dimensional compressible fluid flow, in *Advances in Applied Mechanics* (R. von Mises and T. von Kármán, editors), Academic Press, New York, 1948, pp. 249–285.
25. Faber polynomials in the theory of univalent functions. *Bull. Amer. Math. Soc.* 54 (1948), 503–517.
26. (with P. R. Garabedian) Identities in the theory of conformal mapping. *Trans. Amer. Math. Soc.* 65 (1949), 187–238.
27. (with G. Szegő) Virtual mass and polarization. *Trans. Amer. Math. Soc.* 67 (1949), 130–205.
28. (with A. C. Schaeffer and D. C. Spencer) The coefficient regions of schlicht functions. *Duke Math. J.* 16 (1949), 493–527.
29. (with P. R. Garabedian) On existence theorems of potential theory and conformal mapping. *Ann. of Math.* 52 (1950), 164–187.
30. (with D. C. Spencer) The coefficient problem for multiply-connected domains. *Ann. of Math.* (2) 52 (1950), 362–402.
31. Various types of orthogonalization. *Duke Math. J.* 17 (1950), 329–366.
32. (with S. Bergman) Various kernels in the theory of partial differential equations. *Proc. Nat. Acad. Sci. U.S.A.* 36 (1950), 559–563.
33. (with S. Bergman) Some linear operators in the theory of partial differential equations. *Proc. Nat. Acad. Sci. U.S.A.* 36 (1950), 742–746.
34. Some recent developments in the theory of conformal mapping, in R. Courant, *Dirichlet's Principle, Conformal Mapping, and Minimal Surfaces*, Interscience, New York, 1950, Appendix, pp. 249–323.
35. (with S. Bergman) Kernel functions and conformal mapping. *Compositio Math.* 8 (1951), 205–249.
36. (with S. Bergman) Kernel functions and partial differential equations. I. Boundary value problems in the theory of non-linear partial differential equations of elliptic type. *J. Analyse Math.* 1 (1951), 375–386.
37. (with S. Bergman) A majorant method for non-linear partial differential equations. *Proc. Nat. Acad. Sci. U.S.A.* 37 (1951), 744–749.
38. (with D. C. Spencer) A variational calculus for Riemann surfaces. *Ann. Acad. Sci. Fenn. Ser. A I* 93 (1951), 9 pp.
39. (with D. C. Spencer) On the conformal mapping of one Riemann surface into another. *Ann. Acad. Sci. Fenn. Ser. A I* 94 (1951), 10 pp.
40. Variational methods in the theory of conformal mapping, in *Proceedings of the International Congress of Mathematicians, Cambridge, Mass., 1950*, American Mathematical Society, Providence, R.I., 1952, Vol. 2, pp. 233–240.
41. (with K. S. Miller) On the Green's function of ordinary differential systems. *Proc. Amer. Math. Soc.* 3 (1952), 433–441.
42. (with K. S. Miller) Monotonic properties of the Green's function. *Proc. Amer. Math. Soc.* 3 (1952), 948–956.
43. (with F. S. Bodenheimer) Mathematical studies in animal populations. I. A mathematical study of insect parasitism. *Acta Biotheoretica Ser. A* 10 (1952), 23–56.
44. (with P. R. Garabedian and H. Lewy) Axially symmetric cavitational flow. *Ann. of Math.* (2) 56 (1952), 560–602.
45. (with S. Bergman) Potential-theoretic methods in the theory of functions of two complex variables. *Compositio Math.* 10 (1952), 213–240.
46. (with D. C. Spencer) Some remarks on variational methods applicable to multiply connected domains, in *Construction and Applications of Conformal Maps: Proceedings of a Symposium* (E. F. Beckenbach, editor), National Bureau of Standards, Appl. Math. Series 18, U.S. Government Printing Office, Washington, D.C., 1952, pp. 193–198.
47. (with S. Bergman) Theory of kernel functions in conformal mapping, in *Construction and Applications of Conformal Maps: Proceedings of a Symposium* (E. F. Beckenbach, editor), National Bureau of Standards, Appl. Math. Series 18, U.S. Government Printing Office, Washington, D.C., 1952, pp. 199–206.
48. (with P. R. Garabedian) Variational problems in the theory of elliptic partial differential equations. *J. Rational Mech. Anal.* 2 (1953), 137–171.
49. (with P. R. Garabedian) Convexity of domain functionals. *J. Analyse Math.* 2 (1953), 281–368.

50. Variational methods in the theory of Riemann surfaces, in *Contributions to the Theory of Riemann Surfaces* (L. Ahlfors *et al.*, editors), Annals of Math. Studies 30, Princeton University Press, Princeton, N.J., 1953, pp. 15–30.
51. (with S. Bergman) *Kernel Functions and Elliptic Differential Equations in Mathematical Physics*, Academic Press, New York, 1953.
52. The Applied Mathematics Laboratory at Stanford University, in *Proceedings of a Conference on Training in Applied Mathematics* (1953), pp. 37–40.
53. (with G. Pólya; appendix by H. Helfenstein) Convexity of functionals by transplantation. *J. Analyse Math.* 3 (1954), 245–346.
54. Variation of domain functionals. *Bull. Amer. Math. Soc.* 60 (1954), 303–328.
55. (with D. C. Spencer) *Functionals of Finite Riemann Surfaces*, Princeton University Press, Princeton, N.J., 1954.
56. (with P. R. Garabedian) On estimation of electrostatic capacity. *Proc. Amer. Math. Soc.* 5 (1954), 206–211.
57. (with S. Bergman) Properties of solutions of a system of partial differential equations, in *Studies in Mathematics and Mechanics Presented to Richard von Mises*, Academic Press, New York, 1954, pp. 79–87.
58. (with P. R. Garabedian) On a double integral variational problem. *Canad. J. Math.* 6 (1954), 441–446.
59. (with P. R. Garabedian) A coefficient inequality for schlicht functions. *Ann. of Math.* (2) 61 (1955), 116–136.
60. (with P. R. Garabedian) A proof of the Bieberbach conjecture for the fourth coefficient. *J. Rational Mech. Anal.* 4 (1955), 427–465.
61. Partial differential equations of elliptic type, in *Modern Mathematics for the Engineer* (E. F. Beckenbach, editor), McGraw-Hill, New York, 1956, pp. 110–144.
62. The Fredholm eigen values of plane domains. *Pacific J. Math.* 7 (1957), 1187–1225.
63. Sur les rapports entre les solutions des problèmes intérieurs et celles des problèmes extérieurs. *C. R. Acad. Sci. Paris* 244 (1957), 2680–2683.
64. Sur la polarisation et la masse virtuelle. *C. R. Acad. Sci. Paris* 244 (1957), 3118–3121.
65. Problèmes aux limites et fonctions propres de l'équation intégrale de Poincaré et de Fredholm. *C. R. Acad. Sci. Paris* 245 (1957), 18–21.
66. Partial differential equations of the elliptic type, in *Lecture Series of the Symposium on Partial Differential Equations, University of California, Berkeley, 1955* (N. Aronszajn and C. B. Morrey, Jr., editors), University of Kansas Press, Lawrence, Kansas, 1957, pp. 97–149.
67. Applications of variational methods in the theory of conformal mapping, in *Calculus of Variations and its Applications*, Proc. Symposia Appl. Math. 8 (L. M. Graves, editor), McGraw-Hill, New York, 1958, pp. 93–113.
68. Fredholm eigen values of multiply-connected domains. *Pacific J. Math.* 9 (1959), 211–269.
69. (with G. Pólya) Sur la représentation conforme de l'extérieur d'une courbe fermée convexe. *C. R. Acad. Sci. Paris* 248 (1959), 2837–2839.
70. Extremum problems and variational methods in conformal mapping, in *Proceedings of the International Congress of Mathematicians, Edinburgh, 1958*, Cambridge University Press, New York, 1960, pp. 211–231.
71. Analytical theory of subsonic and supersonic flows, in *Handbuch der Physik*, Springer-Verlag, Berlin, 1960, Vol. 9, Part 3, pp. 1–161.
72. (with Z. Charzyński) A new proof of the Bieberbach conjecture for the fourth coefficient. *Arch. Rational Mech. Anal.* 5 (1960), 187–193.
73. (with Z. Charzyński) A geometric proof of the Bieberbach conjecture for the fourth coefficient. *Scripta Math.* 25 (1960), 173–181.
74. (with N. S. Hawley) Connections and conformal mapping. *Acta Math.* 107 (1962), 175–274.
75. (with P. L. Duren) A variational method for functions schlicht in an annulus. *Arch. Rational Mech. Anal.* 9 (1962), 260–272.
76. (with J. Siciak) Transfinite diameter and analytic continuation of functions of two complex variables, in *Studies in Mathematical Analysis and Related Topics; Essays in Honor of George Pólya* (G. Szegő, editor), Stanford University Press, Stanford, California, 1962, pp. 341–358.
77. (with P. L. Duren) The theory of the second variation in extremum problems for univalent functions. *J. Analyse Math.* 10 (1962/63), 193–252.

78. Fredholm eigenvalues and conformal mapping. *Rend. Mat.* (5) 22 (1963), 447–468.
79. (with E. Reich) Estimates for the transfinite diameter of a continuum. *Math. Z.* 85 (1964), 91–106.
80. (with R. Adler and M. Bazin) *Introduction to General Relativity*, McGraw-Hill, New York, 1965.
81. (with B. Epstein) On the mean-value property of harmonic functions. *J. Analyse Math.* 14 (1965), 109–111.
82. (with G. Springer) Fredholm eigenvalues and conformal mapping of multiply connected domains. *J. Analyse Math.* 14 (1965), 337–378.
83. (with O. Tammi) On the fourth coefficient of bounded univalent functions. *Trans. Amer. Math. Soc.* 119 (1965), 67–78.
84. (with O. Tammi) The fourth coefficient of a bounded real univalent function. *Ann. Acad. Sci. Fenn. Ser. A I* 354 (1965), 32 pp.
85. (with L. Sario and M. Glasner) The span and principal functions in Riemannian spaces. *J. Analyse Math.* 15 (1965), 115–134.
86. (with P. R. Garabedian and G. G. Ross) On the Bieberbach conjecture for even *n*. *J. Math. Mech.* 14 (1965), 975–989.
87. (with N. S. Hawley) Half-order differentials on Riemann surfaces. *Acta Math.* 115 (1966), 199–236.
88. Half-order differentials on Riemann surfaces. *SIAM J. Appl. Math.* 14 (1966), 922–934.
89. (with O. Tammi) A method of variations for functions with bounded boundary rotation. *J. Analyse Math.* 17 (1966), 109–144.
90. A variational method for univalent quasiconformal mappings. *Duke Math. J.* 33 (1966), 395–411.
91. (with P. R. Garabedian) The local maximum theorem for the coefficients of univalent functions. *Arch. Rational Mech. Anal.* 26 (1967), 1–32.
92. (with N. S. Hawley) Riemann surfaces which are doubles of plane domains. *Pacific J. Math.* 20 (1967), 217–222.
93. (with O. Tammi) On the fourth coefficient of univalent functions with bounded boundary rotation. *Ann. Acad. Sci. Fenn. Ser. A I* 396 (1967), 26 pp.
94. Univalent functions whose *n* first coefficients are real. *J. Analyse Math.* 18 (1967), 329–349.
95. On the coefficient problem for univalent functions. *Trans. Amer. Math. Soc.* 134 (1968), 95–101.
96. (with O. Tammi) On bounded univalent functions which are close to identity. *Ann. Acad. Sci. Fenn. Ser. A I* 435 (1968), 26 pp.
97. (with O. Tammi) On the coefficient problem for bounded univalent functions. *Trans. Amer. Math. Soc.* 140 (1969), 461–474.
98. (with J. A. Hummel) Coefficient inequalities for Bieberbach-Eilenberg functions. *Arch. Rational Mech. Anal.* 32 (1969), 87–99.
99. Some distortion theorems in the theory of conformal mapping. *Atti Accad. Naz. Lincei Mem. Cl. Sci. Fis. Mat. Natur. Sez. Ia* (8) 10 (1970), 1–19.
100. (with R. N. Pederson) Further generalizations of the Grunsky inequalities. *J. Analyse Math.* 23 (1970), 353–380.
101. (with H. G. Schmidt) A new set of coefficient inequalities for univalent functions. *Arch. Rational Mech. Anal.* 42 (1971), 346–368.
102. (with O. Tammi) A Green's inequality for the power matrix. *Ann. Acad. Sci. Fenn. Ser. A I* 501 (1971), 15 pp.
103. (with G. Schober) An extremal problem for the Fredholm eigenvalues. *Arch. Rational Mech. Anal.* 44 (1971/72), 83–92.
104. (with R. N. Pederson) A proof of the Bieberbach conjecture for the fifth coefficient. *Arch. Rational Mech. Anal.* 45 (1972), 161–193.
105. Inequalities in the theory of univalent functions, in *Inequalities–III* (O. Shisha, editor), Academic Press, New York, 1972, pp. 311–319.
106. (with G. Schober) A remark on the paper "An extremal problem for the Fredholm eigenvalues." *Arch. Rational Mech. Anal.* 46 (1972), 394.
107. (with R. J. Adler, J. Mark, and C. Sheffield) Kerr geometry as complexified Schwarzschild geometry. *J. Mathematical Phys.* 14 (1973), 52–56.
108. Article in *How to Write Mathematics*, American Mathematical Society, Providence, R.I., 1973, pp. 49–61.

109. (with J. Hersch and L. E. Payne) Some inequalities for Stekloff eigenvalues. *Arch. Rational Mech. Anal.* 57 (1975), 99–114.
110. (with R. Osserman) Doubly-connected minimal surfaces. *Arch. Rational Mech. Anal.* 58 (1975), 285–307.
111. (with G. Schober) Coefficient problems and generalized Grunsky inequalities for schlicht functions with quasiconformal extensions. *Arch. Rational Mech. Anal.* 60 (1975/76), 205–228.
112. (with G. Schober) A distortion theorem for quasiconformal mappings, in *Advances in Complex Function Theory, Maryland 1973/74* (W. E. Kirwan and L. Zalcman, editors), Lecture Notes in Math. 505, Springer-Verlag, Berlin, 1976, pp. 138–147.
113. (with S. Friedland) Global results in control theory with applications to univalent functions. *Bull. Amer. Math. Soc.* 82 (1976), 913–915.
114. (with G. Schober) Representation of fundamental solutions for generalized Cauchy-Riemann equations by quasiconformal mappings. *Ann. Acad. Sci. Fenn. Ser. A I Math.* 2 (1976), 501–531.
115. (with S. Friedland) On coefficient regions of univalent functions. *J. Analyse Math.* 31 (1977), 125–168.
116. (with J. A. Hummel) Variational methods for Bieberbach-Eilenberg functions and for pairs. *Ann. Acad. Sci. Fenn. Ser. A I Math.* 3 (1977), 3–42.
117. (with G. Schober) A variational method for general families of quasiconformal mappings. *J. Analyse Math.* 34 (1978), 240–264.
118. (with H. Samelson) Dedicated to the memory of Stefan Bergman. *Applicable Anal.* 8 (1978/79), 195–199 (1 plate).
119. (with E. Netanyahu) On the monotonicity of some functionals in the family of univalent functions. *Israel J. Math.* 32 (1979), 14–26.
120. (with G. Schober) An application of the calculus of variations for general families of quasiconformal mappings, in *Complex Analysis, Joensuu 1978* (I. Laine, O. Lehto, and T. Sorvali, editors), Lecture Notes in Math. 747, Springer-Verlag, Berlin, 1979, pp. 349–357.
121. (with J. A. Hummel and B. Pinchuk) Bounded univalent functions which cover a fixed disc. *J. Analyse Math.* 36 (1979), 118–138.
122. (with G. Schober) The dielectric Green's function and quasiconformal mapping. *J. Analyse Math.* 36 (1979), 233–243.
123. (with A. Chang and G. Schober) On the second variation for univalent functions. *J. Analyse Math.* 40 (1981), 203–238.
124. Stefan Bergman (1895–1977): in memoriam. *Ann. Polon. Math.* 39 (1981), 5–9 (1 plate).
125. Fredholm eigenvalues and Grunsky matrices. *Ann. Polon. Math.* 39 (1981), 149–164.
126. (with D. Aharonov and L. Zalcman) Potato kugel. *Israel J. Math.* 40 (1981), 331–339.
127. (with P. L. Duren and Y. J. Leung) Support points with maximum radial angle. *Complex Variables Theory Appl.* 1 (1982/83), 263–277.
128. (with L. Bowden) *The Role of Mathematics in Science*, Mathematical Association of America (New Mathematical Library, Vol. 30), Washington, D. C., 1984.
129. George Pólya (1887–1985). *Math. Mag.* 60 (1987), 268–270.
130. Pólya's contributions in mathematical physics, *Bull. London Math. Soc.* 19 (1987), 591–594. [Part of obituary of George Pólya, *ibid.*, pp. 559–608.]
131. (with P. L. Duren) Grunsky inequalities for univalent functions with prescribed Hayman index. *Pacific J. Math.* 131 (1988), 105–117.
132. (with P. L. Duren) Conformal mappings onto nonoverlapping regions, in *Complex Analysis* (J. Hersch and A. Huber, editors), Birkhäuser Verlag, Basel, 1988, pp. 27–39.
133. (with P. L. Duren) Sharpened forms of the Grunsky inequalities. *J. Analyse Math.* 55 (1990), 96–116.
134. (with P. L. Duren) Goluzin inequalities and minimum energy for mappings onto nonoverlapping regions. *Ann. Acad. Sci. Fenn. Ser. A I Math.* 15 (1990), 133–150.
135. (with P. L. Duren) Univalent functions which map onto regions of given transfinite diameter. *Trans. Amer. Math. Soc.* 323 (1991), 413–428.
136. (with P. L. Duren) Robin functions and energy functionals of multiply connected domains. *Pacific J. Math.* 148 (1991), 251–273.

137. (with P. L. Duren) Robin functions and distortion of capacity under conformal mapping. *Complex Variables Theory Appl.* 21 (1993), 189–196.
138. Issai Schur: Some personal reminiscences, in *Mathematik in Berlin*, Vol. II, (H. Begehr, editor), Shaker Verlag, Aachen, 1998, pp. 177–181.

Editors' Note. Two items, which appear in lists of Schiffer's publications prepared in the 1940s but are absent from later lists, have been omitted:

Lösung der Aufgabe 128, *Jber. Deutsch. Math.-Verein.* **42** (1933), Abt. 2, 118–119.

(with S. Sambursky) Static universe and nebular red shift. II, *Phys. Rev.* (2) **53** (1938), 256–263.

[62] The Fredholm eigen values of plane domains

[62] The Fredholm eigen values of plane domains. *Pacific J. Math.* **7** (1957), 1187–1225.

THE FREDHOLM EIGEN VALUES OF PLANE DOMAINS

M. SCHIFFER

Introduction. The method of linear integral equations is an important tool in the theory of conformal mapping of plane simply-connected domains and in the boundary value problems of two-dimensional potential theory, in general. It yields a simple existence proof for solutions of such boundary value problems and leads to an effective construction of the required solution in terms of geometrically convergent Neumann-Liouville series. The convergence quality of these series is of considerable practical importance and has been discussed by various authors [4, 5, 6, 7]. It depends on the numerical value of the lowest nontrivial eigen value of the corresponding homogeneous integral equation which is an important functional of the boundary curve of the domain in question. Ahlfors [1, 10] gave an interesting estimate for this eigen value in terms of the extreme quasi-conformal mapping of the interior of this curve onto its exterior. Warschawski [15] gave a very useful estimate for it in terms of the corresponding eigen value of a nearby curve which allows often a good estimate of the desired eigen value in terms of a well-known one. This method is particularly valuable for special domains, for example, nearly-circular or convex ones.

It is the aim of the present paper to study the eigen functions and eigen values of the homogeneous Fredholm equation which is connected with the boundary value problem of two-dimensional potential theory. In particular, we want to obtain a sharp estimate for the lowest nontrivial eigen value in terms of function theoretic quantities connected with the curve considered. The steps of our investigation might become easier to understand by the following brief outline of our paper.

In § 1 we define the eigen values and eigen functions considered and transform the basic integral equation into a form which exhibits more clearly the interrelation with analytic function theory and extend the eigen functions as harmonic functions into the interior and the exterior of the curve. The boundary relations between these harmonic extensions are discussed and utilized to provide an example of a set of eigen functions and eigen values for the case of ellipses.

In § 2 we show the significance of the eigen value problem for the theory of the dielectric Green's function which depends on a positive parameter ε and is defined in the interior as well as the exterior of the curve. This Green's function has an immediate electrostatic interpretation and its theoretical value consists in the fact that it permits a

Received June 29, 1956. Prepared under contract Nonr-225 (11), (NR-041-086) for Office of Naval Research.

continuous transition from the Green's function of a domain to its Neumann's function. All dielectric Green's functions permit simple series developments in terms of the eigen functions and eigen values studied and the possible applications of these series developments to inequalities in potential theory are briefly indicated. Finally, it is shown that analytic completion of the dielectric Green's functions leads ultimately to univalent analytic functions in the interior as well as the exterior domain. This will lead, on the one hand, to interesting information on potential theoretical questions by use of the numerous distortion theorems of conformal mapping. On the other hand, we obtain in this way a one-parameter family of conformal maps of the domains which start with the identity and end up with the normalized mapping onto a circle. This parametrization is of importance in the theory of univalent functions; it is entirely different from the Löwner parametrization of univalent functions [8].

In §§ 3 and 4 we derive formulas for the variation of the various eigen values and dielectric Green's functions. We use at first interior variations and are thus able to derive precise variational formulas with uniform estimates for the error terms. By superposition of interior variations and simple transformations we can easily derive variational formulas of the Hadamard type. It is seen that the variational formula for the dielectric Green's function is surprisingly similar to that for the ordinary Green's function. It is seen that the circle is a curve for which all nontrivial eigen values are infinite. Thus, the circle leads to a homogeneous integral equation with an eigen value of infinite degeneracy and the usual perturbation theory cannot be applied. We show, therefore, by a special argument how eigen values for nearly-circular domains can be obtained.

Finally, we apply in § 5 the variational formula for the eigen values to a simple extremum problem for the lowest one which leads ultimately to the desired inequality. A characteristic difficulty, however, has to be overcome in this problem. It appears that the eigen values are only continuous functionals of the curve if the curve is deformed in such a way that normals in corresponding points are turned very little. Such a side condition is hard to preserve under general variations. We introduce, therefore, the concept of uniformly analytic curves which is closely related to the theory of univalent functions. Extremum problems within the class of uniformly analytic curves are easy to handle and the problem of the existence of extremum curves is likewise of very elementary nature. As the end result of our study an inequality then appears which estimates the lowest nontrivial Fredholm eigen value from below in terms of the modulus of uniform analyticity. This quantity is, however, easy to determine if a specific analytic curve is prescribed. It seems that the concept of uniform analyticity may play a useful role in many further extremum problems and variational investigations. As a side result of our study we obtain a new class of plane curves for which the Fredholm eigen functions and eigen values can be computed explicitly.

1. The Fredholm eigen values. Let C be a closed curve in the complex z-plane which is three times continuously differentiable; we denote the interior of C by D and its exterior by $\tilde{D}$. The kernel

$$k(z,\zeta)=\frac{\partial}{\partial n_\zeta}\log\frac{1}{|z-\zeta|} \tag{1}$$

is a continuous function of both argument points as these vary on the curve C only. We understand by $\frac{\partial}{\partial n_\zeta}$ the differentiation in direction of the normal at ζ on C pointing into D.

The first boundary value problem of potential theory with respect to the domain D can be reduced to the inhomogeneous integral equation

$$f(z)=\phi(z)+\frac{1}{\pi}\int_C k(z,\zeta)\phi(\zeta)ds_\zeta\,, \qquad z\in C\,, \tag{2}$$

while the second boundary value problem can be solved by reduction to an integral equation with transposed kernel

$$f(z)=\psi(z)-\frac{1}{\pi}\int_C k(\zeta,z)\psi(\zeta)ds_\zeta\,, \qquad z\in C\,. \tag{2'}$$

In view of the Fredholm alternative in the theory of integral equations one is then led naturally to discuss the eigen values and eigen functions of the corresponding homogeneous integral equation

$$\phi_\nu(z)=\frac{\lambda_\nu}{\pi}\int_C k(z,\zeta)\phi_\nu(\zeta)ds_\zeta\,. \tag{3}$$

These functionals of C play an important role in the potential theory of the domains D and $\tilde{D}$ as shall be seen in the following considerations. In this section we shall give a brief survey of their theory and various transformations of the integral equation (3) which will be used later.

We introduce the harmonic function

$$h_\nu(z)=\frac{\lambda_\nu}{\pi}\int_C k(z,\zeta)\phi_\nu(\zeta)ds_\zeta \tag{4}$$

which is defined in D and $\tilde{D}$; for the sake of clarity, we shall denote it by $\tilde{h}(z)$ if its argument point lies in $\tilde{D}$. By the well-known discontinuity behavior of the kernel (1), we have the limit relations, valid for an arbitrary point $z_0 \in C$:

$$(5) \qquad \lim_{z \to z_0} h_\nu(z) = (1+\lambda_\nu)\phi_\nu(z_0), \quad \lim_{z \to z_0} \tilde{h}_\nu(z) = (1-\lambda_\nu)\phi_\nu(z_0) .$$

On the other hand, the normal derivative of a double-layer potential goes continuously through the curve C which carries the charge and, hence,

$$(6) \qquad \frac{\partial}{\partial n} h_\nu(z) = -\frac{\partial}{\partial \tilde{n}} \tilde{h}_\nu(z) , \qquad \text{for } z \in C ,$$

where $\frac{\partial}{\partial \tilde{n}}$ denotes normal differentiation into $\tilde{D}$.

By Green's identity, we have

$$(7) \qquad \frac{1}{2\pi}\int_C \left[h_\nu(\zeta)\frac{\partial}{\partial n_\zeta}\log\frac{1}{|\zeta - z|} - \log\frac{1}{|\zeta - z|}\frac{\partial}{\partial n} h_\nu(\zeta)\right] ds_\zeta = h_\nu(z)\delta(z)$$

and

$$(7') \qquad \frac{1}{2\pi}\int_C \left[\tilde{h}_\nu(\zeta)\frac{\partial}{\partial \tilde{n}_\zeta}\log\frac{1}{|\zeta - z|} - \log\frac{1}{|\zeta - z|}\frac{\partial}{\partial \tilde{n}} \tilde{h}_\nu(\zeta)\right] ds_\zeta = \tilde{h}_\nu(z)\tilde{\delta}(z)$$

where $\delta(z)$ and $\tilde{\delta}(z)$ are the characteristic functions of D and $\tilde{D}$, that is,

$$(8) \qquad \delta(z) = \begin{cases} 1 & \text{if } z \in D \\ 0 & \text{if } z \in \tilde{D} \end{cases} , \quad \tilde{\delta}(z) = 1 - \delta(z) .$$

Combining (7) with (7′) and observing the boundary relations (5) and (6) between $h_\nu(z)$ and $\tilde{h}_\nu(z)$, we obtain

$$(9) \qquad -\frac{\lambda_\nu}{\pi(\lambda_\nu - 1)}\int_C \log\frac{1}{|\zeta - z|}\frac{\partial}{\partial n} h_\nu(\zeta) ds_\zeta = h_\nu(z) , \qquad z \in D$$

$$(10) \qquad -\frac{\lambda_\nu}{\pi(\lambda_\nu + 1)}\int_C \log\frac{1}{|\zeta - z|}\frac{\partial}{\partial \tilde{n}} \tilde{h}_\nu(\zeta) ds_\zeta = \tilde{h}_\nu(z) , \qquad z \in \tilde{D} .$$

Define two analytic functions in D and $\tilde{D}$, respectively, by the formulas

$$(11) \qquad v_\nu(z) = \frac{\partial}{\partial z} h_\nu(z) , \quad \tilde{v}_\nu(z) = \frac{\partial}{\partial z} \tilde{h}_\nu(z) .$$

Differentiating (9) and (10) with respect to z, one obtains easily

$$v_\nu(z)=\frac{\lambda_\nu}{2\pi i}\int_C\frac{\overline{(v_\nu(\zeta)d\zeta)}}{\zeta-z}\,,\quad \tilde{v}_\nu(z)=\frac{\lambda_\nu}{2\pi i}\int_C\frac{\overline{(\tilde{v}_\nu(\zeta)d\zeta)}}{\zeta-z}\,. \tag{12}$$

These are elegant integral equations for the analytic functions $v_\nu(z)$ and $\tilde{v}_\nu(z)$ which are valid for $z\in D$ and $z\in\tilde{D}$, respectively.

We can also bring (12) into the form

$$v_\nu(z)=\frac{\lambda_\nu}{\pi}\iint_D\frac{\overline{v_\nu(\zeta)}}{(\zeta-z)^2}d\tau \qquad \text{for } z\in D \tag{13}$$

and

$$\tilde{v}_\nu(z)=\frac{\lambda_\nu}{\pi}\iint_{\tilde{D}}\frac{\overline{\tilde{v}_\nu(\zeta)}}{(\zeta-z)^2}d\tau \qquad \text{for } z\in\tilde{D} \tag{13'}$$

which expresses $v_\nu(z)$ and $\tilde{v}_\nu(z)$ as solutions of integral equations with improper kernels. The integrals involved have to be understood in the Cauchy principal value sense.

In the transition from the integral equations (3) defined on C to the integral equations (13), (13′) defined in D and $\tilde{D}$, we have lost one particular eigen function. Indeed, if $h(z)=\text{const.}$ were one of our eigen functions $h_\nu(z)$ it would have been cancelled out in the differentiation (11). But by (5), $h_\nu(z)=\text{const.}$ implies $\phi_\nu(z)=\text{const.}$ on C and from (3) and the identity

$$\frac{1}{\pi}\int_C k(z,\ \zeta)ds_\zeta=1\,, \qquad z\in C\,, \tag{14}$$

follows, in fact, that each constant is an eigen function of (3) with the eigen value $\lambda=+1$. The eigen value $\lambda=+1$ plays an exceptional role in the entire Poincaré-Fredholm theory of the boundary value problem; the fact that the equations (13), (13′) lead to all other eigen values and their corresponding eigen functions and eliminate $\lambda=+1$ represents, therefore, a strong argument in favor of this transition.

Let $z(s)$ be the parametric representation of C in terms of its length parameter s. Then

$$z'=\frac{dz}{ds} \tag{15}$$

will be the unit vector in tangential direction to C. The boundary relations (5) and (6) for h_ν and $\tilde{h}_\nu$ go over into the equations on C:

$$\Re\{v_\nu(z)z'\}=\frac{1+\lambda_\nu}{1-\lambda_\nu}\Re\{\tilde{v}_\nu(z)z'\} \tag{16}$$

$$\Im\{v_\nu(z)z'\} = \Im\{\tilde{v}_\nu(z)z'\} \tag{17}$$

which can be combined into the one complex equation

$$v_\nu(z)z' = \frac{1}{1-\lambda_\nu}\tilde{v}_\nu(z)z' + \frac{\lambda_\nu}{1-\lambda_\nu}\overline{\tilde{v}_\nu(z)z'}\,. \tag{18}$$

This relation combined with (12) throws an interesting light on the connection between $v_\nu(z)$ and $\tilde{v}_\nu(z)$. In fact, if we insert (18) into the first equation (12) and apply Cauchy's theorem, we find

$$v_\nu(z) = \frac{1}{2\pi i}\frac{\lambda_\nu}{1-\lambda_\nu}\int_C \frac{\overline{(\tilde{v}_\nu(\zeta)d\zeta)}}{\zeta - z}\,, \qquad z \in D\,. \tag{19}$$

Observe that the second formula (12) yields $\tilde{v}_\nu(z)$ for $z \in \tilde{D}$; now we see that the same expression yields $v_\nu(z)$ for $z \in D$, except for the factor $1-\lambda_\nu$. Similarly, one shows easily

$$\tilde{v}_\nu(z) = \frac{1}{2\pi i}\frac{\lambda_\nu}{1+\lambda_\nu}\int_C \frac{\overline{(v_\nu(\zeta)d\zeta)}}{\zeta - z}\,, \qquad z \in \tilde{D}\,. \tag{20}$$

If $f(z)$ is an arbitrary complex-valued function in the entire z-plane of the class $\mathscr{L}^2$, the equation

$$F(z) = \frac{1}{\pi}\iint \frac{\overline{f(\zeta)}}{(\zeta - z)^2}d\tau\,, \tag{21}$$

defines a new function in $\mathscr{L}^2$, its Hilbert transform. It is well-known [2] that the Hilbert transformation is norm-preserving, that is,

$$\iint |F|^2 d\tau = \iint |f|^2 d\tau\,. \tag{22}$$

Our formulas (13), (13′) and (19), (20) imply that the functions

$$f_\nu(z) = \begin{cases} v_\nu(z) & \text{in } D \\ 0 & \text{in } \tilde{D} \end{cases} \quad \text{and} \quad \tilde{f}_\nu(z) = \begin{cases} 0 & \text{in } D \\ \tilde{v}_\nu(z) & \text{in } \tilde{D} \end{cases} \tag{23}$$

have the Hilbert transforms

$$F_\nu(z) = \begin{cases} \dfrac{1}{\lambda_\nu}v_\nu(z) & \text{in } D \\[2ex] \left(\dfrac{1}{\lambda_\nu}+1\right)\tilde{v}_\nu(z) & \text{in } \tilde{D} \end{cases} \quad \text{and} \quad \tilde{F}_\nu(z) = \begin{cases} \left(\dfrac{1}{\lambda_\nu}-1\right)v_\nu(z) & \text{in } D \\[2ex] \dfrac{1}{\lambda_\nu}\tilde{v}_\nu(z) & \text{in } \tilde{D}\,. \end{cases} \tag{24}$$

Hence (22) yields

$$(\lambda_\nu - 1)\iint_D |v_\nu|^2 d\tau = (\lambda_\nu + 1)\iint_{\tilde{D}} |\tilde{v}_\nu|^2 d\tau . \tag{25}$$

From (25) we conclude easily that

$$|\lambda_\nu| > 1 . \tag{26}$$

For if, for example, $\lambda_\nu = 1$, we would have $\tilde{v}_\nu \equiv 0$ in $\tilde{D}$, $\tilde{h}_\nu(z) = \text{const.}$ and hence by (6) also $h_\nu(z) = \text{const.}$ But this would imply, in turn, $v_\nu(z) \equiv 0$ and no eigen functions would exist.

With each eigen value λ_ν of (4) the eigen value $-\lambda_\nu$ also occurs, except for $\lambda = 1$. In fact, if we denote the conjugate functions of $h_\nu(z)$ and $\tilde{h}_\nu(z)$ by $g_\nu(z)$ and $\tilde{g}_\nu(z)$, we have by the Cauchy-Riemann formulas the relations

$$\frac{\partial}{\partial n} g_\nu(z) = -\frac{1+\lambda_\nu}{1-\lambda_\nu}\frac{\partial}{\partial \tilde{n}}\tilde{g}_\nu(z) , \quad \frac{\partial}{\partial s} g_\nu(z) = \frac{\partial}{\partial s}\tilde{g}_\nu(z) . \tag{27}$$

Hence, putting

$$g_\nu^*(z) = (1-\lambda_\nu) g_\nu(z) , \quad \tilde{g}_\nu^*(z) = (1+\lambda_\nu)\tilde{g}_\nu(z) \tag{28}$$

and adding an appropriate constant we find for $z \in C$:

$$g_\nu^*(z) = \frac{1-\lambda_\nu}{1+\lambda_\nu}\tilde{g}_\nu^*(z) , \quad \frac{\partial}{\partial n} g_\nu^*(z) = -\frac{\partial}{\partial \tilde{n}}\tilde{g}_\nu^*(z) . \tag{29}$$

These are the boundary relations between h_ν and $\tilde{h}_\nu$ but with $-\lambda_\nu$ instead of λ_ν. This proves our assertion.

If we start conversely from the complex integral equations (13) and (13′) and consider any eigen function $v_\nu(z)$ with the eigen value λ_ν, it will be observed that $e^{i\alpha}v_\nu(z)$ is also an eigen function to the eigen value $\lambda_\nu e^{-2i\alpha}$. Hence, if we focus our attention on the integral equations (12) or (13) we may assume without loss of generality that λ_ν is a real positive eigen value. Calculating backward, we can easily see that each such eigen value is also an eigen value of the Fredholm integral equation (4) and so is $-\lambda_\nu$.

It is readily verified that eigen functions $v_\nu(z)$ and $v_\mu(z)$ which belong to different eigen values λ_ν and λ_μ satisfy the orthogonality relation

$$\iint_D v_\nu(z)\overline{v_\mu(z)}\, d\tau = 0 . \tag{30}$$

This condition can be extended to the case of any two linearly independent eigen functions. Similarly

$$\iint_{\tilde{D}} \tilde{v}_\nu \overline{\tilde{v}}_\mu d\tau = 0 \tag{31}$$

for any two different eigen functions $\tilde{v}_\nu$, $\tilde{v}_\mu$. In view of (25), we define

$$w_\nu(z)=\sqrt{\lambda_\nu-1}\ v_\nu(z)\,, \quad \tilde{w}_\nu(z)=i\sqrt{\lambda_\nu+1}\ \tilde{v}_\nu(z)\,. \tag{32}$$

Then we can assume the orthonormality relations

$$\iint_D w_\nu \bar{w}_\mu d\tau=\delta_{\nu\mu}\,, \quad \iint_{\tilde{D}} \tilde{w}_\nu \bar{\tilde{w}}_\mu d\tau=\delta_{\nu\mu}\,. \tag{33}$$

We have in view of (18) the boundary relations on C:

$$w_\nu(z)z'=\frac{i}{\sqrt{\lambda_\nu^2-1}}\tilde{w}_\nu(z)z'-\frac{\lambda_\nu i}{\sqrt{\lambda_\nu^2-1}}\overline{\tilde{w}_\nu(z)z'}\,; \tag{34}$$

from (19) and (20) follows

$$w_\nu(z)=-\frac{\lambda_\nu}{2\pi\sqrt{\lambda_\nu^2-1}}\int_C \frac{\overline{(\tilde{w}_\nu(\zeta)d\zeta)}}{\zeta-z}\,, \qquad z\in D \tag{35}$$

and

$$\tilde{w}_\nu(z)=\frac{\lambda_\nu}{2\pi\sqrt{\lambda_\nu^2-1}}\int_C \frac{\overline{(w_\nu(\zeta)d\zeta)}}{\zeta-z}\,, \qquad z\in \tilde{D}\,. \tag{36}$$

If we were able to guess two functions $w(z)$ and $\tilde{w}(z)$ which are analytic in D and $\tilde{D}$, respectively, and which satisfy on C the relation (34) for a properly chosen λ, we would have obtained a particular solution for the eigen value problems (13) and (13′). It is sometimes possible to construct such pairs of functions and to obtain thus eigen values and eigen functions for the Fredholm integral equation. One possibility of construction is the following: We refer the curve C by conformal mapping to the unit circumference. Let

$$z=f(\zeta) \tag{35}$$

be analytic on and near $|\zeta|=1$ and map it onto C. The condition (34) can now be referred to $|\zeta|=1$ and reads:

$$w[f(\zeta)]f'(\zeta)i\zeta=-\frac{1}{\sqrt{\lambda^2-1}}\tilde{w}[f(\zeta)]f'(\zeta)\zeta-\frac{\lambda}{\sqrt{\lambda^2-1}}\overline{\tilde{w}[f(\zeta)]f'(\zeta)\zeta}\,. \tag{38}$$

Since the conjugation $\bar{\zeta}$ means just $\dfrac{1}{\zeta}$ on $|\zeta|=1$ it is easier to guess solutions in this form.

Let, for example,

$$z=f(\zeta)=\zeta+\frac{\rho^2}{\zeta}\,, \qquad 0<\rho<1\,. \tag{39}$$

This means that C is an ellipse. Let us put

$$w[f(\zeta)]f'(\zeta)\zeta=a\zeta^n+b\zeta^{-n} \tag{40}$$

and

$$\tilde{w}[f(\zeta)]f'(\zeta)\zeta=i\zeta^{-n}. \tag{41}$$

Condition (38) will be fulfilled if we put

$$a=\frac{\lambda}{\sqrt{\lambda^2-1}}, \quad b=\frac{-1}{\sqrt{\lambda^2-1}}. \tag{42}$$

Define $W(f)=\int w\,df$ and $\tilde{W}(f)=\int \tilde{w}\,df$. Then (40) and (41) yield

$$W[f(\zeta)]=\frac{1}{n}(a\zeta^n-b\zeta^{-n})=\frac{\lambda}{n\sqrt{\lambda^2-1}}(\zeta^n+\lambda^{-1}\zeta^{-n}), \tag{43}$$

$$\tilde{W}[f(\zeta)]=-\frac{i}{n}\zeta^{-n}. \tag{44}$$

Now the function (39) is univalent outside of $|\zeta|=\rho$ and, hence, we may consider ζ as a regular analytic function of $z\in\tilde{D}$. Thus, $\tilde{W}(z)$ and $\tilde{w}(z)$ are regular analytic in $\tilde{D}$. In order that $W(z)$, $w(z)$ be analytic in D we must require that

$$\lambda=\rho^{-2n}. \tag{45}$$

In fact, $\zeta^n+\frac{\rho^{2n}}{\zeta^n}$ can be expressed as a Chebysheff polynomial of z. Thus we have guessed an infinity of eigen values and eigen functions for the case of the ellipse. It can be shown that $\lambda^n=\pm\rho^{-2n}$ gives all eigen values of the ellipse for $n=1, 2, \cdots$. Since $\rho=0$ describes a circle, we recognize, in particular, that all eigen values λ_ν for a circle have the value infinity [3].

If we know the eigen values and eigen functions of a given domain D we can find immediately the eigen values and eigen functions of every domain D^* which is obtained from D by a linear transformation

$$z^*=\frac{az+b}{cz+d}=l(z). \tag{46}$$

In fact, let

$$w_\nu^*(z^*)l'(z)=w_\nu(z), \quad \tilde{w}_\nu^*(z^*)l'(z)=\tilde{w}_\nu(z), \quad \lambda_\nu^*=\lambda_\nu. \tag{47}$$

It follows from (33) that the w_ν^* and $\tilde{w}_\nu^*$ form an orthonormal set of analytic functions in D^* and $\tilde{D}^*$, respectively. Since we have on C^*

$$z^{*\prime}(z)=l'(z) \tag{48}$$

it is also obvious that (34) is fulfilled which shows that w_ν^* and $\tilde{w}_\nu^*$ are, in fact, the normalized eigen functions of the domains D^* and $\tilde{D}^*$. In particular, we note that the eigen values λ_ν of a domain are unchanged under linear transformation. Similar domains, for example, have the same set of eigen values.

2. Thc dielectric problem. The consideration of the electrostatic field of a point source at ζ in the presence of a dielectric medium in D with the dielectric constant ε leads to the following heuristic definition of a Green's function $G_\varepsilon(z, \zeta)$:

(a) $G_\varepsilon(z, \zeta)$ is harmonic in D and $\tilde{D}$, except for $z=\zeta$.

(b) $G_\varepsilon(z, \zeta)-\log\dfrac{1}{|z-\zeta|}$ is harmonic at ζ if $\zeta\in\tilde{D}$.

(b′) $G_\varepsilon(z, \zeta)-\varepsilon\log\dfrac{1}{|z-\zeta|}$ is harmonic at ζ if $\zeta\in D$.

(c) $G_\varepsilon(z, \zeta)$ is continuous through C.

(d) $\dfrac{\partial}{\partial n}G_\varepsilon(z, \zeta)+\varepsilon\dfrac{\partial}{\partial\tilde{n}}G_\varepsilon(z, \zeta)=0$ on C for $\zeta\in D$ or $\tilde{D}$.

(e) $\log|z|+G_\varepsilon(z, \zeta)\to 0$ if $z\to\infty$, for $\zeta\in D$ or $\zeta\in\tilde{D}$.

It is easily seen that $G_\varepsilon(z, \zeta)$ is uniquely determined by these conditions and that it satisfies the symmetry condition

$$G_\varepsilon(\zeta, \eta)=G_\varepsilon(\eta, \zeta). \tag{1}$$

We may construct $G_\varepsilon(z, \zeta)$ by means of a line potential as follows. Let $\zeta\in\tilde{D}$ and put

$$G_\varepsilon(z, \zeta)=\log\frac{1}{|z-\zeta|}+\int_C\mu(\eta, \zeta)\log|\eta-z|ds_\eta. \tag{2}$$

This set-up satisfies automatically conditions (a), (b) and (c); we can fulfill condition (e) by the requirement

$$\int_C\mu(\eta, \zeta)ds_\eta=0 \tag{3}$$

and finally (d) by solving the integral equation

$$-\frac{1-\varepsilon}{1+\varepsilon}\cdot\frac{1}{\pi}\frac{\partial}{\partial n_z}\log\frac{1}{|z-\zeta|}=\mu(z, \zeta)-\frac{1-\varepsilon}{1+\varepsilon}\frac{1}{\pi}\int_C\mu(\eta, \zeta)\frac{\partial}{\partial n_z}\log\frac{1}{|\eta-z|}ds_\eta. \tag{4}$$

As long as $\varepsilon > 0$ this equation can be solved in a unique way since $\left|\frac{1-\varepsilon}{1+\varepsilon}\right| < 1$ and all eigen values of the corresponding homogeneous integral equation are larger or equal to one in absolute value. One verifies also from (4) that condition (3) is automatically fulfilled. In a similar way we proceed for $\zeta \in D$.

The integral equation (4) indicates already the close relation between the Green's function $G_\varepsilon(z, \zeta)$ and the Fredholm eigen functions. We obtain a further insight from the Dirichlet identities:

$$(5) \quad \iint_D \nabla G_\varepsilon(z, \zeta) \nabla h_\nu(z) d\tau = -\int_C G_\varepsilon(z, \zeta) \frac{\partial}{\partial n} h_\nu(z) ds_z$$

$$= 2\pi\varepsilon h_\nu(\zeta)\delta(\zeta) - \int_C h_\nu(z) \frac{\partial}{\partial n} G_\varepsilon(z, \zeta) ds_z$$

and

$$(6) \quad \iint_{\tilde{D}} \nabla G_\varepsilon(z, \zeta) \cdot \nabla \tilde{h}_\nu(z)\, d\tau = -\int_C G_\varepsilon(z, \zeta) \frac{\partial}{\partial \tilde{n}} \tilde{h}_\nu(z) ds_z$$

$$= 2\pi \tilde{h}_\nu(\zeta)\tilde{\delta}(\zeta) - \int_C \tilde{h}_\nu(z) \frac{\partial}{\partial \tilde{n}} G_\varepsilon(z, \zeta) ds_z \, .$$

Here we use $\delta(\zeta)$ and $\tilde{\delta}(\zeta)$ as defined in (1.8). Identity (6) is valid in spite of the logarithmic pole of G_ε at infinity since $\tilde{h}_\nu(z)$ vanishes there.

Adding (5) and (6) and using (1.6), we obtain

$$(7) \quad \iint_{\tilde{D}} \nabla G_\varepsilon(z, \zeta) \nabla \tilde{h}_\nu(z) d\tau = -\iint_D \nabla G_\varepsilon(z, \zeta) \nabla h_\nu(z)\, d\tau \, .$$

Putting

$$(8) \quad \rho_\nu = \frac{\lambda_\nu + 1}{\lambda_\nu - 1}$$

and using the boundary relations (1.5) and (d), we find:

$$(9) \quad -\varepsilon\rho_\nu \iint_{\tilde{D}} \nabla G_\varepsilon(z, \zeta) \nabla \tilde{h}_\nu d\tau + \iint_D \nabla G_\varepsilon(z, \zeta) \nabla h_\nu d\tau$$

$$= 2\pi\varepsilon[h_\nu(\zeta)\delta(\zeta) - \rho_\nu \tilde{h}_\nu(\zeta)\tilde{\delta}(\zeta)] \, .$$

Thus, finally,

$$(10) \quad \iint_D \nabla G_\varepsilon(z, \zeta) \nabla h_\nu(z) d\tau = \frac{2\pi\varepsilon}{1 + \varepsilon\rho_\nu} [h_\nu(\zeta)\delta(\zeta) - \rho_\nu \tilde{h}_\nu(\zeta)\tilde{\delta}(\zeta)]$$

$$= -\iint_{\tilde{D}} \nabla G_\varepsilon(z, \zeta) \nabla \tilde{h}_\nu d\tau \, .$$

The eigen functions $h_\nu(z)$ connected with the Fredholm equation appear thus as the eigen functions of the integral equation in D:

$$\frac{1+\varepsilon\rho_\nu}{2\pi\varepsilon}\iint_D \nabla G_\varepsilon(z,\ \zeta)\cdot\nabla h_\nu(z)d\tau=h_\nu(\zeta)\ ,\quad \zeta\in D\ . \tag{10'}$$

Let $G(z,\ \zeta)$ be the ordinary Green's function of D; obviously

$$\iint_D \nabla G(z,\ \zeta)\cdot\nabla h_\nu(z)d\tau=0\ . \tag{11}$$

Hence we obtain for $h_\nu(z)$ the integral equation

$$\frac{1+\varepsilon\rho_\nu}{2\pi\varepsilon}\iint_D \nabla K_\varepsilon(z,\ \zeta)\cdot\nabla h_\nu(z)d\tau=h_\nu(\zeta) \tag{12}$$

with the regular harmonic kernel

$$K_\varepsilon(z,\ \zeta)=G_\varepsilon(z,\ \zeta)-\varepsilon\, G(z,\ \zeta)\ . \tag{13}$$

Let

$$\tilde{G}(z,\ \infty)=\log|z|+\tilde{\gamma}+O\left(\frac{1}{|z|}\right) \tag{14}$$

represent the Green's function of $\tilde{D}$ with the source point at infinity. By (1.5) we have obviously

$$\int_C h_\nu(z)\frac{\partial\tilde{G}(z,\ \infty)}{\partial\tilde{n}}ds=\frac{1+\lambda_\nu}{1-\lambda_\nu}\int_C \tilde{h}_\nu(z)\frac{\partial\tilde{G}(z,\ \infty)}{\partial\tilde{n}}=2\pi\frac{1+\lambda_\nu}{1-\lambda_\nu}\tilde{h}_\nu(\infty)=0\ . \tag{15}$$

We now define the linear space Σ consisting of all functions $h(z)$ which are harmonic in D, have a finite Dirichlet integral there and satisfy the linear homogeneous condition

$$\int_C h(z)\frac{\partial\tilde{G}(z,\ \infty)}{\partial\tilde{n}}ds=0\ . \tag{16}$$

Observe that the only constant element in Σ is the function $h\equiv 0$. All $h_\nu(z)$ lie in Σ; in view of (12) and the symmetry of $K_\varepsilon(z,\ \zeta)$ we may assume that they are orthonormalized by the conditions

$$\iint_D \nabla h_\nu\cdot\nabla h_\mu d\tau=\delta_{\nu\mu} \tag{17}$$

and it is easily seen that they form a complete orthonormal set in Σ [3].

If we use the conditions (c) and (e) in the definition of $G_\varepsilon(z,\ \zeta)$, we can show that the function

(18) $$h(z)=K_\varepsilon(z,\ \zeta)-\tilde{\gamma}$$

lies in Σ. Hence we have for it the following series development

(19) $$G_\varepsilon(z,\ \zeta)-\varepsilon G(z,\ \zeta)=\tilde{\gamma}+\sum_{\nu=1}^{\infty}\frac{h_\nu(z)h_\nu(\zeta)}{1+\varepsilon\rho_\nu}\cdot 2\pi\varepsilon\ , \qquad \zeta\in D\ .$$

The Fourier coefficients in this development have been calculated from (12); the series converges uniformly in each closed subdomain of D.

Suppose next $\zeta\in\tilde{D}$ and consider the harmonic function

(20) $$h(z)=G_\varepsilon(z,\ \zeta)+\tilde{G}(\zeta,\ \infty)-\tilde{\gamma}\ .$$

It is easily seen that $h(z)\in\Sigma$. Hence we may develop $h(z)$ into a series in the complete orthonormal system $h_\nu(z)$. Using (10), we find

(21) $$G_\varepsilon(z,\ \zeta)=\tilde{\gamma}-\tilde{G}(\zeta,\ \infty)-2\pi\varepsilon\sum_{\nu=1}^{\infty}\frac{\rho_\nu}{1+\varepsilon\rho_\nu}h_\nu(z)\tilde{h}_\nu(\zeta)\ .$$

This series converges for $\zeta\in\tilde{D}$ and z in a closed subdomain of D.

Observe that by definition of $G_\varepsilon(z,\ \zeta)$ we have for $\varepsilon=1$

(22) $$G_1(z,\ \zeta)=\log\frac{1}{|z-\zeta|}\ .$$

Hence (19) contains the following series representation for the ordinary Green's function of D:

(23) $$G(z,\ \zeta)=\log\frac{1}{|z-\zeta|}-\tilde{\gamma}-2\pi\sum_{\nu=1}^{\infty}\frac{h_\nu(z)h_\nu(\zeta)}{1+\rho_\nu}\ .$$

On the other hand, (21) reduces for $\varepsilon=1$ to

(24) $$\tilde{G}(\zeta,\ \infty)=\log|z-\zeta|+\tilde{\gamma}-2\pi\sum_{\nu=1}^{\infty}\frac{\rho_\nu}{1+\rho_\nu}h_\nu(z)\tilde{h}_\nu(\zeta)\ .$$

In a similar way we can derive series developments for $G_\varepsilon(z,\ \zeta)$ in the exterior $\tilde{D}$ of C. Observe that in view of the boundary conditions (1.5) and (1.6) the normalization (17) of the $h_\nu(z)$ implies

(25) $$\iint_{\tilde{D}}\nabla\tilde{h}_\nu\cdot\nabla\tilde{h}_\mu d\tau=\rho_\nu^{-1}\delta_{\mu\nu}\ .$$

Let $\tilde{\Sigma}$ be the linear function space consisting of all functions $\tilde{h}(z)$ which are harmonic in $\tilde{D}$, have a finite Dirichlet integral there and which vanish at infinity. Clearly the $\{\rho_\nu^{1/2}\tilde{h}_\nu(z)\}$ form a complete orthonormal set in $\tilde{\Sigma}$.

On the other hand, let $\tilde{G}(z, \zeta)$ be the Green's function of $\tilde{D}$. Then it is easily verified that by condition (e) and (14)

$$\tilde{h}(z)=G_\varepsilon(z, \zeta)-\tilde{G}(z, \zeta)+\tilde{G}(\zeta, \infty)+\tilde{G}(z, \infty)-\tilde{\gamma}, \qquad z, \zeta \in \tilde{D}, \tag{26}$$

lies in $\tilde{\Sigma}$. Again using the Dirichlet formula (10), we find

$$\begin{aligned} G_\varepsilon(z, \zeta)-\tilde{G}(z, \zeta)+\tilde{G}(\zeta, \infty)+\tilde{G}(z, \infty)-\tilde{\gamma} \\ =2\pi\varepsilon \sum_{\nu=1}^{\infty} \frac{\tilde{h}_\nu(z)\tilde{h}_\nu(\zeta)\rho_\nu^2}{1+\varepsilon\rho_\nu}. \end{aligned} \tag{27}$$

Putting, in particular, $\varepsilon=1$, we obtain by virtue of (22):

$$\tilde{G}(z, \zeta)-\tilde{G}(\zeta, \infty)-\tilde{G}(z, \infty)=\log\frac{1}{|z-\zeta|}-\tilde{\gamma}-2\pi\sum_{\nu=1}^{\infty}\frac{\tilde{h}_\nu(z)\tilde{h}_\nu(\zeta)\rho_\nu^2}{1+\rho_\nu}. \tag{28}$$

We have thus shown that all dielectric Green's functions can be constructed simultaneously and in D as well as in $\tilde{D}$, once the system of eigen functions $h_\nu(z)$ and the corresponding eigen values λ_ν are known.

Numerous inequalities can be drawn from these representations. We shall restrict ourselves to one single example. Denote, for $\zeta \in D$,

$$G_\varepsilon(z, \zeta)-\varepsilon\log\frac{1}{|z-\zeta|}=g_\varepsilon(z, \zeta), \quad G(z, \zeta)-\log\frac{1}{|z-\zeta|}=g(z, \zeta). \tag{29}$$

The functions $g_\varepsilon(z, \zeta)$ and $g(z, \zeta)$ represent the potentials induced by a unit pole at ζ in the presence of the dielectric in D, and in the presence of the grounded conductor C, respectively. We find from (19)

$$g_\varepsilon(\zeta, \zeta)-\varepsilon g(\zeta, \zeta)\geqq\tilde{\gamma}. \tag{30}$$

Since $e^{-\tilde{\gamma}}$ represents the electrostatic capacity of the conductor C, we obtain an interesting estimate for the dielectric reaction potential in terms of capacity constants connected with the conductor surface C. For $\varepsilon=1$, we have $g_1(\zeta, \zeta)=0$ and hence

$$\tilde{\gamma}\leqq -g(\zeta, \zeta). \tag{31}$$

This is an inequality connecting the inner and the outer Green's function of C; in the case that C is a circumference and ζ is its center, this inequality becomes an equality.

Up to this point we stressed the connection between the Green's function $G(z, \zeta)$ and the eigen functions $h_\nu(z)$. Since the Fredholm eigen functions appear also in the theory of the second boundary value problem, we should also expect some relations between the $h_\nu(z)$ and the Neumann's function of the domain D.

The Neumann's function is usually defined by its constant normal derivative on C

$$\frac{\partial N(z,\ \zeta)}{\partial n_z}=\frac{2\pi}{L}, \qquad z\in C,\ \zeta\in D,\ L=\text{length of } C, \tag{32}$$

and by the linear homogeneous side condition

$$\int_C N(z,\ \zeta)ds_z=0\,, \qquad \zeta\in D. \tag{33}$$

In order to operate within the class Σ, characterized by (16), we introduce the functions

$$\alpha(z)=\frac{1}{2\pi}\int_C N(t,\ z)\left[\frac{\partial\tilde{G}(t,\ \infty)}{\partial n}+\frac{2\pi}{L}\right]ds_t \tag{34}$$

and

$$h(z)=N(z,\ \zeta)-G(z,\ \zeta)+\alpha(z)+\alpha(\zeta) \tag{35}$$

$$+\frac{1}{4\pi^2}\iint_{CC}N(t,\ \tau)\frac{\partial\tilde{G}(t,\ \infty)}{\partial n}\frac{\partial\tilde{G}(\tau,\ \infty)}{\partial n}ds_t\,ds_\tau.$$

It is easily verified that $h(z)\in\Sigma$. Since obviously

$$\int_C\left[\frac{\partial\tilde{G}(t,\ \infty)}{\partial n}+\frac{2\pi}{L}\right]ds=0\,, \tag{34'}$$

the function $\alpha(z)$ is harmonic in D and has the normal derivative

$$\frac{\partial\alpha}{\partial n}=-\frac{\partial\tilde{G}(z,\ \infty)}{\partial n}-\frac{2\pi}{L}\,, \qquad z\in C. \tag{34''}$$

Hence, finally, we have for $z\in C$

$$\frac{\partial h}{\partial n}=-\frac{\partial G(z,\ \zeta)}{\partial n}-\frac{\partial\tilde{G}(z,\ \infty)}{\partial n} \tag{35'}$$

and consequently in view of (16) valid for each $h_\nu(z)$:

$$\iint_D \nabla h\cdot\nabla h_\nu d\tau=-\int_C h_\nu\frac{\partial h}{\partial n}ds=2\pi h_\nu(\zeta)\,. \tag{36}$$

Since $h(z)\in\Sigma$ and the $h_\nu(z)$ are a complete orthonormal system in Σ, we have the Fourier development

$$(37)\quad N(z, \zeta)-G(z, \zeta)+\frac{1}{2\pi}\int_C N(t, z)\frac{\partial\tilde{G}(t, \infty)}{\partial n}ds+\frac{1}{2\pi}\int_C N(t, \zeta)\frac{\partial\tilde{G}(t, \infty)}{\partial n}ds$$

$$+\frac{1}{4\pi^2}\iint_{CC} N(t, \tau)\frac{\partial\tilde{G}(t, \infty)}{\partial n}\frac{\partial\tilde{G}(\tau, \infty)}{\partial n}ds_t ds_\tau=\sum_{\nu=1}^{\infty} h_\nu(z)h_\nu(\zeta)\cdot 2\pi .$$

This formula is useful to establish the exact asymptotics of the function $G_\varepsilon(z, \zeta)$ as $\varepsilon\to 0$ as can be seen from formula (19).

The dielectric Green's functions $G_\varepsilon(z, \zeta)$ are closely related to a set of interesting univalent analytic functions. In order to show this connection we complete the harmonic functions $G_\varepsilon(z, \zeta)$ to analytic functions in z. We will obtain, of course, two entirely different functions when z lies in D or $\tilde{D}$. Let us denote the analytic completion of $G_\varepsilon(z, \zeta)$ by $P_\varepsilon(z, \zeta)$ if $z\in D$ and by $\tilde{P}_\varepsilon(z, \zeta)$ if $z\in\tilde{D}$. We want to show that for fixed $\zeta\in D$

$$(38)\qquad f_\varepsilon(z)=e^{-1/\varepsilon P_\varepsilon(z,\zeta)},\quad \tilde{f}_\varepsilon(z)=e^{-\tilde{P}_\varepsilon(z,\zeta)}$$

represent univalent analytic functions in D and $\tilde{D}$, respectively.

For the sake of simplicity, we shall assume in the following consideration that C is an analytic curve. There exists, therefore, an analytic function $z=f(t)$ which maps a neighborhood of a segment of the real axis in the t-plane onto a neighborhood of a given arc of C. The function $G_\varepsilon(z, \zeta)$ becomes a harmonic function $g(t)$ to both sides of the segment. It goes continuously through the segment, but its normal derivatives satisfy the discontinuity law

$$(39)\qquad \frac{\partial g}{\partial n}+\varepsilon\frac{\partial g}{\partial\tilde{n}}=0\quad\text{for real } t.$$

Let

$$(40)\qquad p(t)=P_\varepsilon(f(t), \zeta),\ \tilde{p}(t)=\tilde{P}_\varepsilon(f(t), \zeta).$$

We find easily for t in the segment and in view of the described discontinuity behavior of $g(t)$:

$$(41)\qquad \Re\{p'(t)\}=\Re\{\tilde{p}'(t)\},\ \Im\{p'(t)\}=\varepsilon\Im\{\tilde{p}'(t)\}.$$

We can combine the two relations (41) into the one equation:

$$(42)\qquad p'(t)=\frac{1+\varepsilon}{2}\tilde{p}'(t)+\frac{1-\varepsilon}{2}\overline{\tilde{p}'(\bar{t})}.$$

This formula allows an analytic continuation of $\tilde{p}'(t)$ into the upper halfplane and of $p'(t)$ into the lower. This proves that $p'(t)$ and $\tilde{p}'(t)$ are still analytic on the segment of the real axis in the t-plane. Returning to the z-plane we can infer that the functions

$$P'_\varepsilon(z, \zeta)=\frac{d}{dz}P_\varepsilon(z, \zeta), \quad \tilde{P}'_\varepsilon(z, \zeta)=\frac{d}{dz}\tilde{P}_\varepsilon(z, \zeta) \tag{43}$$

are analytic beyond the curve C. Thus we proved that the two determinations of the Green's functions $G_\varepsilon(z, \zeta)$ are still regular harmonic on C if C is an analytic curve.

We derive from (40) and (42) that

$$\frac{P'_\varepsilon(z, \zeta)}{\tilde{P}'_\varepsilon(z, \zeta)}=\frac{1+\varepsilon}{2}+e^{-2i\alpha}\frac{1-\varepsilon}{2}, \qquad \alpha=\arg \tilde{p}'(t). \tag{44}$$

Since we assume throughout $\varepsilon > 0$, we see that the ratio (44) always lies in the right half of the complex plane. This implies

$$\Delta \arg \tilde{P}'_\varepsilon(z, \zeta)=\Delta \arg P'_\varepsilon(z, \zeta) \tag{45}$$

if z runs through the curve C in the positive sense with respect to D.

But by the argument principle we have

$$\Delta \arg P'_\varepsilon(z, \zeta)=Z-P, \quad \Delta \arg \tilde{P}'_\varepsilon(z, \zeta)=\tilde{P}-\tilde{Z} \tag{46}$$

where P, Z are the numbers of zeros and poles of P'_ε in D and $\tilde{P}$, $\tilde{Z}$ have the same meaning with respect to $\tilde{P}'_\varepsilon$ and $\tilde{D}$. In case some zero of P'_ε should lie on C, we can deform the curve in such a way that it does not contain any zero and draw the same conclusion in view of the analyticity of P'_ε and $\tilde{P}'_\varepsilon$ on C.

We know by definition that if $\zeta \in D$ we have

$$P=1, \ Z\geqq 0; \ \tilde{P}=0, \ \tilde{Z}\geqq 1. \tag{47}$$

Hence, from (45), (46) and (47), we conclude

$$Z-1\leqq -1. \tag{48}$$

This is only possible if

$$\tilde{Z}=1, \ Z=0. \tag{49}$$

Hence we can state that $P'_\varepsilon(z, \zeta)$ and $\tilde{P}'_\varepsilon(z, \zeta)$ do not vanish at any finite point of the z-plane.

Consider now the system of differential equations $(z=x+iy)$

$$\frac{dx}{dt}=-\frac{\partial}{\partial x}G_\varepsilon(z, \zeta), \quad \frac{dy}{dt}=-\frac{\partial}{\partial y}G_\varepsilon(z, \zeta). \tag{50}$$

Along each solution curve $x(t)$, $y(t)$ of this system we have

$$\frac{d}{dt} G_\varepsilon(z(t), \zeta) = -\left[\left(\frac{dx}{dt}\right)^2 + \left(\frac{dy}{dt}\right)^2\right] < 0 . \tag{51}$$

We have just shown that no critical point exists where $\nabla G_\varepsilon = 0$. Hence the net of solution curves covers the entire z-plane in a regular manner. All curves start out from the point $z=\zeta$ and run towards infinity. Each curve possesses the integral

$$\Im\{P(z, \zeta)\} = \text{const. or } \Im\{\tilde{P}_\varepsilon(z, \zeta)\} = \text{const.} , \tag{52}$$

according as it is considered in D or in $\tilde{D}$. From these facts it is evident that the functions (38) have the asserted univalency properties in D and $\tilde{D}$, respectively.

The importance of our result lies in the fact that the numerous distortion theorems of univalent function theory are now at our disposal in order to derive estimates of the various potential theoretical quantities connected with $G_\varepsilon(z, \zeta)$ in terms of the geometry of the curve C.

Let us observe, further, that for $\varepsilon = 1$ the function $\tilde{f}_1(z)$ represents the identity mapping while for $\varepsilon = 0$ we conclude from (21) that

$$\tilde{f}_0(z) = e^{-\tilde{\gamma}} \cdot e^{\tilde{P}(z,\infty)} = z + c_1 + \frac{c_1}{z} + \cdots \tag{53}$$

is the univalent function which maps $\tilde{D}$ onto the exterior of a circle of radius $e^{-\tilde{\gamma}}$ and which has at infinity the derivative one. Thus we can interpolate a continuous sequence of univalent mappings between the identity map of $\tilde{D}$ and its normalized mapping onto the exterior of a circle.

The preceding considerations show clearly the significance of the Fredholm eigen values and eigen functions for the dielectric problem and the general potential theory of the curve C. A generalization of most concepts to the physically more interesting case of three dimensions is easily done.

3. The variation of the eigen values. The variation of the eigen values λ_ν under a variation of the curve C can be determined by using the variational theory of the Green's function and of the various kernel functions connected with it [3]. In this paper we wish to give a straightforward and elementary derivation of the variational formulas.

Let z_0 be an arbitrary fixed point in $\tilde{D}$ and consider the mapping

$$z^* = z + \frac{\alpha}{z - z_0} . \tag{1}$$

For small enough α this will be a univalent mapping of C into a new smooth curve C^*. Let us denote its eigen values by λ_ν^* and its eigen functions by $w_\nu^*(z)$. We have used various eigen function definitions in the domain D; the $w_\nu^*(z)$ shall play the same role with respect to D^* (the domain bounded by C^*) as the $w_\nu(z)$ defined in Section 1 played with respect to D.

We have the integral equation

$$w_\nu^*(z^*)=\frac{\lambda_\nu^*}{2\pi i}\int_{C^*}\frac{\overline{(w_\nu^*(\zeta^*)d\zeta^*)}}{\zeta^*-z^*}\,, \qquad z^*\in D^*\,. \tag{2}$$

Let us define

$$m_\nu(z)=w_\nu^*\Big(z+\frac{\alpha}{z-z_0}\Big)\Big(1-\frac{\alpha}{(z-z_0)^2}\Big)\,. \tag{3}$$

This is a regular analytic function in D since (1) maps D univalently onto D^* where $w_\nu^*(z^*)$ is analytic. Using (3), we can rewrite (2) into the simpler form

$$m_\nu(z)=\Big(1-\frac{\alpha}{(z-z_0)^2}\Big)\cdot\frac{\lambda_\nu^*}{2\pi i}\int_C\Big[1-\frac{\alpha}{(z-z_0)(\zeta-z_0)}\Big]^{-1}\frac{\overline{(m_\nu(\zeta)d\zeta)}}{\zeta-z}\,. \tag{4}$$

We have thus referred all variables back to our original domain D, but λ_ν^* and $m_\nu(z)$ appear now as the eigen values and eigen functions of an integral equation with slightly changed kernel.

We may transform the new integral equation (4) by easy calculations into

$$m_\nu(z)=\frac{\lambda_\nu^*}{\pi}\iint_D\frac{\overline{m_\nu(\zeta)}}{(\zeta-z)^2}d\tau-\alpha\frac{\lambda_\nu^*}{\pi}\iint_D\frac{\overline{m_\nu(\zeta)d\tau}}{[(z-z_0)(\zeta-z_0)-\alpha]^2}\,. \tag{5}$$

Observe that by the definition (3) we have

$$\iint_D|m_\nu(z)|^2d\tau=\iint_{D^*}|w_\nu^*(z^*)|^2d\tau^*=1\,. \tag{6}$$

We have thus to determine the normalized eigen functions $m_\nu(z)$ to the integral equation (5) which differs from our original equation (1.13) by an α-term which can be estimated uniformly in z for $z_0\in\tilde{D}$ fixed.

Let us define the analytic function [3]

$$L(z,\zeta)=-\frac{2}{\pi}\frac{\partial^2}{\partial z\,\partial\zeta}G(z,\zeta)=\frac{1}{\pi(z-\zeta)^2}-l(z,\zeta)\,. \tag{7}$$

It is well-known that for every function $f(z)$ which is analytic in D and for which $\iint_D|f|^2d\tau<\infty$ holds

$$(8) \qquad \iint_D L(z, \zeta)\overline{f(\zeta)}d\tau = 0 .$$

Hence we have the identity, valid for each such $f(z)$,

$$(9) \qquad \frac{1}{\pi}\iint_D \frac{\overline{f(\zeta)}}{(\zeta - z)^2}d\tau = \iint_D l(z, \zeta)\overline{f(\zeta)}d\tau .$$

Under our assumptions about the boundary curve C of D, it can be shown that $l(z, \zeta)$ is continuous in both variables in the closed region $D+C$. Thus (5) can be put into the form:

$$(10) \qquad m_\nu(z) = \lambda_\nu^* \iint_D l(z, \zeta)\overline{m_\nu(\zeta)}d\tau - \alpha\frac{\lambda_\nu^*}{\pi}\iint_D \frac{\overline{m_\nu(\zeta)}}{[(z-z_0)(\zeta - z_0) - \alpha]^2}d\tau ,$$

while $w_\nu(z)$ satisfies the unperturbed integral equation

$$(11) \qquad w_\nu(z) = \lambda_\nu \iint_D l(z, \zeta)\overline{w_\nu(\zeta)}d\tau .$$

Now we can apply the general perturbation theory for regular kernels [9] and state that the eigen functions $m_\nu(z)$ and the eigen values λ_ν^* are analytic functions of the perturbation parameters α and $\bar{\alpha}$ and can be developed in power series in them. For $\alpha=0$, λ_ν^* will coincide with λ_ν while $m_\nu(z)$ will then lie in the linear space spanned by the eigen functions of (11) which belong to the unperturbed eigen value λ_ν.

Let $w_\nu^{(j)}(z)$ $(j=1, \cdots, n)$ denote the eigen functions belonging to λ_ν. We have the developments

$$(12) \qquad \lambda_\nu^* = \lambda_\nu + |\alpha|\kappa_\nu + O(|\alpha|^2)$$

and

$$(13) \qquad m_\nu(z) = \sum_{j=1}^n A_j w_\nu^{(j)}(z) + |\alpha|\omega_\nu(z) + O(|\alpha|^2) .$$

Inserting (12) and (13) into (10) and making use of (11), we find

$$(14) \qquad \sum_{j=1}^n A_j w_\nu^{(j)}(z) = \frac{\lambda_\nu^*}{\lambda_\nu}\sum_{j=1}^n \bar{A}_j w_\nu^{(j)}(z) + |\alpha|\lambda_\nu \iint_D l(z, \zeta)\overline{\omega_\nu(\zeta)}d\tau$$

$$- |\alpha|\omega_\nu(z) - \frac{\alpha\lambda_\nu}{(z-z_0)^2\pi}\sum_{j=1}^n \bar{A}_j \iint_D \frac{\overline{w_\nu^{(j)}(\zeta)}}{(\zeta - z_0)^2}d\tau + O(|\alpha|^2) .$$

We multiply this identity with $w_\nu^{(k)}(z)$ and integrate over D. We use the orthonormality of the $w_\nu^{(k)}(z)$, the symmetry of $l(z, \zeta)$ and the integral equation (11). We also make use of the fact that by (1.36)

$$\frac{1}{\pi}\iint_D \frac{\overline{w_\nu(\zeta)}}{(\zeta-z_0)^2}\,d\tau=\frac{\sqrt{\lambda_\nu^2-1}}{i\lambda_\nu}\tilde{w}_\nu(z_0). \tag{15}$$

Hence we arrive at

$$A_k=\frac{\lambda_\nu^*}{\lambda_\nu}\overline{A}_k+2i|\alpha|\Im\left\{\iint_D w_\nu^{(k)}(z)\overline{\omega_\nu(z)}\,d\tau\right\} \tag{16}$$

$$+\alpha\frac{\pi(\lambda_\nu^2-1)}{\lambda_\nu}\sum_{j=1}^{n}\overline{A}_k\tilde{w}_\nu^{(j)}(z_0)\tilde{w}_\nu^{(k)}(z_0)+O(|\alpha|^2)\ ,\quad k=1,2,\cdots,n\,.$$

Using the development (12) and comparing equal powers of $|\alpha|$ on both sides, we obtain

$$\Im\{A_k\}=0\ ,\qquad A_k=\text{real},\ k=1,2,\cdots,n\,. \tag{17}$$

Taking real parts in (16) and putting

$$\operatorname{sgn}\alpha=\frac{\alpha}{|\alpha|}=e^{i\beta}\,, \tag{18}$$

we find

$$\kappa_\nu A_k+\pi(\lambda_\nu^2-1)\sum_{j=1}^{n}A_j\Re\{e^{i\beta}\tilde{w}_\nu^{(j)}(z_0)\tilde{w}_\nu^{(k)}(z_0)\}=0\ ,\quad k=1,2,\cdots,n\,. \tag{19}$$

Thus the possible values of κ_ν in the development (12) of the perturbed eigen value λ_ν^* are the eigen values of the secular equation

$$\det\|\kappa_\nu\delta_{jk}+\pi(\lambda_\nu^2-1)\Re\{e^{i\beta}\tilde{w}_\nu^{(j)}(z_0)\tilde{w}_\nu^{(k)}(z_0)\}\|=0\,. \tag{20}$$

In particular, if λ_ν is a simple (nondegenerate) eigen value, we have the simple variational formula

$$\delta\lambda_\nu=|\alpha|\cdot\kappa_\nu=-\pi(\lambda_\nu^2-1)\Re\{\alpha\tilde{w}_\nu(z_0)^2\}\,. \tag{21}$$

Let us suppose next that we perform a variation (1) of the curve C but now with $z_0\in D$. Since the mapping (1) is regular and univalent in $\tilde{D}$, we can repeat the entire argument by interchanging the roles of D and $\tilde{D}$. We thus find

$$\det\|\kappa_\nu\delta_{jk}+\pi(\lambda_\nu^2-1)\Re\{e^{i\beta}w_\nu^{(j)}(z_0)w_\nu^{(k)}(z_0)\}\|=0 \tag{22}$$

as the secular equation for the κ_ν-terms and

$$\delta\lambda_\nu=|\alpha|\kappa_\nu=-\pi(\lambda_\nu^2-1)\Re\{\alpha w_\nu(z_0)^2\} \tag{23}$$

in the nondegenerate case. Formulas (21) and (23) exhibit the complete symmetry of our theory with respect to D and $\tilde{D}$.

We used the method of interior variations (1) in order to reduce the variational problem for the λ_ν explicitly to the theory of perturbation in classical integral equation theory. The formulas obtained are also very convenient in various extremum problems regarding the λ_ν as we shall show later. It seems, however, desirable to give also a variational formula for deformations of C which are described by the normal shift δn of each point on C. For this purpose we put

$$\begin{aligned}(24)\qquad &\Re\{\pi(\lambda_\nu^2-1)\alpha w_\nu^{(j)}(z_0)w_\nu^{(k)}(z_0)\}\\ &=\Re\left\{\frac{\alpha}{2i}(\lambda_\nu^2-1)\oint_C\frac{w_\nu^{(j)}(\zeta)w_\nu^{(k)}(\zeta)}{\zeta-z_0}d\zeta\right\}, \qquad z_0\in D.\end{aligned}$$

Applying Cauchy's integral theorem with respect to $\tilde{D}$, we also find

$$(25)\qquad 0=\Re\left\{\frac{\alpha}{2i}(\lambda_\nu^2-1)\oint_C\frac{\tilde{w}_\nu^{(j)}(\zeta)\tilde{w}_\nu^{(k)}(\zeta)}{\zeta-z_0}d\zeta\right\}.$$

Finally, we derive from (1.34) that

$$\begin{aligned}(26)\qquad &(\lambda_\nu^2-1)\{w_\nu^{(j)}(\zeta)w_\nu^{(k)}(\zeta)-\tilde{w}_\nu^{(j)}(\zeta)\tilde{w}_\nu^{(k)}(\zeta)\}\zeta'^2\\ &=2\Re\{\lambda_\nu\tilde{w}_\nu^{(j)}(\zeta)\overline{\tilde{w}_\nu^{(k)}(\zeta)}-\lambda_\nu^2\tilde{w}_\nu^{(j)}(\zeta)\tilde{w}_\nu^{(k)}(\zeta)\zeta'^2\}.\end{aligned}$$

Hence, if we subtract (25) from (24), we obtain

$$\begin{aligned}(27)\qquad &\Re\{\pi(\lambda_\nu^2-1)\alpha w_\nu^{(j)}(z_0)w_\nu^{(k)}(z_0)\}\\ &=\oint_C\Re\{\lambda_\nu\tilde{w}_\nu^{(j)}(\zeta)\overline{\tilde{w}_\nu^{(k)}(\zeta)}-\lambda_\nu^2\tilde{w}^{(j)}(\zeta)\tilde{w}_\nu^{(k)}(\zeta)\zeta'^2\}\,\delta n\, ds\end{aligned}$$

where

$$(28)\qquad \delta n=\Re\left\{\frac{1}{i\zeta'}\frac{\alpha}{\zeta-z_0}\right\}$$

represents the normal shift of C under the deformation (1). Thus the coefficients of the secular equation for $\delta\lambda$ have been expressed in terms of δn.

In particular, we have in the nondegenerate case in view of (23)

$$(29)\qquad \delta\lambda_\nu=\int_C[\lambda_\nu^2\Re\{\tilde{w}_\nu(\zeta)^2\zeta'^2\}-\lambda_\nu|\tilde{w}_\nu(\zeta)|^2]\delta n\, ds.$$

It can easily be verified from (1.34) that on C

$$\begin{aligned}(30)\qquad &\lambda_\nu\Re\{\tilde{w}_\nu^{(j)}\overline{\tilde{w}_\nu^{(k)}}\}-\lambda_\nu^2\Re\{\tilde{w}_\nu^{(j)}\tilde{w}_\nu^{(k)}\zeta'^2\}\\ &=-\lambda_\nu\Re\{w_\nu^{(j)}\overline{w_\nu^{(k)}}\}+\lambda_\nu^2\Re\{w_\nu^{(j)}w_\nu^{(k)}\zeta'^2\}.\end{aligned}$$

Thus we may replace $\tilde{w}$ by w in formulas (27) and (29); since transition from D to $\tilde{D}$ implies also a change of sign of the interior normal, the end result is unchanged. Thus the variational formula of the Hadamard type (29) is entirely symmetric with respect to the two complementary domains considered. If we had chosen $z_0 \in \tilde{D}$, we would have obtained the same end result (29).

We derived (29) in the case of a particular variation of the type (1). But since a variational formula depends linearly and additively on the variation, and since we can approximate general δn-variations by superposition of special variations of the type (1), we can extend (29) to the most general case of a δn-variation.

The value of the variational formula (29) is of heuristic nature; it shows the dependence of λ_ν on the geometry of C. For a precise study of extremum problems it is preferable to apply the variational formulas based on interior variations of the type (1).

We can derive, however, interesting monotonicity results by means of (29). Let, for example, $z=f(u)$ give the conformal mapping of the unit circle $|u|<1$ onto the domain D. Let C_r be the image under this map of the circumference $|u|=r<1$; and let $\lambda(r)$, $w(z, r)$ denote, say, the νth eigen value and eigen function of C_r. We assume, for the sake of simplicity, that $\lambda(r)$ is nondegenerate and then easily derive from (29):

$$\frac{d}{dr}\lambda(r) = -\lambda(r)\oint_{|u|=r} |w(z, r)|^2 |f'(u)|^2 \, ds_u + \lambda(r)^2 \Re\left\{\frac{i}{r}\oint_{|u|=r} w(z, r)^2 f'(u)^2 u \, du\right\}. \tag{29'}$$

The function

$$F_r(u) = w[f(u), r]f'(u)$$

is regular analytic for $|u|\leq r$; hence the second integral in (29′) vanishes by Cauchy's integral theorem and we obtain:

$$\frac{d}{dr}\log\lambda(r) = -\int_{|u|=r} |F_r(u)|^2 \, ds_u < 0. \tag{29''}$$

The eigen values $\lambda(r)$ of the level curves C_r are monotonically decreasing if r increases.

For every function $F(u)$ which is regular analytic for $|u|\leq r$ holds the obvious inequality

$$\int_{|u|=r} |F(u)|^2 \, ds_u \geq \frac{2}{r}\iint_{|u|<r} |F(u)|^2 \, d\tau. \tag{29'''}$$

Observe now that because of the normalization of $w(z, r)$ inside of C_r the function $F_r(u)$ is normalized in the circle $|u|<r$. Hence, combining (29″) with (29‴), we finally obtain

$$\frac{d}{dr}(\lambda r^2) \leqq 0 . \tag{29^{iv}}$$

Since we have the trivial estimate $\lambda(1) \geqq 1$ for every curve C, we then derive from (29^{iv}) the useful estimate

$$\lambda(r) \geqq \frac{1}{r^2} \qquad \text{for } r \leqq 1 . \tag{29^{v}}$$

In order to apply the usual perturbation method of integral equation theory we had to replace the integral equation (1.13) with singular kernel by the integral equation (11) which has the regular symmetric kernel $l(z, \zeta)$. The necessity for this transition becomes clear when we consider the exceptional case that C is a circumference. In this case (and only then), we have $l(z, \zeta)=0$. The original integral equation (1.13) has only the eigen value $\lambda=\infty$ and each function $f(z)$ which is analytic in D is an eigen function.

In fact, suppose for the sake of simplicity that C is the unit circumference $z\cdot\bar{z}=1$. We have

$$\frac{1}{\pi}\iint_{|\zeta|<1}\frac{\overline{f(\zeta)}}{(\zeta-z)^2}d\tau=\frac{1}{2\pi i}\oint_{|\zeta|=1}\frac{\overline{f(\zeta)d\zeta}}{\zeta-z}=\overline{\frac{1}{2\pi i}\oint_{|\zeta|=1}\frac{\zeta f(\zeta)d\zeta}{\zeta\bar{z}-1}} . \tag{31}$$

By means of the residue theorem we conclude therefore

$$\frac{1}{\pi}\iint_{|\zeta|<1}\frac{\overline{f(\zeta)}}{(\zeta-z)^2}d\tau=\begin{cases} 0 & \text{if } |z|<1 \\ \dfrac{1}{z^2}\bar{f}\left(\dfrac{1}{z}\right) & \text{if } |z|>1 . \end{cases} \tag{32}$$

This equation proves our statement that $\lambda=\infty$ is the only eigen value of (1.13) in this case and that it is of infinite degeneracy.

Our variational theory does not work in this exceptional case. However, let $|z_0|>1$ and C^* be the image of $|z|=1$ under the variation (1). We define its eigen function $w_\nu^*(z)$ and by (3) a function $m_\nu(z)$ which is regular analytic in D. It satisfies the integral equation (5) which, in view of (32), can be brought into the simple form

$$m_\nu(z)=-\frac{\alpha\lambda_\nu^*}{(z-z_0)^2}\frac{1}{\eta^2}\overline{m_\nu}\left(\frac{1}{\eta}\right), \qquad \eta=z_0+\frac{\alpha}{z-z_0} . \tag{33}$$

Let

$$m_\nu(z)=\frac{d}{dz}M_\nu(z)\ ; \tag{34}$$

if we choose the right constant of integration in the definition of $M_\nu(z)$, we can integrate (33) to the identity

$$M_\nu(z)=-\lambda_\nu^*\overline{M}_\nu(L(z))\ , \tag{35}$$

where

$$L(z)=\eta^{-1}=\frac{z-z_0}{z_0(z-z_0)+\alpha} \tag{36}$$

is a linear function of z. Thus we obtain a simple functional equation for the eigen functions $w_\nu^*(z)$ and the eigen values λ_ν^* of the varied curve C^*. α must be sufficiently small in order that the mapping (1) be univalent in D; but we have not made any neglection of higher powers of α and (35) will give the precise value of λ_ν^*.

If we iterate (35), we obtain

$$M_\nu(z)=\lambda_\nu^{*2}M_\nu(\Lambda(z))\ , \qquad \Lambda=\overline{L}(L(z))\ . \tag{37}$$

If z_1, z_2 are the fixed points of the linear transformation $Z=\Lambda(z)$, we can write

$$\frac{Z-z_1}{Z-z_2}=\tau^2\frac{z-z_1}{z-z_2} \tag{38}$$

where $|z_1|<1$, $|z_2|>1$. The eigen functions $M_\nu(z)$ are of the form

$$M_\nu(z)=A_\nu\left(\frac{z-z_1}{z-z_2}\right)^\nu, \qquad \nu=1, 2\cdots \tag{39}$$

and belong to the eigen values

$$\lambda_\nu^*=\pm\tau^{-\nu}\ . \tag{40}$$

Thus all eigen functions and eigen values of the curve C^* can be calculated explicitly. An easy computation shows that for small values of ε

$$\tau^{-1}=\frac{(|z_0|^2-1)^2}{|\alpha|}+O(1)\ . \tag{41}$$

An analogous calculation can be performed if the unit circle is transformed by a variation (1) with $|z_0|<1$. If we consider a superposition of variations (1), we can still derive an asymptotic formula for the eigen values λ_ν^* obtained. Thus we have shown that the eigen values for nearly circular domains can be obtained asymptotically in spite of the fact that the circle has an infinitely degenerate eigen value.

We showed at the end of § 1 that the eigen values of an ellipse can be calculated explicitly. This result is a particular case of our preceding investigation since the exterior of the ellipse is obtained from the exterior of the unit circle by a transformation (1) with $z_0=0$ and $|\alpha|<1$.

There are relatively few domains for which the eigen values and eigen functions of the Fredholm integral equation are known. It is, therefore, important to possess at least an asymptotic formula for the eigen values of nearly circular domains which admits many arbitrary parameters. Such formulas are particularly useful when one wishes to test hypotheses with respect to the eigen values of general domains.

4. The variation of the dielectric Green's function. In this section we want to derive the formula for the variation of the dielectric Green's function $G_\varepsilon(z, \zeta)$ defined in § 2. It will appear that it possesses a very simple variational formula which is quite similar to that for the ordinary Green's function of a plane domain. We shall again consider the interior variation

$$z^* = z + \frac{\alpha}{z - z_0} \tag{1}$$

which transforms the curve C into a curve C^* defining the two complementary domains D^* and $\tilde{D}^*$. Let $G_\varepsilon^*(z, \zeta)$ be the corresponding dielectric Green's function to the parameter ε.

If $z_0 \in \tilde{D}$, the mapping (1) will be univalent and regular in D for small enough α; hence the function

$$\Gamma_\varepsilon(z, \zeta) = G_\varepsilon^*\left(z + \frac{\alpha}{z - z_0},\ \zeta + \frac{\alpha}{\zeta - z_0}\right) \tag{2}$$

will be harmonic in D. It will also be harmonic in $\tilde{D}$, except for the interior of a circle of radius $|\alpha|^{1/2}$ around the point z_0. The function $\Gamma_\varepsilon(z, \zeta)$ will have logarithmic poles at infinity and for $z=\zeta$ as follows from the definition of $G_\varepsilon(z, \zeta)$.

We consider now Green's identity:

$$\frac{1}{2\pi}\int_C \left[\Gamma_\varepsilon(t, z)\frac{\partial G_\varepsilon(t, \zeta)}{\partial n} - G_\varepsilon(t, \zeta)\frac{\partial \Gamma_\varepsilon(t, z)}{\partial n}\right] ds \tag{3}$$
$$= \varepsilon\Gamma_\varepsilon(\zeta, z)\delta(\zeta) - \varepsilon G_\varepsilon(z, \zeta)\delta(z)\,.$$

Observe that in view of the conformality of (1) on C the function $\Gamma_\varepsilon(z, \zeta)$ has the same continuity (and discontinuity) property on C as the original function $G_\varepsilon(z, \zeta)$. Hence we may transform (3) into

$$\varepsilon[\Gamma_\varepsilon(\zeta, z)\delta(\zeta)-G_\varepsilon(z, \zeta)\delta(z)]=-\frac{\varepsilon}{2\pi}\int_c\Big[\Gamma_\varepsilon(t, z)\frac{\partial G_\varepsilon(t, \zeta)}{\partial \tilde{n}} - G_\varepsilon(t, \zeta)\frac{\partial \Gamma_\varepsilon(t, z)}{\partial \tilde{n}}\Big]ds. \tag{4}$$

Now we can apply Green's identity with respect to the domain $\tilde{D}$ after removing from it the interior of the circle $|z-z_0|=|\alpha|^{1/2}$ which we denote by c. Let us assume that neither z nor ζ lie inside c; then (4) yields

$$\Gamma_\varepsilon(z, \zeta)-G_\varepsilon(z, \zeta)=-\frac{1}{2\pi}\int_c\Big[\Gamma_\varepsilon(t, z)\frac{\partial G_\varepsilon(t, \zeta)}{\partial n}-G_\varepsilon(t, \zeta)\frac{\partial \Gamma_\varepsilon(t, z)}{\partial n}\Big]ds. \tag{5}$$

We have now fully utilized the boundary behavior of $G_\varepsilon(z, \zeta)$. The evaluation of the c-integral follows exactly the lines of the calculation for the ordinary Green's function. We put for $t \in c$

$$t=z_0+|\alpha|^{1/2}e^{i\phi} \tag{6}$$

and evaluate the right-hand integral in (5) by power series development. We define again two analytic functions of z, namely $p_\varepsilon(z, \zeta)$ and $p_\varepsilon^*(z, \zeta)$, by

$$\Re\{P_\varepsilon(z, \zeta)\}=G_\varepsilon(z, \zeta), \quad \Re\{P_\varepsilon^*(z, \zeta)\}=G_\varepsilon^*(z, \zeta). \tag{7}$$

Further, let

$$P_\varepsilon'(z, \zeta)=\frac{d}{dz}P_\varepsilon(z, \zeta), \quad P_\varepsilon^{*\prime}(z, \zeta)=\frac{d}{dz}P_\varepsilon^*(z, \zeta). \tag{7'}$$

Then the usual calculations yield

$$G_\varepsilon^*(z^*, \zeta^*)-G_\varepsilon(z, \zeta)=\Re\{\alpha P_\varepsilon^{*\prime}(z_0, z^*)P_\varepsilon'(z_0, \zeta)\}+O(|\alpha|^2). \tag{8}$$

Further series development leads to the simple result

$$G_\varepsilon^*(z, \zeta)=G_\varepsilon(z, \zeta)+\Re\Big\{\alpha\Big[P_\varepsilon'(z_0, z)P_\varepsilon'(z_0, \zeta) - \frac{P_\varepsilon'(z, \zeta)}{z-z_0}-\frac{P_\varepsilon'(\zeta, z)}{\zeta-z_0}\Big]\Big\}+O(|\alpha|^2). \tag{9}$$

This is exactly the same variational formula as for the ordinary Green's function [12, 13]. It has been derived for $z_0 \in \tilde{D}$.

If we had chosen $z_0 \in D$ instead of $\tilde{D}$ analogous calculations would have been applicable. We could start with

(10)
$$\frac{1}{2\pi}\int_c \left[\Gamma_\varepsilon(t,z)\frac{\partial G_\varepsilon(t,\zeta)}{\partial \tilde{n}} - G_\varepsilon(t,\zeta)\frac{\partial \Gamma_\varepsilon(t,z)}{\partial \tilde{n}}\right]ds$$
$$=\Gamma_\varepsilon(\zeta,z)\tilde{\delta}(\zeta)-G_\varepsilon(z,\zeta)\tilde{\delta}(z)\ .$$

Using the discontinuity of $\frac{\partial G_\varepsilon}{\partial n}$ on C, we find

(11)
$$\varepsilon[\Gamma_\varepsilon(\zeta,z)\tilde{\delta}(\zeta)-G_\varepsilon(z,\zeta)\tilde{\delta}(z)] = -\frac{1}{2\pi}\int_c \left[\Gamma_\varepsilon(t,z)\frac{\partial G_\varepsilon(t,\zeta)}{\partial n} - G_\varepsilon(t,\zeta)\frac{\partial \Gamma_\varepsilon(t,z)}{\partial n}\right]ds$$

and by means of Green's identity

(12)
$$\varepsilon[\Gamma_\varepsilon(z,\zeta)-G_\varepsilon(z,\zeta)] = -\frac{1}{2\pi}\int_c \left(\Gamma_\varepsilon\frac{\partial G_\varepsilon}{\partial n} - G_\varepsilon\frac{\partial \Gamma_\varepsilon}{\partial n}\right)ds$$

where c denotes again the circle $|z-z_0|=|\alpha|^{1/2}$. In this case the same procedure as before yields the result for $z_0 \in D$:

(13)
$$G_\varepsilon^*(z,\zeta)-G_\varepsilon(z,\zeta)=\Re\left\{\alpha\left[\frac{1}{\varepsilon}P_\varepsilon'(z_0,z)P_\varepsilon'(z_0,\zeta) - \frac{P_\varepsilon'(z,\zeta)}{z-z_0}-\frac{P_\varepsilon'(\zeta,z)}{\zeta-z_0}\right]\right\}+O(|\alpha|^2)\ .$$

Observe the factor $\frac{1}{\varepsilon}$ which is now introduced into (13) and causes a slight change in the variational formula.

We have thus derived a very elegant variational formula for the dielectric Green's function; its significance is seen from the numerous applications of its analogue in the case of the ordinary Green's function [**12, 13, 14**].

As mentioned in § 2, the function $P_\varepsilon(z,\zeta)$ consists in reality of two analytic functions, say, $P_\varepsilon(z,\zeta)$ if $z\in D$ and $\tilde{P}_\varepsilon(z,\zeta)$ if $z\in\tilde{D}$. The boundary behavior of $G_\varepsilon(z,\zeta)$ as described in § 2 implies for $z\in C$

(14)
$$\Re\{P_\varepsilon'(z,\zeta)z'\}=\Re\{\tilde{P}_\varepsilon'(z,\zeta)z'\},\quad \Im\{P_\varepsilon'(z,\zeta)z'\}=\varepsilon\Im\{\tilde{P}_\varepsilon'(z,\zeta)z'\}\ .$$

We can combine the variational formulas (9) and (13) into the integral form:

(15)
$$G_\varepsilon^*(z,\zeta)-G_\varepsilon(z,\zeta)$$
$$=\Re\left\{\frac{\alpha}{2\pi i}\oint_C \frac{\frac{1}{\varepsilon}P_\varepsilon'(t,z)P_\varepsilon'(t,\zeta)-\tilde{P}_\varepsilon(t,z)\tilde{P}_\varepsilon'(t,\zeta)}{t-z_0}dt\right\}+O(|\alpha|^2)\ .$$

By use of (14) this can be simplified to

$$(16)\qquad \delta G_\varepsilon(z,\zeta) = \frac{1}{2\pi}\left(\frac{1}{\varepsilon}-1\right)\int_C\left[\frac{\partial G_\varepsilon(t,z)}{\partial s}\frac{\partial G_\varepsilon(t,\zeta)}{\partial s}-\frac{\partial G_\varepsilon(t,z)}{\partial n}\frac{\partial G_\varepsilon(t,\zeta)}{\partial \tilde{n}}\right]\delta n\,ds$$

with

$$(17)\qquad \delta n = \Re\left\{\frac{1}{it'}\frac{\alpha}{(t-z_0)}\right\}.$$

This is the Hadamard type variational formula for the dielectric Green's function which has been proved in a precise manner through use of our interior variational method.

Since we can also write (16) in the form

$$(18)\qquad \delta G_\varepsilon(z,\zeta) = \frac{1}{2\pi}\left(\frac{1}{\varepsilon}-1\right)\int_C\left[\frac{\partial G_\varepsilon(t,z)}{\partial s}\frac{\partial G_\varepsilon(t,\zeta)}{\partial s}+\frac{1}{\varepsilon}\frac{\partial G_\varepsilon(t,z)}{\partial n}\frac{\partial G_\varepsilon(t,\zeta)}{\partial n}\right]\delta n\,ds$$

it is evident that if $\zeta\in D$ the expression $(G_\varepsilon(z,\zeta)+\varepsilon\log|z-\zeta|)_{z=\zeta}$ depends monotonically upon the domain D while for $\zeta\in\tilde{D}$ the same is true for $(G_\varepsilon(z,\zeta)+\log|z-\zeta|)_{z=\zeta}$. In a similar way many other expressions can be constructed which have a definite factor of $\delta n\,ds$ under the integral sign and which depend, therefore, monotonically upon D. The application of Hadamard's formula in order to obtain inequalities and comparison theorems for functionals connected with $G_\varepsilon(z,\zeta)$ is obvious.

For $\varepsilon=1$, we have $G_\varepsilon(z,\zeta)=-\log|z-\zeta|$ independently of the domain. For this reason the factor $\left(\frac{1}{\varepsilon}-1\right)$ must occur in the variational formulas (16) and (18).

We showed at the end of § 2 that the mapping of a domain onto a circle can be connected with the identical mapping by a one-parameter family of univalent functions which are closely related to the dielectric Green's functions. For this reason it is of interest to compute the derivative of $G_\varepsilon(z,\zeta)$ with respect to ε.

We start with Green's identity and with $\varepsilon>0$, $e>0$:

$$(19)\qquad e\delta(\zeta)G_\varepsilon(\zeta,\eta)-\varepsilon\delta(\eta)G_e(\zeta,\eta) = \frac{1}{2\pi}\int_C\left[G_\varepsilon(z,\eta)\frac{\partial G_e(z,\zeta)}{\partial n}-G_e(z,\zeta)\frac{\partial G_\varepsilon(z,\eta)}{\partial n}\right]ds.$$

Using the boundary relations of G_ε and G_e on C and Green's identity with respect to $\tilde{D}$, we find

$$(20)\qquad \frac{G_\varepsilon(\zeta,\ \eta)-G_e(\zeta,\ \eta)}{\varepsilon-e}=\frac{1}{\varepsilon}G_\varepsilon(\zeta,\ \eta)\delta(\zeta)+\frac{1}{2\pi\varepsilon}\int_C G_\varepsilon(z,\ \eta)\frac{\partial G_e(z,\ \zeta)}{\partial \tilde{n}}ds\,.$$

Passing to the limit $\varepsilon=e$, we then obtain

$$(21)\qquad \frac{\partial}{\partial\varepsilon}G_\varepsilon(\zeta,\ \eta)=\frac{1}{\varepsilon}G_\varepsilon(\zeta,\ \eta)\delta(\zeta)+\frac{1}{2\pi\varepsilon}\int_C G_\varepsilon(z,\ \eta)\frac{\partial G_\varepsilon(z,\ \zeta)}{\partial \tilde{n}}ds\,.$$

The symmetry of this expression is more clearly exhibited in the form

$$(22)\qquad \frac{\partial}{\partial\varepsilon}G_\varepsilon(\zeta,\ \eta)=\frac{1}{2\pi\varepsilon^2}\iint_D \nabla G_\varepsilon(z,\ \zeta)\cdot\nabla G_\varepsilon(z,\ \eta)d\tau\,.$$

This result could also have been obtained by straightforward calculation from (2.19) and its analogues.

It is obvious how numerous monotonicity results can be derived from expression (22) by considering combinations with positive derivative. This formula can also be used in order to develop G_ε in powers of ε. The formula is particularly useful in a more detailed discussion of the mapping functions $f_\varepsilon(z)$, defined in § 2; however, we do not enter into this subject in the present paper.

5. An extremum problem for the Fredholm eigen values. We shall now proceed to apply the variational formulas of § 3 to an important extremum problem for the lowest Fredholm eigen value of a given curve C. In order to explain the formulation of the problem considered we start with the following observation. Let C be a three times continuously differentiable curve as was supposed throughout; if λ_1 is its lowest eigen value we have shown that $\lambda_1>1$. Now let C^* be a continuum which consists of all points of C plus a segment which has one endpoint on C and the other in D; let λ_1^* be its lowest eigen value. It can be shown that $\lambda_1^*=1$ however small the additional segment of C^* is; thus two curves in an arbitrary Fréchet neighborhood can have very different lowest Fredholm eigen values.

The fact that λ_1 depends in this discontinuous way on its defining curve C makes it difficult to frame significant extremum problems for it. The side condition on C of three continuous derivatives is, on the one hand, somewhat unnatural and, on the other hand, hard to preserve under variation. We shall restrict ourselves, therefore, in this section to the consideration of analytic curves, but even in this case λ_1 can come as near as we wish to 1. In fact, formula (1.45) shows that we can find ellipses with λ_1 arbitrarily near 1. We have, therefore, to sharpen the concept of an analytic curve by introducing the concept of *uniform analyticity* of a curve. A curve C is called analytic if it is mapped by a regular univalent function $z=f(\zeta)$ from the unit circumference $|\zeta|=1$. $f(\zeta)$ must be regular and univalent in some circular ring $r<|\zeta|<R$ with $r<1<R$. The class of all curves C which are analytic and belong to functions $f(\zeta)$ which are regular and univalent in a fixed ring (r, R) shall be called the class of uniformly analytic curves with the modulus of analyticity (r, R).

Because of the normality of the family of univalent functions in a fixed region the concept of uniform analyticity lends itself easily to the construction of significant extremum problems. In particular, let us ask for the minimum value of λ_1 within the family of all uniformly analytic curves with modulus (r, R).

We may consider our problem as an extremum problem on univalent functions. Given the class of all functions $f(\zeta)$ which are regular and univalent in $r<|\zeta|<R$, to find one in the class which maps the unit circumference onto a curve C with minimum λ_1. The existence of such a function follows easily from the usual normality arguments and we proceed at once to characterize the extremum function by varying it and comparing it with nearby competing functions.

Since the curve C mapped by the extremum function is analytic and since its λ_1 is obviously finite, the lowest eigen value can have only a degeneracy of finite order. Let $w_1^{(1)}(z), \cdots, w_1^{(n)}(z)$ be a complete and linearly independent set of eigen functions belonging to λ_1 in D, while $\tilde{w}_1^{(1)}(z), \cdots, \tilde{w}_1{}^{(n)}(z)$ are the corresponding eigen functions in $\tilde{D}$. Suppose that the image of $|\zeta|=r$ forms a continuum Γ in D while the image of $|\zeta|=R$ forms the continuum $\tilde{\Gamma}$ in $\tilde{D}$. Let $z_0 \in \tilde{\Gamma}$; there exists an infinity of analytic functions which are univalent outside of the continuum $\tilde{\Gamma}$ and which have a series development [11]

$$z^*=z+\sum_{\nu=1}^{\infty}\frac{a_\nu\rho^{\nu+1}}{(z-z_0)^\nu} \tag{1}$$

which converges for $|z-z_0|>\rho$. The coefficients a_ν of this development are uniformly bounded

$$|a_\nu|\leqq 4^{\nu+1} \tag{2}$$

and ρ is a positive parameter which can be chosen arbitrarily small.

Let us insert the extremum function $z=f(\zeta)$ into (1); we will thus obtain an infinity of competing functions regular and univalent in $r<|\zeta|<R$ of the form

$$f^*(\zeta)=f(\zeta)+\frac{a_1\rho^2}{f(\zeta)-z_0}+o(\rho^2)\,. \tag{3}$$

They define curves C^*, the images of $|\zeta|=1$ by $f^*(\zeta)$. If λ_1^* denotes the lowest eigen value of C^*, it defines a root of the secular equation derived in (3.20):

$$\det \|\delta\lambda_1 \cdot \delta_{jk} + \pi(\lambda_1^2 - 1)\Re\{a_1\rho^2\tilde{w}_1^{(j)}(z_0)\tilde{w}_1^{(k)}(z_0)\}\| = 0 \tag{4}$$

with $\delta\lambda_1 = \lambda_1^* - \lambda_1 + o(\rho^2)$. $\delta\lambda_1$ is the lowest root of (4); on the other hand, we conclude from the minimum property of C that

$$\delta\lambda_1 \geqq o(\rho^2) \tag{5}$$

and this holds, *a fortiori*, for all other roots of (4). Hence we can assert that the quadratic form

$$Q_\rho(t) = \sum_{j,k=1}^{n} \Re\{a_1\rho^2\tilde{w}_1^{(j)}(z_0)\tilde{w}_1^{(k)}(z_0)\}t_j t_k \tag{6}$$

satisfies the inequality

$$Q_\rho(t) \leqq o(\rho^2) \tag{7}$$

for every choice of the unit vector $t_1, \cdots, t_n$. Dividing by ρ^2 and passing to the limit $\rho = 0$, we obtain

$$\Re\left\{a_1 \sum_{j,k=1}^{n} \tilde{w}_1^{(j)}(z_0)\tilde{w}_1^{(k)}(z_0)t_j t_k\right\} \leqq 0 . \tag{8}$$

In particular, we obtain

$$\Re\{a_1\tilde{w}(z_0)^2\} \leqq 0 , \quad \tilde{w}(z_0) = \tilde{w}_1^{(1)}(z_0) . \tag{8'}$$

This inequality holds for every choice of the univalent variation function (1). We now apply the following theorem [11, 14]:

If for every point $z_0 \in \tilde{\Gamma}$ and every univalent function (1) *holds*

$$\Re\{a_1 s(z_0)\} \leqq 0 \tag{9}$$

where $s(z_0)$ is regular analytic on $\tilde{\Gamma}$, then $\tilde{\Gamma}$ itself is an analytic curve $z(t)$ which satisfies the differential equation

$$\left(\frac{dz}{dt}\right)^2 s[z(t)] = 1 . \tag{10}$$

Hence we can deduce from (8′) that $\tilde{\Gamma}$ satisfies the differential equation

$$\left(\frac{dz}{dt}\right)^2 \tilde{w}[z(t)]^2 = 1 . \tag{11}$$

In exactly the same way we prove that the extremum function $f(\zeta)$ maps the circumference $|\zeta| = r$ onto an analytic arc Γ which satisfies the differential equation

$$\left(\frac{dz}{dt}\right)^2 w[z(t)]^2 = 1 . \tag{12}$$

Let us put

$$z(\phi) = f(re^{i\phi}) ; \tag{13}$$

if ϕ runs from 0 to 2π the image point $z(\phi)$ will vary over Γ. We deduce from (12) the inequality

$$\zeta^2 f'(\zeta)^2 w[f(\zeta)]^2 < 0 \qquad \text{for } |\zeta| = r . \tag{14}$$

Similarly, we derive from (11) the inequality

$$\zeta^2 f'(\zeta)^2 \tilde{w}[f(\zeta)]^2 < 0 \qquad \text{for } |\zeta| = R . \tag{15}$$

We introduce the analytic functions

$$A(\zeta) = \zeta f'(\zeta) w[f(\zeta)] ; \quad B(\zeta) = \zeta f'(\zeta) \tilde{w}[f(\zeta)] . \tag{16}$$

Clearly, $A(\zeta)$ is regular analytic in the ring domain $r < |\zeta| < 1$ while $B(\zeta)$ is regular analytic for $1 < |\zeta| < R$. (14) and (15) can be expressed as

$$A(\zeta) = \text{imaginary for } |\zeta| = r \tag{14'}$$

$$B(\zeta) = \text{imaginary for } |\zeta| = R \tag{15'}$$

while equation (1.34) leads to

$$-iA(\zeta) = \frac{1}{\sqrt{\lambda_1^2 - 1}} [B(\zeta) + \lambda_1 \overline{B(\zeta)}] \qquad \text{for } |\zeta| = 1 . \tag{17}$$

We have by the Schwarz' reflection principle in view of (14') and (15'):

$$\overline{A(\zeta)} = -A\left(\frac{r^2}{\bar{\zeta}}\right) , \quad \overline{B(\zeta)} = -B\left(\frac{R^2}{\bar{\zeta}}\right) . \tag{18}$$

Now we can rewrite (17) into the form

$$-iA(\zeta) = (\lambda_1^2 - 1)^{-1/2} [B(\zeta) - \lambda_1 B(R^2 \zeta)] \qquad \text{for } |\zeta| = 1 , \tag{19}$$

since $\bar{\zeta} = \zeta^{-1}$ for $|\zeta| = 1$. By (18) we see that $A(\zeta)$ is analytic in the ring $r^2 < |\zeta| < 1$ while $B(\zeta)$ is analytic for $1 < |\zeta| < R^2$. From (19) we can continue $B(\zeta)$ into the ring $k < |\zeta| < 1$ where $k = \max(r^2, R^{-2})$. By (18) again $B(\zeta)$ is, therefore, analytic in the ring $k < |\zeta| < \frac{R^2}{k}$ and by (19) we may continue $A(\zeta)$ beyond the unit circumference. Thus $A(\zeta)$ and $B(\zeta)$ are certainly analytic for $|\zeta| = 1$. The interrelation between $A(\zeta)$ and $B(\zeta)$ is, however, best understood by the use of Laurent series development.

We put

$$A(\zeta)=i\sum_{n=-\infty}^{\infty} a_n\zeta^n,\quad B(\zeta)=i\sum_{n=-\infty}^{\infty} b_n\zeta^n \tag{20}$$

and are sure that both series have a ring of common convergence which contains the unit circumference. The functional equations (18) are reflected in the coefficient relations

$$a_{-n}=\bar{a}_n r^{2n},\ b_{-n}=\bar{b}_n R^{2n}. \tag{21}$$

On the other hand, a comparison of coefficients in (19) yields

$$-ia_n=(\lambda_1^2-1)^{-1/2}(1-\lambda_1 R^{2n})b_n. \tag{22}$$

If we replace n by $-n$ and apply (21), we also find

$$ia_n=(\lambda_1^2-1)^{-1/2}(R^{2n}-\lambda_1)r^{-2n}b_n. \tag{23}$$

But (22) and (23) lead obviously to the alternative

$$a_n=b_n=0 \quad\text{or}\quad \lambda_1=\frac{r^{2n}+R^{2n}}{1+(rR)^{2n}}. \tag{24}$$

Thus $A(\zeta)$ and $B(\zeta)$ are necessarily rational functions and the possible values of λ_1 are restricted to the various values in (24) for integer n. Observe that $n=0$ is excluded since λ_1 is surely greater than one. It is sufficient to consider only positive values of n since $-n$ yields the same λ_1-value as $+n$. We may put equation (24) into the form

$$\frac{\lambda_1-1}{\lambda_1+1}=\frac{R^{2n}-1}{R^{2n}+1}\,\frac{1-r^{2n}}{1+r^{2n}}. \tag{25}$$

This form makes it evident that the minimum value of λ_1 for fixed r and R is attained for $n=1$. Hence, for the lowest eigen value λ_1 which belongs to a uniformly analytic curve C with the modulus (r, R), we have established the inequality:

$$\lambda_1\geqq\frac{r^2+R^2}{1+(rR)^2}. \tag{26}$$

In order to conclude the investigation we have to show that there exists, in fact, a curve C within the class considered for which equality is attained in (26). This curve can be found by a careful analysis of the variational conditions (11) and (12). At first we shall state the nature of an extremum curve C and compute its λ_1-value from its definition. Later we shall show that C is uniquely determined up to linear transformations.

Let us consider the z-plane slit along the linear segment $-i\mu$, $+i\mu$ of the imaginary axis and along the segments $|x| > 1$ of the real axis. Every circular ring $r \leq |\zeta| \leq R$ can be mapped on such a canonical domain; the real parameter μ depends on the ratio $\dfrac{R}{r}$. For reasons of symmetry we can obtain that the points $\zeta = R$ and $\zeta = -R$ are mapped into $z = 1$ and $z = -1$, respectively, while the points $\zeta = ir$ and $\zeta = -ir$ go into $i\mu$ and $-i\mu$. The mapping function $f(\zeta)$ has the symmetry properties:

$$\text{(27)} \qquad \overline{f(\zeta)} = f(\bar{\zeta}) , \quad -\overline{f(\zeta)} = f(-\bar{\zeta}) ,$$

and is uniquely defined. Let C be the image of the unit circumference $|\zeta| = 1$ under the mapping $z = f(\zeta)$. We want to prove that C is the required extremum curve.

We denote again the interior and exterior of C by D and $\tilde{D}$, respectively. Observe that the functions

$$\text{(28)} \qquad W_n(z) = A_n\left(\zeta^n + \frac{(-r^2)^n}{\zeta^n}\right) , \qquad\qquad \zeta = f^{-1}(z)$$

are regular analytic in the entire domain D while the functions

$$\text{(29)} \qquad \widetilde{W}_n(z) = B_n\left(\zeta^n + \frac{R^{2n}}{\zeta^n}\right) , \qquad\qquad \zeta = f^{-1}(z)$$

are regular analytic in D. Let us define the eigen functions of D and $\tilde{D}$ by

$$\text{(30)} \qquad w_n(z) = \frac{d}{dz} W_n(z) , \quad \tilde{w}_n(z) = \frac{d}{dz} \widetilde{W}_n(z) .$$

Differentiating (28) and (29) with respect to ζ, we find

$$\text{(31)} \qquad w_n[f(\zeta)]f'(\zeta)\zeta = nA_n\left(\zeta^n - \frac{(-r^2)^n}{\zeta^n}\right)$$

and

$$\text{(32)} \qquad \tilde{w}_n[f(\zeta)]f'(\zeta)\zeta = nB_n\left(\zeta^n - \frac{R^{2n}}{\zeta^n}\right) .$$

The boundary conditions (1.34) for the eigen functions of D and $\tilde{D}$ will lead to the requirement

$$\text{(33)} \quad -iA_n\left(\zeta^n - \frac{(-r^2)^n}{\zeta^n}\right) = (\lambda_n^2 - 1)^{-1/2}\left[B_n\left(\zeta^n - \frac{R^{2n}}{\zeta^n}\right) + \lambda_n \bar{B}_n\left(\frac{1}{\zeta^n} - R^{2n}\zeta^n\right)\right]$$

for $|\zeta|=1$. This can indeed be fulfilled by satisfying the conditions

$$(34)\qquad -iA_n(\lambda_n^2-1)^{1/2}=B_n-\lambda_n R^{2n}\bar{B}_n$$

$$-iA_n(-r^2)^n(\lambda_n^2-1)^{1/2}=B_nR^{2n}-\lambda_n\bar{B}_n$$

which is always possible if and only if

$$(35)\qquad \lambda_n=\frac{R^{2n}-(-r^2)^n}{|1-(-r^2R^2)^n|}\ .$$

Conversely, it is evident that the values λ_n determined by (35) for $n=1, 2, \cdots$ lead to actual eigen functions for the domains D and $\tilde{D}$. Observe, in particular, that

$$(36)\qquad \lambda_1=\frac{R^2+r^2}{1+r^2R^2}$$

which verifies that C is indeed an extremum curve and that our estimate (26) is the best possible one.

There remains finally the uniqueness question relative to the extremum curve C. In order to answer it we return to the functions $A(\zeta)$ and $B(\zeta)$ connected with the extremum function $f(\zeta)$. Since we know now that in their Laurent development all coefficients vanish except for a_1, a_{-1} and b_1, b_{-1}, we have by (16), (21) and (22)

$$(36')\qquad \zeta f'(\zeta)w[f(\zeta)]=ia_1\left(\zeta+\frac{r^2}{\zeta}\right)$$

and

$$(37)\qquad \zeta f'(\zeta)\tilde{w}[f(\zeta)]=ib_1\left(\zeta-\frac{R^2}{\zeta}\right)$$

with

$$(38)\qquad ia_1(\lambda_1^2-1)^{1/2}=(\lambda_1R^2-1)b_1\ .$$

We made the unessential assumption that a_1 is real which leads to the consequence that b_1 is pure imaginary.

We integrate (36′) and (37) and find

$$(39)\qquad W[f(\zeta)]=ia_1\left(\zeta-\frac{r^2}{\zeta}\right),\quad \tilde{W}[f(\zeta)]=ib_1\left(\zeta+\frac{R^2}{\zeta}\right)$$

where $W(z)$ and $\tilde{W}(z)$ are properly chosen integrals of $w(z)$ and $\tilde{w}(z)$. The function $W(z)$ is single-valued in D; $f(\zeta)$ is regular analytic on $|\zeta|=r$ and can be continued somewhat beyond this circumference. It will take values near the continuum Γ after this continuation; but these values in the z-plane were already attained for some values ζ in $|\zeta|>r$. Hence $W[f(\zeta)]$ must take the same values for $|\zeta|$ somewhat larger than r and for some $|\zeta|$ less than r. From (39) we recognize that these corresponding ζ-values must be connected by the equation

$$\zeta_1-\frac{r^2}{\zeta_1}=\zeta_2-\frac{r^2}{\zeta_2}\,. \tag{40}$$

Hence we proved the functional equation for $f(\zeta)$:

$$f(\zeta)=f\left(-\frac{r^2}{\zeta}\right). \tag{41}$$

In exactly the same manner we derive from the second formula (39) the functional equation

$$f(\zeta)=f\left(\frac{R^2}{\zeta}\right). \tag{42}$$

We know already that the extremum function $f(\zeta)$ will remain an extremum function after a linear transformation since we showed at the end of § 1 that λ_1 does not change under linear transformations. Hence we may assume without loss of generality that

$$f(r)=0\,,\quad f(R)=1\,,\quad f(iR)=\infty\,. \tag{43}$$

From (41) and (42) conclude then that

$$f(-r)=0\,,\quad f(-iR)=\infty \tag{44}$$

and in view of the univalent character of $f(\zeta)$ in $r<|\zeta|<R$ we conclude that $f(\zeta)$ has simple zeros and simple poles at these points. It is now easy to obtain for $f(\zeta)$ a product representation in terms of its known zeros and poles in the entire ζ-plane and to identify it with the function which maps the ring $r<|\zeta|<R$ on the above described slit domain. This completes the uniqueness argument.

Let us return to the inequality (26). An important special case deals with all uniformly analytic curves with the modulus (r, ∞). This is the class of curves which are images of $|\zeta|=1$ mapped by functions which are regular and univalent for $|\zeta|>r$. We find the estimate

$$\lambda_1\geqq r^{-2} \tag{45}$$

and the extremum curve in this case is the ellipse C which is obtained from $|\zeta|=1$ by the mapping

$$z=\zeta+\frac{r^2}{\zeta}\,. \tag{46}$$

This follows directly from (1.45) as well as from our preceding characterization of the extremum domain. The inequality (45) can also be easily derived from the estimate (3.29^{v}); thus this particular result could have been proved by means of a Hadamard type variational formula.

As for the class of uniformly analytic functions with the modulus $(0, R)$, we have analogously the estimate

$$\lambda_1 \geqq R^2 . \tag{47}$$

The extremal curve C is obtained from the unit circumference by the mapping

$$z = \frac{2R\zeta}{R^2+\zeta^2} . \tag{48}$$

This mapping is best understood if we consider the intermediate step

$$\eta = R^{-2}\zeta + \frac{1}{\zeta} \tag{49}$$

which maps the unit circumference onto an ellipse with $\lambda_1 = R^2$ and the circumference $|\zeta| = R$ onto the linear segment $\left\langle -\frac{2}{R}, \frac{2}{R} \right\rangle$. The additional linear transformation $z = \frac{2}{R\eta}$ does not affect the eigen values and leads to a regular univalent function in $|\zeta| < R$. We could have obtained the mapping (48) also as a special case of the preceding characterization of the extremum curve C.

6. Concluding remarks. We have restricted ourselves in the present paper to the case of simply connected domains. It is possible to extend a considerable amount of the results to the case of multiply-connected domains [3, 10, 14]. The investigation becomes, however, more complicated for two reasons. First, we will have a larger number of complementary domains and, second, we will have additional eigen functions belonging to the eigen value one. In fact, let $C_1, C_2, \cdots, C_n$ denote the n components of the boundary C of the domain D; let $\omega_\nu(z)$ be that harmonic function in D which takes on C_μ the boundary value $\delta_{\nu\mu}$. Then it is easily seen that

$$w_\nu(z) = i\frac{\partial}{\partial z}\omega_\nu(z) \tag{1}$$

will satisfy the integral equation

$$w_\nu(z) = \frac{1}{\pi}\iint_D \frac{\overline{w_\nu(z)}}{(\zeta - z)^2} d\tau . \tag{2}$$

All other eigen functions of the integral equation (1.13) belong, however, to eigen values which are larger than one.

The concept of the dielectric Green's function carries over to the case of higher multiplicity and analogous series developments in terms of the eigen functions of the Fredholm integral equation are possible. Likewise, the different variational formulas can be extended to multiple connectivity. But, clearly, it will be much more difficult to draw simple conclusions from these formulas. One has only to consider the great use made in the preceding section of Laurent series developments in order to appreciate the great simplification introduced by the assumption of a simply connected domain.

REFERENCES

1. L. V. Ahlfors, *Remarks on the Neumann-Poincaré integral equation*, Pacific J. Math., **2** (1952), 271–280.

2. ———, *Conformality with respect to Riemann metrics*, Ann. Acad. Sci. Fenn., Series A 202 (1955).

3. S. Bergman and M. Schiffer, *Kernel functions and conformal mapping*, Compositio Math., **8** (1951), 205–249.

4. G. Birkhoff, D. M. Young and E. H. Zarantanello, *Effective conformal transformation of smooth simply-connected domains*, Proc. Nat. Acad. Sci., **37** (1951), 411–414.

5. ———, *Numerical methods in conformal mapping*, Proc. Symposia in Appl. Math., **IV** (1953), 117–140.

6. G. F. Carrier, *On a conformal mapping technique*, Quart. Appl. Math., **5** (1947), 101–104.

7. S. Gershgorin, *On conformal mapping of a simply-connected region onto a circle*, Math. Sb. **40** (1933), 48–58.

8. K. Löwner, *Untersuchungen über schlichte konforme Abbildungen des Einheitskreises*, Math. Ann., **89** (1923), 103–121.

9. F. Rellich, *Störungstheorie der Spektralzerlegung, I. Mitteilung*, Math. Ann., **113** (1937), 600–619.

10. H. L. Royden, *A modification of the Neumann-Poincaré method for multiply-connected regions*, Pacific J. Math., **2** (1952), 385–394.

11. M. Schiffer, *A method of variation within the family of simple functions*, Proc. London Math. Soc., **44** (1938), 432–449.

12. ———, *Variation of the Green's function and theory of the p-valued functions*, Amer. J. Math., **65** (1943), 341–360.

13. ———, *Hadamard's formula and variation of domain functions*, Amer. J. Math., **68** (1946), 417–448.

13. M. Schiffer and D. C. Spencer, *Functionals of finite Riemann surfaces*, Princeton (1954).

14. S. E. Warschawski, *On the effective determination of conformal maps*, Contribution to the Theory of Riemann Surfaces, Princeton (1953).

STANFORD UNIVERSITY

[68] Fredholm eigen values of multiply-connected domains

---

[68] Fredholm eigen values of multiply-connected domains. *Pacific J. Math.* **9** (1959), 211–269.

FREDHOLM EIGEN VALUES OF MULTIPLY-CONNECTED DOMAINS

M. SCHIFFER

Introduction. The solution of the boundary value problems of potential theory can be reduced, according to Poincaré, to an inhomogeneous integral equation of the second kind. It was the study of this particular problem which led, at the beginning of this century, to the development of the modern integral equation theory at the hands of Fredholm and Hilbert. From the beginning, attention was drawn to the eigen value problem for the homogeneous integral equation with the potential theoretical kernel [10]. The eigen functions of this problem can be extended as harmonic functions into the domain considered as well as extended into the complementary domain and give rise to interesting series developments and to a theory relating solutions of the interior and exterior boundary value problems of a closed curve or surface.

In a preceding paper [17], these Fredholm eigen functions were applied to problems of conformal mapping of simply-connected plane domains. Their connection with the dielectric Green's function of such domains was discussed and we showed the possibility of obtaining univalent functions by means of the dielectric Green's function. A variational formula for the Fredholm eigen values was established and an extremum problem for the latter was solved which permitted one to estimate the convergence of the Neumann-Liouville series solving the Dirichlet and Neumann boundary value problems.

In the present paper, the Fredholm eigen value problem is studied in the case of multiply-connected plane domains. Various new difficulties arise in this case. The complementary region of a multiply-connected domain is a domain set and the number of trivial solutions of the problem with the eigen values $|\lambda| = 1$ increases. This fact necessitates a brief restatement of the basic definitions and concepts in § 1. A certain repetition and overlap of material with the preceding paper could not be avoided; but, on the other hand, the presentation of this section makes the paper self-contained and should facilitate the understanding of it.

In § 2, the dielectric Green's functions $g_\varepsilon(z, \zeta)$ of a multiply-connected domain are discussed and their Fourier development in terms of the Fredholm eigen functions is given. The functions g_ε are of geometric-physical significance by themselves; moreover, they represent a one-parameter $(0 < \varepsilon < \infty)$ family of harmonic positive-definite kernels which

Received October 1, 1958. Prepared under contract Nonr-225 (11) for Office of Naval Research.

have also the Fredholm functions as eigen functions. For $\varepsilon = 1$, $g_\varepsilon(z, \zeta)$ reduces to the fundamental singularity $-\log|z - \zeta|$ and leads to the classical kernel of potential theory. A power series development of the dielectric Green's function in terms of $(\varepsilon - 1)/(\varepsilon + 1)$ is given; the coefficient kernels are elementary and can be calculated explicitly by integration of simple functions over the boundary curve system.

The role of the one-parameter family $g_\varepsilon(z, \zeta)$ becomes particularly interesting when one studies the limit cases $\varepsilon = 0$ and $\varepsilon = \infty$. This is done in § 3. It appears that this function family interpolates between two well-known harmonic functions which determine two important canonical mappings of the domain considered; namely the radial-slit mapping and the circular-slit mapping.

In § 4 it is proved that not only the limit cases $\varepsilon = 0$ and $\varepsilon = \infty$ of $g_\varepsilon(z, \zeta)$ give rise to univalent functions in the domain but that each dielectric Green's function does so. We obtain one-parameter families of univalent functions which connect the radial-slit mapping function continuously with the circular-slit mapping function via any prescribed univalent function in the domain. This result is applied to give a new proof for the extremum properties which characterize the above two canonical slit mappings. Another type of one-parameter sets of univalent functions is constructed which interpolates between the canonical parallel-slit mappings.

In § 5, we use the dielectric Green's functions in order to define various norms and scalar products. These are quadratic and bilinear functionals defined for harmonic functions in the multiply-connected domain $\tilde{D}$ as well as for functions harmonic in the complementary domain set D. If one pair of argument functions is defined in $\tilde{D}$, the other pair in D, and if relations between their boundary values on the separating curve system are assumed, equations between the various scalar products are obtained. It is shown that these identities yield estimates and Ritz procedures for solution of boundary value problems in $\tilde{D}$ if the corresponding boundary value problems for the complementary set D are already solved. In the special case $\varepsilon = 1$ the procedure becomes, of course, particularly easy to apply since the dielectric Green's function becomes trivial. It has, indeed, already been used in this form in order to prove interesting isoperimetric inequalities for polarization and for virtual mass [18–20]. The extension of the method to the case of general ε should increase its flexibility and clarify its significance. The various quadratic forms are used, finally, in order to characterize each Fredholm eigen value $|\lambda| > 1$ by the solution of a simple maximum problem without side conditions. This result lays the groundwork for proving the variational formula for the Fredholm eigen values in the next section. The extremum definition is also used in order to prove that all positive

Fredholm eigen values of a subsystem of curves are never less than the corresponding positive eigen values of the full curve system.

In § 6, we derive the variational formula for the dielectric Green's functions under a small deformation of the domain. Through the maximum definition of the Fredholm eigen values, we can derive from this result also the variational formula for the Fredholm eigen values under the same deformation, This formula could also have been obtained immediately from the general perturbation theory of operators. But it seems of methodological interest to utilize fully the maximum property of each eigen value in order to give an elementary proof for this formula.

In order to avoid a discussion of possible degeneration of eigen values it is convenient to deal with symmetric functions of all eigen values and their variation, instead of considering individual eigen values. For this purpose, we define in § 7 the Fredholm determinant of a domain; this concept is rather natural when one comes from the general theory of integral equations. The variational formula for the Fredholm determinant is easily expressed in terms of a complex kernal closely connected with the dielectric Green's function which possesses, moreover, as limit case a kernel well-known in the theory of conformal mapping. Indeed, the variation of the Fredholm determinant for the particular value 1 of the argument is described by this classical kernel itself.

In § 8, at last, we apply the results of the preceding section in order to solve an extremum problem for univalent functions in a multiply-connected domain and involving the Fredholm determinant. This solution gives a new proof for the possibility to map every domain conformally onto a domain bounded by circumferences and characterizes this canonical domain as an extremum domain of a simple variational problem. The treatment of the variational problem for the Fredholm determinant seems also of interest from the methodological point of view and for the general theory of variations of domain functions. In general, one knows from the theory of normal families that a solution of an extremum problem for the family of functions, univalent in a given domain and with specified normalization, does exist; the method of variations has only the task to characterize the extremum domain. In our present problem, we had to restrict ourselves to univalent functions which are analytic in the closed domain in order to be sure of the existence of the Fredholm determinant. In this case, the theory of normal families does not guarantee the existence of an extremum function of equal character. We do not characterize, therefore, the extremum function by our variations, but rather an extremum sequence within the function class, considered. We prove from the very extremum property of the sequence that its limit function does, indeed, belong to the same class and has, moreover, certain characterizing properties. This procedure is very general and may have numerous analogous applications.

1. The Fredholm eigen value problem. Let $\tilde{D}$ be a domain in the complex z-plane containing the point at infinity; let its boundary C consist of N closed curves C_j each of which is three times continuously differentiable. We denote the interior of each C_j by D_j and the union of the N domains D_j by D.

We define the kernel

$$k(z, \zeta) = \frac{\partial}{\partial n_\zeta} \log \frac{1}{|z-\zeta|} \qquad \zeta \in C \tag{1}$$

where $\boldsymbol{n}_\zeta$ denotes the normal of C at ζ pointing into D. It is well known that, under our assumptions about C, the kernel $k(z, \zeta)$ is continuous in both its arguments as long as they are restricted to C.

We want to discuss the eigen value problem

$$\varphi_\nu(z) = \frac{\lambda_\nu}{\pi}\int_C k(z, \zeta)\varphi_\nu(\zeta)ds_\zeta\,, \qquad z \in C \tag{2}$$

which plays an important role in many boundary value problems of potential theory with respect to the multiply-connected domain $\tilde{D}$. The $\varphi_\nu(z)$ and the λ_ν are called the Fredholm eigen functions and the Fredholm eigen values, respectively, of the curve system C. The study of the Fredholm eigen value problem is facilitated by the fact that the kernel $k(z, \zeta)$ is, for fixed $\zeta \in C$, defined and harmonic for all values $z \neq \zeta$ in the complex plane. The integral in (2) represents, therefore, a harmonic function in $\tilde{D}$ and a set of different harmonic functions in D. We shall use the notation

$$\frac{\lambda_\nu}{\pi}\int_C k(z, \zeta)\varphi_\nu(\zeta)ds_\zeta = \begin{cases} h_\nu(z) & \text{for } z \in D \\ \tilde{h}_\nu(z) & \text{for } z \in \tilde{D}\,. \end{cases} \tag{3}$$

The set of harmonic functions $\tilde{h}_\nu(z)$ and $h_\nu(z)$ can be interpreted as the potential due to a double layer of logarithmic charges, spread along C with the density $(\lambda_\nu/\pi)\varphi_\nu(\zeta)$. Hence, the well known discontinuity character of such potentials leads to the boundary relations at each point

$$\lim_{z\to z_0} h_\nu(z) = (1+\lambda_\nu)\varphi_\nu(z_0)\,, \quad \lim_{z\to z_0} \tilde{h}_\nu(z) = (1-\lambda_\nu)\varphi_\nu(z_0)\,, \tag{4}$$

and

$$\frac{\partial}{\partial n}h_\nu(z_0) = -\frac{\partial}{\partial \tilde{n}}\tilde{h}_\nu(z_0)\,, \tag{4'}$$

where $\tilde{n}$ denotes the normal of C pointing into $\tilde{D}$.

The Fredholm eigen value problem may thus be formulated as the following question of potential theory which is of interest by itself:

To determine a harmonic function $\tilde{h}$ in $\tilde{D}$ and a set of harmonic functions h in D which have equal normal derivatives and proportional boundary values on C!. It is easily seen that the two problems are completely equivalent and that the possible factors of proportionality in the second problem are simple functions of the Fredholm eigen values λ_ν.

Instead of the harmonic functions h_ν and $\tilde{h}_\nu$, we may consider their complex derivatives, i.e., the analytic functions

$$v_\nu(z) = \frac{\partial}{\partial z} h_\nu(z) \,, \quad \tilde{v}_\nu(z) = \frac{\partial}{\partial z} \tilde{h}_\nu(z) \,. \tag{5}$$

In view of definition (3) and by our assumption on C it can be asserted that v_ν and $\tilde{v}_\nu$ are continuous in $D + C$ and $\tilde{D} + C$, respectively. In order to translate the relations (4) and (4′) into terms involving v_ν and $\tilde{v}_\nu$, we use the parametric representation $z = z(s)$ of C by means of the arc length s and introduce

$$z' = \frac{dz}{ds}, \tag{6}$$

the unit vector at $z(s)$ in direction of the tangent of C. We can then write (4) and (4′) in the form

$$\Re\left\{\frac{\partial h_\nu}{\partial z} z'\right\} = \frac{1 + \lambda_\nu}{1 - \lambda_\nu} \Re\left\{\frac{\partial \tilde{h}_\nu}{\partial z} z'\right\} \,, \quad \Im\left\{\frac{\partial h_\nu}{\partial z} z'\right\} = \Im\left\{\frac{\partial \tilde{h}_\nu}{\partial z} z'\right\} \,, \tag{7}$$

and combine these two equation into the one complex equation

$$v_\nu(z) z' = \frac{1}{1 - \lambda_\nu} \tilde{v}_\nu(z) z' + \frac{\lambda_\nu}{1 - \lambda_\nu} \overline{\tilde{v}_\nu(z) z'} \,, \qquad z = z(s) \,. \tag{8}$$

Introducing (8) into the Cauchy identity. We obtain for $\zeta \in D$

$$v_\nu(\zeta) = \frac{1}{2\pi i} \oint_C \frac{v_\nu(z)}{z - \zeta} dz = \frac{\lambda_\nu}{1 - \lambda_\nu} \frac{1}{2\pi i} \oint_C \frac{\overline{(\tilde{v}_\nu(z) dz)}}{z - \zeta} \tag{9}$$

while the use of the equation conjugate to (8) leads to

$$\frac{1}{2\pi i} \oint_C \frac{\overline{(v_\nu(z) dz)}}{z - \zeta} = \frac{1}{1 - \lambda_\nu} \frac{1}{2\pi i} \oint_C \frac{\overline{(\tilde{v}_\nu(z) dz)}}{z - \zeta} \,, \qquad \zeta \in D \,. \tag{10}$$

Combining (9) and (10), we arrive thus at the following integral equation for v_ν:

$$v_\nu(\zeta) = \frac{\lambda_\nu}{2\pi i} \oint_C \frac{\overline{(v_\nu(z) dz)}}{z - \zeta} \,. \tag{11}$$

In the same way we prove the analogous equations

(9′) $$\tilde{v}_\nu(\zeta) = \frac{\lambda_\nu}{1-\lambda_\nu}\cdot\frac{1}{2\pi i}\oint_C \frac{\overline{(v_\nu(z)dz)}}{z-\zeta}$$

and

(11′) $$\tilde{v}_\nu(\zeta) = \frac{\lambda_\nu}{2\pi i}\oint_C \frac{\overline{(\tilde{v}_\nu(z)dz)}}{z-\zeta}\,.$$

In all these formulas the integration over the curve system C has to be performed in the positive sense with respect to D.

The line integrals in (9), (9′) and (11), (11′) can be transformed into area integrals and the integral equations take the forms

(12) $$\frac{\lambda_\nu}{\pi}\iint_D \frac{\overline{v_\nu(z)}}{(z-\zeta)^2}d\tau_z = \begin{cases} v_\nu(\zeta) & \text{for } \zeta\in D \\ (1+\lambda_\nu)\tilde{v}_\nu(\zeta) & \text{for } \zeta\in\tilde{D}\end{cases}$$

and

(13) $$-\frac{\lambda_\nu}{\pi}\iint_{\tilde{D}} \frac{\overline{\tilde{v}_\nu(z)}}{(z-\zeta)^2}d\tau_z = \begin{cases} (1-\lambda_\nu)v_\nu(\zeta) & \text{for } \zeta\in D \\ \tilde{v}_\nu(\zeta) & \text{for } \zeta\in\tilde{D}\,.\end{cases}$$

In both integrals $d\tau_z$ denotes the area element with respect to the variable z and the integrals have to be interpreted in the Cauchy principal sense whenever they become improper.

The transformation

(14) $$F(\zeta) = \frac{1}{\pi}\iint_E \frac{\overline{f(z)}}{(z-\zeta)^2}d\tau_z$$

carries every L^2-integrable function $f(z)$ defined in the complex plane E into a new function $F(z)$ of the same class and with the same norm:

(15) $$\iint_E |F(z)|^2 d\tau = \iint_E |f(z)|^2 d\tau\,.$$

This functional transformation plays a role in many problems of function theory [1, 3, 4] and is called the "Hilbert integral transformation". The integral equations (12) and (13) show the close connection between the theories of the Fredholm eigen functions and of the Hilbert transforms of analytic functions.

We introduce next the Green's functions of the domain $\tilde{D}$ and of the set of domains D. While the Green's function $\tilde{g}(z,\zeta)$ of $\tilde{D}$ is defined as usual, the Green's function $g(z,\zeta)$ of D is given by the equation

(16) $$g(z,\zeta) = \begin{cases} g_j(z,\zeta) & \text{for } z,\zeta\in D_j \\ 0 & \text{for } z\in D_j\,\zeta\in D_l,\ l\neq j\,.\end{cases}$$

Here, $g_j(z, \zeta)$ is the usual Green's function of the domain D_j. By complex differentiation, we derive from $g(z, \zeta)$ the analytic function

$$L(z, \zeta) = -\frac{2}{\pi}\frac{\partial^2}{\partial z \partial \zeta} g(z, \zeta) = \frac{1}{\pi(z-\zeta)^2} - l(z, \zeta) . \tag{17}$$

The kernels $L(z, \zeta)$ and $l(z, \zeta)$ are well known in the case that D is a domain [3, 16]. We observe that our generalized kernel $l(z, \zeta)$ still preserves the following important property: If $f(z)$ is regular analytic in D, then

$$\frac{1}{\pi}\iint_D \overline{\frac{f(z)}{(z-\zeta)^2}} d\tau = \iint_D l(z, \zeta)\overline{f(z)} d\tau . \tag{18}$$

In fact, if $\zeta \in D_j$ then $l(z, \zeta) = l_j(z, \zeta)$ for $z \in D_j$ and $l(z, \zeta) = [\pi(z-\zeta)^2]^{-1}$ for $z \in D_l$, $l \neq j$. The identity (18) follows, therefore, directly from the corresponding property of the kernel $l_j(z, \zeta)$.

In particular, we may formulate the integral equations (12) and (13) for $v_\nu(z)$ and $\tilde{v}_\nu(z)$ as follows:

$$\lambda_\nu \iint_D l(z, \zeta)\overline{v_\nu(z)} d\tau = v_\nu(\zeta) , \qquad \zeta \in D \tag{19}$$

and

$$-\lambda_\nu \iint_{\tilde{D}} \tilde{l}(z, \zeta)\overline{\tilde{v}_\nu(\zeta)} d\tau = \tilde{v}_\nu(\zeta) , \qquad \zeta \in \tilde{D} . \tag{20}$$

From the symmetry of the kernels $l(z, \zeta)$ and $\tilde{l}(z, \zeta)$ we can conclude

$$\iint_D v_\nu \bar{v}_\mu d\tau = 0 \quad \text{if} \quad \lambda_\nu \neq \lambda_\mu \tag{21}$$

$$\iint_{\tilde{D}} \tilde{v}_\nu \bar{\tilde{v}}_\mu d\tau = 0 \quad \text{if} \quad \lambda_\nu \neq \lambda_\mu . \tag{21'}$$

Thus, using a familiar argument from theory of integral equation we may assume that any pair of different eigen functions v_ν, v_μ (or $\tilde{v}_\nu$. $\tilde{v}_\mu$) are orthogonal upon each other:

$$\iint_D v_\nu \bar{v}_\mu d\tau = 0 ; \iint_{\tilde{D}} \tilde{v}_\nu \bar{\tilde{v}}_\mu d\tau = 0 \quad \text{for} \quad \nu \neq \mu . \tag{21''}$$

There remains the question of normalizing the v_ν and the $\tilde{v}_\nu$. We have obviously the free choice of a real multiplicator in the definition of v_ν; however, this choice will already determine the function $\tilde{v}_\nu$ in a unique way, for example through equation (12). The relation between the norms of v_ν and $\tilde{v}_\nu$ is best understood by returning to the harmonic functions $h_\nu(z)$ and $\tilde{h}_\nu(z)$ and to their boundary relations (4) and (4′). In fact, we have

$$\iint_D |v_\nu|^2 d\tau = \frac{1}{4}\iint_D |\nabla h_\nu|^2 d\tau = -\frac{1}{4}\oint_C h_\nu \frac{\partial h_\nu}{\partial n} ds \tag{22}$$

$$= \frac{1}{4}\frac{1+\lambda_\nu}{1-\lambda_\nu}\oint_C \tilde{h}_\nu \frac{\partial \tilde{h}_\nu}{\partial \tilde{n}} ds = \frac{\lambda_\nu + 1}{\lambda_\nu - 1}\iint_{\tilde{D}} |\tilde{v}_\nu|^2 d\tau .$$

We can conclude first from (22) that

$$|\lambda_\nu| \geq 1 . \tag{23}$$

Let us consider the limit cases $\lambda_\nu = \pm 1$. For $\lambda_\nu = 1$ we have necessarily $\tilde{v}_\nu(z) \equiv 0$; the second equation (7) yields

$$\mathfrak{J}\{v_\nu(z)z'\} = 0 \quad \text{for} \quad z \in C . \tag{24}$$

Thus, the eigen function $v_\nu(z)$ is a real differential for each component domain D_j. But a simply-connected domain D_j cannot have such real differentials ; hence also $v_\nu(z) \equiv 0$. Thus, as far as the integral equation for v_ν and $\tilde{v}_\nu$ are concerned, $\lambda_\nu = 1$ cannot occur as an eigen value. The situation is, however, different when we return to the original integral equation (2) and to the harmonic functions h_ν and $\tilde{h}_\nu$. To $\lambda_\nu = 1$ must correspond

$$h_\nu(z) \equiv 2c_j \text{ in } D_j , \ \tilde{h}_\nu(z) \equiv 0 \tag{25}$$

and

$$\varphi_\nu(z) = c_j \quad \text{on} \quad C_j . \tag{25'}$$

In fact, it is immediately verified that for arbitrary choice of the constants c_j the function $\varphi(z) = c_j$ on C_j is a solution of the Fredholm eigen value problem (2) to the eigen value $\lambda_\nu = 1$. There exist thus N linearly independent solutions of (2) to the eigen value $\lambda = 1$. These solutions disappear when we replace the original integral equation (2) by the integral equations for v_ν and $\tilde{v}_\nu$, say, by (12) and (13). It is easy to show that the eigen value $\lambda = 1$ is the only one lost in this transition.

We consider next the case $\lambda_\nu = -1$. We conclude now from (22) that $v_\nu(z) \equiv 0$. We find therefore, in view of (8)

$$\mathfrak{J}\{\tilde{v}_\nu(z)z'\} = 0 \quad \text{for} \quad z \in C , \tag{26}$$

i.e., $\tilde{v}_\nu(z)$ is a real differential of $\tilde{D}$. There are $N-1$ linearly independent differentials of this type in $\tilde{D}$ and we can construct a basis for them as follows. Let $\omega_j(z)$ be harmonic in $\tilde{D}$ and satisfy on C the boundary condition

(27) $$\omega_j(z) = \delta_{jl} \quad \text{for} \quad z \in C_l \,.$$

$\omega_j(z)$ is called the harmonic measure of C_j with respect to z of $\tilde{D}$. Clearly, each function

(28) $$\tilde{w}_j(z) = i\frac{\partial \omega_j}{\partial z}$$

is a real differential in $\tilde{D}$. Since $\sum_{j=1}^{N} \omega_j \equiv 1$, we have $\sum_{j=1}^{N} \tilde{w}_j(z) \equiv 0$. But it is easily seen that apart from this relation no other linear condition between the w_j does exist. Thus, we can select any $N-1$ of the $w_j(z)$ as a basis for all real differentials in $\tilde{D}$.

It is clear that each real differential in $\tilde{D}$ satisfies indeed the integral equations (12) and (13). However, there exists no corresponding single valued harmonic function $\tilde{h}_\nu(z)$ connected with the original Fredholm equation (2) which has this real differential as its complex derivative. Indeed, in view of (26) such function would have to satisfy the boundary condition

(29) $$\frac{\partial \tilde{h}_\nu}{\partial n} \equiv 0 \quad \text{on} \quad C$$

which admits only the solution $\tilde{h}_\nu =$ const. and could not lead to a non-vanishing differential. Thus, while we lost in the transition to (12) and (13) the N eigen functions to the eigen value $\lambda = +1$, we have obtained $N-1$ new eigen functions to the eigen value $\lambda = -1$ which have no counterpart in the original Fredholm equation.

After discussing the exceptional cases $\lambda_\nu = \pm 1$, we consider now the eigen functions $v_\nu(z)$ and $\tilde{v}_\nu(z)$ which belong to eigen values $|\lambda_\nu| > 1$. Each such pair is obtained by complex differentiation from a pair of harmonic functions $h_\nu(z), \tilde{h}_\nu(z)$ connected with the original Fredholm problem. Since $h_\nu(z)$ is harmonic in each of the simply-connected domains D_j, it can be completed to a set of single-valued analytic functions in the set of domains D_j:

(30) $$V_\nu(z) = h_\nu(z) + ik_\nu(z) \,.$$

Similarly, we may complete $\tilde{h}_\nu$ in $\tilde{D}$ and define

(31) $$\tilde{V}_\nu(z) = \tilde{h}_\nu(z) + i\tilde{k}_\nu(z) \,.$$

From the boundary conditions (4) and (4′) and from the Cauchy-Riemann equations we derive the boundary conditions for the k_ν:

(32) $$\tilde{k}_\nu(z) = k_\nu(z), \quad \frac{\partial}{\partial n}k_\nu(z) = \frac{1+\lambda_\nu}{1-\lambda_\nu}\frac{\partial}{\partial n}\tilde{k}_\nu(z) \,, \qquad z \in C \,.$$

Equations (32) guarantee that $\tilde{k}_\nu(z)$ is single-valued in $\tilde{D}$ since $k_\nu(z)$ is single-valued in each D_j. We may characterize the single-valued analytic functions $V_\nu(z)$ and $\tilde{V}_\nu(z)$ as follows: Their real parts have equal normal derivatives on C while their boundary values are proportional in the ratio $(1+\lambda_\nu)/(1-\lambda_\nu)$. Their imaginary parts are equal on C but their normal derivatives are proportional with the same ratio.

Let us write $k_\nu^{(1)} = (1-\lambda_\nu)k_\nu$ and $\tilde{k}_\nu^{(1)} = (1+\lambda_\nu)\tilde{k}_\nu$; we have on C

$$(32') \qquad k_\nu^{(1)}(z) = \frac{1-\lambda_\nu}{1+\lambda_\nu}\tilde{k}_\nu^{(1)}(z), \quad \frac{\partial k_\nu^{(1)}}{\partial n} = -\frac{\partial \tilde{k}_\nu^{(1)}}{\partial \tilde{n}}.$$

Thus, $k_\nu^{(1)}$ and $\tilde{k}_\nu^{(1)}$ may be conceived as a pair of k-functions belonging to eigen functions of the Fredholm problem (2) with the eigen value $-\lambda_\nu$. With each eigen value λ_ν with $|\lambda_\nu| > 1$ there occurs also its negative $-\lambda_\nu$ as an eigen value. Their corresponding h-functions are, up to a factor, conjugate harmonic functions.

Finally, we introduce the analytic functions

$$(33) \qquad u_\nu(z) = \sqrt{\lambda_\nu - 1}\, v_\nu(z), \quad \tilde{u}_\nu(z) = i\sqrt{\lambda_\nu + 1}\, \tilde{v}_\nu(z).$$

By virtue of (21″) and (22), we may assume that these functions form orthonormalized sets in D and $\tilde{D}$; that is

$$(34) \qquad \iint_D u_\nu \bar{u}_\mu d\tau = \delta_{\nu\mu}, \quad \iint_{\tilde{D}} \tilde{u}_\nu \bar{\tilde{u}}_\mu d\tau = \delta_{\nu\mu}.$$

Since the u-functions will be frequently used in this paper, we note here some formulas which follow immediately from the corresponding results for the v-functions. From (8) we derive the boundary relation

$$(35) \qquad u_\nu(z)z' = \frac{i}{\sqrt{\lambda_\nu^2 - 1}}\tilde{u}_\nu(z)z' - \frac{\lambda_\nu i}{\sqrt{\lambda_\nu^2 - 1}}\overline{(\tilde{u}_\nu(z)z')}.$$

Equations (9), (9′) and (11), (11′) take on the form

$$(36) \qquad \frac{\lambda_\nu}{2\pi i}\oint_C \frac{\overline{(\tilde{u}_\nu dz)}}{z-\zeta} = \begin{cases} i\sqrt{\lambda_\nu^2 - 1}\, u_\nu(\zeta) & \text{for } \zeta \in D \\ -\tilde{u}_\nu(\zeta) & \text{for } \zeta \in \tilde{D}. \end{cases}$$

and

$$(37) \qquad \frac{\lambda_\nu}{2\pi i}\oint_C \frac{\overline{(u_\nu dz)}}{z-\zeta} = \begin{cases} u_\nu(\zeta) & \text{for } \zeta \in D \\ -i\sqrt{\lambda_\nu^2 - 1}\, \tilde{u}_\nu(\zeta) & \text{for } \zeta \in \tilde{D}. \end{cases}$$

From their connection with the Fredholm integral equation it can be shown that the $u_\nu(z)$ form a complete system of analytic functions in D, in the sense that every function $f(z)$ which is analytic in D and for which $\iint_D |f|^2 d\tau < \infty$ can be represented in the form

$$f(z) = \sum_{\nu=1}^{\infty} a_\nu u_\nu(z),\ a_\nu = \iint_D f\bar{u}_\nu d\tau\ . \tag{38}$$

The series converges uniformly in each closed subdomain of D. In the same sense, the functions $\tilde{u}_\nu(z)$ form a complete orthonormal system within the class of all functions which are analytic in $\tilde{D}$, have a finite norm in $\tilde{D}$ and possess a finite single-valued integral in this multiply-connected domain. If we add to the $\{\tilde{u}_\nu\}$-set any $N-1$ linearly independent real differentials of $\tilde{D}$ we obtain a complete system for all analytic functions in $\tilde{D}$ with finite norm and vanishing at infinity [3, 21].

2. **The dielectric Green's function.** The theory of the Green's function of the domain $\tilde{D}$ is connected with the electrostatic problem of a point charge at a source point ζ in the presence of the system of grounded conductors C_j. We may consider also the problem to determine the electrostatic potential induced by the same point charge at ζ in the presence of N isotropic dielectric media which are spread over the domains D_j and have the dielectric constant ε. The corresponding potential $g_\varepsilon(z, \zeta)$ will now be defined in D as well as in $\tilde{D}$ and will be characterized by the following properties:

(a) $g_\varepsilon(z, \zeta)$ is a harmonic function of z in D and in $\tilde{D}$, except for $z = \zeta$ and for $z = \infty$.

(b) If $\zeta \in \tilde{D}$, the function $g_\varepsilon(z, \zeta) + \log |z - \zeta|$ is harmonic at ζ.

(b′) It $\zeta \in D$, the function $g_\varepsilon(z, \zeta) + \varepsilon \log |z - \zeta|$ is harmonic at ζ.

(c) $g_\varepsilon(z, \zeta)$ is continuous through C.

(d) $\frac{\partial}{\partial n_z} g_\varepsilon(z, \zeta) + \varepsilon \frac{\partial}{\partial \tilde{n}_z} g_\varepsilon(z, \zeta) = 0$ for $z \in C$, ζ in D or in $\tilde{D}$.

(e) $g_\varepsilon(z, \zeta) + \log |z| \to 0$ as $z \to \infty$ for ζ fixed.

If such a function $g_\varepsilon(z, \zeta)$ exists it must be unique and symmetric in its two arguments, as is shown by the standard argument of potential theory based on the second Green's identity. In order to construct the Green's function, we set it up in the form

$$g_\varepsilon(z, \zeta) = \log \frac{1}{|z - \zeta|} + \int_C \mu(\eta, \zeta) \log |\eta - z|\, ds_\eta,\ \zeta \in \tilde{D} \tag{1}$$

and try to determine $\mu(\eta, \zeta)$ in such a way that the above requirements are fulfilled. We proceed analogously, if $\zeta \in D$; only the singularity term on the right side of (1) will now be $-\varepsilon \log |z - \zeta|$. By this formal set up, we have already fulfilled conditions (a) to (c). Condition (e) is satisfied if we require

$$\int_C \mu(\eta, \zeta) ds_\eta = \begin{cases} \varepsilon - 1 & \text{for } \zeta \in D \\ 0 & \text{for } \zeta \in \tilde{D}\,. \end{cases} \tag{2}$$

Finally, we can satisfy (d) by choosing the density function μ of the line potential as solution of the integral equation

$$(3)\qquad \mu(z,\zeta)+\frac{\varepsilon-1}{\varepsilon+1}\frac{1}{\pi}\int_C \mu(\eta,\zeta)k(\eta,z)ds_\eta=\begin{cases}\dfrac{\varepsilon-1}{\varepsilon+1}\cdot\dfrac{\varepsilon}{\pi}k(\zeta,z) & \text{for } \zeta\in D\\[2ex] \dfrac{\varepsilon-1}{\varepsilon+1}\,\dfrac{1}{\pi}k(\zeta,z) & \text{for } \zeta\in\tilde{D}\,.\end{cases}$$

Here $k(\zeta, z)$ is defined by equation (1.1). We observe that

$$(4)\qquad \int_C k(\eta,z)ds_z=\begin{cases}0 & \text{for } \eta\in\tilde{D}\\ \pi & \text{for } \eta\in C\\ 2\pi & \text{for } \eta\in D\,.\end{cases}$$

Hence, if $\mu(z, \zeta)$ is a solution of the integral equation (3) we may integrate this equation with respect to z over C and verify that condition (2) is fulfilled automatically. It is sufficient, therefore, to concentrate upon the inhomogeneous integral equation (3).

For physical reasons, we shall assume $\varepsilon > 0$. In this case, we always have

$$(3')\qquad \left|\frac{\varepsilon-1}{\varepsilon+1}\right|<1\,.$$

Since we showed in § 1 that all eigen values of the kernel $k(z, \zeta)$ have absolute values ≥ 1, it follows that integral equation (3) can always be solved by the usual process of iteration and that the solution can be represented by a Liouville-Neumann series. The convergence of this series will be the better, the nearer ε will be to 1. We observe that

$$(5)\qquad g_1(z,\zeta)=\log\frac{1}{|z-\zeta|}$$

is trivially known.

The function

$$(6)\qquad \gamma_\varepsilon(z,\zeta)=g_\varepsilon(z,\zeta)-\log\frac{1}{|z-\zeta|}$$

is (for $\zeta\in D$ or for $\zeta\in\tilde{D}$) a regular harmonic function of z in $\tilde{D}$, vanishes if z tends to infinity and possesses a single-valued conjugate harmonic function in $\tilde{D}$. This last fact follows from the boundary condition (d) on the dielectric Green's function and the fact that each complementary domain D_j is simply-connected. Let $\tilde{\Sigma}$ be the class of all functions $\tilde{h}(s)$ which are harmonic in $\tilde{D}$, vanish at infinity and have a single-valued conjugate harmonic function. It is easy to show that the harmonic functions $\tilde{h}_\nu(s)$ which belong to eigen values $|\lambda_\nu| > 1$ of the Fredholm problem in § 1 form a basis in the linear space $\tilde{\sum}$. By virtue of (1.5) and (1.21′), we have

$$\iint_{\tilde{D}} \nabla\tilde{h}_\nu \cdot \nabla\tilde{h}_\mu d\tau = 4\Re\left\{\iint_D \tilde{v}_\nu \bar{\tilde{v}}_\mu d\tau\right\} = 0 \qquad \text{for } \nu \neq \mu . \tag{7}$$

By a trivial renormalization we can then achieve that

$$\iint_{\tilde{D}} \nabla\tilde{h}_\nu \cdot \nabla\tilde{h}_\mu d\tau = \delta_{\nu\mu} . \tag{8}$$

We wish now to develop $\gamma_\varepsilon(z, \zeta)$ in terms of the complete orthonormal set $\{\tilde{h}_\nu\}$. In order to determine the Fourier coefficients or $\gamma_\varepsilon(z, \zeta)$, we consider the Dirichlet integrals

$$j_\nu(\zeta) = \iint_D \nabla g_\varepsilon(z, \zeta) \cdot \nabla h_\nu(z) d\tau_z + \iint_{\tilde{D}} \nabla g_\varepsilon(z, \zeta) \cdot \nabla\tilde{h}_\nu(z) d\tau_z . \tag{9}$$

We integrate first by parts with respect to $g_\varepsilon(z, \zeta)$ and use the continuity of this function across C as well as the relation (1.4′) for the normal derivatives of h_ν and $\tilde{h}_\nu$ on C. We find

$$j_\nu(\zeta) \equiv 0 . \tag{10}$$

Next, we integrate by parts with respect to $h_\nu(z)$ and $\tilde{h}_\nu(z)$; we use (1.4) and the condition (d) on $g_\varepsilon(z, \zeta)$. We obtain the equations

$$j_\nu(\zeta) = 2\pi\varepsilon h_\nu(\zeta) - (1 + \varepsilon\rho_\nu) \int_C \frac{\partial g_\varepsilon(z, \zeta)}{\partial \tilde{n}_z} \tilde{h}_\nu(z) ds_z \quad \text{for} \quad \zeta \in D \tag{11}$$

and

$$j_\nu(\zeta) = 2\pi\tilde{h}_\nu(\zeta) - \frac{1 + \varepsilon\rho_\nu}{\varepsilon\rho_\nu} \int_C \frac{\partial g_\varepsilon(z, \zeta)}{\partial n_z} h_\nu(z) ds_z \quad \text{for} \quad \zeta \in \tilde{D} . \tag{11′}$$

Here, we have introduced the abbreviation

$$\rho_\nu = \frac{\lambda_\nu + 1}{\lambda_\nu - 1} ; \tag{12}$$

this simple function of λ_ν will occur frequently in our developments.

From (10), (11) and (11′) we deduce immediately

$$\iint_{\tilde{D}} \nabla g_\varepsilon(z, \zeta) \cdot \nabla\tilde{h}_\nu(z) d\tau_z = -\frac{2\pi\varepsilon}{1 + \varepsilon\rho_\nu} h_\nu(\zeta) \quad \text{for} \quad \zeta \in D \tag{13}$$

and

$$(13')\qquad \iint_{\tilde D} \nabla g_\varepsilon(z,\zeta)\cdot\nabla\tilde h_\nu(z)d\tau_z = \frac{2\pi\varepsilon\rho_\nu}{1+\varepsilon\rho_\nu}\tilde h_\nu(\zeta) \quad \text{for} \quad \zeta\in\tilde D.$$

When we specialize $\varepsilon = 1$, we obtain because of (5) the values of the left-hand integrals with g_ε replaced by $\log 1/|z-\zeta|$. Hence, we obtain finally by subtraction

$$(14)\qquad \iint_{\tilde D} \nabla\gamma_\varepsilon(z,\zeta)\cdot\Delta\tilde h_\nu(z)d\tau_z = \frac{2\pi(1-\varepsilon)}{(1+\rho_\nu)(1+\varepsilon\rho_\nu)}h_\nu(\zeta) \quad \text{for} \quad \zeta\in D.$$

and

$$(14')\qquad \iint_{\tilde D} \nabla\delta_\varepsilon(z,\zeta)\nabla\tilde h_\nu(z)d\tau_z = \frac{2\pi(\varepsilon-1)}{(1+\rho_\nu)(1+\varepsilon\rho_\nu)}\tilde h_\nu(\zeta) \quad \text{for} \quad \zeta\in\tilde D.$$

Having expressed by (14) and (14′) the Fourier coefficients of $\gamma_\varepsilon(z,\zeta)$ with respect to the complete orthonormal system in $\sum$, we obtain thus the two series development for $z\in\tilde D$;

$$(15)\qquad g_\varepsilon(z,\zeta) = \log\frac{1}{|z-\zeta|} + 2\pi(1-\varepsilon)\sum_{\nu=1}^{\infty}\frac{\tilde h_\nu(z)h_\nu(\zeta)}{(1+\rho_\nu)(1+\varepsilon\rho_\nu)} \quad \text{for} \quad \zeta\in D$$

$$(16)\qquad g_\varepsilon(z,\zeta) = \log\frac{1}{|z-\zeta|} + 2\pi(\varepsilon-1)\sum_{\nu=1}^{\infty}\frac{\rho_\nu\tilde h_\nu(z)\tilde h_\nu(\zeta)}{(1+\rho_\nu)(1+\varepsilon\rho_\nu)} \quad \text{for} \quad \zeta\in\tilde D.$$

Both series converge uniformly in each closed subdomain of $\tilde D$.

We wish next to expand analogously $g_\varepsilon(z,\zeta)$ for $z\in D$ in terms of the functions $h_\nu(z)$. By (1.4), (1.4′) and the normalization (8), we have

$$(17)\qquad \iint_D \nabla h_\nu\cdot\nabla h_\mu d\tau = \rho_\nu\delta_{\nu\mu}.$$

Let $\omega_j(z)$ and $\tilde g(z,\infty)$ denote again the j-th harmonic measure and the Green's function with pole at infinity of $\tilde D$. We clearly have

$$(18)\qquad \int_C h_\nu\frac{\partial\omega_j}{\partial n}ds = 0,\quad \int_C h_\nu\frac{\partial\tilde g(z,\infty)}{\partial u}ds = 0.$$

Indeed, because of (1.4) these linear conditions are equivalent to those with $\tilde h_\nu$ and these in turn follow from the fact that all $\tilde h_\nu$ have single-valued harmonic conjugates in $\tilde D$ and that they all vanish at infinity.

Let $\sum$ be the linear space of functions $h(z)$ which are regular hamonic in D and which satisfy the N linear conditions (18). Observe that $\sum$ does not contain any function $h_0(z)$ which has a constant value c_j in each D_j, except for $h_0(z)\equiv 0$. Indeed, the conditions (18) would yield for such a function $h_0(z)$

$$(18')\qquad \sum_{j=1}^{n} c_j p_{ij} = 0,\quad \sum_{j=1}^{n} c_j\omega_j(\infty) = 0$$

where

$$p_{ij} = \frac{1}{2\pi}\int_{C_j} \frac{\partial \omega_i}{\partial n} ds \tag{18''}$$

denotes the period matrix connected with the harmonic measures. But the first system of linear equations (18') implies clearly [5, 15] $c_1 = c_2 = \cdots = c_N = c$ and the last equation yields

$$c \sum_{j=1}^{N} \omega_j(\infty) = c = 0 . \tag{19}$$

Thus, only the trivial function $h_0(z) \equiv 0$ of this type lies in $\sum$.

From this fact and the considerations of § 1, it follows that the functions $\{\rho_\nu^{1/2} h_\nu(z)\}$ form a complete orthonormal set in $\sum$. The function $\gamma_\varepsilon(z, \zeta)$ lies in $\sum$ if $\zeta \in \tilde{D}$; this follows at once from the conditions (c), (d) and (e) on the dielectric Green's function. If $\zeta \in D$, it is seen that $\gamma_\varepsilon(z, \zeta) + (1 - \varepsilon)g(z, \zeta)$ lies in $\sum$ where $g(z, \zeta)$ is the Green's function of D defined by (1.16). The Fourier coefficients of $\gamma_\varepsilon(z, \zeta)$ are easily determined from (9), (10), (13) and (13'). Observe that for $\zeta \in D_j$

$$\iint_D \nabla g(z, \zeta) \cdot \nabla h_\nu(z) d\tau_z = - \int_{C_j} \frac{\partial h_\nu(z)}{\partial n} g(z, \zeta) ds_z = 0 \tag{20}$$

such that the correction term $(1 - \varepsilon)g(z, \zeta)$ does not affect the Fourier coefficients at all. We find without difficulty

$$\begin{aligned} g_\varepsilon(z, \zeta) = \log \frac{1}{|z - \zeta|} &+ (\varepsilon - 1)g(z, \zeta) \\ &+ 2\pi(\varepsilon - 1) \sum_{\nu=1}^{\infty} \frac{h_\nu(z)h_\nu(\zeta)}{\rho_\nu(1 + \rho_\nu)(1 + \varepsilon\rho_\nu)} \qquad \text{for } \zeta \in D \end{aligned} \tag{21}$$

$$g_\varepsilon(z, \zeta) = \log \frac{1}{|z - \zeta|} + 2\pi(1 - \varepsilon) \sum_{\nu=1}^{\infty} \frac{h_\nu(z)\tilde{h}_\nu(\zeta)}{(1 + \rho_\nu)(1 + \varepsilon\rho_\nu)} \qquad \text{for } \zeta \in \tilde{D} . \tag{22}$$

These series also converge uniformly in each closed subdomain of D. Equation (22) could have been derived from (15) and the property of symmetry of the dielectric Green's function in dependence of its two arguments.

The various series developments for $g_\varepsilon(z, \zeta)$ given so far are of theoretical interest and allow the derivation of numerous identities. They help little in the actual determination of the dielectric Green's function of a given domain since we know all Fredholm eigen functions and eigen values only in very few cases. In order to utilize the preceding formulas for actual calculations, we have to add the following considerations.

From the definition of the dielectric Green's functions and from Green's identity, one can derive the identity

(23) $$\frac{1}{e}\iint_D \nabla g_\varepsilon(z,\zeta)\cdot\nabla g_e(z,\eta)d\tau_z + \iint_{\tilde{D}} \nabla g_\varepsilon(z,\zeta)\cdot\nabla g_e(z,\eta)d\tau_z = 2\pi g_\varepsilon(\zeta,\eta)\ .$$

Interchanging ε and e in (23) and subtracting the new identitity, we obtain

(24) $$2\pi[g_\varepsilon(\zeta,\eta) - g_e(\zeta,\eta)] = \left(\frac{1}{e} - \frac{1}{\varepsilon}\right)\iint_D \nabla g_\varepsilon(z,\zeta)\cdot\nabla g_e(z,\eta)d\tau_z\ .$$

In particular, passing to the limit $e \to \varepsilon$, we find

(25) $$\frac{\partial}{\partial\varepsilon} g_\varepsilon(\zeta,\eta) = \frac{1}{\varepsilon^2}\cdot\frac{1}{2\pi}\iint_D \nabla g_\varepsilon(z,\zeta)\nabla g_\varepsilon(z,\eta)d\tau_z\ .$$

We introduce the expression

(26) $$\Gamma(\zeta,\eta) = \frac{1}{2\pi}\iint_D\left(\nabla_z \log\frac{1}{|z-\zeta|}\cdot\nabla_z\log\frac{1}{|z-\eta|}\right)d\tau_z$$

which is a "geometric" functional of D, i.e., can be calculated from elementary functions by a simple process of integration and not by solving any boundary value problem of potential theory. Passing in (25) to the limit $\varepsilon = 1$, we find in view of (5)

(27) $$\frac{\partial}{\partial\varepsilon} g_\varepsilon(\zeta,\eta)\Big|_{\varepsilon=1} = \Gamma(\zeta,\eta)\ .$$

On the other hand, we can calculate this same ε-derivative directly from formulas (15), (16) and (21). Comparing results, we obtain

(28) $$\Gamma(\zeta,\eta) = -2\pi\sum_{\nu=1}^{\infty}\frac{h_\nu(\xi)\tilde{h}_\nu(\xi)}{(1+\rho_\nu)^2} \quad \text{for} \quad \zeta,\eta\in\tilde{D}$$

(28′) $$\Gamma(\zeta,\eta) = 2\pi\sum_{\nu=1}^{\infty}\frac{\rho_\nu\tilde{h}_\nu(\zeta)\tilde{h}_\nu(\eta)}{(1+\rho_\nu)^2} \quad \text{for} \quad \zeta\in D,\eta\in\tilde{D}$$

(28″) $$\Gamma(\zeta,\eta) = g(\zeta,\eta) + 2\pi\sum_{\nu=1}^{\infty}\frac{h_\nu(\zeta)h_\nu(\eta)}{\rho_\nu(1+\rho_\nu)^2} \quad \text{for} \quad \zeta,\eta\in D\ .$$

The fact that these particular series in the h-functions have relatively elementary sums is of considerable interest. It leads to series developments for the dielectric Green's functions in terms of geometric expressions.

Let us define recursively

(29) $$\Gamma^{(n)}(z,\zeta) = \frac{1}{2\pi}\iint_D\left(\nabla_\eta\Gamma^{(n-1)}(\eta,z)\cdot\Delta_\eta\log\frac{1}{|\eta-\zeta|}\right)d\tau_\eta,\ \Gamma^{(1)}\equiv\Gamma\ .$$

Using equations (9), (10) and the Fourier formulas (13), (13′), we derive the series developments

$$\Gamma^{(n)}(z,\zeta) = -2\pi \sum_{\nu=1}^{\infty} \frac{\tilde{h}_\nu(z) h_\nu(\zeta)}{(1+\rho_\nu)^{n+1}} \quad \text{for} \quad z \in \tilde{D}, \zeta \in D \tag{30}$$

$$\Gamma^{(n)}(z,\zeta) = 2\pi \sum_{\nu=1}^{\infty} \frac{\rho_\nu \tilde{h}_\nu(z) \tilde{h}_\nu(\zeta)}{(1+\rho_\nu)^{n+1}} \quad \text{for} \quad z \in \tilde{D}, \zeta \in \tilde{D} \tag{31}$$

$$\Gamma^{(n)}(z,\zeta) = g(z,\zeta) + 2\pi \sum_{\nu=1}^{\infty} \frac{h_\nu(z) h_\nu(\zeta)}{\rho_\nu (1+\rho_\nu)^{n+1}} \quad \text{for} \quad z \in D, \zeta \in D\,. \tag{32}$$

We return now to the formulas (15), (16) and (21) for $g_\varepsilon(z,\zeta)$. We use the series development

$$\frac{\varepsilon - 1}{1+\varepsilon\rho_\nu} = 2\sum_{k=0}^{\infty} \left(\frac{\varepsilon-1}{\varepsilon+1}\right)^{k+1} \frac{(1-\rho_\nu)^k}{(1+\rho_\nu)^{k+1}} = \frac{2}{1-\rho_\nu} \sum_{k=0}^{\infty} \left(\frac{1}{\lambda_\nu}\frac{1-\varepsilon}{1+\varepsilon}\right)^{k+1} \tag{33}$$

which converges absolutely since $\varepsilon > 0$ and $|\lambda_\nu| > 1$. We insert this series into the above formulas for $g_\varepsilon(z,\zeta)$; interchanging the order of summation, we obtain in each case the representation:

$$g_\varepsilon(z,\zeta) = \log \frac{1}{|z-\zeta|} + \sum_{k=0}^{\infty} \left(\frac{\varepsilon-1}{\varepsilon+1}\right)^{k+1} M_k(z,\zeta)\,. \tag{34}$$

The kernels $M_k(z,\zeta)$ are defined as follows:

$$M_k(z,\zeta) = -4\pi \sum_{\nu=1}^{\infty} \frac{(1-\rho_\nu)^k}{(1+\rho_\nu)^{k+2}} \tilde{h}_\nu(z) h_\nu(\zeta) \quad \text{for} \quad z \in \tilde{D}, \zeta \in \tilde{D} \tag{35}$$

$$M_k(z,\zeta) = -4\pi \sum_{\nu=1}^{\infty} \frac{(1-\rho_\nu)^k \rho_\nu}{(1+\rho_\nu)^{k+2}} \tilde{h}_\nu(z) \tilde{h}_\nu(\zeta) \quad \text{for} \quad z \in \tilde{D}, \zeta \in \tilde{D} \tag{36}$$

$$M_k(z,\zeta) = 2g(z,\zeta) + 4\pi \sum_{\nu=1}^{\infty} \frac{(1-\rho_\nu)^k}{\rho_\nu(1+\rho_\nu)^{k+2}} h_\nu(z) h_\nu(\zeta) \quad \text{for} \quad z \in D, \zeta \in D\,. \tag{37}$$

By use of the geometric terms (30), (31) and (32), we can express $M_k(z,\zeta)$ in a uniform way, independently of the location of their arguments. We find

$$M_k(z,\zeta) = \sum_{\sigma=1}^{\infty} (-1)^k \binom{k}{\sigma} 2^{k-\sigma+1} \Gamma^{(k-\sigma+1)}(z,\zeta)\,. \tag{38}$$

Formulas (34) and (38) allow a series development for all dielectric Green's functions in the entire plane in terms of the known iterated Dirichlet integrals $\Gamma^{(n)}(z,\zeta)$. They are closely related to similar developments for the classical Green's function of a multiply-connected domain in terms of geometric expressions [3, 21]. The formulas are convenient for $|\varepsilon - 1|$ small. Observe also that the geometrical terms $M_k(z,\zeta)$ are independent of ε and may be defined as the coefficients of the Taylor's series for $g_\varepsilon(z,\zeta)$ in terms of $(\varepsilon-1)/(\varepsilon+1)$.

3. Limit values of the dielectric Green's function. From the series developments for the dielectric Green's function, given in the preceding section, we can determine the limit values of $g_\varepsilon(z, \zeta)$ as ε converges to zero or to infinity. For this purpose, we have to introduce additional functions of the classes $\sum$ and $\tilde{\sum}$ and to develop them into series of the h-functions.

(a) We suppose $\zeta \in \tilde{D}$ and consider the analytic function $\tilde{\varphi}(z, \zeta)$ of z in $\tilde{D}$ which has a simple pole at $z = \zeta$, vanishes at infinity such that

$$\lim_{z\to\infty} z\tilde{\varphi}(z, \zeta) = 1 \tag{1}$$

and which maps $\tilde{D}$ in a one-to-one manner upon the complex plane slit along concentric circular arcs around the origin. These requirements determine $\tilde{\varphi}(z, \zeta)$ in a unique way.

Let now

$$\tilde{G}(z, \zeta) = \log |\tilde{\varphi}(z, \zeta)| . \tag{2}$$

The function $\tilde{G}(z, \zeta) + \log | z - \zeta |$ is harmonic in $\tilde{D}$, has a single-valued harmonic conjugate there and vanishes as $|z| \to \infty$. Hence, this function lies in the class $\tilde{\sum}$.

We can construct $\tilde{G}(z, \zeta)$ explicitly in terms of the Green's function $\tilde{g}(z, \zeta)$ of $\tilde{D}$. In fact, it is evident that

$$\tilde{G}(z, \zeta) = \tilde{g}(z, \zeta) - \tilde{g}(z, \infty) - \tilde{g}(\zeta, \infty) + \tilde{\gamma} - \sum_{j,k=1}^{\infty} \alpha_{jk}(\omega_j(z) - \omega_j(\infty))(\omega_k(\zeta) - \omega_k(\infty)) , \tag{3}$$

with

$$\tilde{\gamma} = \lim_{z\to\infty} (\tilde{g}(z, \infty) - \log |z|) . \tag{3'}$$

The coefficient matrix α_{jk} has to be chosen in such a way as to make the conjugate of $\tilde{G}$ single-valued along each boundary curve C_l. Hence. we obtain for it the linear equations

$$\omega_l(\zeta) - \omega_l(\infty) = \sum_{j,k=1}^{N-1} \alpha_{jk} p_{jl}[\omega_k(\zeta) - \omega_k(\infty)] \tag{4}$$

where the p_{jl} are the elements of the period matrix defined in (2.18″). Hence, we conclude

$$\sum_{j=1}^{N-1} \alpha_{jk} p_{jl} = \delta_{kl} , \tag{5}$$

i.e., the α-matrix is the inverse of the period matrix of rank $N-1$.

We can develop $\tilde{G}(z,\zeta)+\log|z-\zeta|$ in terms of the complete orthonormal system $\{\tilde{h}_\nu\}$ in $\sum$. Since $\tilde{G}(z,\zeta)$ takes on each curve C_l a constant boundary value

$$\tilde{G}(z,\zeta)=c_l(\zeta) \quad \text{for} \quad z\in C_l,\ \zeta\in\tilde{D}\,, \tag{6}$$

we have

$$\iint_{\tilde{D}}\nabla\tilde{G}(z,\zeta)\cdot\nabla\tilde{h}_\nu(z)d\tau_z=\sum_{l=1}^{N}c_l(\zeta)\int_{C_l}\frac{\partial\tilde{h}_\nu}{\partial n}ds=0\,. \tag{7}$$

Thus, combining (7) with (2.13′) for $\varepsilon=1$, we obtain

$$\iint_{\tilde{D}}\nabla[\tilde{G}(z\,\zeta)+\log|z-\zeta|]\cdot\nabla\tilde{h}_\nu(z)d\tau_z=-\frac{2\pi\rho_\nu}{1+\rho_\nu}\tilde{h}_\nu(\zeta)\,. \tag{8}$$

Consequently, we arrive at the following series development for $\tilde{G}(z,\zeta)$:

$$\tilde{G}(z,\zeta)=\log\frac{1}{|z-\zeta|}-2\pi\sum_{\nu=1}^{\infty}\frac{\rho_\nu}{1+\rho_\nu}\tilde{h}_\nu(z)\tilde{h}_\nu(\zeta)\,. \tag{9}$$

We may now cast (2.16) into the form

$$g_\varepsilon(z,\zeta)-\tilde{G}(z,\zeta)=2\pi\sum_{\nu=1}^{\infty}\frac{\varepsilon\rho_\nu}{1+\varepsilon\rho_\nu}\tilde{h}_\nu(z)\tilde{h}_\nu(\zeta)\,. \tag{10}$$

We recognize, in particular, that

$$\lim_{\varepsilon\to0}g_\varepsilon(z,\zeta)=\tilde{G}(z,\zeta)\,. \tag{11}$$

Thus, the logarithm of the important canonical map function $\tilde{\varphi}(z,\zeta)$ is closely related to the limit of the dielectric Green's function as $\varepsilon\to0$.

Let next $\tilde{\psi}(z,\zeta)$ be analytic for $z\in\tilde{D}$ except for a simple pole at $z=\zeta\in\tilde{D}$, vanish at infinity such that

$$\lim_{z\to\infty}z\tilde{\psi}(z,\zeta)=1 \tag{1′}$$

and map $\tilde{D}$ univalently onto the entire plane slit along rectilinear segments which are all directed towards the origin. $\tilde{\psi}(z,\zeta)$ is uniquely determined and might be constructed explicitly in terms of the Neumann's function of $\tilde{D}$.

Let

$$\tilde{N}(z,\zeta)=\log|\tilde{\psi}(z,\zeta)|\,. \tag{12}$$

Obviously, the function $\tilde{N}(z,\zeta)+\log|z-\zeta|$ lies in the class $\sum$. Since $\tilde{N}(z,\zeta)$ has, by its definition, vanishing normal derivatives on C, we have

$$\iint_{\tilde{D}} \nabla \tilde{N}(z, \zeta) \cdot \nabla \tilde{h}_\nu(z) d\tau_z = 2\pi \tilde{h}_\nu(\zeta) ; \tag{13}$$

therefore, in view of (2.13′) for $\varepsilon = 1$

$$\iint_{\tilde{D}} \nabla[\tilde{N}(z, \zeta) + \log |z - \zeta|] \cdot \nabla \tilde{h}_\nu(z) d\tau_z = \frac{2\pi}{1 + \rho_\nu} \tilde{h}_\nu(\zeta) . \tag{14}$$

Thus, we arrive at the series development

$$\tilde{N}(z, \zeta) = \log \frac{1}{|z - \zeta|} + 2\pi \sum_{\nu=1}^{\infty} \frac{1}{1 + \rho_\nu} \tilde{h}_\nu(z) \tilde{h}_\nu(\zeta) . \tag{15}$$

We can transform (2.16) into

$$\tilde{N}(z, \zeta) - g_\varepsilon(z, \zeta) = \sum_{\nu=1}^{\infty} \frac{\tilde{h}_\nu(z) \tilde{h}_\nu(\zeta)}{1 + \varepsilon \rho_\nu} \tag{16}$$

and read off the limit relation

$$\lim_{\varepsilon \to \infty} g_\varepsilon(z, \zeta) = \tilde{N}(z, \zeta) . \tag{17}$$

The dielectric Green's function $g_\varepsilon(z, \zeta)$ yields thus in $\tilde{D}$ a continuous interpolation between the logarithms of two canonical map functions. The result is the more significant since we shall prove in the next section that each $g_\varepsilon(z, \zeta)$ is analogously related to a univalent function in $\tilde{D}$.

(b) From the fact that the function $\tilde{G}(z, \zeta) + \log |z - \zeta|$ lies in $\tilde{\sum}$, i.e., that it has a single-valued conjugate and that it vanishes at infinity, it follows by virtue of (6) that

$$\sum_{l=1}^{N} c_l(\zeta) \int_{C_l} \frac{\partial \omega_j(z)}{\partial n} ds = \int_C \log \frac{1}{|z - \zeta|} \frac{\partial \omega_j}{\partial n} ds \tag{18}$$

and

$$\sum_{l=1}^{N} c_l(\zeta) \int_{C_l} \frac{\partial \tilde{g}(z, \infty)}{\partial n} ds = \int_C \log \frac{1}{|z - \zeta|} \frac{\partial \tilde{g}(z, \infty)}{\partial n} ds . \tag{18′}$$

We define now for fixed $\zeta \in \tilde{D}$ the harmonic function $c(z, \zeta)$ of z in D by putting

$$c(z, \zeta) = c_l(\zeta) \quad \text{for} \quad z \in D_l . \tag{19}$$

By (18), (18′) and the definition (2.18) of the class Σ, the function $-\log |z - \zeta| - c(z, \zeta)$ lies in this linear space. We may develop it, therefore, into a series of the $h_\nu(z)$. By use of (2.10) and (2.13′), we obtain

$$\log\frac{1}{|z-\zeta|}=c(z,\zeta)-2\pi\sum_{\nu=1}^{\infty}\frac{h_\nu(z)\tilde{h}_\nu(\zeta)}{1+\rho_\nu} \qquad z\in D,\ \zeta\in\tilde{D}. \tag{20}$$

We may combine (20) with (2.22) and find

$$g_\varepsilon(z,\zeta)=c(z,\zeta)-2\pi\varepsilon\sum_{\nu=1}^{\infty}\frac{1}{1+\varepsilon\rho_\nu}h_\nu(z)\tilde{h}_\nu(\zeta). \tag{21}$$

This leads to the limit relation

$$\lim_{\varepsilon\to 0} g_\varepsilon(z,\zeta)=c(z,\zeta) \quad \text{for} \quad z\in D,\ \zeta\in\tilde{D}. \tag{22}$$

The limit of $g_\varepsilon(z,\zeta)$ as $\varepsilon\to\infty$ does not seem to admit a simple geometric interpretation.

(c) Consider next the case $\zeta\in D$, say $\zeta\in D_\iota$. We define now the regular analytic functions $\tilde{\varphi}_\iota(z)$ which map $\tilde{D}$ univalently into a full circle around the origin which is slit along concentric circular arcs, such that $z=\infty$ goes into the center and that

$$\lim_{z\to\infty} z\tilde{\varphi}_\iota(z)=1. \tag{1''}$$

The function $\tilde{\varphi}_\iota(z)$ is uniquely determined by the additional requirement that the special boundary curve C_ι shall correspond to the outer circumference.

Since the function

$$\tilde{G}_\iota(z)=\log|\tilde{\varphi}_\iota(z)| \tag{23}$$

is harmonic in $\tilde{D}$ except for a simple logarithmic pole at infinity and since

$$\tilde{G}_\iota(z)=c_{\iota j} \quad \text{for} \quad z\in C_j, \tag{24}$$

it is evident that $\tilde{G}_\iota(z)$ may again be expressed explicitly in terms of the Green's function $\tilde{g}(z,\zeta)$ of $\tilde{D}$ [5].

Since we assumed $\zeta\in D_\iota$, the function $\tilde{G}_\iota(z)+\log|z-\zeta|$ lies in the class $\tilde{\Sigma}$. We can develop it into a Fourier series of the system $\{\tilde{h}_\nu\}$. The same calculations as before lead to

$$\tilde{G}_\iota(z)=\log\frac{1}{|z-\zeta|}+2\pi\sum_{\nu=1}^{\infty}\frac{\tilde{h}_\nu(z)h_\nu(\zeta)}{1+\rho_\nu},\ z\in\tilde{D},\ \zeta\in D_\iota. \tag{25}$$

From (2.15) we obtain

$$g_\varepsilon(z,\zeta)-\tilde{G}_\iota(z)=-2\pi\varepsilon\sum_{\nu=1}^{\infty}\frac{1}{1+\varepsilon\rho_\nu}\tilde{h}_\nu(z)h_\nu(\zeta); \tag{26}$$

hence

$$\lim_{\varepsilon \to 0} g_\varepsilon(z, \zeta) = \tilde{G}_l(z) \quad \text{for} \quad z \in \tilde{D}, \zeta \in D_l \,. \tag{27}$$

We obtain again interesting canonical mappings from the dielectric Green's function by passing to the limit $\varepsilon = 0$.

(d) The expression $\tilde{G}_l(z) + \log|z - \zeta|$ satisfies the linear relations (2.18) if $\zeta \in D_l$. It has on C the same boundary values as the function $g(z, \zeta) + \log|z - \zeta| + c_l(z)$ which is harmonic in D, with

$$c_l(z) = c_{lj} \quad \text{for} \quad z \in D_j \,. \tag{28}$$

Thus, the new combination will belong to the class Σ and can, therefore, be developed into a Fourier series in the $\{h_\nu\}$-system. An easy calculation leads to

$$g(z, \zeta) = \log \frac{1}{|z - \zeta|} - c_l(z) - 2\pi \sum_{\nu=1}^{\infty} \frac{h_\nu(z) h_\nu(\zeta)}{\rho_\nu(1 + \rho_\nu)}, \; z \in D, \zeta \in D_l \,. \tag{29}$$

From (29) and (2.21) follows

$$g_\varepsilon(z, \zeta) - \varepsilon g(z, \zeta) = c_l(z) + 2\pi\varepsilon \sum_{\nu=1}^{\infty} \frac{1}{\rho_\nu(1 + \varepsilon\rho_\nu)} h_\nu(z) h_\nu(\zeta) \,. \tag{30}$$

Thus, we find the limit formulas, valid for $z \in D, \zeta \in D_l$:

$$\lim_{\varepsilon \to 0} g_\varepsilon(z, \zeta) = c_l(z) \,, \quad \lim_{\varepsilon \to \infty} \frac{1}{\varepsilon} g_\varepsilon(z, \zeta) = g(z, \zeta) \,. \tag{31}$$

4. Dielectric Green's functions and conformal mapping. In this section, we shall show that the dielectric Green's function $g_\varepsilon(z, \zeta)$ leads to a univalent analytic function in $\tilde{D}$ and to a set of univalent analytic functions in D. Let us suppose, for the sake of definiteness, that the source point ζ lies in $\tilde{D}$. Let $p_\varepsilon(z, \zeta)$ be the analytic completion of $g_\varepsilon(z, \zeta)$ for z in $\tilde{D}$; that is, $p_\varepsilon(z, \zeta)$ is analytic for $z \in D$ and we have

$$g_\varepsilon(z, \zeta) = \Re\{p_\varepsilon(z, \zeta)\} \,. \tag{1}$$

$p_\varepsilon(z, \zeta)$ is regular analytic except for the two logarithmic poles at ζ and at ∞. The function has no periods with respect to the boundary curves C_j. Hence

$$\tilde{f}_\varepsilon(z, \zeta) = \exp[-p_\varepsilon(z, \zeta)] \,, \qquad z \in \tilde{D}, \zeta \in \tilde{D} \tag{2}$$

is a single-valued analytic function of $z \in \tilde{D}$ and regular in this domain except for the simple pole at infinity. Since the analytic completion of a harmonic function is only determined up to an additive imaginary constant, we may choose $\tilde{f}_\varepsilon$ in such a way that

$$\tilde{f}_\varepsilon'(\infty, \zeta) = 1 \,, \quad \tilde{f}_\varepsilon(\zeta, \zeta) = 0 \,. \tag{2'}$$

We may similarly complete $g_\varepsilon(z, \zeta)$ to analytic functions of z in D. In order to determine the additive constants for the disjoint domains D_j we proceed as follows. By condition (c) of § 2 on $g_\varepsilon(z, \zeta)$ and because of the Cauchy-Riemann equations, we have, whatever the analytic completion $p_\varepsilon(z, \zeta)$ of $g_\varepsilon(z, \zeta)$ in D:

$$\Im\{p_\varepsilon(z, \zeta)\} = \varepsilon\Im\{\tilde{p}_\varepsilon(z, \zeta)\} + k_j \quad \text{for} \quad z \in C_j\,. \tag{3}$$

Here $\tilde{p}_\varepsilon$ and p_ε shall denote the limits of p_ε from $\tilde{D}$ and D, respectively; we shall use this more specific notation whenever discussing boundary relations. We dispose now of the additive constants in the domains D_j by requiring $k_j = 0$. This convention fixes $p_\varepsilon(z, \zeta)$ in D in a unique way.

In analogy to (2), we define

$$f_\varepsilon(z, \zeta) = \exp\left[-\frac{1}{\varepsilon}p_\varepsilon(z, \zeta)\right] \quad \text{for} \quad z \in D, \zeta \in \tilde{D}\,. \tag{4}$$

We shall prove the

THEOREM. *The function $\tilde{f}_\varepsilon(z, \zeta)$ is univalent in $\tilde{D}$ and the set of functions $f_\varepsilon(z, \zeta)$ is univalent in D in the sense that*

$$f_\varepsilon(z_1, \zeta) = f_\varepsilon(z_2, \zeta) \text{ and } z_1, z_2 \in D \text{ implies } z_1 = z_2\,. \tag{5}$$

In order to prove this theorem, we start with the

LEMMA. *The dielectric Green's function has no critical points. That is, the equation $p'_\varepsilon(z, \zeta) = 0$ is only satisfied at $z = \infty$ and this point is a pole of the Green's function. The dash denotes differentiation of $p_\varepsilon(z, \zeta)$ with respect to its analytic argument z.*

Proof. We denote again, more precisely, the analytic completion of $g_\varepsilon(z, \zeta)$ by $\tilde{p}_\varepsilon$ or by p_ε according to the location of z in $\tilde{D}$ or D, respectively. We combine the boundary conditions (c) and (d) of § 2 on the dielectric Green's function $g_\varepsilon(z, \zeta)$ into the one complex equation

$$p'_\varepsilon(z, \zeta)z' = \frac{1+\varepsilon}{2}\tilde{p}'_\varepsilon(z, \zeta)z' + \frac{1-\varepsilon}{2}\overline{\tilde{p}'_\varepsilon(z, \zeta)z'}\,. \tag{6}$$

Since we assume throughout this paper $\varepsilon > 0$, equation (6) yields

$$\Re\{p'_\varepsilon(z, \zeta)/\tilde{p}'_\varepsilon(z, \zeta)\} > 0 \quad \text{for } z \in C\,. \tag{7}$$

This inequality implies, in particular:

$$\oint_C d\arg p'_\varepsilon(z, \zeta) = \oint_C d\arg \tilde{p}'_\varepsilon(z, \zeta)\,. \tag{8}$$

The statement is evident if p'_ε and $\tilde{p}'_\varepsilon$ are non-zero on C but it can be upheld in the usual way even in the case that these two functions have common zeros on C

Let Z, P and $\tilde{Z}, \tilde{P}$ denote the number of zeros and poles of p'_ε and $\tilde{p}'_\varepsilon$ respectively, in their domains of definition. By the argument principle, we have

$$(9) \qquad \oint_C d \arg p'_\varepsilon(z, \zeta) = Z - P , \quad \oint_C d \arg \tilde{p}'_\varepsilon(z, \zeta) = \tilde{P} - \tilde{Z}$$

if z runs through C in the positive sense with respect to D. Combining (8) and (9), we obtain

$$(10) \qquad Z + \tilde{Z} = P + \tilde{P} .$$

But all poles of p'_ε and $\tilde{p}'_\varepsilon$ are known ; clearly $P = 0, \tilde{P} = 1$ and $Z \geq 1$. Hence, we conclude from (10) :

$$(11) \qquad Z = 0 , \quad \tilde{Z} = 1 .$$

This proves our lemma.

In order to prove the theorem, we consider the lines defined by

$$(12) \qquad \Im\{\tilde{p}_\varepsilon(z, \zeta)\} = \alpha \text{ for } z \in \tilde{D} , \quad \Im\left\{\frac{1}{\varepsilon} p_\varepsilon(z, \zeta)\right\} = \alpha \text{ for } z \in D .$$

Each such line starts from the logarithmic pole ζ and runs to ∞. By virtue of our convention on the analytic completion of $g_\varepsilon(z, \zeta)$ these lines are continuous in the entire plane and, except on C, they are even analytic. Because of our lemma, there is no intersection between different lines except at ζ and ∞. The lines have the physical interpretation as lines of force for the corresponding electrostatic problem and the lemma asserts that there are no points of equilibrium in the field. The lines form for $0 \leq \alpha < 2\pi$ a non-intersecting system which covers the entire complex plane. Along each line, $g_\varepsilon(z, \zeta)$ decreases monotonically when we pass from ζ to ∞. These facts guarantee obviously that the analytic functions $\tilde{f}_\varepsilon(z, \zeta)$ and $f_\varepsilon(z, \zeta)$ have the above stated univalency properties. Thus, the theorem is proved.

Let us assume without loss of generality that $\zeta = 0$. Using the limit theorems of § 3, we can assert :

$$(13) \qquad \tilde{f}_0(z, 0) = \tilde{\varphi}(z, 0)^{-1} , \quad \tilde{f}_1(z, 0) = z , \quad \tilde{f}_\infty(z, 0) = \tilde{\psi}(z, 0)^{-1} .$$

We have thus found a one-parameter family of univalent functions which connects continuously the circular slit mapping through the identity mapping with the radial slit mapping.

In order to illustrate the significance of this result, we calculate from (2.16) that

$$\log|\tilde{f}'_\varepsilon(\zeta,\zeta)| = 2\pi(\varepsilon-1)\sum_{\nu=1}^{\infty}\frac{\rho_\nu}{(1+\rho_\nu)(1+\varepsilon\rho_\nu)}\tilde{h}_\nu(\zeta)^2 . \tag{14}$$

Since all $\rho_\nu > 0$, this is a monotonically increasing function of ε in the interval $[0, \infty)$; it is negative for $0 \leq \varepsilon < 1$ and positive for $1 < \varepsilon$. In particular:

$$|\tilde{f}'_0(\zeta,\zeta)| < 1 \quad |\tilde{f}'_\infty(\zeta,\zeta)| > 1 . \tag{15}$$

We define the family $\mathscr{F}_\zeta$ of all functions $\tilde{f}(z)$ which are analytic and univalent in $\tilde{D}$ and normalized by the requirements

$$\tilde{f}'(\infty) = 1 \qquad \tilde{f}(\zeta) = 0 . \tag{16}$$

Through the mapping $w = \tilde{f}(z)$ we obtain the new domain $\tilde{D}_w$; applying the inequalities (15) in this domain and returning to the original domain $\tilde{D}$, we obtain the inequality

$$|\tilde{f}'_0(\zeta,\zeta)| \leq |\tilde{f}'(\zeta)| \leq |\tilde{f}'_\infty(\zeta,\zeta)| \tag{17}$$

valid for each $f \in \mathscr{F}_\zeta$.

Inequality (16) asserts an extremum property of the canonical slit functions $\tilde{f}_0$ and $\tilde{f}_\infty$ which is well-known [13, 15]. It is, however, not obvious that all real values between the extrema are also possible values for $|f'(\zeta)|$ in $\mathscr{F}_\zeta$. We have now explicity constructed a one-parameter family in $\mathscr{F}_\zeta$ which interpolates between the two extremum values.

There are various other possibilities to obtain from the dielectric Green's function one-parameter families of univalent functions. Consider, for example, the analytic functions

$$A_\varepsilon(z,\zeta) = \frac{\partial}{\partial\xi}p_\varepsilon(z,\zeta) , \quad B_\varepsilon(z,\zeta) = \frac{1}{i}\frac{\partial}{\partial\eta'}p_\varepsilon(z,\zeta) \tag{18}$$

with $\zeta = \xi + i\eta$. Both functions are single-valued in $\tilde{D}$ and in D; they have for $z = \zeta$ simple poles with residue 1 and are else regular in D and in D. We obtain from the identity (6) by differentiation

$$A'_\varepsilon(z,\zeta)z' = \frac{1+\varepsilon}{2}\tilde{A}'_\varepsilon(z,\zeta)z' + \frac{1-\varepsilon}{2}\overline{(\tilde{A}'_\varepsilon(z,\zeta)z')} \tag{19}$$

$$B'_\varepsilon(z,\zeta)z' = \frac{1+\varepsilon}{2}\tilde{B}'_\varepsilon(z,\zeta)z' - \frac{1-\varepsilon}{2}\overline{(\tilde{B}'_\varepsilon(z,\zeta)z')} . \tag{19'}$$

Let a be an arbitrary point on C; integrating (19) along C from a to $z \in C$, we find

$$A_\varepsilon(z,\zeta) - A_\varepsilon(a,\zeta) = \frac{1+\varepsilon}{2}[\tilde{A}_\varepsilon(z,\zeta) - \tilde{A}_\varepsilon(a,\zeta)] + \frac{1-\varepsilon}{2}\overline{[\tilde{A}_\varepsilon(z,\zeta) - \tilde{A}_\varepsilon(a,\zeta)]} . \tag{20}$$

Hence, we have

$$\Re\left\{\frac{A_\varepsilon(z,\zeta)-A_\varepsilon(a,\zeta)}{\tilde{A}_\varepsilon(z,\zeta)-\tilde{A}_\varepsilon(a,\zeta)}\right\}>0 \quad \text{for} \quad z\in C\,. \tag{21}$$

Reasoning as before we can conclude by means of the argument principle that $A_\varepsilon(z,\zeta)$ takes the value $A_\varepsilon(a,\zeta)$ precisely once in $D+C$ and that $\tilde{A}_\varepsilon(z,\zeta)$, likewise, takes every boundary value precisely once. Thus, $A_\varepsilon(z,\zeta)$ and $\tilde{A}_\varepsilon(z,\zeta)$ are univalent in their respective domains of definition. The same reasoning applies to $B_\varepsilon(z,\zeta)$ and $\tilde{B}_\varepsilon(z,\zeta)$.

It is known, and easily verified, that

$$\tilde{A}_0(z,\zeta)=\frac{\partial}{\partial\xi}\log\tilde{\varphi}(z,\zeta)\,,\quad \tilde{B}_0(z,\zeta)=\frac{1}{i}\frac{\partial}{\partial\eta}\log\tilde{\varphi}(z,\zeta) \tag{22}$$

are univalent functions in $\tilde{D}$ with a simple pole at $z=\zeta$ and that they map $\tilde{D}$ onto the entire complex plane, slit along rectilinear segments parallel to the imaginary and the real axis, respectively [16]. Similarly, the analytic functions

$$\tilde{A}_\infty(z,\zeta)=\frac{\partial}{\partial\xi}\log\tilde{\psi}(z,\zeta)\,,\quad \tilde{B}_\infty(z,\zeta)=\frac{1}{i}\frac{\partial}{\partial\eta}\log\tilde{\psi}(z,\zeta) \tag{23}$$

are univalent in $\tilde{D}$ with the same singularity and map the domain onto the entire complex plane, slit along segments parallel to the real and the imaginary axis, respectively. Hence, by the uniqueness theorems on the canonical mappings of a domain, we must have

$$\tilde{A}_\infty(z,\zeta)=\tilde{B}_0(z,\zeta)+\kappa(\zeta)\,;\ \tilde{B}_\infty(z,\zeta)=\tilde{A}_0(z,\zeta)+\lambda(\zeta)\,. \tag{24}$$

Finally, clearly

$$\tilde{A}_1(z,\zeta)=\tilde{B}_1(z,\zeta)=\frac{1}{z-\zeta}\,. \tag{25}$$

Hence, $\tilde{A}_\varepsilon(z,\zeta)$ and $\tilde{B}_\varepsilon(z,\zeta)$ interpolate between the two parallel slit mappings through the simple rational mapping (25).

Using the series development (2.16) for $g_\varepsilon(z,\zeta)$, $\zeta\in\tilde{D}$, we may prove the well-known extremum properties of the canonical slit mappings in the same way, as we did above for the circular and the radial slit mapping.

We do not enter into a more detailed discussion of these families of univalent functions. We want to remark, however, that the dielectric Green's function is not, like the ordinary Green's function, a conformal invariant. By auxiliary mappings of $\tilde{D}$ into a domain $\tilde{D}_w$, one may obtain very different one-parameter families of univalent functions which interpolate between the canonical slit mappings.

5. Dielectric Green's functions and norms in function spaces. With each dielectric Green's function $g_\varepsilon(z, \zeta)$ we can connect a positive-definite quadratic form which may be interpreted as a norm in the linear function spaces Σ and $\tilde{\Sigma}$, defined in § 2. This norm has remarkable properties for function pairs $h \in \Sigma$ and $\tilde{h} \in \tilde{\Sigma}$ which have on C equal boundary values or equal normal derivatives. Useful inequalities and identities can be established which facilitate the solution of the boundary value problem in potential theory by utilizing auxiliary solutions in complementary domains. One can characterize the Fredholm eigen values λ_ν as solutions of certain extremum problems involving these quadratic forms. This characterization, in turn, will lead later to elegant variational formulas for the λ_ν under infinitesimal deformation of the curve system C.

Let h and $\tilde{h}$ be two arbitrary functions of the classes Σ and $\tilde{\Sigma}$, respectively. We have the Fourier developments

$$(1) \qquad h(z) = \sum_{\nu=1}^{\infty} x_\nu h_\nu(z) , \quad \tilde{h}(z) = \sum_{\nu=1}^{\infty} \tilde{x}_\nu \tilde{h}_\nu(z)$$

in terms of the complete orthonomal sets $\{\rho_\nu^{-1/2} h_\nu(z)\}$ and $\{\tilde{h}_\nu(z)\}$ of these linear spaces. The Fourier coefficients are given by

$$(2) \qquad x_\nu = \frac{1}{\rho_\nu} D(h, h_\nu) , \quad \tilde{x}_\nu = \tilde{D}(\tilde{h}, \tilde{h}_\nu)$$

where D and $\tilde{D}$ denote the Dirichlet integral in Σ and $\tilde{\Sigma}$:

$$(3) \qquad D(h, H) = \iint_D \nabla h \cdot \nabla H d\tau , \quad \tilde{D}(\tilde{h}, \tilde{\mathrm{H}}) = \iint_{\tilde{D}} \nabla \tilde{h} \cdot \nabla \tilde{H} d\tau .$$

Let us consider now the particular case that

$$(4) \qquad h(z) = \tilde{h}(z) \quad \text{on} \quad C .$$

By Green's identity and (1.4'), we have obviously

$$(5) \qquad D(h, h_\nu) = - \int_C h \frac{\partial h_\nu}{\partial n} ds = - \int_C \tilde{h} \frac{\partial \tilde{h}_\nu}{\partial n} ds = - \tilde{D}(\tilde{h}, \tilde{h}_\nu)$$

which gives

$$(6) \qquad x_\nu = - \frac{1}{\rho_\nu} \tilde{x}_\nu .$$

We proceed analogously for two function $h \in \Sigma$ and $\tilde{h} \in \tilde{\Sigma}$ which satisfy on C the relation

$$(7) \qquad \frac{\partial h}{\partial n} = \frac{\partial \tilde{h}}{\partial n} .$$

Now, Green's identity and (1.4) yield

$$D(h, h_\nu) = -\int_C h_\nu \frac{\partial h}{\partial n} ds = \rho_\nu \int_C \tilde{h}_\nu \frac{\partial \tilde{h}}{\partial n} ds = \rho_\nu \tilde{D}(\tilde{h}, \tilde{h}_\nu) \tag{8}$$

and, consequently

$$x_\nu = \tilde{x}_\nu \,. \tag{9}$$

Thus, both boundary relations (4) and (7) reflect themselves in a very simple manner in the relations (6) and (9) between the Fourier coefficients.

We define next the bilinear form

$$\pi_\varepsilon(h, H) = \frac{1}{2\pi}\int_C\int_C g_\varepsilon(z, \zeta) \frac{\partial h(z)}{\partial n} \frac{\partial H(\zeta)}{\partial n} ds_z ds_\zeta \tag{10}$$

for any two elements of Σ and in precisely the same manner we define the bilinear from $\tilde{\pi}_\varepsilon(\tilde{h}, \tilde{H})$ for any two elements in $\tilde{\Sigma}$.

By use of the Fourier type formulas (2.13) and (2.13′) we may express the bilinear forms in terms of the Fourier coefficients of the functions involved. Let us denote the Fourier coefficients of $h, \tilde{h}$ by $x_\nu, \tilde{x}_\nu$ and of $H, \tilde{H}$ by $y_\nu, \tilde{y}_\nu$; then a straightforward calculation shows that

$$\pi_\varepsilon(h, H) = \sum_{\nu=1}^{\infty} \frac{\varepsilon\rho_\nu}{1+\varepsilon\rho_\nu} x_\nu y_\nu,\ \tilde{\pi}_\varepsilon(\tilde{h}, \tilde{H}) = \sum_{\nu=1}^{\infty} \frac{\varepsilon\rho_\nu}{1+\varepsilon\rho_\nu} \tilde{x}_\nu \tilde{y}_\nu \,. \tag{11}$$

We verify, first, from (11) that the quadratic forms $\pi_\varepsilon(h, h)$ and $\tilde{\pi}_\varepsilon(\tilde{h}, \tilde{h})$ are positive-definite. This fact allows us to interpret them, indeed, as norms in their corresponding function spaces.

On the other hand, we have because of the normalizations (2.8) and (2.17)

$$D(h, H) = \sum_{\nu=1}^{\infty} \rho_\nu x_\nu y_\nu,\quad \tilde{D}(\tilde{h}, \tilde{H}) = \sum_{\nu=1}^{\infty} x_\nu y_\nu \,. \tag{12}$$

We define further the bilinear forms

$$\Gamma_\varepsilon(h, H) = D(h, H) - \frac{1}{\varepsilon}\pi_\varepsilon(h, H)\,,\quad \tilde{\Gamma}_\varepsilon(\tilde{h}, \tilde{H}) = \tilde{D}(\tilde{h}, \tilde{H}) - \tilde{\pi}_\varepsilon(\tilde{h}, \tilde{H}) \tag{13}$$

and obtain for them the explicit representations:

$$\Gamma_\varepsilon(h, H) = \sum_{\nu=1}^{\infty} \frac{\varepsilon\rho_\nu^2}{1+\varepsilon\rho_\nu} x_\nu y_\nu\,,\quad \tilde{\Gamma}_\varepsilon(\tilde{h}, \tilde{H}) = \sum_{\nu=1}^{\infty} \frac{1}{1+\varepsilon\rho_\nu} \tilde{x}_\nu \tilde{y}_\nu \,. \tag{14}$$

These formulas show that Γ_ε and $\tilde{\Gamma}_\varepsilon$, too, are positive-definite and lead to norms in Σ and $\tilde{\Sigma}$. We have the estimates:

$$0 \leq \frac{1}{\varepsilon}\pi_\varepsilon(h, H) \leq D(h, h)\,; \quad 0 \leq \tilde{\pi}_\varepsilon(\tilde{h}, \tilde{h}) \leq \tilde{D}(\tilde{h}, \tilde{h})\,. \tag{15}$$

By the very definition of π_ε and $\tilde{\pi}_\varepsilon$, we have

THEOREM I. *If*

$$\frac{\partial h}{\partial n} = \frac{\partial \tilde{h}}{\partial n} \quad \textit{and} \quad \frac{\partial H}{\partial n} = \frac{\partial \tilde{H}}{\partial n} \quad \text{on} \quad C \tag{16}$$

we have

$$\pi_\varepsilon(h, H) = \tilde{\pi}_\varepsilon(\tilde{h}, \tilde{H})\,. \tag{17}$$

From (4), (6) and (14), we derive :

THEOREM II. *If*

$$h = \tilde{h} \text{ and } H = \tilde{H} \text{ on } C\,, \tag{18}$$

we have

$$\Gamma_\varepsilon(h, H) = \varepsilon\tilde{\Gamma}_\varepsilon(\tilde{h}, \tilde{H})\,. \tag{19}$$

Finally, we verify from the explicit representations for the bilinear forms

THEOREM III. *If*

$$h = \tilde{h} \text{ and } \frac{\partial H}{\partial n} = \frac{\partial \tilde{H}}{\partial n} \text{ on } C\,, \tag{20}$$

we have

$$D(h, H) = -\tilde{D}(\tilde{h}, \tilde{H}) \tag{21}$$

and

$$\pi_\varepsilon(h, H) = -\varepsilon\tilde{\Gamma}_\varepsilon(\tilde{h}, \tilde{H}), \quad \Gamma_\varepsilon(h, H) = -\tilde{\pi}_\varepsilon(\tilde{h}, \tilde{H})\,. \tag{22}$$

Theorems I–III show a very symmetric interrelation between the various bilinear forms for elements with matching boundary data on C.

The significance of the preceding theorems lies in the fact that one has often to solve a boundary value problem, say in $\tilde{D}$, which is much easier to solve in the complementary domain D. In this case, the above theorems provide valuable information. Let us illustrate the method by the following applications.

(a) Given a function $h \in \Sigma$, to determine the function $\tilde{h} \in \tilde{\Sigma}$ which has on C the same boundary values as h. In particular, we ask for the Dirichlet integral $\tilde{D}(\tilde{h}, \tilde{h})$.

This problem arises, for example, in two-dimensional electrostatics in connection with the question of polarization of a set of conductors in a homogeneous field [19, 22].

We derive inequalities for the Dirichlet integral in question by applying Theorems I–III. We start from the fact that π_ε and Γ_ε have definite quadratic forms and that they satisfy, therefore, the Schwarz inequalities

$$(23)\quad \pi_\varepsilon(h, H)^2 \leq \pi_\varepsilon(h, h) \cdot \pi_\varepsilon(H, H)\,; \quad \Gamma_\varepsilon(h, H)^2 \leq \Gamma_\varepsilon(h, h) \cdot \Gamma_\varepsilon(H, H)\,.$$

We select a pair of test functions $H \in \Sigma$ and $\tilde{H} \in \tilde{\Sigma}$ which have equal normal derivatives on C and obtain from Theorem III and from (23)

$$(24)\qquad \Gamma_\varepsilon(h, H)^2 \leq \tilde{\pi}_\varepsilon(\tilde{h}, \tilde{h}) \cdot \tilde{\pi}_\varepsilon(\tilde{H}, \tilde{H})\,.$$

Using the definition (13) of $\tilde{\Gamma}_\varepsilon$ and Theorems I, II, we can transform (24) into

$$(25)\qquad \Gamma_\varepsilon(h, H)^2 \leq [\tilde{D}(\tilde{h}, \tilde{h}) - \frac{1}{\varepsilon}\Gamma_\varepsilon(h, h)]\pi_\varepsilon(H, H)\,.$$

This inequality contains the sought Dirichlet integral $\tilde{D}(\tilde{h}, \tilde{h})$ and else only the known function of $h \in \Sigma$ and the arbitrary test function $H \in \Sigma$. Thus:

$$(26)\qquad \tilde{D}(\tilde{h}, \tilde{h}) \geq \frac{\Gamma_\varepsilon(h, H)^2}{\pi_\varepsilon(H, H)} + \frac{1}{\varepsilon}\Gamma_\varepsilon(h, h)\,.$$

It is easily seen from our derivation that the inequality (26) is sharp if H is chosen as that function in Σ which has on C the same normal derivative as $\tilde{h}$; in fact, in this case, the Schwarz inequality leading to (24) becomes an equality. Thus, we can express (26) as follows:

$$(26')\qquad \tilde{D}(\tilde{h}, \tilde{h}) = \max \frac{\Gamma_\varepsilon(h, H)^2}{\pi_\varepsilon(H, H)} + \frac{1}{\varepsilon}\Gamma_\varepsilon(h, h) \text{ for all } H \in \Sigma\,.$$

This representation permits us to determine the desired Dirichlet integral by a Ritz procedure in Σ.

It is sometimes more convenient to renounce a precise equation in order to obtain a simple and applicable estimate. We may select, for this purpose, the test function $H(z)$ as equal to the given function $h(z)$; in this case, we have by (13) and (26)

$$(27)\qquad \tilde{D}(\tilde{h}, \tilde{h}) \geq \frac{\Gamma_\varepsilon(h, h)}{\pi_\varepsilon(h, h)} D(h, h)\,.$$

This inequality holds for all pairs of functions $h \in \Sigma, \tilde{h} \in \tilde{\Sigma}$ which have equal boundary values at the same points of C.

In order to understand better the important inequality (27), we express it in terms of the corresponding Fourier coefficients. If we denote again by x_ν the coefficients of $h(z)$, we have by (6) the values $-\rho_\nu x_\nu = \tilde{x}_\nu$ for the Fourier coefficients of $\tilde{h}(z)$. Hence, using the explicit representations (11), (12) and (14) for the quadratic forms, we may write (27) as follows:

$$\sum_{\nu=1}^{\infty} \rho_\nu^2 x_\nu^2 \cdot \sum_{\nu=1}^{\infty} \frac{\varepsilon\rho_\nu}{1+\varepsilon\rho_\nu} x_\nu^2 \geq \sum_{\nu=1}^{\infty} \rho_\nu x_\nu^2 \cdot \sum_{\nu=1} \frac{\varepsilon\rho_\nu^2}{1+\varepsilon\rho_\nu} x_\nu^2 . \tag{27'}$$

We rearrange (27′) into the from

$$\sum_{\nu,\mu=1}^{\infty} \frac{\varepsilon^2\rho_\nu\rho_\mu(\rho_\nu-\rho_\mu)^2}{(1+\varepsilon\rho_\nu)(1+\varepsilon\rho_\mu)} x_\nu^2 x_\mu^2 \geq 0 . \tag{27''}$$

Now the inequality has become evident; but, what is more important, we recognize that equality in (27″) and, hence in (27), holds if and only if all x_ν vanish except for those which belong to a fixed eigen value λ_μ. Thus, equality in (27) holds for

$$h(z) = h_\nu(z) \text{ and } \tilde{h}(z) = -\rho_\nu \tilde{h}_\nu(z) , \qquad \nu = 1, 2, \cdots, \tag{28}$$

and only for these functions.

It is interesting that the inequality (27) becomes precise infinitely often, namely for all functions of the sets $\{h_\nu\}$, $\{\tilde{h}_\nu\}$, which are complete in Σ and $\tilde{\Sigma}$. On the other hand, this fact leads to a new characterization of the Fredholm eigen functions

(b) We deal next with the analogous question: given a function $h \in \Sigma$, to determine the function $\tilde{h} \in \tilde{\Sigma}$ which has at corresponding points of C the same normal derivative as h. In particular, to determine the Dirichlet integral of $\tilde{h}$.

This problem occurs in the theory of a steady incompressible and irrotational fluid flow in the plane around the set of obstacles C. The sought Dirichlet integral, in this case, is the virtual mass of the curve system C [19, 22].

We select now a pair of test functions $H \in \Sigma$, $\tilde{H} \in \tilde{\Sigma}$ which have equal boundary values on C. Starting again with the Schwarz inequality (23) and Theorem III, we have

$$\pi_\varepsilon(h, H)^2 \leq \varepsilon^2 \tilde{\Gamma}_\varepsilon(\tilde{h}, \tilde{h}) \cdot \tilde{\Gamma}_\varepsilon(\tilde{H}, \tilde{H}) . \tag{29}$$

We apply equation (13), make use of Theorems I and II and find

$$\pi_\varepsilon(h, H)^2 \leq \varepsilon\Gamma_\varepsilon(H, H)[\tilde{D}(\tilde{h}, \tilde{h}) - \pi_\varepsilon(h, h)] . \tag{30}$$

Thus finally

$$(31)\qquad \tilde{D}(\tilde{h}, \tilde{h}) \geq \frac{\pi_\varepsilon(h, H)^2}{\varepsilon \Gamma_\varepsilon(H, H)} + \pi_\varepsilon(h, h) .$$

We obtained thus again a lower bound for the Dirichlet integral in terms of the given function h and the arbitrary test function H. If H has on C the same boundary values as $\tilde{h}$, the inequality (31) becomes an equality. This fact allows us again to approximate arbitrarily the Dirichlet integral from below by a Ritz sequence of test functions.

When we choose, on the other hand, $H(z) = h(z)$, we obtain

$$(32)\qquad \tilde{D}(\tilde{h}, \tilde{h}) \geq \frac{\pi_\varepsilon(h, h)}{\Gamma_\varepsilon(h, h)} D(h, h) .$$

This inequality holds for every pair of functions $h \in \Sigma$, $\tilde{h} \in \tilde{\Sigma}$ with equal normal derivatives on C.

This inequality can be verified by means of the explicit Fourier representations (11), (12) and (14) as we did in the case of the inequality (27). We can further show as before that equality in (32) can hold if and only if

$$(33)\qquad h(z) = h_\nu(z) , \quad \tilde{h}(z) = \tilde{h}_\nu(z) , \qquad \nu = 1, 2, \cdots .$$

Thus, inequality (32) leads to another characterization of the Fredholm eigen functions.

We obtain corresponding inequalities when we interchange the role of D and $\tilde{D}$; the Dirichlet integral of a function $h \in \Sigma$ can then be estimated in terms of a function $\tilde{h} \in \tilde{\Sigma}$ which has on C either the same boundary values or the same normal derivative as h.

The most convenient form in which the preceding theory can be applied is obtained by using $\varepsilon = 1$. For, in this case, the dielectric Green's function reduces to the elementary function $-\log|z - \zeta|$ and the bilinear forms can be easily evaluated. Indeed, the general method was first applied to obtain isoperimetric inequalities for polarization and virtual mass with this particular choice of ε [18, 19, 20]. However, the flexibility of the method is obviously increased by considering arbitrary positive ε-values and the significance of the procedure is clarified in this way.

We shall now utilize the quadratic forms in order to obtain estimates for the Fredholm eigen values λ_ν. Let λ_1 be the lowest positive Fredholm eigen value > 1. We have shown in § 1 that with λ_1 also $-\lambda_1$ is an eigen value. We denote $\lambda_2 = -\lambda_1$. By definition (2.12) of the ρ_ν, we have obviously

$$(34)\qquad \frac{1}{\rho_1} = \rho_2 \leq \rho_\nu \leq \rho_1 , \qquad \nu = 1, 2, 3, \cdots .$$

Using now the developments (11), (12) and (14) of the various bilinear forms, we verify by inspection the following theorems:

THEOREM IV. *For every function* $h \in \Sigma$ *the inequalities*

$$\frac{\varepsilon}{1+\varepsilon\rho_1} \leq \frac{\pi_\varepsilon(h, h)}{D(h, h)} \leq \frac{\varepsilon\rho_1}{\rho_1+\varepsilon} \tag{35}$$

hold. The first equality sign holds only for those function $h_\nu \in \Sigma$ *which belong to the eigen value* λ_1*; the second equality sign holds only for functions* $h_\nu \in \Sigma$ *which belong to the eigen value* λ_2.

THEOREM V. *Every function* $\tilde{h} \in \tilde{\Sigma}$ *satisfies the inequalities*

$$\frac{\varepsilon}{\rho_1+\varepsilon} \leq \frac{\tilde{\pi}_\varepsilon(\tilde{h}, \tilde{h})}{\tilde{D}(\tilde{h}, \tilde{h})} \leq \frac{\varepsilon\rho_1}{1+\varepsilon\rho_1}\,. \tag{36}$$

Equality holds only if $\tilde{h} = \tilde{h}_\nu$ *where* $\tilde{h}_\nu$ *belongs to the eigen values* λ_2 *and* λ_1*, respectively.*

We have thus characterized the lowest positive and non-trivial Fredholm eigen value λ_1 by a minimum and a maximum problem in Σ and in $\tilde{\Sigma}$ for the ratio of two positive-definite quadratic forms. This characterization makes it possible to estimate this eigen value by the use of test functions in Σ and in $\tilde{\Sigma}$. The most convenient case for applications is, of course, the case $\varepsilon = 1$.

It is clearly desirable to find analogous extremum problems which characterize the higher eigen values λ_ν. For this purpose, we introduce the bilinear form

$$\Gamma_{\varepsilon,e}(h, H) = \frac{\Gamma_\varepsilon(h, H) - \Gamma_e(h, H)}{\varepsilon - e}\,, \qquad \varepsilon > 0,\, e > 0 \tag{37}$$

in Σ and the bilinear form

$$\tilde{\pi}_{\varepsilon,e}(\tilde{h}, \tilde{H}) = \frac{\tilde{\pi}_\varepsilon(\tilde{h}, \tilde{H}) - \tilde{\pi}_e(\tilde{h}, \tilde{H})}{\varepsilon - e}\,, \qquad \varepsilon > 0,\, e > 0 \tag{37'}$$

in $\tilde{\Sigma}$. From (11) and (14), we obtain the Fourier representations

$$\Gamma_{\varepsilon,e}(h, H) = \sum_{\nu=1}^{\infty} \frac{\rho_\nu^2 x_\nu y_\nu}{(1+\varepsilon\rho_\nu)(1+e\rho_\nu)}\,; \quad \tilde{\pi}_{\varepsilon,e}(\tilde{h}, \tilde{H}) = \sum_{\nu=1}^{\infty} \frac{\rho_\nu \tilde{x}_\nu \tilde{y}_\nu}{(1+\varepsilon\rho_\nu)(1+e\rho_\nu)} \tag{38}$$

The quadratic forms $\Gamma_{\varepsilon,e}(h, h)$ and $\tilde{\pi}_{\varepsilon,e}(\tilde{h}, \tilde{h})$ are evidently positive-definite.

We observe that the function

$$f(x) = \frac{x}{(1+\varepsilon x)(1+e x)} \tag{39}$$

takes in the interval $0 \leq x < \infty$ its maximum value at the point

$$X_m = \frac{1}{\sqrt{\varepsilon e}} . \tag{40}$$

Hence, (38) yields the following theorems:

THEOREM VI. *Every function $h \in \Sigma$ satisfies the inequality*

$$\frac{\Gamma_{\varepsilon,e}(h, h)}{D(h, h)} \leq f(\rho_m) \tag{41}$$

where ρ_m is a value in the sequence of the ρ_ν which gives the largest value of f. Equality holds only for such h_ν which belong to such a value ρ_m.

THEOREM VII. *For every function $\tilde{h} \in \tilde{\Sigma}$, the inequality*

$$\frac{\tilde{\pi}_{\varepsilon,e}(\tilde{h}, \tilde{h})}{\tilde{D}(\tilde{h}, \tilde{h})} \leq f(\rho_m) \tag{42}$$

holds where ρ_m is a value in the sequence of the ρ_ν which gives the largest possible value of $f(\rho)$. Equality holds only for such $\tilde{h}_\nu$ which belong to such a ρ_m.

Given any specific ρ_ν, we can always choose $\sqrt{\varepsilon e} = \rho_\nu^{-1}$ and the corresponding maximum problem will pick out this particular eigen value. We can apply Theorems VI and VII in order to obtain estimates for the location of ρ_ν-values near any given point x_m by the use of test functions in Σ and in $\tilde{\Sigma}$. It is easily seen that Theorems IV and V are contained in Theorems VI and VII as limit cases.

We specialize in Theorem IV $\varepsilon = 1$ and obtain the particular result

$$\frac{1}{2\pi}\int_\sigma\int_\sigma \log\frac{1}{|z-\zeta|}\frac{\partial h(z)}{\partial n}\frac{\partial h(\zeta)}{\partial n}ds_z ds_\zeta \leq \frac{\rho_1}{1+\rho_1}D(h, h) \tag{35'}$$

for every $h \in \Sigma$; equality holds only if $h = h_\nu$ and h_ν belongs to λ_2.

This result permits the following application. Consider the system of curves C^* which consists of the subset $C_1, C_2, \cdots, C_{N^*}$ of C with $N^* < N$. This system of boundaries determines a connected exterior $\tilde{D}^* \supset \tilde{D}$ and the set D^* of the domains D_j, $j = 1, \cdots, N^*$. Let Σ^* be the function class in D^* which is analogous to the class Σ in D and let $h_2^*(z)$ correspond to the largest non-trivial negative eigen value λ_2^* of C^*. We determine a function $h(z) \in \Sigma$ such that

$$\frac{\partial h}{\partial n} = \frac{\partial h_2^*}{\partial n} \text{ on } C^*, \quad \frac{\partial h}{\partial n} = 0 \text{ on } C - C^* . \tag{43}$$

Since the boundary conditions (43) determine $h(z)$ in each D_j only up to an additive constant, we may adjust these constants in such a way that $h(z)$ satisfies the N conditions (2.18) and thus belongs indeed to Σ. Observe that the Dirichlet integral of h coincides in each D_j, $j \leq N^*$ with the corresponding Dirichlet integral of h_2^*, since h and h_2^* differ only by a constant in these domains. In each D_j with $j > N^*$, $h(z)$ is a constant and has the Dirichlet integral zero. Hence:

$$(44) \qquad D^*(h_2^*, h_2^*) = D(h, h) .$$

By (35′) we have

$$(45) \qquad \frac{\rho_1^*}{1+\rho_1^*} D^*(h_2^*, h_2^*) = \frac{1}{2\pi}\int_{C^*}\int_{C^*} \log\frac{1}{|z-\zeta|}\frac{\partial h_2^*(z)}{\partial n}\frac{\partial h_2^*(\zeta)}{\partial n} ds_z ds_\zeta$$

$$= \frac{1}{2\pi}\int_C\int_C \log\frac{1}{|z-\zeta|}\frac{\partial h(z)}{\partial n}\frac{\partial h(\zeta)}{\partial n} ds_z ds_\zeta \leq \frac{\rho_1}{1+\rho_1} D(h, h) .$$

By virtue of (44), we conclude finally

$$(46) \qquad \frac{\rho_1^*}{1+\rho_1^*} \leq \frac{\rho_1}{1+\rho_1}, \quad \rho_1^* \leq \rho_1 .$$

Thus, we proved:

THEOREM VIII. *The lowest positive and non-trivial eigen value λ_1 of a curve system C is never larger than the corresponding eigen value λ_1^* of any subsystem C^* of C.*

Suppose all positive eigen values of C arranged in increasing order, say $\lambda_{\nu'}$, such that $\nu' < \nu''$ implies $\lambda_{\nu'} \leq \lambda_{\nu''}$. Let us do the same with the positive eigen values $\lambda_{\nu'}^*$ of the subsystem C^*. By the above reasoning and by use of the standard methods of eigen value theory [cf. 11], it is easily shown that quite generally

$$(47) \qquad \lambda_{\nu'} \leq \lambda_{\nu'}^*$$

will be fulfilled.

We consider finally the bilinear form

$$(48) \qquad B(h, H) = \frac{1}{2\pi}\int_C\int_C \Gamma(\zeta, \eta)\frac{\partial h(\zeta)}{\partial n}\frac{\partial H(\eta)}{\partial n} ds_\zeta ds_\eta$$

where $\Gamma(\zeta, \eta)$ is the geometric kernel defined in (2.26). For $h \in \Sigma$, $H \in \Sigma$ we have, in view of (2.28) the following Fourier representation for B:

$$(49) \qquad B(h, H) = \sum_{\nu=1}^{\infty} \frac{\rho_\nu}{(1+\rho_\nu)^2} x_\nu y_\nu$$

and the same expression is also valid for $\tilde{h} \in \tilde{\Sigma}$, $\tilde{H} \in \tilde{\Sigma}$.

From (11), (38) and (49) follows

$$\pi_1(h, H) - B(h, H) = \sum_{\nu=1}^{\infty} \frac{\rho_\nu^2}{(1+\rho_\nu)^2} x_\nu y_\nu = \Gamma_{1,1}(h, H) , \tag{50}$$

and

$$B(\tilde{h}, \tilde{H}) = \tilde{\pi}_{1,1}(\tilde{h}, \tilde{H}) . \tag{51}$$

The function

$$f(x) = \frac{x}{(1+x)^2} \tag{39'}$$

takes its maximum 1/4 for positive argument at the point $x_m = 1$ and we derive from (41) and (42) the inequalities

$$0 \leq \pi_1(h, h) - B(h, h) \leq \frac{1}{4} D(h, h) , \qquad h \in \Sigma \tag{52}$$

and

$$0 \leq B(\tilde{h}, \tilde{h}) \leq \frac{1}{4} \tilde{D}(\tilde{h}, \tilde{h}) , \qquad \tilde{h} \in \tilde{\Sigma} . \tag{53}$$

These inequalities are interesting since they yield estimates for the Dirichlet integrals of h and $\tilde{h}$ by means of elementary integrations over C which involve only the normal derivatives on C of these functions and geometric terms. On the other hand, given only these normal derivatives, we could calculate the precise Dirichlet integrals only after solving a Neumann boundary value problem for the domains. We gave by inequality (32) another lower bound for $\tilde{D}(\tilde{h}, \tilde{h})$; but in this estimate we have to assume as known the solution of the corresponding boundary value problem for the complimentary region D of $\tilde{D}$. The present inequalities are, therefore, often easier to apply.

The dielectric Green's functions $g_\varepsilon(z, \zeta)$ and $g_e(z, \zeta)$ which are needed in the calculation of $\tilde{\pi}_{\varepsilon,e}$ and $\Gamma_{\varepsilon,e}$ are known only for very few domains if ε and e are different from 1. We may, however, use the series developments (2.34) for these functions and utilize the partial sums in the development together with a simple estimate for the remainder terms in order to obtain estimates for ρ_m. The calculations are clearly quite laborious, but in principle feasible.

6. Variational formulas for the dielectric Green's functions and for the Fredholm eigen values. The properties (a)–(e) ennumerated in § 2 and defining the dielectric Green's functions $g_\varepsilon(z, \zeta)$ are all invariant under a conformal mapping $z^* = F(z)$ which is normalized at infinity such that $|F'(\infty)| = 1$. Unfortunately, the only conformal mapping of this kind which is regular in the entire complex plane has the trivial form $F(z) = az + b$, $|a| = 1$. We may consider, however, functions $F(z)$ which are analytic with isolated singularities. In this way, we are led naturally to a variational theory for the dielectric Green's functions.

The simplest possible choice of $F(z)$ is evidently

$$z^* = F(z) = z + \frac{\alpha}{z - z_0} \tag{1}$$

which has the right normalization at infinity but has a simple pole at $z = z_0$. We will choose z_0 arbitrarily in D or in $\tilde{D}$ but not on the curve system C. Let $E(z_0)$ denote the entire complex plane from which a circle of radius $\sqrt{|\alpha|}$ around the center z_0 has been removed. It is easily seen that $F(z)$ is univalent in $E(z_0)$. Given, therefore, a fixed point z_0 in D or in $\tilde{D}$, we can always choose $|\alpha|$ so small that C lies in $E(z_0)$ and is mapped in a one-to-one manner into a new curve system C^*. Since $F(z)$ is regular analytic in $E(z_0)$ all differentiability properties of C are transferred to C^*. We denote the dielectric Green's functions of the new curve system C^* by $g_\varepsilon^*(z, \zeta)$. Our aim is to connect these new functions with the functions $g_\varepsilon(z, \zeta)$ of the original system C.

We introduce the function

$$d(z, \zeta) = g_\varepsilon^*(F(z), F(\zeta)) - g_\varepsilon(z, \zeta) . \tag{2}$$

By the definition of g_ε^* and of the curve system C^*, the function $d(z, \zeta)$ is symmetric and harmonic for $z, \zeta \in E(z_0)$, except along the curve set C. The function is still continuous across C but its normal derivatives satisfy the discontinuity relation

$$\frac{\partial}{\partial n_z} d(z, \zeta) + \varepsilon \frac{\partial}{\partial \tilde{n}_z} d(z, \zeta) = 0 \quad \text{for} \quad z \in C, \zeta \in E(z_0) - C . \tag{3}$$

Observe that $d(z, \zeta)$ is still regular harmonic for $z = \zeta$ and that

$$\lim_{z \to \infty} d(z, \zeta) = 0 . \tag{4}$$

We consider now the integral

$$J(\zeta, \eta) = \frac{1}{2\pi} \int_C \left[d(z, \zeta) \frac{\partial}{\partial n_z} g_\varepsilon(z, \eta) - g_\varepsilon(z, \eta) \frac{\partial}{\partial n_z} d(z, \zeta) \right] ds_z . \tag{5}$$

We introduce the characteristic function $\delta(z)$ of D, i.e., we define

$$\delta(z) = \begin{cases} 1 & \text{if} \quad z \in D \\ 0 & \text{if} \quad z \notin D . \end{cases} \tag{6}$$

By Green's identity applied to D, we find

$$J(\zeta, \eta) = \varepsilon d(\zeta, \eta)\delta(\eta) + T(\zeta, \eta)\delta(z_0) . \tag{7}$$

Here

$$T(\zeta, \eta) = \frac{1}{2\pi} \int_{c(z_0)} \left[d(z, \zeta)\frac{\partial}{\partial n_z} g_\varepsilon(z, \eta) - g_\varepsilon(z, \eta)\frac{\partial}{\partial n_z} d(z, \zeta) \right] ds_z , \tag{8}$$

where $c(z_0)$ is the circumference of radius $\sqrt{|\alpha|}$ around z_0 and where **n** is its interior normal.

On the other hand, we may apply Green's identity to $J(\zeta, \eta)$ with respect to the complementary domain $\tilde{D}$. Taking notice of (4) and of the known discontinuity behavior of the various terms in the integrand, we find

$$J(\zeta, \eta) = - \varepsilon d(\zeta, \eta)[1 - \delta(\eta)] - \varepsilon T(\zeta, \eta)[1 - \delta(z_0)] . \tag{9}$$

Subtracting (9) from (7), we obtain finally

$$\varepsilon d(\zeta, \eta) = - T(\zeta, \eta)[\varepsilon + (1 - \varepsilon)\delta(z_0)] . \tag{10}$$

The difference function (2) of g_ε^* and g_ε is thus expressed in terms of an integral over the small circle $c(z_0)$ around the singularity point z_0.

A straightfoward calculation of the type usual in such variational problems [15, 21] yields

$$g_\varepsilon^*(\zeta^*, \eta^*) = g_\varepsilon(\zeta, \eta) + \left[1 + \left(\frac{1}{\varepsilon} - 1\right)\delta(z_0)\right]\Re\{\alpha p_\varepsilon'(z_0, \zeta)p_\varepsilon'(z_0, \eta)\} + O(|\alpha|^2) , \tag{11}$$

where $p_\varepsilon(z, \zeta)$ is the analytic function defined in § 4 whose real part is $g_\varepsilon(z, \zeta)$. The error term $O(|\alpha|^2)$ can be estimated uniformly for ζ and η in $E(z_0)$ and for z_0 in any fixed closed domain which does not contain points of C.

We derived in (11) an interior variational formula for the dielectric Green's function which is very similar to the well-known variational formula for the ordinary Green's function of a domain [14, 15]. Observe that in the special case $\varepsilon = 1$ formula (11) reads

$$\log \frac{1}{|\zeta^* - \eta^*|} = \log \frac{1}{|\zeta - \eta|} + \Re\left\{\frac{\alpha}{(z_0 - \zeta)(z_0 - \eta)}\right\} + O(|\alpha|^2) . \tag{11'}$$

In view of the identity

$$\log |\zeta^* - \eta^*| = |\zeta - \eta| + \log\left|1 - \frac{\alpha}{(z_0 - \zeta)(z_0 - \eta)}\right| , \tag{11''}$$

we can verify (11′) directly by means of the logarithmic series.

We shall not enter into the variational theory of the dielectric

Green's functions since it is entirely analogous to that given in the case of simply-connected domains [17]. We wish to utilize (11) in order to derive analogous variational formulas for the eigen values λ_ν. For this purpose, we shall make use of the extremum principles (5.41) and (5.42) and of the method of transplanting the extremum function [**6**, **11**].

Let us suppose that the singular point z_0 of our variation (1) lies in $\tilde{D}$; in this case, the function $F(z)$ is regular and univalent in D. If $h(z)$ is any analytic function in D, we can define by

$$h^*(z^*) = h(z) \tag{12}$$

a regular analytic function h^* in each component D_j^* of the varied domain set D^*. We call the definition (12) the transplantation of the function $h(z)$ from D into D^*.

We define now the ratios

$$R(h) = \frac{\Gamma_{\varepsilon,e}(h, h)}{D(h, h)}, \qquad R^*(h^*) = \frac{\Gamma^*_{\varepsilon,e}(h^*, h^*)}{D^*(h^*, h^*)} \tag{13}$$

which occur in the extremum problem (5.41). In view of the conformal character of the transplantation, we have clearly

$$D(h, h) = D^*(h^*, h^*) \tag{14}$$

and

$$\frac{\partial h^*(z^*)}{\partial n^*} ds^* = \frac{\partial h(z)}{\partial n} ds, \qquad z \in C, z^* \in C^*. \tag{15}$$

It is, therefore, easy to calculate the ratio $R^*(h^*)$ by referring back to the original region D. By the definitions (5.10), (5.13) and (5.37), we find

$$\Gamma^*_{\varepsilon,e}(h^*, h^*) = \frac{1}{\varepsilon - e} \cdot \frac{1}{2\pi} \int_{C^*}\int_{C^*} \left[\frac{1}{e} g_e^*(\zeta^*, \eta^*) - \frac{1}{\varepsilon} g_\varepsilon^*(\zeta^*, \eta^*)\right] \cdot \\ \cdot \frac{\partial h^*(\zeta^*)}{\partial n^*} \frac{\partial h^*(\eta^*)}{\partial n^*} ds_\zeta^* ds_\eta^* . \tag{16}$$

Now, we use (11) and (15) in order to return to the curve system C as the path of integration. We remember that $z_0 \in \tilde{D}$ and obtain

$$\Gamma^*_{\varepsilon,e}(h^*, h^*) = \Gamma^*_{\varepsilon,e}(h, h) + 2\pi\Re\left\{\alpha \frac{e^{-1}q_e(z_0)^2 - \varepsilon^{-1}q_\varepsilon(z_0)^2}{\varepsilon - e}\right\} + O(|\alpha|^2) \tag{17}$$

with

$$q_\varepsilon(z) = \frac{1}{2\pi}\int_C p_\varepsilon'(z, \zeta) \frac{\partial h(\zeta)}{\partial n} ds_\zeta . \tag{17'}$$

Since $z_0 \in \tilde{D}$, we can express $q_\varepsilon(z)$ as a surface integral

$$q_\varepsilon(z_0) = -\frac{1}{\partial z_0}\left[\frac{1}{\pi}\iint_D \nabla g_\zeta(z_0, \zeta)\cdot \nabla h(\zeta) d\tau_\zeta\right]. \tag{18}$$

The error term $O(|\alpha|^2)$ can be estimated uniformly for all functions $h(z)$ with bounded Dirichlet integral and for z_0 in a closed subdomain of $\tilde{D}$. We have to use the known error term in the variational formula (11) for the dielectric Green's function.

As a first result we can conclude that the eigen values of the ratio $R^*(h^*)$ depend continuously on α and converge with $|\alpha| \to 0$ to the corresponding eigen values of $R(h)$. We can, moreover, derive a precise asymptotic formula for these eigen values.

Let indeed ρ_0 be a particular ρ_ν-value of the original curve system C and let the function $f(x)$, defined in (5.39), be chosen in such a way that it takes its maximum at a point x_m which is nearer to ρ_0 than to any other ρ_ν. If $h_0 \in \Sigma$ is an eigen function which belongs to ρ_0, we will have

$$R(h_0) = f(\rho_0) . \tag{19}$$

We may assume as before (see (2.17)) that

$$D(h_0, h_0) = \rho_0 . \tag{19'}$$

If h_0^* is the transplantation of h_0 into D^*, we can use (14) and (17) in order to determine its ratio $R^*(h_0^*)$. But now we can use formulas (2.9), (2.10) and (2.13′) in order to express the analytic function $q_\varepsilon(z)$ by means of the analytic completion of $\tilde{h}_0(z)$ defined in (1.31). We have

$$q_\varepsilon(z_0) = \frac{\varepsilon\rho_0}{1+\varepsilon\rho_0}\tilde{V}_0'(z_0) , \qquad z_0 \in \tilde{D} . \tag{20}$$

We can now combine (14), (17) and (19) in order to express $R^*(h_0^*)$. We make also use of (19) and of the definition (5.39) of $f(x)$; thus, we arrive finally at

$$R^*(h_0^*) = f(\rho_0) - 2\pi\rho_0 f'(\rho_0)\Re\{\alpha \tilde{V}_0'(z_0)^2\} + O(|\alpha|^2) . \tag{21}$$

The function $h_0^*(z^*)$ defined by the transplantation of $h_0(z)$ will not, in general, belong to the class Σ^* defined with respect to D^* by linear conditions analogous to (2.18). However, we can add to every function $h^*(z^*)$ which is analytic in D^* a different constant in each component D_j^* in order to bring it into the class Σ^*. This trivial readjustment does not affect the Dirichlet integral nor the quadratic from $\Gamma_{\varepsilon,e}^*$ which depends only upon the normal derivatives of h^*. Thus, in the theory of the ratio $R^*(h^*)$ the restriction to the class Σ^* is unessential, since easily achieved.

In particular, we may use h_0^* as a competing function for the extremum problem regarding $R^*(h^*)$ and use the identity (21) in order to estimate the extremum values. Let us suppose that the value ρ_0 belongs to k different eigen functions $h_\beta(z)$ of the unperturbed curve system C; we denote their analytic completions by $V_\beta(z)$. We restrict, at first, $h^*(z^*)$ to the linear sub-space spanned by the k transplanted eigen functions $h_\beta^*(z^*)$. In this case, the ratio $R^*(h^*)$ will have precisely the k stationary values

$$\tau_\beta = f(\rho_0) + 2\pi\rho_0 f'(\rho_0)\sigma_\beta + O(|\alpha|)\,, \qquad \beta = 1, 2, \cdots, k \tag{22}$$

where the σ_β are the eigen values of the secular equation

$$\det \| \Re\{\alpha \tilde{V}_i'(z_0)\tilde{V}_j'(z_0)\} + \sigma\delta_{ij} \|_{i,j=1,2\cdots k} = 0\,. \tag{23}$$

Let us arrange the τ_β in decreasing order; likewise, we shall arrange the values $f(\rho_\beta^*)$ in decreasing order. Since the k first values $f(\rho_\beta^*)$ are the largest stationary values of $R^*(h^*)$ for unrestricted argument function h^*, it follows from standard results on quadratic forms that

$$f(\rho_\beta^*) \geq f(\rho_0) + 2\pi\rho_0 f'(\rho_0)\sigma_\beta + O(|\alpha|^2), \qquad \beta = 1, \cdots, k\,. \tag{24}$$

Because of the continuous dependence of the eigen values ρ_γ^* on α there exists a positive constant δ such that for small enough α all eigen values ρ_γ^* have from ρ_0 a distance larger than δ, except for k eigen values ρ_β^* which can be brought arbitrarily near to ρ_0.

Having now chosen $|\alpha|$ sufficiently small, we can select x_m to the left of ρ_0 and the k neighboring ρ_β^* but so near that all other $f(\rho_\gamma^*)$ are less than any of the $f(\rho_\beta^*)$. Since $f'(\rho) < 0$ for ρ_0 and all ρ_β^*, we derive from (24)

$$\rho_\beta^* \leq \rho_0 + 2\pi\rho_0\sigma_\beta + O(|\alpha|^2)\,, \qquad \beta = 1, 2, \cdots, k\,. \tag{24'}$$

Choosing, on the other hand, x_m to the right of ρ_0 and the ρ_β^* but again so near that $f(\rho_\beta^*)$ is still larger than all $f(\rho_\gamma^*)$, we obtain

$$\rho_\beta^* \geq \rho_0 + 2\pi\rho_0\sigma_\beta + O(|\alpha|^2)\,, \qquad \beta = 1, 2, \cdots, k\,. \tag{24''}$$

Thus, we proved:

The variation of an eigen value ρ_0 with degree of degeneracy $k-1$ is characterized by the formula

$$\rho_\beta^* = \rho_0 + 2\pi\rho_0\sigma_\beta + 0(|\alpha|^2) \tag{25}$$

where the σ_β are the eigen values of the secular equation (23).

In the case that only one eigen function $h_\nu \in \Sigma$ belongs to ρ_ν, we obtain the simpler variational formula

$$\delta\rho_\nu = -\Re\{2\pi\alpha\rho_\nu \tilde{V}_\nu'(z_0)^2\}\,. \tag{26}$$

By the relation (2.12) between ρ_ν and the Fredholm eigen value λ_ν, we obtain in this case finally

$$\delta\lambda_\nu = (\lambda_\nu^2 - 1)\pi\Re\{\alpha \tilde{V}'_\nu(z_0)^2\} . \tag{27}$$

We can proceed in analogous fashion in the case that $z_0 \in D$. We will start then with $\tilde{h}_0 \in \tilde{\Sigma}$ which belongs to ρ_0 and which satisfies by (5.42) the equation

$$\tilde{R}(\tilde{h}_0) = \frac{\tilde{\pi}_{\mathfrak{e},e}(\tilde{h}_0, \tilde{h}_0)}{\tilde{D}(\tilde{h}_0, \tilde{h}_0)} = f(\rho_0) . \tag{28}$$

We transplant $\tilde{h}_0$ by an equation (12) into a comparison function $\tilde{h}_0^*$ in $\tilde{D}^*$. We assume the usual normalization

$$\tilde{D}(\tilde{h}_0, \tilde{h}_0) = 1 \tag{29}$$

and have, therefore, also

$$\tilde{D}^*(\tilde{h}_0^*, \tilde{h}_0^*) = 1 . \tag{29'}$$

The same chain of calculations as before leads to the asymptotic formula

$$\tilde{R}^*(\tilde{h}_0^*) = \frac{\tilde{\pi}^*_{\mathfrak{e},e}(\tilde{h}_0^*, \tilde{h}_0^*)}{\tilde{D}^*(\tilde{h}_0^*, \tilde{h}_0^*)} = f(\rho_0) + 2\pi f'(\rho_0)\Re\{\alpha V_0'(z_0)^2\} + O(|\alpha|^2) . \tag{30}$$

Here, $V_0(z)$ is the analytic completion of $h_0(z)$ in D. This formula is very similar to (21); it differs only by the factor $-\rho_0$. We obtain, therefore, the following result:

If ρ_ν is an eigen value of degeneracy $k-1$ it will change according to the formula

$$\rho_\beta^* = \rho_\nu + 2\pi\sigma_\beta + O(|\alpha|^2) \qquad \beta = 1, 2, \cdots, k \tag{31}$$

under a variation (1) of the curve system C. The σ_β are the k eigen values of the secular equation

$$\det \| \Re\{\alpha V_i'(z_0) V_j'(z_0)\} - \sigma\delta_{ij} \|_{i,j=1,\cdots,k} = 0 \tag{32}$$

and the $V_i(z)$ are the k analytic functions whose real parts are the eigen functions $h_i(z)$ which belong to ρ_ν.

In the particular case $k = 1$, i.e., non-degeneracy, we have

$$\delta\rho_\nu = \Re\{2\pi\alpha V_\nu'(z_0)^2\} \tag{32'}$$

and hence

$$\delta\lambda_\nu = -(\lambda_\nu - 1)^2\pi\Re\{\alpha V_\nu'(z_0)^2\} . \tag{33}$$

There is a lack of symmetry between the variational formulas (23), (25), on the one hand, and (31), (32) on the other. This fact is due to the different normalizations

$$\iint_D \left| V_i'(z) \right|^2 d\tau = D(h_i, h_i) = \rho_i \tag{34}$$

and

$$\iint_{\tilde{D}} \left| \tilde{V}_i'(z) \right|^2 d\tau = D(\tilde{h}_i, \tilde{h}_i) = 1 \,. \tag{35}$$

We were led to these normalizations from the theory of the Fredholm eigen functions $\varphi_i(z)$ through the representation (1.3). These normalizations were also used in the series developments of §§ 2 and 3. However, the variational formulas become symmetric when we define

$$u_\nu(z) = \rho_\nu^{-1/2} V_\nu'(z) \,, \quad \tilde{u}_\nu(z) = i \tilde{V}_\nu'(z) \,. \tag{36}$$

From the definition of the $V_\nu(z)$ and $\tilde{V}_\nu(z)$, their normalizations (34) and (35) and from the definitions (1.33), (1.34) it follows at once that the functions (36) are identical with the functions $u_\nu(z)$ and $\tilde{u}_\nu(z)$ defined at the end of § 1 and normalized by (1.34).

By means of the functions $u_\nu(z)$ and $\tilde{u}_\nu(z)$ we can express the law of variations of the eigen values λ_ν as follows:

THEOREM. *Let λ_ν be a Fredholm eigen value of the curve system C and of degeneracy $k-1$; let $u_\beta(z), \tilde{u}_\beta(z) (\beta = 1, 2, \cdots, k)$ be the set of analytic eigen functions to this eigen value. If we subject the system C to a variation* (1), *we have*

$$\frac{\delta \lambda_\nu}{\lambda_\nu^2 - 1} = \pi \sigma_\beta \tag{37}$$

where σ_β is an eigen value of the secular equation

$$\det \| \Re\{ a u_i(z_0) u_j(z_0) \} + \sigma \delta_{ij} \| = 0 \quad \text{if} \quad z_0 \in D \tag{38}$$

or of

$$\det \| \Re\{ a \tilde{u}_i(z_0) \tilde{u}_j(z_0) \} + \sigma \delta_{ij} \| = 0 \quad \text{if} \quad z_0 \in \tilde{D} \,. \tag{39}$$

In particular, we have in the case of non-degeneracy

$$\frac{\delta \lambda_\nu}{\lambda_\nu^2 - 1} = -\pi \Re\{ a u_\nu^2(z_0) \} \quad \text{for} \quad z_0 \in D \tag{40}$$

and

(40′)
$$\frac{\delta\lambda_\nu}{\lambda_\nu^2 - 1} = -\pi\Re\{\alpha\tilde{u}_\nu^2(z_0)\} \quad \text{for} \quad z_0 \in \tilde{D}\,.$$

The preceding variational formulas can also be derived easily from the original integral equation (1.2) by means of the general theory of perturbations [17]. The above derivation is of interest since it allows a more detailed study of the error terms by means of the dielectric Green's function. It is also possible to obtain more precise statements by using the higher variational terms of these Green's functions. It is particularly easy to develop the higher variations for the lowest positive and non-trivial eigen value λ_1. Consider, for example, a variation (1) of the curve system C with $z_0 \in \tilde{D}$. Let $h(z) \in \Sigma$ and h^* its transplantation into D^*. By definition (5.10) and the identity (11″), we have

(41)
$$\pi_1^*(h^*, h^*) = \pi_1(h, h) - \frac{1}{2\pi}\int_C\int_C \log\left|1 - \frac{\alpha}{(z - z_0)(\zeta - z_0)}\right| \frac{\partial h(z)}{\partial n}\frac{\partial h(\zeta)}{\partial n}\, ds_z ds_\zeta\,.$$

Thus, $\pi_1(h, h)$ has a very simple transformation law under transplantation. The Dirichlet integral is invariant under transplantation. Since ρ_1 leads to the extremum values of the ratio (5.35) it is possible to determine the variations of higher order of λ_1 with relatively little labor.

We wish, finally, to add a simple algebraic remark to the variational formulas (37), (38) and (39). If λ_ν is of degeneracy $k - 1$ a variation (1) will, in general, reduce this degeneracy. It is, however, remarkable that the secular equations (38) and (39) have only two different eigen values such that even after the variation a degenerate eigen value can only split into two different eigen values, at least, up to the order $O(|\alpha|^2)$. Indeed, σ is an eigen value, say of (38) if there exist k real numbers t_j such that the linear equations

(42)
$$\sigma t_i + \sum_{j=1}^{k} \Re\{\alpha u_i u_j\} t_j = 0\,, \qquad i = 1, \cdots, k$$

hold while

(42′)
$$\sum_{j=1}^{k} t_j^2 = 1\,.$$

We denote

(43)
$$\sum_{j=1}^{k} u_j t_j = M$$

and reduce (42) to

(44)
$$\sigma t_i + \Re\{\alpha u_i M\} = 0\,, \qquad i = 1, \cdots, k\,.$$

Multiplying the ith equation (44) with u_i and summing over all i-values, we find:

$$\sigma M + \frac{1}{2}\alpha M \sum_{i=1}^{k} u_i^2 + \frac{1}{2}\bar{\alpha}\bar{M} \sum_{i=1}^{k} |u_i|^2 = 0. \tag{45}$$

On the other hand, multiplying (44) with t_i and summing over i, we obtain from (42')

$$\sigma + \Re\{\alpha M^2\} = 0 . \tag{46}$$

From (45) and (46) we derive

$$-\sigma = \Re\{\alpha M^2\} = \frac{1}{2}\alpha \sum_{i=1}^{k} u_i^2 + \frac{1}{2}\frac{|\alpha M|^2}{\alpha M^2} \sum_{i=1}^{k} |u_i|^2 . \tag{47}$$

Let us put

$$\alpha M^2 = pe^{i\gamma} . \tag{48}$$

The real part and imaginary part of (47) are:

$$\begin{aligned} p\cos\gamma &= \frac{1}{2}\Re\left\{\alpha \sum_{i=1}^{k} u_i^2\right\} + \frac{|\alpha|}{2}\cos\gamma \cdot \sum_{i=1}^{k} |u_i|^2 \\ 0 &= \frac{1}{2}\Im\left\{\alpha \sum_{i=1}^{k} u_i^2\right\} - \frac{|\alpha|}{2}\sin\gamma \cdot \sum_{i=1}^{k} |u_i|^2 \end{aligned} \tag{47'}$$

Eliminating $\cos\gamma$ form the first equation by means of the second, we find

$$\sigma = -\frac{1}{2}\Re\left\{\alpha \sum_{i=1}^{k} u_i(z_0)^2\right\} \pm \frac{1}{2}\sqrt{|\alpha|^2\left(\sum_{i=1}^{k} |u_i(z_0)|^2\right)^2 - \left[\Im\left\{\alpha \sum_{i=1}^{k} u_i(z_0)^2\right\}\right]^2} . \tag{49}$$

We see, in particular, that the first variation of each eigen value, whatever its degree of degeneracy, depends only on

$$U(z_0) = \sum_{i=1}^{k} u_i(z_0)^2 \quad \text{and} \quad \Omega(z_0) = \sum_{i=1}^{k} |u_i(z_0)|^2 . \tag{50}$$

Observe that the product of the two possible σ-values (49) is

$$\frac{1}{4}\left|\alpha \sum_{i=1}^{k} u_i^2\right|^2 - \frac{1}{4}|\alpha|^2\left(\sum_{i=1}^{k} |u_i(z_0)|^2\right)^2 \leq 0 \tag{51}$$

such that under a variation (1) at least one component of a split up multiple eigen value is non-increasing. This is the reason why many maximum problems for positive eigen values lead to degenerate eigen values in the extremum case.

7. The L_ε-kernels and the variation of the Fredholm determinants. In this section, we shall discuss certain kernels obtained by complex differentiation of the dielectric Green's functions which will appear in certain variational formulas for important combinations of Fredholm eigen values. The significance of these kernels is best understood by considering the kernel obtained in an analogous way from the ordinary Green's function, say $\tilde{g}(z, \zeta)$ of $\tilde{D}$.

We defined already in (1.17) a kernel $L(z, \zeta)$ with respect to the Green's function $g(z, \zeta)$ of the domain set D and observed its remarkable property (1.18). Analogously, we introduce the kernel

$$\tilde{L}(z, \zeta) = -\frac{2}{\pi}\frac{\partial^2}{\partial z \partial \zeta}\tilde{g}(z, \zeta) = \frac{1}{\pi(z-\zeta)^2} - \tilde{l}(z, \zeta) . \tag{1}$$

$\tilde{l}(z, \zeta)$ is a regular analytic function for z and ζ in $\tilde{D}$. We shall need two important facts about $\tilde{l}(z, \zeta)$ for later applications.

(a) For $\zeta \in C$ and $z \in \tilde{D}$, we have

$$\frac{\partial \tilde{g}(z. \zeta)}{\partial z} \equiv 0 \quad \text{identically in } z \in \tilde{D}, \zeta \in C . \tag{2}$$

This identity remains even valid when z moves onto C but to a point different from ζ. Let now s be the length parameter on C, $\zeta(s)$ its parametric representation and $\zeta' = d\zeta/ds$ the local tangent unit vector. We differentiate the identity (2) with respect to s and find

$$\frac{\partial^2 \tilde{g}(z, \zeta)}{\partial z \partial \zeta}\zeta' + \frac{\partial^2 \tilde{g}(z, \zeta)}{\partial z \partial \bar{\zeta}}\bar{\zeta}' = 0 , \qquad z \in C, \zeta \in C . \tag{3}$$

We multiply this identity by z' and using the symmetry of the first term in z and ζ as well as the hermitian symmetry of the second term, we conclude:

$$\tilde{L}(z, \zeta)z'\zeta' = \text{real} \quad \text{for } z \in C, \zeta \in C . \tag{4}$$

By use of (1), we may express this result also in the form

$$\Im\{\tilde{l}(z, \zeta)z'\zeta'\} = \frac{1}{\pi}\Im\left\{\frac{z'\zeta'}{(z-\zeta)^2}\right\} . \tag{5}$$

This identity is of great interest since the left side expression is a differential depending on the Green's function while the right hand term depends only on the geometry of the curve system C. Moreover, it can be shown that $\tilde{l}(z, \zeta)$ is continuous in both variables in the closed domain $\tilde{D} + C$ [3, 21]. We may pass to the limit $z = \zeta$ on both sides of (5); an easy calculation yields the boundary condition

$$\Im\{\tilde{l}(z, z)z'^2\} = \frac{1}{6\pi}\Im\left\{\frac{z'''}{z'} - \frac{3}{2}\left(\frac{z''}{z'}\right)^2\right\} . \tag{6}$$

Let us denote by $\kappa = \kappa(s)$ the curvature of C at $z(s)$; then (6) obtains the elegant form

$$\Im\{\tilde{l}(z, z)z'^2\} = \frac{1}{6\pi}\frac{d\kappa}{ds}. \tag{7}$$

In particular, we note that (7) and our assumptions on C yield the

THEOREM. *The function $\tilde{l}(z, z)$ is a quadratic differential of D, i.e., satisfies*

$$\tilde{l}(z, z)z'^2 = \text{real} \quad on\ C \tag{7'}$$

if and only if $\tilde{D}$ is a domain bounded by circumferences C_j.

(b) Let $z^* = f(z)$ be a univalent analytic function in $\tilde{D}$ which maps this domain into $\tilde{D}^*$. The conformal invariance of the Green's function is expressed by the identity

$$\tilde{g}^*(z^*, \zeta^*) = \tilde{g}(z, \zeta) \tag{8}$$

which leads by differentiation to

$$\tilde{L}^*(z^*, \zeta^*)f'(z)f'(\zeta) = \tilde{L}(z, \zeta). \tag{9}$$

The $\tilde{l}$-kernel has, therefore, the transformation law

$$\tilde{l}^*(z^*, \zeta^*)f'(z)f'(\zeta) = \tilde{l}(z, \zeta) + \frac{1}{\pi}\left[\frac{f'(z)f'(\zeta)}{(f(z) - f(\zeta))^2} - \frac{1}{(z - \zeta)^2}\right] \tag{10}$$

and, as a simple calculation shows, in particular

$$\tilde{l}^*(z^*, z^*)f'(z)^2 = \tilde{l}(z, z) + \frac{1}{6\pi}\{f, z\} \tag{11}$$

where

$$\{f, z\} = \frac{f'''(z)}{f'(z)} - \frac{3}{2}\left(\frac{f''(z)}{f'(z)}\right)^2 \tag{12}$$

is the Schwarzian derivative of $f(z)$.

After these remarks on the kernel $\tilde{L}(z, \zeta)$, we introduce now a new kernel by the following formula which is modeled after (1):

$$L_\varepsilon(z, \zeta) = -\frac{2}{\pi}\frac{\partial^2 g_\varepsilon(z, \zeta)}{\partial z \partial \zeta}. \tag{13}$$

This kernel is regular analytic and symmetric in both its arguments in D and in $\tilde{D}$, except for a double pole for $z = \zeta$. We define further two kernels which are regular analytic for $z, \zeta \in D$ and for $z, \zeta \in \tilde{D}$, respectively:

$$l_\varepsilon(z, \zeta) = \frac{1}{\pi(z-\zeta)^2} - \frac{1}{\varepsilon} L_\varepsilon(z, \zeta) \quad \text{in } D \tag{14}$$

and

$$\tilde{l}_\varepsilon(z, \zeta) = \frac{1}{\pi(z-\zeta)^2} - L_\varepsilon(z, \zeta) \quad \text{in } \tilde{D}\,. \tag{15}$$

These kernels have elegant developments in terms of the complex eigen functions of the Fredholm integral equation. We start with the Fourier developments (2.16) and (2.21) for $g_\varepsilon(z, \zeta)$ in terms of the harmonic eigen functions $h_\nu(z)$ and $\tilde{h}_\nu(z)$. Using definition (1.17) and (2.21), we obtain by differentiation

$$l_\varepsilon(z, \zeta) = \left(1 - \frac{1}{\varepsilon}\right)\left[l(z, \zeta) + \sum_{\nu=1}^{\infty} \frac{V_\nu'(z) V_\nu'(\zeta)}{\rho_\nu(1+\rho_\nu)(1+\varepsilon\rho_\nu)}\right] \tag{16}$$

where the $V_\nu(z)$ are the analytic functions whose real part is $h_\nu(z)$. As pointed out in the preceding section, all $V_\nu'(z)$ have a different normalization and it is more convenient to introduce the functions $u_\nu(z)$ defined by (6.36) which have all the norm 1. Then (16) transforms to

$$l_\varepsilon(z, \zeta) = \left(1 - \frac{1}{\varepsilon}\right)\left[l(z, \zeta) + \sum_{\nu=1}^{\infty} \frac{u_\nu(z) u_\nu(\zeta)}{(1+\rho_\nu)(1+\varepsilon\rho_\nu)}\right]. \tag{17}$$

We observe next that with each eigen value $\lambda_\nu > 0$ which belongs to $u_\nu(z)$, there occurs also the eigen value $-\lambda_\nu$ and it belongs to the eigen function $iu_\nu(z)$. This assertion can be verified directly from the complex integral equations (1.36) and (1.37); it is also a consequence of the fact, noted in § 1, that if λ_ν belongs to an eigen function $h_\nu(z)$ then $-\lambda_\nu$ will be an eigen value with the conjugate harmonic eigen function $k_\nu(z)$. Thus, in formula (17), each product $u_\nu(z)\, u_\nu(\zeta)$ occurs, therefore, twice; once coupled with ρ_ν and the other time with opposite sign and coupled with $1/\rho_\nu$. We combine these pairs of terms and sum now only over those ν which correspond to the positive eigen values λ_ν. Using (2.12), we obtain finally

$$l_\varepsilon(z, \zeta) = \left(1 - \frac{1}{\varepsilon}\right)\left[l(z, \zeta) - \sum_{\nu=1}^{\infty} \frac{u_\nu(z) u_\nu(\zeta)}{\lambda_\nu}\right] + E^2 \sum_{\nu=1}^{\infty} \frac{\lambda_\nu^2 - 1}{\lambda_\nu^2 - E^2}\, \frac{u_\nu(z) u_\nu(\zeta)}{\lambda_\nu}\,. \tag{18}$$

with the notation

$$E = \frac{\varepsilon - 1}{\varepsilon + 1}\,. \tag{19}$$

Passing to the limit $\varepsilon = 0$ and using the limit relation (3.31), we derive first from (18)

$$l(z,\zeta)=\sum_{\nu=1}^{\infty}\frac{u_\nu(z)u_\nu(\zeta)}{\lambda_\nu} \tag{20}$$

and, hence, (18) simplifies to

$$l_\varepsilon(z,\zeta)=E^2\sum_{\nu=1}^{\infty}\frac{\lambda_\nu^2-1}{\lambda_\nu^2-E^2}\,\frac{u_\nu(z)u_\nu(\zeta)}{\lambda_\nu}\,. \tag{21}$$

Similarly, we transform (15) by differentiation of (2.16) into the identity

$$\tilde{l}_\varepsilon(z,\zeta)=(\varepsilon-1)\sum_{\nu=1}^{\infty}\frac{\rho_\nu\tilde{V}'_\nu(z)\tilde{V}'_\nu(\zeta)}{(1+\rho_\nu)(1+\varepsilon\rho_\nu)} \tag{22}$$

and replacing $\tilde{V}'_\nu(z)$ by $\tilde{u}_\nu(z)$ by means of (6.36), we find

$$\tilde{l}_\varepsilon(z,\zeta)=-(\varepsilon-1)\sum_{\nu=1}^{\infty}\frac{\rho_\nu\tilde{u}_\nu(\zeta)\tilde{u}_\nu(z)}{(1+\rho_\nu)(1+\varepsilon\rho_\nu)}\,. \tag{23}$$

We combine again terms with ρ_ν and with $1/\rho_\nu$ and sum only over the positive eigen values λ_ν; an easy calculation leads to

$$\tilde{l}_\varepsilon(z,\zeta)=E^2\sum_{\nu=1}^{\infty}\frac{\lambda_\nu^2-1}{\lambda_\nu^2-E^2}\,\frac{\tilde{u}_\nu(z)\tilde{u}_\nu(\zeta)}{\lambda_\nu}\,. \tag{24}$$

The complete symmetry between (21) and (24) is evident.

We consider the limit cases $\varepsilon=0$ and $\varepsilon=\infty$ of formula (24) which correspond both to $E^2=1$. From (3.11) and (3.17) follows

$$\begin{aligned}\tilde{\lambda}(z,\zeta)=\sum_{\nu=1}^{\infty}\frac{\tilde{u}_\nu(z)\tilde{u}_\nu(\zeta)}{\lambda_\nu}&=\frac{1}{\pi(z-\zeta)^2}+\frac{2}{\pi}\frac{\partial^2\tilde{G}(z,\zeta)}{\partial z\partial\zeta}\\&=\frac{1}{\pi(z-\zeta)^2}+\frac{2}{\pi}\frac{\partial^2\tilde{N}(z,\zeta)}{\partial z\partial\zeta}\,.\end{aligned} \tag{25}$$

We can, therefore, express $\tilde{\lambda}(z,\zeta)$ by means of (3.3) in the form

$$\tilde{\lambda}(z,\zeta)=\tilde{l}(z,\zeta)-\frac{1}{2\pi}\sum_{j,k=1}^{N-1}\alpha_{jk}w'_j(z)w'_k(\zeta) \tag{26}$$

where $w_j(z)$ denotes the analytic completion of the harmonic measure $\omega_j(z)$. Formula (26) is the counterpart for $\tilde{D}$ of the relation (20) in D. The kernel $\tilde{\lambda}(z,\zeta)$ is composed of functions with single-valued integral in $\tilde{D}$; the kernel $\tilde{l}(z,\zeta)$ differs from it by a kernel which is composed of a basis of $N-1$ functions in $\tilde{D}$ which do not have a single-valued integral and which are orthogonal in the Dirichlet metric to all functions in $\tilde{D}$ with single-valued integral.

For the sake of completeness, we give also the Fourier developments of the kernels

$$K_\varepsilon(z, \bar{\zeta}) = -\frac{2}{\pi\varepsilon}\frac{\partial^2 g_\varepsilon(z, \zeta)}{\partial z \partial \bar{\zeta}} \quad \text{in } D \tag{27}$$

and

$$\tilde{K}_\varepsilon(z, \bar{\zeta}) = -\frac{2}{\pi}\frac{\partial^2 g_\varepsilon(z, \zeta)}{\partial z \partial \bar{\zeta}} \quad \text{in } \tilde{D}\,. \tag{28}$$

Both kernels are analytic and have hermitian symmetry in their arguments. Putting

$$K(z, \bar{\zeta}) = -\frac{2}{\pi}\frac{\partial^2 g(z, \zeta)}{\partial z \partial \bar{\zeta}} \tag{29}$$

we obtain by differentiation of (2.21) after the above combination of terms

$$K_\varepsilon(z, \bar{\zeta}) = \left(1 - \frac{1}{\varepsilon}\right)\left[K(z, \bar{\zeta}) - \sum_{\nu=1}^{\infty} u_\nu(z)\overline{u_\nu(\zeta)}\right] + E\sum_{\nu=1}^{\infty}\frac{\lambda_\nu^2 - 1}{\lambda_\nu^2 - E^2}u_\nu(z)\overline{u_\nu(\zeta)}\,. \tag{30}$$

Again, we obtain by passage to the limit $\varepsilon = 0$ and in view of (3.31)

$$K(z, \bar{\zeta}) = \sum_{\nu=1}^{\infty} u_\nu(z)\overline{u_\nu(\zeta)} \tag{31}$$

which reduces formula (30) to

$$K_\varepsilon(z, \bar{\zeta}) = E\sum_{\nu=1}^{\infty}\frac{\lambda_\nu^2 - 1}{\lambda_\nu^2 - E^2}u_\nu(z)\overline{u_\nu(\zeta)}\,. \tag{32}$$

Similarly, we find by differentiation of (2.16) the identity

$$\tilde{K}_\varepsilon(z, \bar{\zeta}) = E\sum_{\nu=1}^{\infty}\frac{\lambda_\nu^2 - 1}{\lambda_\nu^2 - E^2}\tilde{u}_\nu(z)\overline{\tilde{u}_\nu(\zeta)}\,. \tag{33}$$

Formulas (21), (24), (32) and (33) for the various kernels depend on ε only through E and this simple rational function of ε has the symmetry property $E(1/\varepsilon) = -E(\varepsilon)$. This leads to the interesting identities:

$$\frac{\partial^2 g_\varepsilon(z, \zeta)}{\partial z \partial \zeta} = \frac{\partial^2 g_{1/\varepsilon}(z, \zeta)}{\partial z \partial \zeta}, \quad \frac{\partial^2 g_\varepsilon(z, \zeta)}{\partial z \partial \bar{\zeta}} = -\frac{\partial^2 g_{1/\varepsilon}(z, \zeta)}{\partial z \partial \bar{\zeta}} \tag{34}$$

if $z, \zeta \in \tilde{D}$ and to a similar identity in $z, \zeta \in D$. These relations are known in the limit case $\varepsilon = 0$ where they represent differential relations between the Green's and the Neumann's function [2, 5, 21].

We define next the Fredholm determinant of the basic integral equation (1.2), Observe again that with each positive eigen value λ_ν

occurs also the eigen value $-\lambda_\nu$ in equal multiplicity. We may thus write

$$D(E) = \prod_{\nu=1}^{\infty}\left(1 - \frac{E^2}{\lambda_\nu^2}\right) \tag{35}$$

where the product is to be extended over all positive eigen values $\lambda_\nu > 1$.

By use of the variational formulas (6.38) and (6.39) and of the identities (21) and (24) one can establish readily the

THEOREM. *If the curve system C is varied according to* (6.1) *the Fredholm determinant $D(E)$ changes according to the variational formulas*

$$\delta \log D(E) = -2\pi\Re\{\alpha l_\varepsilon(z_0, z_0)\} \quad \text{for } z_0 \in D \tag{36}$$

and

$$\delta \log D(E) = -2\pi\Re\{\alpha \tilde{l}_\varepsilon(z_0, z_0)\} \quad \text{for } z_0 \in \tilde{D}\,. \tag{37}$$

$E(\varepsilon)$ is the rational function (19) *of ε.*

The elegant and symmetric variational formulas (36) and (37) show the theoretical interest of the Fredholm determinant (35). We observe that, in particular, for $\varepsilon = \infty$ and $E = 1$ we have by (20) and (25);

$$\delta \log D(1) = -2\pi\Re\{\alpha l(z_0, z_0)\} \quad \text{for } z_0 \in D \tag{38}$$

and

$$\delta \log D(1) = -2\pi\Re\{\alpha \tilde{\lambda}(z_0, z_0)\} \quad \text{for } z_0 \in \tilde{D}\,. \tag{38'}$$

The functional (35) is defined only for curve systems C which are sufficiently differentiable. This fact creates difficulties in applications of the above variational formulas to extremum problems for the Fredholm determinant since it is not sure, a priori, that the extremum system C will have the required smoothness. In many problems, however, it can be shown that the very property of being an extremum set guarantees already that the curve system C is analytic. Thus, we may restrict ourselves from the beginning to the class of analytic curve systems C and formulate the extremum problems only within this class. A first result for a general theory of extremum problems for the Fredholm determinants is the fact that $D(E)$ is semi-continuous from above in the class of all analytic curve systems C. In fact, we will prove the

THEOREM. *Let $\tilde{D}_n$ be a sequence of domains, each being bounded by an analytic curve system C_n and with the Fredholm determinant $D_n(E)$. If the domains $\tilde{D}_n$ converge in the Carathéodory sense to a domain $\tilde{D}$ with analytic boundary C and with the Fredholm determinant $D(E)$, then we have for all $E \geq 0$*

$$\overline{\lim}\, D_n(E) \leq D(E)\ . \tag{39}$$

Proof. We define the kernel

$$\tilde{\lambda}^{(2)}(z, \bar{\zeta}) = \iint_{\tilde{D}} \tilde{\lambda}(z, \eta)\overline{\tilde{\lambda}(z, \eta)} d\tau_\eta = \sum_{\nu=1}^{\infty} \frac{\tilde{u}_\nu(z)\overline{\tilde{u}_\nu(\zeta)}}{\lambda_\nu^2} \tag{40}$$

and define then recursively

$$\tilde{\lambda}^{(2j)}(z, \bar{\zeta}) = \iint_{\tilde{D}} \tilde{\lambda}^{(2j-2)}(z, \bar{\eta})\tilde{\lambda}^{(2)}(\eta, \bar{\zeta}) d\tau_\eta = \sum_{\nu=1}^{\infty} \frac{\tilde{u}_\nu(z)\overline{\tilde{u}_\nu(\zeta)}}{\lambda_\nu^{2j}}\ . \tag{41}$$

We remark that

$$\iint_{\tilde{D}} \tilde{\lambda}^{(2j)}(z, \bar{z}) d\tau_z = \sum_{\nu=1}^{\infty} \frac{1}{\lambda_\nu^{2j}} = S^{(2j)}\ . \tag{42}$$

We denote the corresponding expressions referring to the domain $\tilde{D}_n$ by the subscripts n. We assert, at first:

$$\underline{\lim}\, S_n^{(2j)} \geq S^{(2j)} \tag{43}$$

To prove this assertion, we select a number $\delta > 0$ arbitrarily small and determine a closed subdomain $\tilde{\Delta}$ of $\tilde{D}$ such that

$$\iint_{\tilde{\Delta}} \tilde{\lambda}^{(2j)}(z, \bar{z}) d\tau_z > S^{(2j)} - \delta\ . \tag{44}$$

By the definitions (25), (40), (41) and in view of the continuous dependence of the Green's function $\tilde{G}(z, \zeta)$ on its domain $\tilde{D}$, the kernels $\tilde{\lambda}_n^{(2j)}(z, \bar{\zeta})$ converge to $\tilde{\lambda}^{(2j)}(z, \bar{\zeta})$ uniformly in each closed subdomain of $\tilde{D}$, in particular in $\tilde{\Delta}$. Given δ, we can choose $n(\delta)$ such that for $n > n(\delta)$ the domains $\tilde{D}_n$ contain $\tilde{\Delta}$ and that

$$\begin{aligned} S_n^{(2j)} &= \iint_{\tilde{D}_n} \tilde{\lambda}_n^{(2j)}(z, \bar{z}) d\tau_z > \iint_{\tilde{\Delta}} \tilde{\lambda}_n^{(2j)}(z, \bar{z}) d\tau_z \\ &\geq \iint_{\tilde{\Delta}} \tilde{\lambda}^{(2j)}(z, \bar{z}) d\tau_z - \delta > S^{(2j)} - 2\delta\ . \end{aligned} \tag{45}$$

Since δ can be chosen arbitrarily small, these inequalities imply (43).

We observe next that by definition (35)

$$-\log D(E) = \sum_{j=1}^{\infty} \frac{1}{j} E^{2j} S^{(2j)} \tag{46}$$

and a corresponding representation is valid for $-\log D_n(E)$. Hence, from (43) follows immediately the asserted inequality (39) and the theorem is proved.

The significance of this theorem is the following. Let $\mathfrak{A}$ be a family of analytic curve systems C and let us ask for the maximum of $D(E)$ within the family $\mathfrak{A}$, for some fixed value E. We know that by its definition $D(E) \leq 1$ and is thus trivially bounded in $\mathfrak{A}$. Let $U \leq 1$ denote the least upper bound of $D(E)$ in $\mathfrak{A}$; we can select an extremum sequence of curve sets C_n in $\mathfrak{A}$ such that $D_n(E)$ converges to U. If it is possible to select a subsequence C_{n1} of the C_n such that the corresponding domains D_{n1} converge to a domain D_0 with analytic boundary $C_0 \in \mathfrak{A}$, then C_0 is a maximum curve system. For, by our theorem (38), we have $D_0(E) \geq U$ and, hence, $D_0(E) = U$ since no $D(E)$ in $\mathfrak{A}$ can be larger than U. This argument will be applied in the following section to an interesting problem of conformal mapping.

8. An extremum problem for Fredholm determinants and an existence proof for circular mappings. In this section, we shall utilize the variational formulas for the Fredholm determinants in order to solve a specific maximum problem. The extremum domains of this problem will be characterized by the property that their boundary C consists of circumferences. In this way, we will then prove that every plane domain can be mapped conformally upon a canonical domain whose boundaries are circumferences. This canonical mapping will appear as the solution of a simple extremum problem for the family of all univalent mappings of the given domain.

We formulate the following extremum problem:

Let $\tilde{D}$ be a domain in the complex z-plane which contains the point at infinity and which is bounded by N closed analytic curves C. Let $\mathscr{F}$ be the family of all functions $t = f(z)$ which are analytic in $\tilde{D} + C$, normalized at infinity by $f'(\infty) = 1$ and are univalent in $\tilde{D}$. Each $f(z) \in \mathscr{F}$ will map $\tilde{D}$ upon a domain $\tilde{\Delta}$ with analytic boundary Γ and with the Fredholm determinants $\Delta(E)$. We ask for the functions $f(z) \in \mathscr{F}$ which lead to the maximum value of $\Delta(1)$.

The existence of such maximum functions is by no means obvious. We can assert only that all determinants $\Delta(1)$ obtained by mappings of the family $\mathscr{F}$ have a least upper bound $U \leq 1$. Hence, we may select a sequence of mappings $f_n(z) \in \mathscr{F}$ such that

$$\lim_{n \to \infty} \Delta_n(1) = U . \tag{1}$$

Since the $f_n(z)$ are univalent in $\tilde{D}$ we can use the well-known normality properties of these functions and assume without loss of generality that the $f_n(z)$ converge to a limit function $f(z)$, uniformly in each closed subdomain of $\tilde{D}$. The limit function $f(z)$ provides a univalent map of $\tilde{D}$ into a domain $\tilde{\Delta}$ and is normalized at infinity. The image domains $\tilde{\Delta}_n$ converge in the Carathéodory sense to $\tilde{\Delta}$. If we could prove that $\tilde{\Delta}$ has an analytic boundary Γ, we would know that $f(z) \in \mathscr{F}$ and the semi-continuity from above of $\Delta(1)$ would insure $\Delta(1) = U$, i.e., that $f(z)$ is a maximum function.

In order to prove the fact $f(z) \in \mathscr{F}$ we consider the maximum sequence $f_n(z)$ which converges to $f(z)$. We want to characterize this sequence by comparing it with near-by sequences obtained by infinitesimal variations of their image domains $\tilde{\Delta}_n$. However, if we subject a multiply-connected domain $\tilde{\Delta}_n$ to an interior variation (6.1), we will, in general, obtain a domain $\tilde{\Delta}_n^*$ which is not conformally equivalent to $\tilde{\Delta}_n$ and cannot be obtained from $\tilde{D}$ by a mapping of the family $\mathscr{F}$. Let, indeed, $\omega_l(t)$ be the harmonic measure of the boundary component Γ_l of Γ with respect to $\tilde{\Delta}$ and let $((p_{jk}))$ denote the period matrix (2.18″) of this set of harmonic measures. The period matrix $((p_{jk}))$ is a conformal invariant and if we preserve the point at infinity under the conformal mappings, the numbers $\omega_l(\infty)$ must likewise be unchanged. On the other hand, it is well-known [5, 15, 21] that under a variation of the t-plane of the type (6.1) and with the singular point $t_0 \in \tilde{\Delta}$, we have

$$p_{jk}^* = p_{jk} + \Re\{\alpha w_j'(t_0) w_k'(t_0)\} + O(|\alpha|^2) \tag{2}$$

and

$$\omega_l^*(\infty) = \omega_l(\infty) + \Re\{\alpha p'(t_0, \infty) w_l'(t_0)\} + O(|\alpha|^2) \tag{3}$$

where again $w_l(t)$ and $p(t, \tau)$ denote the analytic completions in t of the harmonic functions $\omega_l(t)$ and $g(t, \tau)$ in $\tilde{\Delta}$. We see that, in general, the numbers p_{jk} and $\omega_l(\infty)$ will change under interior variations and that the domain $\tilde{\Delta}^*$ will not be obtained from $\tilde{D}$ by a mapping of the family $\mathscr{F}$.

Consider now m points t_μ in $\tilde{\Delta}$ and the variation

$$t^* = t + \sum_{\mu=1}^{m} \frac{\alpha_\mu}{t - t_\mu} + O(|\alpha|^2), \quad |\alpha| = \max_\mu (|\alpha_\mu|) \tag{4}$$

where the error term is estimated uniformly in $\tilde{\Delta} + \Gamma$. We may choose the α_μ and the correction term $O(|\alpha|^2)$ such that

$$\Re\left\{\sum_{\mu=1}^{m} \alpha_\mu w_j'(t_\mu) w_k'(t_\mu)\right\} = 0 \tag{5}$$

$$\Re\left\{\sum_{\mu=1}^{m} \alpha_\mu p'(t_\mu, \infty) w_l'(t_\mu)\right\} = 0 \tag{6}$$

and

$$p_{jk}^* = p_{jk}, \quad \omega_l^*(\infty) = \omega_l(\infty). \tag{7}$$

It can be shown, indeed, that given such values t_μ and α_μ, the variation (4) can be selected in such a way that $\tilde{\Delta}^*$ is conformally equivalent to $\tilde{\Delta}$ and that the points at infinity correspond [21]. Even now, we cannot assert that $\tilde{D}$ goes into $\tilde{\Delta}^*$ by a mapping of the family $\mathscr{F}$ which is normalized at infinity. However, the Fredholm determinants do not change under a homothetic mapping of a domain and, hence, the insistence on the normalization at infinity is unnecessary in our problem. Thus, the above variations (4) will transform the domains $\tilde{\Delta}_n$ of the extremum sequence into conformally equivalent domains $\tilde{\Delta}_n^*$ whose Fredholm determinants $\Delta_n^*(1)$ may be compared with the maximum sequence $\Delta_n(1)$.

We observe that the functions $w'_j(t)\cdot w'_k(t)$ and $p'(t,\infty)\cdot w'_l(t)$ are quadratic differentials of $\tilde{\Delta}$, i.e., functions $Q_k(t)$ which are regular analytic in $\tilde{\Delta}+\Gamma$ and satisfy on Γ the boundary condition

$$Q_k(t)t'^2 = \text{real} . \tag{8}$$

At infinity all these functions satisfy the asymptotic relation

$$Q_k(t) = O(|t|^{-3}) . \tag{9}$$

All functions with the properties (8) and (9) from a linear space with real coefficients and of the dimension $3N$-3. We suppose that we have chosen from the above $N(N+1)$ quadratic differentials a fixed basis of $3N$-3 elements $Q_k(t)$, $k=1,2,\cdots,3N$-3.

After these preparations, we return to our maximum sequence of domains $\tilde{\Delta}_n$; we denote by $Q_k^{(n)}(t)$ the corresponding basis of quadratic differentials of $\tilde{\Delta}_n$ and by $Q_k(t)$ the basis for their limit domain $\tilde{\Delta}$. Clearly, we can choose the basis in each $\tilde{\Delta}_n$ and in $\tilde{\Delta}$ such that

$$\lim_{n\to\infty} Q_k^{(n)}(t) = Q_k(t) , \tag{10}$$

uniformly in each closed subdomain of $\tilde{\Delta}$. The determinant

$$\det \| \Re\{Q_k(t_l)\} \| , \qquad l, k = 1, 2, \cdots, 3N\text{-}3 \tag{11}$$

does not vanish identically in $\tilde{\Delta}$ because of the supposed real independence of the $Q_k(t)$. Hence, we can determine $3N$-3 points $t_\mu \in \tilde{\Delta}$ such that

$$\det \| \Re\{Q_k^{(n)}(t_\mu)\} \| \neq 0 \qquad k, \mu = 1, 2, \cdots, 3N\text{-}3 \tag{12}$$

for large enough n; we may even assume, without loss of generality, that (12) holds for all integers n.

Let t_0 be an arbitrary point in $\tilde{\Delta}_n$ and $\alpha^{(n)}$ be an arbitrary complex number. We determine $3N$-3 real numbers $x_\mu^{(n)}$ by the linear equations

$$\Re\{\alpha^{(n)}Q_k^{(n)}(t_0)\} = \sum_{\mu=1}^{3N-3} x_\mu^{(n)}\Re\{Q_k^{(n)}(t_\mu)\} , \quad k = 1, 2, \cdots, 3N\text{-}3 \tag{13}$$

which is always posible because of (12). Observe that $x_\mu^{(n)} = O(|\alpha^{(n)}|)$ for small values of $\alpha^{(n)}$. Consider then the interior variation of $\tilde{\Delta}_n$

$$(14) \qquad t^* = t + \frac{\alpha^{(n)}}{t - t_0} \sum_{\mu=1}^{3N-3} \frac{x_\mu^{(n)}}{t - t_\mu} + O(|\alpha^{(n)}|^2) .$$

This variation is of the type (4), but by the choice (13) of the $x_\mu^{(n)}$, we are sure that the equations (5) and (6) will be fulfilled. We can, therefore, adjust the error term $O(|\alpha^{(n)}|^2)$ in such a way that the varied domain $\tilde{\Delta}_n^*$ is conformally equivalent to $\tilde{\Delta}_n$ and such that the points at infinity correspond. Hence, $\tilde{\Delta}_n^*$ may be used as a competing domain sequence to the maximum sequence $\tilde{\Delta}_n$. We apply now the variational formula (7.38′) in order to characterize the limit domain $\tilde{\Delta}$.

We derive from (7.38′) that the variation (14) of $\tilde{\Delta}_n$ yields

$$(15) \qquad \log \Delta_n^*(1) = \log \Delta_n(1) - 2\pi \Re\{\alpha^{(n)} \tilde{\lambda}_n(t_0, t_0)\} + 2\pi \sum_{\mu=1}^{3N-3} x_\mu^{(n)} \Re\{\tilde{\lambda}_n(t_\mu, t_\mu)\} + O(|\alpha^{(n)}|^2) .$$

Here, the $\tilde{\lambda}_n(t, t)$ denote the $\tilde{\lambda}$-kernels of $\tilde{\Delta}_n$. We denote

$$(16) \qquad \delta_n = \log U - \log \Delta_n(1) .$$

By the definition of the maximum sequence, we have $0 < \delta_n \to 0$. Since $\log \Delta_n^*(1) \leq \log U$, we infer from (15) the inequality

$$(17) \qquad \frac{1}{2\pi}\delta_n \geq - \Re\{\alpha^{(n)} \tilde{\lambda}_n(t_0, t_0)\} + \sum_{\mu=1}^{3N-3} x_\mu^{(n)} \Re\{\tilde{\lambda}_n(t_\mu, t_\mu)\} + O(|\alpha^{(n)}|^2) .$$

We choose finally

$$(18) \qquad \alpha^{(n)} = \delta_n r e^{i\tau} , \qquad r > 0$$

and define the real numbers ξ_μ by the system of linear equations

$$(19) \qquad \sum_{\mu=1}^{3N-3} \xi_\mu \Re\{Q_k(t_\mu)\} = \Re\{e^{i\tau} Q_k(t_0)\} , \qquad k = 1, \cdots, 3N\text{-}3 .$$

We divide equations (13) and (17) by δ_n and pass to the limit $n \to \infty$; comparing (13) with (19), we find

$$(20) \qquad \lim_{n\to\infty} \frac{x_\mu^{(n)}}{r\delta_n} = \xi_\mu ;$$

and since at $t_0, t_1, \cdots, t_{3N-3}$ holds

$$(21) \qquad \lim_{n\to\infty} \tilde{\lambda}_n(t_\mu, t_\mu) = \tilde{\lambda}(t_\mu, t_\mu) ,$$

we obtain from (17)

$$(22)\qquad \frac{1}{2\pi r} \geq -\Re\{e^{i\tau}\tilde{\lambda}(t_0, t_0)\} + \sum_{\mu=1}^{3N-3} \xi_\mu \Re\{\tilde{\lambda}(t_\mu, t_\mu)\} .$$

This inequality holds for arbitrary values $r > 0$; hence, sending $r \to \infty$, we find

$$(23)\qquad 0 \geq -\Re\{e^{i\tau}\tilde{\lambda}(t_0, t_0)\} + \sum_{\mu=1}^{3N-3} \xi_\mu \Re\{\tilde{\lambda}(t_\mu, t_\mu)\} .$$

If we replace in (19) the signum $e^{i\tau}$ by $-e^{i\tau}$, the solution vector ξ_μ changes into $-\xi_\mu$. Since $e^{i\tau}$ is entirely arbitrary, the inequality (23) must also hold for inverted sign of the right hand term. Thus, we arrive finally at the equation

$$(24)\qquad \Re\{e^{i\tau}\tilde{\lambda}(t_0, t_0)\} = \sum_{\mu=1}^{3N-3} \xi_\mu \Re\{\tilde{\lambda}(t_\mu, t_\mu)\} .$$

valid for arbitrary choice of the signum $e^{i\tau}$ and the corresponding choice (19) of the ξ_μ. The fact that, for given fixed $t_1, \cdots, t_{3N-3}$ in $\tilde{\Delta}$ and for arbitrary $t_0 \in \tilde{\Delta}$, the linear equations (19) always imply the equation (24) for arbitrary $e^{i\tau}$, guarantees the existence of $3N$-3 real numbers $\beta_\mu(\mu = 1, \cdots, 3N\text{-}3)$ such that

$$(25)\qquad \tilde{\lambda}(t, t) = \sum_{\mu=1}^{3N-3} \beta_\mu Q_\mu(t) .$$

This identity is then the condition which characterizes the limit domain $\tilde{\Delta}$ of an extremum sequence $\tilde{\Delta}_n$.

Since, in view of (7.26), the function $\tilde{\lambda}(t, t)$ coincides with the more fundamental kernel $\tilde{l}(t, t)$ except for a quadratic differential, we may express the result (25) as follows :

THEOREM I. *If $\tilde{\Delta}$ is the limit domain of a maximum sequence $\tilde{\Delta}_n$, its $\tilde{l}$-kernel satisfies the condition*

$$(26)\qquad \tilde{l}(t, t) = Q(t)$$

where $Q(t)$ is a quadratic differential of $\tilde{\Delta}$.

From Theorem I, we can deduce

THEOREM II. *All boundary curves Γ_l of $\tilde{\Delta}$ are analytic.*

Proof. Let us express equation (26) in terms of functionals of the original domain $\tilde{D}$ which is conformally equivalent to $\tilde{\Delta}$. By (7.11) and because of the covariance character of the quadratic differentials under conformal mapping, we can express (26) in the form

$$(27)\qquad \tilde{l}(z, z) + \frac{1}{6\pi}\{f, z\} = Q(z)$$

where $Q(z)$ is the quadratic differential in $\tilde{D}$ which corresponds to $Q(t)$ under the mapping $t = f(z)$ of $\tilde{D}$ into $\tilde{\Delta}$ and $\tilde{l}(z, z)$ denotes the $\tilde{l}$-kernel of $\tilde{D}$. We have assumed that $\tilde{D}$ has analytic boundaries C_ν; hence, we can assert that $\tilde{l}(z, z)$ and $Q(z)$ are analytic in the closed region $\tilde{D} + C$. By (7.12), we may now interpret the equation (27) as a linear differential equation with analytic coefficient in $\tilde{D} + C$:

$$\mu''(z) + 3\pi[Q(z) - \tilde{l}(z, z)]\mu(z) = 0 \tag{28}$$

for the unknown function

$$\mu(z) = [f'(z)]^{-1/2} . \tag{29}$$

From the general theory of ordinary differential equations we obtain that $\mu(z)$ is regular analytic in $\tilde{D} + C$ and can have only finitely many zeros on C. Hence, $f'(z)$ is analytic on C except for poles which are at least of order 2. At such singular points on C, $f(z)$ would have poles too. But $f(z)$ is univalent in $\tilde{D}$ and has already a pole at infinity. It cannot have additional poles on C; hence, $f(z)$ and $f'(z)$ are regular analytic on C and the theorem is proved.

In particular, we have now shown that the limit function $f(z)$ of the maximum sequence $f_n(z)$ belongs also to the family $\mathscr{F}$ considered and is, therefore, a maximum function of our problem.

Since we know now that the boundary curves Γ_ι of $\tilde{\Delta}$ are analytic, we can combine (26) with (7.7) and find:

$$\Im\{\tilde{l}(t, t)t'^2\} = \Im\{Q(t)t'^2\} = \frac{1}{6\pi}\frac{d\kappa}{ds} . \tag{30}$$

But $Q(t)$ is a quadratic differential of $\tilde{\Delta}$; thus we arrive at

$$\frac{d\kappa}{ds} = 0 \text{ on each } \Gamma_\iota . \tag{31}$$

This leads to

THEOREM III. *Each boundary curve Γ_ι of the maximum domain $\tilde{\Delta}$ is a circumference.*

Since in each given domain $\tilde{D}$ there exists at least one maximum sequence $f_n(z) \in \mathscr{F}$, we have given a new proof for the classical theorem [5, 7, 8, 9, 23]:

THEOREM IV. *Every plane domain $\tilde{D}$ can be mapped onto a domain bounded by circumferences.*

Since the domain $\tilde{\Delta}$ is the limit of a maximum sequence of domains

$\tilde{\Delta}_n$ and since it is analytically bounded, the semi-continuity of the Fredholm determinants leads to

THEOREM V. *Among all conformally equivalent domains, the circular domains have the largest value of the Fredholm determinant* $D(1)$.

BIBLIOGRAPHY

1. L. V. Ahlfors, *Conformality with respect to Riemann metrics*, Ann. Acad. Sci. Fenn., Series A **202** (1955).

2. S. Bergman, and M. Schiffer, *A representation of Green's and Neumann's functions in the theory of partial differential equations of second order*, Duke Math. J. **14**, (1947), 609-638.

3. ———, *Kernel functions and conformal mapping*, Comp. Math. **8** (1951), 205-249.

4. I. E. Block, *Kernel functions and class* L^2, Proc. Amer. Math. Soc. **4** (1953) 110-117.

5. R. Courant, *Dirichlet's principle, conformal mapping, and minimal surfaces*, Appendix by M. Schiffer, New York 1950.

6. P. R. Garabedian, and M. Schiffer, *Convexity of domain functionals*, J. Analyse Math. **2** (1953), 281-368.

7. A. Hurwitz, and R. Courant, *Funktionentheorie*, Berlin, 1929.

8. P. Koebe, *Über die konforme Abbildung mehrfach zusammenhängender Bereiche*, Jahresber. D. Math. Ver. **19** (1910), 339-348.

9. L. Lichtenstein, *Neuere Entwicklung der Potentialtheorie, Konforme Abbildung*, Enzycl. d. Math. Wissensch. II, C. 3, (1918)

10. H. Poincaré, *Théorie du Potentiel Newtonien*, Paris, 1899.

11. G. Pólya, and M. Schiffer, *Convexity of functionals by transplantation*, J. Analyse Math. **3** (1954), 245-346.

12. R. de Possel, *Sur quelques problèmes de représentation conforme*, C. R. Paris **194** (1932), 42-44.

13. E. Rengel, *Existenzbeweise für schlichte Abbildungen mehrfachzusammenhängender Bereiche auf gewisse Normalbereiche*, Jahresber. D. Math. Ver. **44** (1934), 51-55.

14. M. Schiffer, *Variation of the Green's function and theory of the p-valued functions*, Amer. J. Math. **65** (1943), 341-360.

15. ———, *Hadamard's formula and variation of domain functions*, Amer. J. Math. **68** (1946), 417-448.

16. ———, *The kernel function of an orthonormal system*, Duke Math. J. **13** (1946), 529-540.

17. ———, *The Fredholm eigen values of plane domain*, Pacific J. Math. **7** (1957), 1187-1225.

18. ———, *Sur les rapports entre les solutions des problèmes intérieurs et celles des problèmes extérieurs*. C. R. Paris **244** (1957), 2680-2683.

19. ———, *Sur la polarization et la masse virtuelle*, C. R. Paris **244** (1957), 3118-3121,

20. ———, *Problèmes aux limites et fonctions propres de l'équation intégrale de Poincaré et de Fredholm*, C. R. Paris **245** (1957), 18-21.

21. M. Schiffer, and D. C. Spencer, *Functionals of finite Riemann surfaces*, Princeton 1954.

22. M. Schiffer and G. Szegö, *Virtual mass and polarization*, Trans. Amer. Math. Soc. **67** (1949), 130-205.

23. E. Schottky, *Über die konforme Abbildung mehrfach zusammenhängender ebener Flächen*, J.f.r.u. angew. Math. **83** (1877) 300-351.

STANFORD UNIVERSITY

[78] Fredholm eigenvalues and conformal mapping

---

[78] Fredholm eigenvalues and conformal mapping. *Rend. Mat. e Appl.* **22** (1963), 447–468.

MENAHEM SCHIFFER
1963
Rendiconti di Matematica
(3-4) Vol. 22, pp. 447-468

Fredholm eigenvalues and conformal mapping(*)

by **MENAHEM SCHIFFER** (Stanford University)

(dedicated to C. Loewner on his 70th birthday)

Introduction.

This paper is mainly of expository nature. Its aim is to give a connected account of various results regarding the Fredholm eigenfunctions and the Fredholm eigenvalues of plane domains. The Fredholm eigenvalues of a curve system are a set of functionals of these curves whose study leads to many useful applications. They are closely related to the boundary value problem for harmonic functions. They are important in the theory of conformal mapping, in the theory of kernel functions and orthonormal series. They play a role in the theory of the Hilbert transform and in the theory of univalent functions. Their dependence on the curve system is displayed by a very elegant and convenient variational formula. Some applications of these results are given to show the significance of the identities and formulas obtained. For a more detailed development of many points the reader is referred to [10], [11] and [12].

1. Boundary Value Problem of Potential Theory and Fredholm Eigenvalues.

Let C be a closed curve in the complex z-plane with interior domain D and exterior domain $\tilde{D}$. We suppose C to be three times continuously differentiable. It possesses at every point $\zeta \in C$ a normal n_ζ, and it is well known that the kernel

$$k(z, \zeta) = \frac{\partial}{\partial n_\zeta} \log \frac{1}{|z - \zeta|} \tag{1}$$

(*) Conferenze tenute nel Ciclo del CIME (Centro Internazionale Matematico Estivo) su *Autovalori e autosoluzioni* (Chieti, 1-9 agosto 1962).

is a continuous function of both argument points z and ζ on C. We shall always assume the normal n_ζ to be directed into D.

The kernel $k(z, \zeta)$ was introduced by Poincaré in order to solve the first boundary value problem of potential theory. Suppose that we are given a continuous function $f(z)$ on the curve C and wish to find a function $u(z)$ harmonic in D which takes on C the boundary values $f(z)$. We try to solve this problem by superposition of elementary harmonic functions and make the « Ansatz »

$$u(z) = \frac{1}{\pi} \oint_C k(z, \zeta)\, \varphi(\zeta)\, ds_\zeta \tag{2}$$

with a weight function $\varphi(\zeta)$ still to be determined for our problem. This is clearly a harmonic function for $z \in D$. As $z \to z_0 \in C$, we have the well-known jump condition

$$\lim_{z \leftarrow z_0} u(z) = \varphi(z_0) + \frac{1}{\pi} \oint_C k(z_0, \zeta)\, \varphi(\zeta)\, ds_\zeta \tag{3}$$

Since we have prescribed the boundary values $f(z)$ of $u(z)$, we obtain for the determination of the weight function $\varphi(\zeta)$ the Fredholm integral equation of the second kind

$$f(z) = \varphi(z) + \frac{1}{\pi} \oint_C k(z, \zeta)\, \varphi(\zeta)\, d\zeta \tag{4}$$

This particular inhomogeneous integral equation with continuous kernel established by Poincaré led Fredholm to his general theory of linear integral equations. He established his celebrated alternative: Either the integral equation (4) possesses à nontrivial solution $\varphi(z)$ for $f \equiv 0$, or it possesses a unique solution $\varphi(z)$ to each given integrable function $f(z)$.

Thus, the entire boundary value theory of harmonic functions in the plane is reduced to the study of the homogeneous integral equation

$$\varphi_\nu(z) = \frac{\lambda_\nu}{\pi} \oint_C k(z, \zeta)\, \varphi_\nu(\zeta)\, ds_\zeta\,. \tag{5}$$

If it could be shown that -1 is not an eigenvalue λ_ν of (5), the unique existence of a harmonic function $u(z)$ with prescribed boundary values $f(z)$ on C would have been established. This was indeed done by Fredholm and the existence problem of the function $u(z)$ was solved.

There remains the interesting question to study the eigenvalues λ_ν connected with this important integral equation and to determine their potential theoretical significance. They are called the Fredholm eigenvalues of the curve C; they may also be considered as functionals of the domain D or of the domain $\tilde{D}$.

It is well known that the lowest eigenvalue (in absolute value) λ_0 has the value $+1$ and belongs to the eigenfunction $\varphi_0(\zeta) = \text{const.}$ All other eigenvalues are real and satisfy $|\lambda_\nu| > 1$. Since we easily can handle the contribution of the trivial eigenvalue λ_0 to the kernel $k(z, \zeta)$, we actually can reduce the numerical solution of the first and second boundary value problem of potential theory to the process of iteration, i. e., the Liouville-Neumann series development. The details of this procedure were developed by Gershgorin [6]. It probably is the most convenient numerical method in two-dimensional potential theory. However, the speed of convergence of the Liouville-Neumann series depends on the absolute value of the lowest nontrivial Fredholm eigenvalue [1, 9, 17]. Thus, we see the significance to estimate the nontrivial Fredholm eigenvalues of a given curve C. We propose to study the values $\lambda_\nu(C)$ as functionals or as « fonctions de ligne » in the sense of Volterra and to solve extremum problems concerning them.

2. The Fredholm Eigenfunctions.

We consider the Fredholm eigenfunctions $\varphi_\nu(\zeta)$ connected with the eigenvalues λ_ν. These functions are defined only on the curve C. We now introduce harmonic functions $h_\nu(z)$ and $\tilde{h}_\nu(z)$ defined in D and $\tilde{D}$, respectively, which are closely related to the $\varphi_\nu(\zeta)$. We define

$$h_\nu(z) = \frac{\lambda_\nu}{\pi} \oint_C k(z, \zeta)\, \varphi_\nu(\zeta)\, ds_\zeta \,. \tag{6}$$

The right-hand side of (6) represents a harmonic function of z if z ranges over D; it also represents a harmonic function if z varies in $\tilde{D}$. But this harmonic function is not the continuation of $h(z)$ defined in D. For the sake of clarity we therefore shall denote it by the letters $\tilde{h}(z)$.

The jump conditions for the kernel $k(z, \zeta)$ imply for each point $z_0 \in C$

$$\lim_{z \to z_0} h_\nu(z) = (1 + \lambda_\nu) \varphi_\nu(z_0), \quad \lim_{z \to z_0} \tilde{h}_\nu(z) = (1 - \lambda_\nu) \varphi_\nu(z_0). \tag{7}$$

These boundary conditions on C determine the harmonic functions $h_\nu(z)$ and $\tilde{h}_\nu(z)$ uniquely. It is also well known from the theory of the double-layer potential that everywhere on C

$$\frac{\partial h_\nu(z_0)}{\partial n} = \frac{\partial \tilde{h}_\nu(z_0)}{\partial n}. \tag{8}$$

We have thus solved the following interesting problem of potential theory: To determine a harmonic function $h(z)$ in D and a harmonic function $\tilde{h}(z)$ in $\tilde{D}$ with the same normal derivatives at the common boundary and having proportional boundary values there.

We can also give another interpretation to this result. Let

$$D[h] = \iint_D (\nabla h)^2 \, d\tau = -\oint_C h \frac{\partial h}{\partial n} \, ds \tag{9}$$

$$D[\tilde{h}] = \iint_{\tilde{D}} (\nabla \tilde{h})^2 \, d\tau = +\oint_C \tilde{h} \frac{\partial \tilde{h}}{\partial n} \, ds$$

denote the Dirichlet integrals of h and $\tilde{h}$. It follows from (7) and (8) that

$$D[h_\nu] = \frac{\lambda_\nu + 1}{\lambda_\nu - 1} D[\tilde{h}_\nu]. \tag{10}$$

Since Dirichlet integrals are obviously nonnegative; we read off from (10) that $|\lambda_\nu| \geq 1$ and that $|\lambda_\nu| = 1$ is only possible for $\varphi_\nu(\zeta)$ const. Finally, it is easy to see that $\varphi_\nu = \text{const}$ implies $\lambda_\nu = +1$. Thus, the identities (7) and (8) decide the essential step in the above mentioned Fredholm alternative.

Consider next an arbitrary continuous function $f(z)$ defined on C; determine the harmonic functions $h(z)$ and $\tilde{h}(z)$ in D and $\tilde{D}$ which have the common boundary values $f(z)$ on C. Let $D[h]$ and $D[\tilde{h}]$ be their Dirichlet integrals and consider the ratio $D[h]/D[\tilde{h}]$ as a functional of $f(z)$. It is easily seen that for all f holds

$$\frac{\lambda_1 - 1}{\lambda_1 + 1} \le \frac{D[h]}{D[\tilde{h}]} \le \frac{\lambda_1 + 1}{\lambda_1 - 1} \tag{11}$$

where λ_1 is the lowest positive nontrivial Fredholm eigenvalue. The values $\lambda_\nu + 1/\lambda_\nu - 1$ are stationary values of this functional. Because of the complete symmetry of D and $\tilde{D}$, the same values must be the stationary values of the reciprocal ratio or, in other words, the values $\lambda_\nu - 1/\lambda_\nu + 1$ must likewise be stationary values for the original ratio. This is, indeed, the case since with each Fredholm eigenvalue $\lambda_\nu \neq -1$ also the Fredholm eigenvalue $-\lambda_\nu$ will occur. This is best seen from equations (7) and (8). Let $g_\nu(z)$ and $\tilde{g}_\nu(z)$ be the conjugate harmonic functions of $h_\nu(z)$ and $\tilde{h}_\nu(z)$. Then (7) and (8) yield by use of the Cauchy-Riemann equations

$$\frac{\partial g_\nu(z_0)}{\partial n} = \frac{1+\lambda_\nu}{1-\lambda_\nu} \frac{\partial \tilde{g}_\nu(z_0)}{\partial n} \tag{7'}$$

and

$$g_\nu(z_0) = \tilde{g}_\nu(z_0) \tag{8'}$$

if the additive constants in g_ν and $\tilde{g}_\nu$ are properly adjusted. Hence, if we define

$$h_\nu^*(z) = \frac{1}{1+\lambda_\nu} g_\nu(z), \qquad \tilde{h}_\nu^*(z) = \frac{1}{1-\lambda_\nu} \tilde{g}_\nu(z) \tag{12}$$

we find

$$h_\nu^*(z) = \frac{1-\lambda_\nu}{1+\lambda_\nu} \tilde{h}_\nu^*(z), \qquad \frac{\partial h_\nu^*}{\partial n} = \frac{\partial h_\nu^*}{\partial n} \tag{13}$$

which coincides with (7) and (8) if we replace λ_ν by $-\lambda_\nu$.

Since Fredholm eigenvalues occur always in pairs, we shall from now on understand by λ_ν the positive eigenvalue of the pair. Likewise, the $h_\nu(z)$ shall be the eigenfunctions belonging to the positive eigenvalues. Their conjugates will then automatically belong to $-\lambda_\nu$.

We introduce now the analytic functions

$$v_\nu(z) = \frac{\partial}{\partial z} h_\nu(z), \qquad \tilde{v}_\nu(z) = \frac{\partial}{\partial z} \tilde{h}_\nu(z) \tag{14}$$

where $\frac{\partial}{\partial z} = \frac{1}{2}\left(\frac{\partial}{\partial x} - i\frac{\partial}{\partial y}\right)$ is the well-known complex differentiator. A simple calculation leads to the following result:

$$(15)\qquad v_\nu(z) = \frac{\lambda_\nu}{\pi}\iint\limits_D \frac{\overline{v_\nu(\zeta)}}{(\zeta - z)^2}\,d\tau, \quad \tilde{v}_\nu(z) = -\frac{\lambda_\nu}{\pi}\iint\limits_{\tilde{D}} \frac{\overline{\tilde{v}_\nu(\zeta)}}{(\zeta - z)^2}\,d\tau$$

Clearly, the trivial eigenvalue λ_0 has to be discarded since its eigenfunction $h_0(z)$ is constant and is annihilated by differentiation. We understand the improper integrals (15) in the sense of a Cauchy principal value: we exclude first in the integration a circle of radius ε around z and then take the limit value of the integral as $\varepsilon \to 0$. With this understanding (15) represents integral equations for the eigenfunctions $v_\nu(z)$, and we may restrict ourselves to the case of positive eigenvalues $\lambda_\nu > 0$.

It is also easily verified that

$$(16)\qquad \tilde{v}_\nu(z) = \frac{\lambda_\nu}{\pi(1+\lambda_\nu)}\iint\limits_D \frac{\overline{v_\nu(\zeta)}}{(\zeta - z)^2}\,d\tau \qquad \text{for} \quad z \in \tilde{D}$$

$$v_\nu(z) = \frac{\lambda_\nu}{\pi(1-\lambda_\nu)}\iint\limits_{\tilde{D}} \frac{\overline{\tilde{v}_\nu(\zeta)}}{(\zeta - z)^2}\,d\tau \qquad \text{for} \quad z \in D$$

These integrals are proper and display clearly the analytic character of the eigenfunctions.

Another great advantage of introducing the harmonic eigenfunctions $h_\nu(z)$ and the analytic eigenfunctions $v_\nu(z)$ comes from the important orthogonality relations which these functions satisfy. Observe that the kernel $k(z, \zeta)$ is not symmetric in its arguments and hence that the eigenfunctions $\varphi_\nu(\zeta)$ defined on C are, in general, not orthogonal. On the other hand, let $h_\nu(z)$ and $h_\mu(z)$ belong to different eigenvalues: $\lambda_\nu \neq \lambda_\mu$. We compute the Dirichlet integral

$$(17)\qquad D[h_\nu, h_\mu] = \iint\limits_D \nabla h_\nu \cdot \nabla h_\mu\,d\tau = -\int\limits_C h_\nu \frac{\partial h_\mu}{\partial n}\,ds$$

Using the boundary relations (7) and (8), we can write

$$D[h_\nu, h_\mu] = \frac{\lambda_\nu + 1}{\lambda_\nu - 1} \int_C \tilde{h}_\nu \frac{\partial \tilde{h}_\mu}{\partial n} ds = \frac{\lambda_\nu + 1}{\lambda_\nu - 1} D[\tilde{h}_\nu, \tilde{h}_\mu]. \tag{18}$$

But both Dirichlet integrals are symmetric in ν and μ while we know by assumption that $\frac{\lambda_\nu + 1}{\lambda_\nu - 1} \neq \frac{\lambda_\mu + 1}{\lambda_\mu - 1}$. Hence, we proved

$$D[h_\nu, h_\mu] = D[\tilde{h}_\nu, \tilde{h}_\mu] = 0 \qquad \text{for} \quad \lambda_\nu \neq \lambda_\mu \tag{19}$$

We may then assume without loss of generality that

$$D[h_\nu, h_\mu] = D[\tilde{h}_\nu, \tilde{h}_\mu] = 0 \qquad \text{for } \nu \neq \mu \tag{19'}$$

From definition (14) we conclude next the orthogonality relations for the analytic eigenfunctions

$$\iint_D v_\nu(z)\, \overline{v_\mu(z)}\, d\tau = \iint_{\tilde{D}} \tilde{v}_\nu(z)\, \overline{\tilde{v}_\mu(z)}\, d\tau = 0 \qquad \text{for } \nu \neq \mu \tag{20}$$

We should like to normalize the eigenfunctions $h_\nu(z)$ in the Dirichlet norm. Clearly, in view of (18) we shall have to multiply $h_\nu(z)$ and $\tilde{h}_\nu(z)$ with different normalizing factors. Correspondingly, the $v_\nu(z)$ and $\tilde{v}_\nu(z)$ go over into analytic functions $w_\nu(z)$ and $\tilde{w}_\nu(z)$ with the following properties:

$$w_\nu(z) = \frac{\lambda_\nu}{\pi} \iint_D \frac{\overline{w_\nu(\zeta)}}{(\zeta - z)^2} d\tau, \qquad \tilde{w}_\nu(z) = \frac{\lambda_\nu i}{\pi \sqrt{\lambda_\nu^2 - 1}} \iint_D \frac{\overline{w_\nu(\zeta)}}{(\zeta - z)^2} d\tau \tag{21}$$

$$\tilde{w}_\nu(z) = \frac{\lambda_\nu}{\pi} \iint_{\tilde{D}} \frac{\overline{\tilde{w}_\nu(\zeta)}}{(\zeta - z)^2} d\tau, \qquad w_\nu(z) = \frac{\lambda_\nu i}{\pi \sqrt{\lambda_\nu^2 - 1}} \iint_{\tilde{D}} \frac{\overline{\tilde{w}_\nu(\zeta)}}{(\zeta - z)^2} d\tau$$

and the orthonormalization

$$\iint_D w_\nu \overline{w}_\mu \, d\tau = \delta_{\nu\mu} \qquad \iint_{\tilde{D}} \tilde{w}_\nu \overline{\tilde{w}}_\mu \, d\tau = \delta_{\nu\mu} \tag{22}$$

3. Fredholm Eigenvalues and Hilbert Transforms.

We can now connect our results with some important general theorems of analysis. Let $f(z)$ be an arbitrary complex-valued function defined in the entire complex z-plane and of class $\mathcal{L}^2$. Define its so-called Hilbert transform

$$F(z) = \frac{1}{\pi} \iint \frac{\overline{f(\zeta)}}{(\zeta - z)^2} d\tau \tag{23}$$

This will be a new function with the same properties as $f(z)$ and with the same norm

$$\| F \|^2 = \iint | F |^2 d\tau = \iint |f|^2 d\tau = \|f\|^2. \tag{24}$$

The Hilbert transformation is an involution, that is, the Hilbert transform of $F(z)$ is again $f(z)$. Finally, wherever $f(z)$ is analytic, its transform $F(z)$ will be analytic too.

Consider the function $f(z)$ defined as $w_\nu(z)$ in D and as 0 in $\tilde{D}$. Clearly, its Hilbert transform in $\frac{1}{\lambda_\nu} w_\nu(z)$ in D and $\frac{\sqrt{\lambda_\nu^2 - 1}}{\lambda} \tilde{w}_\nu(z)$ in $\tilde{D}$. We may interpret the eigenfunctions $w_\nu(z)$ as the eigenfunctions of the Hilbert transformation restricted to D and to the class of analytic functions in D.

Let now $g(z)$ be a real-valued function in D which vanishes on the boundary C of D and whose complex derivative $\frac{\partial g}{\partial z}$ is in D of the class $\mathcal{L}^2$. It is easily verified that for $z \in D$

$$\frac{1}{\pi} \iint_D \left(\overline{\frac{\partial g(\zeta)}{\partial \zeta}} \right) \frac{1}{(\zeta - z)^2} d\tau = \frac{\partial g(z)}{\partial z} \tag{25}$$

such that all such functions $\frac{\partial g}{\partial z}$ are likewise eigenfunctions of the Hilbert transformation with the eigenvalue 1. However, if $v(z)$ is an arbitrary analytic function in D with a finite norm, we have

$$\iint_D v(z) \left(\overline{\frac{\partial g}{\partial z}} \right) d\tau = 0 \tag{26}$$

Hence, the linear space of all analytic functions with finite norm is orthogonal to these eigenfunctions $\frac{\partial g}{\partial \bar{z}}$.

The linear space of all complex valued functions in D of class $\mathcal{L}^2$ can be split into the two complementary subspaces consisting of the $\frac{\partial g}{\partial \bar{z}}$ and of analytic functions. It is evident that the nontrivial part of the theory of Hilbert transforms belongs to the subspace of analytic functions and not to the trivial orthogonal complement where it reduces to the identity transformation. The theory of the Hilbert transform in the subspace of analytic functions was developed by Bergman and Schiffer [3, 5]. The general theory for the $\mathcal{L}^2$-space was first indicated by Beurling [2, 4].

We shall see that the Hilbert transformation in the analytic subspace can be reduced to an integral transformation with a completely continuous kernel.

4. The Green's Function and its Analytical Kernels.

Let $g(z, \zeta)$ be the harmonic Green's function of D. That is, $g(z, \zeta)$ is harmonic in both arguments for $z \neq \zeta$, vanishes if either argument point lies on the boundary C of D and behaves such that $g(z, \zeta) + \log |z - \zeta|$ is regular harmonic as $z \to \zeta$. It is well known that $g(z, \zeta)$ is symmetric in both arguments. We define now the two kernels [3, 15]

$$K(z, \bar{\zeta}) = -\frac{2}{\pi} \frac{\partial^2 g}{\partial z\, \partial \bar{\zeta}}, \qquad L(z, \zeta) = -\frac{2}{\pi} \frac{\partial^2 g}{\partial z\, \partial \zeta} \tag{27}$$

K is hermitian in the two variables, analytic in z and antianalytic in ζ. It is regular even for $z = \zeta$ since the differentiation process which defines K annihilates the singularity of the Green's function. On the other hand, $L(z, \zeta)$ is analytic and symmetric in its variables, but it has a double pole at $z = \zeta$ and can be written as

$$L(z, \zeta) = \frac{1}{\pi (z - \zeta)^2} - l(z, \zeta) \tag{28}$$

Here $l(z, \zeta)$ is regular analytic and symmetric in both variables. It is even continuous in the closure of D. If C is an analytic curve, it is even analytic in $D + C$.

From the boundary behavior of the Green's function a simple integration by parts leads to the following identities valid for every analytic function $\varphi(z)$ with finite norm over D:

$$(29) \qquad \iint_D K(z, \bar{\zeta}) \varphi(\zeta) \, d\tau = \varphi(z), \qquad \iint_D L(z, \zeta) \overline{\varphi(\zeta)} \, d\tau = 0$$

This shows that $K(z, \bar{\zeta})$ is the Bergman kernel function which reproduces every analytic function with finite norm ; $L(z, \zeta)$ annihilates the same function class under the integration considered. We may rephrase the second identity as follows:

$$(30) \qquad \frac{1}{\pi} \iint_D \frac{\overline{\varphi(\zeta)}}{(\zeta - z)^2} \, d\tau = \iint_D l(z, \zeta) \overline{\varphi(\zeta)} \, d\tau$$

On the left side stands the improper integral which defines according to (23) the Hilbert transform of $\varphi(z)$. On the right we have an integral transformation which is completely continuous and coincides with the Hilbert transform on the subspace of all analytic functions in D with finite norm. This new definition of the Hilbert transform on the subspace is, of course, of very great convenience. Let us consider an arbitrary complete orthonormal system $w_\nu(z)$ in the subspace of analytic functions in D. The Bergman kernel $K(z, \bar{\zeta})$ can be developed into a Fourier series in the system, and we have by virtue of (29)

$$(31) \qquad K(z, \bar{\zeta}) = \sum_{\nu=1}^{\infty} w_\nu(z) \overline{w_\nu(\zeta)}$$

This was, indeed, Bergman's original definition of his kernel function. It is easy to see that the Fredholm eigenfunctions $w_\nu(z)$ defined by (21) and (22) form a complete set and may be used in the representation (31).

Moreover, we have for this particular choice of the orthonormal system in view of (21) and (30)

$$(32) \qquad w_\nu(z) = \lambda_\nu \iint_D l(z, \zeta) \overline{w_\nu(\zeta)} \, d\tau.$$

We can express $l(z,\zeta)$ for z fixed as a Fourier series in the $w_\nu(\zeta)$ and (32) yields us the Fourier coefficients. We have then

$$l(z,\zeta)=\sum_{\nu=1}^{\infty}\frac{w_\nu(z)\,w_\nu(\zeta)}{\lambda_\nu}. \tag{33}$$

We are led next to an important and beautiful identity for the l-kernel by using the concept of the Hilbert transform. Let $f(z)$ be analytic in D and of finite norm. We may conceive it as a complex valued function of class $\mathfrak{L}^2$ in the entire plane if we define it as identically zero in $\tilde{D}$. Its Hilbert transform $F(z)$ can be written as follows:

$$F(z)=\iint_D l(z,\zeta)\overline{f(\zeta)}\,d\tau \qquad \text{if} \quad z\in D \tag{34}$$

$$F(z)=\frac{1}{\pi}\iint_D \frac{\overline{f(\zeta)}}{(\zeta-z)^2}\,d\tau \qquad \text{if} \quad z\in \tilde{D}.$$

The identity of norms (24) for Hilbert transforms yield thus

$$\iint_D |f(z)|^2\,d\tau=\iint_D\iint_D K(\zeta,\bar{\eta})\overline{f(\zeta)}f(\eta)\,d\tau_z\,d\tau_\zeta \tag{35}$$

$$=\iint_D\iint_D\iint_D l(z,\zeta)\overline{l(z,\eta)}f(\eta)\overline{f(\zeta)}\,d\tau_z\,d\tau_\zeta\,d\tau_\eta$$

$$+\iint_{\tilde{D}}\iint_D\iint_D \frac{1}{\pi^2}\frac{1}{(z-\zeta)^2}\frac{1}{\overline{(z-\eta)^2}}f(\eta)\overline{f(\zeta)}\,d\tau_z d\tau_\zeta d\tau_\eta$$

A standard argument leads, therefore, to the identity

$$\iint_D l(z,\zeta)\overline{l(z,\eta)}\,d\tau+\frac{1}{\pi^2}\iint_{\tilde{D}}\frac{d\tau}{(z-\zeta)^2\,\overline{(z-\eta)^2}}=K(\zeta,\bar{\eta}). \tag{36}$$

Observe that the second left-hand integral is regular analytic for $\zeta\in D$ and regular anti-analytic for $\eta\in D$. It can be computed by integrations and is, therefore, more elementary than the kernels K and l which depend on the Green's function of the domain, that is, on the solution of a boundary value problem for harmonic functions. We shall call the expression

$$\Gamma(\zeta, \bar{\eta}) = \frac{1}{\pi^2} \iint_{\tilde{D}} \frac{d\tau}{(z-\zeta)^2 \overline{(z-\eta)}^2} \tag{37}$$

a geometric term in contradistinction to the more transcendental kernels K and l. Clearly, Γ is hermitian and a positive definite kernel. If we insert into (36) the Fourier developments (10) and (33), we find the Fourier development for the geometric kernel

$$\Gamma(\zeta, \bar{\eta}) = \sum_{\nu=1}^{\infty} \left(1 - \frac{1}{\lambda_\nu^2}\right) w_\nu(\zeta) \overline{w_\nu(\eta)}. \tag{38}$$

This representation may serve as basis for calculating the kernels l and K. The basic idea is as follows. All numbers $\xi_\nu = \left(1 - \frac{1}{\lambda_\nu^2}\right)$ lie in the interval $0 < \left(1 - \frac{1}{\lambda_1^2}\right) \leq \xi \leq 1$. In this interval $\frac{1}{\xi} = \sum_{n=0}^{\infty} (1-\xi)^n$ converges absolutely and uniformly. Thus,

$$1 = \xi \sum_{n=0}^{\infty} (1-\xi)^n \tag{39}$$

can be approximated uniformly and arbitrarily by polynomials $P_N(\zeta) = \sum_{\nu=0}^{N} a_{N\nu} \zeta^\nu$. If $\Gamma^{(\nu)}(\zeta, \bar{\eta})$ denotes the νth iterate of the kernel $\Gamma(\zeta, \bar{\eta}) = \Gamma^{(1)}(\zeta, \bar{\eta})$ it is clear that

$$\sum_{\nu=0}^{N} a_{N\nu} \Gamma^{(\nu)}(\zeta, \bar{\eta}) \underset{N}{\rightarrow} K(\zeta, \bar{\eta}). \tag{40}$$

Thus, the kernel K and, as is easily seen, the kernel l can be approximated arbitrarily in terms of iterates of the elementary geometric kernel $\Gamma(\zeta, \bar{\eta})$ [3, 15].

5. Fredholm Eigenvalues and Univalent Functions.

Let us suppose that $f(z)$ is analytic and univalent in the unit circle and maps $|z| < 1$ onto a domain D in the w-plane. Because of the conformal invariance of the Green's function we have for the L-kernel of D the identity

$$L_D(w, \omega) f'(z) f'(\zeta) = L(z, \zeta) \tag{41}$$

with $\omega = f(\zeta)$ and $L(z, \zeta)$ being the L-kernel of the unit circle. Since the Green's of a circle is well known, we have

$$L(z, \zeta) = \frac{1}{\pi (z - \zeta)^2} \tag{42}$$

and hence

$$\begin{aligned} l_D(w, \omega) f'(z) f'(\zeta) &= \frac{1}{\pi} \left[\frac{f'(z) f'(\zeta)}{(f(z) - f(\zeta))^2} - \frac{1}{(z - \zeta)^2} \right] \\ &= \frac{1}{\pi} \frac{\partial^2}{\partial z \, \partial \zeta} \log \frac{f(z) - f(\zeta)}{z - \zeta} . \end{aligned} \tag{43}$$

Now, the function

$$\log \frac{f(z) - f(\zeta)}{z - \zeta} = \sum_{\mu, \nu = 0}^{\infty} c_{\mu\nu} z^{\mu} \zeta^{\nu} \tag{44}$$

plays a central role in the theory of univalent functions. Indeed, a necessary and sufficient condition that $f(z)$ be regular and univalent in $|z| < 1$ is the regularity of this function in two complex variables in the product domain $|z| < 1$, $|\zeta| < 1$. This important formulation of univalency was observed a long time ago, but Grunsky [7] was the first to draw important conclusions from it. Using the preceding relations between the l-and K-kernel (in particular (36)), we can derive the Grunsky inequalities

$$\left| \Sigma c_{\mu\nu} x_{\mu} x_{\nu} \right| \leq \Sigma \nu |x_{\nu}|^2 \tag{45}$$

for arbitrary complex vectors x_{ν} as the necessary and sufficient condition for the univalence of $f(z)$. From these conditions many elementary estimates for the coefficients of univalent functions may be obtained; the most startling one is $|a_4| \leq 4$ proved by Charzynski and Schiffer [5].

We can bring every quadratic form $Q(x, x)$ with symmetric matrix Q into the normal form

$$Q(x, x) = \Sigma l_{\nu} y_{\nu}^2 , \qquad l_{\nu} > 0 \tag{46}$$

where the y_ν are obtained from the x_ν by a unitary transformation. In other words, every symmetric matrix Q can be brought into the Schur normal form [16] $Q = U^{\intercal} \Delta U$ where U is a unitary matrix, $U^{\intercal}$ is its transposed and Δ is a diagonal matrix with positive elements l_ν. There arises then the question: Given a univalent function $f(z)$ in $|z| < 1$ and its corresponding symmetric matrix $c_{\mu\nu}$ defined by (44), what are the positive numbers l_ν which appear in its Schur normal form? The answer is: The l_ν are simple expressions in the Fredholm eigenvalues λ_ν of the image domain D. This surprising relation can be read off from (33), (43) and (44). It shows clearly the great significance of the Fredholm eigenvalues for the general theory of analytic functions and conformal mapping.

6. The Variation of Fredholm Eigenvalues.

The most powerful tool in the study of Fredholm eigenvalues are the variational formulas which show how the λ_ν behave as functionals of the curve C and how they vary with continuously changing boundary. To find the functional derivatives of the λ_ν with respect to C we transform the integral equation (21) for $w_\nu(z)$ by partial integration into

$$w_\nu(z) = \frac{\lambda_\nu}{2\pi i} \oint_C \frac{\overline{(w_\nu(\zeta)\, d\,(\zeta)}}{\zeta - z} \tag{47}$$

We select an arbitrary but fixed point $z_0 \in \widetilde{D}$ and consider the variation

$$z^* = z + \frac{\varepsilon}{z - z_0} \tag{48}$$

The mapping $z^*(z)$ is, for small enough $|\varepsilon|$, a regular analytic function of z in $D + C$ and maps D into a new domain D^* with boundary C^*, eigenvalues λ_ν^* and eigenfunctions $w_\nu^*(z)$. If we can give an asymptotic formula for λ_ν^*, we shall have achieved our aim since the most general variation of C may be obtained by superposition of elementary variations (48).

We write down the corresponding integral equation (47) for λ_ν^*. But putting $z^* = z^*(z)$, $\zeta^* = \zeta^*(\zeta)$, we may refer the integration back to the original curve C and obtain an integral equation for the fixed domain D whose kernel now depends on z_0 and ε. We find

$$(49)\qquad w_\nu^*(z^*(z)) = \frac{\lambda_\nu^*}{2\pi i}\oint_C \frac{\overline{\left(w_\nu^*(\zeta^*(\zeta))\left(1-\dfrac{\varepsilon}{(\zeta-z_0)^2}\right)d\zeta\right)}}{(\zeta-z)\left(1-\dfrac{\varepsilon}{(z-z_0)(\zeta-z_0)}\right)}$$

Let us define the new unknown function in D

$$(50)\qquad m_\nu(z) = w_\nu^*(z^*(z))\left(1-\frac{\varepsilon}{(z-z_0)^2}\right)$$

and bring (49) into the much simpler form

$$(51)\quad m_\nu(z) = \frac{\lambda_\nu^*}{2\pi i}\oint_C \overline{(m_\nu(\zeta)\,d\zeta)}\left[\frac{1}{\zeta-z}+\frac{\varepsilon}{(z-z_0)\left[(z-z_0)(\zeta-z_0)-\varepsilon\right]}\right]$$

We thus have for $m_\nu(z)$ an integral equation whose kernel depends analytically on ε. We may also return to the domain integral form

$$(52)\quad m_\nu(z) = \frac{\lambda_\nu^*}{\pi}\iint_D \overline{m_\nu(\zeta)}\left[\frac{1}{(\zeta-z)^2}+\frac{\varepsilon}{(\zeta-z_0)^2(z-z_0)^2}+0(\varepsilon^2)\right]d\tau$$

We may replace the first term $\dfrac{1}{\pi}(\zeta-z)^{-2}$ in the kernel of (52) by $l(z,\zeta)$. Thus, this integral equation falls well into the pattern of the Rellich theory [8] for eigenvalues of variable kernels. We may assert that

$$(53)\qquad \lambda_\nu^* = \lambda_\nu + |\,\varepsilon\,|\, l_\nu + \ldots$$

admits a power series development in ε. For the sake of simplicity, let us assume that λ_ν is nondegenerate, i. e., it has only one eigenfuction $w_\nu(z)$. In this case we also have a series development

$$(54)\qquad m_\nu(z) = w_\nu(z) + |\,\varepsilon\,|\,\omega_\nu(z) + \ldots$$

We now multiply equation (52) by $w_\nu(z)$ end integrate over D. We use integral equation (21) for $w_\nu(z)$ and observe the asymptotics in ε as given in (53) and (54). Thus,

$$\iint_D m_\nu(z)\overline{w_\nu(z)}\,d\tau = \frac{\lambda_\nu^*}{\lambda_\nu}\iint_D \overline{m_\nu(\zeta)}\,w_\nu(\zeta)\,d\tau \tag{55}$$

$$+\varepsilon\frac{\lambda_\nu}{\pi}\iint_D \frac{\overline{w_\nu(\zeta)}}{(\zeta - z_0)^2}\,d\tau\cdot\iint_D \frac{\overline{w_\nu(z)}}{(z - z_0)^2}\,d\tau + 0(\varepsilon^2).$$

By our assumption $z_0 \in \tilde{D}$, and hence using again the corresponding identity (21), we find

$$\lambda_\nu\iint_D m_\nu(z)\overline{w_\nu(z)}\,d\tau = \lambda_\nu^*\iint_D \overline{m_\nu(\zeta)}\,w_\nu(\zeta)\,d\tau + \tag{56}$$

$$+\pi\,\varepsilon\,\tilde{w}_\nu(z_0)^2(\lambda_\nu^2 - 1) + 0(\varepsilon^2).$$

Take the real parts in (56), use (54) and the normalization of $w_\nu(z)$ to find

$$\lambda_\nu^* - \lambda_\nu = -\operatorname{Re}[\pi\,\varepsilon\,(\lambda_\nu^2 - 1)\,\tilde{w}_\nu(z_0)^2] + 0(\varepsilon^2). \tag{57}$$

This is the desired variational formula for λ_ν. A similar formula may be given if λ_ν is a degenerate eigenvalue. Had we taken a variation (48) with $z_0 \in D$, we might have reasoned in the same way by starting with the integral equation for $\tilde{w}_\nu(z)$ over $\tilde{D}$. We would have found the analogous formula

$$\delta\lambda_\nu = -\operatorname{Re}\{\pi\varepsilon(\lambda_\nu^2 - 1)\,w_\nu(z_0)^2\} + 0(\varepsilon^2) \tag{58}$$

for the varation of a nondegenerate eigenvalue.

To illustrate the power of these variational formulas, we quote one extremum problem which has been solved by using them. Let $f(z)$ be univalent and regular in the circular ring $r \leq |z| \leq R$ with $r < 1 < R$. Let C be the image of $|z| = 1$ by this map. We call C a uniformly analytic curve with the modulus (r, R). The concept of uniform analyticity is an obvious sharpening of the usual assumption of analyticity of a curve. We have the theorem :

The lowest nontrivial eigenvalue λ_1 of a uniformly analytic curve with the modulus (r, R) satisfies the inequality

$$\lambda_1 \geq \frac{r^2 + R^2}{1 + r^2 R^2}. \tag{59}$$

This estimate is the best possible. It shows the importance of the concept of uniform analyticity in the numerical procedures of conformal mapping.

7. Fredholm Eigenvalues for Multiply-Connected Domains.

It is easy to extend the preceding considerations to the case that $\tilde{D}$ is a multiply-connected domain which contains the point at infinity and is bounded by $N > 1$ closed curves C_ν, which shall again be three times continuously differentiable. Its complement D will then consist of N disjoint simply-connected finite domains D_ν. Let $C = \Sigma\, C_\nu$ denote the common boundary of $\tilde{D}$ and D. We may then discuss the Fredholm eigenvalue problem (5) with respect to the curve system C.

We may extend again the eigenfunctions of this problem which are defined only on C into harmonic functions in $\tilde{D}$ and in D. As before, we can interpret the eigenvalues λ_ν also as eigenvalues of integral equations for analytic functions of the Hilbert transform type. We can define a function $w_\nu(z)$ analytic in each component D_ν of D such that

$$w_\nu(z) = \frac{\lambda_\nu}{\pi} \iint_D \frac{\overline{w_\nu(\zeta)}}{(\zeta - z)^2}\, d\tau, \qquad z \in D \tag{60}$$

and one single analytic function $\tilde{w}_\nu(z)$ in the connected domain $\tilde{D}$ such that

$$\tilde{w}_\nu(z) = \frac{\lambda_\nu}{\pi} \iint_{\tilde{D}} \frac{\overline{\tilde{w}_\nu(\zeta)}}{(\zeta - z)^2}\, d\tau, \qquad z \in \tilde{D} \tag{61}$$

These functions are related to the Fredholm eigenfunctions $\varphi_\nu(z)$ of (5) in the same way as in the case of simple connectivity. There is, however, one important difference. The eigenvalue $\lambda = 1$ occurs in (5) in $(N-1)^{st}$ order degeneracy. The integral equation (60) does not possess this eigenvalue at all while $\lambda = 1$ is an eigenvalue of (61) of order $N-1$. The corresponding eigenfunctions are the derivatives of the harmonic measures of the multiply-connected domain $\widetilde{D}$. Since $\lambda = 1$ leads to simple and well-known eigenfunctions, we still shall call it the trivial eigenvalue and assume in all subsequent discussions $\lambda_\nu > 1$.

It is easily seen that for $\lambda_\nu > 1$ still

$$\widetilde{w}_\nu(z) = \frac{\lambda_\nu i}{\pi\sqrt{\lambda_\nu^2 - 1}} \iint_D \frac{\overline{w_\nu(\zeta)}}{(\zeta - z)^2}\, d\tau, \qquad z \in \widetilde{D} \tag{62}$$

$$w_\nu(z) = \frac{\lambda_\nu i}{\pi\sqrt{\lambda_\nu^2 - 1}} \iint_{\widetilde{D}} \frac{\overline{\widetilde{w}_\nu(\zeta)}}{(\zeta - z)^2}\, d\tau, \qquad z \in D$$

and that the $w_\nu(z)$ and $\widetilde{w}_\nu(z)$ form orthonormal systems in their respective domains.

Finally, we can extend the entire theory of the Hilbert transform by means of $l(z,\zeta)$ to the case of the connected region $\widetilde{D}$. However, if we wish to do tho same thing for the set of domains D_ν, we first have to give a proper definition of $l(z,\zeta)$. We start with defining a Green's function $g(z,\zeta)$ for the disconnected region D, namely

$$g(z,\zeta) = \begin{cases} g_\nu(z,\zeta) & \text{if } z,\zeta \text{ lie in same } D_\nu \\ 0 & \text{if } z,\zeta \text{ lie in different } D_\nu \end{cases} \tag{63}$$

Here $g_\nu(z,\zeta)$ is the ordinary Green's function of the simply-connected domain D_ν. We define next $L(z,\zeta)$ from $g(z,\zeta)$ by means of (27) and $l(z,\zeta)$ by means of (28). Thus

$$l(z,\zeta) = \begin{cases} l_\nu(z,\zeta) & \text{if } z,\zeta \text{ lie in same } D_\nu \\ \dfrac{1}{\pi(z-\zeta)^2} & \text{if } z,\zeta \text{ lie in different } D_\nu \end{cases} \tag{64}$$

$l_\nu(z,\xi)$ is the l-kernel of the simply-connected domain D_ν. With this definition, (30) remains obviously valid, and the Fourier representation (33) of $l(z, \zeta)$ in terms of the analytic Fredholm eigenfunctions is preserved.

The variational formulas of the preceding section can be carried over without change since we did not use anywhere in our calcutations that C consists of one single curve.

8. Fredholm Determinants and Conformal Mapping.

Having enumerated many definitions and identities, we shall now show their usefulness and interest by particular applications. An important concept in integral equation theory is the Fredholm determinant [12, 14]

$$D(\lambda) = \prod_{\nu=1}^{\infty}\left(1 - \frac{\lambda^2}{\lambda_\nu^2}\right) \tag{65}$$

where the product is extended over all nontrivial eigenvalues λ_ν. We observe that the eigenvalues $+\lambda_\nu$ and $-\lambda_\nu$ occur always in pairs in our problem; this accounts for the quadratic factors.

Consider $D(\lambda)$ for fixed λ as a functional of C and ask for its variation if C is varied by the standard variation (48) with $z_0 \in D$. By virtue of (58) we find

$$\delta \log D(\lambda) = -\operatorname{Re}\left\{2\pi\varepsilon\lambda^2 \sum_{\nu=1}^{\infty} \frac{\lambda_\nu^2 - 1}{\lambda_\nu^2 - \lambda} \frac{w_\nu(z_0)^2}{\lambda_\nu}\right\}. \tag{66}$$

This formula can also be justified if some λ_ν are degenerate eigenvalues. The result simplifies considerably in the case $\lambda = 1$. Indeed, in view of (33) we can write

$$\delta \log D(1) = -\operatorname{Re}\{2\pi\varepsilon\, l(z_0, z_0)\}. \tag{67}$$

Thus, the important function $l(z, z)$ has been identified as the functional derivative of the logarithm of the Fredholm determinant.

A surprising result occurs in the case of a multiple connectivity $N > 1$. We can speak of the eigenvalues of the curve system C and their Fredholm determinant; we may also consider the eigenvalues $\lambda_\nu^{(k)}$ of the single curve C_k and their Fredholm determinant $D^{(k)}(1)$. If $z_0 \in D_k$ we have by (64) the identity

$$\delta \log D(1) = \delta \log D^{(k)}(1). \tag{68}$$

That is, under a standard variation (48) which is regular analytic outside of D_k, the ratio $D^k(1)/D(1)$ has zero variation. By reasoning typical for variational theory, we can then extend this result to arbitrary finite conformal maps in the exterior of D_k.

THEOREM. *Let D be a set of disjoint finite simply-connected domains D_l with the boundary curve system C. Let $D(1)$ and $D^{(l)}(1)$ be the Fredholm determinants of the curve system C and the single curve C_l, respectively. Let $w=f(z)$ be a conformal mapping in the exterior of D_l. It will carry the curve system C into a curve system Γ; let $\Delta(1)$ and $\Delta^{(l)}(1)$ be the corresponding Fredholm determinants of Γ and Γ_l (the image of C_l). Then*

$$\frac{\Delta^{(l)}(1)}{\Delta(1)}=\frac{D^{(l)}(1)}{D(1)}. \tag{69}$$

It seems difficult to prove the conformal invariance of this ratio in a nonvariational manner.

We recall the fact that if C is a circle, all its eigenvalues are infinite and that the Fredholm determinant of each circle has the constant value 1. In every other case the definition (65) clearly indicates that $D(1)<1$. Hence, suppose that we start with an arbitrary curve set C and map the exterior of C_l conformally onto the exterior of a circle. By (69) we can assert for the Fredholm determinant $\Delta(1)$ of the new curve system

$$\Delta(1)=\frac{D(1)}{D^{(l)}(1)}\geq D(1). \tag{70}$$

Equality in (76) holds only if C_l already happened to be a circle.

This remark throws light on a well-known procedure to map a multiply-connected domain onto a circular domain. One starts with the curve C_1 and maps its exterior onto the exterior of a circle. Then one takes the image of C_2 and maps its exterior onto the exterior of a circle. One continues this procedure indefinitely taking care to run through the images of all starting curves in fixed order. The limit of this map transforms all initial curves C_ν into circles. We see that in this procedure the Fredholm determinant $D(1)$ is steadily increased. One can base on this observation a convergence proof for this method of iteration. We also draw the following conclusion:

THEOREM. *Among all conformally equivalent domains the circular domain has the largest Fredholm determinant* $D^{(1)}$.

This theorem was originally proved by variational methods [12]. The present derivation explains more clearly its significance.

9. Conclusion.

The close relation between the Fredholm eigenvalue problem and the theory of analytic functions of one complex variable has been evident throughout the whole exposition. Hence, it will be expected that the potential theory in more than two dimensions will lead to Fredholm eigenvalues with a less elegant and elastic theory. However, many results can be preserved even in this transition. But one very significant result shows the great difference in the nature of the eigenvalues for different dimensions.

THEOREM. *Let D be a domain in space and let λ_1 be its lowest positive nontrivial Fredholm eigenvalue. Then*

$$\lambda_1 \leq 3. \tag{71}$$

Equality holds only in the case that D is a sphere [13].

Thus, the Liouville-Neumann series development, which solves the boundary value problem in three-dimensional potential theory, will never converge better than a geometric series with ratio $\frac{1}{3}$.

Another significant difference comes from the fact that the concept of conjugate harmonic functions fails in more than two dimensions. Hence, we cannot assert that with each Fredholm eigenvalue λ_ν also its negative $-\lambda_\nu$ will occur as an eigenvalue.

The study of Fredholm eigenvalues in more than two dimensions is thus still an open and promising field of research.

BIBLIOGRAPHY

[1] L. V. AHLFORS, *Remarks on the Neumann-Poincaré integral equation*, Pacific J. Math. **2** (1952), 271-280.

[2] L. V. AHLFORS, *Conformality with respect to Riemann metrics.*, Ann. Acad. Fenn., Series A 206 (1955).

[3] S. BERGMAN and M. SCHIFFER, *Kernel functions and conformal mapping*, Compositio Math. **8** (1951), 205-249.

[4] I. E. BLOCK, *Kernel functions and class L^2*, Proc. Amer. Math. Soc. **4** (1953), 110-117.

[5] S. CHARZYNSKI and M. SCHIFFER, *A new proof of the Bieberbach conjecture for the fourth coefficient*. Arch. Rational Mech. Anal. **5** (1960), 187-193.

[6] S. GERSHGORIN, *On conformal mapping of a simply-connected region onto a circle*, Math. Sb. **40** (1933), 48-59.

[7] H. GRUNSKY, *Koeffizientenbedingungen für schlicht abbildende meromorphe Funktionen*, Math. Z. **45** (1939), 29-61.

[8] F. RELLICH, *Störungstheorie der Spektralzerlegung*, I. Mitteilung, Math. Ann. **113** (1937), 600-619.

[9] H. L. ROYDEN, *A modification of the Neumann-Poincaré method for multiply connected regions*, Pacific J. Math. **2** (1952), 385-394.

[10] M. SCHIFFER, *Applications of variational methods in the theory of conformal mapping*, Proc. Symp. Appl. Math. **8** (1958), 93-113.

[11] M. SCHIFFER, *The Fredholm eigenvalues of plane domains*, Pacific J. Math. **7** (1957), 1187-1225.

[12] M. SCHIFFER, *Fredholm eigenvalues of multiply-connected domains*, Pacific J. Math. **9** (1959), 211-264.

[13] M. SCHIEFER, *Problèmes aux limites et fonctions propres de l'équation intégrale de Poincaré et de Fredholm*. C. R. Paris **245** (1957), 18-21.

[14] M. SCHIFFER and N. HAWLEY, *Connections and conformal mapping*, Acta Math. **107** (1962), 175-274.

[15] M. SCHIFFER and D. C. SPENCER, *Functionals of finite Riemann surfaces*, Princeton 1954.

[16] I. SCHUR, *Ein Satz über quadratische Formen mit komplexen Koeffizienten*, Amer. J. Math, **67** (1945), 472-480

[17] S. E. WARSCHAWSKI, *On the effective determination of conformal maps*, Contribution to the theory of Riemann surface, Princeton 1953.

[*Entrata in Redazione il 7 settembre 1963*].

[125] Fredholm eigenvalues and Grunsky matrices

---

[125] Fredholm eigenvalues and Grunsky matrices. *Ann. Polon. Math.* **39** (1981), 149–164.

ANNALES
POLONICI MATHEMATICI
XXXIX (1981)

Fredholm eigenvalues and Grunsky matrices

by MENAHEM SCHIFFER * (Stanford)

Dedicated to the memory of my friend Stefan Bergman

Abstract. The close relation between the Hilbert transform and the Fredholm integral equation is shown and is used to derive the basic properties of the Fredholm eigenfunctions and their eigenvalues. Applications are given to the theory of conformal mapping and the Grunsky inequalities. The sharp bounds for the inequalities are shown to be expressible in terms of the lowest Fredholm eigenvalue of the image domain. An analogous theory is developed for the case of univalent functions which map onto mutually disjoint domains.

Introduction. In this paper we shall show the close connection between coefficient inequalities of the Grunsky type and the theory of the Hilbert transform and the Fredholm eigenvalues of a domain. We arrive at some illuminating insights into the character of such inequalities. Since the change of the Fredholm eigenvalues under quasi-conformal mapping has been well explored [14], [17], our results may be useful in the theory of univalent functions with k-quasi-conformal extensions. In order to make the paper self-contained, I retrace briefly the theory of the Fredholm eigenvalues and of the Hilbert transform, which are developed from a new point of view. I started this theory almost thirty years ago with my friend Stefan Bergman [3], and I dedicate this paper to his memory.

1. Single and double-layer potentials in the plane. Let Γ be a closed curve in the complex z-plane which encloses a bounded domain Δ and let $\tilde{\Delta}$ be its unbounded complement. To avoid unnecessary arguments, we assume that Γ is three times differentiable with respect to its arc length s. We define two continuous functions of $\zeta(s)\in\Gamma$, say $\mu(\zeta)$ and $\nu(\zeta)$ and introduce the integrals

$$(1)\qquad S(z) = \int_\Gamma \mu(\zeta)\log\frac{1}{|\zeta-z|}\,ds,\qquad D(z) = \int_\Gamma \nu(\zeta)\frac{\partial}{\partial n_\zeta}\log\frac{1}{|\zeta-z|}\,ds,$$

with $\int_\mu \mu ds = 0$ and where $\vec{n}_\zeta$ denotes the normal to Γ at ζ pointing into Δ.

* Research supported in part by NSF grant MCS75-23332-A03.

One calls $S(z)$ a single-layer potential with density μ on Γ and $D(z)$ a double-layer potential with density ν on Γ. Both integrals represent harmonic functions in Δ and $\tilde{\Delta}$. It is well known that $S(z)$ is continuous in the entire plane, while $D(z)$ jumps when z crosses Γ. However, $\partial D/\partial n$ has the same value when evaluated in Δ or in $\tilde{\Delta}$.

We define the scalar product typical for potential theory

$$(2) \qquad (\varphi, \psi) = \int_E \nabla\varphi \cdot \nabla\psi \, dxdy, \qquad E = \Delta + \tilde{\Delta}$$

for any two functions $\varphi(z)$, $\psi(z)$ defined in $\Delta+\tilde{\Delta}$. Now, we use the above continuity behavior of $S(z)$ and $D(z)$ to find

$$(3) \qquad (S, D) = \int_\Delta \nabla S \cdot \nabla D \, dxdy + \int_{\tilde{\Delta}} \nabla S \cdot \nabla D \, dxdy$$

$$= -\int_\Gamma S \frac{\partial D}{\partial n} ds + \int_\Gamma S \frac{\partial D}{\partial n} ds = 0.$$

Thus, every single-layer potential is orthogonal in the Dirichlet metric to every double-layer potential.

Let next $h(z)$ be harmonic in Δ and continuously differentiable in the closure $\Delta+\Gamma$. We have the fundamental identity

$$(4) \qquad \frac{1}{2\pi} \int_\Gamma h(\zeta) \frac{\partial}{\partial n_\zeta} \log \frac{1}{|\zeta - z|} ds - \frac{1}{2\pi} \int_\Gamma \frac{\partial h}{\partial n_\zeta} \log \frac{1}{|\zeta - z|} ds = h(z)\chi_\Delta(z),$$

where $\chi_\Delta(z)$ is the characteristic function of Δ. Thus, every such harmonic function can be expressed as the sum of a single-layer and a double-layer potential since $\int_\Gamma \frac{\partial h}{\partial n} ds = 0$.

If $F(z)$ is any real-valued function which is continuously differentiable in $\Delta+\Gamma$, we can find a harmonic function $h(z)$ which is continuously differentiable in $\Delta+\Gamma$ and has the same boundary values as $F(z)$. We can write

$$(5) \qquad F(z) = N(z) + h(z) = N(z) + S(z) + D(z),$$

where $N(z)$ is continuously differentiable in $\Delta+\Gamma$ and vanishes on Γ. For each such null-function $N(z)$, we have the identity, valid for harmonic h,

$$(6) \qquad \iint_\Delta \nabla N \cdot \nabla h \, dxdy = -\int_\Gamma N \frac{\partial h}{\partial n} ds = 0.$$

If we consider the class of single-layer potentials, double-layer potentials and functions $N(z)$ which vanish in $\tilde{\Delta}+\Gamma$, we have three linear function spaces which are orthogonal in the Dirichlet sense and every continuously differentiable function $F(z)$ can be decomposed in a unique way into three elements of these spaces.

To use complex notation define

$$s(z) = \frac{\partial S}{\partial z}, \quad d(z) = \frac{\partial D}{\partial z}, \quad n(z) = \frac{\partial N}{\partial z}. \tag{7}$$

Then every complex-valued function $f(z)$ which is square-integrable over Δ can be split into

$$f(z) = s(z)+d(z)+n(z) \tag{8}$$

and the three components are orthogonal in the sense of the hermitian metric

$$\int \varphi\bar{\psi}\, dxdy = |\varphi, \psi|. \tag{9}$$

Since every $F(z)$ which is continuously differentiable in the entire plane can be written in the form

$$F(z) = F(z)\chi_\Delta(z)+F(z)\chi_{\tilde{\Delta}}(z), \tag{10}$$

we find that the decomposition (8) is also valid and unique for every $f(z)$ which is square-integrable over the entire complex plane.

2. The Hilbert transformation. For every $f(z)$ of the above type we introduce the integral transformation

$$\mathrm{T}f = \frac{1}{\pi}\int_E \frac{\overline{f(\zeta)}}{(\zeta-z)^2}\, d\xi d\eta \tag{11}$$

understood in the Cauchy improper integral sense. It is called the *Hilbert transform of f* [3], [13].

To understand the significance of this transformation we will discuss its effect on a function $n(z)$, $s(z)$ and $d(z)$ separately.

We start with a function $n(\zeta) = \partial N/\partial\zeta$, where $N(z)$ vanishes on Γ. Let Δ_r be the domain Δ minus a disk of radius r around z. Then

$$\mathrm{T}n(z) = \lim_{r\to 0}\frac{1}{\pi}\int_{\Delta_r} \frac{\partial N}{\partial\bar{\zeta}}\frac{1}{(\zeta-z)^2}\, d\xi d\eta. \tag{12}$$

Integration by parts leads to

$$\frac{1}{\pi}\int_{\Delta_r} \frac{\partial N}{\partial\bar{\zeta}}\frac{1}{(\zeta-z)^2}\, d\xi d\eta = \frac{1}{2\pi i}\int_{|\zeta-z|=r} N(\zeta)\frac{d\zeta}{(\zeta-z)^2}. \tag{13}$$

Observe that

$$\frac{1}{2\pi i}\int_{|\zeta-z|=r} N(\zeta)\frac{d\zeta}{(\zeta-z)^2} = \frac{\partial N}{\partial z}+O(r) = n(z)+O(r). \tag{14}$$

Hence, combining (12), (13) and (14) we find

$$\mathrm{T}n = n. \tag{15}$$

The Hilbert transformation on all n-functions is the identity.

Next, let us start with Green's identity for all single-layer potentials

$$\frac{1}{2\pi}\int_E \nabla_\zeta \log\frac{1}{|\zeta-z|}\cdot \nabla_\zeta S(\zeta)\,d\xi d\eta = S(z). \tag{16}$$

We use complex notation and write

$$\nabla A\cdot\nabla B = 4\,\mathrm{Re}\left\{\frac{\partial A}{\partial\zeta}\,\frac{\partial B}{\partial\bar\zeta}\right\} \tag{17}$$

and hence (16) becomes

$$-\frac{1}{2\pi}\int_E \frac{1}{\zeta-z}\left(\overline{\frac{\partial S}{\partial\zeta}}\right)d\xi d\eta - \frac{1}{2\pi}\int_E \frac{1}{\overline{(\zeta-z)}}\,\frac{\partial S}{\partial\zeta}\,d\xi d\eta = S(z). \tag{18}$$

We differentiate in z and use the Laplace–Poisson identity to find

$$-\frac{1}{2\pi}\int_E \frac{1}{(\zeta-z)^2}\left(\overline{\frac{\partial S}{\partial\zeta}}\right)d\xi d\eta + \frac{1}{2}\,\frac{\partial S}{\partial z} = \frac{\partial S}{\partial z}. \tag{19}$$

Hence we proved for every $s(z) = \partial S/\partial s$ the transformation law

$$\mathrm{T}s = -s. \tag{20}$$

Finally, if $D(z)$ is a double-layer potential, we have

$$\frac{1}{2\pi}\int_E \nabla_\zeta \log\frac{1}{|\zeta-z|}\,\nabla_\zeta D(\zeta)\,d\xi d\eta = 0, \tag{21}$$

as follows easily from the continuity of $\partial D/\partial n$ on Γ. Hence the above calculation yields

$$-\frac{1}{2\pi}\int_E \frac{1}{(\zeta-z)^2}\left(\overline{\frac{\partial D}{\partial\zeta}}\right)d\xi d\eta + \frac{1}{2}\,\frac{\partial D}{\partial z} = 0. \tag{22}$$

For all $d(z) = \partial D/\partial z$ follows the transformation law

$$\mathrm{T}d = d. \tag{23}$$

Formulas (15), (20) and (23) give a clear understanding for the significance of the Hilbert transform. The general $f(z)$ can be decomposed uniquely into

$$f = n+d+s \tag{24}$$

and its norm is

$$\|f\|^2 = \|n\|^2+\|d\|^2+\|s\|^2. \tag{25}$$

Its Hilbert transform is

(26) $$\mathrm{T}f = n+d-s$$

with the norm

(27) $$\|\mathrm{T}f\| = \|n\|^2+\|d\|^2+\|s\|^2 = \|f\|^2.$$

Also, we have obviously

(28) $$\mathrm{TT}f = f.$$

This shows that the Hilbert transform is a norm-preserving involution.

This interpretation of the Hilbert transform in the plane suggests also easy generalizations in potential theory for higher dimensions.

3. Fredholm eigenfunctions. Our preceding analysis makes it quite clear that the Hilbert transformation is only of interest within the class of analytic functions since for their othogonal complement of n-functions it is nothing but the identity operation. Now, the remarkable fact appears that in the case of analytic functions the improper kernel of the transformation can be replaced by an analytic regular kernel.

Indeed, let $g(z,\zeta)$ be the Green's function of Δ. We form the two complex-valued kernels [2], [3]

(29) $$K(z,\bar{\zeta}) = -\frac{2}{\pi}\frac{\partial^2 g(z,\zeta)}{\partial z\partial\bar{\zeta}}, \qquad L(z,\zeta) = -\frac{2}{\pi}\frac{\partial^2 g(z,\zeta)}{\partial z\partial\zeta}.$$

The first kernel is the well-known Bergman kernel. It is regular analytic for all z in Δ and anti-analytic for all ζ in Δ. It has hermitian symmetry $\overline{K(z,\bar{\zeta})} = K(\zeta,\bar{z})$ and for all $\varphi(z)$ analytic in Δ it has the reproducing property

(30) $$\int_\Delta K(z,\bar{\zeta})\varphi(\zeta)\,d\xi d\eta = \varphi(z).$$

The kernel $L(z,\zeta)$ is symmetric in both arguments, but it has a double pole for $z=\zeta$. It can be written in the form

(31) $$L(z,\zeta) = \frac{1}{\pi(z-\zeta)^2} - l(z,\zeta),$$

where $l(z,\zeta)$ is regular analytic in both arguments. For any analytic function $\varphi(z)$ in Δ, we have the identity

(32) $$\int_\Delta L(z,\zeta)\overline{\varphi(\zeta)}\,d\xi d\eta \equiv 0.$$

This implies the identity

(33) $$\frac{1}{\pi}\int_\Delta \frac{\overline{\varphi(\zeta)}}{(\zeta-z)^2}\,d\xi d\eta = \int_\Delta l(z,\zeta)\overline{\varphi(\zeta)}\,d\xi d\eta.$$

This shows that we may replace the singular kernel in the Hilbert transformation of analytic functions in Δ and $\tilde{\Delta}$ by the corresponding kernels $l(z,\zeta)$ and $\tilde{l}(z,\zeta)$.

Now we find at our disposal the classical methods of integral equation theory. We may ask for analytic functions $w(z)$ in Δ which satisfy the integral equation

$$w(z) = \frac{\lambda}{\pi} \int_{\Delta} \frac{\overline{w(\zeta)}}{(\zeta - z)^2} d\xi d\eta, \qquad z \in \Delta. \tag{34}$$

Since $w(\zeta)e^{i\alpha}$ satisfies the same integral equation with $\lambda^* = \lambda e^{-2i\alpha}$, we may be more specific and demand λ to be positive. We may also normalize the eigenfunction by the demand

$$\int_{\Delta} |w|^2 d\xi d\eta = 1. \tag{35}$$

The significance of the eigenfunctions $w_\nu(z)$ and the eigenvalues λ_ν for potential theory is obvious. Observe that the eigenfunctions satisfy also the integral equation with analytic kernel

$$w_\nu(z) = \lambda_\nu \int_{\Delta} l(z,\zeta)\overline{w_\nu(\zeta)}\, d\xi d\eta. \tag{36}$$

Eigenfunctions to different eigenvalues are obviously orthogonal to each other, and we may assume them to form an orthonormal set. It can be shown that the system is complete and we have the spectral decomposition of the two derivatives of the Green's function

$$K(z,\bar{\zeta}) = \sum_{\nu=1}^{\infty} w_\nu(z)\overline{w_\nu(\zeta)}, \qquad l(z,\zeta) = \sum_{\nu=1}^{\infty} \frac{w_\nu(z)\, w_\nu(\zeta)}{\lambda_\nu}. \tag{37}$$

The eigenvalues λ_ν occurred early in potential theory. As is well known, Poincaré attacked the first boundary value problem of two-dimensional potential theory by setting up the sought harmonic function as a double-layer potential

$$h(z) = \frac{1}{\pi} \int_{\Gamma} v(\zeta) \frac{\partial}{\partial n_\zeta} \log \frac{1}{|\zeta - z|} ds_\zeta, \qquad z \in \Delta \tag{38}$$

and was led to the boundary condition

$$h(z) = v(z) + \frac{1}{\pi} \int_{\Gamma} v(\zeta) \frac{\partial}{\partial n_\zeta} \log \frac{1}{|\zeta - z|} ds_\zeta, \qquad z \in \Gamma. \tag{39}$$

This is an integral equation for the sought density $v(z)$ in terms of the known boundary values $h(z)$ on Γ.

This problem gave rise to the modern theory of integral equations and the fundamental Fredholm alternative. According to it, one has to consider the eigenvalue problem

(40) $$\varphi(z) = \frac{\lambda}{\pi} \int_{\Gamma} \varphi(\zeta) \frac{\partial}{\partial n_{\zeta}} \log \frac{1}{|\zeta - z|} ds_{\zeta}, \quad z \in \Gamma$$

and to show that $\lambda = -1$ is not an eigenvalue. Now it is easy to see that up to one trivial eigenvalue $\lambda = 1$ all eigenvalues of the Poincaré–Fredholm theory coincide with our eigenvalues λ_{ν}. They are therefore called the *Fredholm eigenvalues* of the curve Γ [10], [11].

Let us return now to the integral equation (34). The eigenfunction $w_{\nu}(z)$ in Δ defines now also an analytic function

(34′) $$\hat{w}_{\nu}(z) = \frac{\lambda_{\nu}}{\pi} \int_{\Delta} \frac{\overline{w_{\nu}(\zeta)}}{(\zeta - z)^2} d\xi d\eta, \quad z \in \tilde{\Delta}.$$

We may say that the pair $w_{\nu}(\zeta)$ in Δ and 0 in $\tilde{\Delta}$ has the Hilbert transform consisting of $\frac{1}{\lambda_{\nu}} w_{\nu}(z)$ in Δ and $\frac{1}{\lambda_{\nu}} \hat{w}_{\nu}(z)$ in $\tilde{\Delta}$. From the norm preservation under the Hilbert transform, we deduce

(41) $$\|w_{\nu}\|^2 = 1 = \frac{1}{\lambda_{\nu}^2} \|w_{\nu}\|^2 + \frac{1}{\lambda_{\nu}^2} \|\hat{w}_{\nu}\|^2 = \frac{1}{\lambda_{\nu}^2} + \frac{1}{\lambda_{\nu}^2} \|\hat{w}_{\nu}\|^2.$$

We conclude that for all ν

(42) $$\lambda_{\nu} \geqslant 1$$

and $\lambda_{\nu} = 1$ is only possible if $\hat{w}_{\nu} \equiv 0$. It is easy to see that this cannot happen if Γ is a smooth curve. We also find

(43) $$\|\hat{w}_{\nu}\| = \sqrt{\lambda_{\nu}^2 - 1}.$$

We define the normalized analytic function in $\tilde{\Delta}$

(44) $$\tilde{w}_{\nu} = \frac{i}{\sqrt{\lambda_{\nu}^2 - 1}} \hat{w}_{\nu}, \quad \|\tilde{w}_{\nu}\| = 1$$

and rewrite (34′) into

(34″) $$\tilde{w}_{\nu}(z) = \frac{\lambda_{\nu} i}{\pi \sqrt{\lambda_{\nu}^2 - 1}} \int_{\Delta} \frac{\overline{w_{\nu}(\zeta)}}{(\zeta - z)^2} d\xi d\eta, \quad z \in \tilde{\Delta}.$$

Next we apply the Hilbert transformation to the functions $\frac{1}{\lambda_{\nu}} w_{\nu}(z)$ in Δ and $\frac{i}{\lambda_{\nu}} \sqrt{\lambda_{\nu}^2 - 1}\, \tilde{w}_{\nu}(z)$ in $\tilde{\Delta}$. By the involutory character of the T-transformation, we must end up with the function pair $w_{\nu}(z)$ in Δ and 0 in $\tilde{\Delta}$.

Thus

$$(45)\quad \frac{1}{\lambda_\nu}\frac{1}{\pi}\int_\Delta \frac{\overline{w_\nu(\zeta)}}{(\zeta-z)^2}\,d\xi d\eta + \frac{i\sqrt{\lambda_\nu^2-1}}{\lambda_\nu}\frac{1}{\pi}\int_{\tilde{\Delta}} \frac{\overline{\tilde{w}_\nu(\zeta)}}{(\zeta-z)^2}\,d\xi d\eta = w_\nu(z)\chi_\Delta(z).$$

Observe that $w_\nu(z)$ satisfies equations (34) in Δ and (34″) in $\tilde{\Delta}$. Hence for $z\in\tilde{\Delta}$ we find

$$(46)\qquad \tilde{w}_\nu(z) = +\frac{\lambda_\nu}{\pi}\int_{\tilde{\Delta}} \frac{\overline{\tilde{w}_\nu(\zeta)}}{(\zeta-z)^2}\,d\xi d\eta, \qquad z\in\tilde{\Delta}$$

and for $z\in\Delta$

$$(47)\qquad w_\nu(z) = \frac{i\lambda_\nu}{\pi\sqrt{\lambda_\nu^2-1}}\int_{\tilde{\Delta}} \frac{\overline{\tilde{w}_\nu(\zeta)}}{(\zeta-z)^2}\,d\xi d\eta, \qquad z\in\Delta.$$

We see that in view of (46) $\tilde{w}_\nu(z)$ is the eigenfunction in $\tilde{\Delta}$ to the same eigenvalue λ_ν and (47) and (34″) exhibit the complete symmetry between the domains Δ, $\tilde{\Delta}$ and their eigenfunctions.

Much of this theory can be extended to multiply-connected domains [12], [16], and we will make some application in this respect in Section 7.

It should be observed that in the unbounded domain $\tilde{\Delta}$ there exists always an eigenfunction $\tilde{w}_0(z)$ to the eigenvalue $\lambda = -1$. Indeed, let $\tilde{g}(z,\infty)$ be the Green's function of $\tilde{\Delta}$ with the logarithmic pole at infinity. Then $\tilde{w}_0(z) = \dfrac{\partial}{\partial z}\tilde{g}(z,\infty)$ will satisfy the integral equation (46), as can be easily verified by integration by parts. However, this eigenfunction does not have a finite norm over $\tilde{\Delta}$. This observation shows the importance of restricting our considerations to functions with finite norms and the corresponding Hilbert spaces when we pass to orthogonal developments of the eigenfunctions.

4. Fredholm eigenfunctions and conformal mapping. Let $z = f(t)$ be univalent and regular analytic in $|t|<1$ and map this unit disk onto Δ. Also, let $z = g(t)$ be univalent in $|t|>1$ and map that region onto $\tilde{\Delta}$ such that the point at infinity goes into itself. We then transplant the integral equations for $w_\nu(z)$ and $\tilde{w}_\nu(z)$ into the t-plane. We have

$$(48)\qquad w_\nu[f(t)]f'(t) = \frac{\lambda_\nu}{\pi}\int_{|\tau|<1} \overline{w_\nu[f(\tau)]f'(\tau)}\,\frac{f'(t)f'(\tau)}{[f(t)-f(\tau)]^2}\,d\alpha d\beta$$

if $\tau = \alpha+i\beta$. We define the analytic function in the unit disk

$$(49)\qquad m_\nu(t) = w_\nu[f(t)]f'(t).$$

The L-kernel for the unit disk is $1/\pi(z-\zeta)^2$. Hence from (32) and (48) follows the integral equation

(50) $$m_\nu(t) = \frac{\lambda_\nu}{\pi} \int_{|\tau|<1} \overline{m_\nu(\tau)} \left\{ \frac{f'(t)f'(\tau)}{[f(t)-f(\tau)]^2} - \frac{1}{(t-\tau)^2} \right\} d\alpha d\beta.$$

Observe that

(51) $$A(t,\tau) = \frac{f'(t)f'(\tau)}{[f(t)-f(\tau)]^2} - \frac{1}{(t-\tau)^2} = \frac{\partial^2}{\partial t \partial \tau} \log \frac{f(t)-f(\tau)}{t-\tau}.$$

The combination $\log \frac{f(t)-f(\tau)}{t-\tau}$ is well known in the theory of univalent functions [6], [9]. A necessary and sufficient condition for $f(z)$ to be regular analytic and univalent in $|t| < 1$ is the regularity of the expression

(52) $$\log \frac{f(t)-f(\tau)}{t-\tau} = \sum_{m,n=0}^{\infty} c_{mn} t^m \tau^n.$$

The infinite matrix $C = ((c_{mn}))$ was introduced by Grunsky [6] and is called the *Grunsky matrix*. Many necessary conditions for univalence can be expressed by means of it.

Next, we define the analytic function in $\tilde{\Delta}$

(53) $$n_\nu(t) = \tilde{w}_\nu[g(t)]g'(t)$$

and the kernel

(54) $$B(t,\tau) = \frac{g'(t)g'(\tau)}{[g(t)-g(\tau)]^2} - \frac{1}{(t-\tau)^2} = \frac{\partial^2}{\partial t \partial \tau} \log \frac{g(t)-g(\tau)}{t-\tau}.$$

The integral equations (34) and (46) become

(55) $$m_\nu(t) = \frac{\lambda_\nu}{\pi} \int_{|\tau|<1} A(t,\tau)\overline{m_\nu(\tau)}\, d\alpha d\beta,$$

(56) $$n_\nu(t) = \frac{\lambda_\nu}{\pi} \int_{|\tau|>1} B(t,\tau)\overline{n_\nu(\tau)}\, d\alpha d\beta.$$

We also introduce the kernel

(57) $$C(t,\tau) = \frac{g'(t)f'(\tau)}{[g(t)-f(\tau)]^2}.$$

We can then express (34″) and (47) in the simple form

(58) $$n_\nu(t) = \frac{i\lambda_\nu}{\pi\sqrt{\lambda_\nu^2-1}} \int_{|\tau|<1} \overline{m_\nu(\tau)}\, C(t,\tau)\, d\alpha d\beta$$

and

(59) $$m_\nu(t) = \frac{i\lambda_\nu}{\pi\sqrt{\lambda_\nu^2-1}} \int_{|\tau|>1} \overline{n_\nu(\tau)}\, C(\tau,t)\, d\alpha d\beta.$$

We choose a complete orthonormal set $\{\varphi_\alpha(t)\}$ in $|t| < 1$ and an analogous set $\{\psi_\alpha(t)\}$ in $|t| > 1$. We can develop

(60) $$m_\nu(t) = \sum x_{\nu\alpha}\varphi_\alpha(t), \quad n_\nu(t) = \sum y_{\nu\alpha}\psi_\alpha(t).$$

Define the matrices $((A_{\alpha\beta}))$, $((B_{\alpha\beta}))$ and $((C_{\alpha\beta}))$ by the developments

(61) $$A(t,\tau) = \sum_1^\infty A_{\alpha\beta}\varphi_\alpha(t)\varphi_\beta(\tau), \quad B(t,\tau) = \sum_1^\infty B_{\alpha\beta}\psi_\alpha(t)\psi_\beta(\tau),$$

$$C(t,\tau) = \sum_1^\infty C_{\alpha\beta}\psi_\alpha(t)\varphi_\beta(\tau).$$

The matrices $((A_{\alpha\beta}))$ and $((B_{\alpha\beta}))$ are obviously symmetric, the matrix $((C_{\alpha\beta}))$ is not. Using the orthogonality properties, we can transform the four integral equations (55), (56), (58) and (59) into the following discrete equations:

(62) $$x_{\nu\alpha} = \frac{\lambda_\nu}{\pi}\sum_{\beta=1}^\infty A_{\alpha\beta}\bar{x}_{\nu\beta}, \qquad y_{\nu\alpha} = \frac{\lambda_\nu}{\pi}\sum_{\beta=1}^\infty B_{\alpha\beta}\bar{y}_{\nu\beta},$$

(63) $$y_{\nu\alpha} = \frac{i\lambda_\nu}{\pi\sqrt{\lambda_\nu^2-1}}\sum_{\beta=1}^\infty C_{\alpha\beta}\bar{x}_{\nu\beta}, \quad x_{\nu\alpha} = \frac{i\lambda_\nu}{\pi\sqrt{\lambda_\nu^2-1}}\sum_{\beta=1}^\infty C_{\beta\alpha}\bar{y}_{\nu\beta}.$$

5. Grunsky matrices and Fredholm eigenvalues. The most obvious choice of an orthonormal function set in the two circular regions would be the set of powers

(64) $$\varphi_\alpha(t) = \sqrt{\alpha/\pi}\, t^{\alpha-1}, \quad \psi_\alpha(t) = \sqrt{\alpha/\pi}\, t^{-\alpha-1}.$$

In view of definitions (51) and (52), we find

(65) $$A_{\alpha\beta} = \pi\sqrt{\alpha\beta}\, c_{\alpha\beta}, \quad B_{\alpha\beta} = \pi\sqrt{\alpha\beta}\, d_{\alpha\beta}$$

if we write

(66) $$\log\frac{g(t)-g(\tau)}{t-\tau} = \sum_{\alpha,\beta=0}^\infty d_{\alpha\beta}t^{-\alpha}\tau^{-\beta}.$$

Finally let

(67) $$\log[g(t)-f(\tau)] = \log t + \sum e_{\alpha\beta}t^{-\alpha}\tau^\beta.$$

This implies

(68) $$C_{\alpha\beta} = \pi\sqrt{\alpha\beta}\, e_{\alpha\beta}.$$

We recognize the relation between the Grunsky matrices of the mapping functions $f(t)$ and $g(t)$ and the Fredholm eigenvalues. Let us define the matrices

(69) $$P = ((\sqrt{\alpha\beta}\, c_{\alpha\beta})), \quad Q = ((\sqrt{\alpha\beta}\, d_{\alpha\beta})), \quad R = ((\sqrt{\alpha\beta}\, e_{\alpha\beta})).$$

Equations (62) and (63) can then be expressed as follows. There are associated vector pairs $\mathfrak{x}_\nu$ and $\mathfrak{y}_\nu$ such that

$$(70)\qquad \begin{aligned} P\bar{\mathfrak{x}}_\nu &= \frac{1}{\lambda_\nu}\mathfrak{x}_\nu, & Q\bar{\mathfrak{y}}_\nu &= \frac{1}{\lambda_\nu}\mathfrak{y}_\nu,\\ R\bar{\mathfrak{x}}_\nu &= \frac{1}{i}\frac{\sqrt{\lambda_\nu^2-1}}{\lambda_\nu}\mathfrak{y}_\nu, & R^T\bar{\mathfrak{y}}_\nu &= \frac{1}{i}\cdot\frac{\sqrt{\lambda_\nu^2-1}}{\lambda_\nu}\mathfrak{x}_\nu. \end{aligned}$$

Each set $\{\mathfrak{x}_\nu\}$ and $\{\mathfrak{y}_\nu\}$ is a complete orthonormal set in its corresponding Hilbert space. An arbitrary vector $\mathfrak{v}$ in it can be written as

$$(71)\qquad \mathfrak{v} = \sum_{\nu=1}^{\infty} k_\nu \mathfrak{x}_\nu$$

and

$$(72)\qquad P\bar{\mathfrak{v}} = \sum_{\nu=1}^{\infty} \frac{1}{\lambda_\nu}\bar{k}_\nu \mathfrak{x}_\nu .$$

Hence, because of the orthonormality of the $\mathfrak{x}_\nu$

$$(73)\qquad (P\bar{\mathfrak{v}},\bar{\mathfrak{v}}) = \sum_{\nu=1}^{\infty}\frac{1}{\lambda_\nu}\bar{k}^2, \qquad \|\mathfrak{v}\|^2 = \sum |k_\nu|^2.$$

If λ_1 is the least Fredholm eigenvalue of Δ, we have the inequality

$$(74)\qquad \Big|\sum_{\alpha,\beta=1}^{\infty} c_{\alpha\beta}\sqrt{\alpha\beta}\, x_\alpha x_\beta\Big| \leqslant \frac{1}{\lambda_1}\sum_{\alpha=1}^{\infty}|x_\alpha|^2.$$

Since we know that all $\lambda_\nu \geqslant 1$, we can state

$$(75)\qquad \Big|\sum_{\alpha,\beta=1}^{\infty} c_{\alpha\beta}\sqrt{\alpha\beta}\, x_\alpha x_\beta\Big| \leqslant \sum_{\alpha=1}^{\infty}|x_\alpha|^2.$$

This is the well-known Grunsky inequality which plays such a role in the coefficient problem for univalent functions. Inequality (74) is a considerable improvement on it.

To show an important application of (74), we quote a theorem due to G. Springer [17]. Let $f(z)$ be a univalent mapping of the complex plane and let $f(z)$ be analytic in a domain D and k-quasi-conformal in its complement $\tilde{D}$. If λ_1 is the lowest Fredholm eigenvalue of D and Λ_1 the lowest Fredholm eigenvalue of its image $f(D)$, we have the inequality

$$(76)\qquad \frac{\lambda_1+1}{\lambda_1-1}\,\frac{1-k}{1+k} \leqslant \frac{\Lambda_1+1}{\Lambda_1-1} \leqslant \frac{\lambda_1+1}{\lambda_1-1}\,\frac{1+k}{1-k}.$$

We choose D to be the unit disk and $f(t)$ to be a continuous univalent function in the whole plane, analytic in $|t| < 1$ and k-quasi-conformal in $|t| > 1$. It will map the unit disk onto a domain Δ with the lowest Fredholm eigenvalue $\Lambda_1 \geq 1/k$. Indeed, the eigenvalues of the unit disk are all infinite since $l(z, \zeta) = 0$ for it. Hence we can state the following theorem first proved by R. Kühnau [7], [15]:

If the univalent continuous mapping $f(t)$ is analytic in $|t| < 1$ and k-quasi-conformal in $|t| > 1$, its Grunsky matrix satisfies the inequality

$$\left|\sum_{\alpha,\beta=1}^{\infty} \sqrt{\alpha\beta}\, c_{\alpha\beta}\, x_\alpha x_\beta\right| \leq k \sum_{\alpha=1}^{\infty} |x_\alpha|^2. \tag{77}$$

These inequalities are sharp as can be shown by variational methods.

6. Identities between complementary Grunsky matrices. Let Δ and $\tilde{\Delta}$ be complementary domains in the complex plane and let $f(t)$ and $g(t)$ map onto them as before. We can ask for relations between such functions which are so closely connected by geometry. Our present results lead indeed to interesting identities for the coefficients of such pairs.

We form the matrix

$$M = \begin{pmatrix} P & R^T \\ R & Q \end{pmatrix} \tag{78}$$

which is composed of the three infinite matrices defined by the formulas (69). Let

$$\mathfrak{v}_\nu^{(1)} = \begin{pmatrix} \mathfrak{x}_\nu \\ 0 \end{pmatrix}, \quad \mathfrak{v}_\nu^{(2)} = \begin{pmatrix} 0 \\ \mathfrak{y}_\nu \end{pmatrix} \tag{79}$$

be corresponding doubly-infinite vectors constructed from the eigenvectors of the above problems. By virtue of equations (70) we find that

$$M\overline{\mathfrak{v}_\nu^{(1)}} = \begin{pmatrix} \dfrac{1}{\lambda_\nu}\mathfrak{x}_\nu \\ \dfrac{1}{i}\sqrt{1-\dfrac{1}{\lambda_\nu^2}}\,\mathfrak{y}_\nu \end{pmatrix}, \quad M\bar{M}\mathfrak{v}_\nu^{(1)} = \mathfrak{v}_\nu^{(1)}. \tag{80}$$

Similarly

$$M\overline{\mathfrak{v}_\nu^{(2)}} = \begin{pmatrix} \dfrac{1}{i}\sqrt{1-\dfrac{1}{\lambda_\nu^2}}\,\mathfrak{x}_\nu \\ \dfrac{1}{\lambda_\nu}\mathfrak{y}_\nu \end{pmatrix}, \quad M\bar{M}\mathfrak{v}_\nu^{(2)} = \mathfrak{v}_\nu^{(2)}. \tag{81}$$

The $\{\mathfrak{x}_\nu\}$ and $\{\mathfrak{y}_\nu\}$ are complete orthonormal systems in their Hilbert spaces and every doubly-infinite vector $\mathfrak{v}$ in the same space can be written as a linear combination of vectors $\mathfrak{v}_\nu^{(1)}$ and $\mathfrak{v}_\nu^{(2)}$. Thus we have the identity

(82) $$M\bar{M}\mathfrak{v} = \mathfrak{v}$$

valid for all vectors $\mathfrak{v}$. This implies the identity

(83) $$M\bar{M} = I.$$

By its definition (78), M is obviously a symmetric matrix. Now (82) shows that M is also unitary. We find the remarkable identities

(84) $$P\bar{P}+R^T\bar{R} = R\bar{R}^T+Q\bar{Q} = I, \quad R\bar{P}+Q\bar{R} = 0.$$

An immediate consequence of (83) is the fact that the vector transformation

(85) $$\hat{\mathfrak{v}} = M\mathfrak{v}$$

is norm-preserving since

(86) $$(\hat{\mathfrak{v}}_1, \hat{\mathfrak{v}}_2) = (M\mathfrak{v}_1, M\mathfrak{v}_2) = (M\bar{M}\mathfrak{v}_1, \mathfrak{v}_2) = (\mathfrak{v}_1, \mathfrak{v}_2).$$

Hence, applying the Schwarz inequality to the hermitian scalar product we find

(87) $$|(M\mathfrak{v}_1, \mathfrak{v}_2)|^2 \leqslant \|M\mathfrak{v}_1\|^2 \cdot \|\mathfrak{v}_2\|^2 = \|\mathfrak{v}_1\|^2 \cdot \|\mathfrak{v}_2\|^2.$$

This is equivalent to the generalized Grunsky inequality

(88) $$\Big|\sum_{\alpha,\beta=1}^{\infty} \sqrt{\alpha\beta}\,[c_{\alpha\beta}\,\xi_\alpha\,x_\beta + e_{\alpha\beta}\,(x_\beta\,\eta_\alpha + \xi_\beta\,y_\alpha) + d_{\alpha\beta}\,\eta_\alpha\,y_\beta]\Big|^2 \leqslant \sum_{\alpha=1}^{\infty} (|x_\alpha|^2+|y_\alpha|^2) \sum_{\alpha=1}^{\infty} (|\xi_\alpha|^2+|\eta_\alpha|^2).$$

If we use only vectors whose components with index larger than k is zero, we obtain estimates for the $k\times k$ submatrices of $((c_{\alpha\beta}))$, $((e_{\alpha\beta}))$ and $((d_{\alpha\beta}))$.

7. Fredholm eigenvalues of multiply-connected domains. Inequalities (88) are rarely applicable since one knows very few curves Γ for which the mapping functions $f(t)$ and $g(t)$ onto their interior and exterior are explicitly given. We can now use the Fredholm eigenvalues of multiply-connected domains, generalize these inequalities to the same matrices $((c_{\alpha\beta}))$, $((e_{\alpha\beta}))$, $((d_{\alpha\beta}))$ constructed as before but with functions $f(t)$ and $g(t)$ which map on disjoint domains Δ, $\tilde{\Delta}$, not necessarily complementary.

We denote the domain set $\Delta+\tilde{\Delta}$ by D and its complement by $\tilde{D}$. We frame the same eigenvalue problem as before; to find solutions $w_\nu(z)$ and eigenvalues l_ν such that

(89) $$w_\nu(z) = \frac{l_\nu}{\pi}\int_D \frac{\overline{w_\nu(\zeta)}}{(\zeta-z)^2}\,d\xi d\eta, \qquad z\in D.$$

Since the kernel $\dfrac{1}{\pi}(\zeta-z)^{-2}$ may be replaced by $l(\zeta, z)$ and $\tilde{l}(\zeta, z)$ respectively, if z and ζ lie in the same component of D, we are again dealing with an integral equation with regular kernel. We obtain a complete set of eigenfunctions $w_\nu(z)$ which can be orthonormalized by the condition

$$\int_D w_\nu(\zeta)\overline{w_\mu(\zeta)}\,d\xi d\eta = \delta_{\nu\mu} \tag{90}$$

and we may specify

$$l_\nu > 0. \tag{91}$$

The same argument regarding the Hilbert transform leads at once to the result

$$l_\nu \geqslant 1. \tag{92}$$

We can now repeat the formalism of Section 5. We assume that the component $\tilde{\Delta}$ contains the point at infinity and that $f(t)$ maps $|t| < 1$ onto Δ, while $g(t)$ maps $|t| > 1$ onto $\tilde{\Delta}$ such that $g(\infty) = \infty$. Denote the restrictions of $w_\nu(z)$ to Δ and $\tilde{\Delta}$ by $w_\nu^{(1)}(z)$ and $w_\nu^{(2)}(z)$ and define

$$m_\nu(t) = w_\nu^{(1)}[f(t)]\,f'(t), \qquad n_\nu(t) = w_\nu^{(2)}[g(t)]\,g'(t) \tag{93}$$

and using definitions (51), (54) and (57) we can express the integral equation (89) as follows:

$$\begin{aligned} m_\nu(t) &= \frac{l_\nu}{\pi}\Big\{\int_{|\tau|<1} A(t,\tau)\overline{m_\nu(\tau)}\,d\alpha d\beta + \int_{|\tau|>1} C(\tau,t)\overline{n_\nu(\tau)}\,d\alpha d\beta\Big\},\\ n_\nu(t) &= \frac{l_\nu}{\pi}\Big\{\int_{|\tau|<1} C(t,\tau)\overline{m_\nu(\tau)}\,d\alpha d\beta + \int_{|\tau|>1} B(t,\tau)\overline{n_\nu(\tau)}\,d\alpha d\beta\Big\}. \end{aligned} \tag{94}$$

The same calculations as before lead to the system of equations

$$\begin{aligned} x_{\nu\alpha} &= l_\nu\Big\{\sum_{\beta=1}^{\infty}\sqrt{\alpha\beta}\,(c_{\alpha\beta}\,\bar{x}_{\nu\beta} + e_{\beta\alpha}\,\bar{y}_{\nu\beta})\Big\},\\ y_{\nu\alpha} &= l_\nu\Big\{\sum_{\beta=1}^{\infty}\sqrt{\alpha\beta}\,(e_{\alpha\beta}\,\bar{x}_{\nu\beta} + d_{\alpha\beta}\,\bar{y}_{\nu\beta})\Big\}. \end{aligned} \tag{95}$$

In notations (69) and (78) we can combine these equations into

$$M\begin{pmatrix}\bar{\mathfrak{x}}_\nu\\ \bar{\mathfrak{y}}_\nu\end{pmatrix} = \frac{1}{l_\nu}\left(\begin{pmatrix}\mathfrak{x}_\nu\\ \mathfrak{y}_\nu\end{pmatrix}\right). \tag{96}$$

The vectors $\mathfrak{v}_\nu = \left(\begin{pmatrix}\mathfrak{x}_\nu\\ \mathfrak{y}_\nu\end{pmatrix}\right)$ form a complete orthonormal set and every vector $\mathfrak{v}$ can be written as a linear combination of them:

$$\mathfrak{v} = \sum_{\nu=1}^{\infty} k_\nu\,\mathfrak{v}_\nu. \tag{97}$$

We have by iteration of (96)

(98) $$M\bar{M}\mathfrak{v}_\nu = \frac{1}{l_\nu^2}\mathfrak{v}_\nu .$$

Hence

(99) $$\|M\mathfrak{v}\|^2 = (M\bar{M}\mathfrak{v}, \mathfrak{v}) = \sum_{\nu=1}^{\infty} |k_\nu|^2 \frac{1}{l_\nu^2} \leqslant \frac{1}{l_1^2}\|\mathfrak{v}\|^2$$

if l_1 is the least eigenvalue of our problem. Again, we find by the Schwarz inequality

(100) $$|(M\mathfrak{v}, \mathfrak{s})|^2 \leqslant \|M\mathfrak{v}\|^2 \cdot \|\mathfrak{s}\|^2 \leqslant \frac{1}{l_1^2}\|\mathfrak{v}\|^2 \cdot \|\mathfrak{s}\|^2 .$$

This is analogous to (87) and leads to the same consequence (88) because of (92). However, we may now improve this estimate by inserting the factor $1/l_1^2$ before the right-hand term.

Inequalities for the coefficients of pairs of univalent functions which map on disjoint domains are well known [1], [4], [5], [8]. What is new in our derivation is the connection with the theory of the Hilbert transform and the Fredholm eigenvalues. One may generalize further and consider in an analogue manner the Fredholm eigenvalues for sets of k disjoint domains. One obtains so coefficient inequalities for sets of univalent functions which have no common values.

Bibliography

[1] J. E. Alenicyn, *On functions without common values in multiply-connected domains*, Trudy Mat. Inst. Steklof 94 (1968), p. 4–18.

[2] S. Bergman, *The kernel function and conformal mapping*, Amer. Math. Soc. Surveys, No. 5, New York 1950.

[3] – and M. Schiffer, *Kernel functions and conformal mapping*, Compositio Math. 8 (1951), p. 205–249.

[4] D. W. DeTemple, *Univalent functions whose ranges do not overlap*, Math. Z. 128 (1972), p. 23–33.

[5] L. L. Gromowa and N. A. Lebedev, *Area theorems for non-overlapping finitely connected regions*, Vestnik Leningrad Univ. 25 (1970), p. 18–29.

[6] H. Grunsky, *Koeffizientenbedingungen für schlicht abbildende meromorphe Funktionen*, Math. Z. 45 (1939), p. 29–61.

[7] R. Kühnau, *Verzerrungssätze und Koeffizientenbedingungen vom Grunskyschen Typ für quasikonforme Abbildungen*, Math. Nachr. 48 (1971), p. 77–105.

[8] Ch. Pommerenke, *Univalent Funktions*, Göttingen 1975.

[9] M. Schiffer, *Faber polynomials in the theory of univalent functions*, Bull. Amer. Math. Soc. 59 (1948), p. 503–517.

[10] – *Applications of variational methods in the theory of conformal mapping*, Proc. Symp. Appl. Math. 8 (1958), p. 93–113.

[11] – *The Fredholm eigenvalues of plane domains*, Pacific J. Math. 7 (1957), p. 1187–1225.

[12] – *Fredholm eigenvalues of multiply-connected domains*, ibidem 9 (1959), p. 211–264.

[13] – *Fredholm eigenvalues and conformal mapping*, Rend. Mat. 22 (1963), p. 447–468.

[14] – and G. Schober, *An extremal problem for the Fredholm eigenvalues*, Arch. Rational Mech. Anal. 44 (1971), p. 83–92.
[15] – and G. Schober, *Coefficient problems and generalized Grunsky inequalities for schlicht functions with quasi-conformal extensions*, ibidem 60 (1976), p. 205–228.
[16] – and G. Springer, *Fredholm eigenvalues and conformal mapping of multiply-connected domains*, J. Analyse Math. 14 (1965), p. 337–378.
[17] G. Springer, *Fredholm eigenvalues and quasi-conformal mapping*, Acta Math. 111 (1964), p. 121–141.

Reçu par la Rédaction le 15. 9. 1978

Commentary on

[62] *The Fredholm eigen values of plane domains*, Pacific J. Math. **7** (1957), 1187–1225.

[68] *Fredholm eigen values of multiply-connected domains*, Pacific J. Math. **9** (1959), 211–269.

[78] *Fredholm eigenvalues and conformal mapping*, Rend. Mat. (5) **22** (1963), 447–468.

[125] *Fredholm eigenvalues and Grunsky matrices*, Ann. Polon. Math. **39** (1981), 149–164.

The Fredholm eigenvalues of a Jordan curve in the complex plane were first studied in the case of smooth curves in connection with the classical integral equation for solving boundary value problems via Neumann series in potential theory; see the introduction in [78] for a good account. Later they also appeared in connection with the numerical calculation of the Riemann mapping function, integral equations, etc.; cf. [G].

The theory of Fredholm eigenvalues was initiated by Bergman and Schiffer in [35]. For smooth curves, it was developed systematically in Schiffer's masterful paper [62] (see also papers of S.E. Warschawski referenced in [G, p. 289]) and extended in [68] to the case of finite systems of smooth disjoint Jordan curves. The connection with the theory of quasiconformal mappings had already been made by Ahlfors [A], who proved that

$$\frac{1}{\lambda} \leq q := \frac{Q-1}{Q+1} \tag{1}$$

for the smallest Fredholm eigenvalue $\lambda > 1$ and the reflection coefficient Q (the smallest possible dilatation bound in the class of all quasiconformal reflections over the smooth Jordan curve C). It later turned out [Küh3] that equality need not always hold in (1); for the general question of equality in (1), cf. [Kru].

There is also a lower bound (going back to Pommerenke) of the form

$$\Phi(q) \leq \frac{1}{\lambda} \tag{2}$$

with an explicit Φ; cf. [Küh6]. However, the corresponding sharp estimate is unknown. This remains a central open problem in the theory.

In [S], Schober proposed the possibility of using the property

$$\frac{1}{\Lambda} \leq \frac{D[h_1]}{D[h_2]} \leq \Lambda := \frac{\lambda+1}{\lambda-1} \tag{3}$$

for all pairs h_1, h_2, where h_1 is harmonic inside C and h_2 outside, with common values on C, as a new *definition* of λ. Specifically, Λ is now defined as the smallest number for which (3) holds for all such pairs h_1, h_2. The essential point is that this also yields a definition of the smallest Fredholm eigenvalue λ for certain nonsmooth Jordan curves C, namely quasicircles (Jordan curves with a finite reflection coefficient). This approach has several other advantages as well, and the whole theory becomes more transparent. Currently, the connection with quasiconformal mappings is the most important aspect of the theory of Fredholm eigenvalues.

In Schiffer's papers, the higher Fredholm eigenvalues also play an important role for series expansions with eigenfunctions, development of the Green's function, the Bergman kernel, etc.; cf. [B, p. 75ff]. Unfortunately, it has not been possible to obtain these eigenvalues via a definition similar to (3). Even in the classical case of smooth C, there remains the question of the role of the higher Fredholm eigenvalues in the theory of quasiconformal mappings. The question of whether it is possible to reconstruct C (up to a Möbius transformation) from its eigenvalues is also open.

In [62], Schiffer derived from the eigenfunctions for a given smooth curve pairs of analytic functions (one inside, the other outside) with a simple boundary connection; in [Küh2, Satz 5], relations of these pairs to the theory of quasiconformal mappings are found. Krzyż proposed [Krz] the use of these pairs ("conjugate holomorphic eigenfunctions") as a new definition of eigenfunctions also in the nonsmooth case.

Here there arises the question of equality with the Fredholm eigenvalues in the nonsmooth case. Indeed, one may even ask whether there *are* any higher eigenvalues (or higher "conjugate holomorphic eigenfunctions") in the nonsmooth case. This remains unclear even in such a simple case as the Jordan curve (on the Riemann sphere) given by

$$C \equiv \{\mathfrak{Im}\, z = 0, 0 \leq \mathfrak{Re}\, z \leq +\infty\}$$
$$\cup \{\mathfrak{Re}\, z = 0, 0 \leq \mathfrak{Im}\, z \leq +\infty\}.$$

In the survey [78], Schiffer sketched an important new aspect of Fredholm eigenvalue theory. After generalizing the Grunsky coefficient inequalities for schlicht functions of class Σ with a Q-quasiconformal extension (cf. [Kru]), he noted that "simple expressions in the Fredholm eigenvalues" appear after a unitary transformation of the Grunsky matrix. In particular, in the Grunsky coefficient inequalities, sharpened for mappings with a quasiconformal extension, instead of $q = \frac{Q-1}{Q+1}$ the generally smaller factor $\frac{1}{\lambda}$ [cf. (1)] can be inserted on the right-hand side. Later, Schiffer gave a proof [125] for the case of a smooth image C of the unit circle. In [Küh3, Küh4], this result was generalized to the case of a general quasicircle C. This illustrates the great advantage of Schober's definition. This connection also yields a new matrix eigenvalue characterization of (and numerical procedure for calculating) λ in the general case [Küh5]. An interesting problem is to generalize this connection between λ and the Grunsky coefficient inequalities to the multiply connected case. A particular case of special interest is given by conformal mappings of an annulus [Küh1].

Finally, we mention a new direction in which Fredholm eigenvalues play an important role. In [Küh7], a system of inequalities of Grunsky type was derived in which the quasisymmetric substitution corresponding to C yields a new characterization of λ. This system goes back to the work of Partyka [P] concerning the reflection coefficient and offers a characterization of λ via a matrix eigenvalue problem as an additional bonus. This is a surprising analogue to the situation with the classical Grunsky inequalities. For more recent developments in this direction, see [Sh1, Sh2].

Schiffer's papers on Fredholm eigenvalues illustrate in impressive fashion his style and manner of thinking. At the very beginning, there are not, as nowadays is all too common, masses of definitions and notation; instead there are problems. The reader is constantly aware of the author's deep knowledge of mathematical physics as a source of inspiration. In this, we observe a remarkable similarity to the papers of H. Grunsky.

References

[A] Lars V. Ahlfors, *Remarks on the Neumann-Poincaré integral equation*, Pacific J. Math. **2** (1952), 271–280.

[B] Stefan Bergman, *The Kernel Function and Conformal Mapping*, second edition, American Mathematical Society, 1970.

[G] Dieter Gaier, *Konstruktive Methoden der konformen Abbildung*, Springer-Verlag, 1964.

[Kru] S. L. Krushkal, *Quasiconformal extensions and reflections*, Handbook of Complex Analysis: Geometric Function Theory, Vol. 2, ed. R. Kühnau, Elsevier, 2005, pp. 507–553.

[Krz] Jan G. Krzyż, *Conjugate holomorphic eigenfunctions and extremal quasiconformal reflection*, Ann. Acad. Sci. Fenn. Ser. A I Math. **10** (1985), 305–311.

[Küh1] R. Kühnau, *Koeffizientenbedingungen für schlicht abbildende Laurentsche Reihen*, Bull. Acad. Polon. Sci. Sér. Math. Astronom. Phys. **20** (1972), 7–10.

[Küh2] Reiner Kühnau, *Eine Integralgleichung in der Theorie der quasikonformen Abbildungen*, Math. Nachr. **76** (1977), 139–152.

[Küh3] Reiner Kühnau, *Zu den Grunskyschen Koeffizientenbedingungen*, Ann. Acad. Sci. Fenn. Ser. A I Math. **6** (1981), 125–130.

[Küh4] Reiner Kühnau, *Quasikonforme Fortsetzbarkeit, Fredholmsche Eigenwerte und Grunskysche Koeffizientenbedingungen*, Ann. Acad. Sci. Fenn. Ser. A I Math. **7** (1982), 383–391.

[Küh5] R. Kühnau, *Zur Berechnung der Fredholmschen Eigenwerte ebener Kurven*, Z. Angew. Math. Mech. **66** (1986), 193–200.

[Küh6] Reiner Kühnau, *Über die Grunskyschen Koeffizientenbedingungen*, Ann. Univ. Mariae Curie-Skłodowska Lublin Sect. A **54** (2000), 53–60.

[Küh7] Reiner Kühnau, *A new matrix characterization of Fredholm eigenvalues of quasicircles*, J. Analyse Math. **99** (2006), 295–307.

[P] Dariusz Partyka, *The Grunsky type inequalities for quasisymmetric automorphisms of the unit circle*, Bull. Soc. Sci. Lett. Łódź Sér. Rech. Déform. **31** (2000), 135–142.

[S] Glenn Schober, *Estimates for Fredholm eigenvalues based on quasiconformal mapping*, Numerische, insbesondere approximationstheoretische Behandlung von Funktionalgleichungen, Lect. Notes Math. **333**, Springer, 1973, pp. 211–217.

[Sh1] Yuliang Shen, *Generalized Fourier coefficients of a quasisymmetric homeomorphism and the Fredholm eigenvalue*, J. Analyse Math. **112** (2010), 33–48.

[Sh2] Yuliang Shen, *Fredholm eigenvalue for a quasicircle and Grunsky functionals*, Ann. Acad. Sci. Fenn. Math. **35** (2010), 581–593.

REINER KÜHNAU

[69] (with G. Pólya) Sur la représentation conforme de l'extérieur d'une courbe fermée convexe

[69] (with G. Pólya) Sur la représentation conforme de l'extérieur d'une courbe fermée convexe. *C. R. Acad. Sci. Paris* **248** (1959), 2837–2839.

ANALYSE MATHÉMATIQUE. — *Sur la représentation conforme de l'extérieur d'une courbe fermée convexe.* Note (*) de MM. **Georges Pólya** et **Menahem Schiffer.**

Soit L le périmètre et $\bar{r}$ le rayon conforme extérieur d'une courbe convexe fermée. On démontre l'inégalité $8\bar{r} \geq L$ en commençant par le cas des polygones convexes.

Trouver le maximum du périmètre d'un polygone convexe, étant donné que le nombre de ses sommets ne dépasse pas m et que son rayon conforme extérieur est égal à 1 (¹).

1. Sur l'extérieur d'un tel polygone, c'est une fonction de la forme

$$(1)\qquad f(z)=\int\prod_{\nu=1}^{m}\left(1-\frac{\gamma_\nu}{z}\right)^{2t_\nu}dz$$

qui effectue la représentation conforme de l'extérieur du cercle unité dans le plan des z; les constantes t_ν et γ_ν ($\nu=1, 2, \ldots, m$) sont sujettes aux conditions

$$(2)\qquad |\gamma_\nu|=1,\qquad t_\nu\geq 0,\qquad \sum_{\nu=1}^{m}t_\nu=1,\qquad \sum_{\nu=1}^{m}\gamma_\nu t_\nu=0.$$

Inversement, au moyen d'une fonction (1), à chaque système t_ν, γ_ν satisfaisant à (2), correspond un polygone convexe à rayon conforme extérieur 1 ayant au plus m sommets. Ainsi notre problème revient à trouver le maximum de l'intégrale

$$(3)\qquad \int_0^{2\pi}\prod_{\nu=1}^{m}|e^{i\varphi}-\gamma_\nu|^{2t_\nu}d\varphi=\mathfrak{I}(t_\nu,\gamma_\nu)$$

qui dépend de $2m$ variables t_ν, γ_ν satisfaisant aux conditions (2); l'existence du maximum est assurée.

Admettons que parmi les m valeurs t_ν pour lesquelles le maximum de (3) est atteint il y ait n valeurs positives, $t_1, t_2, \ldots, t_n$ ($n\leq m$). Choisissons n nombres réels σ_ν tels que

$$(4)\qquad \sum_{\nu=1}^{n}\sigma_\nu=0,\qquad \sum_{\nu=1}^{n}\sigma_\nu\gamma_\nu=0,$$

mettons $\sigma_\nu=0$ si $\nu>n$ et posons $t_\nu^*=t_\nu+\varepsilon\sigma_\nu$ pour $\nu=1, 2, \ldots, m$. Si $|\varepsilon|$ est suffisamment petit, le système t_ν^*, γ_ν est admissible selon le sens des conditions (2). Mais c'est le système t_ν, γ_ν qui atteint le maximum et ainsi

$$(5)\qquad \mathfrak{I}(t_\nu^*,\gamma_\nu)\leq\mathfrak{I}(t_\nu,\gamma_\nu).$$

Observons que

$$(6)\quad J(t_\nu^*, \gamma_\nu) = \int_0^{2\pi} \prod_{\nu=1}^{m} |e^{i\varphi} - \gamma_\nu|^{2t_\nu} \Bigg\{ 1 + 2\varepsilon \sum_{\nu=1}^{n} \sigma_\nu \log |e^{i\varphi} - \gamma_\nu| + 2\varepsilon^2 \left(\sum_{\nu=1}^{n} \sigma_\nu \log |e^{i\varphi} - \gamma_\nu| \right)^2 + o(\varepsilon^2) \Bigg\} d\varphi.$$

Le signe de ε est arbitraire et ainsi (5) exige que, dans le développement (6), le coefficient de ε s'annule et que le coefficient de ε^2 soit non positif. La dernière condition exige que

$$(7)\qquad \sum_{\nu=1}^{n} \sigma_\nu \log |e^{i\varphi} - \gamma_\nu| = 0 \qquad \text{pour} \quad 0 \leqq \varphi \leqq 2\pi,$$

et ainsi $\sigma_\nu = 0$ pour $\nu = 1, \ldots, n$. Mais, si n était supérieur à 3, le système (4) de trois équations linéaires homogènes à coefficients *réels* et à n inconnues réelles σ_ν admettrait une solution non identiquement nulle. Ainsi, en effet, $n \leqq 3$ et nous avons démontré que, quel que soit m, le polygone à périmètre maximal doit être un triangle qui, toutefois, pourrait être dégénéré.

2. Nous considérons un triangle dont le rayon conforme extérieur est 1 et dont les angles sont $\pi\alpha$, $\pi\beta$, $\pi\gamma$,

$$(8)\qquad \alpha \geqq 0, \quad \beta \geqq 0, \quad \gamma \geqq 0, \quad \alpha + \beta + \gamma = 1.$$

Le périmètre L de ce triangle est donné par la formule

$$(9)\qquad L = 2\pi^2 g(\alpha) g(\beta) g(\gamma)$$

qu'on déduit des résultats connus ([2]) où

$$(10)\qquad g(x) = \left[\Gamma(x) \sin \frac{\pi x}{2} \right]^{-1} x^{\frac{x}{2}} (1 - x)^{-\frac{1-x}{2}}.$$

On vérifie aisément les propriétés

$$(11)\qquad g(x) g(1 - x) = \frac{2}{\pi}, \qquad g(1) = 1, \qquad g(0) = \frac{2}{\pi}.$$

Posons

$$\frac{g'(x)}{g(x)} = l(x).$$

On vérifie que $l'(x) > 0$ si $0 < x < 1/2$ et $l(1 - x) = l(x)$.

En vertu de (8) et (9), L est une fonction de α et β dans le domaine Δ fermé où $\alpha \geqq 0$, $\beta \geqq 0$, $\alpha + \beta \leqq 1$. Le long de la frontière de Δ la fonction L prend la valeur constante 8 en vertu de (11). Si, à l'intérieur de Δ, il y a un point stationnaire, les deux dérivées partielles d'ordre 1 de L s'y annulent et l'on a

$$l(\alpha) = l(\beta) = l(\gamma).$$

Les propriétés mentionnées de $l(x)$ montrent que l'unique point stationnaire est $\alpha = \beta = \gamma = 1/3$, où L prend la valeur 7,113. Ainsi, L atteint son minimum pour le triangle équilatéral et son maximum 8 pour le triangle dégénéré en un segment de longueur 4 (de *périmètre* 8).

3. Désignons par L le périmètre et par $\bar{r}$ le rayon conforme extérieur d'une courbe. La première des deux inégalités

$$2\pi\bar{r} \leqq L \leqq 8\bar{r} \tag{12}$$

est valable pour une courbe quelconque et connue depuis longtemps. La seconde inégalité n'a été énoncée jusqu'ici qu'hypothétiquement ([3]); dans ce qui précède, nous l'avons démontrée pour un polygone convexe quelconque et, par conséquent, pour une courbe convexe quelconque. Pour le cas particulier où la courbe possède un centre de symétrie, M. Chr. Pommerenke nous a communiqué une jolie démonstration tout à fait différente de celle qui vient d'être présentée.

Si la courbe est un polygone, convexe ou non, ayant au plus m côtés, on peut se demander si l'inégalité double (12) pourrait être remplacée par la suivante :

$$\frac{2^{1+\frac{2}{m}}\pi^{\frac{1}{2}}\Gamma\left(\frac{1}{2}+\frac{1}{m}\right)}{\Gamma\left(1+\frac{1}{m}\right)}\bar{r} \leqq L \leqq 8E\left(\frac{m}{2}\right)\bar{r}; \tag{13}$$

le premier membre représente le périmètre du polygone régulier; on désigne par $E(x)$ le plus grand entier contenu dans x. L'inégalité double (13) dit peu de chose lorsque $m = 2$; elle vient d'être démontrée pour $m = 3$; la seconde partie est facile à démontrer si m est pair; tous les autres cas contenus en (13) sont hypothétiques.

(*) Séance du 27 avril 1959.

([1]) Pour la définition du rayon conforme extérieur $\bar{r}$, *voir* G. Pólya et G. Szegö, *Isoperimetric Inequalities in Mathematical Physics*, Princeton, 1951, p. 2.

([2]) Pólya et Szegö, *loc. cit.*, p. 273.

([3]) *Voir* Pólya et Szegö, *loc. cit.*, p. 17, n° 12.

Commentary on

[69] Georges Pólya and Menahem Schiffer, *Sur la répresentation conforme de l'extérieur d'une courbe fermée convexe*, C. R. Acad. Sci. Paris **248** (1959), 2837–2839.

When George Pólya was to retire from Stanford, Max Schiffer was appointed in 1952 as his successor. Their collaboration began in 1954 with a major joint paper [53]. (Space limitations did not allow [53] to be reprinted in these Selected Papers, but it does appear in [P2].) In addition to their mathematical expertise, Schiffer and Pólya shared a strong command of classical physics, which motivated much of their work. The short note [69] proves a conjecture made several years earlier by Pólya and Szegő in their book [PS1] on isoperimetric inequalities. For a Jordan curve C in the plane of length L, enclosing a region Ω with transfinite diameter (= exterior mapping radius) d, it had been "known for a long time" that $2\pi d \leq L$, with equality if and only if C is a circle. It is clear that no reverse inequality of this type can hold in general, since a region of given transfinite diameter can be enclosed by an arbitrarily long curve. However, Pólya and Szegő had conjectured that the inequality $L \leq 8d$ holds for all *convex* regions Ω, with equality only when Ω degenerates to a line segment. Their conjecture is verified in [69] by first considering convex polygons with a fixed number of vertices. The Schwarz–Christoffel formula gives a conformal mapping of the exterior of the unit disk onto the exterior of such a polygon, and calculations yield the desired inequality in this special case. The general result is then deduced by approximation.

The original inequality $2\pi d \leq L$ is quite interesting and deserves to be better known. It can be viewed as a sharpened version of the classical isoperimetric inequality $4\pi A \leq L^2$, where A denotes the area of Ω, since an old result of Pólya [P1] says that $A \leq \pi d^2$, with equality if and only if Ω is a circular disk. The inequality $2\pi d \leq L$ is stated in [PS1] with a cryptic indication of a proof, but it appears as a problem in the English edition of the classic book by Pólya and Szegő [PS2, Part IV, Chap. 2, No. 124], with a hint for a relatively simple proof. Here are further details. Let $\varphi(z) = z + b_0 + b_1/z + \dots$ map the region $|z| > d$ conformally onto the exterior of Ω. For $r > d$, let C_r be the image of a circle $\Gamma_r = \{z : |z| = r\}$, and let L_r denote the length of C_r. Let $h(z) = \sqrt{\varphi'(z)} = 1 + c_1/z^2 + c_2/z^3 + \dots$, and observe that

$$1 + \frac{|c_1|^2}{r^4} + \frac{|c_2|^2}{r^6} + \dots = \frac{1}{2\pi}\int_0^{2\pi} |h(re^{i\theta})|^2 \, d\theta$$

$$= \frac{1}{2\pi r}\int_{\Gamma_r} |\varphi'(z)||dz| = \frac{L_r}{2\pi r} \to \frac{L}{2\pi d}$$

as r decreases to d. Therefore, $2\pi d \leq L$ with strict inequality unless $c_1 = c_2 = \dots = 0$. This in turn implies that $\varphi(z) = z + b_0$, so that Ω is a circular disk.

References

[P1] G. Pólya, *Beitrag zur Verallgemeinerung des Verzerrungssatzes auf mehrfach zusammenhängende Gebiete, II*, Sitzungsberichte der Preussischen Akademie der Wissenschaften, 1928, 280–282.

[P2] *George Pólya: Collected Papers*, Vol. III: Analysis (Joseph Hersch and Gian-Carlo Rota, editors), MIT Press, 1984.

[PS1] G. Pólya and G. Szegő, *Isoperimetric Inequalities in Mathematical Physics*, Princeton University Press, 1951.

[PS2] G. Pólya and G. Szegő, *Problems and Theorems in Analysis*, Vol. II, Springer-Verlag, 1976.

PETER DUREN

[70] Extremum problems and variational methods in conformal mapping

[70] Extremum problems and variational methods in conformal mapping, in Proceedings of the International Congress of Mathematicians, Edinburgh, 1958, Cambridge University Press, New York, (1960), 211–231.

EXTREMUM PROBLEMS AND VARIATIONAL METHODS IN CONFORMAL MAPPING

By MENAHEM SCHIFFER

1. Introduction

One fundamental problem in the classical theory of conformal mapping was the study of the various types of canonical domains upon which any domain, arbitrarily given in the complex plane, can be mapped conformally. The first question to be settled was, therefore, the existence of various types of canonical mapping functions. From the beginning, methods of the calculus of variations were applied in order to establish the necessary existence theorems. The role of the Dirichlet principle in the attempted proof of Riemann's mapping theorem for simply connected domains is well known and also the influence of its initial failure upon the critical period of the calculus of variations and upon the development of the powerful modern direct methods in this important branch of analysis. The existence proofs for canonical conformal mappings by means of extremum problems like the Dirichlet principle are so difficult because they characterize the sought mapping function, which is analytic and univalent, as the extremum function in a much wider class of admissible competing functions. The latter class is so large that the main labour in the proof is spent in establishing the existence of an extremum function of the variational problem considered.

The theory of conformal mapping advanced considerably when one started a systematic study of the univalent analytic functions in a given domain; that is, the class of those functions which realize the various conformal mappings of that domain. The main result of this theory is that all univalent functions in a given domain form a normal family. This fact leads easily to the consequence that for each reasonable extremum problem within the family of univalent functions there exists at least one element of the family which attains the extremum considered[20]. On the basis of this theory, very elegant proofs could be derived for the Riemann mapping theorem and for the existence of numerous other canonical mappings. The characteristic difficulty of the new approach, that is to study extremum problems within the family of univalent functions, lies in the fact that the univalent functions form no linear space; hence, it is not at all easy to characterize an extremum function by comparison with its competitors by infinitesimal variation.

In each particular existence proof a special comparison method had to be devised and the essential step of the whole proof was the characterization of the extremum function by this particular variation.

It is possible to develop a systematic infinitesimal calculus within the family of univalent functions. In 1923 Löwner gave a now classical partial differential equation which has as solutions one-parameter families of univalent functions which admit a very simple geometric interpretation[18]. I showed in 1938 that the univalent extremum functions do satisfy in very many cases a first-order differential equation and gave a standard variational procedure for establishing these ordinary differential equations[26]. In the following years, Schaeffer and Spencer applied this variational procedure systematically to the coefficient problem for functions univalent in the unit circle and developed an extensive theory for it[22, 23, 24]. Golusin applied the same variational technique to numerous questions of geometric function theory[7, 8]. The significance of extremum problems for the general theory of conformal mapping is evident. The great number of possible conformal mappings of a given domain precludes the study of all of them; however, important individual mappings can be singled out as solutions of extremum problems and can be described geometrically and analytically just because of their extremum property. The remaining amorphous mass of conformal mappings is subjected to all the inequalities which flow from the solutions of the various extremum problems and is, thus, at least partially characterized.

In the present paper we shall try to give a brief survey of the basic methods of variation within the family of univalent functions. By discussing a few important extremum problems, we will show the flexibility of the technique. It will appear that the variational method provides very often an elegant and useful transformation of the extremum problem but leads sometimes to functional equations whose solution is a deep problem again. It is clear that the field of research described is by no means completely explored and exhausted and that, because of its interest from the point of view of applied as well as of pure mathematics, it deserves the continued attention of mathematicians.

2. Variation of the Green's function

The simplest approach to the calculus of variations for univalent functions seems to lead through the theory of the Green's function of a domain and its variational formula. Let D be a domain in the complex z-plane whose boundary C consists of n closed analytic curves and let

$g(z,\zeta)$ be its Green's function with the source point ζ. We consider the conformal transformation

$$z^*(z) = z + \frac{e^{i\alpha}\rho^2}{z - z_0} \quad (z_0 \in D,\ \rho > 0). \tag{1}$$

This mapping is univalent in the domain $|z - z_0| > \rho$; hence, for small enough ρ it will be univalent on C and transform it into a new set C^* of n closed analytic curves which bounds a new domain D^*. We denote the Green's function of D^* by $g^*(z,\zeta)$ and wish to express it in terms of $g(z,\zeta)$. We observe that $\gamma(z,\zeta) = g^*(z^*(z), \zeta^*(\zeta))$ is a harmonic function in the domain D_ρ which is obtained from D by removal of the circle $|z - z_0| < \rho$. $\gamma(z,\zeta)$ has a pole for $z = \zeta$ and vanishes on the boundary curves C of D_ρ. We choose two fixed points ζ and η in D_ρ and apply Green's identity in the form

$$\frac{1}{2\pi}\int_{C+c}\left[\frac{\partial}{\partial n}g(z,\eta)\gamma(z,\zeta) - g(z,\eta)\frac{\partial}{\partial n}\gamma(z,\zeta)\right]ds = \gamma(\zeta,\eta) - g(\zeta,\eta). \tag{2}$$

Here, c denotes the small circumference $|z - z_0| = \rho$. Observe now that the integration takes place only over the circumference c since both g and γ vanish on C.

In order to simplify (2), we introduce the analytic functions of z whose real parts are $g(z,\eta)$ and $g^*(z,\zeta)$, respectively, and denote them by $p(z,\eta)$ and $p^*(z,\zeta)$. These functions have logarithmic poles at η or ζ and have also imaginary periods when z circulates around a boundary continuum. It is now easy to express (2) in the form

$$g^*(\zeta^*, \eta^*) - g(\zeta,\eta) = \operatorname{Re}\left\{\frac{1}{2\pi i}\oint_c p^*(z^*,\zeta^*)\,dp(z,\eta)\right\}. \tag{3}$$

This integral equation for $g^*(\zeta^*,\eta^*)$ in terms of $g(\zeta,\eta)$ must now hold for the most general domain D which possesses a Green's function at all. Indeed, such a domain D may be approximated arbitrarily by domains D_ν with analytic boundaries C_ν for which the identity (3) is valid. If D and D_ν go under the variation (1) into the domains D^* and D_ν^*, then the D_ν^* will likewise approximate D^*. Since (3) holds for all approximating domains and since at ζ, η and on c the Green's functions of D_ν and D_ν^* converge uniformly to the Green's functions of D and D^*, respectively, the formula (3) must remain valid in the limit and is thus generally proved[31].

We may apply Taylor's theorem in the form

$$p^*(z^*,\zeta^*) = p^*(z,\zeta^*) + p^{*\prime}(z,\zeta^*)\frac{e^{i\alpha}\rho^2}{z - z_0} + O(\rho^4), \tag{4}$$

where the residual term $O(\rho^4)$ can be estimated equally for all domains D which contain a fixed subdomain Δ which, in turn, contains the point ζ and the circle c. Thus, inserting (4) into (3) and using the residue theorem we obtain after an easy transformation

$$g^*(\zeta^*, \eta^*) = g(\zeta, \eta) + \text{Re}\{e^{i\alpha}\rho^2 p'(z_0, \zeta)p'(z_0, \eta)\} + O(\rho^4), \tag{5}$$

where again $O(\rho^4)$ can be estimated equally as above. Finally, using Taylor's theorem again, we can reduce (5) to[27, 28]

$$g^*(\zeta, \eta) = g(\zeta, \eta) + \text{Re}\left\{e^{i\alpha}\rho^2\left[p'(z_0, \zeta)p'(z_0, \eta) - \frac{p'(\zeta, \eta)}{\zeta - z_0} - \frac{p'(\eta, \zeta)}{\eta - z_0}\right]\right\} + O(\rho^4). \tag{6}$$

In the preceding, we have restricted ourselves to the particular variation (1) for the sake of simple exposition. It is clear that a corresponding formula can be established for each variation $z^* = z + \rho^2 v(z)$, where $v(z)$ is analytic on the boundary C of the varied domain. On the other hand, such a general variation can be approximated arbitrarily by superposition of elementary variations of the type (1). Indeed, for most applications the formulas (1) and (6) are entirely sufficient.

A remarkable transformation of (6) is possible if the boundary C of D is a set of smooth curves. Indeed, we may express (6) in the form

$$g^*(\zeta, \eta) - g(\zeta, \eta) = \text{Re}\left\{e^{i\alpha}\rho^2 \frac{1}{2\pi i}\int_C \frac{p'(z, \zeta)p'(z, \eta)}{z - z_0}\,dz\right\} + O(\rho^4). \tag{7}$$

We observe that the real part of $p(z, \zeta)$ is the Green's function $g(z, \zeta)$ and that it vanishes, therefore, on C. Let $z' = dz/ds$ denote the tangent vector to C at the point $z(s)$; it is easy to see that

$$p'(z, \zeta)z' = -i\frac{\partial g(z, \zeta)}{\partial n}$$

and hence (7) may be given the real form

$$g(\zeta, \eta) = -\frac{1}{2\pi}\int_C \frac{\partial g(z, \zeta)}{\partial n_z}\frac{\partial g(z, \eta)}{\partial n_z}\,\delta n\,ds \tag{8}$$

with

$$\delta n = \text{Re}\left\{\frac{1}{iz'}\frac{e^{i\alpha}\rho^2}{z - z_0}\right\}. \tag{9}$$

Clearly, δn denotes the shift along the interior normal of the boundary point $z \in C$ under the variation (1).

By linear superposition of elementary variations (1), formula (8) can be proved for very general δn-variations of the boundary curves C.

This formula was first given by Hadamard in 1908[11] and has been very frequently used in applied mathematics because of the very intuitive and geometric significance of the normal displacement of the boundary points. We may mention, in particular, Lavrentieff's systematic use of boundary deformations in many problems of fluid dynamics and conformal mapping[16,17].

If D is a simply connected domain there exists a close relationship between the Green's function of D and the univalent function $\phi(z)$ which maps the domain D onto the exterior of the unit circle. In fact, we have

$$g(z, \zeta) = \log \left| \frac{1 - \phi(z)\overline{\phi(\zeta)}}{\phi(z) - \phi(\zeta)} \right|. \tag{10}$$

Julia used this interrelation in order to derive from the Hadamard formula (8) a variational formula for univalent functions[15]. This very intuitive and elegant formula, however, cannot be applied directly to the study of extremum problems in the theory of conformal mapping. In fact, one cannot assert *a priori* that the extremum domain D will possess a boundary C which is smooth enough to admit a variation of the Hadamard–Julia type.

3. Infinitesimal variations and extremum problems

We are now in a position to construct, by means of the fundamental formula (6), in any given domain D, univalent mappings which are arbitrarily close to the identity mapping. We have to assume only that the boundary C of D contains a non-degenerate continuum Γ. Let $D(\Gamma)$ denote the domain of the z-plane which contains the domain D and the point at infinity and which is bounded by Γ; let $g(z, \zeta)$ denote now the Green's function of $D(\Gamma)$. We choose an arbitrary but fixed point $z_0 \in D$ and subject $D(\Gamma)$ to a variation (1) which transforms it into the varied domain $D(\Gamma^*)$ with the Green's function $g^*(z, \zeta)$. The relation between $g^*(z, \zeta)$ and $g(z, \zeta)$ is given by the variational formula (6).

Let $w = \phi(z)$ be univalent in $D(\Gamma)$, normalized at $z = \infty$ by the requirement $\phi'(\infty) = 1$, and let it map $D(\Gamma)$ onto the domain $|w| > 1$. Analogously, we define $w = \phi^*(z)$ with respect to the domain $D(\Gamma^*)$. By virtue of the relation (10), we have obviously

$$g(z, \infty) = \log |\phi(z)|, \quad g^*(z, \infty) = \log |\phi^*(z)|; \tag{11}$$

these relations permit us to connect $\phi^*(z)$ with $\phi(z)$ by use of (6).

The function

$$v(z) = \phi^{*-1}[\phi(z)] \tag{12}$$

is analytic and univalent in $D(\Gamma)$ and hence, *a fortiori*, in D. A simple calculation based on (6) and (11) shows that

$$v(z) = z + e^{i\alpha}\rho^2\left[\frac{1}{z-z_0} - \frac{\phi'(z_0)^2\,\phi(z)}{\phi'(z)\,\phi(z_0)\,[\phi(z)-\phi(z_0)]}\right] + e^{-i\alpha}\rho^2\frac{\overline{\phi'(z_0)}^2\,\phi(z)^2}{\phi'(z)\,\overline{\phi(z_0)}\,[1-\overline{\phi(z_0)}\,\phi(z)]} + O(\rho^4). \tag{13}$$

Since ρ can be made arbitrarily small, we have in (13) the representation for a large class of univalent variations of the domain D considered. We will now show that this set of variations is general enough to characterize the extremum domains for a large class of extremum problems relative to the family of univalent functions.

We shall consider extremum problems of the following type. Let T be a domain in the complex t-plane which contains the point at infinity and which is analytically bounded. We denote by F the family of all analytic functions $f(t)$ in T which are univalent there, have a simple pole at $t = \infty$ and which are normalized by the condition $f'(\infty) = 1$. Let $\phi[f]$ be a real-valued functional defined for all analytic functions $f(t)$ in T. We suppose that $\phi[f]$ is differentiable in the sense that for an arbitrary analytic function $g(t)$ defined in T

$$\phi[f+\epsilon g] = \phi[f] + \mathrm{Re}\{\epsilon\psi[f,g]\} + O(\epsilon^2) \tag{14}$$

holds, where ψ is a complex-valued functional of f and g, linear in g. We suppose that the residual term $O(\epsilon^2)$ can be estimated equally for all analytic functions $g(t)$ which are equally bounded in a specified subdomain of T. Thus, we require for $\phi[f]$ the existence of a Gâteaux differential with the above additional specifications.

We assume also that $\phi[f]$ has an upper bound within the family F. Then, in view of the normality of this family, it is easy to show that there must exist functions $f(t) \in T$ for which $\phi[f]$ attains its maximum value within F. We can characterize each extremum function by subjecting it to infinitesimal variations and comparing $\phi[f]$ with the functional values of the varied univalent elements of the family. Indeed, by means of the functions (13) we can construct the competing functions in F

$$f^*(t) = v[f(t)]\,.\,v'(\infty)^{-1}, \tag{15}$$

where $z = f(t)$ maps the domain T onto the extremum domain D in the z-plane. An easy calculation yields

$$\phi[f^*] = \phi[f] + \mathrm{Re}\{e^{i\alpha}\rho^2 A + e^{-i\alpha}\rho^2 B\} + O(\rho^4), \tag{16}$$

with

$$\left.\begin{aligned} A &= \psi\left[z, \frac{1}{z-z_0} - \frac{\phi'(z_0)^2\,\phi(z)}{\phi'(z)\,\phi(z_0)\,[\phi(z)-\phi(z_0)]}\right] \\ B &= \psi\left[z, \frac{\overline{\phi'(z_0)^2}\,\phi(z)^2}{\phi'(z)\,\overline{\phi(z_0)}\,[1-\overline{\phi(z_0)}\,\phi(z)]} + \frac{\overline{\phi'(z_0)^2}}{\overline{\phi(z_0)^2}}\,z\right] \end{aligned}\right\} \quad (z = f(t)). \tag{17}$$

Since the extremum property of f requires $\phi[f^*] \leqslant \phi[f]$ and since ρ and $e^{i\alpha}$ are at our disposal, we can easily conclude $A + \bar{B} = 0$, that is

$$\psi\left[f(t), \frac{1}{f(t)-z_0}\right]\frac{\phi(z_0)^2}{\phi'(z_0)^2} = \overline{\psi\left[z, \frac{\phi(z)}{\phi'(z)} - z\right]}$$
$$+\,\psi\left[z, \frac{\phi(z)\,\phi(z_0)}{\phi'(z)\,[\phi(z)-\phi(z_0)]}\right] + \overline{\psi\left[z, \frac{\phi(z)\,\overline{\phi(z_0)}^{-1}}{\phi'(z)\,[\phi(z)-\overline{\phi(z_0)}^{-1}]}\right]}. \tag{18}$$

Before discussing the consequences of (18), we introduce some more elementary variations in F which will allow us to simplify the result (18). We map the domain $D(\Gamma)$ onto $|w| > 1$ by means of the function $w = \phi(z)$; we then turn this circle into itself by the linear mapping $w_1 = e^{i\epsilon}w$ and return to the z-plane through $\phi^{-1}(w_1)$. Thus, the function

$$v_1(z) = e^{-i\epsilon}\phi^{-1}[e^{i\epsilon}\phi(z)] \tag{19}$$

is univalent in $D(\Gamma)$ and hence in D. For small ϵ, we have the series development in ϵ

$$v_1(z) = z + i\epsilon\left[\frac{\phi(z)}{\phi'(z)} - z\right] + O(\epsilon^2). \tag{20}$$

Since $f^*(t) = v_1[f(t)]$ is an admissible competing function in F, we deduce easily from the extremum property of $f(t)$ and from the freedom in the choice of the real parameter ϵ

$$\psi\left[z, \frac{\phi(z)}{\phi'(z)} - z\right] = \text{real}. \tag{21}$$

Another possible infinitesimal variation is obtained by

$$v_2(z) = (1+\epsilon)^{-1}\phi^{-1}[(1+\epsilon)\phi(z)] \quad (\epsilon > 0). \tag{22}$$

In fact, we may map $D(\Gamma)$ onto $|w| > 1$, magnify the unit circle by a factor $(1+\epsilon)$ and return through $\phi^{-1}(w)$ to the z-plane. The function $f^*(t) = v_2[f(t)]$ lies also in F and from the extremum property of $f(t)$ we deduce by use of (21) the inequality

$$\psi\left[z, \frac{\phi(z)}{\phi'(z)} - z\right] \leqslant 0. \tag{23}$$

We return now to formula (18) and observe that in view of (21)

$$\lim_{z_0\to\Gamma} \psi\left[f(t), \frac{1}{f(t)-z_0}\right]\frac{\phi(z_0)^2}{\phi'(z_0)^2} = \text{real}. \tag{24}$$

In order to simplify the discussion we shall assume that

$$\psi\left[z, \frac{1}{z-z_0}\right] = W(z_0)$$

is a meromorphic function of z_0; this is, indeed, the case in most applications. We put $z = \psi(w)$, where ψ is the inverse function of $w = \phi(z)$, and obtain from (24) the boundary relation

$$\lim_{|w|\to 1} W[\psi(w)]\, w^2\psi'(w)^2 = \text{real}, \tag{25}$$

for the function $\psi(w)$ which is analytic in $|w| > 1$ and maps this circular domain onto $D(\Gamma)$. By the Schwarz reflection principle, the function $W[\psi(w)]\, w^2\psi'(w)^2$ can then be continued analytically into the domain $|w| \leqslant 1$. Thus, $\psi(w)$ satisfies a first-order differential equation with analytic coefficients in the entire w-plane. This fact shows that Γ is composed of analytic arcs and the same holds for the boundary C of the extremum domain D: *C is composed of analytic arcs.*

In order to complete the argument we need a last elementary variation. We again map $D(\Gamma)$ onto the domain $|w| > 1$ by means of $\phi(z)$. The function

$$\omega = p(w) = w + \frac{w_0^2}{w} \quad (|w_0| = 1) \tag{26}$$

maps the circular region $|w| > 1$ onto the ω-plane slit along the segment between the points $-2w_0$ and $+2w_0$. It is then easily seen that, for $\epsilon > 0$,

$$w_1 = p^{-1}[(1+\epsilon)p(w) + 2\epsilon w_0] = w + \epsilon\frac{w(w+w_0)}{w-w_0} + O(\epsilon^2) \tag{27}$$

provides a mapping of $|w| > 1$ onto the same circular region from which a small radial segment issuing from the periphery point w_0 has been removed. The function

$$\begin{aligned} v_3(z) &= (1+\epsilon)^{-1}\phi^{-1}\left[\phi(z) + \epsilon\frac{\phi(z)\,[\phi(z)+\phi(z_0)]}{\phi(z)-\phi(z_0)} + O(\epsilon^2)\right] \\ &= z + \epsilon\left[\frac{\phi(z)}{\phi'(z)}\frac{\phi(z)+\phi(z_0)}{\phi(z)-\phi(z_0)} - z\right] + O(\epsilon^2) \quad (z_0 \in \Gamma) \end{aligned} \tag{28}$$

is then normalized at infinity and univalent in D. Hence, $f^*(t) = v_3[f(t)]$ is again a competing function in our extremum problem, whence

$$\operatorname{Re}\left\{\psi\left[z, \frac{\phi(z)}{\phi'(z)}\frac{\phi(z)+\phi(z_0)}{\phi(z)-\phi(z_0)} - z\right]\right\} \leqslant 0. \tag{29}$$

But observe that the left side of (29) coincides with the right-hand term of (18) since $z_0 \in \Gamma$. Hence, we have proved

$$\psi\left[f(t), \frac{1}{f(t)-z_0}\right] \frac{\phi(z_0)^2}{\phi'(z_0)^2} \leqslant 0 \quad (z_0 \in \Gamma). \tag{30}$$

Since $|\phi(z)| = 1$ for $z \in \Gamma$ we have $\log \phi(z) =$ imaginary on Γ and, consequently, we can write on each analytic arc of Γ

$$z' \frac{\phi'(z)}{\phi(z)} = \text{imaginary}, \quad z' = \frac{dz}{ds}. \tag{31}$$

Thus, we may express (30) also in the form

$$\psi\left[f(t), \frac{1}{f(t)-z}\right] \left(\frac{dz}{ds}\right)^2 \geqslant 0 \quad \text{on} \quad C. \tag{32}$$

In this final form the characterization of the extremum domain has become independent of the choice of the subcontinuum Γ. The boundary arcs of C are determined by a first-order differential equation involving the meromorphic function $W(z)$ defined above.

Under our assumptions made regarding the functional $\psi[z, 1/(z-z_0)]$ it is also easy to prove that the extremum domain cannot possess exterior points. For, suppose z_0 were an exterior point of an extremum domain D. In this case, the mapping (1) itself would be an admissible univalent variation for ρ small enough and the extremum property of $f(t)$ would imply

$$\operatorname{Re}\left\{e^{i\alpha}\rho^2 \psi\left[z, \frac{1}{z-z_0}\right]\right\} + O(\rho^4) \leqslant 0, \tag{33}$$

whence easily

$$\psi\left[z, \frac{1}{z-z_0}\right] = 0. \tag{34}$$

But if, as supposed, ψ is a specific meromorphic function $W(z_0)$, not identically zero, this result is impossible since (34) would imply by analytic continuation that $W(z_0) \equiv 0$[19]. Thus, we have proved the

Theorem. The extremum domain of the extremum problem $\phi[f] = \max$ *within the family* F *is a slit domain bounded by analytic arcs. Each satisfies the differential equation*:

$$\psi\left[f(t), \frac{1}{f(t)-z(\tau)}\right] \left(\frac{dz}{d\tau}\right)^2 = 1, \tag{35}$$

where τ *is a properly chosen real curve parameter.*

This theorem was proved originally[26] by means of rather deep theorems of measure theory. It can be derived in elementary manner from the variational formula for the Green's function as shown here. It permits now a systematic and unified treatment of numerous extremum problems of conformal mapping. The extremum domain can be determined either by integrating the differential equation (35) for the boundary slits or by solving the differential equation implied by (25) for the functions $\psi(w)$ which map the circular domain $|w| > 1$ onto the domains $D(\Gamma)$. The latter procedure is particularly convenient in the case that the original domain T is simply connected.

4. The coefficient problem

The best studied extremum problem in conformal mapping is without any doubt the coefficient problem for the functions univalent in the unit circle. We consider all power series

$$f(z) = z + a_2 z^2 + \dots + a_n z^n + \dots, \tag{36}$$

which converge for $|z| < 1$ and which represent univalent functions. Bieberbach stated the conjecture

$$|a_n| \leqslant n. \tag{37}$$

Since the 'Koebe function'

$$\frac{z}{(1-z)^2} = z + 2z^2 + \dots + nz^n + \dots \tag{38}$$

is indeed such a univalent power series, this function would seem to be the solution of an infinity of extremum problems. Because of its simple formulation the conjecture (37) has attracted the attention of many analysts. Bieberbach himself proved (37) in 1916 for $n = 2$[2]; Löwner proved the case $n = 3$ in 1923[18] and Garabedian and Schiffer proved the case $n = 4$ in 1955[5]. These proofs are to be considered as tests for our technique in handling extremum problems of conformal mapping and the main significance of the coefficient problem is indeed that it raises a challenge to our various methods in this field. We want to give a brief survey of variational methods applied in this problem.

We define a sequence of polynomials $P_n(x)$ of degree $(n-1)$ by means of the generating function

$$\frac{f(z)}{1-xf(z)} = \sum_{n=1}^{\infty} [a_n + P_n(x)] z^n \quad (P_1(x) = 0). \tag{39}$$

We note down the first few polynomials

$$P_2(x) = x, \quad P_3(x) = 2a_2 x + x^2, \quad P_4(x) = (2a_3 + a_2^2)\, x + 3a_2 x^2 + x^3. \tag{39'}$$

A simple application of the reasoning in the preceding section leads to the following result. Let $f(z)$ be a univalent function which maximizes $|a_n|$; we can make the permissible assumption $a_n > 0$. Then $f(z)$ satisfies the differential equation[27]

$$\frac{z^2 f'(z)^2}{f(z)^2} P_n\left[\frac{1}{f(z)}\right] = \frac{1}{z^{n-1}} + \frac{2a_2}{z^{n-2}} + \frac{3a_3}{z^{n-3}} + \ldots + \frac{(n-1)\,a_{n-1}}{z}$$
$$+ (n-1)\,a_n + (n-1)\,\bar{a}_{n-1} z + \ldots + 3\bar{a}_3 z^{n-3} + 2\bar{a}_2 z^{n-2} + z^{n-1}. \quad (40)$$

The right side as well as the polynomial $P_n(x)$ depends on the coefficients of the unknown function $f(z)$; hence, (40) represents a rather complicated functional equation for the extremum function sought which has been solved until now only in the cases $n \leqslant 4$.

We may attack the functional equation (40) as follows. It is easily shown in all cases $n \leqslant 4$ that the extremum function $w = f(z)$ maps the domain $|z| < 1$ onto the entire w-plane slit along a single analytic arc Γ which runs out to infinity. We consider then the analytic functions

$$w = f(z,t) = e^t[z + a_2(t)\,z^2 + \ldots + a_n(t)\,z^n + \ldots], \quad (41)$$

which map $|z| < 1$ onto the w-plane slit along infinite subarcs Γ_t of Γ. We can read off from (40) that Γ satisfies the differential equation

$$\frac{w'(\tau)^2}{w(\tau)^2} P_n\left[\frac{1}{w(\tau)}\right] + 1 = 0 \quad (\tau = \text{real parameter}), \quad (42)$$

and evidently the subarcs Γ_t satisfy precisely the same equation. Using next the Schwarz reflection principle, we can show that the functions $f(z,t)$ satisfy differential equations which are very similar to (40); namely

$$\frac{z^2 f'(z,t)^2}{f(z,t)^2} P_n\left[\frac{1}{f(z,t)}\right] = \sum_{\nu=-(n-1)}^{n-1} A_\nu(t)\,z^\nu = q(z,t),\ A_{-\nu}(t) = \overline{A_\nu(t)}. \quad (43)$$

We may transform (43) into

$$\int_{f(z_0,t)}^{f(z,t)} \sqrt{\left[P_n\left(\frac{1}{w}\right)\right]}\frac{dw}{w} = \int_{z_0}^{z} \sqrt{[q(z,t)]}\frac{dz}{z}. \quad (44)$$

Löwner has shown[18] that the functions $f(z,t)$ which represent the unit circle on a family of slit domains with growing boundary slits Γ_t of the above type satisfy the partial differential equation

$$\frac{\partial f(z,t)}{\partial t} = z\frac{1+\kappa(t)\,z}{1-\kappa(t)\,z}\frac{\partial f(z,t)}{\partial z} \quad (\kappa(t) \text{ continuous, } |\kappa| = 1). \quad (45)$$

Thus, differentiating (44) with respect to t and using (43) and (44) we find

$$\sqrt{[q(z,t)]}\frac{1+\kappa z}{1-\kappa z}-\sqrt{[q(z_0,t)]}\frac{1+\kappa z_0}{1-\kappa z_0}=\frac{1}{2}\int_{z_0}^{z}\frac{\partial q(z,t)}{\partial t}\frac{1}{\sqrt{[q(z,t)]}}\frac{dz}{z}. \tag{46}$$

Differentiating (46) again with respect to z, we find after simple rearrangement

$$\frac{\partial q(z,t)}{\partial t}=z\frac{1+\kappa z}{1-\kappa z}\frac{\partial q(z,t)}{\partial z}+\frac{4\kappa z}{(1-\kappa z)^2}q(z,t). \tag{47}$$

On the other hand, $q(z,t)$ is a simple rational function of z as is seen from its definition (43). When we insert its expression into (47) and compare the coefficients of equal powers of z on both sides, we obtain

$$\frac{dA_\nu(t)}{dt}=\nu A_\nu(t)+2\sum_{\mu=-(n-1)}^{\nu-1}(2\nu-\mu)A_\mu\kappa^{\nu-\mu}. \tag{48}$$

In order that $A_\nu(t)\equiv 0$ for all $\nu\geqslant n$ it is necessary and sufficient that

$$\sum_{\mu=-(n-1)}^{n-1}A_\mu\kappa^{-\mu}\equiv 0,\qquad \sum_{\mu=-(n-1)}^{n-1}\mu A_\mu\kappa^{-\mu}\equiv 0\quad\text{identically in } t. \tag{49}$$

These conditions guarantee also that $A_{-\nu}\equiv\bar{A}_\nu$ is fulfilled for all values of t.

We observe that the equations (48) for $\nu=-1,-2,\ldots,-(n-1)$ give $(n-1)$ differential equations for the corresponding functions $A_\nu(t)$; their coefficients depend in a very simple manner on $\kappa(t)$. The function $\kappa(t)$, in turn, can be determined from the $A_\nu(t)$ by means of the second equation (49), which can be written in the form

$$\operatorname{Im}\left\{\sum_{\mu=-(n-1)}^{-1}\mu A_\mu\kappa^{-\mu}\right\}=0. \tag{50}$$

Thus, $A_{-1},A_{-2},\ldots,A_{-n-1}$ and κ satisfy a well-determined system of ordinary differential equations.

Let us start with the case $n=3$. The differential system to be considered is

$$\left.\begin{aligned}\frac{dA_{-2}(t)}{dt}=-2A_{-2}(t),\quad \frac{dA_{-1}(t)}{dt}=-A_{-1}(t),\\ \operatorname{Im}\{2A_{-2}\kappa^2+A_{-1}\kappa\}=0.\end{aligned}\right\} \tag{51}$$

We can integrate immediately and find

$$A_{-2}(t)=\alpha_2e^{-2t},\quad A_{-1}(t)=\alpha_1e^{-t}. \tag{52}$$

Since for $t=0$ the function $q(z,0)$ coincides with the right side of (40) for $n=3$, we determine the constants of integration as follows: $\alpha_2=1$, $\alpha_1=2a_2$. Thus, $\kappa(t)$ satisfies the equation

$$e^{-2t}\kappa^2+a_2e^{-t}\kappa=\text{real}. \tag{53}$$

From the general Löwner theory it is well-known that

$$a_2 = -2\int_0^\infty \kappa\, e^{-t}\, dt. \tag{54}$$

We have to utilize now the inequality $|a_3 - a_2^2| \leqslant 1$ which follows from the elementary area theorem. Since we assume $a_3 \geqslant 3$ we can assert $\mathrm{Re}\{a_2^2\} \geqslant 2$ and see that the left side of (53) cannot vanish for $0 \leqslant t < \infty$.

We wish to show next that a_2^2 = real in consequence of (53) and (54). Indeed, if a_2^2 were not real, equation (53) would exclude the possibility $\kappa = \pm \operatorname{sgn} a_2$ and the expression $\mathrm{Im}\{\overline{\kappa(t)}\, a_2\}$ could never change its sign. Consequently

$$\int_0^\infty \mathrm{Im}\{\overline{2\kappa(t)}\, a_2\}\, e^{-t} dt = -\mathrm{Im}\{|a_2|^2\} \tag{55}$$

could not be zero, which yields a contradiction. Thus, a_2^2 = real and in consequence of the area theorem even $a_2^2 > 2$ holds; hence, we conclude a_2 = real. From (53) follows then easily that κ must be real throughout and it can be shown that $a_3 = 3$.

The above proof for $|a_3| \leqslant 3$ is more complicated than Löwner's original proof which made use only of the formula (45). It can, however, be generalized to the problem of a_4 though it becomes in this case still more complicated. The differential system becomes now

$$\left.\begin{aligned} \frac{dA_{-3}}{dt} = -3A_{-3},\quad \frac{dA_{-2}}{dt} = -2A_{-2} - 2A_{-3}\kappa,\quad \frac{dA_{-1}}{dt} = -A_{-1} + 2A_{-3}\kappa^2, \\ \mathrm{Im}\{3A_{-3}\kappa^3 + 2A_{-2}\kappa^2 + A_{-1}\kappa\} = 0. \end{aligned}\right\} \tag{56}$$

We find $A_{-3} = \alpha_3 e^{-3t}$ and, since $A_{-3}(0) = 1$, we have $A_{-3} = e^{-3t}$. We set up

$$A_{-2}(t) = \alpha_2(e^{-t})\, e^{-2t}, \quad A_{-1}(t) = \alpha_1(e^{-t})\, e^{-t}; \tag{57}$$

inserting into (56) and putting $\sigma = e^{-t}$, we arrive at the differential system

$$\left.\begin{aligned} \frac{d\alpha_2(\sigma)}{d\sigma} = 2\kappa,\quad \frac{d\alpha_1(\sigma)}{d\sigma} = -2\kappa^2\sigma \quad (0 \leqslant \sigma \leqslant 1), \\ \mathrm{Im}\{3\kappa^3\sigma^3 + 2\alpha_2(\sigma)\,\kappa^2\sigma^2 + \alpha_1(\sigma)\,\kappa\sigma\} = 0. \end{aligned}\right\} \tag{58}$$

A simple calculation leads to the boundary conditions

$$\left.\begin{aligned} \alpha_2(0) = 3a_2,\quad \alpha_1(0) = 2a_3 + a_2^2, \\ \alpha_2(1) = 2a_2,\quad \alpha_1(1) = 3a_3. \end{aligned}\right\} \tag{59}$$

Those for $\sigma = 1$, $t = 0$ are obvious; those for $\sigma = 0$, $t = \infty$ follow by comparison of coefficients of powers of e^{-t} in (43) and by passage to the limit $t = \infty$.

The differential system (58), together with the boundary conditions (59), represents a typical Sturm–Liouville boundary value problem. We have to start integration of (58) with such initial values $\alpha_1(0)$ and $\alpha_2(0)$ that we end up at the other end of the interval considered with

$$\alpha_1(1) = \tfrac{3}{2}[\alpha_1(0) - \tfrac{1}{9}\alpha_2(0)^2], \quad \alpha_2(1) = \tfrac{2}{3}\alpha_2(0). \tag{60}$$

The difficulty of the problem lies in the non-linear character of the equations and of the boundary conditions. Each possible set $\alpha_1(0)$, $\alpha_2(0)$ determines a set of possible values a_2, a_3. Clearly, $a_2 = 2$, $a_3 = 3$ and $\kappa(\sigma) \equiv -1$ is an admissible solution which leads to the Koebe function (38), the conjectured extremum function.

The question arises now whether the corresponding special values $\alpha_1(0)$, $\alpha_2(0)$ connected with the conjectured extremum function might not be imbedded into a one-parameter family of initial values such that all of them lead to the boundary relations (60). For this purpose, we have to study the variational equations of the system (58) and of the boundary conditions (60). If we denote the derivatives of α_1, α_2 and κ with respect to the parameter by β_1, β_2 and $i\lambda$, we find easily

$$\left.\begin{aligned} &\frac{d\beta_1}{d\sigma} = 4i\lambda\sigma, \quad \frac{d\beta_2}{d\sigma} = 2i\lambda, \quad \lambda = \frac{1}{2p(\sigma)}\operatorname{Im}\{\beta_1 - 2\beta_2\sigma\}, \\ &\beta_1(1) = \tfrac{3}{2}[\beta_1(0) - \tfrac{4}{3}\beta_2(0)], \quad \beta_2(1) = \tfrac{2}{3}\beta_2(0); \quad p(\sigma) = 8\sigma^2 - 12\sigma + 5. \end{aligned}\right\} \tag{61}$$

We are thus led to a linear differential system with linear boundary conditions which can be treated by the standard Sturm–Liouville methods.

It is immediately seen from (61) that λ is real and that $\beta_1(\sigma)$ and $\beta_2(\sigma)$ must be pure imaginary. When we introduce the new unknowns

$$u(\sigma) = \operatorname{Im}\{\beta_1 - 2\sigma\beta_2\}, \quad v(\sigma) = \operatorname{Im}\{\beta_2\}$$

the system (61) simplifies to

$$\frac{du}{d\sigma} = -2v, \quad \frac{dv}{d\sigma} = \frac{1}{p(\sigma)}u, \tag{62}$$

with the boundary conditions

$$u(1) = \tfrac{3}{2}u(0) - \tfrac{10}{3}v(0), \quad v(1) = \tfrac{2}{3}v(0). \tag{63}$$

From the differential system we derive by integration by parts the equality

$$2\int_0^1 v^2 d\sigma = \int_0^1 pv'^2 d\sigma + \tfrac{20}{9}v(0)^2; \quad v(1) = \tfrac{2}{3}v(0). \tag{64}$$

We may now apply the calculus of variations in order to estimate the ratio

$$\left[\int_0^1 pv'^2 d\sigma + \tfrac{20}{9}v(0)^2\right] \cdot \left[\int_0^1 v^2 d\sigma\right]^{-1} = R[v] \tag{65}$$

under the given boundary condition on $v(\sigma)$. Even when we replace in (65) the polynomial $p(\sigma)$ by a piecewise constant function which is nowhere larger than $p(\sigma)$ the minimum value of the new ratio, which can now be computed explicitly, comes out to be larger than 2. Hence, *a fortiori*, we can assert that $R[v] > 2$ for all admissible $v(\sigma)$ and that (64) is impossible. We have thus shown that the solution $a_2 = 2$, $a_3 = 3$, $\kappa(\sigma) \equiv -1$ cannot be imbedded into a one-parameter family of solutions which can be differentiated continuously with respect to this parameter.

By a more careful analysis we may now treat differences of solution systems $\alpha_1(\sigma)$, $\alpha_2(\sigma)$, $\kappa(\sigma)$ instead of differentials. We can then delimit an entire neighborhood of the point $a_2 = 2$, $a_3 = 3$, $\kappa = -1$ in which no other solution point could be located. On the other hand, one can combine the area theorem with various relations between the coefficients of the extremum function which arise from the differential equation (40), in order to estimate the values $|a_2 - 2|$ and $|a_3 - 3|$ in the extremum case. It can be seen by elementary if very tedious calculations that the point a_2, a_3, $\kappa(1)$ must lie precisely in the neighborhood in which 2, 3, -1 is the only solution point. This proves that the Koebe function (38) is, indeed, the extremum function and establishes the inequality $|a_4| \leqslant 4$ for all univalent functions (36).

The actual labor in the proof sketched here lies in the very extensive elementary estimations and could probably be reduced considerably by extending the uniqueness neighborhood through greater attention to the theory of the differential system (58), (59).

It may be remarked, finally, that the Koebe function (38) satisfies the functional equation (40) which characterizes the extremum function for every $n \geqslant 2$. This fact tends, of course, to strengthen the evidence for the Bieberbach conjecture. The following fact should be mentioned, however, in order to caution against too great reliance on this evidence. One may consider the family of functions

$$f(z) = z + b_0 + \frac{b_1}{z} + \frac{b_2}{z^2} + \ldots + \frac{b_n}{z^n} + \ldots, \tag{66}$$

which are univalent in the outside $|z| > 1$ of the unit circle and one may ask for $\max |b_n|$. The same variational technique as above yields for the extremum functions $f_n(z)$ of this 'exterior' problem a differential-functional equation which is analogous to (40). It is easy to show that the functions

$$F_n(z) = \left[z^{n+1} + 2 + \frac{1}{z^{n+1}}\right]^{1/(n+1)} = z + \frac{2}{n+1}\frac{1}{z^n} + \dots \qquad (67)$$

belong to the family considered and satisfy the extremum condition for the corresponding $f_n(z)$. For $n = 1$ and $n = 2$ these functions are, indeed, the extremum functions of the exterior coefficient problem. The estimate $|b_1| \leqslant 1$ was discovered together with the area theorem[2] and $|b_2| \leqslant \frac{2}{3}$ was established in 1938 by Golusin[6] and myself[25]. It was conjectured that $|b_n| \leqslant 2/(n+1)$ was the best possible estimate for the nth coefficient for all values of n. However, in 1955 Garabedian and I[4] showed that the precise value of the maximum for $|b_3|$ is not $\frac{1}{2}$ as expected but $\frac{1}{2}+e^{-6}$. Thus, in spite of the fact that the function $F_3(z)$, defined in (67), satisfies the rather restrictive extremum condition, it is not the extremum function $f_3(z)$. Since e^{-6} is a small number, this example shows also how little empirical numerical evidence can be trusted in problems of this kind. Recently, Waadeland[36] has shown that quite generally

$$\max |b_{2k-1}| \geqslant \frac{1}{k}(1 + 2e^{-2[(k+1)/(k-1)]}), \qquad (67')$$

while for $n = 2k$ no counter example to the conjecture $|b_n| \leqslant 2/(n+1)$ seems to be known.

There are, of course, numerous cross-relations between the coefficient problem for univalent functions and the general theory of conformal mapping. Two examples may serve as illustrations. There is a well-known problem in the theory of conformal mapping: given n points in the complex plane, to find a continuum which contains these points and has minimum capacity[10]. From the topology of the extremum continuum, one can derive by an elementary variation the coefficient inequality $|b_2| \leqslant \frac{2}{3}$[25]. Here, the general theory of conformal mapping helped to solve a coefficient problem. Conversely, de Possel[21] formulated a simple extremum problem for the coefficients of univalent functions in a multiply-connected domain and showed that the extremum functions mapped the domain onto a parallel slit domain. Since the existence of an extremum function is assured, an elegant existence proof for an important canonical mapping was thus established.

5. Fredholm eigenvalues

The problem of conformally mapping a given plane domain D can often be reduced to a boundary value problem for the functions harmonic in D. If the boundary C of D is sufficiently smooth, the latter problem can be attacked through the Poincaré–Fredholm integral equation

$$m(z) = \mu(z) + \frac{1}{\pi}\int_C \frac{\partial}{\partial n_\zeta}\left(\log\frac{1}{|z-\zeta|}\right)\mu(\zeta)\,ds_\zeta \quad (z \in C). \tag{68}$$

In order to solve this fundamental integral equation of two-dimensional potential theory one has to consider the corresponding homogeneous integral equation

$$\phi_\nu(z) = \frac{\lambda_\nu}{\pi}\int_C \frac{\partial}{\partial n_\zeta}\left(\log\frac{1}{|z-\zeta|}\right)\phi_\nu(\zeta)\,ds_\zeta \quad (z \in C), \tag{69}$$

its eigenfunctions $\phi_\nu(z)$ and its eigenvalues λ_ν. The eigenvalue $\lambda = 1$ occurs always and has as eigenfunctions a set of easily described functions on C; we shall call this eigenvalue the trivial eigenvalue of the domain. The non-trivial eigenvalues λ_ν satisfy the inequality $|\lambda_\nu| > 1$. It is easily seen that with each non-trivial eigenvalue λ_ν also the value $-\lambda_\nu$ will occur as eigenvalue of (69) with the same multiplicity. We shall restrict ourselves, therefore, to the positive non-trivial eigenvalues λ_ν and assume them ordered in increasing magnitude. These eigenvalues λ_ν are called the Fredholm eigenvalues of the domain D and they are of importance for the potential theory and the function theory of the domain considered.

It is, for example, of great interest to obtain a lower bound for the first eigenvalue λ_1 of a given domain. Such information would enable us to estimate the speed of convergence of the Neumann–Liouville series which solves the basic equation (68). The larger λ_1 can be asserted to be, the easier the numerical work for the solution of the boundary value problems in the potential theory for D. Thus, the λ_ν seem to be a set of functionals of D which deserves a careful study.

The λ_ν are also closely related to the theory of the Hilbert transformation

$$F(z) = \frac{1}{\pi}\iint_D \frac{\overline{f(\zeta)}}{(\zeta - z)^2}\,d\tau_\zeta, \tag{70}$$

which carries each analytic function in D into a new analytic function in the same domain. There exists a set of eigenfunctions $w_\nu(z)$ which are analytic in D and which satisfy

$$w_\nu(z) = \frac{\lambda_\nu}{\pi}\iint_D \frac{\overline{w_\nu(\zeta)}}{(\zeta - z)^2}\,d\tau_\zeta \quad (\lambda_\nu > 1). \tag{71}$$

15-2

The eigenvalues λ_ν are precisely the Fredholm eigenvalues defined above. We shall assume the $w_\nu(z)$ to be normalized by the usual convention

$$\iint_D |w_\nu(z)|^2 \, d\tau = 1. \tag{72}$$

The eigenfunctions $w_\nu(z)$ form an orthonormal set of analytic functions in D and play an interesting role in the theory of the kernel function of D[33].

In order to establish a unified theory for the treatment of extremum problems for the functionals λ_ν of D it is necessary to determine the variation of each λ_ν for a variation of the defining domain D. If we assume the variation to be of the special type (1) with $z_0 \in D$ and if λ_ν is non-degenerate, we have

$$\lambda_\nu^* = \lambda_\nu + (1-\lambda_\nu^2)\,\pi\,\mathrm{Re}\{e^{i\alpha}\rho^2 w_\nu(z_0)^2\} + O(\rho^4). \tag{73}$$

An analogous, though slightly more complicated, formula can be given for the variation of degenerate eigenvalues.

When one wishes to apply the variational formula (73) to the solution of extremum problems, one runs immediately into a serious difficulty. The entire theory of the Fredholm eigenvalues has been established under certain smoothness conditions for the boundary and one has to be sure that the extremum domain does possess a boundary of this type. One has to introduce a class of domains which possess admissible boundaries and which is compact; within such a class the calculus of variations based on (73) and the theory of extremum problems become possible.

For this purpose, we introduce the concept of *uniformly analytic curves*. A curve is called analytic if it can be obtained as the image of the unit circumference $|z| = 1$ by means of a function $t(z)$ which is analytic and univalent on $|z| = 1$. A set of curves is said to be uniformly analytic with the modulus of uniformity (r, R) (where $r < 1 < R$) if all of them are obtained by means of mapping functions $f(z)$ which are analytic and univalent in the fixed annulus $r \leqslant |z| \leqslant R$. This concept of uniform analyticity seems to be quite useful in the variational theory of domain functionals.

We can now formulate the theorem:

If a simply connected domain is bounded by a curve which is analytic with the modulus (r, R), *then its lowest Fredholm eigenvalue* λ_1 *satisfies the inequality:*

$$\lambda_1 \geqslant \frac{r^2 + R^2}{1 + r^2 R^2}. \tag{74}$$

This estimate is the best possible for every modulus (r, R).

Frequently, the boundary curve C of a domain is given in a parametric representation from which the modulus (r, R) can be readily deduced. Thus, the estimate (74) is often convenient to predict the convergence of the Neumann–Liouville series which solve the various boundary value problems in the domain.

We may also connect with a given domain D the Fredholm determinant

$$D(\lambda) = \prod_{\nu=1}^{\infty} \left(1 - \frac{\lambda^2}{\lambda_\nu^2}\right) \tag{75}$$

of the integral equation (68) and consider, for fixed λ, $D(\lambda)$ as a functional of the domain D. The following extremum problem suggests itself: Let D_0 be a given multiply-connected domain; consider all smoothly bounded domains D which are conformally equivalent to it and ask for those domains in this equivalence class which yield the maximum value of $D(\lambda)$.

This problem has been solved in the case $\lambda = 1$. The main difficulty in the investigation was again the non-compactness of the class of domains considered. It could be overcome by considering maximum sequences of domains and their limit domain; all domains of the sequence were subjected to the same variation (1) and from the fact that they formed a maximum sequence it could be shown that their limit domain is analytically bounded. Then, the existence of a maximum domain is easily established and it can be shown that it is bounded by circumferences. We obtain thus a new proof of Schottky's famous circular mapping theorem and also a characterization of this canonical mapping by an extremum property. Methodologically, the proof is of interest since the method of variation is not applied to the extremum domain, whose existence is not yet known, but to the extremum sequence. This procedure seems to be of very great applicability.

The solution of the maximum problem for general $D(\lambda)$ is not yet known and well deserves additional study.

The Fredholm eigenvalues represent an instructive example for the flexibility of the variational method in dealing with extremum problems for rather difficult types of domain functionals. The great formal elegance of the variational formula (73) enabled us to overcome the quite serious difficulties which arise from the fact that these functionals are defined only for a restricted and non-compact class of domains.

6. Further applications

We have restricted ourselves to a few fundamental problems in order to exhibit clearly the basic ideas of the variational method. It may, however, be applied to much more general function-theoretic problems. It can be used in problems of mapping of domains on Riemann surfaces[29, 33] and leads there to existence theorems for various canonical realizations of Riemann domains. It can be applied to the theory of multivalent functions in a given domain[27], their coefficient problems and distortion theorems. Some interest has been devoted to the problem of developing a calculus of variations within important subclasses of the family of univalent functions in the unit circle. Golusin[9] described a method of variations for the subclass of star-like univalent mappings and Hummel[12, 13] gave an even simpler method of this kind. Singh[34] gave a theory of variations for real univalent functions, for bounded univalent functions and other interesting subclasses. Finally, the role should be mentioned which the method of variations could play as a useful tool in the theory of quasi-conformal mappings and of extremal metrics[14].

The variational method is, of course, only one of many powerful methods in the theory of conformal mapping and complex function theory. There are many problems where other methods give the answer more easily and directly. It seems to me, however, that the method of variations is one of the most systematic and widely applicable methods which we possess in this field.

REFERENCES

[1] Bernardi, S. D. A survey of the development of the theory of schlicht functions. *Duke Math. J.* 19, 263–287 (1952).

[2] Bieberbach, L. Über die Koeffizienten derjenigen Potenzreihen, welche eine schlichte Abbildung des Einheitskreises vermitteln. *Berl. Ber.* 940–955 (1916).

[3] Courant, R. *Dirichlet's Principle, Conformal Mapping and Minimal Surfaces.* New York, 1950. Appendix by M. Schiffer.

[4] Garabedian, P. R. and Schiffer, M. A coefficient inequality for schlicht functions. *Ann. Math.* (2), 61, 116–136 (1955).

[5] Garabedian, P. R. and Schiffer, M. A proof of the Bieberbach conjecture for the fourth coefficient. *J. Rat. Mech. Anal.* 4, 427–465 (1955).

[6] Golusin, G. M. Einige Koeffizientenabschätzungen für schlichte Funktionen. *Rec. Math.* (*Mat. Sbornik*), 3 (45), 321–330 (1938).

[7] Golusin, G. M. Method of variations in the theory of conformal mapping. *Rec. Math.* (*Mat. Sbornik*), 19 (61), 203–236 (1946); 21 (63), 83–117, 119–132 (1947); 29 (71), 455–468 (1951).

[8] Golusin, G. M. *Geometrical Theory of Functions of a Complex Variable.* Moscow, 1952. (German translation: Berlin, 1957.)

[9] Golusin, G. M. A variational method in the theory of analytic functions. *Leningrad Gos. Univ. Uč. Zap.* 144, *Ser. Mat. Nauk*, 23, 85–101 (1952).

[10] Grötzsch, H. Über ein Variationsproblem der konformen Abbildung. *Leipzig. Ber.* 82, 251–263 (1930).

[11] Hadamard, J. Mémoire sur le problème relatif à l'équilibre des plaques élastiques encastrées. *Acad. Sci. Paris, Mém. Sav. étrangers*, 33 (1908).

[12] Hummel, J. A. The coefficient regions of starlike functions. *Pacif. J. Math.* 7, 1381–1389 (1957).

[13] Hummel, J. A. A variational method for starlike functions. *Proc. Amer. Math. Soc.* 9, 82–87 (1958).

[14] Jenkins, J. A. On the existence of certain general extremal metrics. *Ann. Math.* (2), 66, 440–453 (1957).

[15] Julia, G. Sur une équation aux dérivées fonctionnelles liée à la représentation conforme. *Ann. Éc. Normale* (3), 39, 1–28 (1922).

[16] Lavrentieff, M. A. Sur deux questions extrémales. *Rec. Math.* (*Mat. Sbornik*), 41, 157–165 (1934).

[17] Lavrentieff, M. A. Über eine extremale Aufgabe aus der Tragflügeltheorie. *Cent. Aero-Hydrodyn. Inst.* no. 155, 1–40 (1934).

[18] Löwner, K. Untersuchungen über schlichte konforme Abbildungen des Einheitskreises. *Math. Ann.* 89, 103–121 (1923).

[19] Marty, F. Sur le module des coefficients de Maclaurin d'une fonction univalente. *C.R. Acad. Sci., Paris*, 198, 1569–1571 (1934).

[20] Montel, P. *Leçons sur les fonctions univalentes ou multivalentes.* Paris, 1933.

[21] Possel, R. de. Zum Parallelschlitztheorem unendlichvielfach zusammenhängender Gebiete. *Göttinger Nachr.* 199–202 (1931).

[22] Schaeffer, A. C. and Spencer, D. C. The coefficients of schlicht functions. I. *Duke Math. J.* 10, 611–635 (1943); II. *Duke Math. J.* 12, 107–125 (1945); III. *Proc. Nat. Acad. Sci., Wash.*, 32, 111–116 (1946).

[23] Schaeffer, A. C. and Spencer, D. C. A variational method in conformal mapping. *Duke Math. J.* 14, 949–966 (1947).

[24] Schaeffer, A. C. and Spencer, D. C. *Coefficient Regions for Schlicht Functions.* Amer. Math. Soc. Colloquium Publ., vol. 35 (1950).

[25] Schiffer, M. Sur un problème d'extrémum de la représentation conforme. *Bull. Soc. Math. Fr.* 66, 48–55 (1938).

[26] Schiffer, M. A method of variation within the family of simple functions. *Proc. Lond. Math. Soc.* (2), 44, 432–449 (1938).

[27] Schiffer, M. Variation of the Green function and theory of the p-valued functions. *Amer. J. Math.* 65, 341–360 (1943).

[28] Schiffer, M. Hadamard's formula and variation of domain functions. *Amer. J. Math.* 68, 417–448 (1946).

[29] Schiffer, M. Variational methods in the theory of Riemann surfaces. *Contributions to the Theory of Riemann Surfaces*, pp. 15–30. Princeton, 1953.

[30] Schiffer, M. Variation of domain functionals. *Bull. Amer. Math. Soc.* 60, 303–328 (1954).

[31] Schiffer, M. Application of variational methods in the theory of conformal mappings. *Proc. Symp. appl. Math.* 8, 93–113 (1958).

[32] Schiffer, M. The Fredholm eigenvalues of plane domains. *Pacif. J. Math.* 7, 1187–1225 (1957).

[33] Schiffer, M. and Spencer, D. C. *Functionals of Finite Riemann Surfaces.* Princeton, 1954.

[34] Singh, V. Interior variations and some extremal problems for certain classes of univalent functions. *Pacif. J. Math.* 7, 1485–1504 (1957).

[35] Spencer, D. C. Some problems in conformal mapping. *Bull. Amer. Math. Soc.* 53, 417–439 (1947).

[36] Waadeland, H. Über ein Koeffizientenproblem für schlichte Abbildungen des $|\zeta| > 1$. *Norske Vidensk. Selsk. Forh.* 30, 168–170 (1957).

Commentary on

[40] *Variational methods in the theory of conformal mapping*, Proceedings of the International Congress of Mathematicians, Cambridge, Mass., 1950; American Mathematical Society, Providence, R.I., 1952, Vol. 2, pp. 233–240.

[70] *Extremum problems and variational methods in conformal mapping*, Proceedings of the International Congress of Mathematicians, Edinburgh, 1958; Cambridge University Press, New York, 1960, pp. 211–231.

Max Schiffer was twice honored by invitations to address the International Congress of Mathematicians, at Cambridge (Massachusetts) in 1950 and at Edinburgh in 1958. The articles [40, 70] are written versions of his lectures. Both of the articles are devoted to the variational methods for which Schiffer was best known, but they are rather different in character. The paper [40] offers a broad discussion of variational methods and their relative merits, and it surveys a variety of applications. The paper [70], on the other hand, focuses more narrowly on applications to particular extremal problems in function theory such as coefficient problems, and it gives a fairly detailed account of technical advances in the use of variational methods. In this respect, [70] is somewhat dated, since the Bieberbach conjecture has now been proved (see commentaries on [5, 6, 72]) and the conjecture $|b_n| \leq \frac{2}{n+1}$ for functions of class Σ has been disproved for all $n \geq 3$ (see commentaries on [4, 59, 123]). The paper [70] also gives an extended description of Schiffer's method for obtaining a lower bound for the first nontrivial Fredholm eigenvalue of a simply connected domain with analytic boundary curve. Schiffer and others returned repeatedly to this problem (see Kühnau's commentary on [62, 68, 78, 125]). The lower bound is important numerically, because it allows an estimate on the rate at which the Neumann series converges to the solution of the classical Poincaré–Fredholm integral equation associated with the solution of a Dirichlet problem. In his expository paper [78], Schiffer gives a clear and detailed account of this beautiful circle of ideas.

PETER DUREN

[72] (with Z. Charzyński) A new proof of the Bieberbach conjecture for the fourth coefficient

[72] (with Z. Charzyński) A new proof of the Bieberbach conjecture for the fourth coefficient. *Arch. Rational Mech. Anal.* **5** (1960), 187–193.

A New Proof of the Bieberbach Conjecture for the Fourth Coefficient

Z. CHARZYNSKI & M. SCHIFFER

Introduction

In 1955 GARABEDIAN & SCHIFFER [1] proved that the fourth coefficient in the Taylor series of a function $f(z)$, univalent in the unit circle, satisfies the inequality $|a_4| \leqq 4$. The argument consisted in two basic assertions: a) It was shown by means of the calculus of variations and by use of LOEWNER'S differential equation that the problem of finding an extremum function for the Bieberbach problem is equivalent to a nonlinear boundary-value problem for a system of ordinary differential equations. It was proved also that the Koebe function led to a solution of this boundary value problem and that no other solution could lie in a very small neighborhood of this particular solution. b) There remained then the task of showing that outside of this critical neighborhood there could exist no solution leading to an extremum function of the Bieberbach problem. This was achieved by a laborious and complicated set of estimates and inequalities.

While the approach of that paper shows in principle a general method of attack upon coefficient problems for schlicht functions, the amount of labor involved in this relatively simple question seemed discouraging, and it was doubtful whether higher coefficients should be dealt with in a corresponding manner. The purpose of the present paper is to give a great simplification of the proof that $|a_4| \leqq 4$ and to show at the same time that more elementary methods are sufficient to establish the theorem. We start from an inequality implicit in a paper of GRUNSKY [2] of 1939. This inequality can be derived by elementary means, using the Schwarz inequality and standard methods of contour integration. Below we give a new proof of the Grunsky inequality based on variational methods, both in order to make this paper self-consistent, and also since the new proof seems somewhat shorter than GRUNSKY'S. No use of LOEWNER'S differential equation is necessary. We hope that our present proof for the Bieberbach conjecture for a_4 will encourage the attack upon higher coefficients and stimulate new interest in this beautiful field of function theory.

§ 1. The fundamental inequalities

1. We shall be dealing with extremum problems within the family $\mathfrak{S}$ of all functions $f(z)$ which are regular and univalent for $|z| < 1$ and which possess a Taylor series

$$f(z) = z + a_2 z^2 + a_3 z^3 + \cdots + a_n z^n + \cdots. \tag{1}$$

Let $w = f(z)$ yield the image domain D of the unit circle, and let w_0 be an arbitrary point in the w-plane but not in D. There exist for arbitrarily small values of $\varrho > 0$ infinitely many members of $\mathfrak{S}$ and of the form [3]

$$f^*(z) = f(z) + \frac{a\,\varrho^2 f(z)^2}{(f(z) - w_0)\, w_0} + O(\varrho^3), \qquad |a| < 1, \tag{2}$$

where the correction term $O(\varrho^3)$ can be estimated uniformly in each closed subdomain $|z| \leq r < 1$. These functions $f^*(z)$ may then serve as comparison functions with the original member $f(z)$ in variational problems regarding the family $\mathfrak{S}$.

We shall be interested in maximum problems concerning the first three coefficients a_2, a_3, and a_4 in (1). Therefore it is necessary to express the change of these coefficients under the variation (2). An easy calculation yields

$$\begin{aligned} a_2^* &= a_2 - \frac{a\,\varrho^2}{w_0^3} + O(\varrho^3) \\ a_3^* &= a_3 - \frac{a\,\varrho^2}{w_0^3}\left(2a_2 + \frac{1}{w_0}\right) + O(\varrho^3) \\ a_4^* &= a_4 - \frac{a\,\varrho^2}{w_0^3}\left(2a_3 + a_2^2 + \frac{3a_2}{w_0} + \frac{1}{w_0^2}\right) + O(\varrho^3). \end{aligned} \tag{3}$$

Now let $F = F(a_2, a_3, a_4)$ be an analytic function of the three complex variables a_2, a_3, and a_4. F is then a functional of $f \in \mathfrak{S}$ with the corresponding variational formula

$$\begin{aligned} F^* = F - \frac{a\,\varrho^2}{w_0^3}\Big\{ & \frac{\partial F}{\partial a_4}(2a_3 + a_2^2) + \frac{\partial F}{\partial a_3} 2a_2 + \frac{\partial F}{\partial a_2} + \\ & + \left(\frac{\partial F}{\partial a_4} 3a_2 + \frac{\partial F}{\partial a_3}\right)\frac{1}{w_0} + \frac{\partial F}{\partial a_4} \cdot \frac{1}{w_0^2}\Big\} + O(\varrho^3). \end{aligned} \tag{4}$$

Having chosen a fixed functional $F(a_2, a_3, a_4)$, we may ask for the solution of the extremum problem

$$\operatorname{Re}\{F\} = \max \qquad \text{within the family } \mathfrak{S}. \tag{5}$$

The existence of extremizing functions in $\mathfrak{S}$ follows from the compactness of the family. We can characterize the extremum functions $f(z)$ by the fact that under all variations (2) the inequality $\operatorname{Re}\{F^*\} \leq \operatorname{Re}\{F\}$ must hold. In other words,

$$\begin{aligned} &\operatorname{Re}\left\{\frac{a\,\varrho^2}{w_0^3}\left[A + \frac{B}{w_0} + \frac{C}{w_0^2}\right]\right\} \geq O(\varrho^3) \\ &A = (2a_3 + a_2^2)\frac{\partial F}{\partial a_4} + 2a_2\frac{\partial F}{\partial a_3} + \frac{\partial F}{\partial a_2}, \\ &B = 3a_2 \frac{\partial F}{\partial a_4} + \frac{\partial F}{\partial a_3}, \qquad C = \frac{\partial F}{\partial a_4} \end{aligned} \tag{6}$$

must hold for every admissible variation (2).

The basic lemma of the calculus of variations within the class $\mathfrak{S}$ leads now to the conclusion that D is a slit domain bounded by analytic arcs Γ with a parametric representation $w = w(\tau)$ satisfying the differential equation

$$\left(\frac{dw}{d\tau}\right)^2 \cdot \frac{1}{w^3}\left[A + \frac{B}{w} + \frac{C}{w^2}\right] < 0. \tag{7}$$

So far, every extremum problem (5) can be brought within the characterization (7). It is, however, extremely difficult to draw decisive conclusions from (7) and to determine precisely the extremum function sought. Indeed, the coefficient problem of univalent functions is contained in this particular problem. Since it has not been possible to find a satisfactory general solution of (7), we may now ask for particular functionals F for which the complete answer can be given. It is clear from the form of (7) that the integration of the differential equation will be simplified if the term in brackets is a perfect square. We ask, therefore, for those functions $F(a_2, a_3, a_4)$ for which

$$(8)\quad B^2 = 4AC,\ \text{i.e.,}\ \left(3a^2\frac{\partial F}{\partial a_4} + \frac{\partial F}{\partial a_3}\right)^2 = 4\frac{\partial F}{\partial a_4}\left[(2a_3 + a_2^2)\frac{\partial F}{\partial a_4} + 2a_2\frac{\partial F}{\partial a_3} + \frac{\partial F}{\partial a_2}\right].$$

This condition is a simple first order partial differential equation for F. We can give immediately a particular integral of it which is in polynomial form and which contains an arbitrary constant of integration l. Namely,

$$(9)\qquad F(a_2, a_3, a_4) = a_4 - 2a_2a_3 + \tfrac{13}{12}a_2^3 + 2l(a_3 - \tfrac{3}{4}a_2^2) + l^2 a_2.$$

Conversely, if we start with the functional (9) and ask for the solution of the extremum problem (5), we arrive at the differential equation for the boundary curve Γ:

$$(10)\qquad \left(\frac{dw}{d\tau}\right)^2 \cdot \frac{1}{w^3}\left[l + \frac{1}{2}a_2 + \frac{1}{w}\right]^2 < 0.$$

Let $w = f(e^{i\tau})$ be a particular parametric representation for Γ. Now

$$(11)\qquad \frac{dw}{d\tau} = i e^{i\tau} f'(e^{i\tau}).$$

Hence, if we denote

$$(12)\qquad \frac{z^2 f'(z)^2}{f(z)^3}\left[l + \frac{1}{2}a_2 + \frac{1}{f(z)}\right]^2 = q(z),$$

we obtain in $q(z)$ an analytic function of z in the unit circle which is regular there except for a pole of order 3 at the origin and which has non-negative values on the boundary. Hence, by SCHWARZ' reflection principle, $q(z)$ is rational in the entire z-plane and has the form

$$(13)\qquad q(z) = \frac{1}{z^3} + \frac{\alpha}{z^2} + \frac{\beta}{z} + \gamma + \bar{\beta}z + \bar{\alpha}z^2 + z^3, \qquad \gamma \text{ real.}$$

We can easily factorize $q(z)$. It can vanish for $|z| < 1$ only at the point z_0 for which

$$(14)\qquad f(z_0)\,(l + \tfrac{1}{2}a_2) = -1.$$

If we put $z_0 = \varrho e^{i\varphi}$, the point $\bar{z}_0^{-1} = \frac{1}{\varrho}e^{i\varphi}$ will likewise be a zero of $q(z)$. Obviously, both points are double zeros of $q(z)$. Each root of $q(z)$ on the unit circle must be double, since $q(z)$ does not change sign on this curve. Hence

$$(15)\qquad z^3 q(z) = \hat{A}^2(z - e^{i\psi})^2(z - \varrho e^{i\varphi})^2\left(z - \frac{1}{\varrho}e^{i\varphi}\right)^2$$

if a root z_0 of (14) lies in the unit circle, or

$$(15') \qquad z^3 q(z) = A^2 (z - e^{i\varphi})^2 (z - e^{i\varphi_1})^2 (z - e^{i\varphi_2})^2$$

if all roots of $q(z)$ lie on $|z| = 1$.

In any case, in view of (13) we can assert that

$$(16) \qquad \sqrt{q(z)} = \frac{\varkappa z^3 + \mu z^2 + \nu z + 1}{z^{\frac{3}{2}}}$$

and that this expression is real for $|z| = 1$. Hence we may conclude that $\varkappa = 1$, $\mu = \bar{\nu}$, and we may put (12) into the form

$$(17) \qquad \frac{z^{\frac{3}{2}} f'(z)}{f(z)^{\frac{3}{2}}} \left[\left(l + \frac{1}{2} a_2 \right) f(z) + 1 \right] = z^3 + \mu z^2 + \bar{\mu} z + 1 .$$

We insert into the left-hand side of (17) the series development (1) of $f(z)$ and compute the development of this term at the origin. We find that

$$(18) \qquad \begin{aligned} \frac{z^{\frac{3}{2}}}{f(z)^{\frac{3}{2}}} f'(z) \left[1 + \left(l + \frac{1}{2} a_2 \right) f(z) \right] &= 1 + l z + \left(\frac{1}{2} a_3 - \frac{3}{8} a_2^2 + \frac{1}{2} l a_2 \right) z^2 + \\ &+ \left[\frac{3}{2} a_4 - 3 a_2 a_3 + \frac{13}{8} a_2^3 + l \left(\frac{3}{2} a_3 - \frac{9}{8} a_2^2 \right) \right] z^3 + \cdots . \end{aligned}$$

Comparing coefficients in (17) and (18), we obtain

$$(19) \qquad \begin{aligned} l &= \bar{\mu} \\ (a_3 - \tfrac{3}{4} a_2^2) + l a_2 &= 2\mu \\ a_4 - 2 a_2 a_3 + \tfrac{13}{12} a_2^3 + l (a_3 - \tfrac{3}{4} a_2^2) &= \tfrac{2}{3} . \end{aligned}$$

We multiply the second equality (19) by $l = \bar{\mu}$ and add it to the third equality. We arrive finally at the result

$$(20) \qquad F = a_4 - 2 a_2 a_3 + \tfrac{13}{12} a_2^3 + 2 l (a_3 - \tfrac{3}{4} a_2^2) + l^2 a_2 = \tfrac{2}{3} + 2 |l|^2 .$$

The right-hand side of (20) depends only on the parameter l, which we can choose arbitrarily at the beginning of the inquiry. The left-hand side is the value of the functional in the extremum case. Hence we find the general coefficient inequality, valid for all functions of the family $\mathfrak{S}$:

$$(21) \qquad \mathrm{Re}\{a_4 - 2 a_2 a_3 + \tfrac{13}{12} a_2^3 + 2 l (a_3 - \tfrac{3}{4} a_2^2) + l^2 a_2\} \leqq \tfrac{2}{3} + 2 |l|^2 .$$

This inequality is fundamental for our further applications. It asserts a relation between a quadratic expression on the left and a hermitian expression in l on the right. It thus reminds one of the important inequalities due to GRUNSKY [2] for coefficients of univalent functions. It can, indeed, be derived from the Grunsky inequalities. This fact is of great methodological interest since GRUNSKY'S inequalities are derived without use of variational methods and are closely related to the classical area theorem. We gave the above direct proof of (21) in order to show the significance of the Grunsky inequalities in the theory of variations, where they are closely related to the problem of differential equations containing perfect squares [4].

2. We shall later need a simple inequality for the coefficients in $\mathfrak{S}$ which follows from the area theorem. Consider

$$g(z) = \frac{1}{\sqrt{f\left(\frac{1}{z^2}\right)}} = z + \frac{b_1}{z} + \frac{b_3}{z^3} + \cdots. \tag{22}$$

Clearly, $g(z)$ is regular and univalent for $|z| > 1$, and an easy computation yields

$$b_1 = -\tfrac{1}{2} a_2, \qquad b_3 = -\tfrac{1}{2}\left(a_3 - \tfrac{3}{4} a_2^2\right). \tag{23}$$

We have the area theorem:

$$1 \geqq |b_1|^2 + 3|b_3|^2 + 5|b_5|^2 + \cdots. \tag{24}$$

The combination

$$\lambda = a_3 - \tfrac{3}{4} a_2^2 \tag{25}$$

will occur frequently in our calculations. We derive for it from (23) and (24) the inequality

$$3|\lambda|^2 \leqq 4 - |a_2|^2. \tag{26}$$

§ 2. Proof that $|a_4| \leqq 4$

1. We wish to show now that the inequalities (1.21) and (1.26) imply the Bieberbach inequality for the fourth coefficient. We may assume without loss of generality that

$$a_4 \geqq 0, \qquad \operatorname{Re}\{a_2\} \geqq 0, \tag{1}$$

since this case can always be achieved by replacing $f(z)$ by the new member of $\mathfrak{S}$: $e^{-i\alpha} f(e^{i\alpha} z)$.

We set

$$\lambda = \xi + i\eta, \qquad a_2 = 2 - x + iy, \qquad 0 \leqq x \leqq 2. \tag{2}$$

We write (1.21) with l as a real parameter at our disposal:

$$a_4 \leqq \tfrac{2}{3} + \operatorname{Re}\{2a_2\lambda\} + \tfrac{5}{12}\operatorname{Re}\{a_2^3\} - 2l\operatorname{Re}\{\lambda\} + l^2(2 - \operatorname{Re}\{a_2\}). \tag{3}$$

We specialiez by selecting l so as to give the minimum value to the right-hand side:

$$l = \frac{\operatorname{Re}\{\lambda\}}{2 - \operatorname{Re}\{a_2\}}. \tag{4}$$

Now (3) becomes

$$a_4 \leqq \frac{2}{3} + \operatorname{Re}\{2a_2\lambda\} + \frac{5}{12}\operatorname{Re}\{a_2^3\} - \frac{\operatorname{Re}\{\lambda\}^2}{2 - \operatorname{Re}\{a_2\}}. \tag{5}$$

We insert here the real and imaginary parts according to the definition (2):

$$a_4 \leqq \frac{2}{3} + 2\{(2-x)\xi - y\eta\} + \frac{5}{12}\{(2-x)^3 - 3(2-x)y^2\} - \frac{\xi^2}{x}. \tag{6}$$

Observe that

$$2(2-x)\xi - \frac{\xi^2}{x} \leqq (2-x)^2 x. \tag{7}$$

Hence

(8) $$a_4 \leqq 4 + (2-x)^2 x - 5x + \tfrac{5}{2}x^2 - \tfrac{5}{12}x^3 - (\tfrac{5}{2} - \tfrac{5}{4}x)\, y^2 - 2y\eta.$$

By using the identity

(9) $$2y\eta = \alpha\left(y + \frac{1}{\alpha}\eta\right)^2 - \alpha y^2 - \frac{1}{\alpha}\eta^2, \qquad \alpha > 0,$$

we may transform (8) into

(10) $$a_4 - 4 \leqq -x - \frac{3}{2}x^2 + \frac{7}{12}x^3 - \left(\frac{5}{2} - \frac{5}{4}x\right) y^2 + \alpha y^2 + \frac{1}{\alpha}\eta^2.$$

We use next the inequality (1.26), from which, expressed in the notation (2), follows the estimate

(11) $$\eta^2 \leqq \tfrac{4}{3}x - \tfrac{1}{3}(x^2 + y^2).$$

Hence, we can eliminate η from (10) and obtain

(12) $$a_4 - 4 \leqq \left(\frac{4}{3\alpha} - 1\right)x - \left(\frac{3}{2} + \frac{1}{3\alpha}\right)x^2 + \frac{7}{12}x^3 - \left(\frac{5}{2} + \frac{1}{3\alpha} - \alpha - \frac{5}{4}x\right)y^2.$$

The final proof of the inequality $a_4 \leqq 4$ is achieved by two estimates. We first choose

(13) $$\alpha = \tfrac{4}{3}$$

and find that

(14) $$a_4 - 4 \leqq -\frac{7}{4}x^2\left(1 - \frac{x}{3}\right) - \left(\frac{17}{12} - \frac{5}{4}x\right)y^2.$$

We can infer from this inequality that

(15) $$a_4 \leqq 4 \quad \text{when} \quad 0 \leqq x \leqq \tfrac{17}{15}.$$

Next, we select $\alpha = \frac{1}{2}$ and obtain

(16) $$a_4 - 4 \leqq \tfrac{5}{3}x - \tfrac{13}{6}x^2 + \tfrac{7}{12}x^3 - (\tfrac{8}{3} - \tfrac{5}{4}x)\, y^2.$$

The coefficient of y^2 is negative for all values $0 \leqq x \leqq 2$; hence, everything depends on the sign of

(17) $$r(x) = 5 - \tfrac{13}{2}x + \tfrac{7}{4}x^2.$$

One verifies directly that

(18) $$r(2) = -1, \qquad r(\tfrac{17}{15}) = -\tfrac{107}{900}.$$

Thus $r(x)$ must be negative in the entire interval $\frac{17}{15} \leqq x \leqq 2$. Hence we have proved that

(19) $$a_4 \leqq 4 \quad \text{when} \quad \tfrac{17}{15} \leqq x \leqq 2.$$

This completes the proof of the Bieberbach inequality

(20) $$|a_4| \leqq 4.$$

2. For the sake of completeness we wish to indicate briefly how the fundamental inequality (1.21) follows from GRUNSKY's inequality. Let

(21) $$g(z) = z + \sum_{\nu=1}^{\infty} \frac{b_\nu}{z^\nu}$$

be univalent in the circular domain $|z|>1$. Define the nth Faber polynomial of $g(z)$ as the polynomial of degree n of $g(z)$ for which we have the series development

$$G_n[g] = z^n + \sum_{\nu=1}^{\infty} \frac{b_{n\nu}}{z^\nu}. \tag{22}$$

It is easily seen that $G_n(t)$ is uniquely determined by this condition. GRUNSKY proved that the set of inequalities between quadratic and hermitian forms,

$$\left|\sum_{n,m=1}^{N} m\, b_{nm}\, x_n\, x_m\right| \leq \sum_{n=1}^{N} n\,|x_n|^2, \qquad N=1,2,\ldots, \tag{23}$$

are necessary and sufficient conditions that the function (21) be univalent.

We can apply this result to the particular function (1.22), choose $x_1=l$, $x_2=0$, $x_3=\frac{1}{3}$, and find precisely the required inequality (1.21).

This work was performed partially under an Office of Naval Research contract, Nonr-225(11), with Stanford University.

References

[1] GARABEDIAN, P. R., & M. SCHIFFER: A proof of the Bieberbach conjecture for the fourth coefficient. J. Rational Mech. Anal. **4**, 427–465 (1955).

[2] GRUNSKY, H.: Koeffizientenbedingungen für schlicht abbildende meromorphe Funktionen. Math. Z. **45**, 29–61 (1939).

[3] SCHIFFER, M.: A method of variation within the family of simple functions. Proc. London Math. Soc. **44**, 432–449 (1938).

[4] SCHIFFER, M.: Faber polynomials in the theory of univalent functions. Bull. Amer. Math. Soc. **54**, 503–517 (1948).

Stanford University
California

(Received March 7, 1960)

Commentary on

[72] Z. Charzyński and M. Schiffer, *A new proof of the Bieberbach conjecture for the fourth coefficient*, Arch. Rational Mech. Anal. **5** (1960), 187–193.

The Bieberbach conjecture said that for all functions $f(z) = z + a_2z^2 + \dots$ analytic and univalent in the unit disk, the bounds $|a_n| \leq n$ hold, with strict inequality for all n unless f is a rotation of the Koebe function $k(z) = z(1-z)^{-2} = \sum_{n=1}^{\infty} na_nz^n$. This famous problem lay open for many years and attracted the efforts of distinguished analysts, who devised ingenious new techniques in attempts to resolve it. The conjecture dated back to 1916, when Bieberbach proved $|a_2| \leq 2$, leading to sharp versions of Koebe's distortion theorem and related geometric results. In 1923, Loewner developed a method which enabled him to prove $|a_3| \leq 3$ but did not seem to capture the result for higher coefficients. The inequality $|a_4| \leq 4$ resisted all assaults until 1955, when Garabedian and Schiffer [60] finally managed to prove it. Their paper was a *tour de force* that applied both Loewner's method and Schiffer's variational method, then resorted to a long series of laborious estimates to arrive at the desired result.

Five years later, the paper of Charzyński and Schiffer [72] appeared, giving a relatively short, entirely elementary proof of $|a_4| \leq 4$. Their proof appealed only to the Grunsky inequalities [Gr], a simple refinement of the area principle. Needless to say, the paper caused quite a sensation. I well remember the air of excitement and anticipation at Stanford in 1960, when I arrived to take a postdoctoral position. There was general optimism that the same method would soon yield a full proof of the Bieberbach conjecture.

That optimism faded in later years, however, as the new approach was tried on the next few coefficients. Ozawa [O1] and Pederson [P] did manage to prove $|a_6| \leq 6$ with the Grunsky inequalities; but their arguments were much more involved than that of [72], and some nonelementary results were invoked. Ozawa [O2] was later able to remove the extraneous elements from his argument and give a purely elementary proof of $|a_6| \leq 6$, but the details were still discouragingly complicated. The method applied more readily to coefficients of even order, because of the effect of a square-root transformation. That insight motivated a paper by Garabedian, Ross, and Schiffer [86], which Pederson applied in his proof of $|a_6| \leq 6$. Pederson and Schiffer [104] eventually succeeded in modifying the approach of [72] by borrowing from [60, 100] to devise a nonelementary and rather difficult proof that $|a_5| \leq 5$. Some studies of a_8 followed, but all became moot in 1984, when de Branges [Br] used a totally different method to prove a conjecture of Milin (see [D, Chap. 5]) that was known to imply the Bieberbach conjecture. Then FitzGerald and Pommerenke [FP] translated de Branges's proof into classical terms, reducing it to a combination of Loewner's method with an inequality of Askey and Gasper involving hypergeometric functions. (To be sure, de Branges's proof had also used the Askey–Gasper inequality and a variant of Loewner's classical method, among other ideas.)

After [72] appeared, G. V. Kuzmina gave a simplified version of the Charzyński–Schiffer proof in the supplement to the second (posthumous) edition of Goluzin's book [Go] (adapted in [D, Sect. 4.6]). It was apparent from the proofs of $|a_4| \leq 4$ in [60, 72] that $|a_4|$ is much smaller than 4 when $|a_2|$ is near zero. In fact, it is a corollary of Kuzmina's version of the proof that $|a_4| \leq \frac{2}{3}$ when $a_2 = 0$, and the function $\{k(z^3)\}^{1/3}$ shows that the bound is sharp. Ahlfors [A] used the Grunsky inequalities to improve the estimate $|a_4| \leq 4$ to

$$|a_4| \leq 4\left\{1 - \tfrac{1}{15}(4 - |a_2|^2)\right\}^{1/2},$$

and Baranova [Ba] made some further improvements. Among stronger results, she showed for instance that $|a_4| \leq \frac{4}{15}(11 + 2|a_2|)$.

Following their proof of $|a_4| \leq 4$ via the Grunsky inequalities in [72], Charzyński and Schiffer gave a different proof, also relatively short, by a variational method [73]. There they analyzed the resulting quadratic differential by a method similar to that which Garabedian and Schiffer [60] had used to give a new proof of $|a_3| \leq 3$.

References

[A] Lars V. Ahlfors, *Conformal Invariants: Topics in Geometric Function Theory*, McGraw-Hill, 1973; second edition, American Mathematical Society, 2010.

[Ba] V. A. Baranova, *An estimate of the coefficient c_4 of univalent functions depending on $|c_2|$*, Mat. Zametki **12** (1972), 127–130 (in Russian); English transl., Math. Notes **12** (1972), 510–512.

[Br] Louis de Branges, *A proof of the Bieberbach conjecture*, Acta Math. **154** (1985), 137–152.

[D] Peter L. Duren, *Univalent Functions*, Springer-Verlag, 1983.

[FP] Carl H. FitzGerald and Ch. Pommerenke, *The de Branges theorem on univalent functions*, Trans. Amer. Math. Soc. **290** (1985), 683–690.

[Go] G. M. Goluzin, *Geometric Theory of Functions of a Complex Variable*, second edition, Izdat. "Nauka", Moscow 1966; English transl., American Mathematical Society, 1969.

[Gr] Helmut Grunsky, *Koeffizientenbedingungen für schlicht abbildende meromorphe Funktionen*, Math. Z. **45** (1939), 29–61.

[O1] Mitsuro Ozawa, *On the sixth coefficient of univalent function*, Kōdai Math. Sem. Rep. **17** (1965), 1–9.

[O2] Mitsuro Ozawa, *An elementary proof of the Bieberbach conjecture for the sixth coefficient*, Kōdai Math. Sem. Rep. **21** (1969), 129–132.

[P] Roger N. Pederson, *A proof of the Bieberbach conjecture for the sixth coefficient*, Arch. Rational Mech. Anal. **31** (1968/69), 331–351.

PETER DUREN

[75] (with P. L. Duren) A variational method for functions schlicht in an annulus

[75] (with P. L. Duren) A variational method for functions schlicht in an annulus. *Arch. Rational Mech. Anal.* **9** (1962), 260–272.

A Variational Method for Functions Schlicht in an Annulus

P. L. DUREN & M. SCHIFFER

1. Introduction

In his investigations on numerical construction of conformal mappings, D. GAIER [1] has been led to consider the family of functions $f(z)$ regular analytic and schlicht in the annulus $R: r<|z|<1$, and satisfying the following three conditions:

$$(1) \qquad |f(z)|<1 \quad \text{in } R, \quad \text{while} \quad |f(z)|=1 \quad \text{on} \quad |z|=1;$$

that is, $f(z)$ maps R onto the unit disk minus some continuum Γ;

$$(2) \qquad f(z)\neq 0 \quad \text{in } R \quad (\Gamma \text{ contains the origin});$$

$$(3) \qquad f(1)=1 \qquad \text{(a normalization).}$$

We shall denote this family by $\mathfrak{F}$. We shall also consider the families $\mathfrak{F}_0$ and $\mathfrak{F}_1$ of schlicht analytic functions satisfying only (1) and (2), and only (1), respectively. Of course, $\mathfrak{F}\subset\mathfrak{F}_0\subset\mathfrak{F}_1$.

GAIER raised the question of the maximum of $|f(z)-z|$ for all $f\in\mathfrak{F}$ and $z\in R$. This and other extremal problems will be solved in the present paper using a specific method of variation within the families $\mathfrak{F}$, $\mathfrak{F}_0$, and $\mathfrak{F}_1$. The variation leads from every given $f(z)$ to a large set of comparison functions within the family, and the method seems to be of wide applicability to extremum problems in the given families.

In view of the maximum modulus principle, GAIER's distortion problem

$$(4) \qquad \max_{f\in\mathfrak{F},\, z\in R} |f(z)-z|$$

will be solved if $\max_{f\in\mathfrak{F}} |f(b)-b|$ can be found for each fixed b either on the inner or on the outer boundary of R. The problem for the inner boundary will be treated in §3, the outer boundary §§5, 6. The estimate on the inner boundary is derived by solving the extremal problem

$$\max_{f\in\mathfrak{F}} |f(b)|, \quad \text{where} \quad |b|=r_1, \quad r<r_1<1,$$

a result which is of interest in itself. The use of our variational method is further illustrated in §4, where it is applied to find the maximum diameter of the continuum Γ (see (1)) for all functions in the family $\mathfrak{F}$.

Each function in $\mathfrak{F}_1$ maps the unit circumference onto itself and hence may be continued by Schwarz reflection to a function analytic in $r<|z|<\frac{1}{r}$. No such continuation is in general possible through the inner boundary, and the expression $f(z)$ may be meaningless for $|z|=r$. Existence almost everywhere of a radial limit is guaranteed by FATOU's theorem, but we shall make no use of this.

Another important remark is that $\overline{f(\bar{z})}\in\mathfrak{F}$ whenever $f(z)\in\mathfrak{F}$, and similarly for $\mathfrak{F}_0$ and $\mathfrak{F}_1$.

The existence of solutions to most of the extremal problems we shall consider is based on the fact that $\mathfrak{F}$, $\mathfrak{F}_0$, and $\mathfrak{F}_1$ are compact normal families. Normality is a consequence of the uniform bound $|f(z)|\leqq 1$. Property (1) and the fact that $f(z)$ is schlicht in R ensure that

$$\int\limits_{|z|=1} d\arg f(z) = 2\pi$$

for every $f(z)\in\mathfrak{F}_1$. This rules out constant limits within the normal families considered since the limit function must satisfy the same equality. HURWITZ' theorem shows that the uniform limit of a sequence $f_n\in\mathfrak{F}_0$ must again satisfy (2).

Each of the families depends, of course, upon the basic annulus R and hence upon the inner radius r. To emphasize this dependence, we might use the notation $\mathfrak{F}(r)$ instead of $\mathfrak{F}$. It is obvious that $\mathfrak{F}(r_1)\subset\mathfrak{F}(r_2)$ if $r_1<r_2$. Moreover, $\mathfrak{F}(0)$ consists only of the identity function $f(z)\equiv z$, since any $f(z)\in\mathfrak{F}(0)$ defines a conformal mapping of the unit disk onto itself for which $f(0)=0$ and $f(1)=1$. This is a linear fractional transformation which reduces to the identity because 0 and 1 are fixed. Hence, for $r=0$, $|f(z)-z|\equiv 0$, and it is not surprising that $|f(z)-z|=O(r)$, as GAIER has pointed out.

The following well known result illustrates the importance of the normalization (3) of $\mathfrak{F}$. Since it will be used repeatedly, we include a short proof.

Uniqueness Lemma. *If two functions $f_1(z), f_2(z)\in\mathfrak{F}$ have the same range, then $f_1(z)\equiv f_2(z)$.*

Proof. $F(z)=f_1^{-1}(f_2(z))$ maps R onto itself, and $F(1)=1$. Consider the function $\log\left|\frac{F(z)}{z}\right|$, which is harmonic in R and vanishes on both boundaries. By the maximum principle, $\log\left|\frac{F(z)}{z}\right|\equiv 0$ in R, and $\log F(z)\equiv\log z+i\alpha$ (α a real constant). Thus $F(z)\equiv e^{i\alpha}z\equiv z$ because $F(1)=1$.

2. Construction of the Variational Formulas

Corresponding to each function $w=f(z)$ in the family $\mathfrak{F}$ we shall develop a class of variational functions $V_\varrho(w)$ analytic and schlicht in the range $\Delta(f)$ of f having the properties

$$(1)\qquad \begin{aligned} &\text{(i)} && \lim_{\varrho\to 0} V_\varrho(w)\equiv w,\\ &\text{(ii)} && |V_\varrho(w)|=1 \quad\text{for}\quad |w|=1,\\ &\text{(iii)} && V_\varrho(w)\neq 0 \quad\text{for}\quad w \text{ in } \Delta(f),\\ &\text{(iv)} && V_\varrho(1)=1.\end{aligned}$$

Properties (ii), (iii), (iv) ensure that $V_\varrho(f(z)) \in \mathfrak{F}$ whenever $f(z) \in \mathfrak{F}$. If we drop requirement (iv), we obtain variation functions $V_\varrho^{(0)}(w)$ for the class $\mathfrak{F}_0$. Similarly, variation functions $V_\varrho^{(1)}(w)$ for $\mathfrak{F}_1$ result if (iii) and (iv) are eliminated.

Let $w = f(z)$ be a given function in $\mathfrak{F}$, and let Γ denote the complement within the open unit disk of the range of f. Let $w_0 \neq 0$ be a fixed point on the continuum Γ. Corresponding to each such choice of w_0, it is known [4] that there exist functions of the form

$$F(w) = w + \frac{a\,\varrho^2}{w - w_0} + O(\varrho^3) \tag{2}$$

analytic and univalent in the domain D_ϱ which consists of all points exterior either to Γ or to the circle $|w - w_0| = \varrho$. Here the constant a depends on the small parameter ϱ; however, $|a| = |a(\varrho)| \leqq 1$. Furthermore, the error term $O(\varrho^3)$ can be estimated uniformly in each closed subdomain of D_ϱ. The normalized function

$$F_0(w) = F(w) - F(0) = w + \frac{a\,\varrho^2 w}{(w - w_0)\,w_0} + O(\varrho^3) \tag{3}$$

has the same properties, and, in addition, $F_0(0) = 0$. Finally, since $w^2/1 - \overline{w}_0 w$ satisfies a Lipschitz condition in each disk $|w| \leqq c$ $(c < 1/|w_0|)$, it can easily be seen that the function

$$W^* = W - \frac{\bar{a}\,\varrho^2 W^2}{(1 - \overline{w}_0 W)\,\overline{w}_0}$$

is, for each sufficiently small ϱ, analytic and univalent in $|W| \leqq c$ with $1 < c < 1/|w_0|$. Inserting here $W = F_0(w)$, we obtain the function

$$\begin{aligned} G(w) &= F_0(w) - \frac{\bar{a}\,\varrho^2 w^2}{(1 - \overline{w}_0 w)\,\overline{w}_0} + O(\varrho^3) \\ &= w\left[1 + \frac{a\,\varrho^2}{(w - w_0)\,w_0} - \frac{\bar{a}\,\varrho^2}{\left(\frac{1}{w} - \overline{w}_0\right)\overline{w}_0}\right] + O(\varrho^3), \end{aligned} \tag{4}$$

which is, for small enough ϱ, univalent in $\Delta(f)$.

Besides preserving the origin, G preserves the periphery of the unit circle up to order ϱ^3. To see this, note that for $|w| = 1$, $G(w)$ is of the form

$$G(w) = w\,[1 + i\,\varrho^2 A] + O(\varrho^3), \tag{5}$$

where A is a real quantity. Hence,

$$|G(w)| = 1 + O(\varrho^3) \quad \text{for} \quad |w| = 1. \tag{6}$$

From G we now construct our variation function $V_\varrho(w)$, which will preserve the unit circumference precisely. For each small ϱ, $\eta = G(w)$ maps the circumference $|w| = 1$ onto a nearby simple closed curve C_ϱ in the η-plane. Let

$$\zeta = \chi(\eta), \qquad \eta = \psi(\zeta)$$

be the conformal mapping of the interior of C_ϱ onto $|\zeta| < 1$, satisfying

$$\psi(0) = 0, \qquad \psi'(0) > 0. \tag{7}$$

The function

$$V_\varrho(w) = \frac{\chi(G(w))}{\chi(G(1))} \tag{8}$$

now has the four desired properties (1).

It is important to know how $V_\varrho(w)$ depends on ϱ. POISSON's formula gives

$$\log\left(\frac{\psi(\zeta)}{\zeta}\right)=\frac{1}{2\pi}\int_0^{2\pi}\log\left|\frac{\psi(e^{i\vartheta})}{e^{i\vartheta}}\right|\frac{e^{i\vartheta}+\zeta}{e^{i\vartheta}-\zeta}\,d\vartheta.$$

The additive imaginary constant disappears because of (7). But the curve C_ϱ always stays at a distance $1+O(\varrho^3)$ from the origin, by (6); hence $|\psi(e^{i\vartheta})|=1+O(\varrho^3)$. Thus $\log\left(\frac{\psi(\zeta)}{\zeta}\right)=O(\varrho^3)$, so that $\psi(\zeta)=\zeta+O(\varrho^3)$. Consequently, the inverse function has the form

(9) $$\chi(\eta)=\eta+O(\varrho^3),$$

and an easy calculation from (8) and (4) yields

(10) $$V_\varrho(w)=w\left[1+\frac{a\,\varrho^2(1-w)}{(w-w_0)(1-w_0)\,w_0}+\frac{\bar a\,\varrho^2(1-w)}{(1-\bar w_0\,w)(1-\bar w_0)\,\bar w_0}\right]+O(\varrho^3).$$

For the family $\mathfrak{F}_0$ an admissible variation is given by $V_\varrho^{(0)}(w)=\chi(G(w))$, where G and χ are as above. Using (4) and (9), we have

(11) $$V_\varrho^{(0)}(w)=w\left[1+\frac{a\,\varrho^2}{(w-w_0)\,w_0}-\frac{\bar a\,\varrho^2\,w}{(1-w\,\bar w_0)\,\bar w_0}\right]+O(\varrho^3).$$

To construct a variational function for the family $\mathfrak{F}_1$, we need not make the normalization (1.3). Instead, starting with the variation (2), we form the univalent function

$$H(w)=F(w)-\frac{\bar a\,\varrho^2\,w}{\frac{1}{w}\left(\frac{1}{w}-\bar w_0\right)}+O(\varrho^3),$$

which again has the structure $H(w)=w[1+i\varrho^2B]+O(\varrho^3)$ on $|w|=1$, where B is real. Defining $V_\varrho^{(1)}(w)=\chi(H(w))$, we find by the same reasoning as before

(12) $$V_\varrho^{(1)}(w)=w\left[1+\frac{a\,\varrho^2}{w(w-w_0)}-\frac{\bar a\,\varrho^2\,w^2}{(1-w\,\bar w_0)}\right]+O(\varrho^3).$$

Clearly, the variation within the narrower class will also be an admissible variation within the wider class and may be applied there for solving extremum problems.

3. Maximum Modulus on the Inner Boundary

It is the eventual aim of this section to prove

(1) $$\limsup_{|z|\to r}|f(z)-z|\leqq 5r$$

for all $f(z)\in\mathfrak{F}$. Our technique is first to discover which functions in $\mathfrak{F}$ take the largest absolute value on a given circle $|z|=r_1$ $(r<r_1<1)$, a result which is of interest in itself. Specifically, the extremum problem to be considered is

(2) $$\max_{f\in\mathfrak{F}}|f(b)|,\quad |b|=r_1\quad (r<r_1<1),$$

for fixed b. GRÖTZSCH [*2*] previously solved this problem using the method of extremal length, and GAIER [*1*, p. 151] used this result to obtain a bound for (1), but we treat it here as an illustration of our systematic approach. The solution to (2) is used to show that the continuum Γ (1.1) cannot extend farther than $4r$ from the origin. The estimate (1) then follows.

Since the normalization (1.3) has no essential bearing on (2), it is enough to consider the problem

$$\max_{f\in\mathfrak{F}_0} |f(b)|, \quad |b| = r_1 \quad (r < r_1 < 1). \tag{3}$$

The advantage is that the variation function (2.11) for $\mathfrak{F}_0$ is simpler than that for $\mathfrak{F}$.

The existence of an extremal function $f_0(z)$ is assured because $\mathfrak{F}_0$ is normal and compact. The variation $V_\varrho^{(0)}(w)$, given in equation (2.11), may be applied to f_0, resulting in a function $V_\varrho^{(0)}(f_0(z)) \in \mathfrak{F}_0$. Since f_0 is extremal, we have

$$|V_\varrho^{(0)}(f_0(b))| \leqq |f_0(b)|.$$

Upon introducing (2.11) and the notation $B = f_0(b)$, this yields

$$\left|1 + \frac{a\varrho^2}{(B-w_0)w_0} - \frac{\bar{a}\varrho^2 B}{(1-B\bar{w}_0)\bar{w}_0} + O(\varrho^3)\right| \leqq 1.$$

Thus, using the identity $\operatorname{Re}\{\bar{\alpha}\} = \operatorname{Re}\{\alpha\}$,

$$\operatorname{Re}\left\{\frac{a\varrho^2}{w_0}\left[\frac{1}{B-w_0} - \frac{\bar{B}}{1-w_0\bar{B}}\right] + O(\varrho^3)\right\} \leqq 0.$$

Finally, since $|B| < 1$, an easy calculation leads to

$$\operatorname{Re}\left\{\frac{a\varrho^2}{w_0(B-w_0)(1-w_0\bar{B})} + O(\varrho^3)\right\} \leqq 0. \tag{4}$$

We remark parenthetically that (4) also results if the more restrictive variation $V_\varrho(w)$ of (2.10) is used.

We apply now the main lemma of the method of boundary variation for univalent functions, which may be formulated as follows [*4*]:

Let Γ be a continuum in the w-plane and $s(w) \not\equiv 0$ a regular analytic function on Γ. If the inequality

$$\operatorname{Re}\{a\varrho^2 s(w_0) + O(\varrho^3)\} \leqq 0$$

holds for every univalent variation (2.2) of Γ, then Γ is an analytic curve $w(t)$ satisfying the differential equation

$$w'(t)^2 s(w(t)) > 0.$$

In our particular case we conclude from this lemma and inequality (4) that $f_0(z)$ maps the domain R onto the unit disk slit along an analytic arc Γ: $w = w(t)$ satisfying the differential equation

$$\frac{(w')^2}{w(B-w)(1-w\bar{B})} > 0. \tag{5}$$

The slit Γ must pass through the origin because $f_0(z) \neq 0$ in R, by (1.2). It can be shown with the aid of the transformation $\omega = \sqrt{w}$ that (5) has a unique solution (apart from choice of parameter t) passing through the origin; in fact, the origin must be an endpoint of the slit.

It is easily verified that $w(t) = Bt$ $(0 \leqq t \leqq 1)$ satisfies (5); hence, it is the only solution passing through the origin. Therefore, each extremal function $f_0(z)$ for problem (3) must map R onto the unit disk slit along a part of the straight segment from 0 to $B = f_0(b)$. This geometric characterization of $f_0(z)$ determines the extremal function up to a rotation. The length M (less than $|B|$) of the slit is determined by the modulus of the ring. The value of M and the sought maximum value $|B|$ can be expressed explicitly in terms of elliptic functions.

We shall now normalize $f_0(z)$ in order to identify it with a known mapping function. Set $B=|B|\,e^{i\beta}$, and let α be defined by $f_0(e^{i\alpha})=e^{i\beta}$. The function $\Phi(z)=e^{-i\beta}f_0(e^{i\alpha}z)$ belongs to the family $\mathfrak{F}$ and maps R onto the unit disk slit along the real axis from 0 to M. But by our uniqueness lemma (§1) there is only one such function; it is a classical function expressible in terms of Jacobian elliptic functions.* The length M of the slit depends upon the modulus of the ring according to the formula*

$$M=2L(1+L^2)^{-1}, \quad \text{where } L=2r\prod_{n=1}^{\infty}\left(\frac{1+r^{8n}}{1+r^{8n-4}}\right)^2. \tag{6}$$

Hence

$$M\leqq 2L=4r\prod_{n=1}^{\infty}\left(\frac{1+r^{8n}}{1+r^{8n-4}}\right)^2\leqq 4r, \tag{7}$$

each factor of the infinite product being less than unity. The constant 4 is best possible because $M\sim 4r$ as $r\to 0$.

The solution to (2) may now be applied to solve the problem

$$\max_{f\in\mathfrak{F}}|f(b)|, \quad |b|=r, \tag{8}$$

which is to be interpreted as requesting the maximum distance from the origin attainable on a continuum Γ. It is not immediately clear that a maximum in (8) is attained, since the normality of $\mathfrak{F}$ implies uniform convergence only in each closed subdomain of R. We are about to show, however, that the maximum in question is precisely M, the length of the slit as given in (6) above.**

Indeed, each of the problems (2) is solved by the same function $\Phi(z)$ mapping R onto the unit disk slit radially from 0 to M. The value of the maximum on the circle $|z|=r_1$ $(r<r_1<1)$ is B $(M<B<1)$. As r_1 approaches r, B tends to M. Hence, if there were any function $h(z)\in\mathfrak{F}$ having a continuum Γ with a point of absolute value $M_1>M$, it would also possess the property

$$\max_{|z|=r_1}|h(z)|>M_1 \qquad (r<r_1<1),$$

and we need only choose r_1 sufficiently close to r to arrive at a contradiction.

This proves $|f(z)|\leqq M$ for z on the inner boundary of R. Using (7), we have

$$\limsup_{|z|\to r}|f(z)-z|\leqq 5r \quad \text{for all } f\in\mathfrak{F}. \tag{9}$$

4. Maximum Diameter

A function $f(z)\in\mathfrak{F}$ maps the annulus R onto the unit disk from which a continuum Γ, containing the origin, is removed. In §3 we saw that Γ extends farthest from the origin when it is stretched into a straight segment. We shall

* See Nehari [3], where the function is constructed which maps R onto the unit disk slit from $-L$ to L. Our function $\Phi(z)$ is obtained from this by application of the linear transformation $(w+L)(1+Lw)^{-1}$. In particular, $M=2L(1+L^2)^{-1}$.

** That the straight slit mappings are the only possible solutions to (8) may be verified by the variational procedure used to solve (2).

now digress from our main problem to explore the question of the maximum diameter of Γ. An intuitive guess would be that the diameter is a maximum when Γ is a straight segment with center at the origin; and, indeed, this will be the answer.

With the notation $d(\Gamma)$ for the diameter of Γ, our extremum problem is

$$\max_{f \in \mathfrak{F}} d(\Gamma). \tag{1}$$

It is convenient to solve this problem within the larger class $\mathfrak{F}_1$ of competing functions. It will turn out that the maximum diameter for $f \in \mathfrak{F}_1$ is actually attained within the subfamily $\mathfrak{F}$. The modified problem is then

$$\max_{f \in \mathfrak{F}_1} d(\Gamma). \tag{2}$$

The existence proof for a solution to (2) presents a difficulty, because it is not immediately clear that $d(\Gamma)$ is a continuous functional, and thus the normality of $\mathfrak{F}_1$ cannot be applied. To prove existence, let

$$\bar{d} = \sup_{f \in \mathfrak{F}_1} d(\Gamma),$$

and let $f_n \in \mathfrak{F}_1$ be an extremal sequence of functions in the sense that $d(\Gamma_n) \to \bar{d}$. Let α_n and β_n be points on Γ_n such that $|\alpha_n - \beta_n| = d(\Gamma_n)$. Because of the normality of $\mathfrak{F}_1$ we may suppose, after passing to a subsequence if necessary, that $f_n(z)$ tends uniformly in each closed subdomain to a limit $f(z) \in \mathfrak{F}_1$. We may further suppose that $\alpha_n \to \alpha$ and $\beta_n \to \beta$ because $\{\alpha_n\}$ and $\{\beta_n\}$ are bounded sets. The remainder of the proof consists in showing that the limit function f cannot assume either of the values α, β. Indeed, suppose $f(z_0) = \alpha$ for some $z_0 \in R$. Let $\varrho > 0$ be so small that the disk $|z - z_0| \leq \varrho$ is in R, and let Δ be the f-image of this disk. Now choose $\varepsilon > 0$ so small that the disk $|w - \alpha| \leq \varepsilon$ is interior to Δ and therefore has a positive distance δ from the boundary of Δ. Let ω be any fixed number such that $|\omega - \alpha| < \varepsilon$. We write

$$f_n(z) - \omega = [f_n(z) - f(z)] + [f(z) - \omega].$$

Now for all z on the circumference $|z - z_0| = \varrho$ we have $|f(z) - \omega| > \delta$ by construction. And for $n > N$ sufficiently large, the uniform convergence gives $|f_n(z) - f(z)| < \delta$ on $|z - z_0| = \varrho$. Hence, for all $n > N$ we conclude from ROUCHÉ's theorem that $[f_n(z) - \omega]$ has precisely as many zeros in the disk $|z - z_0| < \varrho$ as does $[f(z) - \omega]$, namely, one. Consequently, the entire disk $|w - \alpha| < \varepsilon$ is in the range of $f_n(z)$ for all $n > N$. But this is impossible, since α_n is not in the range of f_n and $\alpha_n \to \alpha$.

Thus α and β belong to the continuum Γ which corresponds to the limit function $f(z)$. Therefore,

$$\bar{d} = |\alpha - \beta| \leqq d(\Gamma) \leqq \bar{d},$$

so $d(\Gamma) = \bar{d}$, and the extremal problem (2) has a solution.

For variation within $\mathfrak{F}_1$ we shall use not $V_\varrho^{(1)}(w)$ but the more special variational function $V_\varrho^{(0)}$ given in (2.11), which, surprisingly, is more convenient. Let $f_0(z)$ be an extremal function for (2), and let $f_0^*(z) = V_\varrho^{(0)}(f_0(z))$. Let Γ_0 and Γ_0^* be the corresponding continua. Let

$$|\alpha - \beta| = d(\Gamma_0), \qquad \alpha, \beta \in \Gamma_0.$$

For ϱ sufficiently small $V_\varrho^{(0)}(w)$ is defined and univalent on the union D_ϱ of the set $\{|w-w_0|>\varrho\}$ and the range of f_0 (see §2). Here $w_0 \in \Gamma_0$ is a fixed point upon which $V_\varrho^{(0)}(w)$ depends. If $w_0 \neq \alpha, \beta$, then for ϱ sufficiently small $\alpha, \beta \in D_\varrho$. We claim that $\alpha^* = V_\varrho^{(0)}(\alpha)$ and $\beta^* = V_\varrho^{(0)}(\beta)$ belong to Γ_0^*. For if $f^*(z_0)=\alpha^*$ for some $z_0 \in R$, we have only to apply the inverse of $V_\varrho^{(0)}$ to conclude $f(z_0)=\alpha$, which is absurd. Consequently, it follows from the extremal property of Γ_0 that

$$|\alpha^* - \beta^*| \leqq d(\Gamma_0^*) \leqq d(\Gamma_0) = |\alpha - \beta|. \tag{3}$$

Using (2.11) and simplifying, we find from (3) that

$$\left|1 - \frac{a\varrho^2}{(\alpha-w_0)(\beta-w_0)} - \frac{\bar{a}\varrho^2(\alpha+\beta-\alpha\beta\bar{w}_0)}{\bar{w}_0(1-\alpha\bar{w}_0)(1-\beta\bar{w}_0)} + O(\varrho^3)\right| \leqq 1.$$

Hence, we may conclude by the usual argument

$$\operatorname{Re}\left\{\frac{a\varrho^2}{w_0}\left[\frac{w_0}{(\alpha-w_0)(\beta-w_0)} + \frac{\bar{\alpha}+\bar{\beta}-\bar{\alpha}\bar{\beta}w_0}{(1-\bar{\alpha}w_0)(1-\bar{\beta}w_0)}\right] + O(\varrho^3)\right\} \geqq 0. \tag{4}$$

Before discussing the consequences of this inequality, we wish to derive a relation between the constants α and β by an additional elementary variation.

We consider a fixed number ω $(|\omega|<1)$ and form the function

$$f_1(z) = \frac{f_0(z)-\omega}{1-\bar{\omega}f_0(z)} \in \mathfrak{F}_1.$$

By the extremal property of $|\alpha-\beta|$,

$$\left|\frac{\alpha-\omega}{1-\bar{\omega}\alpha} - \frac{\beta-\omega}{1-\bar{\omega}\beta}\right| = \frac{|\alpha-\beta|(1-|\omega|^2)}{|1-\bar{\omega}\alpha||1-\bar{\omega}\beta|} \leqq |\alpha-\beta|.$$

Setting $\omega = \varepsilon e^{i\sigma}$, $\varepsilon>0$,

$$(1-\varepsilon^2)\,|1+\varepsilon(\alpha+\beta)e^{-i\sigma}+O(\varepsilon^2)| \leqq 1,$$

so

$$\operatorname{Re}\{\varepsilon(\alpha+\beta)e^{-i\sigma}+O(\varepsilon^2)\} \leqq 0.$$

Letting $\varepsilon \to 0$ and observing that σ is arbitrary, we have proved

$$\alpha+\beta=0. \tag{5}$$

Using (5) and noting that $|\alpha|<1$, we reduce inequality (4) to

$$\operatorname{Re}\left\{\frac{a\varrho^2}{(w_0^2-\alpha^2)(1-\bar{\alpha}^2w_0^2)} + O(\varrho^3)\right\} \geqq 0.$$

Consequently, using again our fundamental lemma [1], we find that Γ_0 is an analytic curve $w=w(t)$ satisfying the differential equation

$$\frac{(w')^2}{(w^2-\alpha^2)(1-\bar{\alpha}^2w^2)} < 0. \tag{6}$$

Equation (6) has a unique solution passing through $w=\alpha$; it is immediately verified that this solution is $w=t\alpha$, $-1<t<1$. Thus, Γ_0 is the segment joining $-\alpha$ and α. In particular, Γ_0 passes through the origin, so $f_0(z)\in\mathfrak{F}_0$. To construct from f_0 an explicit extremal function in $\mathfrak{F}$, let

$$\alpha = |\alpha|\,e^{i\tau}, \qquad f_0(e^{i\sigma}) = e^{i\tau}.$$

Then
$$g_0(z) = e^{-i\tau} f_0(e^{i\sigma} z) \in \mathfrak{F}$$
maps R onto the unit disk slit along the real axis from $-|\alpha|$ to $|\alpha|$. By our uniqueness theorem (§1), $g_0(z)$ must therefore be the classical mapping [3] constructible in terms of elliptic functions. The length of the slit is
$$\bar{d} = 2L,$$
where L is given in (3.6). This solves (1).

5. Maximum Displacement on the Outer Boundary

In §3 we proved that for each $f \in \mathfrak{F}$
$$\limsup_{|z| \to r} |f(z) - z| \leqq 5r.$$
We shall now estimate $|f(z) - z|$ for $|z| = 1$. The maximum modulus principle will allows us to combine these two results to obtain a uniform bound for $|f(z) - z|$ for all z in the annulus R.

For fixed $b \neq 1$ on the unit circle, we pose the problem
$$\mu_b = \max_{f \in \mathfrak{F}} |f(b) - b|, \qquad |b| = 1. \tag{1}$$
Actually, we are interested not in μ_b but in the *maximum maximorum*; namely,
$$\mu = \max_{|b|=1} \mu_b. \tag{2}$$
For this purpose, it is sufficient to solve the problem
$$\vartheta_b = \max_{f \in \mathfrak{F}} \arg\left\{\frac{f(b)}{b}\right\}, \qquad |b| = 1. \tag{3}$$
Indeed, the quantity
$$\bar{\vartheta} = \max_{|b|=1} \vartheta_b \tag{4}$$
is positive, since $\overline{f(\bar{z})} \in \mathfrak{F}$ whenever $f(z) \in \mathfrak{F}$. If $\bar{\vartheta} \leqq \pi$, elementary geometry provides the relation
$$\mu = 2 \sin \tfrac{1}{2} \bar{\vartheta}. \tag{5}$$

If $\bar{\vartheta} > \pi$, there is a function $g \in \mathfrak{F}$ such that the expression $\arg\{g(z)/z\}$ exceeds π somewhere on $|z| = 1$. Since this expression equals 0 at $z = 1$ and is continuous on $|z| = 1$, it must somewhere be precisely equal to π. In this case, the trivial estimate $|f(z) - z| \leqq 2$ will be sharp, and $\mu = 2$ if $\bar{\vartheta} > \pi$. Combining this with (5), we have
$$\mu = 2 \sin\left(\tfrac{1}{2} \min[\bar{\vartheta}, \pi]\right). \tag{6}$$

We shall solve (3) in an equivalent form by asking for
$$\max_{f \in \mathfrak{F}} \operatorname{Im}\{\log f(b)\}, \qquad |b| = 1. \tag{7}$$
(That branch of the logarithm is chosen for which $\log 1 = 0$.) This problem has a solution $f_0(z)$ because $\mathfrak{F}$ is a compact normal family in $r < |z| < \frac{1}{r}$. If the

variation function $V_\varrho(w)$ of (2.10) is applied, we obtain another function $V_\varrho(f_0(z))$ in $\mathfrak{F}$. The inequality

$$\operatorname{Im}\{\log V_\varrho(f_0(b))\} \leqq \operatorname{Im}\{\log f_0(b)\}$$

reduces to

$$\operatorname{Im}\left\{\frac{a\varrho^2(1-B)}{(B-w_0)(1-w_0)w_0} + \frac{\bar{a}\varrho^2(1-B)}{(1-\overline{w}_0 B)(1-\overline{w}_0)\overline{w}_0} + O(\varrho^3)\right\} \leqq 0, \tag{8}$$

where $B=f_0(b)$. Applying the identity $\operatorname{Im}\{\bar{\alpha}\} = -\operatorname{Im}\{\alpha\}$ and noting that $\overline{B} = \frac{1}{B}$, we transform (8) into

$$\operatorname{Im}\left\{\frac{2a\varrho^2(1-B)}{(B-w_0)(1-w_0)w_0} + O(\varrho^3)\right\} \leqq 0;$$

i.e.,

$$\operatorname{Re}\left\{\frac{a\varrho^2(1-B)}{i(B-w_0)(1-w_0)w_0} + O(\varrho^3)\right\} \leqq 0. \tag{9}$$

From (9) we conclude by the same method as before that $f_0(z)$ maps R onto the unit disk slit along an arc $w=w(t)$ satisfying

$$\frac{(w')^2(1-B)}{i(B-w)(1-w)w} > 0. \tag{10}$$

Since $f_0(z) \in \mathfrak{F}$, the arc $w(t)$ must pass through the origin. There is a unique solution curve to the differential equation (10) which satisfies this initial condition.

With $B=e^{i\beta}$ $(0<\beta<2\pi)$, we shall verify that

$$w(t) = -e^{i\frac{\beta}{2}}t, \qquad t>0, \tag{11}$$

does satisfy (10), hence is the unique solution. Indeed, substitution of (11) into the left-hand side of (10) gives

$$\frac{e^{i\beta}(1-e^{i\beta})}{-i(e^{i\beta}+e^{i\beta/2}t)(1+e^{i\beta/2}t)e^{i\beta/2}t} = \frac{-e^{-i\beta/2}+e^{i\beta/2}}{i|1+e^{i\beta/2}t|^2 t}$$
$$= \frac{2\sin\beta/2}{t|1+e^{i\beta/2}t|^2} > 0 \quad \text{for} \quad 0<\beta<2\pi.$$

Thus the extremal function $f_0(z)$ maps R onto $|w|<1$ slit radially from 0 to $-Me^{i\beta/2}$, where the length M is again given by (3.6).

We now turn our attention to the evaluation of the quantity μ defined in (2). Our first object is to show that μ may be expressed in terms only of the function $\Phi(z) \in \mathfrak{F}$ whose range is the unit disk with the *real* slit from 0 to M. Let

$$\nu = \max_{0 \leqq t \leqq 2\pi} \arg\left\{\frac{\Phi(e^{it})}{e^{it}}\right\}. \tag{12}$$

With $\bar{\vartheta}$ as in (4), we shall prove

$$\bar{\vartheta} = 2\nu. \tag{13}$$

Our investigations have shown that an extremal function $f_0(z) = f_0(z;b)$ for (3) maps R onto $|w|<1$ slit from 0 to $-Me^{i\beta/2}$, where $f_0(b;b)=e^{i\beta}$. Therefore, defining γ by the relation $\Phi(e^{i\gamma}) = -e^{-i\beta/2}$, we have by the uniqueness lemma of §1

$$f_0(z;b) = \frac{\Phi(e^{i\gamma}z)}{\Phi(e^{i\gamma})}. \tag{14}$$

Setting $z=b=e^{i\alpha}$ in (14) gives $\Phi(e^{i(\gamma+\alpha)}) = -e^{i\beta/2}$.

Since the range of $\Phi(z)$ is symmetric with respect to the real axis, we can infer from the uniqueness lemma

$$\overline{\Phi(\bar{z})} = \Phi(z), \tag{15}$$

and in particular

$$\overline{\Phi(e^{i\gamma})} = \Phi(e^{-i\gamma}) = -e^{i\frac{\beta}{2}}.$$

Hence, since $\Phi(z)$ is univalent, we find

$$e^{i(\gamma+\alpha)} = e^{-i\gamma}. \tag{16}$$

Thus, by (14), (15), and (16),

$$\frac{f_0(b;b)}{b} = \frac{f_0(e^{i\alpha};e^{i\alpha})}{e^{i\alpha}} = \frac{\Phi(e^{i(\gamma+\alpha)})}{e^{i\alpha}\,\Phi(e^{i\gamma})} = \frac{[\Phi(e^{-i\gamma})]^2}{e^{i\alpha}} = \left[\frac{\Phi(e^{-i\gamma})}{e^{-i\gamma}}\right]^2,$$

which implies $\bar{\vartheta} \leq 2\nu$.

Conversely, let γ be such that $\arg\left\{\frac{\Phi(e^{-i\gamma})}{e^{-i\gamma}}\right\} = \nu$, and define α by (16). Then

$$\frac{\Phi(e^{i(\gamma+\alpha)})}{e^{i\alpha}\,\Phi(e^{i\gamma})} = \left[\frac{\Phi(e^{-i\gamma})}{e^{-i\gamma}}\right]^2.$$

Since $\frac{\Phi(e^{i\gamma}z)}{\Phi(e^{i\gamma})} \in \mathfrak{F}$, this shows $2\nu \leq \bar{\vartheta}$, and the proof of (13) is complete. Equation (13) may be combined with (6) to give

$$\mu = 2\sin\left(\min\left[\nu, \frac{\pi}{2}\right]\right). \tag{17}$$

6. The Maximum Value of $|f(z) - z|$

We found in the last section that

$$\mu = \max_{f\in\mathfrak{F},\ |z|=1} |f(z) - z|$$

may be expressed by (5.17) in terms of the function $\Phi(z) \in \mathfrak{F}$ which maps R onto the unit disk slit along $0 \leq w \leq M$. Our remaining task is to evaluate the quantity

$$\nu = \max_{0\leq t\leq 2\pi} \arg\left\{\frac{\Phi(e^{it})}{e^{it}}\right\}$$

which occurs in (5.17).

Although $\Phi(z)$ is representable [3] by elliptic functions, we find it more advantageous to construct the inverse function $z = \psi(w)$ in terms of elliptic integrals. By Schwarz reflection ψ maps the entire plane slit from 0 to M and from $\frac{1}{M}$ to ∞ onto the annulus $r < |z| < \frac{1}{r}$. We shall now construct ψ explicitly.

The function

$$S(w) = \int_0^w \frac{dw}{\sqrt{w(M-w)(1-Mw)}} \tag{1}$$

maps the half-plane $\operatorname{Im}\{w\} > 0$ onto a rectangle with vertices $S(0) = 0$, $S(M)$, $S(1/M)$, and $S(\infty)$. Here $S(M)$ lies on the positive real axis, $S(\infty)$ on the positive imaginary axis. Therefore,

$$\sigma(w) = -\exp\left\{-\pi i \frac{S(w)}{S(M)}\right\}$$

maps $\operatorname{Im}\{w\} > 0$ onto the half annulus $1 < |z| < \varrho$, $\operatorname{Im}\{z\} > 0$, where $\varrho = \exp[\pi \operatorname{Im}\{S(1/M)\}/S(M)]$. Since $\sigma(w)$ is real in the real intervals $(-\infty, 0)$ and $(M, 1/M)$, it may be extended by reflection to a mapping of the entire exterior of the slits $0 \leqq w \leqq M$, $\frac{1}{M} \leqq w \leqq \infty$ onto the full annulus $1 < |z| < \varrho$. Thus the mapping by $\sigma(w)$ as well as that by the function $\psi(w)$ carries the same slit domain into a circular ring. The modulus being a conformal invariant, we have necessarily $\varrho = r^{-2}$.

The function $r\sigma(w)$ maps the doubly slit plane onto $r < |z| < \frac{1}{r}$; so does $\psi(w)$. To see that the two functions are identical, consider $F(z) = r\sigma(\psi^{-1}(z))$, a mapping of the annulus $r < |z| < \frac{1}{r}$ onto itself. Under $F(z)$ the inner boundary $|z| = r$ corresponds to itself, and $F(1) = r\sigma(1) > 0$. An argument similar to the proof of our uniqueness lemma (§1) now shows that $F(z) \equiv z$. Hence $\psi(w) = r\sigma(w)$; that is,

$$\psi(w) = -r\exp\left\{-\pi i \frac{S(w)}{S(M)}\right\}. \tag{2}$$

If $e^{i\varphi} = \Phi(e^{i\alpha})$, then $e^{i\alpha} = \psi(e^{i\varphi})$, and

$$\arg\left\{\frac{\Phi(e^{i\alpha})}{e^{i\alpha}}\right\} = -\arg\left\{\frac{\psi(e^{i\varphi})}{e^{i\varphi}}\right\}.$$

By (2),

$$\alpha = \arg\{\psi(e^{i\varphi})\} = \pi\left[1 - \frac{\operatorname{Re}\{S(e^{i\varphi})\}}{S(M)}\right]. \tag{3}$$

But we find from (1)

$$S(e^{i\varphi}) = S(1) - \int_0^{\varphi} \frac{dt}{|M - e^{it}|}. \tag{4}$$

Since $\operatorname{Re}\{S(1)\} = S(M)$, (3) and (4) yield

$$\alpha = \alpha(\varphi) = \frac{\pi}{S(M)} \int_0^{\varphi} \frac{dt}{|M - e^{it}|}.$$

Consequently, using the fact $\alpha(\pi) = \pi$ for the determination of $S(M)$, we find

$$\nu = \max_{0 \leqq \varphi \leqq \pi} [\alpha(\varphi) - \varphi] = \max_{0 \leqq \varphi \leqq \pi} D(\varphi), \tag{5}$$

where

$$D(\varphi) = \pi \frac{\int_0^{\varphi} \frac{dt}{|M - e^{it}|}}{\int_0^{\pi} \frac{dt}{|M - e^{it}|}} - \varphi. \tag{6}$$

The maximum of $D(\varphi)$ will occur at a value φ_0 for which $D'(\varphi_0)=0$, which gives

(7) $$\frac{\pi}{|M-e^{i\varphi_0}|}=\int_0^\pi \frac{dt}{|M-e^{it}|}.$$

By (5), (6), and (7) we have solved our extremum problem in principle and reduced it to a numerical calculation. It is of interest to make the following observations in this respect. From (6), (7), and (3.6) one finds after trivial but somewhat lengthy calculation

$$\nu = D(\varphi_0) = 4r - \tfrac{10}{3}r^3 + O(r^4).$$

Thus

(8) $$2\sin\nu = 8r - 28r^3 + O(r^4).$$

In view of (5.17), this last result shows $\mu \leqq 8r$ for sufficiently small r. But by (3.9) we have already established the estimate $5r$ for the inner boundary, so by the maximum modulus principle

(9) $$|f(z)-z| \leqq 8r \quad \text{for all} \quad f\in\mathfrak{F},\ \ z\in R,$$

provided r is sufficiently small. Since (9) is trivially true for $r\geqq\frac{1}{4}$, we are led to conjecture that the estimate holds uniformly for $0\leqq r\leqq 1$. This could be checked numerically. Suffice it to note that GAIER proved the existence of a constant C such that $|f(z)-z|\leqq Cr$ and showed $4\leqq C_0\leqq 12.6$, where C_0 is the smallest possible constant. We have now improved this estimate to $8\leqq C_0\leqq 12.6$.

This work was supported in part by Office of Naval Research Contract Nonr-225(11) at Stanford University.

References

[1] GAIER, D.: Untersuchungen zur Durchführung der konformen Abbildung mehrfach zusammenhängender Gebiete. Arch. Rational Mech. Anal. **3**, 149–178 (1959).

[2] GRÖTZSCH, H.: Über einige Extremalprobleme der konformen Abbildung. Leipzig Berichte **80**, 367–376 (1928).

[3] NEHARI, Z.: Conformal Mapping. New York: McGraw-Hill 1952.

[4] SCHIFFER, M.: A method of variation within the family of simple functions. Proc. London Math. Soc. **44**, 432–449 (1938).

Department of Mathematics
Stanford University
California

(Received September 18, 1961)

Commentary on

[75] P. L. Duren and M. Schiffer, *A variational method for functions schlicht in an annulus*, Arch. Rational Mech. Anal. **9** (1962), 260–272.

Max Schiffer always emphasized the importance of *methods*, which are often more valuable than isolated results. The paper [75] offers a case in point. The problem it treats was later solved more completely by simpler methods, but the variational method developed there has given results not accessible by other means.

When Dieter Gaier visited Stanford to give a Colloquium talk in 1961, he raised the problem that led to this paper, to find the maximum value of $|f(z)-z|$ for all points z in the annulus $R = \{z : r < |z| < 1\}$ and all functions $f \in \mathscr{F}$. Here $0 < r < 1$ and $\mathscr{F}$ is the family of conformal mappings of R into the unit disk with the properties $f(z) \neq 0$, $|f(z)| = 1$ for $|z| = 1$, and $f(1) = 1$. Gaier [Ga1] had shown that $|f(z)-z| \leq Cr$ for $C = 12.6$, but he asked for the best value of the constant C. The constant was important because it governed the speed of convergence of Gaier's iterative method for numerical construction of conformal mappings of multiply connected domains (see also [Ga2]). Applying the variational method introduced in [75], Schiffer and I determined the maximum value of $|f(z)-z|$ in terms of elliptic integrals, and we found it to be less than $8r$ for small r, the constant 8 being the best possible. We also conjectured that the bound $8r$ remains valid for all $r < 1$. This was confirmed by Gaier and Huckemann [GaHu], and the same result was obtained independently by Gehring and af Hällström [GeHa]. Huckemann [Hu] then found a simpler solution to the problem by the method of extremal length.

Gaier's problem could be solved by simpler methods because the extremal function maps the annulus onto a radially slit disk, but the variational method of [75] can be applied to solve problems with more complicated extremal functions. For instance, the method was used in [D] to find the maximum distortion $|f'(z_0)|$ for functions $f \in \mathscr{F}$ at a prescribed point $z_0 \in R$. Here the extremal function maps onto a radially slit disk when $|z_0|$ is near 1, but for $|z_0|$ near r the slit has a bifurcation. McLaughlin [McL1, McL2] later applied the variational method to other extremal problems for the family $\mathscr{F}$, and Menke [Me1, Me2] solved some similar problems. Kühnau [K1] used the method of [75] in his solution of an extremal problem for capacity raised by Gaier. Meanwhile, White [W] adapted the method of [90] to develop a quasiconformal analogue of the method of [75].

More recently, Kühnau [K2] has developed an elliptic generalization of the variational method introduced in [75]. The larger annulus

$$\tilde{R} = \{z : r < |z| < 1/r\}$$

is *diametrically symmetric* in the sense that the antipodal point $z^* = 1/\bar{z} \in \tilde{R}$ whenever $z \in \tilde{R}$, and each mapping $f \in \mathscr{F}$ extends by Schwarz reflection to a diametrically symmetric conformal mapping of $\tilde{R}$, since $f(z^*) = f(z)^*$ for every $z \in \tilde{R}$. On the other hand, each $f \in \mathscr{F}$ is *elliptically schlicht* in the sense that $f(z_1) \neq f(z_2)^*$ for all pairs of points $z_1, z_2 \in R$. Kühnau discovered that the variational method, while developed in [75] only for the special case of conformal mappings of an annulus R that belong to the family $\mathscr{F}$, can be applied (with suitable modifications) to arbitrary families of diametrically symmetric or elliptically schlicht conformal mappings.

References

[D] P. L. Duren, *Distortion in certain conformal mappings of an annulus*, Michigan Math. J. **10** (1963), 431–441; **11** (1964), 95.

[Ga1] Dieter Gaier, *Untersuchungen zur Durchführung der konformen Abbildung mehrfach zusammenhängender Gebiete*, Arch. Rational Mech. Anal. **3** (1959), 149–178.

[Ga2] Dieter Gaier, *Konstruktive Methoden der konformen Abbildung*, Springer-Verlag, 1964.

[GaHu] Dieter Gaier and Friedrich Huckemann, *Extremal problems for functions schlicht in an annulus*, Arch. Rational Mech. Anal. **9** (1962), 415–421.

[GeHa] F. W. Gehring and Gunnar af Hällström, *A distortion theorem for functions univalent in an annulus*, Ann. Acad. Sci. Fenn. Ser. A I **325** (1963).

[Hu] Friedrich Huckemann, *Über einige Extremalprobleme bei konformer Abbildung eines Kreisringes*, Math. Z. **80** (1962), 200–208.

[K1] Reiner Kühnau, *Über ein Modul-Kapazitätsproblem von D. Gaier*, Mitt. Math. Sem. Giessen **198** (1990), 85–119.

[K2] Reiner Kühnau, *Variation of diametrically symmetric or elliptically schlicht conformal mappings* J. Analyse Math. **89** (2003), 303–316.

[McL1] Renate McLaughlin, *Extremalprobleme für eine Familie schlichter Funktionen*, Math. Z. **118** (1970), 320–330.

[McL2] Renate McLaughlin, *Some extremal problems for functions univalent in an annulus*, Math. Scand. **28** (1971), 129–138.

[Me1] Klaus Menke, *Some properties of functions schlicht in an annulus*, Complex Variables Theory Appl. **5** (1985), 77–86.

[Me2] Klaus Menke, *Distortion theorems for functions schlicht in an annulus*, J. Reine Angew. Math. **375/376** (1987), 346–361.

[W] Alvin M. White, *An extremal problem for quasiconformal mappings in an annulus*, Proc. Amer. Math. Soc. **71** (1978), 267–274.

PETER DUREN

[81] (with B. Epstein) On the mean-value property of harmonic functions

[81] (with B. Epstein) On the mean-value property of harmonic functions. *J. Analyse Math.* **14** (1965), 109–111.

ON THE MEAN-VALUE PROPERTY OF HARMONIC FUNCTIONS

By

BERNARD EPSTEIN and M. M. SCHIFFER
in Stanford, California, U.S.A.

In a brief note with the same title [1], one of the authors has proven the following theorem: "Let D be a plane simply connected domain of finite area and t a point of D such that, for every function u harmonic in D and integrable over D, the mean value of u over the area of D equals $u(t)$. Then D is a disc and t is its center." The restriction to plane domains was necessitated in our proof by the use of analytic functions, while the restriction to a simply connected domain arose from the fact that the proof exploited the relationship between the "reproducing kernel" of such a domain and the functions which map it conformally onto a disc. It appears plausible, however, that these two restrictions are related only to the method of proof, and do not reflect the actual state of affairs. Indeed, we are now able to announce the following result:

Theorem: *Let D be any domain in Euclidean space, E^n, possessing finite measure, and let the complement of D possess non-empty interior. Suppose that there exists a point P_0 in D such that, for every function u harmonic in D and integrable over D, the mean value of u over D equals $u(P_0)$. Then D is a sphere with center at P_0.*

Proof. First we restrict consideration to the case that the domain D is bounded. For simplicity in exposition we consider tha case $n=3$, the modifications for other values of n being quite obvious. Also, we normalize the measure of D to unity. The integral $\iiint_D \text{grad}_Q\,(1/PQ)dV_Q$ converges for all points P in E^3 and defines a continuous vector-valued function $\overrightarrow{u(P)}$. For each point P in the complement of D the rectangular components of the integrand are harmonic functions of Q $(Q \in D)$; from the assumed mean-value property we then obtain the equality

$$\overrightarrow{u(P)} = \frac{1}{(P_0P)^3}\overrightarrow{P_0P}, \qquad P \notin D. \tag{1}$$

In order to analyze the behavior of $\overrightarrow{u(P)}$ for P in D, we construct a sphere S, with center at P_0, so large that $D \subset S$, and we write

$$\overrightarrow{u(P)} = \iiint_S \operatorname{grad}_Q\left(\frac{1}{PQ}\right)dV_Q - \iiint_{S-D} \operatorname{grad}_Q\left(\frac{1}{PQ}\right)dV_Q. \tag{2}$$

Now, the first integral in (2) works out to $\frac{4}{3}\pi\,\overrightarrow{P_0P}$, and so, for $P \in D$, we conclude that

$$\overrightarrow{u(P)} = \frac{4}{3}\pi\,\overrightarrow{P_0P} + \overrightarrow{h(P)} \tag{3}$$

where each component of $\overrightarrow{h(P)}$ is harmonic. Comparing (1) with (3) and taking account of the continuity of $\vec{u}$, we conclude that each component of the outer product $\overrightarrow{P_0P} \times \overrightarrow{h(P)}$ must vanish everywhere on the boundary of D. Since each component of this outer product is shown by a trivial computation to be harmonic, if follows, by the maximum principle, that the outer product vanishes everywhere in D. Therefore, $\overrightarrow{h(P)}$ is always parallel to $\overrightarrow{P_0P}$, and it is both solenoidal and irrotational. Any easy calculation shows that these three properties imply that $\overrightarrow{h(P)}$ vanishes identically. Thus, (3) simplifies to

$$\overrightarrow{u(P)} = \frac{4}{3}\pi\,\overrightarrow{P_0P}. \tag{4}$$

Since $\vec{u}$ remains continuous on the boundary of D, we conclude frcm (1) and (4) that $(P_0P)^3$ equals $3/4\pi$ everywhere on the boundary of D. Thus, the boundary of D lies entirely on the spherical surface Σ defined by the equality $P_0P = (3/4\pi)^{1/3}$. Since D possesses finite content, it is evident that it coincides with the interior of Σ. Thus, the proof is complete for *bounded* domains

Turning to the case that D is unbounded, it is readily seen that we may define the vector-valued function $\overrightarrow{u(P)}$ as before, and that both (1) and (3) continue to hold (While (2) now is not meaningful, we can, for any point P in D, employ a sphere S with center at P_0 which contains P in its interior and obtain (3); the vector $\overrightarrow{h(P)}$ is independent of the size of S, since the factor $\frac{4}{3}\pi$ appearing in (3) does not involve S.) We may conclude, as before, that the outer product $\overrightarrow{P_0P} \times \overrightarrow{h(P)}$ vanishes at all (finite) boundary points of D. The same is therefore true for $\overrightarrow{P_0P} \times \overrightarrow{u(P)}$. Since each component of $\overrightarrow{u(P)}$ is harmonic in D (as we see from (3)), we conclude as before that each component of $\overrightarrow{P_0P} \times \overrightarrow{u(P)}$ is harmonic in D. Simple estimates based on the integral used to define $\overrightarrow{u(P)}$ show that the quantities $\frac{1}{P_0P} u_{ik}(P)$, where $u_{ik}(P)$ denotes any component of $\overrightarrow{P_0P} \times \overrightarrow{u(P)}$, are bounded at infinity. If we now perform an inversion in any sphere which lies entirely in the complement of D, we obtain from the functions $u_{ik}(P)$ new harmonic functions $\tilde{u}_{ik}$ which vanish at every boundary point of $\tilde{D}$ (the image of D) except perhaps at the center of the sphere, which is the image of infinity. The *boundedness* of the functions $\frac{1}{P_0P} u_{ik}$ enables us to conclude that the functions $\tilde{u}_{ik}$ vanish identically in $\tilde{D}$, and so the functions u_{ik} must vanish identically in D. We may now argue, as in the bounded case, that the boundary of D lies on the sphere Σ defined previously, and so we conclude that D is indeed a sphere with center at P_0.

The possibility still remains that there exists a domain D which possesses finite content and is dense in E^n for which the mean-value property stated in the theorem holds. Our use of inversion prevents us from ruling out this possibility, but it appears likely that a more delicate argument may succeed.

Reference

1. B. Epstein, On the Mean-Value Property of Harmonic Functions, Amer. Math. Soc. 13 (1962), p. 830.

Department of Mathematics
Stanford University
Stanford, California, U.S.A.

(Received June 1st, 1964)

Commentary on

[81] Bernard Epstein and M.M. Schiffer, *On the mean-value property of harmonic functions*, J. Analyse Math. 14 (1965), 109–111.

In [E], Epstein had shown that if D is a simply connected domain of finite area in the plane and $P_0 \in D$ is a point such that for every function u harmonic in D and integrable over D, the mean value of u over the area of D equals $u(P_0)$, then D must be a disc centered at P_0. In [81], Epstein and Schiffer generalize this result to the case in which D is a domain in euclidean space $\mathbb{R}^n$ ($n \geq 2$) whose complement contains an open set. A few years later, in another very short paper [K] bearing the very same title as its predecessors, Kuran gave an extremely simple proof of this result, assuming only that D is a domain in $\mathbb{R}^n$. In fact, as noticed by Netuka [N, p. 405], even the condition that D be connected turns out to be unnecessary. Thus, the final result may be stated as follows.

Theorem. *Let D be an open set of finite volume in $\mathbb{R}^n$ ($n \geq 2$). Suppose there exists a point $P_0 \in D$ such that for every function u harmonic in D and integrable over D, the (volume) mean value of u over D equals $u(P_0)$. Then D is an open ball with center at P_0.*

Proof. We may assume $P_0 = 0$. Let $y \in \mathbb{R}^n \setminus D$ have minimal distance r from 0 and denote by B and $\overline{B}$, respectively, the open and closed balls of radius r about 0; then, clearly, $B \subset D$. By the mean-value theorem for harmonic functions and the hypothesis on D, it follows that $\int_{D\setminus\overline{B}} u\,dx = 0$ whenever u is an integrable harmonic function on D satisfying $u(0) = 0$. Now put

$$K(x) = (|x|^2 - |y|^2)/|x-y|^n$$

for $x \in \mathbb{R}^n \setminus \{y\}$ and set $u = K - K(0)$. Then u is harmonic on D and integrable there, and $u(0) = 0$; so $\int_{D\setminus\overline{B}} u\,dx = 0$. Since $u(x) > 0$ for $x \in D \setminus \overline{B}$, the open set $D \setminus \overline{B}$ must have zero measure. Thus $B \subset D \subset \overline{B}$, and so $D = B$.

Further generalizations are also possible. For instance, Armitage and Goldstein [AG] showed that the above theorem holds in case u is restricted to the class of *positive* harmonic functions in $L^p(D)$ for any fixed $0 < p < n/(n-2)$. They also prove that if the mean-value property is assumed only for bounded harmonic functions, then $D = B \setminus E$, where B is an open ball and E is a polar set. For further results on these and related matters, see the extremely comprehensive survey [NV].

Later, Aharonov, Schiffer, and Zalcman took up a closely related problem in potential theory [126]; see the comment on that paper in this volume.

The results discussed above are also intimately connected with the subject of *quadrature domains*, for which, see [GS] and the other papers in the same volume.

References

[AG] D.H. Armitage and M. Goldstein, *The volume mean-value property of harmonic functions*, Complex Variables Theory Appl. **13** (1990), 185–193.

[E] Bernard Epstein, *On the mean-value property of harmonic functions*, Proc. Amer. Math. Soc. **13** (1962), 80.

[GS] B. Gustafsson and H.S. Shapiro, *What is a quadrature domain?*, Quadrature Domains and Their Applications, Birkhäuser Verlag, 2005, pp. 1–25.

[K] Ü. Kuran, *On the mean-value property of harmonic functions*, Bull. London Math. Soc. **4** (1972), 311–312.

[N] Ivan Netuka, *Harmonic functions and mean value theorems*, Časopis Pešt. Mat. **100** (1975), 391–409.

[NV] Ivan Netuka and Jiří Veselý, *Mean value property and harmonic functions*, Classical and Modern Potential Theory and Applications, Kluwer Acad. Publ., 1994, pp. 359–398.

LAWRENCE ZALCMAN

[87] (with N. S. Hawley) Half-order differentials on Riemann surfaces

[87] (with N. S. Hawley) Half-order differentials on Riemann surfaces. *Acta Math.* **115** (1966), 199–236.

HALF-ORDER DIFFERENTIALS ON RIEMANN SURFACES

BY

N. S. HAWLEY and M. SCHIFFER

Stanford University, Stanford, Calif., U.S.A. (1)

Dedicated to Professor R. Nevanlinna on the occasion of his 70th birthday

Introduction

In this paper we wish to exhibit the utility of differentials of half integer order in the theory of Riemann surfaces. We have found that differentials of order $\frac{1}{2}$ and order $-\frac{1}{2}$ have been involved implicitly in numerous earlier investigations, e.g., Poincaré's work on Fuchsian functions and differential equations on Riemann surfaces. But the explicit recognition of these differentials as entities to be studied for their own worth seems to be new. We believe that such a study will have a considerable unifying effect on various aspects of the theory of Riemann surfaces, and we wish to show, by means of examples and applications, how some parts of this theory are clarified and brought together through investigating these half-order differentials.

A strong underlying reason for dealing with half-order differentials comes from the general technique of contour integration; already introduced by Riemann. In the standard theory one integrates a differential (linear) against an Abelian integral (additive function) and uses period relations and the residue theorem to arrive at identities. As we shall demonstrate, one can do an analogous thing by multiplying two differentials of order $\frac{1}{2}$ and using the same techniques of contour integration.

As often happens, when one discovers a new (at least to him) entity and starts looking around to see where it occurs naturally, one is stunned to find so many of its hiding places —and all so near the surface.

Our current point of view concerning the study of Riemann surfaces has evolved from an earlier one in which we introduced the notion of a *meromorphic connection* in analogy with classical notions in real differential geometry; we now view the theory of connections

(1) This work was supported in part by National Science Foundation grant GP 4069 and Air Force contract AF 49(638) 1345 at Stanford University.

on Riemann surfaces as being the theory of first order linear differential equations. The present paper is concerned with the next step—second order linear differential equations. The remarkable difference between first and second order linear differential equations on a closed Riemann surface of genus greater than one is that there exist everywhere regular second order equations, although each first order equation must be singular.

I. The Schwarzian differential parameter and related connections

1. In the theory of conformal mappings and univalent functions the following expression plays a central role. If $f(z)$ is an analytic function in the plane domain D, the function [5, 8, 15, 20]

$$F(z,\zeta)=\log\frac{f(z)-f(\zeta)}{z-\zeta}\tag{1}$$

is analytic in the Cartesian product domain $D\times D$ except for logarithmic poles. A necessary and sufficient condition for $f(z)$ to be univalent in D is the regularity of $F(z,\zeta)$ in $D\times D$. Since a linear transformation

$$f^*(z)=\frac{af(z)+b}{cf(z)+d}\tag{2}$$

does not affect the univalence of the function, it is to be expected that the corresponding function in two variables $F^*(z,\zeta)$ stands in a simple relation to $F(z,\zeta)$. Indeed, we find

$$F^*(z,\zeta)=F(z,\zeta)+\log\frac{ad-bc}{(cf(z)+d)\,(cf(\zeta)+d)}.\tag{3}$$

Thus it seems useful to define

$$U(z,\zeta)=\frac{\partial^2}{\partial z\,\partial\zeta}F(z,\zeta)=\frac{f'(z)f'(\zeta)}{[f(z)-f(\zeta)]^2}-\frac{1}{(z-\zeta)^2}\tag{4}$$

which is in view of (3) invariant under linear transformation. Clearly, the univalence of $f(z)$ in D is still equivalent to the regularity of $U(z,\zeta)$ in $D\times D$.

Let $w=f(z)$ and $\omega=f(\zeta)$ and define

$$[w,\omega;\,z,\zeta]=\frac{f'(z)f'(\zeta)}{[f(z)-f(\zeta)]^2}-\frac{1}{(z-\zeta)^2}.\tag{5}$$

If we consider $z=g(t)$, $\zeta=g(\tau)$, we can form the analogous expression

$$[z,\zeta;\,t,\tau]=\frac{g'(t)g'(\tau)}{[g(t)-g(\tau)]^2}-\frac{1}{(t-\tau)^2}.\tag{6}$$

It is easy to verify that for $w=f\{g(t)\}$, $\omega=f\{g(\tau)\}$ we have

$$[w, \omega;\, z, \zeta]dz d\zeta + [z, \zeta;\, t, \tau]dt d\tau = [w, \omega;\, t, \tau]dt d\tau. \tag{7}$$

This additive law for $[w, \omega;\, z, \zeta]dz d\zeta$ under composition of mappings is of significant value in the theory of conformal mappings.

If we identify the two arguments in $U(z, \zeta)$, we obtain

$$U(z, z) = [w, w;\, z, z] = -\frac{1}{6}\left[\left(\frac{f''(z)}{f'(z)}\right)' - \frac{1}{2}\left(\frac{f''(z)}{f'(z)}\right)^2\right]. \tag{8}$$

We are thus led in a natural way to the Schwarz differential parameter

$$\{w, z\} = \left(\frac{f''(z)}{f'(z)}\right)' - \frac{1}{2}\left(\frac{f''(z)}{f'(z)}\right)^2 \tag{9}$$

in terms of which (8) can be expressed as

$$U(z, z) = [w, w;\, z, z] = -\tfrac{1}{6}\{w, z\}. \tag{8'}$$

From the preceding properties of $U(z, \zeta)$ follow then the well-known properties of the Schwarzian differential parameter:

(a) $\{w, z\}$ is invariant under a linear transformation of w.

(b) Under the composition $w=f(z)$ and $z=g(t)$ we have

$$\{w, z\}dz^2 + \{z, t\}dt^2 = \{w, t\}dt^2. \tag{10}$$

From (a) follows that for the linear transformation

$$z = \frac{\alpha t + \beta}{\gamma t + \delta} \tag{11}$$

holds identically $\{z, t\}=0$. Hence we infer from (b) in this case:

$$\{w, t\}dt^2 = \{w, z\}dz^2 \tag{12}$$

that is, the Schwarzian differential parameter transforms like a quadratic differential under a linear transformation of the independent variable.

Finally, let in (10) $t=w$ and use the fact that in this case again $\{w, t\}=0$. Thus

$$\{w, z\}dz^2 = -\{z, w\}dw^2 \tag{13}$$

which determines the Schwarzian differential parameter of inverse functions.

We may consider w as a function on the domain D and the independent variable z as local coordinate. Then the mapping $z = g(t)$ can be conceived as a change of coordinates and the transformation law (10) shows that the Schwarzian differential parameter $\{w, z\}$ transforms under such change of coordinates according to a linear inhomogeneous law of transformations. The expression behaves almost like a quadratic differential; however, we have to add the inhomogeneity $\{z, t\}dt^2$ which does not depend on the function w considered, but only on the transformation law from z to t. We may call $\{w, z\}$ a connection in analogy to the corresponding concept in differential geometry [18].

2. Naturally we are now led to consider connections analogous to the Schwarzian differential parameter on Riemann surfaces. We ask for quantities S_z defined in terms of the local uniformizer z, which transform under a change of uniformizer $z = g(t)$ according to the law

$$S_t(t)\,dt^2 = S_z(z)\,dz^2 + \{z, t\}\,dt^2 \tag{14}$$

Clearly, it is enough to construct on a given Riemann surface only one such Schwarzian connection. For, if S_z and $\hat{S}_z$ are two such expressions with the law of transformation (14), their difference would be a quadratic differential on the surface and this class is well known and completely understood.

It is now remarkable that on every Riemann surface $\mathfrak{R}$ there does indeed exist a regular analytic Schwarzian connection S. In order to construct it, we introduce the Abelian integral of the third kind [10, 14, 22] $w(p; r, s)$ which is analytic in p, has logarithmic poles with residues $+1$ and -1, respectively, at the two given points $r \in \mathfrak{R}$ and $s \in \mathfrak{R}$ and which is normalized to have the periods zero with respect to the cross cuts $\mathfrak{A}_\nu$ of a canonical cut system $\{\mathfrak{A}_\nu, \mathfrak{B}_\nu\}$. The analytic dependence of the Abelian integral so defined upon its parameters is best understood by means of the fundamental theorem that for every quadruple $p, q; r, s$ on $\mathfrak{R}$ the combination

$$W(p, q; r, s) = w(p; r, s) - w(q; r, s) \tag{15}$$

is symmetric in the pair p, q of arguments and r, s of parameters. In particular, we see that

$$\frac{\partial^2 w(p; r, s)}{\partial p\, \partial r} = \frac{\partial^2 w(r; p, q)}{\partial r\, \partial p} = \frac{\partial^2 W(p, q; r, s)}{\partial p\, \partial r} \tag{16}$$

depends analytically on p and r, is symmetric in p and r and is independent of s and q. This expression has a singularity if $p = r$, and to study it we introduce a local uniformizer z such that p has the coordinate z and r has the coordinate ζ. We then find that

$$\frac{\partial^2 W(p, q; r, s)}{\partial z\, \partial \zeta} = \frac{1}{(z-\zeta)^2} + l_z(z, \zeta), \tag{17}$$

where $l_z(z, \zeta)$ is symmetric in both variables and regular analytic in the uniformizer neighborhood. If we change uniformizers by the analytic relations $z = g(t)$, $\zeta = g(\tau)$, we find by use of (6)

$$l_t(t, \tau)\,dt\,d\tau = l_z(z, \zeta)\,dz\,d\zeta + [z, \zeta; t, \tau]\,dt\,d\tau. \tag{18}$$

Let us then define the expression

$$S_z(z) = -6\,l_z(z, z) = \lim_{\zeta \to z}\left[-6\left(\frac{\partial^2 W(p, q; r, s)}{\partial z\, \partial \zeta} - \frac{1}{(z-\zeta)^2}\right)\right]. \tag{19}$$

This is an analytic function in each uniformizer neigborhood which depends, however, upon the choice of uniformizer. In view of (8′) and (18) we have

$$S_t(t)\,dt^2 = S_z(z)\,dz^2 + \{z, t\}\,dt^2. \tag{20}$$

A comparison of this transformation law with (14) shows that S satisfies the proper transformation rule of a connection.

We obtain all Schwarzian connections on $\mathfrak{R}$ by adding to the expression $S_z(z)$ constructed explicitly any quadratic differential on $\mathfrak{R}$. However, we may construct another Schwarzian connection in a different way and arrive at a remarkable identity. Let $\varphi(p)$ be the analytic function which is defined on the universal covering surface $\mathfrak{R}_u$ of $\mathfrak{R}$ and maps $\mathfrak{R}_u$ onto the unit disk. It is well known that the $\varphi(p)$ is polymorphic on $\mathfrak{R}$, that is, at points of $\mathfrak{R}_u$ over the same point of $\mathfrak{R}$ the function $\varphi(p)$ has different determinations which are related by a linear transformation. Thus, if $\hat{\varphi}(p)$ and $\varphi(p)$ are two such determinations, then

$$\hat{\varphi}(p) = L[\varphi(p)], \tag{21}$$

where L transforms the unit disk onto itself. If we introduce a local uniformizer z at p, we see that in view of (21)

$$\{\varphi, z\} = \{\hat{\varphi}, z\} \tag{22}$$

that is, the Schwarzian differential parameter of the polymorphic function $\varphi(p)$ is the same for all branches of this function; it is single-valued and analytic in each uniformizer neighborhood on $\mathfrak{R}$. It depends, however, on the choice of the uniformizer; if we replace z by t through the analytic relation $z = g(t)$, we find by (10)

$$\{\varphi, t\}\,dt^2 = \{\varphi, z\}\,dz^2 + \{z, t\}\,dt^2. \tag{23}$$

Thus $\{\varphi, z\}$ has the same transformation law of a Schwarzian connection as $S_z(z)$. Hence we have

$$\{\varphi, z\} = -6l_z(z, z) + Q(z), \tag{24}$$

where $Q(z)$ is a regular quadratic differential on $\mathfrak{R}$.

Since the Abelian integrals and quadratic differentials of a surface $\mathfrak{R}$ are of a more elementary character than the uniformizing transcendatal function $\varphi(z)$, we may consider (24) as a useful differential equation for $\varphi(z)$ in terms of the easier accessible expressions $l_z(z, z)$ and $Q(z)$.

II. Schwarz' differential equation and half-order differentials

1. We return to the case of planar domains D. Suppose that a function $S(z)$ is given in D and that we wish to find the solution of the differential equation of the third order in $w(z)$,

$$\{w, z\} = S(z). \tag{1}$$

Schwarz showed that the solution of this nonlinear differential equation can be reduced to the simpler problem of finding two independent solutions of the linear second order equation

$$u''(z) + \tfrac{1}{2} S(z)\, u(z) = 0. \tag{2}$$

Indeed, if $u_1(z)$ and $u_2(z)$ are independent solutions of (2), their Wronskian

$$u_1(z)\, u_2'(z) - u_2(z)\, u_2'(z) = W(z) \tag{3}$$

will be not identically zero. On the other hand, we see from (2) at once that $W(z)$ must be a constant; thus we may choose $W(z) \equiv 1$. In any case, as a simple calculation shows, the ratio

$$w(z) = \frac{u_1(z)}{u_2(z)} \tag{4}$$

will satisfy the differential equation (1).

Let us now change the independent variable by a conformal mapping $z = g(t)$ and refer to the independent variable t in a domain Δ. Clearly, $w = w[g(t)]$ is defined in Δ by composition and we have

$$\{w, t\}\, dt^2 = S(z)\, dz^2 + \{z, t\}\, dt^2. \tag{5}$$

To find w as a function of t, we might also consider the corresponding linear second order equation

$$v''(t) + \tfrac{1}{2} \{w, t\}\, v(t) = 0 \tag{6}$$

and express w as the ratio of two independent solutions of this equation.

However, the natural question arises whether there is any relation between the solutions $u_\nu(z)$ of (2) and $v_\nu(t)$ of (6). An easy calculation shows that

$$v_\nu(t) = u_\nu(g(t))[g'(t)]^{-\frac{1}{2}}, \quad \nu = 1, 2 \tag{7}$$

yields a system of two independent solutions of (6) whose Wronskian has also the value 1.

We have thus the remarkable fact: If S transforms like a Schwarzian connection

$$S_t(t)\,dt^2 = S_z(z)\,dz^2 + \{z, t\}\,dt^2 \tag{8}$$

the solutions of the differential equation

$$u''(z) + \tfrac{1}{2} S_z(z)\,u(z) = 0 \tag{9}$$

transform like differentials of order $-\frac{1}{2}$, i.e.,

$$u_t(t)\,dt^{-\frac{1}{2}} = u_z(z)\,dz^{-\frac{1}{2}}. \tag{10}$$

2. To show the usefulness of this covariance of the auxiliary functions $u_\nu(z)$, we rederive an interesting theorem of Nehari [13] which connects the univalence of a function $f(z)$ analytic in the unit disk with the growth of its Schwarzian differential parameter $S(z) = \{f, z\}$.

Suppose that $f(z)$ is not univalent in $|z| < 1$. There would be two different points in the disk, say a and b, such that $f(a) = f(b)$. By a linear transformation of the independent variable we can achieve that $a = 0$, $b = r > 0$, and by a linear transformation of $f(z)$ we can achieve that $f(0) = f(r) = 0$. We introduce now the solutions $u_1(z)$ and $u_2(z)$ of the differential equation (2) and express $f(z)$ as their ratio

$$f(z) = \frac{u_1(z)}{u_2(z)}. \tag{11}$$

The non-univalence of $f(z)$ leads to the conclusion

$$u_1(0) = u_1(r) = 0. \tag{12}$$

To utilize this equation we map the unit disk onto the strip

$$-\tfrac{1}{4}\pi < \operatorname{Im}\{t\} < \tfrac{1}{4}\pi \tag{13}$$

by means of the function

$$z = g(t) = \tanh t, \qquad t = \tfrac{1}{2}\log\frac{1+z}{1-z}. \tag{14}$$

The points $z = 0$ and $z = r$ go into the points $t = 0$, $t = \frac{1}{2}\log((1+r)/(1-r)) = \varrho > 0$ on the real axis. By virtue of (10) we know that the functions

$$v_\nu(t) = u_\nu[g(t)]g'(t)^{-\frac{1}{2}} \tag{15}$$

are defined in the strip (13) and satisfy the differential equation

$$v_\nu''(t) + \tfrac{1}{2} S_t(t) v_\nu(t) = 0, \tag{16}$$

where $S_t(t)$ is obtained from $S(z)$ by the transformation (8). Equation (12) implies

$$v_1(0) = v_1(\varrho) = 0. \tag{17}$$

Consider equation (16) for $\nu = 1$, multiply it by $\overline{v_1(t)}$ and integrate the result along the real t-axis from 0 to ϱ. Integration by parts and the boundary conditions (17) lead to

$$\frac{1}{2}\int_0^\varrho S_t(t)\,|v_1(t)|^2\,dt = \int_0^\varrho |v_1'(t)|^2\,dt \geqslant 0. \tag{18}$$

We compute now the factor $S_t(t)$ by means of (8) and (14). We find by an easy computation

$$S_t(t) = S(z)\cdot(1-z^2)^2 - 2. \tag{19}$$

In particular, if on the real axis

$$|S(z)| < \frac{2}{(1-z^2)^2} \tag{20}$$

we clearly have $\operatorname{Re}\{S_t(t)\} < 0$ for real t and the inequality (18) is impossible. On the other hand, the inequality

$$|S(z)|\,|dz|^2 < \frac{2\,|dz|^2}{(1-|z|^2)^2} \tag{21}$$

is invariant under linear transformations in view of (1.12) and the invariance of the non-Euclidean line element in the unit disk. Thus, if the Schwarzian differential parameter $S(z) = \{f, z\}$ satisfies the inequality

$$|\{f, z\}| < \frac{2}{(1-|z|^2)^2}, \quad |z| < 1 \tag{21'}$$

we can assert that $f(z)$ cannot take the same value at two different points in the unit disk and $f(z)$ must be univalent.

(21′) is Nehari's sufficient condition for the univalence of $f(z)$. It is also known that the inequality [5, 13]

$$|\{f, z\}| \leqslant \frac{6}{(1-|z|^2)^2} \tag{22}$$

is necessary for univalence. However, the gap between the two conditions (21′) and (22) cannot be narrowed since Hille showed that [11]

$$f(z) = A\left(\frac{1-z}{1+z}\right)^{\delta}, \qquad \{f, z\} = \frac{2(1-\delta^2)}{(1-z^2)^2} \tag{23}$$

is univalent for real δ, but non-univalent for arbitrarily small imaginary δ. Thus, the constant 2 in the sufficient condition (21′) cannot be replaced by any larger one.

We gave the above derivation of (21′) to show how useful the covariance of the auxiliary functions $u_\nu(z)$ can be. It allows us a great freedom in simplifying transformations and a clear understanding of the meaning of (21′). The reader may deduce from (18) many other (though not so elegant) conditions on $\{f, z\}$ which would ensure the univalence of $f(z)$.

3. As we have shown in Section I.2, we have Schwarzian conections on every Riemann surface $\mathfrak{R}$. We may thus consider the second order differential equation

$$\frac{d^2 u_z}{dz^2} + \frac{1}{2} S_z(z)\, u_z(z) = 0 \tag{24}$$

in each neighborhood with a uniformizing parameter z and continue the differential equation into adjacent neighborhoods with $z = g(t)$ through the transformation laws

$$\left.\begin{aligned} u_t(t) &= u_z[g(t)]\,[g'(t)]^{-\frac{1}{2}}, \\ S_t(t)\,dt^2 &= S_z(z)\,dz^2 + \{z, t\}\,dt^2. \end{aligned}\right\} \tag{25}$$

Thus we can express the differential equation (24) in an invariant manner as

$$\frac{d^2 u(p)}{dp^2} + \frac{1}{2} S(p)\, u(p) = 0 \tag{24′}$$

for all points $p \in \mathfrak{R}$.

We start with an arbitrary but fixed pair $u_\nu(p)$, $\nu = 1, 2$, of independent solutions of (24′) and continue them analytically along a closed path Γ on the surface. By the principle of permanence of analytic relations the functions will remain solutions of (24′) under this continuation, and on returning to the starting point on Γ we will arrive with new solution functions $u_\nu^{(\Gamma)}(p)$. However, these new determinations must be linear combinations of the original solution set $u_\nu(p)$. Thus we have

$$u_\nu^{(\Gamma)}(p) = A_{\nu 1}(\Gamma)\, u_1(p) + A_{\nu 2}(\Gamma)\, u_2(p). \tag{26}$$

The matrix

$$A(\Gamma) = \begin{pmatrix} A_{11}(\Gamma) & A_{12}(\Gamma) \\ A_{21}(\Gamma) & A_{22}(\Gamma) \end{pmatrix} \tag{26′}$$

is complex-valued and depends on the cycle Γ described.

We consider the Wronskian

$$W(p) = u_1(p)\,u_2'(p) - u_2(p)\,u_1'(p). \tag{27}$$

It is easily seen that in view of the law of transformation (25) this expression is independent of the choice of uniformizers and that because of the differential equation (24) it is a constant. We may assume without loss of generality that $W(p) \equiv 1$. It then follows from (26) and (27) that

$$W^{(\Gamma)}(p) = \|A(\Gamma)\| \cdot W(p), \quad \|A(\Gamma)\| = 1. \tag{28}$$

The transformation matrices for the fundamental system $u_\nu(p)$ are unimodular for every cycle Γ. Thus there exists one constraint among the four elements of the matrix $A(\Gamma)$, and we find that each such matrix depends on three independent complex parameters.

If $\mathfrak{R}$ is of genus g, we can select a canonical set of cross cuts $\{\mathfrak{A}_\nu, \mathfrak{B}_\nu\}$, $\nu = 1, \ldots, g$, and express each homotopy class of curves on $\mathfrak{R}$ in terms of the homotopy classes of these cross cuts, i.e., we take the homotopy classes of the cross cuts as generators of the fundamental group of $\mathfrak{R}$. If Γ is a closed curve on $\mathfrak{R}$, then $A(\Gamma)$ depends only on the homotopy class of Γ and not on Γ itself. It is therefore sufficient to study the $2g$ unimodular matrices $A(\mathfrak{A}_\nu)$ and $A(\mathfrak{B}_\nu)$. We note that the transformation matrices of different cross cuts do not necessarily commute, and the matrix $A(\Gamma)$ of curve Γ depends on the *homotopy class* of Γ and not merely on the *homology class* of the cycle which Γ gives rise to. The study of the various matrices $A(\Gamma)$ leads therefore to a deeper theory of the moduli of a Riemann surface than that of the period matrices of Abelian integrals.

The $2g$ matrices $A(\mathfrak{A}_\nu)$, $A(\mathfrak{B}_\nu)$ depend on $6g$ complex parameters. However, we have a certain freedom in the choice of the solutions $u_\nu(p)$ whose transformations they represent. A change of the fundamental system $u_\nu(p)$ leads to a similarity transformation

$$\hat{A}(\Gamma) = M^{-1}A(\Gamma)M \tag{29}$$

of the corresponding matrices. It is easily seen that M contains three essential complex parameters such that the $2g$ matrices depend on $6g-3$ complex numbers.

We may count, on the other hand, the freedom in the choice of Schwarzian connections. Let $S(p)$ be the specific connection constructed in (1.19) from the Abelian integral of the third kind. Then the most general Schwarzian connection which is regular on $\mathfrak{R}$ is of the form

$$S(p;\lambda_\nu) = S(p) + \sum_{\nu=1}^{3g-3} \lambda_\nu Q_\nu(p), \tag{30}$$

where the $Q_\nu(p)$ are a base for all quadratic differentials, regular on $\mathfrak{R}$. We see that the Schwarzian connections form a linear manifold depending on $3g-3$ complex parameters.

It would be more precise to denote the matrices A as $A(\Gamma;\lambda_\nu)$ to describe the cycle as well as the specific differential equation from which they arise. Finally, it should be observed that the transformation matrices $A(\Gamma;\lambda_\nu)$ depend also on the moduli of the Riemann surface $\mathfrak{R}$ considered. The number of these moduli is well known to be $3g-3$. Thus the $6g-3$ complex parameters which determine the transformation matrices depend on the following $6g-6$ complex parameters: The $3g-3$ moduli of the surface and the $3g-3$ accessory parameters λ_ν.

One would therefore expect that three further relations should hold between the parameters determining the $2g$ matrices $A(\Gamma;\lambda_\nu)$. This is indeed the case since the totality of cross cuts transforms $\mathfrak{R}$ into a simply-connected domain and the continuation of each $u_\nu(p)$ along the boundary of this domain must return each function to its initial value. Thus, under proper numeration of the cross cuts, we must have the relation

$$\prod_{\alpha=1}^{g} A(\mathfrak{A}_\alpha;\lambda_\nu)A(\mathfrak{B}_\alpha;\lambda_\nu)A^{-1}(\mathfrak{A}_\alpha;\lambda_\nu)A^{-1}(\mathfrak{B}_\alpha;\lambda_\nu)=I \tag{28'}$$

which represents the sought additional three constraints on the elements of the unimodular matrices.

The enumeration of parameters at our disposal and of the essential parameters in the transformation matrix set shows that the set $A(\mathfrak{A}_\nu;\lambda_\mu)$, $A(\mathfrak{B}_\nu;\lambda_\mu)$ may be considered as a possible set of moduli for the surface $\mathfrak{R}$. On the other hand, we are led to the interesting problem of determining those coefficient vectors λ_ν which lead to important classes of transformation groups. For example, the question arises how to determine those λ_ν whose corresponding solution set $u_\nu(p)$ has as ratio the polymorphic function $\varphi(p)$, discussed in Section I.2, which maps the universal covering surface of $\mathfrak{R}$ onto the unit disk.

4. Let us consider a domain D in the complex plane and let $g(z,\zeta)$ be its Green's function. We form the analytic kernel [5, 8, 20]

$$L(z,\zeta)=-\frac{2}{\pi}\frac{\partial^2 g(z,\zeta)}{\partial z\,\partial\zeta}=\frac{1}{\pi(z-\zeta)^2}-l(z,\zeta) \tag{31}$$

which plays a central role in the theory of the Bergman kernel function. $l(z,\zeta)$ is regular analytic in D and $L(z,\zeta)$ has a double pole at $z=\zeta$, as is explicitly displayed in (31).

From the conformal invariance of the Green's function follows that under a mapping $z=g(t)$, $\zeta=g(\tau)$ holds

$$L(z,\zeta)\,dz\,d\zeta=\hat{L}(t,\tau)\,dt\,d\tau. \tag{32}$$

In terms of $l(z,\zeta)$ this leads to the transformation law

$$l(z, \zeta)\, dz\, d\zeta = l(t, \tau)\, dt\, d\tau + \frac{1}{\pi}\,[z, \zeta; t, \tau]\, dt\, d\tau, \tag{33}$$

where $[z, \zeta; t, \tau]$ is defined in (1.6). In particular, by (1.8′) we have

$$l(z, z)\, dz^2 = l(t, t)\, dt^2 - \frac{1}{6\pi}\,\{z, t\}\, dt^2. \tag{34}$$

We recognize that

$$S(z) = 6\pi\, l(z, z) \tag{35}$$

transforms as a Schwarzian connection.

Let us suppose that the boundary of the domain is analytic and admits a parametrization $z = z(s)$. Denoting

$$\dot{z} = \frac{dz}{ds} \tag{36}$$

we can easily derive from the boundary behavior of the Green's function that

$$L(z, \zeta)\, \dot{z}\dot{\zeta} = \text{real}, \quad z, \zeta \in \partial D,\ z \neq \zeta. \tag{37}$$

Hence,

$$\operatorname{Im}\{l(z, \zeta)\, \dot{z}\dot{\zeta}\} = \frac{1}{\pi} \operatorname{Im}\left\{\frac{\dot{z}\dot{\zeta}}{(z-\zeta)^2}\right\}. \tag{38}$$

The right-hand side depends here in an elementary way on the geometry of ∂D, while the left-hand term is an expression involving the much deeper Green's function of the domain. In particular, letting $z = \zeta$ and making an elementary computation, we find

$$\operatorname{Im}\{l(z, z)\, \dot{z}^2\} = \frac{1}{6\pi} \frac{d}{ds}\, \varkappa(s), \quad z = z(s), \tag{39}$$

where $\varkappa(s)$ is the curvature of the boundary curve at $z(s)$.

We are now able to understand the significance of the differential equation

$$u''(z) + [3\pi\, l(z, z) + \Sigma\, \lambda_\nu\, Q_\nu(z)]\, u(z) = 0 \tag{40}$$

with real λ_ν and where the $Q_\nu(z)$ are a basis for all real quadratic differentials of D, i.e., of all $Q(z)$ which satisfy on ∂D the condition

$$Q(z)\, \dot{z}^2 = \text{real}. \tag{41}$$

Indeed, let C_α be a component curve of ∂D. We may assume without loss of generality that C_α is the real axis since this can always be achieved by a conformal mapping and since the covariance of the solutions $u(z)$ of (40) is given by (25). By virtue of (39) and (41) we see that $u(z)$ satisfies on the real axis a second-order differential equation with real coefficients. We may therefore choose a basic solution set $u_\nu(z)$ which is real on the real axis and find as the general solution

$$u(z) = A\, u_1(z) + B\, u_2(z) \tag{42}$$

with arbitrary complex constants A and B. The ratio of two independent solutions of (40) is by (25) a conformal invariant. If we denote

$$\frac{u_1(z)}{u_2(z)} = r_\alpha(z) = \text{real} \quad \text{for } z \in C_\alpha \tag{43}$$

we find for any two solutions $u(z)$ and $v(z)$ of (40) the ratio

$$R(z) = \frac{u(z)}{v(z)} = \frac{A^{(\alpha)} r_\alpha(z) + B^{(\alpha)}}{C^{(\alpha)} r_\alpha(z) + D^{(\alpha)}} \quad \text{on } C_\alpha. \tag{44}$$

The values of the ratio $R(z)$ on C_α lie therefore always on a circle.

There are various canonical mappings of a domain D which transform boundary curves into circles. All of them may be obtained by solving the second-order differential equation (40). We are led again to the problem of determining those λ_ν which lead to a univalent mapping of D on the canonical circular domain. The significance of differentials of order $-\frac{1}{2}$ in the theory of such canonical mappings is evident.

5. It should be pointed out that the concept of differentials of half-integer order is implicit in the general theory of the Schwarzian derivative. Indeed, let us consider the third-order nonlinear differential equation

$$\{w, z\} = S(z) \tag{45}$$

for given $S(z)$ and unknown $w = f(z)$. If $f(z)$ solves (45), one defines

$$u_1(z) = f'(z)^{-\frac{1}{2}}, \qquad u_2(z) = f(z)\, f'(z)^{-\frac{1}{2}} \tag{46}$$

and proves that both functions $u_\nu(z)$ satisfy the same linear second-order differential equation [2, p. 311]

$$u''(z) + \tfrac{1}{2} S(z)\, u(z) = 0. \tag{47}$$

Likewise our considerations regarding the group of linear transformations of the $u_\nu(z)$ under homotopy classes of paths of the Riemann surface $\mathfrak{R}$ considered are closely related to the approach of Poincaré and Klein in the study of the uniformization problem. The coefficients λ_ν in the Schwarzian connection (30) are the well-known accessory parameters in this theory. Instead of dealing with the linear differential equation (24), one considers usually the equivalent nonlinear differential equation of the third order for the uniformizing function $f(p)$. The parameters λ_ν in the Schwarzian connection must then be adjusted in such a way that the linear transformations of the uniformizing function under the various cycles form a Fuchsian group.

It is of interest to show that this requirement simplifies if we use the half-order differentials $u_\nu(p)$. Indeed, the condition on the linear transformations is that they preserve the unit circle, and this engenders the requirement that the non-Euclidean line element

$$\frac{|dw|}{1-|w|^2}=ds \tag{48}$$

be unchanged if $w=f(p)$ undergoes its linear transformations for any closed trajectory. Since by (4) we may express $w(p)$ as the ratio of the $u_\nu(p)$, we find

$$ds=\frac{|dp|}{|u_2(p)|^2-|u_1(p)|^2} \tag{49}$$

if we assume the system of solutions normalized by the condition that its Wronskian have the value 1. Thus, the group of linear transformations must have the invariant

$$Q(u_1, u_2)=|u_1|^2-|u_2|^2. \tag{50}$$

The use of the $u_\nu(p)$ leads to a very short proof of a theorem of Poincaré's which we may paraphrase: The groups of linear transformations engendered by the solution vectors $u_\nu(p)$ and $v_\nu(p)$ of the differential equations

$$u''+\tfrac{1}{2}Su=0, \qquad v''+\tfrac{1}{2}Tv=0 \tag{51}$$

coincide only if the Schwarzian connections S and T are identical.

Indeed, suppose that for given Schwarzian connections S and T we have solution systems $u_\nu(p)$ and $v_\nu(p)$ with the same unimodular transformation matrices $A(\mathfrak{A}_\alpha)$, $A(\mathfrak{B}_\alpha)$. We form the determinant

$$\delta(p)=u_1v_2-u_2v_1 \tag{52}$$

which is a reciprocal differential on $\mathfrak{R}$. Under any cycle Γ on $\mathfrak{R}$ we find $\delta(p)$ unchanged since the vectors $u_\nu(p)$ and $v_\nu(p)$ transform cogrediently and since the determinant of each linear transformation is by construction exactly 1. Hence we have constructed a regular and single-valued reciprocal differential on $\mathfrak{R}$. This is clearly impossible by the Riemann-

Roch theorem if the genus g is >1. Thus, necessarily $\delta(p)\equiv 0$, the vectors $u_\nu(p)$ and $v_\nu(p)$ are linearly dependent and the Schwarzian derivatives of their ratios are equal. This proves $S\equiv T$ as asserted.

A proof can also be given for the case $g=1$. We omit this step since our main purpose was to show the usefulness of the $u_\nu(p)$. Again we have only paraphrased a standard proof of the Poincaré theorem [2], but a comparison of the arguments will show how the explicit use of half-order differentials has been illuminating.

6. The limitation of our consideration to second-order equations of the form of (24) or, as invariantly expressed, (24′)—may seem at first to be an arbitrary restriction. This is not the case, as we shall show. In fact any second-order, linear, homogeneous, differential equation which is invariantly defined and *everywhere regular* on a compact Riemann surface must be of the form

$$\frac{d^2u}{dp^2}+\frac{1}{2}S(p)\,u=0, \tag{53}$$

where $S(p)$ transforms as

$$S_t=S_z\left(\frac{dz}{dt}\right)^2+\{z,t\}; \tag{54}$$

here, as before, S_t is the representation of $S(p)$ in terms of the local uniformizer t, and S_z is the representation of $S(p)$ in terms of the local uniformizer z.

To be more specific, consider a general second-order, linear, homogeneous, differential equation on the Riemann surface. Let its representation in terms of t be

$$v''+pv'+qv=0, \tag{55}$$

and its representation in terms of z be

$$u''+Pu'+Qu=0 \tag{55′}$$

where ′ denotes differentiation with respect to the obvious argument. We require that the coefficients in the differential equation transform according to a linear inhomogeneous law under change of uniformizer

$$p=\alpha P+\beta,\qquad q=\gamma Q+\delta \tag{56}$$

with coefficients which depend on the relation between z and t only. The dependent variable shall transform according to a linear homogeneous rule

$$v(t)=\Phi(t)\,u[z(t)]. \tag{57}$$

Thus

$$v'=\Phi u'z'+\Phi' u \tag{58}$$

14 – 662945 *Acta mathematica*. 115. Imprimé le 11 mars 1966.

which means
$$\frac{dv(t)}{dt}=\Phi(t)\,\frac{du(z)}{dz}\,\frac{dz}{dt}+\Phi'(t)\,u(z). \tag{58'}$$

We also have
$$v''=\Phi z'^2u''+(2\Phi' z'+\Phi z'')\,u'+\Phi'' u. \tag{59}$$

Upon substituting from (58) and (59) into (55) we get

$$\Phi z'^2u''+(2\Phi' z'+\Phi z''+p\Phi z')\,u'+(\Phi''+\Phi' p+\Phi q)\,u=0. \tag{60}$$

Since the differential equation is invariant, we can immediately conclude, upon comparing (56) and (60), that
$$\Phi''+\Phi' p+\Phi q=\Phi z'^2Q \tag{61}$$

and
$$2\Phi' z'+\Phi z''+p\Phi z'=\Phi z'^2P. \tag{62}$$

Since $Q(z)$ depends on t and q, but certainly not on p, we may conclude from (61) that either

(i) $\Phi\equiv$ constant (so $\Phi'\equiv 0$)

or (ii) $p\equiv 0$.

Let us examine the consequences of (i) first. In this case (61) reduces to

$$q(t)=Q(z)\,z'^2; \tag{61 i}$$

thus q, i.e., the last coefficient in the differential equation, is a *quadratic differential*. Also, equation (62) becomes
$$p(t)=P(z)\,z'-\frac{z''}{z'}\,. \tag{62 i}$$

This means that p, i.e., the coefficient of the first order term, is a connection, see [18, p. 251].

But it is known (see [18, p. 252]) that the sum of the residues of a connection on a compact Riemann surface of genus g is $2-2g$; therefore, if $g\neq 1$ the connection must have singularities. Thus the differential equation (55) must be singular unless either $g=1$, the case of elliptic function theory, or $p\equiv 0$, which brings us to case (ii). Before we consider case (ii), let us mention that since $\Phi\equiv$ const, we can take $\Phi\equiv 1$, which means that the solutions of (55) are *functions*, at least locally, i.e., they transform like functions.

In considering case (ii) we again return to equation (61) which now becomes

$$\frac{\Phi''}{\Phi}+q=Qz'^2. \tag{61 ii}$$

And equation (62) becomes
$$2\Phi' z'+\Phi z''=0. \tag{62 ii}$$

This means that $$\Phi = kz'^{-\frac{1}{2}}, \tag{63}$$

where k is a constant which we choose to be 1. In this case an easy calculation shows that

$$\frac{\Phi''}{\Phi} = -\frac{1}{2}\{z, t\}$$

so that (61 ii) becomes

$$q = Q \cdot z'^2 + \tfrac{1}{2}\{z, t\}, \tag{61 ii$'$}$$

and setting $$q = \tfrac{1}{2} S_t \quad \text{and} \quad Q = \tfrac{1}{2} S_z$$

we have established our assertion. In case the equation (55) is regular, i.e., $p \equiv 0$ (in all coordinates), then $\Phi = (z')^{-\frac{1}{2}}$, so the solutions of (55) transform like reciprocal half-order differentials, i.e., differentials of order $-\frac{1}{2}$.

Finally, we should mention that the case $g = 1$ has not been overlooked (in fact, we consider it a very important testing ground for general theories—as it has been for more than a century!). Although one nonsingular connection does exist in this case, we may use a uniformizer on the surface which makes it equal to zero everywhere, which brings us back to the case (ii) again. All other connections in the case $g = 1$ are singular, so our assertion is established with complete generality.

7. In this section we wish to make precise the concept of analytic continuation of differentials of order $\pm\frac{1}{2}$ over the Riemann surface. This consideration is necessary in order to define clearly the meaning of the transformation matrices $A(\Gamma)$ introduced in Section 3 of this chapter.

If we consider the Riemann surface $\mathfrak{R}$ realized as a covering over the complex z-plane, we might define the half-order differentials locally as analytic functions of z and their continuation over $\mathfrak{R}$ as explicit analytic continuation. However, for the general theory of abstract Riemann surfaces the following argument may be more appropriate.

In order to save space, we adopt the notations and definitions given in [18, pp. 249–51]. Thus, by a differential of order $\frac{1}{2}$ we mean a collection of variables $\{\psi_\alpha\}$ which satisfy the transformation laws

$$\psi_\alpha = \psi_\beta \left(\frac{dz_\beta}{dz_\alpha}\right)^{\frac{1}{2}} \quad \text{in } U_\alpha \cap U_\beta. \tag{64}$$

By $(dz_\beta/dz_\alpha)^{\frac{1}{2}}$ we mean an analytic function in $U_\alpha \cap U_\beta$ whose square is dz_β/dz_α. Of course, this function is not unique (there are two choices in each case) and we must show that we can choose them consistently, i.e., such that

$$\left(\frac{dz_\beta}{dz_\alpha}\right)^{\frac{1}{2}} \left(\frac{dz_\gamma}{dz_\beta}\right)^{\frac{1}{2}} \left(\frac{dz_\alpha}{dz_\gamma}\right)^{\frac{1}{2}} \equiv 1 \quad \text{in } U_\alpha \cap U_\beta \cap U_\gamma. \tag{65}$$

In order to show that we can make such a choice, let $\theta_{\alpha\beta}$ be any collection of analytic functions chosen on the sets $U_\alpha \cap U_\beta$ such that

$$\theta_{\alpha\beta}^2 = \frac{dz_\beta}{dz_\alpha} \quad \text{in } U_\alpha \cap U_\beta.$$

Then

$$\theta_{\alpha\beta}\,\theta_{\beta\gamma}\,\theta_{\gamma\alpha} \equiv s_{\alpha\beta\gamma} = \pm 1 \quad \text{for } U_\alpha \cap U_\beta \cap U_\gamma \neq \emptyset. \tag{66}$$

Let $N(\mathfrak{U})$ be the nerve of covering $\mathfrak{U} = \{U_\alpha\}$, and $\mathfrak{U}$ chosen as in [18, p. 255]. Then $s_{\alpha\beta\gamma}$ determines a 2-cocycle on $N(\mathfrak{U})$ given by

$$s[\sigma_{\alpha\beta\gamma}] = s_{\alpha\beta\gamma} \tag{67}$$

which in turn determines an element of $H^2(\mathfrak{R}, G)$, the two-dimensional cohomology group of $\mathfrak{R}$ with coefficients in G, the multiplicative group consisting of the two elements 1 and -1.

Clearly, we have

$$\prod_{\sigma^2} s[\sigma^2] = 1, \tag{68}$$

where the product is taken over all the positively oriented σ^2 in $N(\mathfrak{U})$. Indeed $\prod_\sigma s[\sigma]$ as given in (68) represents the product of all the $s_{\alpha\beta\gamma}$. But each $\sigma_{\alpha\beta}$ occurs as the face of exactly two two-simplexes, say $\sigma_{\alpha\beta\gamma}$ and $\sigma_{\beta\alpha\delta}$ (see illustration [18, p. 255]). Since $\theta_{\alpha\beta}$ and $\theta_{\beta\alpha} = 1/\theta_{\alpha\beta}$ each occurs once in the product, equation (68) holds. But this equation means that s is cohomologous to the identity (upon using the fact that $\mathfrak{R}$ is an orientable, two-dimensional manifold), i.e.,

$$s_{\alpha\beta\gamma} = s_{\alpha\beta} s_{\beta\gamma} s_{\gamma\alpha}, \quad U_\alpha \cap U_\beta \cap U_\gamma \neq \emptyset, \tag{69}$$

where $s_{\alpha\beta} = \pm 1$, etc.

Now define

$$\left(\frac{dz_\beta}{dz_\alpha}\right)^{\frac{1}{2}} = s_{\alpha\beta}\,\theta_{\alpha\beta},$$

then clearly

$$\left(\frac{dz_\beta}{dz_\alpha}\right)^{\frac{1}{2}} \left(\frac{dz_\gamma}{dz_\beta}\right)^{\frac{1}{2}} \left(\frac{dz_\alpha}{dz_\gamma}\right)^{\frac{1}{2}} \equiv 1 \quad \text{in } U_\alpha \cap U_\beta \cap U_\gamma \tag{65}$$

since

$$s_{\alpha\beta}\,\theta_{\alpha\beta}\,s_{\beta\gamma}\,\theta_{\beta\gamma}\,s_{\gamma\alpha}\,\theta_{\gamma\alpha} \equiv s_{\alpha\beta\gamma}\,s_{\alpha\beta\gamma} = 1.$$

We have thus defined a coherent set of expressions $(dz_\beta/dz_\alpha)^{\frac{1}{2}}$. By defining

$$u_\alpha = u_\beta \left(\frac{dz_\beta}{dz_\alpha}\right)^{\frac{1}{2}} \tag{70}$$

we can now give an unambiguous definition to differentials of order $\frac{1}{2}$ on $\mathfrak{R}$.

The differentials of order $-\frac{1}{2}$ can be defined by the same procedure.

III. The Szegö kernel of a Riemann surface

1. Let D be a planar domain bounded by a finite set of analytic curves. In the boundary value problems for analytic functions in D, as well as in the theory of conformal mapping, one can very successfully apply the theory of orthogonal analytic functions and their kernels. The most important norms used in such theories are those involving area integrals

$$(\varphi, \psi) = \iint_D \varphi(z)\overline{\psi(z)}\,dx\,dy, \quad z = x + iy, \tag{1}$$

and boundary line integrals

$$((\varphi, \psi)) = \int_{\partial D} \varphi(z)\overline{\psi(z)}\,ds. \tag{2}$$

If we have a complete orthonormal set of analytic functions $\varphi_\nu(z)$ in D, we can form their kernel [3, 4, 8, 20]

$$K(z, \zeta) = \sum_{\nu=1}^{\infty} \varphi_\nu(z)\overline{\varphi_\nu(\zeta)}, \tag{3}$$

which converges in both cases almost uniformly in D and is independent of the particular choice of the complete orthonormal set. For normalization (1) one obtains the Bergman kernel which is closely related to the Green's function of D and which has many applications in the theory of conformal mapping. In the case of normalization (2) one obtains a kernel which was first introduced by Szegö [21]. Garabedian [9] showed the close connection between the Bergman and the Szegö kernels in the case of planar domains.

We can characterize the Bergman kernel by the reproducing property

$$\iint_D K_B(z, \zeta)\,f(\zeta)\,dx\,dy = f(z) \tag{4}$$

and the Szegö kernel by the analogous equation

$$\int_{\partial D} K_S(z, \zeta)\,f(\zeta)\,ds = f(z). \tag{5}$$

In order to study the behavior of the kernels under conformal mapping and with the aim of extending the theory to domains on Riemann surfaces we shall write (4) and (5) in the form

$$\iint_D K_B(z, \zeta)\,f'(\zeta)\,dx\,dy = f'(z) \tag{4'}$$

and

$$\int_{\partial D} K_S(z, \zeta)\sqrt{f'(\zeta)}\,ds = \sqrt{f'(z)}, \tag{5'}$$

where we assume that $\sqrt{f'(z)}$ is a single-valued analytic function in D. If we now define the transformation laws

$$\hat{K}_B(w, \bar{\omega})\, dw\, d\bar{\omega} = K_B(z, \zeta)\, dz\, d\bar{\zeta} \tag{6}$$

and

$$\hat{K}_S(w, \bar{\omega})\, dw^{\frac{1}{2}}\, d\bar{\omega}^{\frac{1}{2}} = K_S(z, \zeta)\, dz^{\frac{1}{2}}\, d\bar{\zeta}^{\frac{1}{2}} \tag{7}$$

it can easily be seen that the reproducing properties (4′) and (5′) are preserved under conformal mapping. We see, in particular, that a more penetrating theory of the Szegö kernel leads necessarily to the consideration of differentials of half-integer order.

2. It is well known that the function theory of planar domains can be easier understood if we complete them to compact Riemann surfaces by adding to them their double. In particular, the Green's function and the Bergman kernel can be expressed in terms of certain Abelian integrals and differentials of the symmetric Riemann surface so obtained. We therefore shall start with an arbitrary closed Riemann surface $\mathfrak{R}$ and consider there such expressions which for the case of symmetric surfaces will reduce to differentials like the Bergman and the Szegö kernel.

We start with the symmetric Abelian integral $W(p, r; q, s)$ of the surface $\mathfrak{R}$, as we defined in Section I.2, and form the double differential

$$L(p, q) = \frac{\partial^2 W(p, r; q, s)}{\partial p\, \partial q} \tag{8}$$

which is independent of r, s and symmetric in p, q. It is regular for $p, q \in \mathfrak{R}$, except for the case $p=q$ when we have a double pole as indicated in (1.17). This double differential is closely related to the Bergman kernel in the case of planar domains; we may therefore refer to it as the Bergman kernel of $\mathfrak{R}$.

We wish now to construct correspondingly a Szegö kernel for the Riemann surface which shall be a half-order differential in each variable, have a simple pole if both arguments coincide and which is anti-symmetric in p and q. Let us denote it by $\Lambda(p, q)$. Clearly, $\Lambda(p, q)^2$ will be a double differential on $\mathfrak{R}$ with all the regularity and symmetry properties of the Bergman kernel $L(p, q)$. If it exists at all, it must have the form

$$\Lambda(p, q)^2 = L(p, q) + \sum_{i, k=1}^{g} a_{ik}\, w_i'(p)\, w_k'(q), \tag{9}$$

where the $w_j'(p)$ are the normalized Abelian differentials of the first kind of $\mathfrak{R}$ and where the a_{ik} form a symmetric matrix. Since $L(p, q)$ and the Abelian differentials $w_i'(p)$ are well studied, we are led to the algebraic problem to form a combination of $L(p, q)$ and the g differential $w_i'(p)$, which has only double zeros on $\mathfrak{R}$.

An algebraic approach to this problem is as follows. Let p_k $(k=1, 2, \dots, g)$ be a set of points on $\mathfrak{R}$ and form the determinant

$$D(p; q; p_j) = \begin{Vmatrix} L(p, q) & w_i'(p) \\ L(p_k; q) & w_i'(p_k) \end{Vmatrix} \tag{10}$$

of $g+1$ rows and $g+1$ columns. This expression represents for fixed q and p_k a differential of the first order on $\mathfrak{R}$ with the g zeros p_k and the double pole at q. Since a differential of the first order with a double pole has precisely $2g$ zeros, we see that $D(p;\, q,\, p_j)$ has another set of g zeros on $\mathfrak{R}$. But we have still the freedom in the choice of the p_j to achieve that each zero p_j is a double zero. For this purpose, we must fulfill the g conditions

$$D'(p_\varrho; q, p_j) = \begin{Vmatrix} L'(p_\varrho; q) & w_i''(p_\varrho) \\ L(p_k; q) & w_i'(p_k) \end{Vmatrix} = 0, \quad \varrho = 1, \dots, g. \tag{11}$$

We thus have g equations for the g zeros p_j, which shows that the problem is hopeful.

But in order to avoid the theory of elimination for algebraic functions, we shall construct $\Lambda(p, q)$ through the deeper theory of Abelian integrals and by use of the classical results on the Jacobi inversion problem. We follow here an analogous approach as was used by Garabedian in constructing the Szegö kernel of plane domains. We use the Abelian integral of the third kind $w(p; r, s)$ defined in Section I.2. Since it is normalized to have the periods zero around each cross cut $\mathfrak{A}_\nu$ of an arbitrary but fixed system of canonical cross cuts, we know that it has the following periods along a cross cut $\mathfrak{B}_\nu$:

$$\int_{\mathfrak{B}_\nu} dw(p; r\, s) = 2\pi i[w_\nu(r) - w_\nu(s)]. \tag{12}$$

The integrals of the first kind $w_\nu(p)$ are normalized with respect to the same canonical cut system such that

$$\int_{\mathfrak{A}_\mu} dw_\nu = \delta_{\nu\mu}. \tag{13}$$

We select an arbitrary but fixed differential of the first kind $v'(p)$ on $\mathfrak{R}$ and denote its zeros by $p_1, p_2, \dots, p_\rho$; $\varrho = 2g-2$. We pick another set of g points $q_1, \dots, q_g$ on $\mathfrak{R}$ and form the expression

$$E(p; q, q_\nu) = \log v'(p) + \sum_{\nu=1}^{g-1} [w(p; q_\nu, p_{2\nu-1}) + w(p; q_\nu, p_{2\nu})] + 2w(p; q_g, q) + 4\pi i \sum_{\alpha=1}^{g} a_\alpha w_\alpha(p). \tag{14}$$

On the right-hand side we have cancelled out all logarithmic poles at the zeros p_ν ($\nu = 1, \ldots, 2g-2$) of the given differential $v'(p)$. We have, however, logarithmic poles at the points q_ν ($\nu = 1, \ldots, g$) with the residue $+2$ and a logarithmic pole at q with the residue -2.

The expression $E(p; q, q_\nu)$ is not single valued on $\mathfrak{R}$. If we describe a cycle $\mathfrak{A}_\beta$, the $w(p; q_\nu, p_\mu)$ do not change, but $\log v'(p)$ may change by $2\pi i m_\beta$ ($m_\beta =$ integer) and the last sum increases by $4\pi i a_\beta$. Under a cycle $\mathfrak{B}_\beta$ we have in view of (12) the period

$$\int_{\mathfrak{B}_\beta} dE(p; q, q_\nu) = 2\pi i \Big\{ n_\beta + \sum_{\nu=1}^{g-1} [2 w_\beta(q_\nu) - w_\beta(p_{2\nu-1}) - w_\beta(p_{2\nu})] + 2(w_\beta(q_g) - w_\beta(q)) + 2 \sum_{\alpha=1}^{g} a_\alpha P_{\alpha\beta} \Big\} \tag{15}$$

if we introduce the Riemann matrix of periods for the normalized Abelian integrals of the first kind,

$$P_{\alpha\beta} = \int_{\mathfrak{B}_\beta} dw_\alpha. \tag{16}$$

n_β is again an integer.

We apply now the existence theorem for the Jacobi inversion problem. Given any set of g complex numbers ξ_β, we can always find g points q_ν on $\mathfrak{R}$ such that

$$\sum_{\nu=1}^{g} w_\beta(q_\nu) = \xi_\beta + k_\beta + \sum_{\alpha=1}^{g} l_\alpha P_{\alpha\beta}, \quad \beta = 1, 2, \ldots, g, \tag{17}$$

with integers k_β and l_α. That is, the left-hand sum differs from the ξ_β only by a period of $w_\beta(p)$ [12].

Given an arbitrary but fixed $q \in \mathfrak{R}$, we determine q_ν in such a way that

$$\sum_{\nu=1}^{g} w_\beta(q_\nu) = w_\beta(q) + \tfrac{1}{2} \sum_{\varrho=1}^{2g-2} w_\beta(p_\nu) + k_\beta + \sum_{\alpha=1}^{g} l_\alpha P_{\alpha\beta}. \tag{18}$$

This is always possible by the inversion theorem. Finally, we determine the coefficients a_α as the integers

$$a_\alpha = -l_\alpha. \tag{19}$$

With these choices of parameters we find that

$$\int_{\mathfrak{B}_\beta} dE(p; q, q_\nu) = 2\pi i (n_\beta + 2 k_\beta) \tag{20}$$

and

$$\int_{\mathfrak{A}_\beta} dE(p; q, q_\nu) = 2\pi i (m_\beta - 2 l_\beta). \tag{21}$$

All periods of $E(p; q, q_\nu)$ on $\mathfrak{R}$ are integer multiples of $2\pi i$. Hence

$$\lambda(p, q) = \exp\{E(p; q, q_\nu)\} \tag{22}$$

is single valued on $\mathfrak{R}$ and has a double pole at the chosen point q and double zeros at the points q_ν determined by it. Since $\lambda(p, q)$ has the factor $v'(p)$ and is else expressed in terms of Abelian integrals only, it is a differential of the first order on $\mathfrak{R}$ with a double pole and g double zeros.

Finally, we construct the expression

$$\Lambda(p, q) = \sqrt{\lambda(p, q)}. \tag{23}$$

It is a half-order differential in p which is regular everywhere on $\mathfrak{R}$, except for a simple pole at q. It is determined only up to a $\pm$ sign and can change its determination if we continue it over a closed cycle on $\mathfrak{R}$. It therefore will, in general, be single valued only on a proper two-sheeted covering of the surface. It is easily seen that the indeterminacy of sign comes solely from the behavior of $\sqrt{v'(p)}$. Indeed, the change of argument of $\Lambda(p, q)$ over $\mathfrak{A}_\beta$ and $\mathfrak{B}_\beta$ is $\frac{1}{2}m_\beta - l_\beta$ and $\frac{1}{2}n_\beta + k_\beta$, respectively, as can be seen from (20) and (21). Thus, only the parity of the periods m_β, n_β depending on $v'(p)$ decides the changes of sign in $\Lambda(p, q)$.

Hence, if we construct $\Lambda(p, q)$ for different values of q but with the same differential of the first kind $v'(p)$, the two-sheeted covering of $\mathfrak{R}$ will always be the same.

3. We normalize the half-order differential $\Lambda(p, q)$ by the requirement that in a uniformizer neighborhood of q we have

$$\Lambda(p, q)\, dp^{\frac{1}{2}}\, dq^{\frac{1}{2}} = \frac{\sqrt{dz\, d\zeta}}{z - \zeta} + \text{regular differential.} \tag{24}$$

It is easily seen that this normalization is independent of the specific parameter used.

Let now $\Lambda(p, q)$ and $\Lambda(p, q_1)$ be any two half-order differentials on $\mathfrak{R}$ with simple poles at q and q_1, respectively, the normalization (24) and both single valued on the same two-sheeted covering of $\mathfrak{R}$. In this case the product $\Lambda(p, q)\, \Lambda(p, q_1)$ is a single-valued differential on $\mathfrak{R}$ with two simple poles at q and q_1. Hence the sum of its residues must equal to zero and we find

$$\Lambda(q_1, q) + \Lambda(q, q_1) = 0. \tag{25}$$

We thus proved

$$\Lambda(q, q_1) = -\Lambda(q_1, q). \tag{25'}$$

The half-order differential $\Lambda(p, q)$ in p is antisymmetric in both its arguments. It is therefore also a half-order differential in q. It is single valued in dependence on q on the same two-sheeted covering of $\mathfrak{R}$. Our argument shows further that $\Lambda(p, q)$ is uniquely determined by the fact that it is a half-order differential in p with a simple pole at q and the normalization (24), provided that its sign changes on cycles on $\mathfrak{R}$ are prescribed. We shall call $\Lambda(p, q)$ the Szegö kernel of $\mathfrak{R}$ with respect to the two-sheeted covering considered.

We can construct interesting combinations of Szegö kernels which are single valued on $\mathfrak{R}$. Consider, for example, the expression

$$U(p; r, s) = \frac{\Lambda(p, r)\, \Lambda(p, s)}{\Lambda(r, s)}, \quad r \neq s. \tag{26}$$

It is a single-valued differential of first order in p and a single-valued function in r and s on the surface $\mathfrak{R}$. Indeed, if any variable changes on $\mathfrak{R}$, the corresponding sign changes occur always in pairs in the product (26). The differential in p has two simple poles at r and s with the residues $+1$ and -1, respectively. It is thus an Abelian differential of the third kind, analogous to $w'(p; r, s)$ used before. The new differential has, however, a remarkable factorization. The identity (26) indicates the significance of the half-order differentials as building blocks for the classical single-valued differentials on a Riemann surface.

4. Given a Szegö kernel $\Lambda(p, q)$, which is single valued on a specified two-sheeted covering of $\mathfrak{R}$, we shall call two points m and n on $\mathfrak{R}$ associated if they satisfy the equation

$$\Lambda(m, n) = 0. \tag{27}$$

Because of the antisymmetry of the Szegö kernel this relation is a symmetric one. Each given point $q \in \mathfrak{R}$ has precisely g associated points q_ν $(\nu = 1, 2, \ldots, g)$ if we count them by their multiplicity.

The construction of $\Lambda(p, q)$ suggests a close relation between the set of associated points q_ν and the Jacobi inversion problem. This relation can be made more explicit by the following consideration. Let $u'(p)$ and $v'(p)$ be two differentials of the first kind on $\mathfrak{R}$. Then their ratio will be a function on $\mathfrak{R}$ and the integral

$$\frac{1}{2\pi i} \sum_{\nu} \int_{\mathfrak{A}_\nu + \mathfrak{B}_\nu} w_k(p)\, d\left[\log \frac{u'(p)}{v'(p)}\right] = I_k \tag{28}$$

will be defined for every normalized integral of the first kind $w_k(p)$. The standard method of contour integration shows that I_k is a period of $w_k(p)$:

$$I_k = n_k + \sum_{\nu=1}^{g} m_\nu P_{\nu k}, \quad n_k, m_\nu \text{ integers}. \tag{29}$$

On the other hand, let α_ρ be the set of zeros of $u'(p)$ and β_ρ the corresponding set of $v'(p)$. Then the residue theorem yields

$$I_k = \sum_{\varrho=1}^{2g-2} w_k(\alpha_\varrho) - \sum_{\varrho=1}^{2g-2} w_k(\beta_\varrho), \quad k=1,2,\ldots,g. \tag{29'}$$

Thus the sum of each $w_k(p)$ extended over the set of zeros of any differential of the first kind is the same except for a period of $w_k(p)$.

Each integral of the first kind $w_k(p)$ is defined only up to an additive constant. We may normalize these integrals further as follows. Let $v'(p)$ be the special integral of the first kind used in Section 2 to construct the Szegö kernel; we demand

$$\sum_{\varrho=1}^{2g-2} w_k(p_\varrho) = 0, \tag{30}$$

where the sum is extended over all zeros p_ϱ of $v'(p)$. This implies

$$\sum_{\varrho=1}^{2g-2} w_k(\alpha_\varrho) \equiv 0 \quad (\text{mod. period of } w_k(p)), \tag{30'}$$

where α_ρ is the set of zeros of any differential of the first kind.

In view of (18) we find the following characterization of the set q_ν of points associated to q:

$$\sum_{\nu=1}^{g} w_\beta(q_\nu) \equiv w_\beta(q) \quad (\text{mod. period of } w_\beta(p)). \tag{31}$$

The associated point set q_ν of q solves a special Jacobi inversion probleml

The normalization (30) is obviously only possible if the genus of $\Re$ satisfies $g>1$ since for $g=1$ the differential of the first kind has no zeros. We shall consider this special case briefly in the following section.

We may use the concept of associated points to construct the following differentials; let q_ν be one point associated to q and form

$$v_\nu'(p) = \Lambda(p,q)\,\Lambda(p,q_\nu). \tag{32}$$

It is easily seen that $v_\nu'(p)$ is a differential of first order in p, single valued on $\Re$ and regular everywhere. Indeed, the poles of the factors are just cancelled out because of the relation $\Lambda(q,q_\nu)=0$. Thus we can construct g differentials of the first kind on $\Re$ by choosing $\nu=1$, 2, ..., g. Since $v_\nu'(p)$ vanishes at all associated points q_μ of q except for q_ν, we see that the g differentials of the first kind are linearly independent and form a basis for all differentials of the first kind. If $w'(p)$ is a differential of the first kind, we have the development

$$w'(p) = \sum_{\nu=1}^{g} \frac{\Lambda(p,q_\nu)\,\Lambda(p,q)}{\Lambda'(q_\nu,q)}\, w'(q_\nu). \tag{33}$$

4. Let us illustrate the general theory of the Szegö kernel by considering the special case of genus 1. Here we may visualize the Riemann surface in the complex u-plane in the form of a parallelogram generated by the vectors $2\omega_1$ and $2\omega_2$ in which opposite sides are identified. We interpret the effect of a closed cycle $\mathfrak{A}$ as a parallel displacement by the vector $2\omega_1$ and interpret a parallel displacement by the vector $2\omega_2$ as the outcome of a $\mathfrak{B}$-cycle. We have at our disposal the elliptic functions with the periods $2\omega_1$, $2\omega_2$ in order to construct the various Abelian integrals, and we shall use the notations of the Weierstrass theory.

We find at once that

$$w(p) = \frac{1}{2\,\omega_1} u \tag{34}$$

is the normalized integral of the first kind and that the Riemann matrix reduces to

$$P = \int_0^{2\omega_2} dw = \frac{\omega_2}{\omega_1}, \tag{35}$$

the modulus of the parallelogram. The function

$$t(p; q) = \zeta(u - v) - \frac{\eta_1}{\omega_1} u, \quad v = u(q), \tag{36}$$

is clearly a normalized integral of the second kind since it has a simple pole at $p = q$ and the period zero over the $\mathfrak{A}$-cycle. The expression

$$L(p, q) = -\frac{d}{dp} t(p; q) = \wp(u - v) + \frac{\eta_1}{\omega_1} \tag{37}$$

is a differential in p; since it is symmetric in p and q, it is a double differential in both variables. It has a double pole for $p = q$ and has the residue zero. We verify that

$$\int_{\mathfrak{A}} L(p, q)\, dp = 0 \tag{38}$$

from which it follows that this kernel coincides with the Bergman kernel (8) of the Riemann surface.

According to (9) and because of (34) we find for the Szegö kernel the representation

$$\Lambda(p, q)^2 = \wp(u - v) + \frac{\eta_1}{\omega_1} + a \cdot \frac{1}{4\,\omega_1^2}. \tag{39}$$

The constant a has to be chosen such that the right-hand side has a double zero. Since

$$\wp'(u)^2 = 4(\wp(u) - e_1)(\wp(u) - e_2)(\wp(u) - e_3) \tag{40}$$

we may choose a in three different ways:

$$\Lambda(p, q)^2 = \wp(u - v) - e_\varrho, \quad \varrho = 1, 2, 3, \tag{41}$$

and

$$a_\varrho = -4\omega_1^2\left[e_\varrho + \frac{\eta_1}{\omega_1}\right]. \tag{41'}$$

The functions

$$\sqrt{\wp(u) - e_\varrho} = \frac{\sigma_\varrho(u)}{\sigma(u)} \tag{42}$$

are very familiar in the theory of elliptic functions and of theta functions. Their significance is now explained by the role as the Szegö kernels of the surface. We have three different Szegö kernels for the surface $\mathfrak{R}$ since it possesses precisely three different two-sheeted covering surfaces. These are obtained by assigning independently the sign $+1$ or -1 to the effect of a cycle $\mathfrak{A}$ or $\mathfrak{B}$ and omitting the combination $+1$, $+1$ which corresponds to the original Riemann surface itself.

Since

$$\wp(\omega_\varrho) = e_\varrho \tag{43}$$

we see that the condition for associated points $\Lambda(q^*, q) = 0$ leads to the equation

$$v^* = v + \omega_\varrho. \tag{44}$$

The well-known equation

$$\sqrt{\wp(u) - e_1}\sqrt{\wp(u + \omega_1) - e_1} = -\sqrt{(e_1 - e_2)(e_1 - e_3)} \tag{45}$$

is therefore nothing but the special case of (32) since the only differential of the first kind is a constant in our choice of uniformizer.

In the following chapter we shall discuss some relations between the Riemann period matrix $((P_{\alpha\beta}))$ and the matrix $((a_{\alpha\beta}))$ which connects the Bergman and the Szegö kernel through the identity (9). While some interesting results will be obtained, the explicit form of the coefficient a_ρ given in (41') shows already that the relation between the two $g \times g$ matrices is by no means elementary.

IV. The variational formula for the Szegö kernel

1. In the preceding chapter we have shown that the Szegö kernel $\Lambda(p, q)$ is uniquely determined by the Riemann surface $\mathfrak{R}$ and the two-sheeted covering on which it is single valued. If we change $\mathfrak{R}$ continuously, we can deform simultaneously the covering in a corresponding fashion and are thus led to a continuous variation of the Szegö kernel. A study of this variation will then disclose the functional dependence of $\Lambda(p, q)$ upon the Riemann surface $\mathfrak{R}$ and its moduli.

Since there exist many ways of describing $\mathfrak{R}$ in terms of moduli, we shall use a somewhat special but very intuitive kinematics to deform $\mathfrak{R}$. We select an arbitrary but fixed point $r_0 \in \mathfrak{R}$ and introduce at r_0 the uniformizer $z(p)$ such that $z(r_0) = 0$. A neighborhood $\mathfrak{N} \subset \mathfrak{R}$ of r_0 corresponds to a domain Δ in the complex z-plane which contains the origin. Consider now the conformal mapping [14, 17, 19, 20]

$$z^* = z + \frac{e^{2i\alpha}\varrho^2}{z}, \quad \varrho > 0, \ \alpha = \text{real}, \tag{1}$$

in the entire z-plane. We assume that the disk $|z| < \varrho$ lies in Δ. Outside of this disk the mapping $z \to z^*$ is one-to-one and regular analytic for all finite values z. The circumference $|z| = \varrho$ is mapped onto the rectilinear segment $\langle -2\varrho e^{i\alpha}, 2\varrho e^{i\alpha} \rangle$ such that the points $z_1 = \varrho e^{i\varphi}$ and $z_2 = \varrho e^{i(2\alpha - \varphi)}$ go into the same point $z^* = 2\varrho e^{i\alpha} \cos(\varphi - \alpha)$. If we divide the circumference $|z| = \varrho$ into two arcs by drawing the diameter $z = \tau e^{i\alpha}$, $-\varrho \leqslant \tau \leqslant \varrho$, we see that points on this circumference and symmetric to the diameter go into the same points z^*.

We are now able to define a rather radical deformation of the complex z-plane. We cut from it the disk $|z| < \varrho$ and identify points $z_1 = \varrho e^{i\varphi}$ and $z_2 = \varrho e^{i(2\alpha - \varphi)}$, which removes all boundary points of the cut domain and makes it to a new Riemann domain. We may still use the parameter z as a uniformizer on the new Riemann domain, but a function in this domain will be considered analytic only if it is an analytic function of $z^* = z + e^{2i\alpha}\varrho^2/z$.

The deformation of the complex z-plane just defined determines a deformation of the Riemann surface $\mathfrak{R}$ as follows. We delete from $\mathfrak{R}$ all points in $\mathfrak{N}$ which correspond to the disk $|z| < \varrho$ in the uniformizer neighborhood of z and identify points p_1 and p_2 which correspond to points z_1 and z_2 with the same value z^*. This leaves us with a new Riemann surface $\mathfrak{R}^*$ of the same genus as $\mathfrak{R}$. If ϱ is small, the Riemann surface $\mathfrak{R}^*$ is near to $\mathfrak{R}$. This means that corresponding normalized Abelian integrals differ numerically arbitrarily little at corresponding points if ϱ is small enough.

We now wish to give an asymptotic formula for the Szegö kernel $\Lambda^*(p, q)$ of the deformed surface $\mathfrak{R}^*$ in terms of the Szegö kernel $\Lambda(p, q)$ of the original surface $\mathfrak{R}$. For this purpose we introduce a canonical set of cross cuts $\{\mathfrak{A}_\nu, \mathfrak{B}_\nu\}$ for $\mathfrak{R}$ and take care that none of its loops passes through the neighborhood $\mathfrak{N}$ of the point r_0 at which we perform the variation. Under this assumption the same set may also serve as canonical cross-cut system for $\mathfrak{R}^*$. We construct the Szegö kernel $\Lambda^*(p, q)$ of $\mathfrak{R}^*$ by the procedure of Section III.2, using the corresponding differential of the first kind $v^{*\prime}(p)$ on $\mathfrak{R}^*$. Clearly, the kernel

$\Lambda^*(p, q)$ so obtained will be single valued on the corresponding two-sheeted covering of $\mathfrak{R}^*$. In particular, we come to the fundamental conclusion that

$$\Pi(p; q_1, q_2) = \Lambda^*(p, q_1)\Lambda(p, q_2) \tag{2}$$

is a differential of order 1 for $p \in \mathfrak{R} - \mathfrak{N}$ which is single valued on the residual Riemann surface. It has simple poles with the residues $\Lambda(q_1, q_2)$ at q_1 and $\Lambda^*(q_2, q_1)$ at q_2 if we assume that q_1 and q_2 lie also in $\mathfrak{R} - \mathfrak{N}$. Let $|z| = r$ $(\varrho < r)$ be a fixed circumference in the uniformizer neighborhood $\mathfrak{N}$ and let Γ be its corresponding image in $\mathfrak{R}$. We apply the residue theorem to $\Pi(p; q_1, q_2)$ with respect to the part of $\mathfrak{R}$ outside of Γ. Using further the antisymmetry (III.25) of the Szegö kernel, we find

$$\frac{1}{2\pi i}\oint_{|z|=r} \Lambda^*(z, q_1)\Lambda(z, q_2)\, dz = \Lambda^*(q_1, q_2) - \Lambda(q_1, q_2) \tag{3}$$

if we run over the circumference in the positive sense.

Observe that $\Lambda(z, q_2)$ may be developed into a convergent power series in z since it is analytic on $\mathfrak{N}$. This is not the case for the Szegö kernel $\Lambda^*(z, q_1)$ of $\mathfrak{R}^*$ whose development proceeds in powers of z^*. Since $\Lambda^*(p, q)$ is a differential of order $\frac{1}{2}$, we have

$$\Lambda^*(z, q_1) = \mathfrak{F}(z^*, q_1)\left(\frac{dz^*}{dz}\right)^{\frac{1}{2}}, \tag{4}$$

where $\mathfrak{F}(z^*, q_1)$ is a power series in z^*. Using the relation (1) between z and z^*, we thus obtain

$$\Lambda^*(z, q_1) = \mathfrak{F}(z, q_1) + e^{2i\alpha}\varrho^2\left[\frac{\mathfrak{F}'(z, q_1)}{z} - \frac{\mathfrak{F}(z, q_1)}{2z^2}\right] + O(\varrho^4). \tag{5}$$

Inserting (5) into (3), we may now apply the residue theorem with respect to $|z| < r$ since $\mathfrak{F}(z, q_1)$ is analytic there. We find the asymptotic formula

$$\Lambda^*(q_1, q_2) - \Lambda(q_1, q_2) = \tfrac{1}{2}e^{2i\alpha}\varrho^2[\mathfrak{F}'(0, q_1)\Lambda(0, q_2) - \Lambda'(0, q_2)\mathfrak{F}(0, q_1)] + O(\varrho^4). \tag{6}$$

Since $\Lambda^*(z, q_1)$ depends on ϱ so does $\mathfrak{F}(z^*, q_1)$. But it is evident that this analytic function remains bounded as $\varrho \to 0$; we therefore infer from (6) that $\Lambda^*(q_1, q_2) - \Lambda(q_1, q_2) = O(\varrho^2)$ uniformly in each closed region in $\mathfrak{R}$ which does not contain the point r_0 at which we deform the surface. From this fact and (5) we can obtain

$$\Lambda(z, q_1) - \mathfrak{F}(z, q_1) = O(\varrho^2) \quad \text{for } |z| = r. \tag{7}$$

Since the left hand of (7) is analytic for $|z| < r$, we infer the same asymptotic formula for all $|z| < r$ and, in particular,

$$\Lambda(0, q_1) = \mathfrak{F}(0, q_1) + O(\varrho^2), \quad \Lambda'(0, q_1) = \mathfrak{F}'(0, q_1) + O(\varrho^2). \tag{8}$$

Hence, finally (6) takes the symmetric form

$$\Lambda^*(q_1, q_2) - \Lambda(q_1, q_2) = \tfrac{1}{2} e^{2i\alpha} \varrho^2 [\Lambda'(0, q_1)\Lambda(0, q_2) - \Lambda'(0, q_2)\Lambda(0, q_1)] + O(\varrho^4). \tag{9}$$

This formula allows an asymptotic estimate of the new Szegö kernel $\Lambda^*(p, q)$ in terms of the original known Szegö kernel $\Lambda(p, q)$ and its first derivative. The error term can be estimated uniformly in each closed region of $\mathfrak{R}$ which does not contain r_0.

We summarize the result of this section as follows. If we perform a variation of $\mathfrak{R}$ at a point r_0, which in terms of the local uniformizer z has the form (1), we have

$$\delta\Lambda(p, q) = \tfrac{1}{2} e^{2i\alpha} \varrho^2 [\Lambda'(r_0, p)\Lambda(r_0, q) - \Lambda(r_0, p)\Lambda'(r_0, q)], \tag{10}$$

where

$$\Lambda'(r_0, q) = \frac{\partial}{\partial z} \Lambda(r, q)\big|_{z=0}. \tag{10'}$$

In the case of plane domains the variational formula for the Szegö kernel was derived in [16].

2. The variational formula for the Szegö kernel $\Lambda(p, q)$ leads us to the interesting combination

$$H_0(r; p, q) = \Lambda'(r, p)\Lambda(r, q) - \Lambda(r, p)\Lambda'(r, q). \tag{11}$$

It is antisymmetric in p and q and clearly a differential of order $\frac{1}{2}$ in each variable. It is easily verified that it is a quadratic differential in r; indeed, we can write

$$H_0(r; p, q) = \Lambda(r, q)^2 \frac{\partial}{\partial r} \left(\frac{\Lambda(r, p)}{\Lambda(r, q)}\right) \tag{12}$$

which displays clearly the covariant character of H_0.

It is clear that H_0 in dependence on r has two singularities, namely at points p and q. If we introduce at p a local uniformizer $\zeta(r)$ such that $\zeta(p) = 0$, we find by (III.24)

$$H_0(r; p, q) = -\frac{1}{\zeta^2} \Lambda(p, q) - \frac{2}{\zeta} \Lambda'(p, q) + \text{regular terms} \tag{13}$$

as the series development near p.

We construct next the expression

$$H_1(r; p, q) = -\Lambda(p, q)\{\Lambda'(r, p)\Lambda(r, q) - \Lambda(r, p)\Lambda'(r, q)\}. \tag{14}$$

It is now symmetric in p and q and is a quadratic differential in r, a linear differential in p and q. We have thus succeeded in constructing a differential of integer order in all three variables. For r near p, we have by (13) the development in terms of the local uniformizer $\zeta(r)$:

$$H_1(r; p, q) = \frac{1}{\zeta^2}\Lambda(p, q)^2 + \frac{1}{\zeta} 2\Lambda'(p, q)\Lambda(p, q) + \dots. \tag{15}$$

We observe that the simpler expression $\Lambda^2(r, p)\Lambda^2(r, q)$ is also symmetric in p and q, is a quadratic differential in r, and a linear differential in p and q and has for r near q precisely the same principal part (15). Indeed, by (III.24) we have for $\Lambda(r, p)$ the series development

$$\Lambda(r, p) = \frac{1}{\zeta} + \alpha_1\zeta + \alpha_3\zeta^3 + \dots \tag{16}$$

since it is antisymmetric in r and p and must be an odd power series. Thus we can assert that

$$H_1(r; p, q) - \Lambda^2(r, p)\Lambda^2(r, q) = H_2(r; p, q) \tag{17}$$

is symmetric in p, q, a linear differential in each of them, a quadratic differential in r, and regular analytic on $\mathfrak{R}$ in all its variables.

To understand more clearly the significance of this term

$$-\Lambda(p, q)\{\Lambda'(r, p)\Lambda(r, q) - \Lambda(r, p)\Lambda'(r, q)\} - \Lambda^2(r, p)\Lambda^2(r, q) \tag{18}$$

we shall identify it as the limit case of a more general expression which involves four variables but is of particularly simple structure. We define

$$N(r, s; p, q) = \Lambda(r, s)\Lambda(p, q)\{\Lambda(r, p)\Lambda(s, q) - \Lambda(r, q)\Lambda(s, p)\} + \Lambda(r, p)\Lambda(r, q)\Lambda(s, p)\Lambda(s, q) \tag{19}$$

This is a linear differential in all four variables. It has the symmetries

$$N(r, s; p, q) = N(s, r; p, q) = N(r, s; q, p) = N(p, q; r, s) = N(p, s; r, q). \tag{20}$$

We easily verify that it remains finite in each variable on $\mathfrak{R}$. Hence we can express it in terms of the Abelian differentials of the first kind and obtain the multilinear representation

$$N(r, s; p, q) = \sum_{i,k,l,m=1}^{g} c_{iklm}\, w_i'(r)\, w_k'(s)\, w_l'(p)\, w_m'(q). \tag{21}$$

The symmetries (20) express themselves in terms of the coefficients as

$$c_{iklm} = c_{kilm} = c_{ikml} = c_{lmik} = c_{lkim}. \tag{22}$$

The coefficients are thus completely symmetric in all four indices.

Let us write the identity (21) by means of the definition (19) in the form

$$-\Lambda(r,s)\Lambda(p,q)\{\Lambda(r,p)\Lambda(s,q)-\Lambda(r,q)\Lambda(s,p)\}$$
$$=\Lambda(r,p)\Lambda(r,q)\Lambda(s,p)\Lambda(s,q)-\sum_{i,k,l,m=1}^{g} c_{iklm}\, w_i'(r)\, w_k'(s)\, w_l'(p)\, w_m'(q). \tag{23}$$

Pass here to the limit $s=r$ and find $H_1(r;p,q)$:

$$-\Lambda(p,q)\{\Lambda'(r,p)\Lambda(r,q)-\Lambda'(r,q)\Lambda(r,p)\}$$
$$=\Lambda(r,p)^2\Lambda(r,q)^2-\sum_{i,k,l,m=1}^{g} c_{iklm}\, w_i'(r)\, w_k'(r)\, w_l'(p)\, w_m'(q). \tag{24}$$

We have thus found a simple and highly symmetric expression for the important variational terms (11), (14), and (18) by means of differentials of the first kind. The coefficient set c_{iklm} is a set of possible moduli for the surface $\mathfrak{R}$. We shall show its importance in the general theory of moduli on a Riemann surface.

3. We have derived in Section 1 a variational formula for the Szegö kernel and obtained in Section 2 remarkable identities for the variational expressions which will facilitate its applications. We wish to show now to what use the entire variational theory can be put.

We return to the identity (II.9) which must be fulfilled by the Szegö kernel

$$\Lambda(p,q)^2=L(p,q)+\sum_{i,k=1}^{g} a_{ik}\, w_i'(p)\, w_k'(q). \tag{25}$$

The coefficient matrix $((a_{ik}))$ is symmetric and uniquely determined by the Riemann surface $\mathfrak{R}$ and the two-sheeted covering on which $\Lambda(p,q)$ is single valued. The a_{ik} may thus be considered as a set of moduli for $\mathfrak{R}$. A very similar symmetric $g\times g$ matrix of moduli is given by the period matrix $((P_{ik}))$ of the integrals of the first kind as defined in (III.16). This matrix has been extensively studied and its importance in the moduli problem is well known. The question arises whether the two matrices $((a_{ik}))$ and $((P_{ik}))$ stand in any simple relation.

Let Γ_1 and Γ_2 be two closed curves on $\mathfrak{R}$. Integrating the identity (25) with $p\in\Gamma_1$, $q\in\Gamma_2$, we obtain the equation

$$\int_{\Gamma_1}\int_{\Gamma_2}\Lambda(p,q)^2\,dp\,dq=\int_{\Gamma_1}\int_{\Gamma_2}L(p,q)\,dp\,dq+\sum_{i,k=1}^{g} a_{ik}\int_{\Gamma_1}dw_i\int_{\Gamma_2}dw_k. \tag{26}$$

By definition (III: 8) of the L-kernel and in view of (I: 15) we have

$$\int_{\Gamma}L(p,q)\,dp=\int_{\Gamma}\frac{\partial^2 w(p;q,s)}{\partial p\,\partial q}\,dp=\frac{\partial}{\partial q}\int_{\Gamma}dw(p;q,s). \tag{27}$$

Because of the normalization of the Abelian integrals of the third kind we then obtain

$$\int_{\mathfrak{A}_\alpha} L(p,q)\,dp = 0, \qquad \int_{\mathfrak{B}_\alpha} L(p,q)\,dp = 2\pi i\, w_\alpha'(q), \tag{28}$$

and consequently

$$\int_{\mathfrak{A}_\alpha}\int_{\mathfrak{A}_\beta} L(p,q)\,dp\,dq = \int_{\mathfrak{B}_\alpha}\int_{\mathfrak{A}_\beta} L(p,q)\,dp\,dq = 0, \tag{29}$$

$$\int_{\mathfrak{A}_\alpha}\int_{\mathfrak{B}_\beta} L(p,q)\,dp\,dq = 2\pi i\,\delta_{\alpha\beta} \tag{29'}$$

and

$$\int_{\mathfrak{B}_\alpha}\int_{\mathfrak{B}_\beta} L(p,q)\,dp\,dq = 2\pi i\,P_{\alpha\beta}. \tag{29''}$$

Observe the asymmetry of the integrals extended over $\mathfrak{A}_\alpha$ and $\mathfrak{B}_\alpha$. It is due to the fact that the integral is in this case improper and therefore, in spite of the symmetry of its kernel, takes different values for a different order of integration.

The equations (26), (29), (29′) and (29″) lead to the period formulas

$$\int_{\mathfrak{A}_\alpha}\int_{\mathfrak{A}_\beta} \Lambda(p,q)^2\,dp\,dq = a_{\alpha\beta}, \tag{30}$$

$$\int_{\mathfrak{B}_\alpha}\int_{\mathfrak{A}_\beta} \Lambda(p,q)^2\,dp\,dq = \sum_{k=1}^{g} P_{\alpha k}\,a_{k\beta}, \tag{30'}$$

$$\int_{\mathfrak{A}_\alpha}\int_{\mathfrak{B}_\beta} \Lambda(p,q)^2\,dp\,dq = 2\pi i\,\delta_{\alpha\beta} + \sum_{k=1}^{g} a_{\alpha k}\,P_{k\beta}, \tag{30''}$$

$$\int_{\mathfrak{B}_\alpha}\int_{\mathfrak{B}_\beta} \Lambda(p,q)^2\,dp\,dq = 2\pi i\,P_{\alpha\beta} + \sum_{i,k=1}^{g} P_{\alpha i}\,a_{ik}\,P_{k\beta}. \tag{30'''}$$

The symmetric matrix $((a_{\alpha\beta}))$ of coefficients in (25) has thus been identified as the period matrix of $\Lambda(p,q)^2$ with respect to the cycles $\mathfrak{A}_\alpha$:

$$A = ((a_{\alpha\beta})) = \left(\left(\int_{\mathfrak{A}_\alpha}\int_{\mathfrak{A}_\beta} \Lambda(p,q)^2\,dp\,dq\right)\right). \tag{31}$$

It is therefore very analogous to the Riemann matrix $((P_{\alpha\beta}))$ which is the period matrix of the L-kernel with respect to the cycles $\mathfrak{B}_\alpha$:

$$P = ((2\pi i P_{\alpha\beta})) = \left(\left(\int_{\mathfrak{B}_\alpha}\int_{\mathfrak{B}_\beta} L(p,q)\,dp\,dq\right)\right). \tag{32}$$

We define next the matrices

$$K = ((k_{\alpha\beta})) = \left(\left(\int_{\mathfrak{B}_\alpha}\int_{\mathfrak{A}_\beta} \Lambda(p,q)^2\, dp\, dq\right)\right) \tag{33}$$

and

$$\hat{K} = ((\hat{k}_{\alpha\beta})) = \left(\left(\int_{\mathfrak{A}_\alpha}\int_{\mathfrak{B}_\beta} \Lambda(p,q)^2\, dp\, dq\right)\right). \tag{33'}$$

While A and P are symmetric matrices, K and $\hat{K}$ are not. By (30′) and (30″) we have

$$\hat{K} = 2\pi i I + K^T, \tag{33''}$$

where I is the unit matrix and K^T is the transposed matrix of K. We can now bring the equations (30′) and (30‴) into matrix form if we also define the period matrix of $\Lambda(p,q)^2$ with respect to the $\mathfrak{B}$-cycles:

$$B = ((b_{\alpha\beta})) = \left(\left(\int_{\mathfrak{B}_\alpha}\int_{\mathfrak{B}_\beta} \Lambda(p,q)^2\, dp\, dq\right)\right). \tag{34}$$

We find

$$K = \frac{1}{2\pi i} P \cdot A; \qquad B = P - \frac{1}{4\pi^2} PAP. \tag{35}$$

We can condense the matrix relations into one single equation if we introduce the symmetric matrix

$$S = \begin{pmatrix} A & K^T \\ K & B \end{pmatrix} \tag{36}$$

which depends only on the periods of $\Lambda^2(p,q)$ and the matrix

$$\Sigma = \begin{pmatrix} I & 0 \\ \frac{1}{2\pi i} P & I \end{pmatrix} \tag{37}$$

which depends only on the periods of $L(p,q)$. The equations (35) can then be combined in the matrix equation

$$S = \Sigma \begin{pmatrix} A & 0 \\ 0 & P \end{pmatrix} \Sigma^T. \tag{38}$$

4. After these formal considerations we are now ready to study the dependence of the various period matrices upon the Riemann surface $\mathfrak{R}$ for which they are defined. By virtue of the identities (35) or (38) it is sufficient to know how the matrices P and A change under a deformation of the surface $\mathfrak{R}$ in order to compute the change of the remaining periods under the same variation.

It is known that under a variation of $\mathfrak{R}$ at r_0 according to (1) we have the asymptotic formula [14, 20]

$$\delta P_{\mu\nu} = -2\pi i\, e^{2i\alpha} \varrho^2 w_\mu'(r_0)\, w_\nu'(r_0). \tag{39}$$

To find the variation of the matrix A we have to use the identity (30) and the known variational formula for the Szegö kernel. By use of (10) and (14) we find

$$\delta a_{\mu\nu} = -e^{2i\alpha} \varrho^2 \int_{\mathfrak{A}_\mu} \int_{\mathfrak{A}_\nu} H_1(r_0; p, q)\, dp\, dq. \tag{40}$$

We simplify considerably by representing $H_1(r_0; p, q)$ by means of (24). Indeed,

$$\delta a_{\mu\nu} = -e^{2i\alpha} \varrho^2 \left\{ \int_{\mathfrak{A}_\mu} \Lambda(r_0, p)^2\, dp \int_{\mathfrak{A}_\nu} \Lambda(r_0, q)^2\, dq - \sum_{j,k=1}^{g} c_{jk\mu\nu}\, w_j'(r_0)\, w_k'(r_0) \right\}. \tag{41}$$

We reduce the formulas further by use of (25) and (28) which yield

$$\int_{\mathfrak{A}_\mu} \Lambda(r_0, p)^2\, dp = \sum_{j=1}^{g} a_{j\mu}\, w_j'(r_0). \tag{42}$$

We therefore arrive at the final result:

$$\delta a_{\mu\nu} = e^{2i\alpha} \varrho^2 \left\{ \sum_{j,k=1}^{g} c_{jk\mu\nu}\, w_j'(r_0)\, w_k'(r_0) - \sum_{j,k=1}^{g} a_{j\mu}\, a_{k\nu}\, w_j'(r_0)\, w_k'(r_0) \right\}. \tag{43}$$

If we combine the variational formulas (39) and (43), we obtain the elegant equation

$$2\pi i \delta a_{\mu\nu} = \sum_{j,k=1}^{g} (a_{j\mu}\, a_{k\nu} - c_{jk\mu\nu})\, \delta P_{jk}. \tag{44}$$

Let us introduce a set of moduli m_α of which the P_{jk} and a_{jk} are real analytic functions. We then find

$$2\pi i \frac{\partial a_{\mu\nu}}{\partial m_\alpha} = \sum_{j,k=1}^{g} (a_{j\mu}\, a_{k\nu} - c_{jk\mu\nu}) \frac{\partial P_{jk}}{\partial m_\alpha}. \tag{45}$$

Observe that because of (22) we have

$$a_{j\mu}\, a_{k\nu} - c_{jk\mu\nu} = a_{\mu j}\, a_{\nu k} - c_{\mu\nu jk}. \tag{46}$$

This implies

$$\sum_{\mu,\nu=1}^{g} \left(\frac{\partial a_{\mu\nu}}{\partial m_\alpha} \frac{\partial P_{\mu\nu}}{\partial m_\beta} - \frac{\partial a_{\mu\nu}}{\partial m_\beta} \frac{\partial P_{\mu\nu}}{\partial m_\alpha} \right) = 0 \tag{47}$$

for any pair of indices α and β. But (47) is the well-known integrability condition which guarantees that the integral

$$\int \sum_{\mu,\nu=1}^{g} a_{\mu\nu}\, dP_{\mu\nu} = J \tag{48}$$

is unchanged under continuous deformation of the path in the space of moduli provided the end points are kept fixed. Thus we can define a function $J(m_\alpha)$ in the space of moduli such that

$$dJ = \sum_{\mu,\nu=1}^{g} a_{\mu\nu}\, dP_{\mu\nu}. \tag{48'}$$

These identities and theorems illustrate the value and the significance of the variational formulas for the various differentials on a Riemann surface $\mathfrak{R}$. The amount of new identities and suggestive relations involving the Szegö kernel show the usefulness of the new concept for the general theory of Abelian integrals.

5. We introduced the coefficient scheme $c_{jk\mu\nu}$ by the definition (21), and we may also characterize them in view of (44) as differential coefficients of the matrix $((a_{\mu\nu}))$ with respect to the matrix $((P_{jk}))$. Another interesting role for this set of coefficients can be deduced from identity (23) if we specialize the point p in this formula to be a point q_ν associated to q according to the relation

$$\Lambda(q_\nu, q) = 0. \tag{49}$$

In this case, (23) reduces to

$$\sum_{i,k,l,m=1}^{g} c_{iklm}\, w_i'(r)\, w_k'(s)\, w_l'(q_\nu)\, w_m'(q) = \Lambda(r, q_\nu)\, \Lambda(r, q)\, \Lambda(s, q_\nu)\, \Lambda(s, q). \tag{50}$$

On the other hand, we showed in Section III.3 that $v_\nu'(p) = \Lambda(p, q)\Lambda(p, q_\nu)$ is a differential of the first kind if q and q_ν are associated points. We may express each $v_\nu'(p)$ in terms of the canonical basis for such differentials and write

$$v_\nu'(p) = \Lambda(p, q)\, \Lambda(p, q_\nu) = \sum_{\varrho=1}^{g} c_{\nu\varrho}(q)\, w_\varrho'(p). \tag{51}$$

With the coefficient matrix $((c_{\nu\varrho}(q)))$ so defined and in view of the linear independence of all $w_\nu'(p)$ we derive from (50) the identity

$$\sum_{l,m=1}^{g} c_{iklm}\, w_l'(q_\nu)\, w_m'(q) = c_{\nu i}(q)\, c_{\nu k}(q). \tag{52}$$

We can eliminate from this identity the $c_{\nu\rho}(q)$ entirely and bring (52) into the form

$$\sum_{l,m,\lambda,\mu=1}^{g} (c_{iklm}\, c_{ik\lambda\mu} - c_{iilm}\, c_{kk\lambda\mu})\, w_l'(q_\nu)\, w_\lambda'(q_\nu)\, w_m'(q)\, w_\mu'(q) = 0. \tag{53}$$

It is convenient to introduce the bilinear forms of quadratic differentials

$$B_{ik}(p,q)=\sum_{l,m,\lambda,\mu=1}^{g}(c_{iklm}\,c_{ik\lambda\mu}-c_{illm}\,c_{kk\lambda\mu})\,w_l'(p)\,w_\lambda'(p)\,w_m'(q)\,w_\mu'(q) \tag{54}$$

which is based on the coefficient scheme c_{iklm}. We then see by (53) that

$$B_{ik}(q_\nu,q)=0,\quad 1\leqslant i<k\leqslant g, \tag{55}$$

for any pair of associated points q_ν, q on $\mathfrak{R}$.

We may also express the property (55) by writing

$$B_{ik}(p,q)=\Lambda(p,q)\,T_{ik}(p,q), \tag{56}$$

where the kernels $T_{ik}(p,q)$ are antisymmetric in p and q and 3/2-order differentials in each variable. To represent the $T_{ik}(p,q)$ in a simple manner, we have to investigate the class of all regular 3/2-order differentials on $\mathfrak{R}$ which are single valued on the same two-sheeted covering of $\mathfrak{R}$ as $\Lambda(p,q)$.

We easily see that

$$T(p)=\Lambda(p,q)\left\{\sum_{\nu=1}^{g}C_\nu w'(p;q_\nu,q_0)+\sum_{\nu=1}^{g}c_\nu w_\nu'(p)\right\} \tag{57}$$

will be the most general, regular 3/2-order differential if the coefficients C_ν and c_ν satisfy the two linear, homogeneous conditions

$$\sum_{\nu=1}^{g}C_\nu=0,\qquad \sum_{\nu=1}^{g}C_\nu w'(q;q_\nu,q_0)+\sum_{\nu=1}^{g}c_\nu w_\nu'(q)=0. \tag{57'}$$

The point $q_0\in\mathfrak{R}$ can be chosen arbitrarily except for being different from q. There are precisely $2(g-1)$ linearly independent regular 3/2-order differentials on $\mathfrak{R}$. We choose a basis $T_\alpha(p)$ $(\alpha=1,\dots 2(g-1))$ of such differentials and can then write

$$T_{ik}(p,q)=\sum_{\alpha,\beta=1}^{2g-2}d_{ik,\alpha\beta}\,T_\alpha(p)\,T_\beta(q) \tag{58}$$

with $d_{ik,\alpha\beta}$ being antisymmetric in the last pair of indices.

Observe now that there are $3g-3$ linearly independent quadratic differentials. Hence the most general symmetric bilinear form of quadratic differentials depends on $\binom{3g-2}{2}$ independent coefficients. On the other hand, the antisymmetric bilinear forms of differentials of order 3/2 depend only upon $\binom{2g-2}{2}$ independent parameters. Thus, each of the coefficient sets of the forms $B_{ik}(p,q)$ has to satisfy a large number of constraints.

These set up numerous conditions in the matrix $((c_{iklm}))$ which may be interpreted as differential relations for the $((a_{ik}))$ in their dependence on the $((P_{lm}))$.

The significance of the variational formula for the Szegö kernel in the problem of the moduli of a Riemann surface and the Teichmüller spaces is evident. These problems have been treated extensively and successfully in recent years [1, 6, 7]. We hope to be able to contribute to these questions by the present developments and techniques.

Bibliography

[1]. AHLFORS, L. V., Teichmüller spaces. *Proc. Intern. Congr. Math., Stockholm* 1962, 3–9.
[2]. APPELL, P. & GOURSAT, E., *Théorie des fonctions algébriques*, t. II. Paris 1930.
[3]. BERGMAN, S., Über die Entwicklung der harmonischen Funktionen der Ebene und des Raumes nach Orthogonalfunktionen. *Math. Ann.*, 86 (1922), 238–271.
[4]. —— *The kernel function and conformal mapping.* Amer. Math. Soc. Survey, New York 1950.
[5]. BERGMAN, S. & SCHIFFER, M., Kernel functions and conformal mapping. *Compositio Math.*, 8 (1951), 205–249.
[6]. BERS, L., Spaces of Riemann surfaces. *Proc. Intern. Congr. Math., Edinburgh* 1958, 349–361.
[7]. —— Automorphic forms and general Teichmüller spaces. *Proc. Conf. Complex Anal., Minneapolis* 1964, 109–113.
[8]. COURANT, R., *Dirichlet's principle, conformal mapping and minimal surfaces.* Appendix by M. Schiffer. New York 1950.
[9]. GARABEDIAN, P. R., Schwarz's lemma and the Szegö kernel function. *Trans. Amer. Math. Soc.*, 67 (1949), 1–35.
[10]. HENSEL, K. & LANDSBERG, G., *Theorie der algebraischen Funktionen einer Variabeln.* Leipzig 1902.
[11]. HILLE, E., Remarks on a paper by Zeev Nehari. *Bull. Amer. Math. Soc.*, 55 (1949), 552–553.
[12]. KRAZER, A., *Lehrbuch der Thetafunktionen.* Leipzig 1903.
[13]. NEHARI, Z., The Schwarzian derivative and schlicht functions. *Bull. Amer. Math. Soc.*, 55 (1949), 545–551.
[14]. SCHIFFER, M., Hadamard's formula and variation of domain functions. *Amer. J. Math.*, 68 (1946), 417–448.
[15]. —— Faber polynomials in the theory of univalent functions. *Bull. Amer. Math. Soc.*, 54 (1948), 503–517.
[16]. —— Various types of orthogonalization. *Duke Math. J.*, 17 (1950), 329–366.
[17]. —— Variational methods in the theory of Rieman surfaces. *Contributions to the theory of Riemann surfaces*, Princeton 1953, 15–30.
[18]. SCHIFFER, M. & HAWLEY, N. S., Connections and conformal mapping. *Acta Math.*, 107 (1962), 175–274.
[19]. SCHIFFER, M. & SPENCER, D. C., A variational calculus for Riemann surfaces. *Ann. Acad. Scient. Fenn. Ser. A I*, 93, Helsinki 1951.
[20]. —— *Functionals of finite Riemann surfaces.* Princeton 1954.
[21]. SZEGÖ, G., Über orthogonale Polynome, die zu einer gegebenen Kurve der komplexen Ebene gehören. *Math. Z.*, 9 (1921), 218–270.
[22]. WEYL, H., *Die Idee der Riemannschen Fläche.* 3. Aufl., Stuttgart 1955.

Received July 31, 1965

Commentary on

[87] N. S. Hawley and M. Schiffer, *Half-order differentials on Riemann surfaces*, Acta Math. 115 (1966), 199–236.

This paper marks the first appearance of the Szegő kernel $\Lambda(p,q)$ for a bundle of half-order differentials on a compact Riemann surface R of genus $g \geq 1$. As a classical reproducing kernel [S], $\Lambda(p,q)$ had played an important role in extremal problems (Schwarz Lemma) for bounded analytic functions on planar domains D [G]. In [87], Hawley and Schiffer construct a differential $\Lambda^2(p,q)$ on R with double zeros and pole along $p = q$ which gives this kernel when R is the double of D.

For a standard (A,B)-marking of R, there is also a symmetric bilinear differential $L(p,q)$ (related to the Bergman kernel for the double of D) with zero A-periods and double pole along $p = q$. One then has a relation

$$\Lambda^2(p,q) = L(p,q) + \sum_{1\leq i,k\leq g} a_{ik} w_i'(p) w_k'(q), \qquad p,q \in R,\ a_{ik} \in \mathbf{C}, \tag{1}$$

where w_i' are the A-normalized differentials on R with period matrix (τ_{ik}). From his variational method applied to $\Lambda(p,q)$, Schiffer is able to show that $\sum_{1\leq i,k\leq g} a_{ik} d\tau_{ik} = dJ$ for some function J on Torelli space.

The paper [87] appeared during the period of renewed study of Riemann's theta function initiated by Harry Rauch. Classically, the holomorphic half-order differentials had been constructed from the gradients of the odd theta functions; and in his construction of $\Lambda(p,q)$, Schiffer's Jacobi-variety condition on the zeros of $\Lambda(p,q)$ for fixed q was exactly that for the divisor of zeros of $\theta(w(p) - w(q))$. Subsequently, Schiffer's student Dennis Hejhal gave a complete theta-function explanation of (1) and the function J in his thesis [H]. Thus, (1) and also (IV-23) in [87] anticipated key relations between theta functions and abelian differentials on R.

More generally, the Szegő kernels for bundles of half-order differentials on R with scalar or matrix multipliers have appeared in such diverse contexts as integrable systems (K-P hierarchy) and conformal field theory [KNTY], analytic torsion (determinant of spin-Laplacian) [DS], and non-abelian theta functions [F, BB]. These basic kernels all have $\Lambda(p,q)$ as their common ancestor.

The paper also includes several observations clarifying the role of half-order differentials and projective connections in conformal mapping and Schwarzian equations defining projective structures on R.

References

[BB] David Ben-Zvi and Indranil Biswas, *Theta functions and Szegő kernels*, Int. Math. Res. Not. **2003**, 1305–1340.

[DS] Michael J. Dugan and Hidenori Sonoda, *Functional determinants on Riemann surfaces*, Nuclear Phys. B **269** (1987), 227–252.

[F] John Fay, *The nonabelian Szegő kernel and theta-divisor*, Curves, Jacobians and Abelian Varieties, Contemp. Math. **136**, Amer. Math. Soc., 1992, pp. 171–183.

[G] P. R. Garabedian, *Schwarz's lemma and the Szegő kernel function*, Trans. Amer. Math. Soc. **67** (1949), 1–35.

[H] Dennis A. Hejhal, *Theta Functions, Kernel Functions and Abelian Integrals*, Memoirs Amer. Math. Soc. **129** (1972).

[KNTY] Noburo Kawamoto, Yukihiko Namikawa, Akihiro Tsuchiya and Yasuhiko Yamada, *Geometric realization of conformal field theory on Riemann surfaces*, Comm. Math. Phys. **116** (1988), 247–308.

[S] G. Szegő, *Über orthogonale Polynome, die zu einer gegebenen Kurve der komplexen Ebene gehören*, Math. Z. **9** (1921), 218–270.

JOHN FAY

[91] (with P. R. Garabedian) The local maximum theorem for the coefficients of univalent functions

[91] (with P. R. Garabedian) The local maximum theorem for the coefficients of univalent functions. *Arch. Rational Mech. Anal.* **26** (1967), 1–32.

The Local Maximum Theorem for the Coefficients of Univalent Functions

P. R. GARABEDIAN & M. SCHIFFER

Introduction

The purpose of this paper is to prove the local version of the Bieberbach conjecture regarding the coefficients of a univalent power series

$$f(z)=z+a_2 z^2+\cdots+a_n z^n+\cdots \tag{0.1}$$

which converges for $|z|<1$. In 1916 BIEBERBACH [2] conjectured that a necessary condition for univalence of (0.1) in the unit disk was the set of inequalities

$$\operatorname{Re}\{a_n\}\leqq n \tag{0.2}$$

for all values of n. The "Koebe function"

$$\frac{z}{(1-z)^2}=\sum_{\nu=1}^{\infty}\nu z^{\nu} \tag{0.3}$$

is of the form (0.1), is univalent in $|z|<1$, and would thus be an extremum function in (0.2) for every $n>1$. The Bieberbach inequality (0.2) has been proved up to $n=4$ and for a large number of important subclasses of univalent functions, but the general case still remains a challenge to analysts.

To deal with this problem, which is evidently a maximum problem within the class of univalent functions (0.1), variational methods have been developed in order to characterize the extremum functions. A study of the first variation leads to an important and rather restrictive differential-functional equation which must be satisfied by the extremum functions. It is significant that the Koebe function (0.3) satisfies these necessary conditions for all values of n [14]. To continue the investigation by the standard procedure of the calculus of variations, one is next led to consider the second variation in order to ensure the actual maximum character of the Koebe function. Such an attack was carried out by DUREN-SCHIFFER [5] and led to a certain inequality which guarantees that the Koebe function is, at least, the extremum function among all univalent functions (0.1) which are obtained from the Koebe function by a sufficiently small analytic variation. This inequality, which is of independent interest in the theory of trigonometric series, could be verified for $n\leqq 9$.

The approach through the second variation thus leads naturally to the local Bieberbach conjecture, which would state that the Koebe function yields the maximum of $\operatorname{Re}\{a_n\}$ at least for those normalized univalent functions which are close enough to it in some appropriate topology. Curiously enough, it was not a frontal attack on the Bieberbach inequality by means of the second variation which led to the solution of this more restricted problem. Beforehand one had to

1 Arch. Rational Mech. Anal., Vol. 26

solve a somewhat different but more accessible problem which leads to an inequality with a large number of free parameters. This is the Grunsky inequality, which gives numerous estimates for the coefficients of univalent functions [10]. It was applied by CHARZYNSKI-SCHIFFER [4] to give a simple proof of the inequality (0.2) for $n=4$ and was utilized by GARABEDIAN-ROSS-SCHIFFER [9] to prove the local Bieberbach conjecture for the case of even index $n=2m$.

To understand the meaning of the detour via the Grunsky inequalities, one has to observe that maximum problems for univalent functions lead always to differential equations of the form

$$f'(z)^2 E\left(\frac{1}{f}\right)=R(z), \tag{0.4}$$

where $E(t)$ is a polynomial and $R(z)$ a rational function. One is thus in general led to equations between hyperelliptic integrals which are exceedingly difficult to discuss. The extremum problem which leads to the Grunsky inequality is distinguished by the fact that its corresponding $E(t)$ is a perfect square of another polynomial. This allows an immediate integration of the differential equation and a complete solution of the problem. Now it is also easy to integrate (0.4) if $E(t)=(t-\alpha)\,P(t)^2$, where $P(t)$ is a polynomial. Thus it was natural to look for coefficient maximum problems leading to such left-hand sides in (0.4). GARABEDIAN [6] constructed such an expression and integrated the corresponding differential equation to prove that the local Bieberbach conjecture is true for $n=5$. It thus became clear that a new set of inequalities had to be discovered through differential equations of the almost quadratic type for all odd values $n=2m+1$. An outline of this approach was given by GARABEDIAN in [7].

In the present paper we derive a generalization of the Grunsky inequality which is obtained through a differential equation (0.4) with almost quadratic $E(t)$ and use it to prove the local Bieberbach conjecture in the general case. In Chapter I we start out with an inequality which involves free parameters, coefficients of the univalent function $f(z)$ considered, and values w which are not taken by $f(z)$. This inequality is of interest by itself since it throws light on the relation between conformal mapping into the w-plane and the coefficients of the mapping function. But the introduction of an exceptional w-value complicates the study of the coefficient problem. We show in Chapter II that we can replace the exceptional value w by a complex value u which satisfies an algebraic equation whose coefficients depend only on the parameters and the a_ν; the differential equation for this extremum problem is still of the desired form. However, in order to be able to integrate it completely, we have to restrict the parameters properly. An admissible set of parameters is obtained by an interesting reasoning based on the theory of univalent mappings by means of solutions of elliptic partial differential equations.

In Chapter III the inequalities obtained in Chapter II are used to prove the local Bieberbach conjecture for $n=2m+1$ and to complete thus the work started in [9]. It is of methodological interest to observe that again one has to consider the combination $a_{n+2}-a_n$ rather than a_n. This device has been used in the proof of the Bieberbach conjecture in the case of real a_n [13], has been necessary in the proof of the local theorem for even index, and is applied now in the odd case, too. This observation may be useful when the full Bieberbach conjecture is approached.

Our present result seems to promise further progress in the classical coefficient problem (0.2). Indeed, if one could bound off a finite and well-defined neighborhood around the Koebe function in which $\mathrm{Re}\{a_n\}<n$, except for the Koebe function, one might deal with the remaining coefficient region by less delicate estimates. Such a procedure was used by GARABEDIAN-SCHIFFER [8] in the first proof of (0.2) for $n=4$. There is a good chance that at least the cases $n=5$ and $n=6$ might be settled in this way. A similar approach has been followed by BOMBIERI, who has also obtained significant results in the local Bieberbach problem [3].

Finally, we should mention that a special case of our general inequality was used by JENKINS to prove (0.2) in the case $n=3$ and was derived by him by means of his "General Coefficient Theorem" [11].

I. A Generating Function for Generalized Grunsky Coefficients

1.1.

In this section we shall give a brief summary of the calculus of variations for univalent functions. Let $\mathscr{F}$ be the family of analytic functions which are regular and univalent in a domain D of the complex z-plane. Let $\Phi[f]$ be a real-valued functional which is defined for all analytic functions $f(z)$ in D. We suppose that $\Phi[f]$ satisfies the asymptotic equation

$$\Phi[f+\varepsilon g]=\Phi[f]+\mathrm{Re}\{\varepsilon\psi[f,g]\}+o(\varepsilon) \tag{1.1}$$

for any pair of analytic functions $f(z)$ and $g(z)$ in D. Here $\psi[f,g]$ is a complex-valued functional of $f(z)$ and $g(z)$ which is linear in $g(z)$. The remainder term $o(\varepsilon)$ is supposed to be equibounded for all functions $g(z)$ which are equibounded in a specified subdomain of D. Clearly, $\psi[f,g]$ is the Fréchet derivative of the functional $\Phi[f]$.

We may now pose the problem, to find functions $f_0(z)\in\mathscr{F}$ such that

$$\Phi[f]\leqq\Phi[f_0], \qquad f(z)\in\mathscr{F}. \tag{1.2}$$

It is easy to see that if $f(z)\in\mathscr{F}$ does not take the value w for any $z\in D$, there exist functions $f^*(z)\in\mathscr{F}$ of the asymptotic form

$$f^*(z)=f(z)+\frac{\varepsilon}{f(z)-w}+o(\varepsilon), \tag{1.3}$$

for $|\varepsilon|$ arbitrarily small, such that the residual term $o(\varepsilon)$ can be estimated uniformly in each closed subdomain of D. From the maximum property (1.2) of $f_0(z)$ and the local behavior (1.1) under infinitesimal variations we deduce the inequality

$$\Phi[f_0^*]=\Phi[f_0]+\mathrm{Re}\left\{\varepsilon\psi\left[f_0,\frac{1}{f_0-w}\right]\right\}+o(\varepsilon)\leqq\Phi[f_0] \tag{1.4}$$

and find the condition

$$\mathrm{Re}\left\{\varepsilon\psi\left[f_0,\frac{1}{f_0-w}\right]\right\}+o(\varepsilon)\leqq 0 \tag{1.4'}$$

for every admissible variation of the form (1.3).

1*

Let us introduce the analytic function

$$s(w)=\psi\left[f_0, \frac{1}{f_0-w}\right] \tag{1.5}$$

and apply the following fundamental lemma of the calculus of variations for univalent functions [*14*, *16*–*18*]:

Lemma. *Let $w=f(z)$ map D univalently onto a domain Δ; let $s(w)$ be analytic on the boundary of Δ and suppose that for every admissible variation* (1.3) *of $f(z)$*

$$\operatorname{Re}\{\varepsilon\, s(w)\}+o(\varepsilon)\leqq 0 \tag{1.6}$$

holds. Then Δ consists of the entire w-plane slit along analytic arcs $w(\tau)$, each of which satisfies the differential equation

$$\left(\frac{dw}{d\tau}\right)^2 s(w)=1 \tag{1.7}$$

for proper choice of the real curve parameter τ.

In view of this lemma we can assert that the function $f_0(z)\in\mathscr{F}$ which yields the maximum value of $\Phi[f]$ within the family maps the domain D onto a slit domain Δ in the w-plane whose analytic boundary arcs satisfy the differential equation

$$\psi\left[f_0, \frac{1}{f_0-w}\right]\left(\frac{dw}{d\tau}\right)^2=1. \tag{1.8}$$

For a more detailed exposition of the variational algorithm we refer the reader to [*14*, *16*–*18*].

1.2.

In studying the extremum problem $\Phi[f]=\max$, we have to discuss two questions. First, we have to make sure that an element $f_0(z)\in\mathscr{F}$ exists which yields the maximum value of the functional. Secondly, we have to use the characterization by the differential equation (1.8) to determine the extremum function and the maximum value $\Phi[f_0]$.

To guarantee the existence of an extremum function, one uses the fact that the univalent functions in a given domain D form a normal family. Hence they comprise a compact function set if one normalizes them at an arbitrary but fixed point $\zeta\in D$ by the requirement $f(\zeta)=0$, $f'(\zeta)=1$. Since now each maximum sequence has limit functions, the existence of an extremum function within the class of normalized univalent functions in $\mathscr{F}$ is ensured for any reasonable functional $\Phi[f]$.

Observe that each function $f(z)\in\mathscr{F}$ can be normalized by a linear transformation $\hat{f}(z)=af(z)+b$. Hence, if the functional $\Phi[f]$ considered is invariant under such linear transformations, we may drop the restriction to normalized functions in $\mathscr{F}$ and can still be sure that an extremum function exists.

We consider the following important example of such a functional. Let z_j $(j=1, 2, \ldots, n)$ be n arbitrary but fixed points in D and attach to each such point a complex number x_j. We impose the single restriction

(1.9) $$\sum_{j=1}^{n} x_j = 0$$

and define the functional

(1.10) $$\Phi[f] = \operatorname{Re}\left\{ \sum_{j,k=1}^{n} x_j x_k \log \frac{f(z_j)-f(z_k)}{z_j - z_k} \right\};$$

in the case $j=k$ we interpret the logarithmic term as $\log f'(z_j)$. It is easily seen that this functional satisfies all assumptions of the variational calculus and that it is invariant under a linear transformation of the above type. Hence there exists at least one extremum function $f_0(z)$ which yields the maximum value of this functional.

To determine the extremum function by means of the differential equation (1.8), we have first to calculate the Fréchet derivative

$$\psi\left[f_0, \frac{1}{f_0 - w}\right]$$

of our functional. This is obviously achieved by replacing the function $f_0(z)$ by its modification

(1.11) $$f_0^*(z) = f_0(z) + \frac{\varepsilon}{f_0(z) - w} + o(\varepsilon) = f_0(z) + \delta f_0(z) + o(\varepsilon)$$

and calculating the increment of the functional to the first oder in ε. We find

(1.12) $$\log \frac{f_0^*(z) - f_0^*(\zeta)}{z - \zeta} = \log \frac{f_0(z) - f_0(\zeta)}{z - \zeta} + \\ + \log\left[1 - \frac{\varepsilon}{(f_0(z) - w)(f_0(\zeta) - w)}\right] + o(\varepsilon),$$

i.e.,

(1.12′) $$\delta \log \frac{f_0(z) - f_0(\zeta)}{z - \zeta} = -\frac{\varepsilon}{[f_0(z) - w][f_0(\zeta) - w]},$$

whence by definitions (1) and (10)

(1.13) $$\psi\left[f_0, \frac{1}{f_0 - w}\right] = -\sum_{j,k=1}^{n} \frac{x_j x_k}{[f_0(z_j) - w][f_0(z_k) - w]} = -\left[\sum_{j=1}^{n} \frac{x_j}{f_0(z_j) - w}\right]^2.$$

The extremum function $f_0(z)$ therefore maps the domain D onto a slit domain Δ in the w-plane whose boundary arcs $w(\tau)$ satisfy the differential equation

(1.14) $$\left(\frac{dw}{d\tau}\right)^2 \left[\sum_{j=1}^{n} \frac{x_j}{f_0(z_j) - w}\right]^2 + 1 = 0.$$

This condition can be used in the case of quite general domains D to solve the maximum problem and to establish important inequalities for univalent functions. In this paper we shall be concerned with the problem for functions in the unit disk $|z| < 1$; hence we shall now simplify matters by considering that special case only. It is then more convenient to parametrize the boundary arcs in the form

$$(1.15)\qquad w(\varphi)=f_0(e^{i\varphi}),\qquad \frac{dw}{d\varphi}=i\,e^{i\varphi}f_0'(e^{i\varphi})$$

and to deduce from (1.14) the condition

$$(1.16)\qquad z^2 f_0'(z)^2\left[\sum_{j=1}^{n}\frac{x_j}{f_0(z_j)-f_0(z)}\right]^2>0,\qquad |z|=1.$$

The particular functional (1.10) has the remarkable property that the left side of the differential inequality (1.16) is a perfect square. This fact will enable us to solve the extremum problem in an elegant and convenient way.

We replace (1.16) by the equivalent and simpler equation

$$(1.16')\qquad z f_0'(z)\sum_{j=1}^{n}\frac{x_j}{f_0(z)-f_0(z_j)}=\text{real},\qquad |z|=1.$$

The left side of this equation is meromorphic in $|z|<1$ and has simple poles at the distinguished points z_j, with residues $x_j z_j$. Hence it can be extended into the entire complex plane by means of the Schwarz reflection principle; it is a rational function of the form

$$(1.17)\qquad z f_0'(z)\sum_{j=1}^{n}\frac{x_j}{f_0(z)-f_0(z_j)}=\sum_{j=1}^{n}\left\{\frac{z x_j}{z-z_j}+\frac{\bar{x}_j}{1-\bar{z}_j z}\right\}.$$

Here we have made use of the fact that the sum of all x_j is zero and we have determined an additive real constant by comparing both sides of (1.17) at the point $z=0$. Subtract on the right side the term $\sum\bar{x}_j=0$, divide by z and arrive at

$$(1.18)\qquad f_0'(z)\sum_{j=1}^{n}\frac{x_j}{f_0(z)-f_0(z_j)}=\sum_{j=1}^{n}\left\{\frac{x_j}{z-z_j}+\frac{\bar{z}_j\bar{x}_j}{1-\bar{z}_j z}\right\}.$$

We integrate and find

$$(1.19)\qquad \sum_{j=1}^{n}x_j\log\frac{f_0(z)-f_0(z_j)}{z-z_j}=\sum_{j=1}^{n}\left\{x_j\log\frac{f_0(z_j)}{z_j}+\bar{x}_j\log\frac{1}{1-\bar{z}_j z}\right\}.$$

We have obtained an equation for the extremum function $f_0(z)$ which, however, still contains the n unknown values $f_0(z_j)$. Nevertheless, we can calculate the maximum value $\Phi[f_0]$ for our problem. We specialize (1.19) by choosing $z=z_k$; the equation so obtained is multiplied by x_k and then we add the results for $k=1,2,\ldots,n$. Using again the restriction (1.9) for the x_k, we finally arrive at the result

$$(1.20)\qquad \begin{aligned}\Phi[f_0]&=\max_{f\in\mathscr{F}}\operatorname{Re}\left\{\sum_{j,k=1}^{n}x_j x_k\log\frac{f(z_j)-f(z_k)}{z_j-z_k}\right\}\\&=\operatorname{Re}\left\{\sum_{j,k=1}^{n}x_j\bar{x}_k\log\frac{1}{1-z_j\bar{z}_k}\right\}.\end{aligned}$$

In a completely analogous way we may treat the functional

$$\Phi[f]=\mathrm{Re}\left\{-\frac{1}{4\pi^2}\iint_{\Gamma\Gamma}\mu(z)\,\mu(\zeta)\log\frac{f(z)-f(\zeta)}{z-\zeta}\,dz\,d\zeta\right\}, \tag{1.21}$$

where Γ is an arbitrary curve in D and $\mu(z)$ is a complex-valued weight function satisfying

$$\int_\Gamma \mu(z)\,dz=0. \tag{1.21'}$$

We find for every regular univalent function $f(z)$ in $|z|<1$ the inequality

$$\begin{aligned}&\mathrm{Re}\left\{-\frac{1}{4\pi^2}\iint_{\Gamma\Gamma}\mu(z)\,\mu(\zeta)\log\frac{f(z)-f(\zeta)}{z-\zeta}\,dz\,d\zeta\right\}\\ &\leqq \mathrm{Re}\left\{\frac{1}{4\pi^2}\iint_{\Gamma\Gamma}\mu(z)\,\overline{\mu(\zeta)}\log\frac{1}{1-z\bar{\zeta}}\,dz\,d\bar{\zeta}\right\}.\end{aligned} \tag{1.22}$$

Clearly, this result can also be obtained from (1.20) by a suitable limit procedure.

The most important application of inequality (1.22) is obtained when the curve Γ is closed around the origin and we choose

$$\mu(z)=\sum_{j=1}^{n}\frac{\lambda_j}{z^{j+1}}. \tag{1.23}$$

This choice is compatible with condition (1.21'). Let us define the coefficient matrix $((c_{jk}))$ by the equation

$$\log\frac{f(z)-f(\zeta)}{z-\zeta}=\sum_{j,k=0}^{\infty}c_{jk}\,z^j\zeta^k. \tag{1.24}$$

The c_{jk} are the "Grunsky coefficients" connected with the function $f(z)$. Using the residue theorem in (1.22) with respect to the weight function (1.23), we obtain the important Grunsky inequality [*10*, *15*]

$$\mathrm{Re}\left\{\sum_{j,k=1}^{n}c_{jk}\lambda_j\lambda_k\right\}\leqq\sum_{j=1}^{n}\frac{1}{j}|\lambda_j|^2. \tag{1.25}$$

This inequality has interesting applications to the coefficient problem for univalent functions in the unit disk. By means of it one can verify the Bieberbach conjecture for the fourth coefficient in an easy and elementary way. The same method enabled us to prove that the Koebe function yields a local maximum for all coefficients with even index [*9*].

Up to now we covered known ground in order to prepare for a generalization of the Grunsky inequality which will play a central role in further developments of the coefficient problem. We wish to stress the essential features of our derivation. We have been dealing with a functional which is invariant under linear transformations $\hat{f}=af+b$ and that leads to a differential equation for the extremum function that contains a perfect square. The latter fact is based on an important property of the functional

$$\Phi[f]=\mathrm{Re}\{\log[f(z)-f(\zeta)]\}, \tag{1.26}$$

whose Fréchet derivative has the special form

$$(1.26') \qquad \psi\left[f, \frac{1}{f-w}\right] = -\frac{\varepsilon}{[f(z)-w][f(\zeta)-w]};$$

that is, it splits into a product of a function of z with the same function of the variable ζ. In the next section we shall construct a more general functional (or generating function) with the same properties.

1.3.

The most general one-to-one conformal mapping of the complex plane onto itself has the form

$$(1.27) \qquad \hat{w} = \frac{a w + b}{c w + d}.$$

It is therefore natural to consider functionals $\Phi[f]$ which are invariant under these general linear transformations. Extremum problems for such functionals will refer to the class of all univalent meromorphic functions in the domain considered.

If the functional depends specifically upon values $f(z_j)$ of the function at distinguished points z_j in the domain, it must in reality depend upon the various cross ratios of the function values, since only these remain unchanged under all transformations (1.27). Let us therefore start with four points z, ζ, ω, η in the domain and define the cross ratio

$$(1.28) \qquad R[f; z, \zeta; \omega, \eta] = \frac{f(z)-f(\omega)}{f(z)-f(\eta)} \cdot \frac{f(\zeta)-f(\eta)}{f(\zeta)-f(\omega)}.$$

Having chosen these four points once and for all, we ask now for a functional

$$(1.29) \qquad \Phi[f] = \mathrm{Re}\{m(R[f; z, \zeta; \omega, \eta])\},$$

where $m(x)$ is a complex-valued function chosen in such a way that the Fréchet derivative

$$\psi\left[f, \frac{1}{f-w}\right]$$

decomposes into a product of a function of z with the same function of ζ. For this purpose we calculate first how the cross ration (1.28) changes under a variation (1.3) of the function $f(z)$. A straightforward calculation yields

$$(1.30) \qquad \begin{aligned} &R[f^*; z, \zeta; \omega, \eta] = R[f; z, \zeta; \omega, \eta] \times \\ &\times \left\{1 - \varepsilon \frac{[f(z)-f(\zeta)][f(\omega)-f(\eta)]}{[f(z)-w][f(\zeta)-w][f(\omega)-w][f(\eta)-w]} + O(\varepsilon^2)\right\}. \end{aligned}$$

We insert this asymptotic formula into the „Ansatz" (1.29) and find

$$1.31) \qquad \begin{aligned} &\Phi[f^*] = \Phi[f] - \\ &- \mathrm{Re}\left\{\varepsilon R\, m'(R) \frac{[f(z)-f(\zeta)][f(\omega)-f(\eta)]}{[f(z)-w][f(\zeta)-w][f(\omega)-w][f(\eta)-w]}\right\} + o(\varepsilon). \end{aligned}$$

We observe the identity

$$1-R[f;z,\zeta;\omega,\eta]=\frac{[f(z)-f(\zeta)][f(\omega)-f(\eta)]}{[f(z)-f(\eta)][f(\zeta)-f(\omega)]}. \tag{1.32}$$

Hence combining (1.31) with (1.28) and (1.32) and using the definition (1.1) of the Fréchet derivative, we find

$$\begin{aligned}\psi\left[f,\frac{1}{f-w}\right]&=-\sqrt{R}(1-R)\,m'(R)\times\\&\times\frac{\sqrt{[f(z)-f(\eta)][f(z)-f(\omega)]}\sqrt{[f(\zeta)-f(\eta)][f(\zeta)-f(\omega)]}}{[f(z)-w][f(\zeta)-w][f(\omega)-w][f(\eta)-w]}.\end{aligned} \tag{1.33}$$

The fractional factor on the right in (1.33) is indeed split into a product of a function of z with the same function of ζ. Since R is not symmetric in z and ζ, we eliminate it from equation (1.33) by requiring the function $m(R)$ to satisfy the differential equation

$$\sqrt{R}(1-R)\,m'(R)=-1. \tag{1.34}$$

Thus we find

$$m(R)=\log\frac{1-\sqrt{R}}{1+\sqrt{R}} \tag{1.35}$$

and arrive at the functional

$$\Phi[f]=\mathrm{Re}\left\{\log\frac{1-\sqrt{R[f;z,\zeta;\omega,\eta]}}{1+\sqrt{R[f;z,\zeta;\omega,\eta]}}\right\}. \tag{1.36}$$

Its Fréchet derivative has the very useful form

$$\psi\left[f,\frac{1}{f-w}\right]=\frac{\sqrt{[f(z)-f(\eta)][f(z)-f(\omega)]}\sqrt{[f(\zeta)-f(\eta)][f(\zeta)-f(\omega)]}}{[f(z)-w][f(\zeta)-w][f(\omega)-w][f(\eta)-w]}. \tag{1.37}$$

We can also write

$$\Phi[f]=\mathrm{Re}\{N(z,\zeta;\omega,\eta)\}, \tag{1.38}$$

where $N(z, \zeta; \omega, \eta)$ depends analytically upon its four variables as follows:

$$\begin{aligned}N(z,\zeta;\omega,\eta)\\=\log\frac{\sqrt{[f(z)-f(\eta)][f(\zeta)-f(\omega)]}-\sqrt{[f(z)-f(\omega)][f(\zeta)-f(\eta)]}}{\sqrt{[f(z)-f(\eta)][f(\zeta)-f(\omega)]}+\sqrt{[f(z)-f(\omega)][f(\zeta)-f(\eta)]}}.\end{aligned} \tag{1.39}$$

An easy calculation leads to the useful result

$$\frac{\partial}{\partial z}N(z,\zeta;\omega,\eta)=\frac{f'(z)}{f(z)-f(\zeta)}\,\frac{\sqrt{[f(\zeta)-f(\eta)][f(\zeta)-f(\omega)]}}{\sqrt{[f(z)-f(\eta)][f(z)-f(\omega)]}}. \tag{1.40}$$

If we set

$$u=f(\omega)^{-1},\quad v=f(\eta)^{-1}, \tag{1.41}$$

we are led to consider the following generating function:

$$(1.42)\quad \begin{aligned} &M(z,\zeta;u,v)\\ &=\log\frac{\sqrt{[1-uf(z)][1-vf(\zeta)]}-\sqrt{[1-uf(\zeta)][1-vf(z)]}}{\left(\sqrt{[1-uf(z)][1-vf(\zeta)]}+\sqrt{[1-uf(\zeta)][1-vf(z)]}\right)(z-\zeta)}.\end{aligned}$$

An easy transformation of (1.40) yields

$$(1.43)\quad \frac{\partial}{\partial z}M(z,\zeta;u,v)=\frac{f'(z)}{f(z)-f(\zeta)}\frac{\sqrt{[1-uf(\zeta)][1-vf(\zeta)]}}{\sqrt{[1-uf(z)][1-vf(z)]}}-\frac{1}{z-\zeta},$$

while an analogous calculation leads to

$$(1.44)\quad \frac{\partial}{\partial u}M(z,\zeta;u,v)=\frac{1}{u-v}\sqrt{\frac{1-vf(z)}{1-uf(z)}}\sqrt{\frac{1-vf(\zeta)}{1-uf(\zeta)}}.$$

We may assume without loss of generality that $f(0)=0$ and $f'(0)=1$. Near the origin, the function $M(z,\zeta;u,v)$ is analytic in z and ζ, for u, v fixed, and admits a power series development

$$(1.45)\quad M(z,\zeta;u,v)=\sum_{j,k=0}^{\infty}c_{jk}(u,v)\,z^j\zeta^k.$$

Here we have

$$(1.46)\quad c_{00}(u,v)=\log\frac{v-u}{4},$$

while all other $c_{jk}(u,v)$ are polynomials in u and v. They are also simple polynomials in the coefficients a_k of the power series for $f(z)$. The function $M(z,\zeta;u,v)$ appears thus as the generating function of an interesting class of polynomials and it is for these polynomials that we will be able to obtain generalizations of the Grunsky inequality.

The close relationship between the coefficient matrix $c_{jk}(u,v)$ and the Grunsky matrix c_{jk} defined in (1.24) is a consequence of the limit formula

$$(1.47)\quad \lim_{u,v\to 0}\left[M(z,\zeta;u,v)-\log\frac{v-u}{4}\right]=\log\frac{f(z)-f(\zeta)}{z-\zeta},$$

which implies

$$(1.47')\quad c_{jk}(0,0)=c_{jk},\qquad (j,k)\neq(0,0).$$

In analogy to the functional (1.21) we consider now the functional

$$(1.48)\quad \Phi[f]=\mathrm{Re}\left\{-\frac{1}{4\pi^2}\iint_{\Gamma\Gamma}\mu(z)\,\mu(\zeta)\,M(z,\zeta;u,v)\,dz\,d\zeta\right\}$$

based on a curve Γ in the unit disk and the complex-valued weight function $\mu(z)$. The Fréchet derivative of $\Phi[f]$ can easily be calculated by means of (1.37) and (1.41); we find

$$\psi\left[f,\frac{1}{f-w}\right] = \left\{\frac{1}{\sqrt{(1-uw)(1-vw)}}\,\frac{1}{2\pi i}\int_\Gamma \mu(z)\frac{\sqrt{[1-uf(z)][1-vf(z)]}}{f(z)-w}\,dz\right\}^2. \tag{1.49}$$

We have again a perfect square in the variational formula, except for the square-root factor in the denominator, which can easily be handled, as will be seen in the sequel. It is this factor which allows the treatment of coefficients with odd index in a way analogous to that with even index. Consider the function

$$p(u,v) = -\frac{1}{4\pi^2}\iint_{\Gamma\Gamma} \mu(z)\,\mu(\zeta)\,M(z,\zeta;u,v)\,dz\,d\zeta \tag{1.50}$$

for given $f(z)$, and fixed Γ and $\mu(z)$. In view of (1.44), we find the equation

$$\frac{\partial p(u,v)}{\partial u} = \frac{1}{u-v}\left\{\frac{1}{2\pi i}\int_\Gamma \frac{\sqrt{1-vf(z)}}{\sqrt{1-uf(z)}}\,\mu(z)\,dz\right\}^2, \tag{1.51}$$

which will play an important role in our further considerations.

We obtain interesting expressions for $\Phi[f]$ and $p(u,v)$ if we choose Γ to be a closed curve around the origin and if we define

$$Q(z) = \sum_{\nu=-m}^{m} \lambda_m z^{-m}, \qquad \lambda_{-m} = \bar{\lambda}_m, \tag{1.52}$$

and set

$$\mu(z) = \frac{1}{z}\,Q(z). \tag{1.53}$$

In this case we obtain from (1.45) and (1.50) by means of the residue theorem

$$p(u,v) = \sum_{j,k=0}^{m} c_{jk}(u,v)\,\lambda_j\,\lambda_k. \tag{1.54}$$

We will now study the extremum problem of maximizing the functional $\Phi[f] = \mathrm{Re}\{p(u,v)\}$, which generalizes the maximum problem (1.25) that we solved in Section 1.2.

1.4.

In dealing with extremum problems for the functional

$$\Phi[f] = \mathrm{Re}\left\{\sum_{j,k=0}^{m} c_{jk}(u,v)\,\lambda_j\,\lambda_k\right\} = \mathrm{Re}\{p(u,v)\} \tag{1.55}$$

it must be born in mind that the $c_{jk}(u,v)$ vary under change of $f(z)$ for two different reasons. First of all, the coefficients a_k of the defining function $f(z)$ are changed; secondly, the argument points u and v vary also cogrediently. Because of definition (1.41) we have under a variation (1.3) the following variation of u and v:

$$\delta u = -\frac{\varepsilon u^3}{1-uw}, \qquad \delta v = -\frac{\varepsilon v^3}{1-vw}. \tag{1.56}$$

We can pose three types of extremum problems which lead to the variational formula (1.49) for the functional (1.55):

a) We may prescribe two arbitrary points ω and η in the unit disk and ask for the univalent function $f(z)$ for which

$$\operatorname{Re}\left\{p\left(\frac{1}{f(\omega)}, \frac{1}{f(\eta)}\right)\right\} = \max. \tag{1.57}$$

By our general theory and in view of the form (1.49) of the Fréchet derivative we can assert that the extremum function maps the disk onto a slit domain whose boundary arcs satisfy the differential equation

$$\left\{\frac{dw}{d\tau} \frac{1}{\sqrt{\left(1-\frac{w}{f(\omega)}\right)\left(1-\frac{w}{f(\eta)}\right)}} \frac{1}{2\pi i} \times \oint_{\Gamma} Q(z) \frac{\sqrt{\left(1-\frac{f(z)}{f(\omega)}\right)\left(1-\frac{f(z)}{f(\eta)}\right)}}{f(z)-w} \frac{dz}{z}\right\}^2 + 1 = 0. \tag{1.58}$$

b) We may prescribe one arbitrary point ω in the unit disk and ask for a univalent function $f(z)$ for which

$$\operatorname{Re}\left\{p\left(\frac{1}{f(\omega)}, \frac{1}{w_0}\right)\right\} = \max, \tag{1.59}$$

where w_0 is a value which is not taken by $f(z)$ in the disk. In this problem both $f(z)$ and w_0 have to be varied in order to determine the maximum value. It is clear that our variational method applies in this case, too. For suppose that $f_0(z)$ and w_0 are the optimal choice. Under a variation (1.3) of $f_0(z)$ we are sure that the point

$$w_0^* = w_0 + \frac{\varepsilon}{w_0 - w} + o(\varepsilon) \tag{1.60}$$

will be a point not taken by $f_0^*(z)$ in $|z|<1$. Hence, if we vary $f_0(z)$ according to (1.3) and w_0 cogrediently according to (1.60), the value of the functional (1.59) cannot increase. Hence we can apply the fundamental lemma to formula (1.49) and conclude that the extremum function maps the unit disk onto a slit domain whose boundary arcs satisfy the differential equation

$$\left\{\frac{dw}{d\tau}\left(1-\frac{w}{w_0}\right)^{-\frac{1}{2}}\left(1-\frac{w}{f(\omega)}\right)^{-\frac{1}{2}} \frac{1}{2\pi i} \times \oint_{\Gamma} Q(z) \frac{\sqrt{\left(1-\frac{f(z)}{f(\omega)}\right)\left(1-\frac{f(z)}{w_0}\right)}}{f(z)-w} \frac{dz}{z}\right\}^2 + 1 = 0. \tag{1.58'}$$

Similarly, we may ask for the maximum value of the functional

$$\Phi[f] = \operatorname{Re}\left\{p\left(\frac{1}{w_0}, \frac{1}{w_1}\right)\right\}, \tag{1.61}$$

where both w_0 and w_1 are values not taken by $f(z)$ in the disk. The reasoning in this case is precisely as before.

c) Let us choose $v=0$; this is surely the reciprocal of an exceptional value w_0 and satisfies the variational equation (1.56). We determine u as the solution of the algebraic equation

$$\frac{\partial p(u,0)}{\partial u}=\frac{1}{u}\left\{\frac{1}{2\pi i}\oint_{\Gamma}\frac{Q(z)}{\sqrt{1-uf(z)}}\frac{dz}{z}\right\}^2=0. \tag{1.62}$$

According to (1.56), u will in general not vary cogrediently with $f(z)$. However, the variation of u is now unessential since we insured by the definition (1.62) of u that the functional be insensitive to an infinitesimal variation of u. Hence we may still apply the variational formula (1.49) to determine the boundary of the extremum domain. Many further extremum problems can be discussed by combining these three types of conditions on u and v. The special problem *c)* posed here is particularly important for the coefficient problem and will be treated in detail in Chapter II.

1.5.

All extremum problems indicated in the preceding section can be treated and completely solved. However, the formulas sometimes become quite forbidding; a better insight into the general theory may be obtained from the discussion of important special cases. We shall deal here with the case of type *b)* in which u^{-1} and v^{-1} are both values not taken by $f(z)$ in the unit disk.

We thus ask for those functions $f(z)$ for which $\mathrm{Re}\{p(u,v)\}$ is maximal. In view of the normalization of $f(z)$ we know that $|u|\leq 4$, $|v|\leq 4$, and since all $c_{jk}(u,v)$ are polynomials in u and v if $(j,k)\neq(0,0)$, we only have to consider the exceptional term $c_{00}(u,v)$ in order to see whether the functional is bounded. However, we see from (1.46) that $\mathrm{Re}\{c_{00}(u,v)\}$ becomes negatively infinite if u and v coincide; likewise, $\mathrm{Im}\{c_{00}(u,v)\}$ is not necessarily bounded. We assume therefore that λ_0 is real. In this case $\Phi[f]$ is bounded from above for all normalized univalent functions and, since the class is compact, the existence of a maximum function is certain.

We may now apply the calculus of variations to characterize an extremum function $f_0(z)$ for our problem. Such a function maps the unit disk onto the w-plane slit along analytic arcs satisfying the differential equation

$$\frac{dw}{d\tau}\frac{1}{\sqrt{(1-uw)(1-vw)}}\frac{1}{2\pi i}\oint_{\Gamma}Q(t)\frac{\sqrt{[1-uf_0(t)][1-vf_0(t)]}}{f_0(t)-w}\frac{dt}{t}=\sqrt{-1}. \tag{1.63}$$

To utilize this characterization, we observe that

$$\frac{1}{2\pi i}\oint_{\Gamma}Q(t)\frac{\sqrt{[1-uf_0(t)][1-vf_0(t)]}}{w-f_0(t)}\frac{dt}{t}=\frac{1}{w}\Pi_m\left(\frac{1}{w}\right) \tag{1.64}$$

is a polynomial of degree $m+1$ in $1/w$ if w lies outside the image of Γ by $f(z)$. This polynomial depends on u, v, the λ_ν, and the coefficients a_k in the development of $f(z)$. The boundary arcs of the extremum domain thus satisfy the differential equation

$$\frac{1}{w}\frac{dw}{d\tau}\Pi_m\left(\frac{1}{w}\right)\frac{1}{\sqrt{(1-uw)(1-vw)}}=\sqrt{-1}. \tag{1.63'}$$

With the parametrization $w=f_0(e^{i\varphi})$ we may bring (1.63) into the form

$$H(z)=\frac{z f_0'(z)}{\sqrt{[1-uf_0(z)][1-vf_0(z)]}}\frac{1}{2\pi i}\oint_\Gamma Q(t)\frac{\sqrt{[1-uf_0(t)][1-vf_0(t)]}}{f_0(z)-f_0(t)}\frac{dt}{t} \tag{1.65}$$
$$=\text{real}, \qquad |z|=1.$$

By virtue of formula (1.43) we can express $H(z)$ in the elegant form

$$H(z)=z\frac{1}{2\pi i}\oint_\Gamma Q(t)\left[\frac{\partial}{\partial z}M(z,t;u,v)+\frac{1}{z-t}\right]\frac{dt}{t}. \tag{1.66}$$

Next we apply (1.45) and (1.52) and arrive at

$$H(z)=\sum_{\nu=0}^{m}\lambda_\nu z^{-\nu}+\sum_{\nu=1}^{\infty}l_\nu z^\nu;\qquad l_\nu=\nu\sum_{\mu=0}^{m}c_{\nu\mu}(u,v)\lambda_\mu. \tag{1.67}$$

We have shown that $H(z)$ is regular analytic for $|z|<1$, except for a pole at the origin whose principal part is displayed in (1.67), and that $H(z)$ is real for $|z|=1$. It can be continued in a unique way into the region $|z|>1$ by use of the Schwarz reflection principle and is analytic there except for a pole at infinity. On the circumference $|z|=1$, as is easily seen, $H(z)$ is analytic. Indeed, consider the analytic function

$$\frac{1}{i}\int_1^z[H(z)-\lambda_0]\frac{dz}{z}=\mathscr{I}(z), \tag{1.68}$$

which is bounded and real for $|z|=1$. Since it is single-valued for $|z|<1$, it is single-valued by the reflection principle for $|z|>1$. Hence because $\mathscr{I}(z)$ can have at most isolated singularities on $|z|=1$, these are all removable and $\mathscr{I}(z)$ is analytic on the entire circumference. By differentiation the same is true for $H(z)$.

Finally, we can state that in view of the reflection principle

$$H(z)=\sum_{\nu=-m}^{m}\lambda_\nu z^{-\nu}=Q(z). \tag{1.69}$$

Observe how we had to use repeatedly the assumption that λ_0 is real.

We thus proved that the extremum function $f_0(z)$ satisfies the differential equation

$$\frac{z f_0'(z)}{f_0(z)}\frac{\Pi_m\left(\frac{1}{f_0(z)}\right)}{\sqrt{[1-uf_0(z)][1-vf_0(z)]}}=Q(z). \tag{1.70}$$

We can also deduce from (1.67) and (1.69) that

$$\frac{1}{\nu}\bar{\lambda}_\nu=\sum_{\mu=0}^{m} c_{\nu\mu}(u,v)\,\lambda_\mu,\qquad \nu=1,2,\ldots,m, \tag{1.71}$$

$$0=\sum_{\mu=0}^{m} c_{\nu\mu}(u,v)\,\lambda_\mu,\qquad \nu>m. \tag{1.71'}$$

If $\lambda_0=0$, we can rebuild $\Phi[f]$ by means of the equations (1.71) and determine its maximum value. In any other case we have to use an additional step of integration to obtain one more equation.

We start with the identity

$$\frac{1}{2\pi i}\oint_\Gamma Q(t)\,M(z,t;u,v)\frac{dt}{t}=\sum_{\mu=0}^{m}\lambda_\mu\sum_{j=0}^{\infty}c_{j\mu}(u,v)\,z^j, \tag{1.72}$$

which follows from (1.45) and (1.52). Because of (1.71) and (1.71′) this reduces to

$$\frac{1}{2\pi i}\oint_\Gamma Q(t)\,M(z,t;u,v)\frac{dt}{t}=\sum_{\mu=0}^{m}c_{0\mu}(u,v)\,\lambda_\mu+\sum_{j=1}^{m}\frac{\bar{\lambda}_j}{j}z^j. \tag{1.72'}$$

From the nature of the slit mapping provided by the extremum function $f_0(z)$ follows the existence of at least one point η on the circumference $|z|=1$ such that $f(\eta)^{-1}=v$. On the other hand, we read off from definition (1.42) that for this value η

$$M(\eta,\zeta;u,v)=\log\frac{1}{\eta-\zeta}. \tag{1.73}$$

Hence if we let $z\to\eta$ in (1.72′), we obtain

$$\frac{1}{2\pi i}\oint_\Gamma Q(t)\log\frac{1}{\eta-t}\frac{dt}{t}=\sum_{\mu=0}^{m}c_{0\mu}(u,v)\,\lambda_\mu+\sum_{j=1}^{m}\frac{\bar{\lambda}_j}{j}\eta^j. \tag{1.74}$$

On the other hand, it is clear in view of (1.52) that

$$\frac{1}{2\pi i}\oint_\Gamma Q(t)\log\frac{1}{\eta-t}\frac{dt}{t}=\lambda_0\log\frac{1}{\eta}+\sum_{j=1}^{m}\frac{\lambda_j}{j}\eta^{-j}. \tag{1.75}$$

Hence we found because of $|\eta|=1$ the equation

$$\sum_{\mu=0}^{m}c_{0\mu}(u,v)\,\lambda_\mu=\lambda_0\log\frac{1}{\eta}+2i\,\mathrm{Im}\left\{\sum_{j=1}^{m}\frac{\lambda_j}{j}\eta^{-j}\right\}, \tag{1.76}$$

which shows that the left side is pure imaginary.

If we multiply the ν-th equation (1.71) by λ_ν, multiply equation (1.76) by the real number λ_0, take real parts and add, we end up with the equation

$$\mathrm{Re}\left\{\sum_{\mu,\nu=0}^{m}\lambda_\nu\lambda_\mu c_{\nu\mu}(u,v)\right\}=\sum_{\nu=1}^{m}\frac{|\lambda_\nu|^2}{\nu}. \tag{1.77}$$

Since the functional so determined is a maximum value, we can assert the following

Theorem. *For every univalent $f(z)$ in $|z|<1$*

$$\operatorname{Re}\{p(u,v)\}=\operatorname{Re}\left\{\sum_{\mu,\nu=0}^{m}\lambda_\mu\lambda_\nu c_{\mu\nu}(u,v)\right\}\leqq\sum_{\nu=1}^{m}\frac{1}{\nu}|\lambda_\nu|^2 \tag{1.78}$$

holds.

This theorem gives a beautiful generalization of the Grunsky inequality (1.25) and provides ample information about the values u^{-1} and v^{-1} which are left out in the conformal mapping by the univalent function $f(z)$.

The coefficients $c_{jk}(u,v)$ can easily be calculated by use of the formulas (1.44) and (1.47′). We find, for example,

$$\begin{aligned}&c_{00}(u,v)=\log\frac{v-u}{2},\qquad c_{01}(u,v)=c_{10}(u,v)=a_2+\frac{u+v}{2},\\&c_{11}(u,v)=c_{11}+\frac{(u-v)^2}{8}=a_3-a_2^2+\frac{(u-v)^2}{8}\end{aligned} \tag{1.79}$$

if we define the coefficients a_k by the equation

$$f(z)=z+\sum_{n=2}^{\infty}a_n z^n. \tag{1.80}$$

We can therefore state the inequality

$$\lambda_0^2\log\left|\frac{v-u}{4}\right|+2\lambda_0\operatorname{Re}\left\{a_2+\frac{u+v}{2}\right\}+\operatorname{Re}\left\{a_3-a_2^2+\frac{(u-v)^2}{8}\right\}\leqq 1 \tag{1.81}$$

for all real values of λ_0 and any two values u^{-1}, v^{-1} not taken by $f(z)$. If we assume $v=0$, which is always allowed, we can deduce from (1.81) the simpler estimate

$$\lambda_0^2\log\left|\frac{u}{4}\right|+2\lambda_0\operatorname{Re}\left\{a_2+\frac{u}{2}\right\}+\operatorname{Re}\left\{a_3-a_2^2+\frac{u^2}{8}\right\}\leqq 1, \tag{1.81′}$$

which contains the Koebe theorem $|u|\leqq 4$ as a special case.

1.6.

The central role of the generating function $M(z,\zeta;u,v)$ for the theory of univalent functions is best seen through some important specialization, which will only be indicated briefly. We have already shown that in the limit $u\to 0, v\to 0$ we are led back to the generating function for the original Grunsky coefficients.

We may assume $u=f(\omega)^{-1}$ and $v=f(\eta)^{-1}$ and consider

$$\begin{aligned}&\lim_{\omega\to\eta}\left\{M(z,\zeta;f(\omega)^{-1},f(\eta)^{-1})+\log\frac{1}{\omega-\eta}\right\}\\&=m(z,\zeta;\eta)\\&=\log\left(-\frac{f'(\eta)}{4}\right)-\log[f(z)-f(\eta)]-\log[f(\zeta)-f(\eta)]+\log\frac{f(z)-f(\zeta)}{z-\zeta}.\end{aligned} \tag{1.82}$$

We use the definition of the Faber polynomials $\mathcal{F}_n(t)$ by means of the generating function

$$(1.83)\qquad \log\{1-t f(z)\}=\sum_{n=1}^{\infty}\left[\frac{1}{n}\mathcal{F}_n(t)-c_{n0}\right]z^n,$$

where the c_{n0} stand for the old Grunsky coefficients. We can then determine the coefficients of the series development

$$(1.84)\qquad m(z,\zeta;\eta)=\sum_{j,k=0}^{\infty}\gamma_{jk}(\eta)\,z^j\zeta^k$$

as follows:

$$(1.85)\qquad \gamma_{00}(\eta)=\log\left[-\frac{f'(\eta)}{4}\right]+\log f'(0)$$

and

$$(1.85')\qquad \begin{aligned}\gamma_{j0}(\eta)&=\frac{1}{j}\,\mathcal{F}_j\left(\frac{1}{f(\eta)}\right)=\gamma_{0j}(\eta), && j>0,\\ \gamma_{jk}(\eta)&=c_{jk}, && j,k>0.\end{aligned}$$

In view of (1.45) and (1.82) we have thus proved

$$(1.86)\qquad \begin{aligned}&\lim_{u\to v} c_{jk}(u,v)=c_{jk}, && j,k>0,\\ &\lim_{u\to v} c_{j0}(u,v)=\frac{1}{j}\,\mathcal{F}_j(v), && j>0.\end{aligned}$$

We may consider the functional

$$(1.87)\qquad \Phi[f]=\mathrm{Re}\left\{\sum_{j,k=0}^{m}\lambda_j\lambda_k\gamma_{jk}(\eta)\right\}$$

and ask for its maximum value. As can be seen from the differential equation (1.58), the coincidence of ω and η leads in this case to a disappearance of all square-root terms and the treatment of the problem simplifies considerably. We state the final result of the calculation as a

Theorem. *For all normalized univalent functions $f(z)$ in $|z|<1$, the inequality*

$$(1.88)\qquad \begin{aligned}&\mathrm{Re}\left\{\lambda_0^2\log\left[\frac{\eta^2 f'(\eta)f'(0)}{4f(\eta)^2}\right]+2\lambda_0\sum_{j=1}^{m}\frac{\lambda_j}{j}\left[\mathcal{F}_j\left(\frac{1}{f(\eta)}\right)-\frac{1}{\eta^j}\right]+\sum_{j,k=1}^{m}\lambda_j\lambda_k c_{jk}\right\}\\ &\quad\leqq|\lambda_0|^2\log\frac{1}{1-|\eta|^2}+\sum_{j=1}^{m}\frac{1}{j}|\lambda_j|^2-2\,\mathrm{Re}\left\{\lambda_0\sum_{j=1}^{m}\frac{1}{j}\bar{\lambda}_j\eta^j\right\}\end{aligned}$$

holds.

This result was derived in a different form in [*15*]. It was indeed found by searching for a functional which leads to a differential equation with a perfect square term.

Finally, let us make in $M(z,\zeta;u,v)$ the choice $v=f(\eta)^{-1}$ with $\eta=0$. Because of the normalization of $f(z)$ we have to pass to the limit $v=\infty$ in (1.42). We arrive at the generating function

$$(1.89)\qquad M(z,\zeta;u,\infty)=\log\frac{\sqrt{f(\zeta)}\sqrt{1-u f(z)}-\sqrt{f(z)}\sqrt{1-u f(\zeta)}}{[\sqrt{f(\zeta)}\sqrt{1-u f(z)}+\sqrt{f(z)}\sqrt{1-u f(\zeta)}](z-\zeta)}.$$

To avoid a branch point at the origin, we replace z and ζ by z^2 and ζ^2 and define the closely related expression

$$(1.90)\quad \hat{M}(z,\zeta;u)=\log\frac{[\sqrt{f(\zeta^2)}\sqrt{1-u\,f(z^2)}-\sqrt{f(z^2)}\sqrt{1-u\,f(\zeta^2)}](z+\zeta)}{[\sqrt{f(\zeta^2)}\sqrt{1-u\,f(z^2)}+\sqrt{f(z^2)}\sqrt{1-u\,f(\zeta^2)}](z-\zeta)},$$

which leads to a variational formula of the same type.

The particular case $u=0$ leads to the important generating function

$$(1.91)\qquad \hat{M}(z,\zeta;0,\infty)=\log\frac{[\sqrt{f(\zeta^2)}-\sqrt{f(z^2)}](z+\zeta)}{[\sqrt{f(\zeta^2)}+\sqrt{f(z^2)}](z-\zeta)},$$

which plays a decisive role in the proof that the even coefficients of the Koebe function yield local maxima [9].

II. Application of a Lemma of Berg and Lewy

2.1.

In the previous chapter we established a generalization of the Grunsky inequality which involved the coefficients of a univalent function $f(z)$ and the values omitted by it. These inequalities are of interest since they have various implications in the geometry of conformal mapping; however, they yield no new results in the direction of the Bieberbach conjecture. In this chapter we shall develop an alternative type of inequality which is also based on the coefficient matrix $c_{jk}(u,v)$ of the generalized generating function of Chapter I, but where the argument points u, v have been chosen in a different way. These new inequalities will lead us ultimately to a proof of the Bieberbach conjecture for any coefficient with odd index a_{2m+1} of univalent functions close enough to the Koebe function.

We deal now with the extremum problem of Section 1.4 of type *c)*. Since we will choose $v=0$ throughout, we shall now use the following notation. The generating function

$$(2.1)\qquad M(z,\zeta;u)=\log\frac{\sqrt{1-u\,f(z)}-\sqrt{1-u\,f(\zeta)}}{[\sqrt{1-u\,f(z)}+\sqrt{1-u\,f(\zeta)}](z-\zeta)}$$

leads to the coefficient matrix $c_{jk}(u)$ given by

$$(2.2)\qquad M(z,\zeta;u)=\sum_{j,k=0}^{\infty} c_{jk}(u)\,z^j\zeta^k$$

and to the corresponding quadratic form

$$(2.3)\quad p(u)=-\frac{1}{4\pi^2}\oint\oint M(z,\zeta;u)\,Q(z)\,Q(\zeta)\frac{dz}{z}\frac{d\zeta}{\zeta}=\sum_{j,k=0}^{m} c_{jk}(u)\,\lambda_j\lambda_k,$$

which depends on the coefficients of $f(z)$, the choice of u and the coefficients λ_ν of the rational function

$$(2.4)\qquad Q(z)=\sum_{\nu=-m}^{m}\lambda_\nu z^{-\nu},\qquad \bar{\lambda}_\nu=\lambda_{-\nu}.$$

We pose now the extremum problem to find

$$\operatorname{Re}\{p(u)\}=\text{maximum} \tag{2.5}$$

in the family of all normalized univalent functions $f(z)$, with u defined as a root of the polynomial equation (1.62), namely

$$p'(u)=\frac{1}{u}\left\{\frac{1}{2\pi i}\oint\frac{Q(z)}{\sqrt{1-u\,f(z)}}\frac{dz}{z}\right\}^2=0. \tag{2.6}$$

We shall prove the following

Theorem. *If λ_0 is real and if $Q(z)$ has all its roots on the unit circumference, so that*

$$Q(z)=\prod_{k=1}^{2m}\left(\frac{e^{i\theta_k}}{\sqrt{z}}-\frac{\sqrt{z}}{e^{i\theta_k}}\right), \tag{2.7}$$

then we have the generalized Grunsky inequality

$$\operatorname{Re}\{p(u)\}\leqq\sum_{j=1}^{m}\frac{1}{j}|\lambda_j|^2, \tag{2.8}$$

valid for all normalized univalent functions in $|z|<1$.

This is the same inequality as (1.78), but the choice of the argument point u is quite different. To prove our assertion, we shall again use the method of variations. As in Chapter I, we first verify that the extremum problem (2.5), (2.6) has a solution within the family of normalized univalent functions and that the extremum functions map the unit disk conformally onto the w-plane slit along analytic arcs which satisfy an ordinary differential equation of the type (1.8).

The formal calculations for the variation of our new functional are precisely the same as those carried out in Section 1.5, since the variation of u cannot make a contribution to the differential equation for $f(z)$ because of the side condition (2.6) we have imposed. As in Section 1.5, we are led to the function

$$H(z)=\frac{z\,f'(z)\,\Pi_m\left(\dfrac{1}{f(z)}\right)}{f(z)\sqrt{1-u\,f(z)}}, \tag{2.9}$$

where Π_m is a polynomial of degree m. This function is real for $|z|=1$ and has easily determined singularities. Since $f(z)=u^{-1}$ might have a solution for $|z|<1$, we see that $H(z)$ could have an algebraic singularity in $|z|<1$; moreover, it has a pole at the origin and one branch of it behaves there according to (1.67) as follows:

$$H(z)=\sum_{\nu=0}^{m}\lambda_\nu z^{-\nu}+O(|z|)=Q(z)+O(|z|). \tag{2.10}$$

Observe next that (for λ_ν fixed) u depends algebraically on the coefficients $a_2,\dots,a_{2m+1}$ of $f(z)$ and that $p(u)$ itself depends only on these coefficients. Thus the variational equation (2.9) characterizes an extremum problem with respect to the $2m+1$ first coefficients of a univalent function. It is well known that in this

2*

case the only infinity of the differential equation cccurs at the origin; hence we may conclude that the denominator $\sqrt{1-uf(z)}$ has to cancel out against a corresponding factor

$$\left(\frac{1}{f}-u\right) \quad \text{in} \quad \Pi_m\left(\frac{1}{f}\right).$$

Thus $H(z)^2$ is regular analytic in $|z|<1$, except for a pole at the origin, and is nonnegative for $|z|=1$. It is therefore a rational function in the entire complex plane with poles at $z=0$ and $z=\infty$.

Let

$$\Pi_m\left(\frac{1}{f}\right)=P\left(\frac{1}{f}\right)\left(\frac{1}{f}-u\right). \tag{2.11}$$

Then we find the following differential equation for the extremum function:

$$\frac{z^2 f'(z)^2}{f(z)^3} P\left(\frac{1}{f}\right)^2\left(\frac{1}{f}-u\right)=\sum_{\nu=-2m}^{2m} A_\nu z^{-\nu}, \quad \bar{A}_\nu=A_{-\nu}. \tag{2.12}$$

The right-hand side is nonnegative for $|z|=1$ and has all its roots as double roots, except for one pair of roots symmetric in the unit circumference. Hence we necessarily have

$$\sum_{\nu=-2m}^{2m} A_\nu z^{-\nu}=\left[\sum_{\nu=1}^{m} B_\nu z^{-(\nu-\frac{1}{2})}-\sum_{\nu=1}^{m} \bar{B}_\nu z^{\nu-\frac{1}{2}}\right]^2\left(\frac{1}{z e^{2i\theta}}-2x+z e^{2i\theta}\right), \quad x\geqq 1. \tag{2.12'}$$

Observe that the left side in (2.12') is $H(z)^2$. Hence we conclude from (2.10) and (2.12') that

$$e^{i\theta}\frac{Q(z)+O(|z|)}{\sqrt{1-2xz e^{2i\theta}+z^2 e^{4i\theta}}}=\sum_{\nu=1}^{m} B_\nu z^{-\nu}-\sum_{\nu=1}^{m} \bar{B}_\nu z^{\nu-1}. \tag{2.13}$$

We introduce the development

$$\frac{1}{\sqrt{1-2xz e^{2i\theta}+z^2 e^{4i\theta}}}=\sum_{n=0}^{\infty} P_n(x) z^n e^{2in\theta} \tag{2.14}$$

by means of the Legendre polynomials $P_n(x)$ and find

$$B_1=\sum_{\nu=1}^{m} \lambda_\nu P_{\nu-1}(x) e^{(2\nu-1)i\theta} \tag{2.15}$$

and

$$-\bar{B}_1=\sum_{\nu=0}^{m} \lambda_\nu P_\nu(x) e^{(2\nu+1)i\theta}. \tag{2.15'}$$

Eliminating B_1 from these two relations, we obtain the equation

$$\sum_{\nu=0}^{m} \lambda_\nu P_\nu(x) e^{2i\nu\theta}+\sum_{\nu=1}^{m} \bar{\lambda}_\nu P_{\nu-1}(x) e^{-2i\nu\theta}=0. \tag{2.16}$$

This equation gives us information about the point ω for which $uf(\omega)=1$. If $x=1$ for all roots θ, $x\geqq 1$ of this equation, then any point

$$\omega=r\,e^{-2i\theta}, \quad x=\frac{1}{2}\left(r+\frac{1}{r}\right) \tag{2.17}$$

mapping into u^{-1} must lie on the unit circle $r=1$, so that u^{-1} is an omitted value of the univalent function $f(z)$. In that case the results of Chapter I (cf. (1.78)) yield the asserted sharp upper bound (2.8) on $\mathrm{Re}\{p\}$. We are therefore interested in finding conditions on the parameters $\lambda_0, \ldots, \lambda_m$ that will assure us that (2.16) has no real roots in the range $x>1$.

2.2.

We shall prove that when all roots of the parameter function $Q(z)$ lie on the unit circumference, so that it has the factorization (2.7), then all the real roots θ, $x\geqq 1$ of (2.16) lie on the line $x=1$.

The proof depends on the observation that the function

$$w(x,\theta)=\sum_{\nu=0}^{m}\lambda_\nu P_\nu(x)\,e^{2i\nu\theta}+\sum_{\nu=1}^{m}\bar{\lambda}_\nu P_{\nu-1}(x)\,e^{-2i\nu\theta} \tag{2.18}$$

satisfies the elliptic partial differential equation

$$w_{\varphi\varphi}+w_{tt}+i\,w_\varphi+(\operatorname{cth} t)\,w_t=0, \tag{2.19}$$

where

$$x=\operatorname{ch} t, \quad \varphi=2\theta. \tag{2.20}$$

This is an immediate consequence of the differential equations satisfied by the Legendre polynomials,

$$[(x^2-1)\,P_n'(x)]'=n(n+1)\,P_n(x). \tag{2.19'}$$

We observe that (2.19) has the particular solution

$$e^{-i\theta}P_{-\frac{1}{2}}(x)=\frac{2e^{-i\theta}}{\pi}\int_0^1\frac{d\tau}{\sqrt{(1-\tau^2)\left(1+\frac{x-1}{2}\tau^2\right)}}. \tag{2.21}$$

Define the complex-valued function

$$\zeta(t,\varphi)=\frac{w(x,\theta)}{e^{-i\theta}P_{-\frac{1}{2}}(x)} \tag{2.22}$$

and find that it satisfies the partial differential equation

$$\zeta_{\varphi\varphi}+\zeta_{tt}+(\operatorname{cth} t)\,\zeta_t+2(\operatorname{sh} t)\frac{P'_{-\frac{1}{2}}(\operatorname{ch} t)}{P_{-\frac{1}{2}}(\operatorname{ch} t)}\,\zeta_t=0. \tag{2.23}$$

This is an elliptic partial differential equation with real analytic coefficients.

Our analysis of the location of the roots of (2.16) will be based on the following lemma for equations of the above type, which is implicit in the work of P. Berg [*1*] and H. Lewy [*12*]:

Lemma. *If a complex-valued solution $\zeta=\xi+i\eta$ of the analytic real linear elliptic partial differential equation*

$$\zeta_{\varphi\varphi}+\zeta_{tt}+a\,\zeta_{\varphi}+b\,\zeta_{t}=0 \tag{2.24}$$

defines a univalent mapping of some region of the (φ, t)-plane, then its Jacobian cannot vanish anywhere in that region.

To prove this lemma, we may assume without loss of generality that ζ maps the origin into itself, since we may add to ζ any constant and still preserve the equation (2.24). Thus we have only to show that

$$J=\frac{\partial(\xi,\eta)}{\partial(\varphi,t)}\neq 0 \tag{2.25}$$

at the origin. Let $\sigma=\varphi+it$ and note that $\zeta(\varphi, t)$ has a Taylor series expansion around the origin of the form

$$\zeta=\alpha\,\sigma^{n}+\beta\,\bar{\sigma}^{n}+O(|\sigma|^{n+1}), \tag{2.26}$$

where α and β are constants and n is a positive integer. Indeed, the sum of the terms of lowest degree must be harmonic. Since we assumed ζ to be univalent at $\sigma=0$, we can assert that $n=1$; for otherwise the level curves of, say, ξ would cross at the origin. Clearly, we then have for $\sigma=0$

$$J=|\alpha|^{2}-|\beta|^{2}, \tag{2.27}$$

and if J were zero, we would have $\beta=e^{2i\gamma}\alpha$ and hence

$$\zeta=\alpha\,e^{i\gamma}(e^{-i\gamma}\sigma+e^{i\gamma}\bar{\sigma})+O(|\sigma|^{2}). \tag{2.26'}$$

Thus the level curves of $\operatorname{Im}\{\alpha^{-1}e^{-i\gamma}\zeta\}$ would have to cross at the origin, which again is excluded by our assumption of univalence. The lemma is proved by this contradiction.

We will study the solution $w(x, \theta)$ in the fundamental region

$$x>1, \qquad 0\leqq\theta<2\pi \tag{2.28}$$

and, in particular, on its boundary $x=1$. Since

$$P_{n}(1)=1, \qquad n=1,2,\ldots, \tag{2.29}$$

we see by comparison of (2.4) and (2.18) that

$$w(1,\theta)=Q(e^{-2i\theta}). \tag{2.29'}$$

Thus $w(x, \theta)$ is real for $x=1$.

Let us next calculate the Jacobian

$$J=\begin{vmatrix}\xi_x & \eta_x\\ \xi_\theta & \eta_\theta\end{vmatrix}=\frac{i}{2}\begin{vmatrix}\zeta_x & \bar{\zeta}_x\\ \zeta_\theta & \bar{\zeta}_\theta\end{vmatrix} \tag{2.30}$$

for $x=1$. Observe that $x=1$ corresponds to $t=0$ and that

$$\zeta_{tt}+(\operatorname{cth} t)\,\zeta_{t}=(x^{2}-1)\,\zeta_{xx}+2x\,\zeta_{x}.$$

Hence the differential equation (2.23) reduces in the case $x=1$ to the relation

$$2\zeta_x+\tfrac{1}{4}\zeta_{\theta\theta}=0. \tag{2.31}$$

Therefore we find from (2.31) and (2.22)

$$\begin{aligned} J&=-\operatorname{Im}\{\zeta_x\bar{\zeta}_\theta\}=\frac{1}{8}\operatorname{Im}\{\bar{\zeta}_\theta\zeta_{\theta\theta}\}=\frac{|\zeta_\theta|^2}{8}\frac{\partial}{\partial\theta}\arg\zeta_\theta \\ &=\frac{1}{8}\operatorname{Im}\{(w_\theta-i\,w)(w_{\theta\theta}+2i\,w_\theta-w)\}=\frac{w^2+2w_\theta^2-w\,w_{\theta\theta}}{8}, \end{aligned} \tag{2.32}$$

since $w(1,\theta)$ is real.

Up to now we have not used the special form (2.7) of the parameter function $Q(z)$. If we make here this assumption and also suppose that all roots of $Q(e^{-2i\theta})=w(1,\theta)$ are disjoint, we see that the trigonometric polynomials w and w_θ each have at $x=1$ precisely $4m$ interlacing roots. Thus the quotient w/w_θ has $4m$ simple poles with negative residues at the maxima and minima of w. Between adjacent poles it takes on every finite real value just once, since any linear combination of w and w_θ is a trigonometric polynomial with no more than $4m$ roots. It follows that w/w_θ increases monotonically between its poles, and

$$0<\frac{\partial}{\partial\theta}\frac{w}{w_\theta}=\frac{w_\theta^2-w\,w_{\theta\theta}}{w_\theta^2}. \tag{2.33}$$

Consequently on the boundary $x=1$ of the region (2.28) we have

$$J>\frac{w^2+w_\theta^2}{8}>0. \tag{2.34}$$

Moreover, formula (2.32) also shows that the image of the line $x=1$ by the mapping ζ is a self-intersecting locally convex curve C which passes $4m$ times through the origin.

If we knew that $J>0$ in the entire region (2.28), we could conclude from these results that ζ maps this region univalently onto a piece of a Riemann surface with $2m+1$ sheets bounded by the locally convex curve C and possessing at infinity a branch point of order $2m$. The lemma of Berg and Lewy enables us to draw the desired conclusion under the hypothesis (2.7) in general, provided we can verify it in just one particular case of a parameter function $Q(z)$.

Indeed, consider the manifold of sets of roots of Q on the unit circumference for which the corresponding ζ-maps are univalent. This manifold is closed because $\zeta(x,\theta)$ solves an analytic elliptic partial differential equation. On the other hand, it is also open because of the positive lower bound on the Jacobian J implied by the lemma and our estimate (2.34). Thus if the manifold is not empty, it has to contain all the possible combinations of roots of Q on the unit circumference. Finally, it follows in this case from the geometry of the ζ-mapping that $\zeta\neq0$ except at the roots of Q on the line $x=1$. This is a most important consequence of our considerations, as can be seen from the remarks at the end of Section 2.1.

To complete the argument, we have to exhibit a special choice of the parameter function Q for which $J>0$ in the entire region (2.28). To that end let

(2.35) $$\lambda_m=1, \quad \lambda_{m-1}=\cdots=\lambda_0=0.$$

In this case a direct calculation gives

(2.36) $$J=2R_m(R'_m+R'_{m-1})+(2m-1)(R_m-R_{m-1})(R'_m+R'_{m-1})+2y\sin^2 2m\theta,$$

where

(2.37) $$R_m=R_m(x)=\frac{P_m(x)}{P_{-\frac{1}{2}}(x)}, \quad R'_m=R'_m(x)$$

and

(2.38) $$y(x)=(2m-1)R_{m-1}R'_m-(2m+1)R_m R'_{m-1}.$$

Since all $R_m(x)$ increase with their index and their argument for $x\geqq 1$, we see that the first two terms on the right of (2.36) are positive. To show that $y(x)$ is positive, too, we use the ordinary differential equation

(2.39) $$(x^2-1)R''_m+\left[2x+2(x^2-1)\frac{P'_{-\frac{1}{2}}}{P_{-\frac{1}{2}}}\right]R'_m-(m+\tfrac{1}{2})^2 R_m=0$$

for R_m, which follows directly from the Legendre equations satisfied by the $P_\nu(x)$. From (2.39) we derive the following equation for $y(x)$:

(2.40) $$\frac{d}{dx}[(x^2-1)y]+2(x^2-1)\frac{P'_{-\frac{1}{2}}}{P_{-\frac{1}{2}}}y = \tfrac{1}{2}(4m^2-1)R_{m-1}R_m-2(x^2-1)R'_{m-1}R'_m.$$

We use the inequality

(2.41) $$xP_{-\frac{1}{2}}(x)+2(x^2-1)P'_{-\frac{1}{2}}(x)=\frac{1}{\pi}\int_0^1\frac{2x-x\tau^2+\tau^2}{\sqrt{(1-\tau^2)\left(1+\frac{x-1}{2}\tau^2\right)^3}}d\tau>0$$

to obtain

(2.42) $$\frac{d}{dx}[(m+\tfrac{1}{2})^2R_m^2-(x^2-1)R'^2_m]=2R'^2_m\left[x+2(x^2-1)\frac{P'_{-\frac{1}{2}}}{P_{-\frac{1}{2}}}\right]>0.$$

Integrating this inequality from $x=1$ to values $x>1$, we find

(2.43) $$(m+\tfrac{1}{2})R_m^2>(x^2-1)R'^2_m.$$

Thus the right-hand side of (2.40) is clearly positive and

(2.44) $$\frac{d}{dx}[(x^2-1)P^2_{-\frac{1}{2}}y(x)]\geqq 0.$$

Thus $y(x)\geqq 0$ and $J>0$ follows. We have therefore shown one case of Q for which $J>0$ in the entire region and completed our argument. The theorem (2.7), (2.8) announced in Section 2.1 is herewith proved.

2.3.

For later applications we mention an important extension of the theorem just proved. The coefficient inequality (2.8) has meaning even when $\lambda_l \neq 0$ is fixed and $\lambda_{l+1}, \dots, \lambda_m$ are small, while the λ_ν with $\nu < l$ are bounded. In this case it makes sense to separate the m roots u of (2.6) into one set of l finite points and another set of $m-l$ points which are unbounded and tend to infinity as the parameters $\lambda_{l+1}, \dots, \lambda_m$ shrink to zero. If u is chosen to be one of the l smaller roots of the equation (2.6), we maintain that the inequality (2.8) remains true as long as $\lambda_0, \dots, \lambda_l$ are chosen such that

$$\sum_{\nu=-l}^{l} \lambda_\nu z^{-\nu} = \prod_{k=1}^{2l} \left(\frac{e^{i\theta_k}}{\sqrt{z}} - \frac{\sqrt{z}}{e^{i\theta_k}} \right). \tag{2.45}$$

For the proof, note that our analysis still implies that $J > 0$ in any finite interval $1 < x < X$ for $\lambda_{l+1}, \dots, \lambda_m$ small enough. Consequently (2.16) is seen to have no roots in such a range. On the other hand, the roots of (2.16) with $x \geqq X$ would (for large enough X) correspond under the extremal univalent mapping to the large roots u of (2.6), which we have expressly excluded. Thus the relevant root of (2.16) lies on the line $x = 1$, so that u^{-1} is again an omitted value of $f(z)$ and (2.8) follows from the results of Chapter I.

III. The Local Theorem

Our aim in this chapter is to establish the following local

Theorem. *If the normalized univalent function $f(z)$ is close enough to the Koebe function*

$$\frac{z}{(1-z)^2} = z + 2z^2 + 3z^3 + \cdots, \tag{3.1}$$

for example, if

$$|a_2 - 2| < \varepsilon_m \tag{3.2}$$

for an appropriate small number $\varepsilon_m > 0$, then the Bieberbach conjecture is true for the coefficient a_{2m+1} of $f(z)$, in other words

$$\operatorname{Re}\{a_{2m+1}\} \leqq 2m+1, \tag{3.3}$$

with equality holding only for the Koebe function.

The proof of the theorem will be achieved by suitably combining Taylor series expansions of cases of the inequality (2.8) for which the Koebe function is extremal. Let us put

$$f(z) = \frac{z}{(1-z)^2} + \delta f(z) = \sum_{n=1}^{\infty} (n + \delta^1 a_n + \delta^2 a_n) z^n, \tag{3.4}$$

where $\delta^1 a_n$ is supposed to be pure imaginary and $\delta^2 a_n$ is supposed to be real. We shall set up an estimate of $\delta^2 a_{2m+1}$ in terms of a positive-definite quadratic form in the quantities $\delta^1 a_n$ and a linear expression in the variables $\delta^2 a_n$ with small coefficients whose signs can be specified at will. To this end it will suffice to retain only the first and second order terms in our Taylor series expansion of (2.8).

It is therefore of the utmost importance to observe that the equation (2.6) defining u involves a perfect square, so that wherever $p'(u)$ vanishes all its first partial derivatives vanish, too. Thus in the analysis to follow it will be permissible to fix u at the extremal value

$$u=-4, \tag{3.5}$$

which will simplify our calculations in an essential way.

In order to expand $p(u)$ in a Taylor series, we first develop the generating function (2.1) in powers of δf. Since

$$\begin{aligned}\sqrt{1-uf(z)}&=\sqrt{\left(\frac{1+z}{1-z}\right)^2+4\delta f(z)}\\&=\frac{1+z}{1-z}+2\frac{1-z}{1+z}\delta f-2\left(\frac{1-z}{1+z}\right)^3\delta f^2+\cdots,\end{aligned} \tag{3.6}$$

we have

$$\begin{aligned}\delta M(z,\zeta;u)=&\frac{(1-z)(1-\zeta)}{z-\zeta}\left[\frac{1-z}{1+z}\delta f(z)-\frac{1-\zeta}{1+\zeta}\delta f(\zeta)\right]-\\&-\frac{(1-z)(1-\zeta)}{1-z\zeta}\left[\frac{1-z}{1+z}\delta f(z)+\frac{1-\zeta}{1+\zeta}\delta f(\zeta)\right]-\\&-\frac{(1-z)^2(1-\zeta)^2}{2(z-\zeta)^2}\left[\frac{1-z}{1+z}\delta f(z)-\frac{1-\zeta}{1+\zeta}\delta f(\zeta)\right]^2+\\&+\frac{(1-z)^2(1-\zeta)^2}{2(1-z\zeta)^2}\left[\frac{1-z}{1+z}\delta f(z)+\frac{1-\zeta}{1+\zeta}\delta f(\zeta)\right]^2-\\&-\frac{(1-z)(1-\zeta)}{z-\zeta}\left[\left(\frac{1-z}{1+z}\right)^3\delta f(z)^2-\left(\frac{1-\zeta}{1+\zeta}\right)^3\delta f(\zeta)^2\right]+\\&+\frac{(1-z)(1-\zeta)}{1-z\zeta}\left[\left(\frac{1-z}{1+z}\right)^3\delta f(z)^2+\left(\frac{1-\zeta}{1+\zeta}\right)^3\delta f(\zeta)^2\right]+\\&+\cdots.\end{aligned} \tag{3.7}$$

Multiplying both sides of (3.7) by $Q(z)/z$ and by $Q(\zeta)/\zeta$ and taking (2.3) into account, we find after appropriate contour integration that

$$\begin{aligned}\delta p(u)=&\frac{1}{2\pi i}\oint\delta f(z)\frac{(1-z)^3}{1+z}\frac{Q(z)^2}{z^2}dz+\frac{1}{8\pi i}\oint\delta f(z)^2 d\left[\frac{(1-z)^6}{(1+z)^2}\frac{Q(z)^2}{z^2}\right]+\\&+\sum_{k=1}^{m-1}\frac{k}{(2\pi i)^2}\oint\delta f(z)\frac{(1-z)^3}{1+z}\frac{Q(z)}{z}\frac{z^k}{z}dz\times\\&\times\oint\delta f(\zeta)\frac{(1-\zeta)^3}{1+\zeta}\frac{Q(\zeta)}{\zeta}\frac{\zeta^{-k}+\zeta^k}{\zeta}d\zeta+\cdots,\end{aligned} \tag{3.8}$$

where it has been assumed that Q is a real function with the property $Q(1/z)=Q(z)$. To this we want next to add a contribution obtained by varying in an optimal way the parameters λ_k that occur in the coefficient inequality (2.8).

In the case of the Koebe function we have

$$M(z,\zeta;-4)=\log\frac{1}{1-z\zeta}=\sum_{j=1}^{\infty}\frac{1}{j}z^j\zeta^j, \tag{3.6'}$$

and in view of (2.6),

$$p'(-4)=-\frac{1}{4}\left\{\frac{1}{2\pi i}\oint\frac{1-z}{1+z}\frac{Q(z)}{z}dz\right\}^2=-\frac{1}{4}\left[\sum_{\nu=-m}^{m}(-1)^\nu\lambda_\nu\right]^2. \tag{3.6''}$$

Thus the Koebe function solves the maximum problem (2.5), (2.6) with $u=-4$ if all λ_k are real and $Q(-1)=0$. We shall require that Q have a double root at $z=-1$. We shall confine our attention to the case where all the other roots of Q, which have to lie on the unit circle, are distinct. It is then feasible to vary the imaginary parts of the λ_k infinitesimally in any desired fashion, since this keeps

$$\delta Q(-1)=0 \tag{3.9}$$

and consequently none of the roots of Q can leave the unit circle. The choice of the imaginary perturbations $\delta^1\lambda_k$ of the parameters λ_k that makes the inequality (2.8) strongest is clearly

$$\delta^1\lambda_k=-\frac{k}{2}\sum_{j=0}^{m}\lambda_j\delta^1 c_{jk}, \tag{3.10}$$

where $\delta^1 c_{jk}$ stands for the imaginary part of the deviation of c_{jk} from its extremal value

$$c_{jk}=\frac{\delta_{jk}}{k}. \tag{3.11}$$

Our closing remarks in Chapter 2 show that it is permissible to use (3.10) over the full range $k=1,\dots,m$ even when only the coefficients $\lambda_0,\dots,\lambda_l$ of Q differ from zero, and $l<m$. Then, too, we can assert the validity of the inequality (2.8). The perturbations (3.10) result in the appearance on the left in (2.8) of a positive contribution

$$\begin{aligned}&-\tfrac{1}{2}\sum_{k=1}^{m}k\left(\sum_{j=0}^{m}\lambda_j\delta^1 c_{jk}\right)^2\\ &=-\tfrac{1}{2}\sum_{k=1}^{m}k\left(\frac{1}{2\pi i}\oint\delta f(z)\frac{(1-z)^3}{1+z}\frac{Q(z)}{z}\frac{z^{-k}+z^k}{z}dz\right)^2.\end{aligned} \tag{3.12}$$

Adding (3.12) to (3.8) and substituting into (2.8), we arrive at the final estimate

$$\begin{aligned}\text{Re}&\left\{\frac{1}{2\pi i}\oint\delta f(z)\frac{(1-z)^3}{1+z}\frac{Q(z)^2}{z^2}dz+\right.\\ &+\frac{1}{8\pi i}\oint\delta f(z)^2 d\left[\frac{(1-z)^6}{(1+z)^2}\frac{Q(z)^2}{z^2}\right]+\\ &\left.+\tfrac{1}{2}\sum_{k=-m}^{m}k\left(\frac{1}{2\pi i}\oint\delta f(z)\frac{(1-z)^3}{1+z}\frac{Q(z)}{z}\frac{z^k}{z}dz\right)^2+\cdots\right\}\leq 0.\end{aligned} \tag{3.13}$$

Our problem is to average (3.13) over various possibilities for the parameter function Q so as to be left with the desired inequality

$$\delta^2 a_{2m+1} < 0. \tag{3.14}$$

In choosing Q we shall be motivated both by the ordinary differential equation for the extremal function $f(z)$ maximizing $\operatorname{Re}\{a_{2m+1}\}$ and by the known proof of the Bieberbach conjecture for real schlicht functions. To start with note that because

$$\sum_{k=-m}^{m} k z^k \zeta^k = -(z\zeta)^{-m}\left[\frac{m+1}{1-z\zeta} - \frac{1}{(1-z\zeta)^2}\right] + \frac{m(z\zeta)^{m+2} - (m+1)(z\zeta)^{m+1}}{(1-z\zeta)^2} \tag{3.15}$$

we can write (3.13) in the form

$$\begin{aligned} \operatorname{Re}\Bigg\{ & \frac{1}{2\pi i}\oint \delta f(z)\frac{(1-z^2)^3 T(z)^2}{z^4}\,dz + \\ & + \frac{1}{8\pi i}\oint \delta f(z)^2\, d\left[\frac{(1-z)^6(1+z)^2 T(z)^2}{z^4}\right] - \\ & - \frac{1}{2}\frac{1}{(2\pi i)^2}\oint\oint \delta f(z)\,\delta f(\zeta)\frac{(1-z)^2(1-\zeta)^2(1-z^2)(1-\zeta^2)}{(z\zeta)^{m+3}} \times \\ & \times T(z)\,T(\zeta)\left[\frac{m+1}{1-z\zeta} - \frac{1}{(1-z\zeta)^2}\right] dz\, d\zeta + \cdots \Bigg\} \leqq 0, \end{aligned} \tag{3.16}$$

where

$$Q(z) = \left(\frac{1}{z} + 2 + z\right) T(z). \tag{3.17}$$

We now insert for T the Tchebycheff polynomial

$$T(z) = T_l(z) = \frac{z^{-l} - z^l}{z^{-1} - z} \tag{3.18}$$

to obtain

$$\begin{aligned} \delta^2 a_{2l+1} - \delta^2 a_{2l-1} + \operatorname{Re}\Bigg\{ & \frac{1}{8\pi i}\oint \delta f(z)^2\, d\left[\frac{(1-z)^6(1+z)^2 T_l(z)^2}{z^4}\right] - \\ & - \frac{1}{2}\frac{1}{(2\pi i)^2}\oint\oint \delta f(z)\,\delta f(\zeta)\frac{(1-z)^2(1-\zeta)^2(1-z^2)(1-\zeta^2)}{(z\zeta)^{m+3}} \times \\ & \times T_l(z)\,T_l(\zeta)\left[\frac{m+1}{1-z\zeta} - \frac{1}{(1-z\zeta)^2}\right] dz\, d\zeta + \cdots \Bigg\} \leqq 0. \end{aligned} \tag{3.19}$$

Our intention is to sum this inequality over the index l from 1 to m in order to derive (3.14).

It is easy to establish the explicit representation

$$\begin{aligned} k(z,\zeta) &= \sum_{l=1}^{m} T_l(z)\,T_l(\zeta) \\ &= \frac{z\zeta}{(1-z^2)(1-\zeta^2)}\left[\frac{(z\zeta)^{-m} - (z\zeta)^{m+1}}{1-z\zeta} - \frac{z^{m+1}\zeta^{-m} - z^{-m}\zeta^{m+1}}{z-\zeta}\right] \end{aligned} \tag{3.20}$$

for the truncated kernel $k(z, \zeta)$ of the Tchebycheff polynomials. The diagonal case

$$k(z, z)=\frac{z^3}{(1-z^2)^3}\left[\frac{1}{z^{2m+1}}-\frac{2m+1}{z}+(2m+1)z-z^{2m+1}\right] \tag{3.21}$$

of (3.20) is of special interest because it is connected with the right-hand side of the ordinary differential equation

$$\begin{aligned}\frac{z^2 f'^2}{f^3} F\left(\frac{1}{f}\right)=\frac{1}{z^{2m}}+\frac{2a_2}{z^{2m-1}}+\cdots+\frac{2m\,a_{2m}}{z}+2m\,a_{2m+1}+\\ +2m\,\bar{a}_{2m}z+\cdots+2\bar{a}_2 z^{2m-1}+z^{2m}\end{aligned} \tag{3.22}$$

for the extremal function maximizing $\operatorname{Re}\{a_{2m+1}\}$. Summing (3.19) on l and using (3.20) and (3.21) to simplify the result, we find after suitable contour integration that

$$\begin{aligned}\operatorname{Re}\Bigg\{\frac{1}{2\pi i}\oint\delta f(z)\frac{(1-z^2)^3 k(z, z)}{z^4}\,dz-\frac{1}{4\pi i}\oint\delta f(z)^2\frac{(1-z)^2}{z^{2m+3}}\,dz-\\ -\frac{1}{2}\frac{1}{(2\pi i)^2}\oint\oint\delta f(z)\,\delta f(\zeta)\frac{(1-z)^2(1-\zeta)^2}{(z\zeta)^{2m+2}}\times\\ \times\left[\frac{m+1}{(1-z\zeta)^2}-\frac{1}{(1-z\zeta)^3}\right]dz\,d\zeta+\cdots\Bigg\}\leqq 0.\end{aligned} \tag{3.23}$$

Although the first term on the left is identical with $\delta^2 a_{2m+1}$, it is important to observe that through slight alteration of the real parameters λ_k we have chosen it is possible to replace that term by $\delta^2 a_{2m+1}$ plus a linear combination of

$$\delta^2 a_2, \ldots, \delta^2 a_{2m}$$

with coefficients of arbitrary sign, except that the coefficient of $\delta^2 a_2$ comes out negative. Thus we are assured that $\delta^2 a_2, \ldots, \delta^2 a_{2m}$ will make a negative contribution to the upper bound on $\delta^2 a_{2m+1}$ we shall obtain from (3.23).

No generality is lost in our discussion of the local theorem if we rotate $f(z)$ so that

$$\delta^1 a_{2m+1}=0. \tag{3.24}$$

Moreover, let us put

$$(1-z)\,\delta f(z)=\sum_{n=2}^{\infty}(\delta^1 a_n-\delta^1 a_{n-1})z^n+\cdots=\sum_{n=2}^{\infty} b_n z^n+\cdots, \tag{3.25}$$

where the b_n are supposed to be pure imaginary. Then calculation of the residues from the contour integrals in (3.23) gives

$$\begin{aligned}\delta^2 a_{2m+1}&\leqq\tfrac{1}{2}\sum_{j+k=2m+2} b_j b_k+\tfrac{1}{4}\sum_{n=0}^{2m-1}(n+1)(2m-n)(b_{2m-n+1}-b_{2m-n})^2\\ &=\tfrac{1}{4}\sum_{k=0}^{m-2}\left[4b_{k+2}b_{2m-k}+2(m-k)\sum_{n=k}^{2m-k-1}(b_{2m-n+1}-b_{2m-n})^2\right]+\\ &\quad+\tfrac{1}{2}b_{m+1}^2+\tfrac{1}{2}(b_{m+2}-b_{m+1})^2+\tfrac{1}{2}(b_{m+1}-b_m)^2+\cdots.\end{aligned} \tag{3.26}$$

Completing the square appropriately, we conclude that

$$\begin{aligned}\delta^2 a_{2m+1} \leqq \tfrac{1}{2} \sum_{k=0}^{m-2} \Bigg\{ (m-k) \sum_{n=k+1}^{2m-k-2} \left(b_{2m-n+1} - b_{2m-n} + \frac{b_{2m-k}}{m-k} \right)^2 + \\ + \frac{2 b_{2m-k}^2}{m-k} + (m-k) \left[(b_{2m-k+1} - b_{2m-k})^2 + (b_{k+2} - b_{k+1})^2 \right] \Bigg\} + \\ + \tfrac{1}{2} b_{m+1}^2 + \tfrac{1}{2} (b_{m+2} - b_{m+1})^2 + \tfrac{1}{2} (b_{m+1} - b_m)^2 + \cdots,\end{aligned} \tag{3.27}$$

where the quadratic form on the right in the pure imaginary variables

$$\delta^1 a_2, \ldots, \delta^1 a_{2m}$$

is obviously positive-definite. This finishes our proof of the local theorem, for it shows that (3.14) holds whenever any of the quantities $\delta^1 a_2, \ldots, \delta^1 a_{2m}$ or $\delta^2 a_2, \ldots, \delta^2 a_{2m}$ differ from zero.

Our proof in an earlier publication [9] of the local theorem

$$\delta^2 a_{2m} < 0 \tag{3.28}$$

for the even coefficients a_{2m} looks superficially different from the present work. However, we can prove (3.28) in almost the same way that we establish (3.14) if we simply replace the generating function (2.1) throughout our argument by the expression

$$\hat{M}(z, \zeta; \infty, 0) = \log \frac{[\sqrt{f(z^2)} - \sqrt{f(\zeta^2)}](z+\zeta)}{[\sqrt{f(z^2)} + \sqrt{f(\zeta^2)}](z-\zeta)}, \tag{3.29}$$

which was already mentioned in Chapter I. With other choices of the generating function doubtless further results of the kind we have presented here could be obtained. Finally, the implication should not be lost on the reader that the extended Grunsky inequality (2.8) may well suffice to prove the Bieberbach conjecture (3.3) for the odd coefficients globally.

In order to learn whether the new coefficient inequality (2.8) we have discovered is strong enough to settle the Bieberbach conjecture in the large, it is of interest to investigate the example of a_5, which is the lowest still remaining open. In a previous publication [6] we proved the local theorem (3.14) for a_5 by essentially the same method we have described here, with the significant qualification that a different set of orthogonal polynomials was used in the representation (3.20) of the kernel $k(z, \zeta)$. A reasonable initial step toward the global discussion of the fifth coefficient would be to refine that proof so as to obtain a specific numerical estimate of the size ε_2 of the neighborhood (3.2) to which the local theorem is applicable. In such a manner one might hope to build up step by step a proof of the global result

$$|a_5| \leqq 5 \tag{3.30}$$

based on a division of cases according to the location of the second coefficient a_2.

As a start in this direction we indicate how to establish the local theorem for a_5 by a version of our method in which the parameters λ_k, instead of being specified in an optimal way, are selected so they depend on a_2 only, while at the same time they serve to eliminate a_4 from our upper bound on $\operatorname{Re}\{a_5\}$ altogether. Such a set of choices is given by

$$(3.31)\quad \lambda_0=2-\varepsilon\sqrt{-\delta^2 a_2},\qquad \lambda_1=1,\qquad \lambda_2=0;$$

$$(3.32)\quad \lambda_0=\varepsilon[2+3\sqrt{-\delta^2 a_2}],\quad \lambda_1=\frac{a_2+\bar{a}_2}{4}+\varepsilon\frac{a_2}{\bar{a}_2}[1+\sqrt{-\delta^2 a_2}],\quad \lambda_2=\frac{\bar{a}_2}{a_2},$$

and

$$(3.33)\quad \lambda_0=4-\varepsilon[2+\sqrt{-\delta^2 a_2}],\ \lambda_1=3\,\frac{a_2+\bar{a}_2}{4}-\varepsilon\frac{\bar{a}_2}{a_2}[1+\sqrt{-\delta^2 a_2}],\ \lambda_2=\frac{a_2}{\bar{a}_2},$$

where $\varepsilon>0$ is a sufficiently small fixed number. The success of these cases of the parameters in proving that

$$(3.34)\qquad \delta^2 a_5 \leqq -E\,|\delta^2 a_2|^{\frac{3}{2}}<0$$

for some $E>0$ is verified by substituting them directly into the variational formula (3.8) and averaging. Observe, however, that the range of values of a_2 for which (3.33) can be used is narrowly limited by our admissibility criterion (2.7).

This work was supported in part by Office of Naval Research Contract Nonr-285(55) at New York University and by Air Force Contract AF 49(638)1345 at Stanford University.

Bibliography

[1] Berg, P. W., On univalent mappings by solutions of linear elliptic partial differential equations. Trans. Amer. Math. Soc. **84**, 310–318 (1957).

[2] Bieberbach, L., Über die Koeffizienten derjenigen Potenzreihen, welche eine schlichte Abbildung des Einheitskreises vermitteln. Preuss. Akad. Wiss. Berlin, Sitzungsberichte 1916, S. 940–955.

[3] Bombieri, E., Sul problema di Bieberbach per le funzioni univalenti. Atti Accad. Naz. Lincei Rend. Cl. Sci. Fis. Mat. Natur. **35**, 469–471 (1963).

[4] Charzynski, Z., & M. Schiffer, A new proof of the Bieberbach conjecture for the fourth coefficient. Arch. Rational Mech. Anal. **5**, 187–193 (1960).

[5] Duren, P. L., & M. Schiffer, The theory of the second variation in extremum problems for univalent functions. J. Analyse Math. **10**, 193–252 (1962/63).

[6] Garabedian, P. R., Inequalities for the fifth coefficient. Comm. Pure Appl. Math. **19**, 199–214 (1966).

[7] Garabedian, P. R., An extension of Grunsky's inequalities bearing on the Bieberbach conjecture. J. Analyse Math. **18**, 81–97 (1967).

[8] Garabedian, P. R., & M. Schiffer, A proof of the Bieberbach conjecture for the fourth coefficient. J. Rational Mech. Anal. **4**, 427–465 (1955).

[9] Garabedian, P. R., G. G. Ross, & M. Schiffer, On the Bieberbach conjecture for even *n*. J. Math. Mech. **14**, 975–989 (1965).

[10] Grunsky, H., Koeffizientenbedingungen für schlicht abbildende meromorphe Funktionen. Math. Z. **45**, 29–61 (1939).

[11] Jenkins, J. A., On certain coefficients of univalent functions. Analytic functions, Princeton 1960, pp. 159–194.

[12] Lewy, H., On the non-vanishing of the Jacobian in certain one-to-one mappings. Bull. Amer. Math. Soc. **42**, 689–692 (1936).

[13] Rogosinski, W., Über positive harmonische Entwicklungen und typisch-reelle Potenzreihen. Math. Z. **35**, 93–121 (1932).

[14] Schiffer, M., A method of variation within the family of simple functions. Proc. London Math. Soc. **44**, 432–449 (1938).

[15] SCHIFFER, M., Faber polynomials in the theory of univalent functions. Bull. Amer. Math. Soc. **54**, 503–517 (1948).
[16] SCHIFFER, M., Some recent developments in the theory of conformal mapping. Appendix to R. COURANT: Dirichlet's principle, conformal mapping and minimal surfaces, pp. 249–323. New York 1950.
[17] SCHIFFER, M., Applications of variational methods in the theory of conformal mapping. Proc. Symp. in Appl. Math. **8**, 93–113 (1956).
[18] SCHIFFER, M., Extremum problems and variational methods in conformal mapping. Proc. Internat. Congress Math. 1958, pp. 211–231.

Courant Institute of Mathematical Sciences
New York University
and
Stanford University
Stanford, California

(Received March 22, 1967)

Commentary on

[91] P. R. Garabedian and M. M. Schiffer, *The local maximum theorem for the coefficients of univalent functions*, Arch. Rational Mech. Anal. **26** (1967), 1–32.

In 1960, the paper of Charzyński and Schiffer [72] appeared, giving an elementary proof of the Bieberbach conjecture for the fourth coefficient using nothing but the Grunsky inequalities [Gr]. It was a sensational development, and it drew a lot of attention to the Grunsky inequalities, which then seemed likely to hold the key to a general proof. Although the full conjecture was verified some years later by a totally different method, the approach of Charzyński and Schiffer was successfully adapted to give proofs for $n = 5$ and 6. (See the commentary on [72] for more complete discussion.) The paper [91] develops a new generalization of the Grunsky inequalities, now known as the *Garabedian–Schiffer inequalities*, which enables the authors to prove the local Bieberbach conjecture: for each given n, the inequality $|a_n| \leq n$ holds for all functions $f \in S$ with $|a_2 - 2|$ sufficiently small. As they explain, the local form of the theorem was a crucial part of their original proof in [55] that $|a_4| \leq 4$.

Previously, Garabedian, Ross, and Schiffer [86] had used a generalized form of the Grunsky inequalities to prove the local Bieberbach conjecture for even n, but the case of odd n required new ideas. Garabedian [Ga] then outlined the method that the authors implement in [91]. Their proof proceeds by Schiffer's method of boundary variation [5]. In an extended introduction to [91], the authors explain the rationale for the special choice of functional to be maximized. A few years later, Pommerenke [P] reformulated the Garabedian–Schiffer inequalities and gave an elementary proof based upon the theory of Bieberbach–Eilenberg functions. He then showed how the result can be applied to prove Loewner's theorem that $|a_3| \leq 3$ and the sharp inequality $|b_3| \leq \frac{1}{2} + e^{-6}$ obtained in [59] by Garabedian and Schiffer for functions $g(z) = z + b_0 + b_1/z + \ldots$ of class Σ.

We now prepare to state the Garabedian–Schiffer inequalities in Pommerenke's form. Given a function $g \in \Sigma$ and distinct complex parameters u and v, let

$$\tilde{g}(z) = \left\{ \frac{g(z) - u}{g(z) - v} \right\}^{1/2} = 1 + \tfrac{1}{2}(v - u)/z + \ldots$$

for z in some neighborhood of ∞. We define the *Garabedian–Schiffer coefficients* $c_{jk}(u,v)$ by the generating relation

$$\log \frac{\tilde{g}(z) - \tilde{g}(\zeta)}{(z^{-1} - \zeta^{-1})(\tilde{g}(z) + \tilde{g}(\zeta))} = -\sum_{j=0}^{\infty}\sum_{k=0}^{\infty} c_{jk}(u,v) z^{-j} \zeta^{-k}.$$

A set of complex numbers $\lambda_0, \lambda_1, \ldots, \lambda_m$ is said to be *admissible* if λ_0 is real and the rational function $\lambda_0 + \sum_{k=1}^{m}(\lambda_k z^k + \overline{\lambda_k} z^{-k})$ has all $2m$ of its zeros on the unit circle. Define the function

$$p(u,v) = \sum_{j=0}^{m}\sum_{k=0}^{m} c_{jk}(u,v)\lambda_j \lambda_k + \sum_{k=1}^{m} \frac{1}{k} |\lambda_k|^2.$$

Finally, let E denote the complement of the range of g. Then the Garabedian–Schiffer inequalities state that $\mathrm{Re}\{p(u,v)\} \geq 0$ if $\lambda_0, \lambda_1, \ldots, \lambda_m$ is an admissible set and the parameters u and v ($u \neq v$) are chosen so that $u \in E$ or $\frac{\partial}{\partial u} p(u,v) = 0$, and also $v \in E$ or $\frac{\partial}{\partial v} p(u,v) = 0$.

It is clear that any attempt to apply the theorem will meet with practical difficulties, including the calculation of Garabedian–Schiffer coefficients and verification of the required constraints in choice of parameters. These issues are addressed in [91] and [P], as well as in [D], where the theorem is applied to obtain, among other results, the sharp inequality $|b_5 + b_1 b_3 + b_2^2| \leq \frac{1}{3}$.

References

[D] Peter L. Duren, *Applications of the Garabedian–Schiffer inequality*, J. Analyse Math. **30** (1976), 141–149.

[Ga] P. R. Garabedian, *An extension of Grunsky's inequalities bearing on the Bieberbach conjecture*, J. Analyse Math. **18** (1967), 81–97.

[Gr] Helmut Grunsky, *Koeffizientenbedingungen für schlicht abbildende meromorphe Funktionen*, Math. Z. **45** (1939), 29–61.

[P] Ch. Pommerenke, *Univalent Functions*, Vandenhoeck & Ruprecht, 1975.

PETER DUREN

[99] Some distortion theorems in the theory of conformal mapping

---

[99] Some distortion theorems in the theory of conformal mapping. *Atti Accad. Naz. Lincei Mem. Cl. Sci. Fis. Mat. Natur. Sez.* Ia (8) **10** (1970), 1–19.

RELAZIONE

letta e approvata nella seduta del 10 gennaio 1970, sulla Memoria di MENAHEM SCHIFFER, presentata nella seduta del 15 novembre 1969 dal Corrisp. G. FICHERA, intitolata: *Some distortion theorems in the theory of conformal mapping.*

La presente Memoria di Menahem Schiffer si inquadra in un campo di ricerche nel quale l'Autore ha, con la sua ben nota produzione scientifica, recato contributi di grande interesse. Tale campo è relativo allo studio delle variazioni subite da certe funzioni di insieme (quali la lunghezza di un arco di curva, l'area di una regione piana, il diametro transfinito di un insieme di punti del piano, etc.) in conseguenza di una trasformazione conforme (teoremi di distorsione), nonché alla determinazione di quelle fra tali trasformazioni che estremizzano le dette variazioni.

Nel lavoro presente viene considerata una particolare funzione di insieme Φ (D), costruita mediante il diametro transfinito di un dominio pluriconnesso D, mediante i periodi di certe funzioni analitiche connesse alla funzione di Green di D e mediante i valori all'infinito delle misure armoniche relative ai contorni del dominio D. Le proprietà variazionali ed estremali, rispetto alle trasformazioni conformi, della funzione Φ (D) vengono esaurientemente studiate ed interessanti conseguenze vengono tratte dai risultati ottenuti.

La Commissione è ben lieta di proporre che la Memoria, appartenente ad uno dei campi di ricerca più elevati nella teoria delle funzioni di variabile complessa, sia integralmente accolta negli Atti Accademici.

GAETANO FICHERA
CARLO MIRANDA
BENIAMINO SEGRE

Some distortion theorems in the theory of conformal mapping

Memoria (*) di MENAHEM SCHIFFER

RIASSUNTO. — Nel presente lavoro viene considerata una particolare funzione di insieme $\Phi(D)$ costruita fondandosi sulla conoscenza del diametro transfinito del dominio pluriconnesso D, dei periodi di certe funzioni analitiche connesse alla funzione di Green del dominio D e dei valori all'infinito delle misure armoniche relative ai contorni del dominio D.

INTRODUCTION.

The theory of conformal mappings leads to the problem of distortion. That is, the question how much a given set in the domain of the function can be changed under the mapping. The most intuitive distortion results would express the change of simple measures like length, area, etc., under the mapping. However, it is often convenient to introduce more refined set functions which can be studied in a convenient way. The transfinite diameter of a point set has been used in various applications [2], [4] of this kind since it is closely related to potential theoretical tools. In selecting the appropriate set functions, we may be guided by the following consideration. To find the maximum change of a set function under conformal mapping, we may apply the calculus of variations and characterize the extremum mapping. Hence we should study such set functions which possess a simple variational derivative. It is well known that the differential equations for the extremum function can be integrated in a particularly simple way if the variational derivative is a perfect square. Thus, we should try to introduce set functions which have such a variational derivative.

In the present paper we consider a large class of such set functions and arrive at some quite specific answers to their extremal values under normalized conformal mappings. The interest of these results in enhanced by the fact that they can easily be extended to problems of quasi-conformal mappings [7], and that they focus the attention on certain domain functions which may be of use in a further development of the theory.

(*) This work was supported in part by contract AF 49(638)1345 at Stanford University.

1. Definitions and Preliminaries.

We shall collect in this section the definitions of the various functionals which will be used in the distortion theorems of this paper. Let D be a domain in the complex z-plane which contains the point $z = \infty$ and whose boundary ∂D consists of N proper continua C_ν. We denote by $g(z, \zeta)$ the Green's function of D with pole at $z = \zeta$ and write, in particular,

$$g(z, \infty) = g(z). \tag{1}$$

The Green's function $g(z, \zeta)$ is the real part of an analytic function $p(z, \zeta)$

$$g(z, \zeta) = \operatorname{Re}\{p(z, \zeta)\}. \tag{2}$$

This function has a logarithmic pole at ζ and has periods

$$-2\pi i\, \omega_\nu(\zeta) = \oint_{\tilde{C}_\nu} dp(z, \zeta) = -i \oint_{\tilde{C}_\nu} \frac{\partial g(z, \zeta)}{\partial n}\, ds_z \tag{3}$$

if z runs in the positive sense through a closed curve $\tilde{C}_\nu$ which is homotopic to C_ν. The harmonic function $\omega_\nu(\zeta)$ is the harmonic measure of the continuum C_ν with respect to D. We shall write for short.

$$\omega_\mu(\infty) = \omega_\mu. \tag{4}$$

Clearly, $\omega_\nu(z)$ is harmonic in D and has the boundary value $\delta_{\nu\mu}$ on C_μ.

Let us next complete the harmonic function $\omega_\nu(z)$ to the analytic function $w_\nu(z)$,

$$\omega_\nu(z) = \operatorname{Re}\{w_\nu(z)\} \tag{5}$$

and define the periods of $w_\nu(z)$ with respect to the continuum C_μ,

$$-2\pi i\, p_{\nu\mu} = \oint_{\tilde{C}_\mu} dw_\nu(z) = -i \oint_{\tilde{C}_\mu} \frac{\partial \omega_\nu}{\partial n}\, ds. \tag{6}$$

It is well known that the matrix $((p_{\nu\mu}))$ is symmetric and negative semi-definite. The quadratic form

$$\sum_{\mu,\nu=1}^{N} p_{\mu\nu}\, x_\mu\, x_\nu = P(x, x) \tag{7}$$

vanishes if and only if all x_ν are equal [1], [4].

Finally, we develop the special Green's function $g(z)$ near the point at infinity

$$g(z) = \log|z| + \log\frac{1}{d} + o\left(\frac{1}{|z|}\right). \tag{8}$$

The quantity d is the so-called transfinite diameter of the set ΣC_ν and plays a very useful role in various problems of complex analysis. It may, of course, also be considered as a functional of D.

The functional which we will discuss in this paper will be constructed from the terms d, ω_ν and $p_{\nu\mu}$. We will select a set of N real constants x_ν and define the functional

$$\Phi[D] = \log\frac{1}{d} + 2\sum_{\nu=1}^{N} x_\nu\,\omega_\nu + \sum_{\nu,\mu=1}^{N} x_\mu\,x_\nu\,p_{\mu\nu}. \tag{9}$$

Next we have to discuss the variation of the above functionals of D under a variation of the domain. We use the following kinematics of variation. Let $z_0 \in D$ and let ρ be a small positive number such that the disk $|z - z_0| < \rho$ lies in D. Consider the mapping

$$z^* = z + \frac{e^{i\alpha}\rho^2}{z - z_0} \tag{10}$$

which is regular and univalent in $|z - z_0| > \rho$. It carries the continua C_ν into new continua C^*_ν which bound a domain D^* of the same type as D. Denoting corresponding functionals of D^* by the same letter as in D but with an asterisk added, we have ([1], [4])

$$g^*(z^*, \zeta^*) = g(z, \zeta) + \mathrm{Re}\{e^{i\alpha}\rho^2 p'(z_0, z)\,p'(z_0, \zeta)\} + o(\rho^2) \tag{11}$$

$$\omega^*_\nu(z^*) = \omega_\nu(z) + \mathrm{Re}\{e^{i\alpha}\rho^2 p'(z_0, z)\,w'_\nu(z_0)\} + o(\rho^2) \tag{12}$$

$$p^*_{\mu\nu} = p_{\mu\nu} + \mathrm{Re}\{e^{i\alpha}\rho^2 w'_\mu(z_0)\,w'_\nu(z_0)\} + o(\rho^2) \tag{13}$$

$$\log\frac{1}{d^*} = \log\frac{1}{d} + \mathrm{Re}\{e^{i\alpha}\rho^2 p'(z_0)^2\} + o(\rho^2), \tag{14}$$

where $p(z)$ is an analytic completion of the particular Green's function $g(z)$. Combining all these variational formulas, we finally arrive at the result

$$\Phi[D^*] = \Phi[D] + \mathrm{Re}\left\{e^{i\alpha}\rho^2\left[p'(z_0) + \sum_{\nu=1}^{N} x_\nu\,w'_\nu(z_0)\right]^2\right\} + o(\rho^2) \tag{15}$$

if the domain D^* arises from D by the variation (9). We see now the usefulness of the particular functional (9). Its variational expression is a perfect square and it is well known that such functionals admit a very simple solution of extremum problems connected with them.

2. A MINIMUM PROBLEM FOR $\Phi[D]$.

Consider the exterior $|\zeta| > 1$ of the unit disk and in it a set of N continua Γ_ν. We map the domain $|\zeta| > 1$ by a univalent normalized conformal mapping

$$z = f(\zeta) = \zeta + a_0 + \frac{a_1}{\zeta} + \cdots \tag{1}$$

which preserves the point at infinity. The continua Γ_ν go over into continua C_ν in the z-plane which determine a domain D. For this domain we can define the functional Φ [D] introduced in Section 1, and we can raise the question to find that univalent function (1) for which Φ [D] is minimal.

The existence of a corresponding function follows from the usual compactness arguments for univalent functions. Our problem is to characterize this extremum function. Let C_0 be the image of $|\zeta| = 1$ under the extremal mapping and D_0 the exterior of C_0. If $z_0 \notin D_0$, we can always find a univalent conformal mapping of D_0 of the form

$$z^* = z + \frac{a\rho^2}{z - z_0} + o(\rho^2) \tag{2}$$

with arbitrarily small ρ and $|a| < 1$ [3], [5], [6]. The remainder term can be estimated uniformly in each compact subdomain of D_0. This fact allows us to construct in $|\zeta| > 1$ a competing function of type (1), namely

$$f^*(\zeta) = f(\zeta) + \frac{a\rho^2}{f(\zeta) - z_0} + o(\rho^2). \tag{3}$$

This function maps the continua Γ_ν onto continua C^*_ν defining a domain D* and the results of Section 1 lead immediately to the conclusion

$$\Phi[D^*] = \Phi[D] + \operatorname{Re}\left\{a\rho^2 \left[p'(z_0) + \sum_{\nu=1}^{N} x_\nu w'_\nu(z_0)\right]^2\right\} + o(\rho^2). \tag{4}$$

We observe that $z_0 \in D$ and that all terms on the right side are well defined.

In view of our minimum requirement we have $\Phi[D^*] \geq \Phi[D]$, i.e., for all admissible variations (3) of $f(\zeta)$ we have

$$\operatorname{Re}\left\{a\rho^2 \left[p'(z_0) + \sum_{\nu=1}^{N} x_\nu w'_\nu(z_0)\right]^2\right\} + o(\rho^2) \geq 0. \tag{5}$$

The basic lemma of the theory of boundary variations in conformal mappings [3], [5], [6] then leads to the conclusion that D_0 covers the entire complex z-plane except for a set of analytic arcs with the differential equation

$$\left(\frac{dz}{dt}\right)^2 \left[p'(z) + \sum_{\nu=1}^{N} x_\nu w'_\nu(z)\right]^2 + 1 = 0 \tag{6}$$

if $z(t)$ is a properly chosen parametric representation of the arc.

We may also express (6) in the form

$$d\left[p(z) + \sum_{\nu=1}^{N} x_\nu w_\nu(z)\right] = \text{imaginary} \tag{7}$$

on the boundary of D_0 or equivalently

$$g(z) + \sum_{\nu=1}^{N} x_\nu \omega_\nu(z) = \text{const} = x_0 \qquad \text{on } C_0. \tag{8}$$

By the very definition of the harmonic functions $g(z)$ and $\omega_\nu(z)$ we also have the equation

$$g(z) + \sum_{\nu=1}^{N} x_\nu \, \omega_\nu(z) = x_\nu \qquad \text{on } C_\nu . \tag{9}$$

We do not know yet the extremum domain D, but we can consider the harmonic function

$$H(\zeta) = g[f(\zeta)] + \sum_{\nu=1}^{N} x_\nu \, \omega_\nu[f(\zeta)] \tag{10}$$

in the domain $\tilde{\Delta}$ bounded by $|\zeta| = 1$ and the N continua Γ_ν. Let us denote by Γ_0 the unit circumference and the corresponding functionals of $\tilde{\Delta}$ by the appropriate letters with a tilde. These expressions may be considered known since they depend only upon the starting configuration but not on the extremum function $f(\zeta)$. We obviously have by (8) and (9)

$$H(\zeta) = \tilde{g}(\zeta) + \sum_{\nu=0}^{N} x_\nu \, \tilde{\omega}_\nu(\zeta) . \tag{11}$$

Let now $\zeta \to \infty$ and find by comparing (10) with (11)

$$\log \frac{1}{d} + \sum_{\nu=1}^{N} x_\nu \, \omega_\nu = \log \frac{1}{\tilde{d}} + \sum_{\nu=0}^{N} x_\nu \, \tilde{\omega}_\nu . \tag{12}$$

Similarly, describing closed contours around each Γ_μ ($\mu = 0, 1, \cdots, N$) and calculating the periods of the conjugate harmonic function, we obtain

$$\omega_\mu + \sum_{\nu=1}^{N} x_\nu \, p_{\mu\nu} = \tilde{\omega}_\mu + \sum_{\nu=0}^{N} x_\nu \, \tilde{p}_{\mu\nu} , \qquad \mu = 1, \cdots, N \tag{13}$$

$$0 = \tilde{\omega}_0 + \sum_{\nu=0}^{N} x_\nu \, \tilde{p}_{0\nu} . \tag{14}$$

We multiply each equation (13) by the corresponding x_μ and add the result on to (12) to find

$$\log \frac{1}{d} + 2 \sum_{\nu=1}^{N} x_\nu \, \omega_\nu + \sum_{\mu,\nu=1}^{N} x_\mu \, x_\nu \, p_{\mu\nu} = \tag{15}$$

$$= \log \frac{1}{\tilde{d}} + 2 \sum_{\nu=1}^{N} x_\nu \, \tilde{\omega}_\nu + \sum_{\mu,\nu=1}^{N} x_\mu \, x_\nu \, \tilde{p}_{\mu\nu} - x_0^2 \, \tilde{p}_{00} .$$

Here the x_ν ($\nu = 1, \cdots, N$) are given by definition and x_0 is determined by (14). We find thus the inequality valid for all admissible mappings (1):

$$\Phi[D] \geq \log \frac{1}{\tilde{d}} + 2 \sum_{\nu=1}^{N} x_\nu \, \tilde{\omega}_\nu + \sum_{\mu,\nu-1}^{N} x_\mu \, x_\nu \, \tilde{p}_{\mu\nu} - \frac{\left[\tilde{\omega}_0 + \sum_{\nu=1}^{N} x_\nu \tilde{p}_{0\nu} \right]^2}{\tilde{p}_{00}} . \tag{16}$$

3. A Maximum Problem for Φ [D].

We now raise the corresponding question to obtain the maximum value for Φ [D] for all admissible univalent mappings (2.1). We may use the same variational technique as in Section 2 and arrive at the analogous differential equation for the boundary slits

$$\left(\frac{dz}{dt}\right)^2 \left[p'(z) + \sum_{\nu=1}^{N} x_\nu w'_\nu(z)\right]^2 = 1 . \tag{1}$$

Indeed, the maximum problem differs from the previous minimum problem only in the sign of the inequality (2.5).

Thus we can assert that on C_0 we have

$$\operatorname{Im}\left\{p(z) + \sum_{\nu=1}^{N} x_\nu w_\nu(z)\right\} = \text{const.} \tag{2}$$

Again we refer back to the ζ–plane and introduce the harmonic function (2.10) in $\tilde{\Delta}$. Then we can state that

$$H(\zeta) = x_\nu \qquad \text{on} \quad \Gamma_\nu , \qquad\qquad \nu = 1, 2, \cdots, N , \tag{3}$$

and

$$\frac{\partial}{\partial n} H(\zeta) = 0 \qquad \text{on} \quad \Gamma_0 . \tag{4}$$

Let us now define the domain $\hat{\Delta}$ in the ζ–plane which is symmetric with respect to the circle $|\zeta| = 1$ and is bounded by the N continua Γ_ν ($\nu = 1, 2, \cdots, N$) and their reflected images $\Gamma_{\nu+N}$ with respect to this circumference. We then may use equation (4) to continue $H(\zeta)$ into $\hat{\Delta}$ by means of the reflection identity

$$H\left(\frac{1}{\bar{\zeta}}\right) = H(\zeta) . \tag{5}$$

Let $\hat{g}(\zeta, \eta)$ be the Green's function of $\hat{\Delta}$ and $\hat{\omega}_\rho(\zeta)$ denote the harmonic measures of its various boundary continua ($\rho = 1, \cdots, 2N$). We clearly have

$$H(\zeta) = \hat{g}(\zeta) + \hat{g}\left(\frac{1}{\bar{\zeta}}\right) + \sum_{\nu=1}^{N} x_\nu \left[\hat{\omega}_\nu(\zeta) + \hat{\omega}_\nu\left(\frac{1}{\bar{\zeta}}\right)\right] . \tag{6}$$

Then let ζ converge to infinity and obtain by comparison of (2.10) and (6)

$$\log\frac{1}{d} + \sum_{\nu=1}^{N} x_\nu \omega_\nu = \log\frac{1}{\hat{d}} + \hat{g}(0) + \sum_{\nu=1}^{N} x_\nu (\hat{\omega}_\nu + \hat{\omega}_\nu(0)) . \tag{7}$$

On the other hand, if ζ describes a closed curve around Γ_μ ($\mu = 1, \cdots, N$), the conjugate of $H(\zeta)$ will have periods which can be calculated by use of

(2.10) or of (6). Comparing the results of the two computations leads to the equation

$$(8)\qquad \omega_\mu + \sum_{\nu=1}^{N} x_\nu p_{\nu\mu} = \hat{\omega}_\mu + \hat{\omega}_{N+\mu} + \sum_{\nu=1}^{N} x_\nu [\hat{p}_{\nu\mu} + \hat{p}_{\nu N+\mu}], \quad \mu = 1, 2, \cdots, N$$

To verify the choice of signs on the right-hand side of (8) one has to remember that, for example,

$$(9)\qquad \int_{\Gamma_\mu} \frac{\partial}{\partial n} \hat{g}\left(\frac{1}{\bar{\zeta}}\right) ds = -\int_{\Gamma_{N+\mu}} \frac{\partial}{\partial n} \hat{g}(\zeta)\, ds,$$

but that under reflection the sense of the circulation has been inverted such that the period of $\hat{g}\left(\frac{1}{\bar{\zeta}}\right)$ is $\hat{\omega}_{N+\mu}$ if ζ runs around Γ_μ in the positive sense.

We combine now (7) with the N equations (8) to find

$$(10)\qquad \Phi[D] = \log\frac{1}{d} + 2\sum_{\nu=1}^{N} x_\nu \omega_\nu + \sum_{\mu,\nu=1}^{N} x_\mu x_\nu p_{\mu\nu}$$

$$= \log\frac{1}{\hat{d}} + \hat{g}(0) + 2\sum_{\nu=1}^{N} x_\nu (\hat{\omega}_\nu + \hat{\omega}_{N+\nu}) + \sum_{\nu,\mu=1}^{N} x_\mu x_\nu (\hat{p}_{\nu\mu} + \hat{p}_{\nu N+\mu}).$$

Here we used the fact that

$$(11)\qquad \hat{\omega}_\nu\left(\frac{1}{\bar{\zeta}}\right) = \hat{\omega}_{N+\nu}(\zeta)$$

and hence $\hat{\omega}_\nu(0) = \hat{\omega}_{N+\nu}$.

Since equation (10) holds in the maximum case, we obtain the following distortion theorem:

Under all normalized univalent mappings (2.1) the functional $\Phi[D]$ of the image continua C_ν satisfies the inequality

$$(12)\qquad \Phi[D] \leq \log\frac{1}{\hat{d}} + \hat{g}(0) + 2\sum_{\nu=1}^{N} x_\nu (\hat{\omega}_\nu + \hat{\omega}_{N+\nu}) + \sum_{\nu,\mu=1}^{N} x_\mu x_\nu (\hat{p}_{\nu\mu} + \hat{p}_{\nu N+\mu}).$$

We have thus given an estimate for the maximum of $\Phi[D]$ under all admissible mappings in terms of the functionals of the known domain $\hat{\Delta}$. In Section 2 the minimum of $\Phi[D]$ was expressed by means of the functionals of another known domain $\tilde{\Delta}$. It is easy, however, to relate the functionals of $\hat{\Delta}$ and $\tilde{\Delta}$. Indeed, one sees at once that

$$(13)\qquad \hat{g}(\zeta) - \hat{g}\left(\frac{1}{\bar{\zeta}}\right) = \tilde{g}(\zeta), \qquad \hat{\omega}_\nu(\zeta) - \hat{\omega}_\nu\left(\frac{1}{\bar{\zeta}}\right) = \tilde{\omega}_\nu(\zeta)$$

for $\nu = 1, 2, \cdots, N$. Hence, also

$$(14)\qquad \log\frac{1}{\hat{d}} - \hat{g}(0) = \log\frac{1}{\tilde{d}}, \qquad \hat{p}_{\nu\mu} - \hat{p}_{\nu N+\mu} = \tilde{p}_{\nu\mu}.$$

Finally, since

$$(15) \qquad \sum_{\nu=0}^{N} \tilde{\omega}_\nu(\zeta) = 1 ,$$

one can express

$$(16) \qquad \tilde{\omega}_0 = 1 - \sum_{\nu=1}^{N} \tilde{\omega}_\nu \quad , \quad \tilde{p}_{0\mu} = - \sum_{\nu=1}^{N} \tilde{p}_{\nu\mu} \quad (\mu = 1, \cdots, N) , \quad \tilde{p}_{00} = \sum_{\nu,\mu=1}^{N} \tilde{p}_{\nu\mu} .$$

Thus the maximum and the minimum of Φ [D] can be expressed in terms of the functionals of $\hat{\Delta}$.

4. An Application of the Distortion Theorem.

Let $\Gamma_{\nu,N+\nu}$ be a closed curve which surrounds the two symmetric continua Γ_ν and $\Gamma_{N+\nu}$, but no other of the continua. We have the idendity

$$(1) \qquad \frac{1}{2\pi} \int_{\Gamma_{\nu,N+\nu}} \frac{\partial \hat{g}(z,\zeta)}{\partial n_z} ds_z = \hat{\omega}_\nu(\zeta) + \hat{\omega}_{N+\nu}(\zeta) ,$$

and if $\Gamma_{\mu,N+\mu}$ is a corresponding curve for the symmetric pair $\Gamma_\mu, \Gamma_{N+\mu}$, we can write

$$(2) \qquad \frac{1}{4\pi^2} \int_{\Gamma_{\nu,N+\nu}} \int_{\Gamma_{\mu,N+\mu}} \frac{\partial^2 \hat{g}(z,\zeta)}{\partial n_z \, \partial z_\zeta} ds_z \, ds_\zeta = \hat{p}_{\nu\mu} + \hat{p}_{N+\nu\mu} + \hat{p}_{\nu N+\mu} + \hat{p}_{N+\nu\, N+\mu} .$$

But in view of (3.11) we have

$$(3) \qquad \hat{p}_{\nu\mu} = \hat{p}_{N+\nu\, N+\mu} \quad , \quad \hat{p}_{\nu N+\mu} = \hat{p}_{N+\nu\mu} .$$

Thus (2) reduces to

$$(4) \qquad \frac{1}{4\pi^2} \int_{\Gamma_{\nu,N+\nu}} \int_{\Gamma_{\mu,N+\mu}} \frac{\partial^2 g(z,\zeta)}{\partial n_z \, \partial n_\zeta} ds_z \, ds_\zeta = 2[\hat{p}_{\nu\mu} + \hat{p}_{\nu N+\mu}] .$$

The continua Γ_ν may now be deformed and brought continuously into arcs on the unit circumference. They will coincide in the limit with their reflected images $\Gamma_{N+\nu}$. Let us denote the Green's function of the domain bounded by the arc system by $G(\zeta, \eta)$; let $\Omega_\nu(\zeta)$ and $P_{\nu\mu}$ denote the corresponding harmonic measures and their periods. We clearly have

$$(5) \qquad G(\zeta, \eta) - G\left(\frac{1}{\bar{\zeta}}, \eta\right) = \log \left| \frac{1 - \bar{\eta}\zeta}{\zeta - \eta} \right|$$

and

$$(5') \qquad G(\zeta) - G\left(\frac{1}{\bar{\zeta}}\right) = \log |\zeta| .$$

Hence if T is the transfinite diameter of the arc system

$$(6) \qquad \log \frac{1}{T} = G(0) .$$

From (1) and (4) we find the limit relations

$$\hat{\omega}_\nu(\zeta) + \hat{\omega}_{N+\nu}(\zeta) = \Omega_\nu(\zeta) \tag{7}$$

$$2[\hat{p}_{\nu\mu} + \hat{p}_{\mu N+\mu}] = P_{\nu\mu}. \tag{8}$$

The estimate (3.12) holds for all choices of the Γ_ν; we can pass to the limit under which $\hat{g}(z, \zeta) \to G(z, \zeta)$ and find the following remarkable result:

Let Γ_ν $(\nu = 1, 2, \cdots, N)$ be a set of arcs on the unit circumference and let C_ν $(\nu = 1, 2, \cdots, N)$ be their images under any normalized mapping (2.1). Then we have the estimate

$$\Phi[D] = \log\frac{1}{d} + 2\sum_{\nu=1}^{N} x_\nu \omega_\nu + \sum_{\mu,\nu=1}^{N} x_\mu x_\nu p_{\mu\nu} \tag{9}$$

$$\leq 2\log\frac{1}{T} + 2\sum_{\nu=1}^{N} x_\nu \Omega_\nu + \sum_{\mu,\nu=1}^{N} x_\mu x_\nu P_{\mu\nu}.$$

This estimate holds for every choice of the real numbers x_ν. The special case

$$d \geq T^2 \tag{10}$$

was first proved in the case of a single arc by Schiffer [4] and generalized by Pommerenke [2] to arbitrary sets on the unit periphery.

We note now the interesting counterpart

$$\left|\sum_{\mu,\nu=1}^{N} x_\mu x_\nu p_{\mu\nu}\right| \geq \frac{1}{2}\left|\sum_{\mu,\nu=1}^{N} x_\mu x_\nu P_{\nu\mu}\right|. \tag{11}$$

The appearance of T^2 in (10) becomes understandable if we return to the general case of boundary curves Γ_ν outside the unit disc. We specialize the inequality (3.12) by putting all $x_\nu = 0$ and find

$$\log\frac{1}{d} \leq \log\frac{1}{\hat{d}} + \hat{g}(0). \tag{12}$$

Hence using (3.14) we find

$$d \geq \frac{\hat{d}^2}{\tilde{d}} \tag{13}$$

for all possible conformal maps $\Delta \to D$. In the case that all Γ_ν lie on $|\zeta| = 1$, we have $\hat{d} = T$ and $\tilde{d} = 1$ and the estimate (10) is recovered.

Since the identity mapping is an admissible competing mapping, we find the interesting relation

$$d \cdot \tilde{d} \geq \hat{d}^2 \tag{14}$$

for the transfinite diameters of the three initial domains Δ, $\tilde{\Delta}$ and $\hat{\Delta}$. However, this inequality is never sharp since in the case that (13) is an equality the image of the unit circle is a slit.

2*

5. A Generalized Minimum Problem for Φ [D].

We may remove the distinguished role which the unit circle played in the minimum problem of Section 2 and consider the following situation. There is a set E_0 of M continua γ_μ which defines a domain Δ_0 in the complex ζ-plane, containing the point at infinity. In Δ_0 lies a set E_1 of N continua Γ_ν. Let S be the class of all univalent functions $f(\zeta)$ which are normalized at infinity by (2.1) and are elsewhere regular analytic in Δ_0. These functions will map E_1 onto a set of continua C_ν in the z-plane and define thus the domain D bounded by the C_ν. We then may ask for those functions $f(\zeta) \in S$ for which the functional $\Phi[D]$, defined as before in (1.9), takes the minimum value.

The same method of variation as used in Section 2 tells us that the extremum function maps the continua γ_μ into analytic arcs c_μ for which

$$g(z) + \sum_{\nu=1}^{N} x_\nu \, \omega_\nu(z) = y_\mu = \text{const} \qquad \text{on} \quad c_\mu . \tag{1}$$

Indeed, in the entire argument of Section 2 leading to (2.8) no use was made of the fact that the basic domain was circular.

We also have the information

$$g(z) + \sum_{\nu=1}^{N} x_\nu \, \omega_\nu(z) = x_\nu \qquad \text{on} \quad C_\nu . \tag{2}$$

Referring back to the domain Δ_0, we obtain the harmonic function

$$H(\zeta) = g(f(\zeta)) + \sum_{\nu=1}^{N} x_\nu \, \omega_\nu(f(\zeta)) \tag{3}$$

which is regular except for a pole at $\zeta = \infty$ in the domain $\tilde{\Delta}$ bounded by the continua γ_μ and Γ_ν. We have

$$H(\zeta) = y_\mu \quad \text{on} \quad \gamma_\mu \quad , \quad H(\zeta) = x_\nu \quad \text{on} \quad \Gamma_\nu . \tag{4}$$

Hence, in terms of the functionals $\tilde{g}(\zeta)$, $\tilde{\omega}_\nu(\zeta)$, $\tilde{p}_{\nu\mu}$ of $\tilde{\Delta}$ we may express $H(\zeta)$ as follows:

$$H(\zeta) = \tilde{g}(\zeta) + \sum_{\nu=1}^{N} x_\nu \, \tilde{\omega}_\nu(\zeta) + \sum_{\mu=1}^{M} y_\mu \, \tilde{\omega}_{N+\mu}(\zeta) \tag{5}$$

if we assign to γ_μ the harmonic measure $\tilde{\omega}_{N+\mu}(\zeta)$ and count the periods $\tilde{p}_{\nu\mu}$ accordingly.

Again, the functionals of $\tilde{\Delta}$ are to be considered known. The x_ν are given from the definition of $\Phi[D]$. But the constants y_μ still have to be determined from the data of the problem. For this purpose we observe that the conjugate function of $g(z)$ has no periods if we describe a closed curve around the images c_μ of the γ_μ. The same is therefore true for $H(\zeta)$ if ζ runs around a loop about γ_μ. Hence by (5) we obtain the condition

$$(6) \qquad 0 = \tilde{\omega}_{N+\mu} + \sum_{\nu=1}^{N} x_\nu \tilde{p}_{\nu N+\mu} + \sum_{\varrho=1}^{M} y_\varrho \tilde{p}_{N+\varrho N+\mu}, \qquad \mu = 1, \cdots, M.$$

Thus we found M linear equations for the unknowns y_ϱ and the definite character of the matrix $((\tilde{p}_{\nu\mu}))$ guarantees the nonvanishing of the determinant of this equation system.

Now we can proceed as before. Letting $\zeta \to \infty$ yields

$$(7) \qquad \log\frac{1}{d} + \sum_{\nu=1}^{N} x_\nu \omega_\nu = \log\frac{1}{\tilde{d}} + \sum_{\nu=1}^{N} x_\nu \tilde{\omega}_\nu + \sum_{\mu=1}^{M} y_\mu \tilde{\omega}_{N+\mu}.$$

Comparing the periods of $H(\zeta)$ around the Γ_α leads to

$$(8) \qquad \omega_\alpha + \sum_{\nu=1}^{N} x_\nu p_{\nu\alpha} = \tilde{\omega}_\alpha + \sum_{\nu=1}^{N} x_\nu \tilde{p}_{\nu\alpha} + \sum_{\mu=1}^{M} y_\mu \tilde{p}_{N+\mu\alpha}, \qquad \alpha = 1, 2, \cdots, N.$$

Thus we finally obtain

$$(9) \qquad \Phi[D] = \log\frac{1}{d} + 2\sum_{\nu=1}^{N} x_\nu \omega_\nu + \sum_{\nu,\alpha=1}^{N} x_\nu x_\alpha p_{\nu\alpha}$$

$$= \log\frac{1}{\tilde{d}} + 2\sum_{\nu=1}^{N} x_\nu \tilde{\omega}_\nu + \sum_{\nu,\alpha=1}^{N} x_\nu x_\alpha \tilde{p}_{\nu\alpha} + \sum_{\mu=1}^{M} y_\mu \left[\tilde{\omega}_{N+\mu} + \sum_{\alpha=1}^{N} x_\alpha \tilde{p}_{N+\mu\alpha}\right].$$

Since this is the value of $\Phi[D]$ in the minimum case, we can derive from (6) and (9) the following result:

Under all mappings $f(z) \in S$ the functional $\Phi[D]$ satisfies the inequality

$$(10) \qquad \Phi[D] \geq \log\frac{1}{\tilde{d}} + 2\sum_{\nu=1}^{N} x_\nu \tilde{\omega}_\nu + \sum_{\nu,\alpha=1}^{N} x_\nu x_\alpha \tilde{p}_{\nu\alpha} - \sum_{\varrho,\mu=1}^{M} y_\mu y_\varrho \tilde{p}_{N+\varrho N+\mu}$$

where the y_μ are determined by the equations (6).

Since the $((\tilde{p}_{\mu\nu}))$ form a negative semidefinite matrix, we can simplify (10) to

$$(10') \qquad \log\frac{1}{d} + 2\sum_{\nu=1}^{N} x_\nu \omega_\nu + \sum_{\nu,\alpha=1}^{N} x_\nu x_\alpha p_{\nu\alpha} \geq \log\frac{1}{\tilde{d}} + 2\sum_{\nu=1}^{N} x_\nu \tilde{\omega}_\nu + \sum_{\nu,\alpha=1}^{N} x_\nu x_\alpha \tilde{p}_{\nu\alpha}.$$

This worsened inequality may still be sharp if we can find numbers x_ν such that

$$(11) \qquad 0 = \tilde{\omega}_{N+\mu} + \sum_{\nu=1}^{N} x_\nu \tilde{p}_{\nu N+\mu} \qquad \text{for } \mu = 1, 2, \cdots, M.$$

For in this case all y_ν vanish and (10) reduces to (10′).

6. Conformal Radii.

So far we defined the functionals $\Phi[D]$ for a domain D with N independent real parameters x_ν. Now we will specialize these numbers and obtain important new functionals of the domain. We select an arbitrary but fixed index ν, say $\nu = 1$, and ask for the maximum value of $\Phi[D]$ for the given domain D and value $x_1 = 0$, while all other real parameters x_ν remain free. Since the quadratic form

$$Q(x) = \sum_{\mu,\nu=2}^{N} x_\mu x_\nu p_{\mu\nu} \tag{1}$$

is negative definite, we are sure that $\Phi[D]$ is bounded from above and that a maximum vector $(x_2, x_3, \cdots, x_N)$ exists.

We characterize the maximum vector by the rules of elementary calculus and find the conditions

$$\omega_\nu + \sum_{\mu=2}^{N} x_\mu p_{\mu\nu} = 0, \qquad \nu = 2, 3, \cdots, N. \tag{2}$$

This system determines the maximum vector uniquely in terms of the functionals of D, and we may use (2) to express the maximum value as

$$\Phi[D] = \log\frac{1}{d} + \sum_{\nu=2}^{N} x_\nu \omega_\nu . \tag{3}$$

To understand the significance of this expression, we consider the harmonic function

$$g_1(z) = g(z) + \sum_{\mu=2}^{N} x_\mu \omega_\mu(z), \tag{4}$$

where the x_ν are defined by (2). We verify at once that

$$\frac{1}{2\pi}\int_{C_\nu} \frac{\partial g_1(z)}{\partial n}\, ds = \omega_\nu + \sum_{\mu=2}^{N} x_\mu p_{\mu\nu} = \delta_{\nu 1}. \tag{5}$$

Thus, if we complete $g_1(z)$ to the analytic function

$$p_1(z) = \log f_1(z) = g_1(z) + i h_1(z) = \log z + \cdots, \tag{6}$$

we find that $f_1(z)$ is analytic and single valued in D. Its argument does not change along a closed loop around C_ν $(\nu = 2, 3, \cdots, N)$ but increases by 2π if we run around C_1. We also have

$$|f_1(z)| = \begin{cases} 1 & \text{on } C_1 \\ e^{x_\nu} & \text{on } C_\nu, \qquad \nu = 2, 3, \cdots, N. \end{cases} \tag{7}$$

Thus $f_1(z)$ yields the canonical mapping of D onto a domain bounded by the unit circumference and N — 1 concentric circular slits. The point at infinity corresponds to the point at infinity and the continuum C_1 goes into the unit circle. We have

$$(8) \qquad |f_1'(\infty)| = \frac{1}{d} \exp\left(\sum_{\nu=2}^{n} x_\nu \omega_\nu\right).$$

This leads us to define the conformal radius

$$(9) \qquad d_1 = d \exp\left(-\sum_{\nu=2}^{n} x_\nu \omega_\nu\right).$$

It is well known that d_1 is the largest transfinite diameter which an image of C_1 can attain if we consider all univalent normalized functions

$$(10) \qquad f(z) = z + \alpha_0 + \frac{\alpha_1}{z} + \cdots$$

in D and their image domains [4]. We see by comparison of (3) and (9) that the maximum value of $\Phi[D]$ can be expressed as

$$(11) \qquad \Phi[D] = \log \frac{1}{d_1}.$$

Now suppose we subject the basic domain D to a variation (1.10) and thus deform it into the domain D^*. We can find the variation of the functionals d, ω_ν and $p_{\mu\nu}$ by the formulas of Section 1 and calculate the corresponding variation of the maximum vector $(x_2, \cdots, x_N)$ by (2). However, if we define $\Phi[D]$ by the representation (1.9), we may disregard the variation of the x_ν completely. Indeed, it is clear that the x_ν have a variation of order $O(\rho^2)$ at most, while equation (2) expresses the condition

$$(12) \qquad \frac{\partial \Phi[D]}{\partial x_\nu} = 0, \qquad \nu = 2, 3, \cdots, N.$$

Hence the change of the x_ν enters into our formulas at most in the order $O(\rho^4)$. Thus we proved (see (1.15))

$$(13) \qquad \log \frac{1}{d_1^*} = \log \frac{1}{d_1} + \operatorname{Re}\{e^{i\alpha} \rho^2 p_1'(z_0)^2\} + o(\rho^2).$$

The conformal radii have also a variational formula involving a perfect square.

We return now to the assumptions made in Section 2. Consider a set of N continua Γ_ν in the exterior of the unit disc $|\zeta| > 1$ and all mappings (2.1) which will carry the Γ_ν into various sets of continua C_ν ($\nu = 1, 2, \cdots, N$). The following problem may now be posed:

Find those conformal mappings (2.1) for which the conformal radius d_1 takes its largest possible value.

We argue precisely as we did in Section 2, using the complete analogy of the variational formulas. In the extremum mapping the unit circle goes into a continuum C_0 in the z-plane on which

$$(14) \qquad g_1(z) = x_0 = \text{const.} \qquad \text{on } C_0.$$

Since the continuum C_0 lies in the domain D bounded by $C_1, \cdots, C_N$, the harmonic function $g_1(z)$ is regular on C_0 and has an analytic completion $p_1(z)$ which is single valued around C_0. Thus the function

$$f_1(z) = e^{p_1(z)} = \frac{1}{d_1} z + \cdots \tag{15}$$

maps the domain in the z-plane bounded by $C_0, C_1, \cdots, C_N$ onto a domain bounded by the unit periphery and N concentric circular slits, such that C_1 corresponds to the unit circle.

Finally, let

$$F_1(\zeta) = f_1[f(\zeta)] = \frac{1}{\tilde{d}_1} \zeta + \cdots \tag{16}$$

be the corresponding function in the domain $\tilde{\Delta}$. We see that the maximum value d_1 of the domain D coincides with the known value $\tilde{d}_1$ of the domain $\tilde{\Delta}$. We proved the following:

THEOREM. *The conformal radius d_ν of the images C_ν of the set of continua Γ_ν in $|\zeta| > 1$ under all normalized conformal mappings* (2.1) *is always less or equal to the conformal radius $\tilde{d}_\nu$ of the domain $\tilde{\Delta}$ bounded by $\Gamma_0 = (|\zeta| = 1)$ and the Γ_ν. There exists a mapping for which equality is attained.*

7. SOME NEW CANONICAL DOMAINS.

Our technique can obviously be applied to numerous analogous problems. The selection of additional examples should be determined by the elegance of the results which can be obtained. We wish to stress one particular type of questions which leads to an existence proof for a remarkable type of canonical domains.

Let Δ be a domain in the complex ζ-plane bounded by N continua $\Gamma_1, \Gamma_2, \cdots, \Gamma_N$ and containing the point $\zeta = \infty$. We denote by Δ_ν the larger domain which has all boundaries of Δ, except for the continuum Γ_ν. Next we admit all conformal mappings

$$z = f(\zeta) = \zeta + a_0 + \frac{a_1}{\zeta} + \cdots \tag{1}$$

which are univalent in Δ and regular analytic except for the pole at infinity. Each such mapping carries the domains Δ and Δ_ν into domains D and D_ν in the z-plane. Let $d(D_\nu)$ denote the transfinite diameter of the domain D_ν, i.e., of its boundary continua $\sum_{\alpha \neq \nu} C_\alpha$. We then raise the following

PROBLEM. *Find the extremum mapping* (1) *for which*

$$\sum_{\nu=1}^{N} d(D_\nu) = \max . \tag{2}$$

Since the existence of such an extremum mapping is again evident, we proceed at once to characterize it by our variational technique. Let $z_0 \in C_\beta$ and consider the univalent variation

$$z^* = z + \frac{a\rho^2}{z - z_0} + o(\rho^2) \tag{3}$$

which is univalent outside of C_β. Observe now that C_β belongs to the boundary of all D_ν except for D_β. Since the transfinite diameter of a domain is invariant under a normalized univalent mapping, we have

$$d(D_\nu^*) = d(D_\nu) \qquad \text{for} \quad \nu \neq \beta . \tag{4}$$

On the other hand, using the variational formula (1.14) for $d(C_\beta)$, we find

$$\log d(D_\beta^*) = \log d(D_\beta) - \operatorname{Re}\{a\rho^2 p'_{D_\beta}(z_0)^2\} + o(\rho^2), \tag{5}$$

where $p_{D_\beta}(z)$ is the analytic completion of the Green's function $g_{D_\beta}(z)$ of D_β. The assumed maximum property of the domain D leads then to the consequence

$$\operatorname{Re}\{a\rho^2 p'_{D_\beta}(z_0)^2\} + o(\rho^2) \geq 0 \tag{6}$$

for all admissible variations (3). This, in turn, implies that C_β consists of analytic arcs for which

$$z' p'_{D_\beta}(z) = \text{imaginary}. \tag{7}$$

In other words,

$$g_{D_\beta}(z) = \xi_\beta = \text{const.} \qquad \text{on} \quad C_\beta . \tag{8}$$

Since this harmonic function vanishes on all other C_ν, we can write it in the form

$$g_{D_\beta}(z) = g(z) + \xi_\beta\, \omega_\beta(z), \tag{9}$$

where $g(z)$ and $\omega_\nu(z)$ are the Green's function and the harmonic measures of the entire image domain D.

Since $g_{D_\beta}(z)$ is regular harmonic on C_β, we have the period condition

$$0 = \frac{1}{2\pi}\int_{C_\beta}\left[\frac{\partial g(z)}{\partial n} + \xi_\beta \frac{\partial \omega_\beta(z)}{\partial n}\right] ds = \omega_\beta + \xi_\beta\, p_{\beta\beta} . \tag{10}$$

Thus we find

$$g_{D_\beta}(z) = g(z) - \frac{\omega_\beta}{p_{\beta\beta}}\, \omega_\beta(z) . \tag{11}$$

This identity must hold for all values of β since we picked the continuum C_β for the variation at random. Using the asymptotics (1.8) of the Green's function at infinity, we find

$$d(D_\beta) = d \exp\left(\frac{\omega_\beta^2}{p_{\beta\beta}}\right). \tag{12}$$

Observe that the transfinite diameter d, the values ω_β of the β–th harmonic measure at infinity and the period matrix $p_{\mu\nu}$ are conformal invariants and have the same value for the basic domain Δ as for the extremum domain D. Hence the right-hand side of (12) is known.

Had we asked only for the extremum domain D for which the single functional $d(D_\beta)$ is maximal, the same consideration would have been posssible and we would have come to the conclusion that

$$d(C_\beta) \leq d \exp\left(\frac{\omega_\beta^2}{p_{\beta\beta}}\right) \tag{13}$$

under all normalized mappings (1). We thus have proved the following:

THEOREM. *Let Δ be a domain containing the point at infinity and bounded by* N *continua* Γ_ν. *We can map it by a normalized function* (1) *onto a canonical domain* D *bounded by* N *analytic arcs* C_ν *with the following properties*:

1) *Each domain* D_β *has the largest value* (12) *for the transfinite diameter possible under mappings* (1).

2) *The Green's function of* D *can be transformed into a regular harmonic function* $g(z) - \frac{\omega_\beta}{p_{\beta\beta}} \omega_\beta(z)$ *across the* β*–th boundary arc by combination with the* β*–th harmonic measure.*

It seems that this remarkable canonical domain deserves a more detailed study and may prove useful for the general theory of multiply-connected domains. However, even in the very special case of a doubly-connected domain Δ do we obtain a very remarkable consequence. In this case we may refer the domain to a circular ring

$$R_1 < |z| < R_2 \tag{14}$$

and express the harmonic measures in the form

$$\omega_2(z) = \frac{\log \frac{|z|}{R_1}}{\log \frac{R_2}{R_1}} \quad , \quad \omega_1(z) + \omega_2(z) = 1 . \tag{15}$$

We obtain

$$-p_{11} = \frac{1}{\log \frac{R_2}{R_1}} = -p_{22} . \tag{16}$$

Hence in view of the general theorem we easily find

$$\sqrt{\log \frac{d}{d(\Gamma_1)}} + \sqrt{\log \frac{d}{d(\Gamma_2)}} \geq \sqrt{\log \frac{R_2}{R_1}} , \tag{17}$$

an interesting subadditivity relation between the transfinite diameters and the module of the ring domain.

BIBLIOGRAPHY.

[1] COURANT R., *Dirichlet's principle, conformal mapping and minimal surfaces,* New York 1950. Appendix.
[2] POMMERENKE C., *On the logarithmic capacity and conformal mapping,* « Duke Math. J. », *35*, 321–325 (1968).
[3] SCHIFFER M., *A method of variation within the family of simple functions,* « Proc. London Math. Soc. », (2), *44*, 432–449 (1938).
[4] SCHIFFER M., *Hadamard's formula and variation of domain functions,* « Amer. J. Math. », *68*, 417–448 (1946).
[5] SCHIFFER M., *Applications of variational methods in the theory of conformal mapping,* « Proc. Symp. Appl. Math. », *8*, 93–113 (1958)
[6] SCHIFFER M., *Extremum problems and variational methods in conformal mapping* « Proc. International Congress Math. », 211–231, Edinburgh 1958.
[7] SCHIFFER M., *A variational method for univalent quasi-conformal mappings,* « Duke Math. J. », *33*, 395–412 (1966).

Commentary on

[99] *Some distortion theorems in the theory of conformal mapping*, Atti Accad. Naz. Lincei Mem. Cl. Sci. Fis. Mat. Natur. Sez. Ia (8) **10** (1970), 1–19.

Schiffer begins this paper by revealing his simple strategy to get the most out of the variational method: apply it to a functional whose variational derivative is a perfect square. Then, at least in principle, the resulting quadratic differential can be integrated to produce quantitative information about the extremal functions. Of course, it is not easy to devise interesting functionals with this property, but Schiffer had the Midas touch. In [99], he considers a multiply connected domain D containing ∞ and bounded by N proper continua $C_1, C_2, \dots, C_N$. Letting $\omega_k(z)$ denote the harmonic measure of C_k with respect to D, with $\omega_k = \omega_k(\infty)$ and (p_{jk}) the Riemann matrix of periods of the harmonic conjugates, and letting d denote the transfinite diameter of ∂D, he proposes consideration of the improbable domain functional

$$\Phi[D] = \log\frac{1}{d} + 2\sum_{k=1}^{N} x_k\omega_k + \sum_{j=1}^{N}\sum_{k=1}^{N} x_j x_k p_{jk},$$

where the x_k are arbitrarily selected real constants. He then applies his method of interior variation [13,17] to perturb the domain D and calculate the variational derivative of $\Phi[D]$, which turns out to be a perfect square.

After these preliminaries, Schiffer fixes N disjoint proper continua $\Gamma_1, \Gamma_2, \dots, \Gamma_N$ in the domain $\Delta = \{\zeta : |\zeta| > 1\}$ and considers the family of domains D complementary to the set of continua $C_k = f(\Gamma_k)$, where f belongs to the class Σ of conformal mappings of Δ with the form $f(\zeta) = \zeta + b_0 + b_1\zeta^{-1} + \dots$. Applying his "basic lemma" of boundary variation [5], Schiffer is then able to calculate the sharp upper and lower bounds of $\Phi[D]$ for all $f \in \Sigma$. As the continua Γ_k move onto the unit circle, the upper inequality reduces to

$$\Phi[D] \le 2\log\frac{1}{T} + 2\sum_{k=1}^{N} x_k\Omega_k + \frac{1}{2}\sum_{j=1}^{N}\sum_{k=1}^{N} x_j x_k P_{jk},$$

where T is the transfinite diameter of the system of arcs Γ_k, and Ω_k and P_{jk} denote the corresponding harmonic measures and the periods of their harmonic conjugates. (Here there is an apparent misprint in the paper; the constant $\frac{1}{2}$ is missing.) As a special case, with all $x_k = 0$, Schiffer obtains the sharp inequality $d \ge T^2$, established by Pommerenke [P1] via application of the Goluzin inequalities (see also [P2]). In fact, Pommerenke had shown that $d(f(E)) \ge d(E)^2$ for all $f \in \Sigma$ and for every closed subset E of the unit circle. Previously, Schiffer [17] had given a variational proof in the special case where E is a single arc. A more direct proof of Pommerenke's theorem was found later in [137], where the result is explained in terms of Robin capacity.

The paper [99] has an unusual feature: it is preceded by a kind of public referees' report, prepared by a committee of three distinguished Italian mathematicians, summarizing the paper and recommending its publication in the proceedings of the Italian National Academy. Their remarks begin with a laudatory description of the author's previous work, suggestive of introducing an invited speaker. Written in the report, and again at the top of the paper, is the phrase "Memoria di Menahem Schiffer." Although this has the simple translation "Memoir by ...," I suspect it was the source of a disturbing rumor that circulated around the time the paper was published, a rumor that Schiffer had died! Fortunately, the reports of his death were greatly exaggerated; he flourished for another 25 years.

References

[P1] Ch. Pommerenke, *On the logarithmic capacity and conformal mapping*, Duke Math. J. **35** (1968), 321–325.

[P2] Ch. Pommerenke, *Univalent Functions*, Vandenhoeck & Ruprecht, 1975.

PETER DUREN

[103] (with G. Schober) An extremal problem for the Fredholm eigenvalues

[103] (with G. Schober) An extremal problem for the Fredholm eigenvalues. *Arch. Rational Mech. Anal.* **44** (1971/72), 83–92.

An Extremal Problem for the Fredholm Eigenvalues

M. SCHIFFER & G. SCHOBER

1. Introduction

The Dirichlet problem of two-dimensional potential theory for a domain bounded by a smooth Jordan curve C can be reduced to solving the integral equation

$$\phi(z)+\int_C k(z,t)\,\phi(t)\,ds_t=u(z), \qquad z,t\in C \tag{1}$$

with double layer kernel

$$k(z,t)=\frac{1}{\pi}\,\frac{\partial}{\partial n_t}\log\frac{1}{|z-t|}. \tag{2}$$

Here n_t denotes the inner normal to C at t, and the integration is with respect to arc length. The integral equation (1) with a sign change is useful in constructive methods of conformal mapping [11], and with transposed kernel it leads to a solution of the Neumann boundary value problem.

It can be shown [3] that the eigenvalues of the kernel $k(z,t)$ are all real, are symmetric with respect to the origin, and lie outside the unit circle. The only exception is the trivial eigenvalue 1 corresponding to constant eigenfunctions.

The integral equation (1) can be solved by successive approximations as follows. Let $\phi_0=u$ and

$$\phi_n(z)=-\int_C k(z,t)\,\phi_{n-1}(t)\,ds_t+u(z), \qquad n=1,2,3,\ldots. \tag{3}$$

Then $(\phi_n+\phi_{n-1})/2$ converges uniformly to a solution of (1). The convergence is geometric of order $1/\lambda^n$ where λ is the smallest eigenvalue larger than the trivial one [11]. Therefore it is important to study the eigenvalue λ and to obtain estimates, especially lower bounds, for it.

2. An Extremal Problem

Let $\mathcal{F}$ be the family of plane homeomorphisms f which are analytic in $\mathcal{A}=\{r<|\zeta|<R\}$, are K-quasiconformal in $\Delta=\{|\zeta|<r\}$, and are $\tilde{K}$-quasiconformal in $\tilde{\Delta}=\{|\zeta|>R\}$. The quasiconformality means that relative to each standard rectangle in $\Delta\cup\tilde{\Delta}$, f is absolutely continuous on a.e. horizontal and vertical line and satisfies the dilatation condition

$$D(f)=\frac{|f_\zeta|+|f_{\bar{\zeta}}|}{|f_\zeta|-|f_{\bar{\zeta}}|}\leqq Q \qquad \text{a.e.} \tag{4}$$

We denote by Q the function which is identically K in Δ and $\tilde{K}$ in $\tilde{\Delta}$.

If $f \in \mathscr{F}$ and $r<1<R$, then $C=f(|\zeta|=1)$ is an analytic Jordan curve, and we may consider the associated eigenvalue λ as a functional on $\mathscr{F}$. Consider the problem of extremizing λ, which we shall refer to as *the Fredholm eigenvalue*, over the family $\mathscr{F}$. Since $\lambda=\infty$ is associated with the unit circle itself [2], the interesting problem is to *minimize* λ *over the family* $\mathscr{F}$.

In Section 3 of this article we shall discuss a general variational procedure for quasiconformal mappings (cf. [7]). In Section 4 we shall apply it to our extremal problem. We shall show that the minimum

$$\lambda(f)=\frac{R^2+k\tilde{k}r^2}{\tilde{k}+kr^2R^2}, \qquad k=\frac{K-1}{K+1}, \qquad \tilde{k}=\frac{\tilde{K}-1}{\tilde{K}+1}. \tag{5}$$

In Section 5 we shall actually construct an extremal function.

Based on (5) we have the following sharp estimate.

Theorem. *If* λ *is the Fredholm eigenvalue of a Jordan curve which is the image of* $|\zeta|=1$ *under a mapping which is analytic for* $r<|\zeta|<R$, *K-quasiconformal for* $|\zeta|<r<1$, *and* $\tilde{K}$*-quasiconformal for* $|\zeta|>R>1$, *then*

$$\lambda \geqq \frac{R^2+k\tilde{k}r^2}{\tilde{k}+kr^2R^2} \tag{6}$$

where $k=(K-1)/(K+1)$ *and* $\tilde{k}=(\tilde{K}-1)/(\tilde{K}+1)$.

The estimate (6) is an extension of the estimate

$$\lambda \geqq \frac{R^2+r^2}{1+r^2R^2} \tag{7}$$

obtained earlier in [5] for the class of "uniformly analytic" curves, corresponding to $K=\tilde{K}=\infty$. The estimate (6) also relates to the estimate

$$\lambda \geqq \frac{1+k\tilde{k}}{\tilde{k}+k} \tag{8}$$

obtained by SPRINGER [10] for the class of "quasicircles", corresponding to $r=R=1$.

3. The Method of Variations

Suppose that f is a quasiconformal mapping of a simply connected domain Ω and that in a subdomain $\Omega_0 \subseteq \Omega$

$$D(f) \leqq K_0 \quad \text{a.e. in } \Omega_0. \tag{9}$$

Suppose also that f maximizes a functional χ over the family $\mathscr{F}_0$ of all quasiconformal mappings g of Ω that satisfy

$$D(g) \leqq K_0 \quad \text{a.e. in } \Omega_0 \tag{10}$$

and

$$D(g)=D(f) \quad \text{a.e. in } \Omega-\Omega_0. \tag{11}$$

We may approximate [1, Theorem 9] f by quasiconformal mappings $f_n \in \mathscr{F}_0$ which are of class $C^{(1)}$ in Ω_0 in such a way that f_n converges locally uniformly to f, and the respective partial derivatives $u_n = (f_n)_\zeta$ and $v_n = (f_n)_{\bar\zeta}$ converge to f_ζ and $f_{\bar\zeta}$ locally in $L^2(\Omega)$. *We assume that χ is sufficiently continuous for the validity of*

$$\lim_{n\to\infty} \chi(f_n) = \chi(f). \tag{12}$$

Suppose that the disks $|z - z_{n\nu}| < \rho_{n\nu}$, $\nu = 1, \ldots, N$, are disjoint and belong to $f_n(\Omega_0)$. Define

$$\varphi_{n\nu}(z) = \begin{cases} \rho_{n\nu}^2/(z - z_{n\nu}) & \text{for } |z - z_{n\nu}| \geqq \rho_{n\nu} \\ (\overline{z - z_{n\nu}})(2 - \rho_{n\nu}^{-2}|z - z_{n\nu}|^2) & \text{for } |z - z_{n\nu}| < \rho_{n\nu} \end{cases} \tag{13}$$

and

$$\Phi_n(z) = z + \sum_{\nu=1}^{N} a_{n\nu}\varphi_{n\nu}(z), \qquad |a_{n\nu}| < \varepsilon. \tag{14}$$

Then Φ_n is of class $C^{(1)}$, and since

$$|(\Phi_n)_{\bar z}| \leqq \frac{2\varepsilon}{1-\varepsilon} |(\Phi_n)_z|, \tag{15}$$

Φ_n is a quasiconformal function for $\varepsilon < \frac{1}{3}$. Such functions are light and open [4, Kapitel VI]. Therefore since Φ_n is univalent in a neighborhood of infinity, Φ_n is a plane quasiconformal homeomorphism for $\varepsilon < \frac{1}{3}$.

The composition $f_n^* = \Phi_n \circ f_n$ is quasiconformal in Ω and of class $C^{(1)}$ in Ω_0. Since Φ_n is analytic outside $U = \bigcup_{\nu=1}^{N} \{z : |z - z_{n\nu}| \leqq \rho_{n\nu}\}$, the function f_n^* satisfies both (11) and (10) in $\Omega_0 - U$. In U it is convenient to consider the function

$$R(f_n) = \tfrac{1}{2}[D(f_n) + D(f_n)^{-1}]. \tag{16}$$

Note that the following are equivalent:

$$D(f_n) \leqq K_0, \tag{17}$$

$$R(f_n) \leqq \tfrac{1}{2}(K_0 + K_0^{-1}), \tag{18}$$

$$\frac{|u_n v_n|}{|u_n|^2 - |v_n|^2} \leqq \frac{1}{4}(K_0 - K_0^{-1}). \tag{19}$$

From the identity

$$R(f_n^*) = R(f_n)\frac{|(\Phi_n)_z|^2 + |(\Phi_n)_{\bar z}|^2}{|(\Phi_n)_z|^2 - |(\Phi_n)_{\bar z}|^2} + \frac{4\,\mathrm{Re}\{(\Phi_n)_z \overline{(\Phi_n)_z}\, u_n v_n\}}{(|(\Phi_n)_z|^2 - |(\Phi_n)_{\bar z}|^2)(|u_n|^2 - |v_n|^2)} \tag{20}$$

we obtain the asymptotic form

$$R(f_n^*) = R(f_n) + 8\left(1 - \frac{|f_n - z_{n\nu}|^2}{\rho_{n\nu}^2}\right)\frac{\mathrm{Re}\{\bar a_{n\nu} u_n v_n\}}{|u_n|^2 - |v_n|^2} + O(\varepsilon^2) \tag{21}$$

corresponding to the disk $|z - z_{n\nu}| < \rho_{n\nu}$. Because of (15), (18), and (19) there is a uniform bound for $O(\varepsilon^2)/\varepsilon^2$ depending only on K_0.

Let $\eta>0$ and define

(22) $$\Sigma_n(\eta)=\{\zeta\in\Omega_0:\ D(f_n)<K_0-\eta\}.$$

Since f_n is a class $C^{(1)}$ homeomorphism, $f_n(\Sigma_n(\eta))$ is an open set. Assume that the disks $|z-z_{n\nu}|<\rho_{n\nu}$, $\nu=1,\ldots,N$, all belong to $f_n(\Sigma_n(\eta))$. Then (21) implies that f_n^* belongs to $\mathscr{F}_0$ for ε sufficiently small, independent of n.

We assume now that there exist functions A_n analytic in $f_n(\Omega_0)$ and converging locally uniformly in $f(\Omega_0)$ to $A\not\equiv 0$ such that

(23) $$\chi(f_n^*)\geqq\chi(f_n)+\sum_{\nu=1}^{N}\pi\rho_{n\nu}^2\operatorname{Re}\{a_{n\nu}A_n(z_{n\nu})\}+O(\varepsilon^2)$$

where $O(\varepsilon^2)/\varepsilon^2$ is uniformly bounded independent of n and N.

Since f maximizes χ over $\mathscr{F}_0$, it follows that

(24) $$\sum_{\nu=1}^{N}\pi\rho_{n\nu}^2\operatorname{Re}\{a_{n\nu}A_n(z_{n\nu})\}\leqq\chi(f)-\chi(f_n)-O(\varepsilon^2)$$

for ε sufficiently small. We may choose

(25) $$a_{n\nu}=\frac{|A_n(z_{n\nu})|}{A_n(z_{n\nu})}\,\varepsilon_n,\qquad \varepsilon_n=\sqrt{\chi(f)-\chi(f_n)}$$

since $\varepsilon_n\to 0$ as $n\to\infty$. Then (24) becomes

(26) $$\sum_{\nu=1}^{N}\pi\rho_{n\nu}^2|A_n(z_{n\nu})|\leqq O(\varepsilon_n).$$

Let G be any open set whose closure is compact in Ω_0 and disjoint from the zeros of $A\circ f$. By the Vitali covering theorem we may exhaust almost all of $f_n(G\cap\Sigma_n(\eta))$ by disjoint disks. We may then interpret (26) to say that the area of $f_n(G\cap\Sigma_n(\eta))$ tends to zero as $n\to\infty$. Under the mappings f_n^{-1}, two dimensional measure satisfies the Hölder relation [4, pp. 226–227]

(27) $$m(f_n^{-1}(S))\leqq B\,m(S)^{\delta}$$

for some $\delta>0$ and B which can be chosen uniformly for the sequence $\{f_n^{-1}\}$ relative to any compact subset of $f(\Omega)$. Consequently, the area of $G\cap\Sigma_n(\eta)$ tends to zero as $n\to\infty$. In other words, the dilatation functions $D(f_n)$ converge to K_0 in measure relative to $G\cap\Omega_0$. Due to the arbitrary nature of G we may, by passing to the limit, conclude that

(28) $$D(f)=\frac{|f_z|+|f_{\bar z}|}{|f_z|-|f_{\bar z}|}=K_0\quad\text{ a.e. in }\Omega_0.$$

Suppose now that $\zeta_{n\nu}$ ($\nu=1,\ldots,N$) are points in the open set

(29) $$T_n(\eta)=\{\zeta\in\Omega_0:\ 0<\eta<\arg[u_n v_n(A_n\circ f_n)]<2\pi-\eta\},$$

and let

(30) $$\delta_{n\nu}=\arg[u_n(\zeta_{n\nu})\,v_n(\zeta_{n\nu})],\qquad \tau_{n\nu}=-\arg[A_n\circ f_n(\zeta_{n\nu})].$$

Because of (29) we may assume that

$$\eta < \delta_{n\nu} - \tau_{n\nu} < 2\pi - \eta . \tag{31}$$

Define

$$a_{n\nu} = -i\varepsilon_n \exp\{\tfrac{1}{2} i(\delta_{n\nu} + \tau_{n\nu})\}, \quad \varepsilon_n = \sqrt{\chi(f) - \chi(f_n)} . \tag{32}$$

Then at $\zeta_{n\nu}$

$$\operatorname{Re}\{\bar{a}_{n\nu} u_n v_n\} = -\varepsilon_n |u_n v_n| \sin\left(\frac{\delta_{n\nu} - \tau_{n\nu}}{2}\right) < -\varepsilon_n |u_n v_n| \sin \frac{\eta}{2} < 0, \tag{33}$$

and by continuity we may find disjoint neighborhoods of the $\zeta_{n\nu}$'s in $T_n(\eta)$ such that

$$\operatorname{Re}\{\bar{a}_{n\nu} u_n v_n\} < 0 \tag{34}$$

whenever $|f_n(\zeta) - f_n(\zeta_{n\nu})| < \rho_{n\nu}$. From (21) and (34) we see that we can again apply the variation (14) to obtain (24). The choice of $a_{n\nu}$ in (32) then gives

$$\sum_{\nu=1}^{N} \pi \rho_{n\nu}^2 \operatorname{Im}\{\exp[\tfrac{1}{2} i(\delta_{n\nu} + \tau_{n\nu})] A_n \circ f_n(\zeta_{n\nu})\} \leqq O(\varepsilon_n) . \tag{35}$$

Hence from (30) and (31) we have

$$\sum_{\nu=1}^{N} \pi \rho_{n\nu}^2 (\sin \tfrac{1}{2}\eta) |A_n \circ f_n(\zeta_{n\nu})| \leqq O(\varepsilon_n) . \tag{36}$$

Just as in the argument following (26) we may conclude that

$$\arg[f_\zeta f_{\bar{\zeta}} (A \circ f)] = 0 \quad \text{a.e. in } \Omega_0 . \tag{37}$$

Define

$$J(z) = \int \sqrt{A(z)}\, dz \quad \text{and} \quad k_0 = (K_0 - 1)/(K_0 + 1) . \tag{38}$$

Then using (28) and (37), we have

$$\begin{aligned} \frac{\partial}{\partial \bar{\zeta}} [J \circ f - k_0 \overline{J \circ f}] &= \sqrt{A} f_{\bar{\zeta}} - k_0 \overline{\sqrt{A} f_\zeta} \\ &= \sqrt{A} f_{\bar{\zeta}} - \left| \frac{f_{\bar{\zeta}}}{f_\zeta} \right| \overline{\sqrt{A} f_\zeta} \\ &= \frac{1}{\sqrt{A} f_\zeta} [A f_\zeta f_{\bar{\zeta}} - |A f_\zeta f_{\bar{\zeta}}|] \\ &= 0 \quad \text{a.e. in } \Omega_0 . \end{aligned} \tag{39}$$

It follows directly as in [7] and also from the theory of quasiconformal functions [4, Kapitel VI] that

$$J \circ f - k_0 \overline{J \circ f} \tag{40}$$

is analytic in Ω_0. We view (28), (37), and the analyticity of (40) as important necessary conditions for an extremal function f.

4. Minimization of the Fredholm Eigenvalue

We return to the extremal problem posed in Section 2.

Suppose $g_k \in \mathscr{F}$ is a minimizing sequence, i.e.,

$$\lim_{k\to\infty} \lambda(g_k) = \inf_{\mathscr{F}} \lambda. \tag{41}$$

Since λ is invariant if g_k is followed by a Möbius transformation [2], we may assume by a normality argument that g_k converges locally uniformly to a function $f \in \mathscr{F}$. Moreover, derivatives of g_k converge locally uniformly to derivatives of f on $\mathscr{A}$. Under this convergence the functional λ is continuous [8, Theorem 7] so that

$$\lambda(f) = \inf_{\mathscr{F}} \lambda, \tag{42}$$

and hence a minimizing function f exists within $\mathscr{F}$.

We shall apply the variational procedure of Section 3 to the extremal quasiconformal mapping f of $\Omega = \mathbb{C}$ relative to the functional $\chi = -\lambda$ and both subdomains $\Omega_0 = \Delta$ and $\Omega_0 = \tilde{\Delta}$. The continuity condition (12) is satisfied, again by [8, Theorem 7].

To examine the condition (23) we note that under the variation (14) the functional λ has the asymptotic form [5, p. 1207]

$$\lambda(f_n^*) = \lambda(f_n) - [\lambda(f_n)^2 - 1] \sum_{\nu=1}^{N} \pi \rho_{n\nu}^2 \sigma_{n\nu} + O(\varepsilon^2) \tag{43}$$

where $\sigma_{n\nu}$ is the largest characteristic value of the symmetric matrix

$$[\operatorname{Re}\{a_{n\nu} w_{ni}(z_{n\nu}) w_{nj}(z_{n\nu})\}]_{i,j=1}^{m} \tag{44}$$

and the functions $w_{n1}, \dots, w_{nm}$ are obtained as follows. Suppose that $\phi_1, \dots, \phi_m$ form a basis for the space of eigenfunctions of the homogeneous equation

$$\phi(z) = \lambda(f_n) \int_{C_n} k(z,t)\,\phi(t)\,ds_t, \qquad C_n = f_n(|\zeta| = 1). \tag{45}$$

Then

$$w_{nj}(z) = \begin{cases} \lambda(f_n)\sqrt{\lambda(f_n) - 1} \displaystyle\int_{C_n} \frac{\partial}{\partial z} k(z,t)\,\phi_j(t)\,ds_t & \text{for } z \in D_n = f_n(|\zeta| < 1) \\ i\lambda(f_n)\sqrt{\lambda(f_n) + 1} \displaystyle\int_{C_n} \frac{\partial}{\partial z} k(z,t)\,\phi_j(t)\,ds_t & \text{for } z \in \tilde{D}_n = f_n(|\zeta| > 1). \end{cases} \tag{46}$$

It follows from (46) and Green's theorem that

$$w_{nj}(z) = \frac{\lambda(f_n)}{\pi} \iint_{D_n} \frac{\overline{w_{nj}(\zeta)}}{(\zeta - z)^2}\, d\xi\, d\eta \quad \text{for } z \in D_n, \tag{47}$$

$$w_{nj}(z) = \frac{\lambda(f_n)}{\pi} \iint_{\tilde{D}_n} \frac{\overline{w_{nj}(\zeta)}}{(\zeta - z)^2}\, d\xi\, d\eta \quad \text{for } z \in \tilde{D}_n. \tag{48}$$

That is, the restrictions of w_{nj} to D_n and $\tilde{D}_n$ are eigenfunctions of the complex Hilbert transform [2]. Another consequence of (46) is that

$$\iint_{D_n} |w_{nj}(z)|^2 \, dx\, dy = \iint_{\tilde{D}_n} |w_{nj}(z)|^2 \, dx\, dy, \tag{49}$$

and we may normalize ϕ_j so that their norms equal one.

Since $\sigma_{n\nu}$ is the largest characteristic value of (44), the quadratic form

$$\sum_{i,j=1}^{m} \mathrm{Re}\{a_{n\nu} w_{ni}(z_{n\nu}) w_{nj}(z_{n\nu})\} t_i t_j \leqq \sigma_{n\nu} \tag{50}$$

for every choice of unit vector $(t_1, \ldots, t_m)$. In particular, for $(1, 0, \ldots, 0)$ we have

$$\mathrm{Re}\{a_{n\nu}[w_{n1}(z_{n\nu})]^2\} \leqq \sigma_{n\nu}. \tag{51}$$

Consequently, (43) and (51) give the inequality (23) with $A_n = [\lambda(f_n)^2 - 1] w_{n1}^2$. Since $\lambda(f_n) \to \lambda(f) > 1$ as $n \to \infty$, convergence of A_n to a non-zero limit hinges on the functions w_{n1}.

Because the integrals (49) are normalized, we may by passing to a subsequence assume that w_{n1} converges locally uniformly on $D = f(|\zeta| < 1)$ and $\tilde{D} = f(|\zeta| > 1)$ as $n \to \infty$ to an analytic function w. If T denotes the complex Hilbert transform and $\|\cdot\|_S$ denotes the norm in $L^2(S)$, then

$$\|Tw\|_D = \lim_{n \to \infty} \|Tw_{n1}\|_{D_n} = \lim_{n \to \infty} 1/\lambda(f_n) = 1/\lambda(f) \tag{52}$$

and

$$\|Tw\|_{\tilde{D}} = \lim_{n \to \infty} \|Tw_{n1}\|_{\tilde{D}_n} = \lim_{n \to \infty} 1/\lambda(f_n) = 1/\lambda(f). \tag{53}$$

Consequently, w does not vanish identically in either D or $\tilde{D}$. Moreover, the restrictions of w to D and $\tilde{D}$ are eigenfunctions of the complex Hilbert transform with eigenvalue $\lambda(f)$.

Having verified the assumptions in Section 3, we may conclude that the dilatation

$$D(f) = Q \quad \text{a.e. in } \Delta \cup \tilde{\Delta}. \tag{54}$$

Furthermore, if we define $q = (Q-1)/(Q+1)$ and $W = \mathscr{W} \circ f$ where $\mathscr{W}_z = w$ in $D \cup \tilde{D}$, then

$$W - q\overline{W} \tag{55}$$

is analytic in $\Delta \cup \tilde{\Delta}$.

Since f is analytic in $r < |\zeta| < R$, we may expand

$$W(\zeta) = \begin{cases} \sum_{\nu=-\infty}^{\infty} b_\nu \zeta^\nu & \text{for } r < |\zeta| < 1 \\ \sum_{\nu=-\infty}^{\infty} c_\nu \zeta^\nu & \text{for } 1 < |\zeta| < R. \end{cases} \tag{56}$$

Then the discontinuity properties of the kernel (2) imply [5, p. 1194] the relation

$$(57)\qquad \sum_{\nu=-\infty}^{\infty} b_\nu \zeta^\nu = \frac{i}{\sqrt{\lambda(f)^2-1}} \sum_{\nu=-\infty}^{\infty} c_\nu \zeta^\nu - \frac{\lambda(f)\, i}{\sqrt{\lambda(f)^2-1}} \overline{\sum_{\nu=-\infty}^{\infty} c_\nu \zeta^\nu} + \text{constant}$$

on $|\zeta|=1$. Since $\bar{\zeta}=1/\zeta$, we have by equating coefficients

$$(58)\qquad b_\nu = \frac{i}{\sqrt{\lambda(f)^2-1}}\,[c_\nu - \lambda(f)\,\bar{c}_{-\nu}] \qquad \forall \nu \neq 0.$$

Let $k=(K-1)/(K+1)$ and $\tilde{k}=(\tilde{K}-1)/(\tilde{K}+1)$. Then since W is continuous across $|\zeta|=r$ and $|\zeta|=R$, we may conclude from (55) that

$$(59)\qquad b_\nu - k r^{-2\nu}\,\bar{b}_{-\nu}=0 \qquad \forall \nu<0$$

and

$$(60)\qquad c_\nu - \tilde{k} R^{-2\nu}\,\bar{c}_{-\nu}=0 \qquad \forall \nu>0.$$

From (58) and (59) we see that

$$(61)\qquad \begin{aligned} c_\nu - \lambda(f)\,\bar{c}_{-\nu} &= \frac{1}{i}\sqrt{\lambda(f)^2-1}\, b_\nu = \frac{1}{i}\sqrt{\lambda(f)^2-1}\, k r^{-2\nu}\,\bar{b}_{-\nu} \\ &= -k r^{-2\nu}[\bar{c}_{-\nu} - \lambda(f)\, c_\nu] \qquad \forall \nu<0. \end{aligned}$$

So

$$(62)\qquad c_\nu\left(1-\lambda(f)\, k r^{-2\nu}\right) = \bar{c}_{-\nu}\left(\lambda(f) - k r^{-2\nu}\right) \qquad \forall \nu<0,$$

and (60) implies that

$$(63)\qquad c_\nu\left(1-\lambda(f)\, k r^{-2\nu}\right) = c_\nu \tilde{k} R^{2\nu}\left(\lambda(f) - k r^{-2\nu}\right) \qquad \forall \nu<0.$$

For each $\nu<0$ we conclude that

$$(64)\qquad \lambda(f) = \frac{R^{-2\nu} + k\tilde{k} r^{-2\nu}}{\tilde{k} + k r^{-2\nu} R^{-2\nu}}$$

or $c_\nu=0$. Not all c_ν, for $\nu<0$, can be zero, for then (60) and (56) would imply that $w=0$ in $\tilde{D}$. We note that the right side of (64) is least when $\nu=-1$. Consequently, we have the lower bound

$$(65)\qquad \lambda(f) \geqq \frac{R^2 + k\tilde{k} r^2}{\tilde{k} + k r^2 R^2}.$$

We shall deduce equality in (65) by constructing a function f_0 in $\mathscr{F}$ for which equality holds.

5. An Extremal Function

Consider the z-plane slit along the segment σ from $-i\mu$ to $i\mu$ and along the real rays $x \geqq 1$ and $x \leqq -1$. Let $z=\varphi(\zeta)$ map $\sqrt{\tilde{k}}\, r<|\zeta|<R/\sqrt{\tilde{k}}$ conformally onto such a domain. The ratio $R/r\sqrt{k\tilde{k}}$ determines μ. Because of the inherent symmetry we may normalize φ so that

$$(66)\qquad \varphi(\bar{\zeta}) = \overline{\varphi(\zeta)}, \qquad \varphi(i\bar{\zeta}) = \overline{-\varphi(i\zeta)}.$$

Let $C_0=\varphi(|\zeta|=1)$, $D_0=\text{Int}\, C_0$, $\tilde{D}_0=\text{Ext}\, C_0$.

Now

$$\mathscr{W}(z)=i\sqrt{R^4-\tilde{k}^2}(kr^2\zeta^{-1}-\zeta), \quad \zeta=\varphi^{-1}(z), \tag{67}$$

is analytic in $D_0-\sigma$, and since $kr^2\zeta^{-1}-\zeta$ is purely imaginary on $|\zeta|=\sqrt{k}r$, we may (by the symmetry of φ) extend $\mathscr{W}$ analytically across σ to all of D_0. Similarly,

$$\mathscr{W}(z)=\sqrt{1-k^2r^4}(R^2\zeta^{-1}+\tilde{k}\zeta), \quad \zeta=\varphi^{-1}(z), \tag{68}$$

is analytic in $\tilde{D}_0-\{|x|\geqq 1\}$, and since $R^2\zeta^{-1}+\tilde{k}\zeta$ is real on $|\zeta|=R/\sqrt{\tilde{k}}$, we may extend $\mathscr{W}$ analytically across $|x|\geqq 1$ to all of $\tilde{D}_0$. The function $\mathscr{W}$ is then analytic in D_0 and $\tilde{D}_0$, and across C_0 it has been constructed to possess the discontinuity behavior (57) relative to

$$\lambda_0=\frac{R^2+k\tilde{k}r^2}{\tilde{k}+kr^2R^2}. \tag{69}$$

However, the relation (57) is not only necessary, but also sufficient ([5, p. 1194]) for $\mathscr{W}_z$ to be an eigenfunction in D_0 and $\tilde{D}_0$ with eigenvalue λ_0. It remains to associate λ_0 with a function in $\mathscr{F}$.

Define

$$h(\rho e^{i\theta})=\begin{cases} \frac{1}{2}\left[(K+1)\rho-(K-1)\frac{r^2}{\rho}\right]\cos\theta \\ \quad +\frac{i}{2K}\left[(K+1)\rho+(K-1)\frac{r^2}{\rho}\right]\sin\theta & \text{for } r\sqrt{k}<\rho<r \\ \rho e^{i\theta} & \text{for } r\leqq\rho\leqq R \\ \left\{\frac{1}{2\tilde{K}}\left[(\tilde{K}-1)\frac{\rho}{R^2}+(\tilde{K}+1)\frac{1}{\rho}\right]\cos\theta \right. \\ \quad \left. +\frac{i}{2}\left[(\tilde{K}-1)\frac{\rho}{R^2}-(\tilde{K}+1)\frac{1}{\rho}\right]\sin\theta\right\}^{-1} & \text{for } R<\rho<R/\sqrt{\tilde{k}}. \end{cases} \tag{70}$$

Then h maps $\sqrt{k}r<|\zeta|<R/\sqrt{\tilde{k}}$ quasiconformally onto the whole plane with a vertical slit from $-\frac{ir}{K}\sqrt{K^2-1}$ to $\frac{ir}{K}\sqrt{K^2-1}$ and horizontal slits from $\pm\frac{\tilde{K}R}{\sqrt{\tilde{K}^2-1}}$ to ∞. Furthermore, h has piecewise constant dilatation

$$D(h)=\begin{cases} K & \text{in } \sqrt{k}r<|\zeta|<r \\ 1 & \text{in } r<|\zeta|<R \\ \tilde{K} & \text{in } R<|\zeta|<R/\sqrt{\tilde{k}}. \end{cases} \tag{71}$$

Finally, the function $f_0=\varphi\circ(h^{-1})$ has associated eigenvalue

$$\lambda(f_0)=\lambda_0=\frac{R^2+k\tilde{k}r^2}{\tilde{k}+kr^2R^2} \tag{72}$$

because $f_0(|\zeta|=1)=C_0$. Since the slits are removable, f_0 extends to a plane homeomorphism which is analytic in $r<|\zeta|<R$, K-quasiconformal in Δ, and $\tilde{K}$-quasiconformal in $\tilde{\Delta}$. Therefore f_0 is our desired extremal function in $\mathscr{F}$.

This work was supported in part by Air Force contract F 44620-69-C-0106 and National Science Foundation grants GP 16115 and GP 19694.

References

1. AHLFORS, L. V., & L. BERS, Riemann's mapping theorem for variable metrics. Ann. of Math. **72**, 385–404 (1960).
2. BERGMAN, S., & M. SCHIFFER, Kernel functions and conformal mapping. Compositio Math. **8**, 205–249 (1951).
3. BLUMENFELD, J., & W. MAYER, Über Poincarésche Fundamentalfunktionen. Sitz. Wien. Akad. Wiss., Math.-Nat. Klasse **122**, Abt. IIa, 2011–2047 (1914).
4. LEHTO, O., & K. I. VIRTANEN, Quasikonforme Abbildungen. Berlin-Heidelberg-New York: Springer 1965.
5. SCHIFFER, M., The Fredholm eigenvalues of plane domains. Pacific J. Math. **7**, 1187–1225 (1957).
6. SCHIFFER, M., Fredholm eigenvalues and conformal mapping. Rend. Mat. **22**, 447–468 (1963).
7. SCHIFFER, M., A variational method for univalent quasiconformal mappings. Duke Math. J. **33**, 395–412 (1966).
8. SCHOBER, G., Continuity of curve functionals and a technique involving quasiconformal mapping. Arch. Rational Mech. Anal. **29**, 378–389 (1968).
9. SCHOBER, G., Semicontinuity of curve functionals. Arch. Rational Mech. Anal. **33**, 374–376 (1969).
10. SPRINGER, G., Fredholm eigenvalues and quasiconformal mapping. Acta. Math. **111**, 121–141 (1964).
11. WARSCHAWSKI, S. E., On the solution of the Lichtenstein-Gershgorin integral equation in conformal mapping: I. Theory. Nat. Bur. Standards Appl. Math. Ser. **42**, 7–29 (1955).

Stanford University
and
Indiana University

(Received June 29, 1971)

[106] (with G. Schober) A remark on the paper "An extremal problem for the Fredholm eigenvalues"

[106] (with G. Schober) A remark on the paper "An extremal problem for the Fredholm eigenvalues," *Arch. Rational Mech. Anal.* **46** (1972), 394.

M. SCHIFFER & G. SCHOBER

A Remark on the Paper "An Extremal Problem for the Fredholm Eigenvalues"

Vol. 44, pp. 83–92 (1971)

H. RENELT in a recent Ph. D. thesis has given an alternative derivation of our variational method and has pointed out to us that the maximal dilatation of the variation in § 3 is not independent of N as asserted in formula (15). However, this is easily remedied by adding terms of order $O(\varepsilon^2)$. Indeed, suppose $|z-z_{n\nu}|<\rho_{n\nu}$, $\nu=1, 2, \ldots, N$, are disjoint disks in a fixed compact set S,

$$\kappa_n=\begin{cases} a_{n\nu}(\varphi_{n\nu})_{\bar z}[1+a_{n\nu}(\varphi_{n\nu})_z]^{-1} & \text{in } |z-z_{n\nu}|<\rho_n, \quad \nu=1, 2, \ldots, N \\ 0 & \text{otherwise} \end{cases}$$

where $\varphi_{n\nu}$ is defined in formula (13), and $|a_{n\nu}|\leqq\varepsilon<\frac{1}{3}$. Since κ_n is Lipschitz continuous and $|\kappa_n|\leqq 2\varepsilon/(1-\varepsilon)<1$, the solution of the Beltrami equation

$$(\Phi_n)_{\bar z}=\kappa_n(\Phi_n)_z$$

with $\Phi_n(z)=z+o(1)$ near ∞ is a class $C^{(1)}$ plane quasiconformal mapping. Moreover,

$$\Phi_n(z)=z+\sum_{\nu=1}^{N} a_{n\nu}\varphi_{n\nu}(z)+O(\varepsilon^2),$$

where $O(\varepsilon^2)/\varepsilon^2$ is uniformly bounded independent of n, N, and z.

The rest of the procedure remains unchanged.

References

LEHTO, O., & K. I. VIRTANEN, Quasikonforme Abbildungen. Berlin-Heidelberg-New York: Springer 1965.

RENELT, H., Behandlung von Abbildungsproblemen der quasikonformen Abbildung mittels direkter Variationsmethoden. Ph. D. Thesis, Martin-Luther-Universität, Halle-Wittenberg, May, 1971.

Department of Mathematics
Stanford University
Stanford, Calif.
and
Department of Mathematics
Indiana University
Bloomington

(Received March 29, 1972)

[117] (with G. Schober) A variational method for general families of quasiconformal mappings

[117] (with G. Schober) A variational method for general families of quasiconformal mappings. *J. Analyse Math.* **34** (1978), 240–264.

A VARIATIONAL METHOD FOR GENERAL FAMILIES OF QUASICONFORMAL MAPPINGS

By

M. SCHIFFER AND G. SCHOBER[†]

1. Introduction

Classically, the calculus of variations has been valuable from at least two points of view. On the one hand, it can be used to solve extremal problems for functionals over specified domains, say over some family of functions. In general, the procedure leads to a relation, such as an Euler equation, which is satisfied by functions for which the extremum is attained. If this relation is sufficiently restrictive and simple enough, one can determine the extremal functions or, at least, the extreme value of the functional. This leads then to estimates for the value of the functional over its entire domain and to useful inequalities.

On the other hand, by constructing appropriate functionals over compact families of functions one can obtain existence theorems for solutions of those equations or relations that arise from the variational procedure.

In this article we shall illustrate both points of view, perhaps with an emphasis on the latter. However, our framework will not be a standard one. We shall consider extremal problems over compact families of quasiconformal (q.c.) mappings. The general variational procedure is developed in Section 2.

In Section 3 we introduce some compact families of q.c. mappings that are important for applications.

Section 4 contains the solution of a coefficient problem for families of schlicht functions with q.c. extensions. In this section it is the extremal problem that is of primary interest.

In Section 5 we pose some extremal problems that yield (known) existence theorems for fundamental solutions of some linear partial differential equations. More importantly, our methods give representations for fundamental solutions in terms of q.c. mappings. Such representations are valuable for studying properties

[†] This work was supported in part by grants MCS-75-23332-A03 and MCS-77-01831-A01 from the National Science Foundation. It originated during the Samuel Neiman Research Year at the Technion in Haifa.

(e.g., stream lines, equipotential lines) of solutions. Additional extremal problems lead to mappings onto canonical domains.

In Section 6 we indicate some extensions of the work in Section 5 to certain nonlinear partial differential equations.

2. General theory

Let D be a domain in the extended complex plane $\bar{\mathbf{C}} = \mathbf{C} \cup \{\infty\}$. A quasiconformal (q.c.) mapping of D is a homeomorphism $f: D \to \bar{\mathbf{C}}$ that is locally absolutely continuous on a.e. horizontal and vertical line in D and whose complex dilatation

$$\mu = f_{\bar{z}} / f_z \tag{2.1}$$

satisfies $\|\mu\|_\infty < 1$. If $\infty \in D$, we shall assume that $f(\infty) = \infty$; otherwise $f(D) \subset \mathbf{C}$ if $\infty \notin D$.

In this section we shall develop a calculus of variations for compact families $\mathcal{F}$ of q.c. mappings $f: D \to \bar{\mathbf{C}}$ whose complex dilatations μ are restricted by a relation of the form

$$F(\mu(z), z) \leqq 0 \qquad \text{a.e.} \tag{2.2}$$

We shall assume that (2.2) implies that $\|\mu\|_\infty < 1$. In addition, in $\{t: |t| < 1\} \times D$ we shall assume that

(i) $F(t, z)$ is measurable in z for each fixed t,

(ii) $F(t, z)$ is continuous in t for a.e. fixed z, and

(iii) for a possible exceptional set N of measure zero, the complex partial derivative $F_t(t, z)$ exists in $\mathcal{D} = \{t: |t| < 1\} \times (D \setminus N)$, is uniformly bounded on compact subsets of $\mathcal{D}$, and

$$F(t_1, z) = F(t, z) + 2\mathrm{Re}\{(t_1 - t) F_t(t, z)\} + o(|t_1 - t|)$$

as $t_1 \to t$, where the term $o(|t_1 - t|)$ is uniform on compact subsets of $\mathcal{D}$.
Of course, condition (iii) implies (ii). We have identified the so-called *Carathéodory conditions* (i) and (ii) which imply ([2]) that $F(\mu(z), z)$ is a measurable function of z for any measurable function μ, so that a relation of the form (2.2) is meaningful.

Let $\lambda: \mathcal{F} \to \mathbf{R}$ be an upper semicontinuous functional, and consider the problem of finding the maximum value of λ over $\mathcal{F}$. Since λ is upper semicontinuous and $\mathcal{F}$ is compact, there will be a function $f \in \mathcal{F}$ for which

$$\lambda(f) = \max_{\mathcal{F}} \lambda. \tag{2.3}$$

We shall assume that λ is *Gâteaux differentiable at f* in the sense that a finite complex measure κ, compactly supported in D, exists such that

$$\lambda(f^*)=\lambda(f)+\varepsilon\,\mathrm{Re}\iint_D g\,d\kappa+o(\varepsilon) \tag{2.4}$$

whenever $f^*\in\mathcal{F}$, $\varepsilon>0$, and $f^*=f+\varepsilon g+o(\varepsilon)$ (the latter $o(\varepsilon)$ is to be uniform on compact subsets of D).

For the purpose of constructing variations, let E be a compact set and $a(w)$ a complex-valued measurable function with $|a(w)|=\chi_E$, the characteristic function of the set E. Then for each ε, $0<\varepsilon<1$, there exists (cf. [11, lemma 13.1]) a q.c. mapping $\Phi:\bar{\mathbf{C}}\to\bar{\mathbf{C}}$ that satisfies the Beltrami equation

$$\Phi_{\bar{w}}=\varepsilon a\Phi_w \quad \text{a.e.} \tag{2.5}$$

and has the normalization $\Phi(w)=w+o(1)$ near ∞. Moreover, the variation Φ has the representation

$$\Phi(w)=w-\frac{\varepsilon}{\pi}\iint_E\frac{a(\zeta)}{\zeta-w}\,dm(\zeta)+O(\varepsilon^2) \tag{2.6}$$

where m denotes planar Lebesgue measure and the term $O(\varepsilon^2)/\varepsilon^2$ is uniformly bounded depending only on E.

We shall say that *variations of the extremal function f can be normalized in $\mathcal{F}$* if for each variation Φ with the support of a contained in $f(D)$ and with $F((\Phi\circ f)_{\bar{z}}/(\Phi\circ f)_z,z)\leqq 0$ a.e., there exists a Möbius transformation τ such that $\tau\circ\Phi\circ f\in\mathcal{F}$. Furthermore, we require that there exists a function $\Psi(\zeta,w)$ (*variational derivative relative to $\mathcal{F}$ of f at ζ*) such that

$$\tau\circ\Phi(w)=w+\frac{\varepsilon}{\pi}\iint_{f(D)}a(\zeta)\Psi(\zeta,w)\,dm(\zeta)+o(\varepsilon),\qquad w\in f(D), \tag{2.7}$$

where $o(\varepsilon)/\varepsilon\to 0$ locally uniformly on $f(D)$ as $\varepsilon\to 0$.

Finally, we assume that $\Psi(\zeta,f(z))$ is locally integrable on $f(D)\times D$ with respect to $dm\times d\kappa$ and

$$A(\zeta)=\iint_D\Psi(\zeta,f(z))\,d\kappa(z)\neq 0 \qquad \text{for a.e. } \zeta\in f(D), \tag{2.8}$$

where κ is the measure associated with the Gâteaux derivative of λ.

Important applications where these hypotheses pertain will be given in Sections 4 and 5.

Theorem 2.1. *Under the hypotheses stated above in this section, the extremal function f for the problem* (2.3) *satisfies the differential relations*

$$F(f_{\bar z}/f_z, z) = 0 \quad \textit{and} \quad \overline{F_t(f_{\bar z}/f_z, z)A(f)(f_z)^2} \geqq 0 \tag{2.9}$$

a.e. in D.

Corollary 2.2. *If, in addition, the function A defined in* (2.8) *is analytic and does not vanish in a neighborhood of a finite point* $w_0 = f(z_0)$, *then f satisfies the differential relations*

$$F((J\circ f)_{\bar z}/(J\circ f)_z, z) = 0 \quad \textit{and} \quad \overline{F_t((J\circ f)_{\bar z}/(J\circ f)_z, z)(J\circ f)_z^2} \geqq 0 \tag{2.10}$$

a.e. in a neighborhood of z_0, *where J denotes any local integral*

$$J(w) = \int^w \sqrt{A(\zeta)}\,d\zeta.$$

Corollary 2.2 is an immediate consequence of Theorem 2.1.

Proof of Theorem 2.1. Let μ be the complex dilatation of the extremal function f for (2.3). For fixed $\delta > 0$ set

$$\Sigma_\delta = \{z \in D : F(\mu(z), z) < -\delta\},$$

and let E be any compact subset of $f(\Sigma_\delta)$. If Φ is a variation with $a(w) = e^{i\alpha}\chi_E(w)$, α real, then

$$\mu^* = (\Phi\circ f)_{\bar z}/(\Phi\circ f)_z = \mu + \varepsilon e^{i\alpha}(1-|\mu|^2)\bar f_{\bar z}/f_z + O(\varepsilon^2) \quad \text{a.e.}$$

if $f(z) \in E$, and $\mu^* = \mu$ otherwise. Here $O(\varepsilon^2)/\varepsilon^2$ is uniformly bounded. Consequently,

$$F(\mu^*, z) = F(\mu, z) + 2\varepsilon(1-|\mu|^2)\operatorname{Re}\{e^{i\alpha}F_t(\mu, z)\overline{f_{\bar z}}/f_z\} + o(\varepsilon)$$

if $f(z) \in E$, where $o(\varepsilon)$ is uniform in z, and $F(\mu^*, z) = F(\mu, z)$ otherwise. Since

$F_t(t, z)$ is uniformly bounded, we have $F(\mu^*, z) \leqq 0$ for all ε sufficiently small. Therefore, by assumption, a Möbius transformation τ exists such that $f^* = \tau \circ \Phi \circ f \in \mathcal{F}$. It follows that

$$0 \geqq \lambda(f^*) - \lambda(f) = \frac{\varepsilon}{\pi} \operatorname{Re} e^{i\alpha} \iint_D \iint_E \Psi(\zeta, f(z)) dm(\zeta) d\kappa(z) + o(\varepsilon).$$

Divide by ε and let $\varepsilon \to 0$. Then

$$\operatorname{Re} e^{i\alpha} \iint_D \iint_E \Psi(\zeta, f(z)) dm(\zeta) d\kappa(z) \leqq 0.$$

Since this is valid for all real α, we have

$$\iint_D \iint_E \Psi(\zeta, f(z)) dm(\zeta) d\kappa(z) = 0.$$

By Fubini's theorem we may interchange the order of integration. Finally, since E is arbitrary, we have

$$A(\zeta) = \iint_D \Psi(\zeta, f(z)) d\kappa(z) = 0$$

for a.e. $\zeta \in f(\Sigma_\delta)$. By (2.8) this is possible only if $f(\Sigma_\delta)$ has measure zero. Since q.c. mappings (and their inverse mappings) preserve null sets, the set Σ_δ has measure zero for all $\delta > 0$. As a consequence, $F(\mu(z), z) = 0$ for a.e. $z \in D$.

Now let $q(z) = \overline{F_t(\mu(z), z)} A(f(z)) f_z^2(z)$ and $p(z) = \sqrt{q(z)}$ where we choose values of the square root in the upper half plane. For $\delta > 0$ define

$$T_\delta = \left\{ z \in D : q(z) \neq 0 \quad \text{and} \quad |F_t(\mu(z), z)| \operatorname{Im} \frac{p(z)}{|p(z)|} > \delta \right\}$$

and, as before, let E be an arbitrary compact subset of $f(T_\delta)$. If Φ is a variation with

$$a(w) = -i \frac{p(z)|A(w)|}{|p(z)| A(w)} \chi_E(w), \qquad z = f^{-1}(w),$$

then

$$\begin{aligned}F(\mu^*, z) &= F(\mu, z) + 2\varepsilon(1 - |\mu|^2)\operatorname{Re}\left\{F_t(\mu, z)\, a\frac{\overline{f_z}}{f_z}\right\} + o(\varepsilon)\\ &= F(\mu, z) - 2\varepsilon(1 - |\mu|^2)|F_t(\mu, z)|\operatorname{Im}\frac{p(z)}{|p(z)|} + o(\varepsilon)\\ &\leqq F(\mu, z) - 2\varepsilon(1 - \|\mu\|_\infty^2)\delta + o(\varepsilon)\end{aligned}$$

if $f(z) \in E$, and $F(\mu^*, z) = F(\mu, z)$ otherwise. Here the $o(\varepsilon)$ terms are uniform in z. Therefore for each ε sufficiently small a Möbius transformation τ exists such that $f^* = \tau \circ \Phi \circ f \in \mathcal{F}$, and we conclude as before that

$$0 \geqq \lambda(f^*) - \lambda(f) = \frac{\varepsilon}{\pi}\operatorname{Re}\iint_D \iint_E a(\zeta)\Psi(\zeta, f(z))\,dm(\zeta)\,d\kappa(z) + o(\varepsilon).$$

Divide by ε, let $\varepsilon \to 0$, and use Fubini's theorem. Then

$$0 \geqq \iint_E \operatorname{Re}\{a(\zeta)A(\zeta)\}\,dm(\zeta) = \iint_E |A|\operatorname{Im}\frac{p}{|p|}\,dm(\zeta) \geqq \delta\iint_E \frac{|A|}{|F_t|}\,dm(\zeta) \geqq 0.$$

Consequently, $A(\zeta) = 0$ a.e. in E, and since E is arbitrary, a.e. in $f(T_\delta)$. We conclude as before that T_δ has measure zero for all $\delta > 0$. Therefore, except for a set of measure zero, either $q = 0$ or $\operatorname{Im} p = 0$. In any case, $q = p^2 \geqq 0$ a.e. in D. This completes the proof of the theorem.

3. Compact families of q.c. mappings

It is important for our development and of independent interest to determine compact families of q.c. mappings that are determined by the restriction (2.2) and convenient normalizations. In our program, each such family will provide a framework in which to pose extremal problems and also to provide solutions for certain partial differential equations.

Let $\{f_n\}$ be a sequence of q.c. mappings of a domain D, that converges uniformly on compact subsets of D to a q.c. mapping f. Let μ_n and μ be the complex dilatations of f_n and f, respectively. It is well known ([8]) that if $|\mu_n| \leqq k$ a.e. for some fixed constant $k < 1$, then $|\mu| \leqq k$ a.e. It is a perhaps less well-known result of K. Strebel [12] that

$$|\mu| \leqq \limsup_{n\to\infty} |\mu_n| \qquad \text{a.e.}$$

Consequently, if k is a fixed measurable function and $|\mu_n| \leqq k$ a.e., then $|\mu| \leqq k$ a.e. also. The following theorems are extensions of these results.

Theorem 3.1. *Let c be a complex function and k a nonnegative function, both measurable in a domain D, with $\| |c| + k \|_\infty < 1$. Let f_n $(n = 1, 2, 3, \cdots)$ be q.c. mappings of D with complex dilatations μ_n that satisfy*

$$|\mu_n - c| \leqq k \qquad \text{a.e. in } D.$$

If f_n converges to a q.c. mapping f as $n \to \infty$, uniformly on compact subsets of D, then the complex dilatation μ of f satisfies

$$|\mu - c| \leqq k \qquad \text{a.e. in } D. \tag{3.1}$$

Proof. If $r = \| |c| + k \|_\infty < 1$, then $\|\mu_n\|_\infty \leqq r$ by the triangle inequality and $\|\mu\|_\infty \leqq r$ follows from the classical result. To show that μ satisfies the much stronger restriction (3.1), we begin with an affine change of independent variable

$$z = z(\zeta) = \zeta + h\bar{\zeta}, \qquad h \text{ constant}, \quad |h| < 1.$$

If $F_n = f_n \circ z$ and $F = f \circ z$, then $F_n \to F$ uniformly on compact subsets of $z^{-1}(D)$ as $n \to \infty$, and the complex dilatations

$$M_n = \frac{(F_n)_{\bar{\zeta}}}{(F_n)_\zeta} = \frac{\mu_n + h}{1 + \bar{h}\mu_n}, \qquad M = \frac{F_{\bar{\zeta}}}{F_\zeta} = \frac{\mu + h}{1 + \bar{h}\mu}.$$

Strebel's result [12] implies that

$$|M| \leqq \limsup_{n \to \infty} |M_n| \tag{3.2}$$

a.e. in $z^{-1}(D)$, for each fixed h. By ignoring a countable union of null sets we also have (3.2) holding for a dense set of h's in $|h| < 1$, except for a fixed null set in $z^{-1}(D)$. Since the sequence $\{(\mu_n + h)/(1 + \bar{h}\mu_n)\}$ is equicontinuous as a function of h, we even have (3.2) holding for all h, $|h| < 1$, a.e. in $z^{-1}(D)$. It is important that the exceptional set in $z^{-1}(D)$ is independent of h. We may then choose h to depend on ζ (or z) in (3.2). The choice

$$h = -\frac{1}{2}\frac{c}{|c|^2}\left[1 + |c|^2 - k^2 - \sqrt{(1 + |c|^2 - k^2)^2 - 4|c|^2}\right] \tag{3.3}$$

makes the disk $|\mu - c| \leqq k$ correspond under the mapping $M = (\mu + h)/(1 + \bar{h}\mu)$ to a disk $|M| \leqq R$,

$$R = \frac{1}{2k}\left[1 - |c|^2 + k^2 - \sqrt{(1 + |c|^2 - k^2)^2 - 4|c|^2}\right] < 1. \tag{3.4}$$

Now since $|M_n| \leqq R$ for all n, we have $|M| \leqq R$ for the limit function F (a.e.). In the z-plane this corresponds to the desired inequality $|\mu - c| \leqq k$ a.e. in D for the limit function f.

Definition. A set $\mathcal{K}$ in $\Delta_r = \{t : |t| \leqq r\}$ is *Δ_r-convex* if it is a (arbitrary) nonempty intersection of closed disks contained in Δ_r.

Consequently, Δ_r-convex sets are compact and (strictly) convex in the usual sense.

Example 1. Evidently, closed disks in $\{|t| < 1\}$ are Δ_r-convex for some $r < 1$.

Example 2. Let $\mathcal{K}$ be a closed convex set containing the origin and having a twice continuously differentiable boundary curve. Let d be the distance from the origin to $\partial\mathcal{K}$ and R ($\leqq \infty$) the maximum radius of curvature of $\partial\mathcal{K}$. Then $\mathcal{K}$ is Δ_{2R-d}-convex.

Example 3. In particular, the ellipse $\mathcal{E} = \{x + iy : x^2/a^2 + y^2/b^2 \leqq 1\}$ in standard position is $\Delta_{2a^2/b-b}$-convex for $0 < b < a$.

One can easily describe other criteria for Δ_r-convexity.

Given any standard normalization, the following theorem leads to very many compact families of q.c. mappings.

Theorem 3.2. *Suppose $F : \{t : |t| < 1\} \times D \to \mathbf{R}$ has the properties:*

(a) *$F(t, z)$ is measurable in z for each fixed t,*

(b) *$F(t, z)$ is continuous in t for a.e. fixed z, and*

(c) *for some r, $0 < r < 1$, the sets $\mathcal{K}(z) = \{t : F(t, z) \leqq 0\}$ are Δ_r-convex for a.e. z.*

Let f_n ($n = 1, 2, 3, \cdots$) be q.c. mappings of the domain D, with complex dilatations μ_n that satisfy

$$F(\mu_n(z), z) \leqq 0 \qquad \text{a.e. in } D.$$

If f_n converges to a q.c. mapping f as $n \to \infty$, uniformly on compact subsets of D, then the complex dilatation μ of f satisfies

$$F(\mu(z), z) \leqq 0 \qquad a.e. \text{ in } D. \tag{3.5}$$

Proof. Let $\{\Delta_m\}$ be an enumeration of all closed disks Δ_m in $\{t: |t| < 1\}$ whose radii r_m are rational and whose centers c_m have rational coordinates. Similarly, let $\{t_\nu\}$ be a denumerable dense subset of $\{t: |t| < 1\}$. Since the set $E_m = \bigcup_{t_\nu \notin \Delta_m} \{z : F(t_\nu, z) \leqq 0\}$ is measurable, the functions

$$c_m(z) = \begin{cases} 0 & \text{if } z \in E_m \\ c_m & \text{if } z \in D \setminus E_m \end{cases} \quad \text{and} \quad k_m(z) = \begin{cases} r & \text{if } z \in E_m \\ r_m & \text{if } z \in D \setminus E_m \end{cases}$$

are measurable. In addition, if $\mathcal{K}_m(z) = \{t : |t - c_m(z)| \leqq k_m(z)\}$, then with this construction we may represent

$$\mathcal{K}(z) = \bigcap_{m=1}^{\infty} \mathcal{K}_m(z).$$

Now apply Theorem 3.1 for each m. We conclude that the complex dilatation μ of the limit function has the property that $\mu(z) \in \mathcal{K}_m(z)$ a.e. for each m. Therefore $\mu(z) \in \mathcal{K}(z)$ for a.e. $z \in D$, and (3.5) is satisfied.

4. Extremal problems

In this section we shall apply the calculus of variations from Section 2 to solve a coefficient problem. We shall consider q.c. homeomorphisms of the plane, whose complex dilatations assume values in a fixed set for $|z| < 1$ and vanish for $|z| > 1$.

Definition. Suppose B is a nonempty compact subset of the open unit disk. Let $\Sigma'(B)$ be the family of all q.c. homeomorphisms $f : \bar{\mathbf{C}} \to \bar{\mathbf{C}}$ whose restriction to $|z| > 1$ is analytic and has an expansion of the form

$$f(z) = z + \sum_{n=1}^{\infty} \frac{b_n}{z^n}, \qquad |z| > 1, \tag{4.1}$$

and whose complex dilatation μ assumes values in B for a.e. z, $|z| < 1$.

If B is the disk $\{t : |t| \leqq k\}$ for some constant k, $0 \leqq k < 1$, then R. Kühnau [5] has proved that $\operatorname{Re}\{e^{i\theta} b_1\} \leqq k$ for all $f \in \Sigma'(|t| \leqq k)$ and all $\theta \in [0, 2\pi)$. The following theorem is a generalization to the case that B is a disk not necessarily centered at the origin.

Theorem 4.1. *Let $B = \{t : |t - c| \leqq k\}$ where $c \in \mathbf{C}$ and $k \geqq 0$ are constants satisfying $|c| + k < 1$. If $f \in \Sigma'(B)$ and $\theta \in [0, 2\pi)$, then*

$$\operatorname{Re}\{e^{i\theta}b_1\} \leqq k + \operatorname{Re}\{e^{i\theta}c\}. \tag{4.2}$$

Equality in (4.2) *occurs only for*

$$f(z) = \begin{cases} z + (c + ke^{-i\theta})\dfrac{1}{z} & |z| \geqq 1, \\ z + (c + ke^{-i\theta})\bar{z} & |z| \leqq 1. \end{cases} \tag{4.3}$$

Proof. It is a consequence of Theorem 3.1 and the normalization imposed by (4.1) that this family $\Sigma'(B)$ is compact. It is of the form studied in Section 3 with

$$F(t, z) = \begin{cases} |t|^2 & \text{for } |z| > 1, \\ |t - c|^2 - k^2 & \text{for } |z| \leqq 1. \end{cases}$$

It admits variations of the form (2.6), and its variational derivative is $\Psi(\zeta, w) = 1/(w - \zeta)$.

The functional $\lambda(f) = \operatorname{Re}\{e^{i\theta}b_1\}$ defined on $\Sigma'(B)$ is continuous and Gâteaux differentiable. We may choose for κ in (2.4) any measure that produces the coefficient of $1/z$ in the expansion of $e^{i\theta}g$ for analytic functions g in $|z| > 1$. Therefore the function in (2.8) is

$$A(\zeta) \equiv e^{i\theta}.$$

If f is an extremal function for the problem

$$\max_{\Sigma'(B)} \operatorname{Re}\{e^{i\theta}b_1\},$$

then by Theorem 2.1 it satisfies the equations

$$\left|\frac{f_{\bar{z}}}{f_z} - c\right| = k \quad \text{and} \quad \left(\frac{f_{\bar{z}}}{f_z} - c\right) e^{i\theta}(f_z)^2 = \left|\frac{f_{\bar{z}}}{f_z} - c\right| |f_z|^2$$

a.e. in $|z| < 1$. Combining these equations, we find that

$$f_{\bar{z}} = cf_z + e^{-i\theta}k\overline{f_z} \qquad \text{a.e. in } |z| < 1. \tag{4.4}$$

Define the function

$$I(\zeta) = e^{i\theta}f(\zeta - b\bar{\zeta}) - \overline{af(\zeta - b\bar{\zeta})}$$

where the constants

$$a = \frac{1}{2k}[1 + k^2 - |c|^2 - \sqrt{(1 + k^2 - |c|^2)^2 - 4k^2}],$$

$$b = \frac{1}{2\bar{c}}[1 - k^2 + |c|^2 - \sqrt{(1 + k^2 - |c|^2)^2 - 4k^2}]$$

are chosen to satisfy the equations $abk - b + c = 0$ and $ab\bar{c} - a + k = 0$. If $k = 0$, define $a = 0$, and if $c = 0$, define $b = 0$. In any case, $0 \leqq a < 1$ and $|b| < 1$. Now the equation (4.4) implies that $I_{\bar{\zeta}} = 0$ for ζ in the ellipse $E = \{\zeta : |\zeta - b\bar{\zeta}| < 1\}$.

Since I is analytic E, it follows that for $n \geqq 0$

$$0 = (1 - |b|^2)^{n+1} \int_{\partial E} I(\zeta)\zeta^n d\zeta$$

$$= \int_{|z|=1} \left[e^{i\theta} f(z) - a\overline{f\left(\frac{1}{\bar{z}}\right)}\right]\left(z + b\frac{1}{z}\right)^n \left(1 - \frac{b}{z^2}\right) dz$$

$$= \int_{|z|=1} \left[e^{i\theta} f(z) - a\overline{f\left(\frac{1}{\bar{z}}\right)}\right]\left[z^n - \frac{b^{n+1}}{z^{n+2}} + \sum_{m=1}^{n}\left\{\binom{n}{m} - \binom{n}{m-1}\right\} b^m z^{n-2m}\right] dz.$$

For $n = 0$ this gives

$$e^{i\theta}(b_1 - b) - a(1 - b\overline{b_1}) = 0 \tag{4.5}$$

or

$$e^{i\theta} b_1 = \frac{a(1 - |b|^2) + e^{i\theta} b(1 - |a|^2)}{1 - |ab|^2} = k + e^{i\theta} c,$$

from which (4.2) follows. For $n \geqq 1$ it gives

$$e^{i\theta} b_{n+1} + ab^{n+1}\overline{b_{n+1}} + \sum_{m=1}^{1+[\frac{1}{2}n]} \left\{\binom{n}{m} - \binom{n}{m-1}\right\}\{e^{i\theta} b^m b_{n-2m+1} + ab^{n-m+1}\overline{b}_{n-2m+1}\} = 0$$

where $b_{-1} = 1$ and $b_0 = 0$. Using (4.5) one sees that terms involving b_{-1} and b_1 in this sum add to zero. Therefore $b_{n+1} = 0$ for $n \geqq 1$ by induction. It follows that (4.3) is the only extremal function.

Theorem 4.1 admits an extension to the case where B is any Δ_r-convex (cf. Section 3) set. The solution will be in terms of a distinguished point on ∂B. If B is

Δ_r-convex (or just strictly convex), then for each $\theta \in [0;2\pi)$ there exists a unique point $t_\theta \in \partial B$ and a closed half-plane H_θ with exterior normal $e^{-i\theta}$ such that $B \subset H_\theta$ and $t_\theta \in \partial B \cap \partial H_\theta$. It is this point t_θ that appears in the following theorem.

Theorem 4.2. *Suppose that the set B is Δ_r-convex for some $r<1$. If $f \in \Sigma'(B)$ and $\theta \in [0,2\pi)$, then*

$$\operatorname{Re}\{e^{i\theta}b_1\} \leqq \operatorname{Re}\{e^{i\theta}t_\theta\}. \tag{4.6}$$

Equality in (4.6) *occurs only for*

$$f(z)=\begin{cases} z+t_\theta \dfrac{1}{z} & |z|\geqq 1,\\ z+t_\theta \bar{z} & |z|<1. \end{cases} \tag{4.7}$$

Proof. Since B is Δ_r-convex, one easily verifies that there is a closed disk S_θ such that $B \subset S_\theta \subset \Delta_r$ and ∂S_θ osculates ∂H_θ at t_θ. If S_θ has center c_θ and radius k_θ, then

$$t_\theta = c_\theta + e^{-i\theta}k_\theta. \tag{4.8}$$

We now make use of Theorem 4.1. If $f \in \Sigma'(B)$, then

$$\operatorname{Re}\{e^{i\theta}b_1\} \leqq k_\theta + \operatorname{Re}\{e^{i\theta}c_\theta\} \tag{4.9}$$

since $\Sigma'(B) \subset \Sigma'(S_\theta)$. The inequality (4.6) follows by substituting (4.8) into (4.9). Since the function (4.7) belongs to $\Sigma'(B)$, the inequality (4.6) is sharp. Moreover, the function (4.7) is the only extremal function since it is the only one in the larger family $\Sigma'(S_\theta)$.

Corollary 1. *Let B be the ellipse $\{x+iy : x^2/a^2+y^2/b^2 \leqq 1\}$ where the constants a and b satisfy $a^2<\frac{1}{2}b(1+b)$ and $0<b<a<1$. If $f \in \Sigma'(B)$ and $\theta \in [0,2\pi)$, then*

$$\operatorname{Re}\{e^{i\theta}b_1\} \leqq \sqrt{a^2\cos^2\theta + b^2\sin^2\theta}. \tag{4.10}$$

Proof. Refer to Example 3 in Section 3. The restrictions on a and b imply that the ellipse is Δ_r-convex for some $r<1$. Since

$$t_\theta = \frac{a^2 \cos\theta - ib^2 \sin\theta}{\sqrt{a^2\cos^2\theta + b^2\sin^2\theta}},$$

the estimate (4.10) follows from Theorem 4.2.

Corollary 2. *Let B be the lens $\{t : |t-c| \leqq k\} \cap \{t : |t+c| \leqq k\}$ where the constants c and k satisfy $0 \leqq c < k$ and $c+k<1$. If $f \in \Sigma'(B)$, then*

$$\operatorname{Re}\{e^{i\theta}b_1\} \leqq \begin{cases} k - c|\cos\theta| & \text{if } |\cos\theta| \geqq c/k, \\ \sqrt{k^2-c^2}\,|\sin\theta| & \text{if } |\cos\theta| < c/k. \end{cases}$$

Proof. In this case

$$t_\theta = \begin{cases} -c\dfrac{\cos\theta}{|\cos\theta|} + ke^{-i\theta} & \text{if } |\cos\theta| \geqq c/k \\ -i\sqrt{k^2-c^2}\dfrac{\sin\theta}{|\sin\theta|} & \text{if } |\cos\theta| < c/k \end{cases}$$

and the estimate follows directly from Theorem 4.2.

In both corollaries the estimates are sharp and equality occurs only for the function (4.7) with the respective choice of t_θ.

5. Fundamental solutions for a linear partial differential system

In this section we shall use the calculus of variations from Section 2 to obtain existence theorems for fundamental solutions of the linear partial differential system

$$\begin{aligned} u_x &= \alpha v_x + \beta v_y \\ -u_y &= \gamma v_x + \alpha v_y \end{aligned} \tag{5.1}$$

and, more importantly, representations for them in terms of q.c. mappings. The results of this section contain some of those of the article [10] as a special case ($\alpha = 0$ and $\beta = \gamma$). Many of the arguments are similar and will not be repeated.

By eliminating v in (5.1), one has the second order equation

$$\left(\frac{\gamma u_x + \alpha u_y}{\beta\gamma - \alpha^2}\right)_x + \left(\frac{\alpha u_x + \beta u_y}{\beta\gamma - \alpha^2}\right)_y = 0 \tag{5.2}$$

for the function u. Since α, β, γ will be assumed only to be measurable, essentially bounded functions, we shall study this equation in the form of the system (5.1).

By a (*weak*) *solution* of the system (5.1) we mean a pair (u, v) of continuous functions that are ACL (absolutely continuous on a.e. horizontal and vertical line), have square integrable first order partial derivatives, and satisfy the system (5.1) a.e. Let $D \subset \bar{\mathbf{C}}$ be a domain and $z_0, z_1 \in D$. If (u, v) is a solution of (5.1) in a neighborhood of each point of $D \setminus \{z_0, z_1\}$ and u has logarithmic singularities at z_0 and z_1, we shall call (u, v) a *fundamental solution* of (5.1) in D (i.e., u is a fundamental solution of (5.2)). Of course, v may be multivalued. The precise nature of the fundamental singularities is determined by the equations (cf. [3, 4]); further information may be obtained from the representations in the theorems that follow.

A measurable function α satisfies a *Dini condition* at z_0 if its essential modulus of continuity

$$\omega(r, z_0, \alpha) = \operatorname*{ess\,sup}_{|z - z_0| \leq r} |\alpha(z) - \alpha(z_0)|$$

has the property that

$$\int_0^{\varepsilon} \frac{\omega(r, z_0, \alpha)}{r}\, dr < \infty \tag{5.3}$$

for some $\varepsilon > 0$. In particular, (5.3) holds if α is Hölder continuous at z_0. By convention, α satisfies a Dini condition at ∞ if $\alpha(1/z)$ satisfies one at the origin.

Theorem 5.1. *Suppose that the functions* $\alpha : \bar{\mathbf{C}} \to (-\infty, \infty)$ *and* $\beta, \gamma : \bar{\mathbf{C}} \to [0, \infty)$ *are measurable, essentially bounded, and satisfy a Dini condition at* z_0 *and* ∞. *Assume either that* $\beta\gamma - \alpha^2 \leqq 1$ *a.e. and* $\operatorname{ess\,inf}(\beta\gamma - \alpha^2) > 0$ *or that* $\beta\gamma - \alpha^2 \geqq 1$ *a.e. Then there exists a q.c. mapping* $f : \bar{\mathbf{C}} \to \bar{\mathbf{C}}$, $f(\infty) = \infty$, *such that*

$$u(z) + iv(z) = -\log[f(z) - f(z_0)] \tag{5.4}$$

is a (*weak*) *fundamental solution of the system* (5.1).

The proofs of theorems in this section will be postponed to the end of the section.

M. A. Lavrent'ev [6], L. Bers [1], and others have obtained homeomorphic solutions $u + iv$ of (5.1) under ellipticity conditions. Our limitations on the coefficients are more restrictive, but we obtain a solution with the stronger property that e^{u+iv} is univalent. The univalence of solutions is an important property, for example, for the understanding of stream lines, equipotential lines, etc., for a variety of problems.

The next two theorems concern fundamental solutions of the first and second kind for domains D with boundary. Suppose that D admits a harmonic Green's function. If f is a q.c. mapping of D, then $\Omega = f(D)$ also admits a harmonic Green's function G_f. That is, as a function of w,

(a) $\Gamma_f(w, \omega) = G_f(w, \omega) + \log|w - \omega|$ is harmonic in Ω, and

(b) $G_f(w, \omega) \to 0$ as $w \to \partial\Omega$.

In addition, G_f is symmetric: $G_f(w, \omega) = G_f(\omega, w)$. We let γ_f be the limit

$$\gamma_f(w) = \Gamma_f(w, w) \tag{5.5}$$

and denote by $P_f(w;\omega)$ an (multivalued) analytic completion of $G_f(w, \omega)$ in $\Omega \setminus \{\omega\}$ as a function of w.

Theorem 5.2. *Let D be a domain that admits a harmonic Green's function. Suppose that the functions $\alpha : D \to (-\infty, \infty)$ and $\beta, \gamma : D \to [0, \infty)$ are measurable, essentially bounded, and satisfy a Dini condition at $z_0 \in D$. Assume either that $\beta\gamma - \alpha^2 \leqq 1$ a.e. and $\operatorname{ess\,inf}(\beta\gamma - \alpha^2) > 0$ or that $\beta\gamma - \alpha^2 \geqq 1$ a.e. Then there exists a q.c. mapping f of D such that*

$$u(z) + iv(z) = P_f(f(z); f(z_0)), \tag{5.6}$$

where P_f is an analytic completion of the harmonic Green's function G_f of $f(D)$, is a (weak) fundamental solution of the system (5.1) *in D, with singularity at z_0 and zero boundary values.*

Theorem 5.3. *Let D be a domain bounded by a finite number of nondegenerate continua. Suppose that the functions $\alpha : D \to (-\infty, \infty)$ and $\beta, \gamma : D \to [0, \infty)$ are measurable, essentially bounded, and satisfy a Dini condition at $z_1, z_2 \in D$. Assume either that $\beta\gamma - \alpha^2 \leqq 1$ a.e. and $\operatorname{ess\,inf}(\beta\gamma - \alpha^2) > 0$ or that $\beta\gamma - \alpha^2 \geqq 1$ a.e. Then there exists a q.c. mapping g of D onto the extended plane with radial slits, such that z_1 and z_2 correspond to 0 and ∞, respectively, and*

$$u + iv = \log g \tag{5.7}$$

is a (weak) fundamental solution of the system (5.1) *in D, with singularities at z_1 and z_2.*

The function u is a fundamental solution of the second kind in the sense that the function v is constant on each boundary component of D. If ∂D is smooth and there is sufficient differentiability, the fact that the tangential derivative $\partial v/\partial s = 0$ translates into the boundary condition $\partial u/\partial \nu = 0$ where ν is direction of the conormal.

The proofs of these theorems will depend on posing appropriate extremal problems over compact families of q.c. mappings that are related to Theorem 3.1.

This will be done in such a way that the relations (2.10) satisfied by an extremal function reduce to the differential system (5.1).

Some terminology will be necessary to normalize our families of q.c. mappings and to define functionals on them.

Let $\mathcal{A}(r; z_0, f)$ denote the area of $f(|z - z_0| < r)$. If Q is a bounded measurable function in a neighborhood of z_0, $Q \geqq 1$, and Q satisfies a Dini condition (5.3) at z_0, and if the dilatation quotient

$$D_f = \frac{|f_z| + |f_{\bar{z}}|}{|f_z| - |f_{\bar{z}}|}$$

satisfies $D_f \leqq Q$ a.e. in a neighborhood of z_0, then define

$$\phi(z_0, Q, f) = \lim_{r \to 0} \frac{\sqrt{\mathcal{A}(r; z_0, f)/\pi}}{r^{1/Q(z_0)}} \tag{5.8}$$

and

$$\psi(z_0, Q, f) = \phi(f(z_0), Q \circ f^{-1}, f^{-1})^{-Q(z_0)} \tag{5.9}$$

$$= \lim_{\rho \to 0} \frac{\rho}{[\sqrt{\mathcal{A}(\rho; f(z_0), f^{-1})/\pi}]^{Q(z_0)}}.$$

The existence of a (finite) limit in (5.8) and a (possibly infinite) limit in (5.9) was proved in [10, lemma 1]. In general, the limits

$$\lim_{z \to z_0} \frac{|f(z) - f(z_0)|}{|z - z_0|^{1/Q(z_0)}} \quad \text{and} \quad \lim_{z \to z_0} \frac{|f(z) - f(z_0)|}{|z - z_0|^{Q(z_0)}}$$

may not exist; however, if either does, then it agrees with the corresponding limit (5.8) or (5.9). Important properties of the functionals (5.8) and (5.9) are that they are upper semicontinuous and lower semicontinuous, respectively, as functions of f ([10, lemma 2]).

If $z_0 = \infty$, then define $\phi(\infty, Q, f) = 1/\phi(0, Q(1/z), 1/f(1/z))$ and $\psi(\infty, Q, f) = 1/\psi(0, Q(1/z), 1/f(1/z))$.

Proof of Theorem 5.1. Suppose $\delta \equiv \beta\gamma - \alpha^2 \leqq 1$ a.e. and $\delta_0 \equiv \operatorname{ess\,inf}(\beta\gamma - \alpha^2) > 0$. Define

$$c = \frac{\beta - \gamma - 2i\alpha}{\beta + \gamma + 1 + \delta} \quad \text{and} \quad k = \frac{1 - \delta}{\beta + \gamma + 1 + \delta}.$$

Then $k \geqq 0$ and

$$\| |c| + k \|_\infty = \left\| \frac{\sqrt{(\beta+\gamma)^2 - 4\delta} + 1 - \delta}{\beta + \gamma + 1 + \delta} \right\|_\infty \leqq \left\| \frac{\beta + \gamma + 1 - \delta}{\beta + \gamma + 1 + \delta} \right\| \leqq \frac{\|\beta + \gamma + 1\|_\infty - \delta_0}{\|\beta + \gamma + 1\|_\infty + \delta_0} < 1.$$

By Theorem 3.1 the family of all q.c. mappings $f : \bar{\mathbf{C}} \to \bar{\mathbf{C}}$, $f(\infty) = \infty$, whose complex dilatations $\mu = f_{\bar{z}} / f_z$ satisfy

$$|\mu - c| \leqq k \qquad \text{a.e.} \tag{5.10}$$

is closed. An additional constraint will be imposed below.

We shall make two separate affine changes of variables, one to specify the additional constraint and the other to pose the extremal problem. Let

$$z^\infty(\zeta) = \zeta + h(\infty)\bar{\zeta} \quad \text{and} \quad z^0(\zeta) = \zeta + h(z_0)\bar{\zeta}$$

where the constants $h(\infty)$ and $h(z_0)$ are defined by the formula (3.3) in terms of $c(\infty)$, $k(\infty)$ and $c(z_0)$, $k(z_0)$, respectively. Set $\zeta_0 = (z^0)^{-1}(z_0)$.

If h denotes either of the constants $h(\infty)$ or $h(z_0)$ and if f satisfies (5.10), then the complex dilatation M of $f(\zeta + h\bar{\zeta})$ satisfies

$$|M - C| \leqq K \qquad \text{a.e.} \tag{5.11}$$

where

$$C(\zeta) = \frac{[h + c(\zeta + h\bar{\zeta})][1 + \overline{hc(\zeta + h\bar{\zeta})}] - hk^2(\zeta + h\bar{\zeta})}{|1 + \bar{h}c(\zeta + h\bar{\zeta})|^2 - |h|^2 k^2(\zeta + h\bar{\zeta})}$$

and

$$K(\zeta) = \frac{(1 - |h|^2)k(\zeta + h\bar{\zeta})}{|1 + \bar{h}c(\zeta + h\bar{\zeta})|^2 - |h|^2 k^2(\zeta + h\bar{\zeta})} \geqq 0.$$

Of course, $\| |C| + K \|_\infty < 1$ remains true. The choice $h = h(\infty)$ has the property that $C(\infty) = 0$, and the choice $h = h(z_0)$ has the property that $C(\zeta_0) = 0$. That is, certain of the disks (5.11) are centered at the origin, and all are contained in

$$|M| \leqq |C| + K.$$

Finally, define

$$Q = \frac{1 + |C| + K}{1 - |C| - K}.$$

We shall append a superscript depending on the choice $h = h(\infty)$ or $h = h(z_0)$.

Since $D_{f\circ z^\infty} \leqq Q^\infty$ and $D_{f\circ z^0} \leqq Q^0$ a.e., the functionals $\phi(\infty, Q^\infty, f\circ z^\infty)$ and $\phi(\zeta_0, Q^0, f\circ z^0)$ are defined.

We are now in a position to further constrain the family of q.c. mappings f that satisfy (5.10). We require that

$$\phi(\infty, Q^\infty, f\circ z^\infty) = 1. \tag{5.12}$$

There is a question whether a q.c. mapping exists that satisfies both (5.10) and (5.12). To show that there is, let g be a q.c. homeomorphism of the plane, with 0 and ∞ fixed, that satisfies the Beltrami equation

$$g_{\bar\omega} = \tilde\mu g_\omega \qquad \text{a.e.}$$

with

$$\tilde\mu(\omega) = \frac{\omega}{\bar\omega}\,\frac{K^\infty(\infty) - K^\infty(1/\zeta(\omega)) + \frac{\omega}{\bar\omega}C^\infty(1/\zeta(\omega))}{1 + K^\infty(\infty)\left[\frac{\omega}{\bar\omega}C^\infty(1/\zeta(\omega)) - K^\infty(1/\zeta(\omega))\right]},$$

and $\zeta(\omega) = |\omega|^{Q^\infty(\infty)}\cdot \omega/|\omega|$. Then $\tilde\mu(0) = 0$ and $\tilde\mu$ satisfies a Dini condition at $\omega = 0$. By theorem V.7.1 of [8] the mapping g is regular and conformal at $\omega = 0$. Therefore

$$0 \neq \lim_{\omega\to 0}\frac{|g(\omega)|}{|\omega|} = \lim_{\zeta\to 0}\frac{|G(\zeta)|}{|\zeta|^{1/Q^\infty(\infty)}}$$

where $g(\omega) = G(\zeta(\omega))$. The complex dilatation of $\gamma(\zeta) = 1/G(1/\zeta)$ is

$$C^\infty(\zeta) - \frac{\zeta}{\bar\zeta}K^\infty(\zeta),$$

which satisfies (5.11). Consequently, $f = \gamma\circ(z^\infty)^{-1}$ satisfies (5.10) and $\phi(\infty, Q^\infty, f\circ z^\infty) \neq 0$. Thus a constant times f satisfies both (5.10) and (5.12). By an analogous modification of the complex dilatation in a neighborhood of z_0, we may also assume that $\phi(\zeta_0, Q^0, f\circ z^0) \neq 0$.

Now consider the problem of finding the maximum of

$$\lambda(f) = \log\phi(\zeta_0, Q^0, f\circ z^0) \tag{5.13}$$

over the family of q.c. mappings f that satisfy (5.10) and (5.12). Since the functional (5.13) is upper semicontinuous and the normalizing functional (5.12) is lower semicontinuous, it follows as in [10] that an extremal function f exists that satisfies (5.10) and (5.12) and provides the maximum of (5.13).

We are now prepared to invoke Theorem 2.1. The variations (2.6) preserve our normalization, so that the variational derivative

$$\Psi(\zeta, w) = \frac{1}{w - \zeta}.$$

Although the functional (5.13) is not Gâteaux differentiable in the sense requested for Theorem 2.1, it is sufficiently well behaved. Under a variation of the form (2.6) the area

$$\mathcal{A}(r; z_0, f^* \circ z^0) = \mathcal{A}(r; z_0, f \circ z^0) - 2\varepsilon \operatorname{Re} \frac{1}{\pi} \iint_{\Omega} \iint_{E} \frac{a(\zeta)}{(\zeta - w)^2} \, dm(\zeta) dm(w) + O(\varepsilon^2)$$

where $\Omega = f \circ z^0 \, (|\zeta - \zeta_0| < r)$ and $E \subset \mathbf{C} \setminus \Omega$. It follows that

$$\lambda(f^*) = \lambda(f) - \varepsilon \operatorname{Re} \frac{1}{\pi} \iint_{E} \frac{a(\zeta) dm(\zeta)}{[\zeta - f(z_0)]^2} + O(\varepsilon^2)$$

under variations of the form (2.6). This expansion is all that is necessary for the proof of Theorem 2.1. Corresponding to formula (2.8), we let

$$A(\zeta) = -1/[\zeta - f(z_0)]^2$$

and

$$J(w) = \int^{w} \sqrt{A(\zeta)} \, d\zeta = i \log[w - f(z_0)].$$

The constraint (5.10) may be described by the function

$$F(t, z) = |t - c(z)|^2 - k(z)^2.$$

Therefore, by Corollary 2.2, the extremal function f satisfies

$$\left| \frac{(J \circ f)_{\bar{z}}}{(J \circ f)_z} - c \right| = k \quad \text{and} \quad \left(\frac{(J \circ f)_{\bar{z}}}{(J \circ f)_z} - c \right) (J \circ f)_z^2 \geqq 0 \qquad \text{a.e.}$$

If $\mathcal{J} = iJ \circ f = -\log[f - f(z_0)]$, then

$$\left(\frac{\mathcal{J}_{\bar z}}{\mathcal{J}_z} - c\right) \mathcal{J}_z^2 = -\left|\frac{\mathcal{J}_{\bar z}}{\mathcal{J}_z} - c\right| |\mathcal{J}_z|^2 = -k|\mathcal{J}_z|^2$$

or

$$\mathcal{J}_{\bar z} - c\mathcal{J}_z + k\overline{\mathcal{J}_z} = 0. \tag{5.14}$$

Therefore $\mathcal{J} = -\log[f - f(z_0)] = u + iv$ satisfies

$$(1+k)u_{\bar z} - cu_z = -i(1-k)v_{\bar z} + icv_z.$$

Elimination of $u_{\bar z}$ leads to

$$u_z = \frac{i(|c|^2 + 1 - k^2)v_z - 2i\bar{c}v_{\bar z}}{(1+k)^2 - |c|^2}$$

$$= \frac{i}{2}(\beta + \gamma)v_z + \left[\alpha - \frac{i}{2}(\beta - \gamma)\right] v_{\bar z},$$

which is equivalent to the system (5.1). Thus the variational method has led to a q.c. mapping with the desired properties.

If $\delta = \beta\gamma - \alpha^2 \geqq 1$ a.e., then a parallel development leads to the same conclusion. One defines c as before and replaces k by its negative. Then $k \geqq 0$ and

$$\| |c| + k \|_\infty \leqq \frac{\|\beta + \gamma + \delta\|_\infty - 1}{\|\beta + \gamma + \delta\|_\infty + 1} < 1.$$

In the constraint (5.12) one replaces ϕ by ψ and then asks for the maximum of the functional

$$\lambda(f) = -\log \psi(\zeta_0, Q^0, f \circ z^0).$$

An extremal function f then satisfies the equation

$$\mathcal{J}_{\bar z} - c\mathcal{J}_z - k\overline{\mathcal{J}_z} = 0$$

with $\mathcal{J} = -\log[f - f(z_0)] = u + iv$, which again is equivalent to the system (5.1).

Proof of Theorem 5.2. The proof is similar to the proof of Theorem 5.1. In the first case one uses the family of q.c. mappings of D that satisfy (5.10) and

$$\phi(\zeta_0, Q^0, f \circ z^0) = 1. \tag{5.15}$$

Then one poses the problem of finding the maximum of

$$\lambda(f) = -\gamma_f(f(z_0))$$

where γ_f is defined by (5.5). Under a variation of the form (2.7) that preserves ∞ and the normalization at z_0, the functional

$$\lambda(f^*) = \lambda(f) - \operatorname{Re} \frac{\varepsilon}{\pi} \iint_{f(D)} a(\zeta)[P'(\zeta; f(z_0))]^2 dm(\zeta) + o(\varepsilon)$$

as in [10]. Then Corollary 2.2 implies that an extremal function satisfies the equation (5.14) with

$$\mathscr{I}(z) = P_f(f(z); f(z_0)) = u(z) + iv(z),$$

and consequently, (5.6) satisfies (5.1).

In the second case one replaces ϕ by ψ in the normalization (5.15) and finds the maximum of

$$\lambda(f) = \gamma_f(f(z_0))$$

in a similar fashion.

Proof of Theorem 5.3. One uses the family of q.c. mappings of D that satisfy (5.10) and asks for the extremes of the same functionals as in theorem 6 of [10]. The details are omitted since they would be repetitious.

Remarks. One important aspect of Theorem 5.3 is that it asserts the existence of a canonical slit mapping. That is, with the indicated hypotheses on α, β, γ a solution $u + iv$ of the system (5.1) exists such that $g = e^{u+iv}$ maps the finitely connected domain D onto the extended plane with radial slits, such that two pre-assigned points correspond to 0 and ∞.

By altering the functionals in the proof of Theorem 5.3 as in [10, theorems 8 and 9] one can obtain other canonical mappings. For example, (with the same hypotheses on α, β, γ) there exists a solution $u + iv$ of the system (5.1) such that $g = e^{u+iv}$ maps the finitely connected domain D onto the unit disk with either concentric circular slits or radial slits. In particular, a simply connected domain (with at least two boundary points) can be mapped onto the unit disk in such a fashion.

Thus, just as in conformal mapping theory, appropriate extremal problems for families of q.c. mappings lead to interesting canonical mappings.

6. Fundamental solutions for nonlinear equations

The program of Section 5 can be carried out to provide existence and a representation in terms of q.c. mappings for fundamental solutions of many more differential equations. In fact, each admissible function $F(t, z)$ used to define a family of q.c. mappings by means of (2.2) gives rise to a differential equation. In this section we simply indicate some nonlinear equations that arise in this way.

In the case of fundamental solutions for the entire plane or for the first or second boundary value problems in a plane domain, let $u + iv$ be defined as in (5.4), (5.6), or (5.7), respectively. If $F(t, z)$ satisfies the hypotheses of Theorem 2.1, then by Corollary 2.2 an extremal function f satisfies

$$(6.1)\qquad F\left(\frac{(u+iv)_{\bar z}}{(u+iv)_z}, z\right) = 0 \quad\text{and}\quad \pm\overline{F_t\left(\frac{(u+iv)_{\bar z}}{(u+iv)_z}, z\right)}\,[(u+iv)_z]^2 \geqq 0.$$

The $\pm$ sign is determined by which extremal problem is posed. It is sometimes convenient to write the latter equation as

$$\pm\overline{F_t\left(\frac{u_{\bar z}+iv_{\bar z}}{u_z+iv_z}, z\right)}\,(u_z+iv_z) = \left|F_t\left(\frac{u_{\bar z}+iv_{\bar z}}{u_z+iv_z}, z\right)\right|\,(u_{\bar z}-iv_{\bar z}).$$

For the purpose of illustration, let

$$F(\sigma+i\tau, z) = \frac{\sigma^2}{a^2} + \frac{\tau^2}{b^2} - 1$$

where a and b are nonnegative measurable functions that are bounded away from 0 and ∞. Assume that there is a constant $r < 1$ such that $2a^2/b - b \leqq r$ at points where $b \leqq a$ and $2b^2/a - a \leqq r$ at points where $a \leqq b$. Then the family of q.c. mappings whose complex dilatations μ satisfy $F(\mu(z), z) \leqq 0$ a.e. is closed by Theorem 3.2 and the preceding example.

For this "ellipse family" the system of differential equations (6.1) becomes

$$\frac{\sigma^2}{a^2} + \frac{\tau^2}{b^2} = 1 \quad\text{and}\quad \pm\left(\frac{\sigma}{a^2} + i\frac{\tau}{b^2}\right)(u_z + iv_z) = \left|\frac{\sigma}{a^2} + i\frac{\tau}{b^2}\right|(u_{\bar z} - iv_{\bar z})$$

where

$$\sigma + i\tau = \frac{(u+iv)_{\bar z}}{(u+iv)_z} = \frac{u_{\bar z}^2 + v_{\bar z}^2}{|u_z + iv_z|^2}.$$

That is

$$4a^2(u_xu_y + v_xv_y)^2 + b^2(u_x^2 + v_x^2 - u_y^2 - v_y^2)^2 = a^2b^2[(u_x + v_y)^2 + (v_x - u_y)^2]^2$$

and

$$\pm[2ia^2(u_xu_y+v_xv_y)+b^2(u_x^2+v_x^2-u_y^2-v_y^2)][(u_x+v_y)+i(v_x-u_y)]^2 \geqq 0.$$

To clarify the meaning of these two equations, let us write

$$v_y=\rho u_x, \qquad v_x=-\sigma u_y$$

and determine factors ρ and σ as functions of u_x and u_y. The first equation becomes

$$(6.2)\, 4a^2(1-\rho\sigma)^2u_x^2u_y^2+b^2[(1-\rho^2)u_x^2-(1-\sigma^2)u_y^2]^2=a^2b^2[(1+\rho)^2u_x^2+(1+\sigma)^2u_y^2]^2.$$

Regarding the second equation, let us express the fact that the imaginary part on the left is zero:

$$a^2(1-\rho\sigma)[(1+\rho)^2u_x^2-(1+\sigma)^2u_y^2]=b^2(1+\rho)(1+\sigma)[(1-\rho^2)u_x^2-(1-\sigma^2)u_y^2].$$

This leads to the simple relation

$$(6.3) \qquad \begin{aligned} &u_x^2(1+\rho)^2[(a^2-b^2)(1-\rho\sigma)+b^2(\rho-\sigma)] \\ &\quad = u_y^2(1+\sigma)^2[(a^2-b^2)(1-\rho\sigma)-b^2(\rho-\sigma)]. \end{aligned}$$

Observe that equation (6.2) depends only on the ratio u_x^2/u_y^2. Hence, if we insert into it the value from (6.3), we find after some rearrangement

$$(6.4) \qquad \frac{(1-\rho\sigma)^2}{b^2}+\frac{(\rho-\sigma)^2}{a^2-b^2}=(1+\rho)^2(1+\sigma)^2.$$

Let us define

$$P=\frac{1-\rho}{1+\rho}, \qquad \Sigma=\frac{1-\sigma}{1+\sigma}.$$

Then we arrive at the two equations

$$(6.5) \qquad \frac{(\Sigma+P)^2}{4b^2}+\frac{(\Sigma-P)^2}{4(a^2-b^2)}=1$$

and

$$(6.6) \qquad \frac{u_x^2}{u_y^2}=\left(\frac{1+P}{1+\Sigma}\right)^2\frac{a^2P+(a^2-2b^2)\Sigma}{(a^2-2b^2)P+a^2\Sigma}.$$

These are two algebraic equations to express Σ, P in terms of the ratio $(u_x/u_y)^2$. Observe that an interchange of ρ and σ interchanges in the defining equations u_x and u_y. Thus, the equations arrived at finally are of the form

$$\left(\rho\left(\frac{u_x}{u_y}\right)u_x\right)_x+\left(\rho\left(\frac{u_y}{u_x}\right)u_y\right)_y=0, \tag{6.7}$$

where $\rho(u_x/u_y)$ can be determined from (6.5) and (6.6). The coefficients a and b in these equations can be quite general functions of z.

To get an idea of the equations for which we have derived the general existence theorem, let us assume

$$a^2=b^2+\varepsilon b^2 \tag{6.8}$$

and that ε is small, so that we can neglect higher orders than the first. An easy calculation leads to the asymptotic result

$$\rho=\frac{1-b}{1+b}-\varepsilon\frac{1}{1-b^2}\frac{(u_y^2+3u_x^2)(u_y^2-u_x^2)}{(u_x^2+u_y^2)^2}+O(\varepsilon^2).$$

The quantities b and ε may vary according to quite arbitrary assignment. The case $\varepsilon=0$ reduces, of course, to the well-known linear differential equation arising from the condition $|\mu(z)|\leqq b(z)$ (cf. [10]).

The nonlinear differential equation in the special case of our problem does not seem to be of particular theoretical interest. However, we see the general method which might be useful in similar and more significant questions.

REFERENCES

1. L. Bers, *Univalent solutions of linear elliptic systems*, Comm. Pure Appl. Math. **6** (1953), 513–526.

2. C. Carathéodory, *Vorlesungen über Reelle Funktionen*, Chelsea Publ. Co., New York, 1948.

3. R. Courant and D. Hilbert, *Methods of Mathematical Physics*, Vol. II, Interscience Publ., New York, 1962.

4. P. R. Garabedian, *Partial Differential Equations*, John Wiley and Sons, New York, 1964.

5. R. Kühnau, *Wertannahmeprobleme bei quasikonformen Abbildungen mit ortsabhängiger Dilatationsbeschränkung*, Math. Nachr. **40** (1969), 1–11.

6. M. A. Lavrent'ev, *The general problem of the theory of quasiconformal mappings of plane regions*, Amer. Math. Soc. Transl. No. 46 (1951).

7. O. Lehto, *Schlicht functions with a quasiconformal extension*, Ann. Acad. Sci. Fenn. Ser. A I 500 (1971).

8. O. Lehto and K. I. Virtanen, *Quasiconformal Mappings in the Plane*, Grundlehren Math. Wiss. **126**, Springer-Verlag, New York–Heidelberg–Berlin, 1973.

9. M. Schiffer, *A variational method for univalent quasiconformal mappings*, Duke Math. J. **33** (1966), 395–412.

10. M. Schiffer and G. Schober, *Representation of fundamental solutions for generalized Cauchy–Riemann equations by quasiconformal mappings*, Ann. Acad. Sci. Fenn. Ser. A I, **2** (1976), 501–531.

11. G. Schober, *Univalent Functions — Selected Topics*, Springer-Verlag Lecture Notes in Math. **478**, 1975.

12. K. Strebel, *Ein Konvergensatz für Folgen quasikonformer Abbildungen*, Comment. Math. Helv. **44** (1969), 469–475.

STANFORD UNIVERSITY
STANFORD, CALIFORNIA 94305 USA

AND

INDIANA UNIVERSITY
BLOOMINGTON, INDIANA 47401 USA

(Received September 20, 1978)

Commentary on

[103] M. Schiffer and G. Schober, *An extremal problem for the Fredholm eigenvalues*, Arch. Rational Mech. Anal. **44** (1971/72), 83–92.

[106] M. Schiffer and G. Schober, *A remark on the paper "An extremal problem for the Fredholm eigenvalues"*, Arch. Rational Mech. Anal. **46** (1972), 394.

[117] M. Schiffer and G. Schober, *A variational method for general families of quasiconformal mappings*, J. Analyse Math. **34** (1978), 240–264.

The variational method for solving extremal problems in classes of quasiconformal mappings was introduced by P.P. Belinskiĭ in the 1950s (cf. [B]) and was extended and used intensively in papers of S.L. Krushkal and others from 1964 onward; see the extensive surveys [K1, K2].

Schiffer's work in this area began in 1966 with [90], in which he derived a new variational formula, based on a somewhat simpler underlying idea. After a hiatus of several years, there followed in relatively rapid succession the eight papers [103, 106, 111, 112, 114, 117, 120, 122], all written in collaboration with Glenn Schober. It was noticed by H. Renelt [R], who gave an alternative derivation of the variational method developed in [90], that the specific variations used in [90, 103] lacked a certain uniformity property used in the procedure; this gap turned out to be easily remedied [106]. Later, A. Pfluger derived another variational procedure for solving extremal problems for quasiconformal mappings; but, unfortunately, this was never published.

Extremal problems for quasiconformal mappings have been usually studied in mapping classes with the complex dilatation $\mu(z)$ bounded by a given fixed function, especially $|\mu(z)| \leq k$ for a fixed constant $k < 1$. In [117, 120], Schiffer and Schober solved extremal problems with more general dilatation restrictions of the form

$$F(\mu(z), z) \leq 0$$

with a given function F. This also yields interesting new existence proofs and insights on fundamental solutions to systems of elliptic differential equations. Of special interest because of the simplicity of the results are restrictions of the form

$$|\mu(z) - c| \leq k.$$

Comments and another approach to this case are in [Kü].

References

[B] P. P. Belinskiĭ, *General Properties of Quasiconformal Mappings*, Nauka Sibirsk. Otdel., 1974 (Russian).

[K1] S. L. Krushkal, *Variational principles in the theory of quasiconformal maps*, Handbook of Complex Analysis: Geometric Function Theory, Vol. 2, ed. R. Kühnau, Elsevier, 2005, pp. 31–98.

[K2] S. L. Krushkal, *Univalent holomorphic functions with quasiconformal extensions (variational approach)*, Handbook of Complex Analysis: Geometric Function Theory, Vol. 2, ed. R. Kühnau, Elsevier, 2005, pp. 165–241.

[Kü] Reiner Kühnau, *Zur Methode der Randintegration bei quasikonformen Abbildungen, II*, Ann. Polon. Math. **43** (1983), 105–110.

[R] Heinrich Renelt, *Modifizierung und Erweiterung einer Schifferschen Variationsmethode für quasikonforme Abbildungen*, Math. Nachr. **55** (1973), 353–379.

REINER KÜHNAU

[109] (with J. Hersch and L. E. Payne) Some inequalities for Stekloff eigenvalues

---

[109] (with J. Hersch and L. E. Payne) Some inequalities for Stekloff eigenvalues. *Arch. Rational Mech. Anal.* **57** (1975), 99–114.

Some Inequalities for Stekloff Eigenvalues

J. HERSCH, L. E. PAYNE & M. M. SCHIFFER

1. Introduction

1.1. In 1902 STEKLOFF introduced an interesting eigenvalue problem for harmonic functions [15]. Since then many authors have discussed this problem and have derived various isoperimetric inequalities [1, 2, 6, 7, 8, 9, 10, 13, 14, 17, 19]. The aim of the present paper is to sharpen some results of WEINSTOCK [19] and of HERSCH & PAYNE [6] and to obtain further inequalities for higher eigenvalues of various generalized Stekloff problems in simply and multiply connected plane domains.

1.2. Let G be a plane domain and let ∂G denote its boundary (of finite length L). Along ∂G let a non-negative function $\rho(s)$ $(0 \leqq s < L)$ be given. For convenience $\rho(s)$ will be called the "mass density" of ∂G and $\oint_{\partial G} \rho(s)\, ds = M$ the "total mass" of ∂G. The domain G may contain the point at infinity.

The Stekloff eigenvalue problem is defined by the conditions

$$\Delta \varphi = 0 \qquad \text{in } G$$

$$\partial \varphi / \partial n = \mu \rho(s) \varphi \qquad \text{on } \partial G,$$

where $\partial/\partial n$ is the outward directed normal derivative on ∂G.

Let $\mu_1 < \mu_2 \leqq \mu_3 \leqq \dots$ be the eigenvalues of this problem and let $\varphi_1, \varphi_2, \varphi_3, \dots$ be the corresponding eigenfunctions. The eigenvalues are the stationary values of the Rayleigh quotient

$$R[u] = D(u) / \oint_{\partial G} \rho u^2\, ds$$

where $D(u) \equiv \iint_G |\text{grad}\, u|^2\, dA$ is the Dirichlet integral; thus $\mu_n = R[\varphi_n] \geqq 0$ for $n \geqq 1$, while $\mu_1 = 0$ and $\varphi_1 = \text{const}$.

2. The Stekloff Problem in a Simply Connected Domain. A Sharpened Form of Weinstock's Inequality

2.1. For a simply connected domain J, WEINSTOCK [19] proved that $\mu_2 M \leqq 2\pi$ (the equality holding for a circle when $\rho = \text{const}$). His proof was inspired by that of G. SZEGÖ ([16]; see also H. F. WEINBERGER [18]) for the corresponding inequality in the free membrane problem. The proofs of those authors made use of a topological fixed point theorem. More recently, PAYNE & HERSCH [6] sharpened WEINSTOCK's inequality to the form $(\mu_2^{-1}+\mu_3^{-1})M^{-1}\geqq\pi^{-1}$ (again with the equality holding for a circle when $\rho=\text{const}$).

2.2. We shall here prove the still sharper inequality

$$\mu_2\mu_3 M^2 \leqq 4\pi^2. \tag{1}$$

Let $m(s)=\int_{t=0}^{s}\rho(t)\,dt$; then $\rho\,ds=dm$. For any function $f(m)$ such that $f(0)=f(M)$ and $\int_{m=0}^{M} f'^2(m)\,dm<\infty$, we define $u(s)=f(m(s))$; then $du/ds=f'(m(s))\rho(s)$. Further let $u(x,y)$ be the harmonic function in J with the boundary values $u(s)$, and let $\tilde{u}(x,y)$ be the conjugate harmonic function, satisfying $\oint\tilde{u}\,dm=0$ (this fixes the choice of the additive constant for $\tilde{u}$). Then

$$D(u)=D(\tilde{u})=\oint\tilde{u}\frac{\partial\tilde{u}}{\partial n}ds=-\oint\tilde{u}\frac{\partial u}{\partial s}ds=-\oint\tilde{u}f'(m)\,dm;$$

hence by Schwarz's inequality $D(u)D(\tilde{u})\leqq\oint\tilde{u}^2dm\cdot\oint f'^2(m)\,dm$, and

$$R[u]\,R[\tilde{u}]\leqq\oint f'^2(m)\,dm/\oint f^2(m)\,dm=R_{\partial J}[f]. \tag{2}$$

The upper bound $R_{\partial J}[f]$ is the Rayleigh quotient of a (periodic) homogeneous string of length M.

In order to insure that $R[u]\geqq\mu_2$ and $R[\tilde{u}]\geqq\mu_2$, we require both of the functions $u(s)$ and $\tilde{u}(s)$ to be orthogonal in the ρ-metric to the first eigenfunction $\varphi_1=\text{const}$ of the Stekloff problem. Now $\tilde{u}$ already satisfies $\oint\tilde{u}\rho\,ds=\oint\tilde{u}\,dm=0$; we must therefore request additionally only that $\oint f(m)\,dm=\oint u\rho\,ds=0$. Under this restriction, the best possible bound in (2) is the second eigenvalue of the "periodic" string, namely $(2\pi/M)^2$ (this and all higher eigenvalues have multiplicity two). The latter bound is achieved for $f(m)=c_1\sin(2\pi m/M)+c_2\cos(2\pi m/M)$ with $c_1^2+c_2^2\neq0$. Clearly it is always possible to choose c_1 and c_2 in such a way that the further condition

$$(u,\varphi_2)=\oint u\varphi_2\rho\,ds=\oint f(m)\,\varphi_2\,dm=0$$

is fulfilled; this then implies that $R[u]\geqq\mu_3$ and $R[\tilde{u}]\geqq\mu_2$ and thus that

$$\mu_2\mu_3\leqq R[u]\,R[\tilde{u}]\leqq R_{\partial J}[f]=(2\pi/M)^2,$$

which proves (1).

2.3. Remarks. (a) Inequality (1) is *sharp*, since equality holds for a circle of radius R when $\rho=\text{constant}=\rho_0$. If the center of the circle is chosen as the origin then $M=2\pi R\rho_0$, $\varphi_2=x$, $\varphi_3=y$ and $\mu_2=\mu_3=1/R\rho_0=2\pi/M$. Moreover we even have equality for any simply connected domain J provided we define the "mass density" along ∂J by means of a conformal mapping of J onto a circle.

(b) Both inequalities mentioned in 2.1 follow immediately from (1), since

$$2\mu_2^{-1}M^{-1}\geqq(\mu_2^{-1}+\mu_3^{-1})M^{-1}\geqq2\sqrt{\mu_2^{-1}\mu_3^{-1}}\,M^{-1}\geqq\pi^{-1}.$$

(c) The proof given in 2.2 does not use the fixed point theorem as in the papers cited earlier.

2.4. An Example. C. Bandle ([1], pp. 633–4) has recently considered the Stekloff problem for a rectangle having two sides of length π, where $\rho=1$, and two sides of length 1.1996, where $\rho=0$ (she chose this length so that $\mu_3=\mu_4$). Then $M=2\pi$ and (1) gives $\mu_2\mu_3 \leqq (2\pi/M)^2=1$. The actual values are $\mu_2 \simeq 0.5369$ and $\mu_3 \simeq 1.6672$, so that $\mu_2\mu_3 \simeq 0.8951$.

Remark. If $\rho=0$ on some boundary arc α, then f and u are constant along α. This in turn gives $\partial\tilde{u}/\partial n=0$ along α.

3. Higher Eigenvalues in Simply Connected Domains

3.1. The homogeneous string of length M with periodicity conditions

$$f''(m)+\nu f(m)=0, \qquad f(0)=f(M), \qquad f'(0)=f'(M)$$

has the following eigenvalues and eigenfunctions:

$$\nu_1=0, \qquad f_1=\text{const}$$

$$\nu_{2n}=\nu_{2n+1}=(2\pi n/M)^2$$

and

$$f_{2n}(m)=\sin(2\pi nm/M), \qquad f_{2n+1}(m)=\cos(2\pi nm/M),$$

where $n=1, 2, 3, \ldots$.

To each of the functions $f_k(m)$ $(k=2, 3, 4, \ldots)$ corresponds a harmonic function $u_k(x, y)$ in J with boundary values $u_k(s)=f_k(m(s))$. Let $\tilde{u}_k(x, y)$ be the conjugate harmonic function satisfying $\oint \tilde{u}_k\, dm=0$. Then u_k and $\tilde{u}_k$ are orthogonal (in the ρ-metric) to the first eigenfunction $\varphi_1=\text{const}$ of the Stekloff problem. The functions u_k moreover are orthogonal to each other in the ρ-metric and therefore are linearly independent. Consequently the functions $\tilde{u}_k$ are also linearly independent. (On the other hand, the functions $\tilde{u}_k$ may depend linearly on u_j: for the circle with $\rho=\text{const}$, we have $\tilde{u}_2=-u_3$ and $\tilde{u}_3=u_2$.)

Let p and q be any two integers greater than 1. We put

$$N=p+q-2, \qquad f(m)=\sum_2^N c_k f_k(m)$$

and

$$u=\sum_2^N c_k u_k, \qquad \tilde{u}=\sum_2^N c_k \tilde{u}_k.$$

There exist $N-1$ constants, $c_2, c_3, \ldots, c_N$ (not all vanishing) satisfying the following $N-2$ linear homogeneous conditions of orthogonality in the ρ-metric;

$$u \perp \varphi_2, \varphi_3, \ldots, \varphi_{p-1}$$

and

$$\tilde{u} \perp \varphi_2, \varphi_3, \ldots, \varphi_{q-1}.$$

For these constants we have $\mu_p \leqq R[u]$ and $\mu_q \leqq R[\tilde{u}]$. By inequality (2) we obtain

$$\mu_p\mu_q \leqq R[u]\,R[\tilde{u}] \leqq R_{\partial J}[f] \leqq R_{\partial J}[f_N]=\nu_N=\begin{cases}[(N-1)\pi/M]^2 & \text{if } N \text{ is odd,}\\ (N\pi/M)^2 & \text{if } N \text{ is even.}\end{cases}$$

Thus

$$\mu_p \mu_q M^2 \leqq (p+q-3)^2 \pi^2 \quad \text{if } p+q \text{ is odd} \tag{3}$$

$$\mu_p \mu_q M^2 \leqq (p+q-2)^2 \pi^2 \quad \text{if } p+q \text{ is even}. \tag{3'}$$

We note that (3) implies (3'). Indeed, if $p+q$ is even, then $(p+1)+q$ is odd and $\mu_p \mu_q M^2 \leqq \mu_{p+1} \mu_q M^2 \leqq (p+q-2)^2 \pi^2$.

The specialization of (3') to the case $p=q$ gives

$$\mu_q M \leqq 2(q-1)\pi. \tag{3''}$$

For $q=2$ this is WEINSTOCK's inequality.

3.2. Remarks. (a) In the particular case $q=2$, $p=3$, inequality (3) reduces to the sharp inequality (1); in the general case, however, (3) is not expected to be sharp. Indeed for a circle of radius R with $\rho=\text{const}=\rho_0$, we have $M=2\pi R\rho_0$, $\mu_{2n}=\mu_{2n+1}=n/R\rho_0$; thus, for $n \geqq 1$ and $k \geqq 1$

$$\mu_{2n}\mu_{2k+1} M^2 = 4nk\pi^2 \leqq [2n+(2k+1)-3]^2 \pi^2 = 4(n+k-1)^2 \pi^2,$$

since $(n+k-1)^2 - nk = (n-1)^2 + (k-1)^2 + nk - 1 \geqq 0$; equality holds only if $n=k=1$.

(b) In the case $p=q+1$, inequality (3) gives $\mu_q \mu_{q+1} M^2 \leqq 4(q-1)^2 \pi^2$, from which we obtain

$$\frac{1}{\mu_q} + \frac{1}{\mu_{q+1}} \geqq \frac{2}{\sqrt{\mu_q \mu_{q+1}}} \geqq \frac{M}{(q-1)\pi} \qquad (q \geqq 2).$$

This inequality is interesting in itself, though for domains which are "symmetric of order q" with respect to a center it is weaker than the sharp estimate

$$\mu_q^{-1} + \mu_{q+1}^{-1} \geqq 2M/q\pi$$

obtained by BANDLE ([1], p. 631).

3.3. An Example. BANDLE ([1]; p. 634) has recently considered a square of side length $2a$ with $\rho=1$; then $M=8a$ and $\mu_2=\mu_3 \simeq 0.6882/a$, $\mu_4=1/a$, and $\mu_5 \simeq 2.3044/a$. Inequality (3) gives

$$\mu_2 \mu_3 \leqq \pi^2/16a^2 \simeq 0.61685/a^2, \quad \text{whereas} \quad \mu_2\mu_3 \simeq 0.4736/a^2,$$

$$\mu_3 \mu_4 \leqq \pi^2/4a^2 \simeq 2.46740/a^2, \quad \text{whereas} \quad \mu_3\mu_4 \simeq 0.6882/a^2,$$

$$\mu_2 \mu_5 \leqq \pi^2/4a^2 \simeq 2.46740/a^2, \quad \text{whereas} \quad \mu_2\mu_5 \simeq 1.5859/a^2,$$

$$\mu_4 \mu_5 \leqq 9\pi^2/16a^2 \simeq 5.55165/a^2, \quad \text{whereas} \quad \mu_4\mu_5 \simeq 2.3044/a^2.$$

4. Stekloff Problems in Doubly Connected Domains

4.1. We consider a doubly connected domain D with boundary curves γ_a and γ_b. Their arc lengths (positively oriented with respect to D) will be designated as s_a and s_b. The "mass density" is given by $\rho_a(s_a)$ $(0 \leqq s_a < L_a)$ on γ_a and $\rho_b(s_b)$

$(0 \leqq s_b < L_b)$ on γ_b. Let

$$m_a(s_a) = \int_{t=0}^{s_a} \rho_a(t)\, dt,$$

$$m_b(s_b) = \int_{t=0}^{s_b} \rho_b(t)\, dt,$$

$$m_a(L_a) = M_a, \qquad m_b(L_b) = M_b.$$

We consider functions $f^{(a)}(m_a)$ $(0 \leqq m_a \leqq M_a)$ with $f^{(a)}(0) = f^{(a)}(M_a)$ and $f^{(b)}(m_b)$ $(0 \leqq m_b \leqq M_b)$ with $f^{(b)}(0) = f^{(b)}(M_b)$. The symbol f will denote the system of two functions $f^{(a)}(m_a(s_a))$ on γ_a and $f^{(b)}(m_b(s_b))$ on γ_b. We write $\gamma_a \cup \gamma_b = \partial D = \Gamma$.

Let $u(x, y)$ be the harmonic function in D with the boundary values f. The requirement that the conjugate harmonic function $\tilde{u}(x, y)$ be single valued in D, that is that $\oint_{\gamma_a} \partial u/\partial n\, ds_a = 0$, yields one linear homogeneous restriction on f. We choose the additive constant in $\tilde{u}$ so that $\oint_\Gamma \tilde{u}\, dm = 0$. Then

$$D(u)\, D(\tilde{u}) \leq \oint_\Gamma \tilde{u}^2\, dm \oint_\Gamma [f'(m)]^2\, dm$$

as in 2.2 and

$$R[u]\, R[\tilde{u}] \leqq \frac{\int_0^{M_a} [f^{(a)\prime}(m_a)]^2\, dm_a + \int_0^{M_b} [f^{(b)\prime}(m_b)]^2\, dm_b}{\int_0^{M_a} f^{(a)2}\, dm_a + \int_0^{M_b} f^{(b)2}\, dm_b} = R_\Gamma[f]. \tag{2'}$$

The upper bound $R_\Gamma[f]$ is the Rayleigh quotient of a system of *two* (independent) homogeneous strings of lengths M_a and M_b. The corresponding eigenvalues (taken separately) are

$$\nu_1^{(a)} = \nu_1^{(b)} = 0; \quad \nu_{2n}^{(a)} = \nu_{2n+1}^{(a)} = (2\pi n/M_a)^2; \quad \nu_{2n}^{(b)} = \nu_{2n+1}^{(b)} = (2\pi n/M_b)^2, \tag{4}$$

$n = 1, 2, 3, \ldots$; let $f_1^{(a)}, f_2^{(a)}, f_3^{(a)}, \ldots, f_1^{(b)}, f_2^{(b)}, f_3^{(b)}, \ldots$ denote the corresponding eigenfunctions.

The two strings system has the eigenvalues $0 = \nu_1 = \nu_2 < \nu_3 = \nu_4 \leqq \nu_5 = \nu_6 \leqq \ldots$ and the corresponding eigenfunctions

$$f_1 = \begin{cases} f_1^{(a)} = \text{const} & \text{on } \gamma_a \\ 0 & \text{on } \gamma_b \end{cases}, \qquad f_2 = \begin{cases} 0 & \text{on } \gamma_a \\ f_1^{(b)} = \text{const} & \text{on } \gamma_b, \end{cases} \quad \text{etc.,}$$

each vanishing either along γ_a or along γ_b. The individual eigenvalues ν_k are precisely the eigenvalues of the series $\nu_i^{(a)}$ and $\nu_i^{(b)}$ arranged in increasing order. If $\nu_k = \nu_i^{(a)}$, for example, then

$$f_k = \begin{cases} f_i^{(a)} & \text{on } \gamma_a \\ 0 & \text{on } \gamma_b. \end{cases}$$

In particular,

$$\nu_3 = \nu_4 = \min(\nu_2^{(a)}, \nu_2^{(b)}) = [2\pi/\max(M_a, M_b)]^2. \tag{4'}$$

4.2. Again, let p and q be any two integers greater than one. In contrast to Section 3.1 we now write $N=p+q-1$. Again

$$f=\sum_2^N c_k f_k, \quad u=\sum_2^N c_k u_k \quad \text{and} \quad \tilde{u}=\sum_2^N c_k \tilde{u}_k .$$

Note that the functions $\tilde{u}_k$ will not, in general, be single valued.

There exist $N-1$ constants $c_2, c_3, \ldots, c_N$ (not all vanishing) such that the following $N-2$ linear homogeneous conditions are satisfied: $\oint_{\gamma_a} \partial u/\partial n \, ds=0$ (that is $\tilde{u}$ must be single valued), $u \perp \varphi_2, \varphi_3, \ldots, \varphi_{p-1}$ and $\tilde{u} \perp \varphi_2, \varphi_3, \ldots, \varphi_{q-1}$. Then $\mu_p \leqq R[u]$ and $\mu_q \leqq R[\tilde{u}]$. Hence by (2')

$$\mu_p \mu_q \leqq R[u] \, R[\tilde{u}] \leqq R_\Gamma[f] \leqq R_\Gamma[f_N] = \nu_N .$$

It follows then that

$$(3''') \qquad \mu_p \mu_q \leqq \nu_{p+q-1} \qquad [\text{see } (4), (4')].$$

Since $\nu_{2n-1}=\nu_{2n}$, $n=1,2,3,4,\ldots$, the inequalities (3''') for $p+q$ odd imply those for $p+q$ even. Indeed, if $p+q$ is even, then $(p+1)+q$ is odd and $\mu_p \mu_q \leqq \mu_{p+1} \mu_q \leqq \nu_{p+q-1}$. For example we have by (4')

$$(3^{\text{iv}}) \qquad \mu_2 \mu_3 \leqq \nu_4 = [2\pi/\max(M_a, M_b)]^2 .$$

4.3. In the particular case when $\rho_b \equiv 0$ (the boundary curve γ_b carries no mass), then $M_b=0$, $M=M_a$ and we obtain inequality (1): $\mu_2 \mu_3 M^2 \leqq 4\pi^2$. A direct proof could be given as in Section 2.2.

Remark. Another direct proof could be given by defining $u=\text{const}$ on γ_b, with the constant so chosen that $\tilde{u}$ is single valued, that is, so that $\oint_{\gamma_b} \partial u/\partial n \, ds=0$. Then $\partial \tilde{u}/\partial n=0$ along γ_b, $D(\tilde{u})=\oint_{\gamma_a} \tilde{u} \, \partial \tilde{u}/\partial n \, ds$, and so on.

Moreover, by modifying the proof of Section 3.1 in an analogous way ($\partial u_k/\partial n=0$ along γ_b), we could prove (3) and (3'). These may be obtained as a limit case as $M_b \to 0$ since $\nu_1^{(b)}=0$ but $\nu_2^{(b)} \to \infty$. Thus, $\nu_1=\nu_2=0$, $\nu_{p+q-1}=\nu_{p+q-2}^{(a)}=[\pi(p+q-3)/M_a]^2$ if $p+q$ is odd, by (4).

4.4. The same method of proof remains valid for the Stekloff problem in *multiply connected domains*, if only *one* boundary curve γ_1 carries all the mass, and shows again the validity of (1), (3), (3'). If γ_1 is the *outer* boundary curve, this follows immediately from the simply connected case (Sections 2 and 3) by the monotony principle: creating holes diminishes the Rayleigh quotient. If γ_1 is an *inner* boundary curve, one considers *the Stekloff problem with respect to its exterior*.

4.5. In a doubly connected domain D with boundary curves γ_a and γ_b, we now consider the harmonic function $\hat{u}$ which assumes a constant value $\hat{a}$ on γ_a and another constant value $\hat{b}$ on γ_b. In order that $\hat{u}$ be orthogonal in the ρ-metric to the first (constant) eigenfunction, we must have $\hat{a} M_a + \hat{b} M_b = 0$. Further, its Dirichlet integral is $D(\hat{u})=(\hat{a}-\hat{b})^2/\mu$, where μ is the conformal modulus of the doubly connected domain D. For instance, for a circular ring $1<r<R$, $\mu=(1/2\pi)\ln R$. (There should be no confusion between the conformal modulus μ

and the eigenvalues μ_k of the Stekloff problem.) We now have

$$R[\hat{u}] = D(\hat{u})/\oint_{\Gamma} \hat{u}^2\, dm = (\hat{a}-\hat{b})^2/\mu(\hat{a}^2 M_a + \hat{b}^2 M_b) = \mu^{-1}(M_a^{-1}+M_b^{-1}).$$

Thus

$$\mu_2 \leqq \mu^{-1}(M_a^{-1}+M_b^{-1}). \tag{5}$$

This inequality is sometimes but not always sharp (depending on the values of M_a, M_b and μ). Its proof is independent of all preceding considerations.

5. Extensions to Multiply Connected Domains

5.1. Stekloff Eigenvalues and the Maxwell Matrix of Multiply Connected Domains. We consider an n-fold connected domain D, bounded by the closed curves $\gamma_1, \gamma_2, \ldots, \gamma_n$, and let $\Gamma = \gamma_1 \cup \gamma_2 \cup \cdots \cup \gamma_n$. Let ω_i be the harmonic measure of γ_i in D, namely the harmonic function in D with boundary values 1 on γ_i, 0 on all other γ. The Maxwell matrix (P_{ij}) of D is defined by

$$2\pi P_{ij} = \oint_{\gamma_j} (\partial\omega_i/\partial n)\, ds$$

(see [2], pp. 159, 164). It is symmetric and positive semi-definite since

$$2\pi P_{ij} = \oint_{\Gamma} \omega_j(\partial\omega_i/\partial n)\, ds = \iint_D \operatorname{grad}\omega_i \cdot \operatorname{grad}\omega_j\, dA = D(\omega_i, \omega_j) = 2\pi P_{ji},$$

and $2\pi \sum_{i,j=1}^{n} P_{ij} c_i c_j = D(c_1\omega_1 + \cdots + c_n\omega_n) \geqq 0$. The latter Dirichlet integral is zero if and only if $c_1\omega_1 + \cdots + c_n\omega_n = \text{const}$, that is, if and only if $c_1 = c_2 = \cdots = c_n$. Thus all eigenvalues of (P_{ij}) are positive except the simple eigenvalue zero corresponding to the eigenvector with equal components.

We now consider the Stekloff problem in D and write $\oint_{\gamma_i} \rho\, ds = M_i$. If we restrict consideration to harmonic functions which are constant on each boundary curve, that is, which have the form $v = c_1\omega_1 + \cdots + c_n\omega_n$, then by the monotony principle upper bounds for the eigenvalues are given by the stationary values of the Rayleigh quotient

$$R[v] = D(v)/\oint_{\Gamma} \rho\, v^2\, ds = 2\pi \sum_{i,j=1}^{n} P_{ij} c_i c_j \Big/ \sum_{1}^{n} M_i c_i^2.$$

Let $0 = \nu_1 \leqq \nu_2 \leqq \cdots \leqq \nu_n$ be the eigenvalues of the matrix $2\pi(P_{ij})$ with respect to the diagonal matrix with elements $M_1, \ldots, M_n$; then

$$\mu_2 \leqq \nu_2;\ \mu_3 \leqq \nu_3;\ \ldots;\ \mu_n \leqq \nu_n. \tag{6}$$

In the special case of a doubly connected domain ($n=2$) of conformal modulus μ, the Maxwell matrix is given by

$$2\pi P_{11} = \oint_{\gamma_1} \frac{\partial\omega_1}{\partial n}\, ds = \oint_{\Gamma} \omega_1 \frac{\partial\omega_1}{\partial n}\, ds = D(\omega_1) = \frac{1}{\mu}; \qquad 2\pi P_{22} = D(\omega_2) = \frac{1}{\mu};$$

$$2\pi P_{21} = 2\pi P_{12} = \int_{\gamma_2} \frac{\partial\omega_1}{\partial n}\, ds = -\oint_{\gamma_1} \frac{\partial\omega_1}{\partial n}\, ds = -D(\omega_1) = -\frac{1}{\mu}.$$

Here v_1 and v_2 are the roots of the secular equation

$$0=\begin{vmatrix} \mu^{-1}-vM_1 & -\mu^{-1} \\ -\mu^{-1} & \mu^{-1}-vM_2 \end{vmatrix} = v[vM_1 M_2 - \mu^{-1}(M_1+M_2)];$$

thus $v_1=0$ and $v_2=\mu^{-1}(M_1^{-1}+M_2^{-1})$. The relation $\mu_2 \leqq v_2$ corresponds to inequality (5) in Section 4.

5.2. The considerations of Sections 4.1–4.2 can of course be extended to multiply connected domains with boundary curves $\gamma_1, \gamma_2, \ldots, \gamma_n$. We again consider the harmonic function $u(x, y)$ with given boundary values f. We require its conjugate harmonic function $\tilde{u}(x, y)$ to be single valued, thus putting $n-1$ linear homogeneous restrictions on f. We therefore put $N=p+q+n-3$ and obtain easily

$$(3^{\mathrm{v}}) \qquad \mu_p \mu_q \leqq v_{p+q+n-3}.$$

Clearly $v_1=v_2=\cdots=v_n=0$ and $v_{n+1}=v_{n+2} \leqq v_{n+3}=v_{n+4} \leqq \ldots$.

6. Simply Connected Domains of Even Symmetry Order q

We consider a simply connected domain J of symmetry order q [11], by which is meant that there exists a "symmetry center" 0 such that J is invariant with respect to a rotation of angle $2\pi/q$ around 0. We suppose further that the mass density has the same symmetry along ∂J, namely $\rho(z)=\rho(e^{2\pi i/q}z)$. In the case of even q, BANDLE [1] proved the sharp inequality $\mu_q^{-1}+\mu_{q+1}^{-1} \geqq 2M/\pi q$; we shall here prove the even sharper result

$$(7) \qquad \mu_q \mu_{q+1} M^2 \leqq q^2 \pi^2.$$

To see this, consider the function $f(m)=c_1 \sin(\pi q m/M)+c_2 \cos(\pi q m/M)$ defined on the boundary ∂J. Let $u(z)$ again be the harmonic function in J with boundary values $f(m)$, and let $\tilde{u}(z)$ be the conjugate harmonic function satisfying $\oint \tilde{u}\, dm=0$. We have $u(z)=-u(e^{2\pi i/q}z)$ and $\tilde{u}(z)=-\tilde{u}(e^{2\pi i/q}z)$. Therefore, as in BANDLE's work ([1], pp. 632-3)

$$\oint u\varphi_j\, dm=\oint \tilde{u}\varphi_j\, dm=0 \qquad \text{for } j=1,2,\ldots,q-1.$$

We can further choose c_1 and c_2 so that also $\oint u\varphi_q\, dm=0$. Then $\mu_{q+1} \leqq R[u]$ and $\mu_q \leqq R[\tilde{u}]$. We proceed once more as in Section 2.2, obtaining $\mu_q \mu_{q+1} \leqq R_{\partial J}[f]=(\pi q/M)^2$ as required.

Example. For a square of side $2a$ with $\rho \equiv 1$, $M=8a$, $q=4$, BANDLE ([1], p. 634) showed that $\mu_4=a^{-1}$ and that $\mu_5 \simeq 2.3044 a^{-1}$. It follows then from (7) that $\mu_4 \mu_5 M^2 q^{-2} \simeq 9.2176 < \pi^2 \simeq 9.8696$.

7. Lower Bounds for Sums of Reciprocal Stekloff Eigenvalues

7.1. Let J be a simply connected domain. Then for any $n \geqq 1$,

$$(8) \qquad \frac{1}{\mu_2}+\frac{1}{\mu_3}+\cdots+\frac{1}{\mu_{2n+1}} \geqq \frac{M}{\pi}\left(1+\frac{1}{2}+\cdots+\frac{1}{n}\right).$$

Proof. Let $u_k(x, y)$ be the harmonic function in J with boundary values $u_k(m) = \sin(2\pi km/M)$ on ∂J $(k=1, 2, \ldots, n)$, and let $\tilde{u}_k(x, y)$ be the conjugate harmonic function satisfying $\oint \tilde{u}_k\, dm = 0$. We have then

$$\begin{aligned} D(u_j, \tilde{u}_k) &= \oint u_j(\partial \tilde{u}_k/\partial n)\, ds = -\oint u_j(\partial u_k/\partial s)\, ds = -\oint u_j(\partial u_k/\partial m)\, dm \\ &= -(2\pi k/M) \oint \sin(2\pi jm/M) \cos(2\pi km/M)\, dm = 0. \end{aligned}$$

(This is trivial for $j=k$, since $\operatorname{grad} u_k \perp \operatorname{grad} \tilde{u}_k$ everywhere.) Furthermore $D(\tilde{u}_j, \tilde{u}_k) = D(u_j, u_k)$ since $\operatorname{grad} \tilde{u}_j \cdot \operatorname{grad} \tilde{u}_k = \operatorname{grad} u_j \cdot \operatorname{grad} u_k$. We now choose

$$v_1 = u_1;$$

$$v_2 = \alpha_{21} u_1 + \alpha_{22} u_2,$$

with constants α_{21} and α_{22} so chosen that $D(v_1, v_2) = 0$;

$$v_3 = \alpha_{31} u_1 + \alpha_{32} u_2 + \alpha_{33} u_3$$

with constants α_{31} and α_{32} and α_{33} so chosen that $D(v_1, v_3) = D(v_2, v_3) = 0$;

$$v_n = \alpha_{n1} u_1 + \alpha_{n2} u_2 + \cdots + \alpha_{nn} u_n$$

with constants $\alpha_{n1}, \ldots, \alpha_{nn}$ so chosen that $D(v_1, v_n) = \cdots = D(v_{n-1}, v_n) = 0$.

We consider the $2n$-dimensional linear space L_{2n} spanned by v_1, $\tilde{v}_1$, v_2, $\tilde{v}_2, \ldots, v_n, \tilde{v}_n$. These functions all satisfy the condition $\oint v\, dm = 0$, and are orthogonal to each other in the Dirichlet metric; therefore, by a variational characterization of sums of reciprocal eigenvalues (see for example [3], p. 399; [4]),

$$\begin{aligned} &\mu_2^{-1} + \mu_3^{-1} + \cdots + \mu_{2n+1}^{-1} \\ (9) \qquad &\geqq TR\,inv[L_{2n}] = R[v_1]^{-1} + R[\tilde{v}_1]^{-1} + \cdots + R[v_n]^{-1} + R[\tilde{v}_n]^{-1} \\ &\geqq 2(R[v_1]\, R[\tilde{v}_1])^{-1/2} + \cdots + 2(R[v_n]\, R[\tilde{v}_n])^{-1/2}. \end{aligned}$$

As in Section 2.2 we obtain therefore

$$R[v_k]\, R[\tilde{v}_k] \leqq \frac{\oint \left(\dfrac{\partial v_k}{\partial m}\right)^2 dm}{\oint v_k^2\, dm} = \frac{\left(\dfrac{2\pi}{M}\right)^2 \dfrac{M}{2} (\alpha_{k1}^2 + 2^2 \alpha_{k2}^2 + \cdots + k^2 \alpha_{kk}^2)}{\dfrac{M}{2} (\alpha_{k1}^2 + \alpha_{k2}^2 + \cdots + \alpha_{kk}^2)} \leqq \left(\frac{2\pi}{M}\right)^2 k^2,$$

which proves (8).

This inequality is *sharp* since equality holds for a circle with constant mass density.

7.2. Remarks. (a) As mentioned in Section 6, BANDLE [1] has obtained the inequality $\mu_q^{-1} + \mu_{q+1}^{-1} \geqq 2M/\pi q$ for domains of even *symmetry* order q. Her result is reminiscent of the inequalities in [11] and [5] for symmetric (homogeneous) membranes, while our inequality corresponds to the following isoperimetric theorem ([12], p. 306): *Among all simply connected membranes, the expression* $(\lambda_1^{-1} + \lambda_2^{-1} + \cdots + \lambda_n^{-1})\, \dot{r}^{-2}$ *is least for the circle.* (Here $\dot{r}$ is the maximum conformal radius of the domain.)

(b) It is of interest to consider the possible inequality

$$\frac{1}{\mu_2\mu_3}+\frac{1}{\mu_4\mu_5}+\cdots+\frac{1}{\mu_{2n}\mu_{2n+1}} \overset{?}{\geqq} \left(\frac{M}{2\pi}\right)^2\left(1+\frac{1}{2^2}+\cdots+\frac{1}{n^2}\right),$$

which extends (1) and is again sharp in the case of the circle. We have verified it for rectangles with $\rho=0$ along one pair of opposite sides and $\rho=1$ along the other pair, a case already considered by BANDLE [1].

7.3. Consider the doubly connected domain D of Section 4. Let $u_k^{(a)}$ be a harmonic function in D with boundary values $u_k^{(a)}=0$ along b and

$$u_k^{(a)}(m_a)=c_1\sin(2\pi k m_a/M_a)+c_2\cos(2\pi k m_a/M_a)$$

along a, with constants c_1 and c_2 determined so that the conjugate harmonic function $\tilde{u}_k^{(a)}$ is single valued in D. We again choose the additive constant for $\tilde{u}_k^{(a)}$ so that $\oint_\Gamma \tilde{u}_k^{(a)}\,dm=0$. Let $u_k^{(b)}$ and $\tilde{u}_k^{(b)}$ be defined in a similar way. Clearly $\partial\tilde{u}_k^{(a)}/\partial n = -\partial u_k^{(a)}/\partial s=0$ along b. Then it is easily verified that

$$D(u_j^{(a)},\tilde{u}_k^{(a)})=D(u_j^{(b)},\tilde{u}_k^{(b)})=0,$$

$$D(u_j^{(a)},\tilde{u}_k^{(b)})=\oint_\Gamma u_j^{(a)}\frac{\partial\tilde{u}_k^{(b)}}{\partial n}\,ds=0$$

(cf. Section 7.1). Now rearrange the quantities

$$1M_a^{-1}, 2M_a^{-1}, 3M_a^{-1}, \ldots \quad \text{and} \quad 1M_b^{-1}, 2M_b^{-1}, 3M_b^{-1}, \ldots$$

in increasing order and call them $\eta_1^{-1}\leqq\eta_2^{-1}\leqq\eta_3^{-1}\leqq\ldots$. Let the functions $u_{k_a}^{(a)}$, $u_{k_b}^{(b)}$ be arranged and numbered correspondingly as $u_1, u_2, u_3, \ldots$. We write $u_k=u_p^{(a)}$ if $\eta_k^{-1}=pM_a^{-1}$.

Define $v_1=u_1$, $v_2=\alpha_{21}u_1+\alpha_{22}u_2, \ldots$ as in Section 7.1, and again consider the linear space L_{2n} spanned by $v_1, \tilde{v}_1, \ldots, v_n, \tilde{v}_n$. These functions satisfy $\oint v\,dm=0$ and are orthogonal to each other in the Dirichlet metric so that (9) holds. Moreover

$$R[v_k]\,R[\tilde{v}_k]\leqq\oint_\Gamma[\partial v_k/\partial m]^2\,dm\Big/\oint_\Gamma v_k^2\,dm.$$

We note that the right hand side here is a weighted mean of the n quantities $4\pi^2(k_a/M_a)^2$ and $4\pi^2(k_b/M_b)^2$ with the indices k_a and k_b corresponding to the $u_1, \ldots, u_k$ as indicated above. Then $R[v_k]\,R[\tilde{v}_k]\leqq(2\pi\eta_k^{-1})^2$. We thus obtain

$$\mu_2^{-1}+\mu_3^{-1}+\cdots+\mu_{2n+1}^{-1}\geqq\frac{1}{\pi}(\eta_1+\eta_2+\cdots+\eta_n). \tag{10}$$

Remark. If $\rho=0$ along the boundary curve b, then $\eta_k^{-1}=kM_a^{-1}$. Thus

$$\mu_2^{-1}+\mu_3^{-1}+\cdots+\mu_{2n+1}^{-1}\geqq(M_a/\pi)(1+2^{-1}+\cdots+n^{-1}).$$

This is just (8) again since now $M_a=M$ (see Sections 4.3 and 4.4).

8. Some "Mixed Problems" of Stekloff Type

8.1. A trilateral is a simply connected domain with three prescribed boundary points. Let T be a trilateral, let a, b, and c be its sides, let $\rho(s)$ be a given non-negative function on the side c, and let $\rho=0$ along the sides a and b. Naturally, $M=\int_c \rho\, ds$. Let $\lambda_1^{(a)}<\lambda_2^{(a)}\leqq\dots$ be the eigenvalues of the mixed Stekloff problem with fixed side a, and let $\varphi_1^{(a)}, \varphi_2^{(a)}, \dots$ be the corresponding eigenfunctions: thus $\Delta\varphi^{(a)}=0$ in T, $\varphi^{(a)}=0$ on a, $\partial\varphi^{(a)}/\partial n=0$ on b, and $\partial\varphi^{(a)}/\partial n=\lambda\rho\varphi^{(a)}$ on c. By Rayleigh's principle

$$\lambda_1^{(a)}=\min_{v=0\text{ along }a} R[v] \qquad \left(R[v]=D(v)/\int_c \rho v^2\, ds\right).$$

Let $\lambda_n^{(b)}$, $\varphi_n^{(b)}$ be defined in the same way, but with the side b being the distinguished part in place of a.

HERSCH & PAYNE [6] obtained the sharp inequality

$$(\lambda_1^{(a)^{-1}}+\lambda_1^{(b)^{-1}})M^{-1}\geqq 4\pi^{-1};$$

we shall prove here the even sharper result

$$\lambda_1^{(a)}\lambda_1^{(b)}M^2\leqq(\pi/2)^2 \tag{11}$$

and moreover shall extend its validity to higher eigenvalues (Section 8.2). Some cases of equality with $\rho=$const along c are known, namely, circular sectors with boundary radii a and b and half-strips with infinite sides a and b.

Proof of (11). Let s denote the arc length along c with $s=0$ at the common end point of a and c $(0\leqq s\leqq L_c)$, and let $m(s)=\int_{t=0}^{s}\rho(t)\,dt$ and $u(s)=\sin[\pi m(s)/2M]$ on c. Let u be the solution of the mixed problem

$$\Delta u=0 \text{ in } T$$

$$u=0 \text{ on } a, \quad u=u(s) \text{ on } c, \quad\text{and}\quad \partial u/\partial n=0 \text{ on } b.$$

Clearly $\lambda_1^{(a)}\leqq R[u]$. Any conjugate harmonic function $\tilde u$ satisfies $\tilde u=$const on b and $\partial u/\partial n=0$ on a. We choose $\tilde u=0$ on b. Then clearly $\lambda_1^{(b)}\leqq R[\tilde u]$.

As in Section 2.2, we now have

$$D(u)=D(\tilde u)=\oint \tilde u\frac{\partial\tilde u}{\partial n}ds=-\int_c \tilde u\frac{\partial u}{\partial s}ds=-\frac{\pi}{2M}\int_{m=0}^{M}\tilde u\cos\frac{\pi m}{2M}dm.$$

By Schwarz's inequality

$$D(u)\cdot D(\tilde u)\leqq\left(\frac{\pi}{2M}\right)^2\int_{m=0}^{M}\cos^2\frac{\pi m}{2M}dm\cdot\int_c \tilde u^2\,dm.$$

But

$$\int_{m=0}^{M}\cos^2\frac{\pi m}{2M}dm=\frac{M}{2}=\int_{m=0}^{M}\sin^2\frac{\pi m}{2M}dm=\int_c u^2\,dm,$$

so that

$$\lambda_1^{(a)}\,\lambda_1^{(b)} \leqq R[u]\,R[\tilde u] \leqq (\pi/2M)^2.$$

Remark. Another proof can be given by means of a conformal mapping of T onto one quarter of an ellipse, in analogy to the procedure used in [6].

8.2. We now look for bounds for higher eigenvalues. We choose a function $f(m)$ $(0 \leqq m \leqq M)$ on c with $f(0)=0$, and define the harmonic function u in T by $u=0$ on a, $u=f(m)$ on c, $\partial u/\partial n=0$ on b. Let $\tilde u$ be the conjugate harmonic function satisfying $\tilde u=0$ on b. As in Section 2.2 we have

$$D(u)\cdot D(\tilde u) \leqq \int_c \tilde u^2\,dm \cdot \int_c f'^2(m)\,dm,$$

which leads to

$$(2'') \qquad R[u]\,R[\tilde u] \leqq \int_0^M f'^2(m)\,dm \Big/ \int_0^M f^2(m)\,dm = R_c[f].$$

We now proceed as in Section 3. The stationary values of $R_c[f]$ under the imposed condition $f(0)=0$ are the eigenvalues of the homogeneous (periodic) string of length M, namely, $\nu_k=[(2k-1)\pi/2M]^2$ and $f_k=\sin[(2k-1)\pi m/2M]$, $k=1,2,3,\ldots$. To each $f_k(m)$ corresponds a harmonic function u_k together with a conjugate harmonic function $\tilde u_k$ vanishing on b. The functions u_k are orthogonal in the ρ-metric and are therefore linearly independent; hence the $\tilde u_k$ are also linearly independent.

Let p and q be any two positive integers. We put

$$N=p+q-1; \quad f(m)=\sum_1^N c_k f_k(m); \quad u=\sum_1^N c_k u_k; \quad \tilde u=\sum_1^N c_k \tilde u_k.$$

There exist N constants $c_1,\ldots,c_N$ (not all vanishing) satisfying the following $N-1$ linear homogeneous conditions of orthogonality in the ρ-metric:

$$u \perp \varphi_1^{(a)},\ldots,\varphi_{p-1}^{(a)} \quad\text{and}\quad \tilde u \perp \varphi_1^{(b)},\ldots,\varphi_{q-1}^{(b)}.$$

For these constants we have $\lambda_p^{(a)} \leqq R[u]$ and $\lambda_q^{(b)} \leqq R[\tilde u]$. We now apply inequality (2″) to the present functions u and $\tilde u$. This gives

$$\lambda_p^{(a)}\,\lambda_q^{(b)} \leqq R[u]\,R[\tilde u] \leqq R_c[f] \leqq R_c[f_N] = \nu_N = [(2N-1)\pi/2M]^2,$$

from which follows

$$(12) \qquad \lambda_p^{(a)}\,\lambda_q^{(b)}\,M^2 \leqq (p+q-\tfrac{3}{2})^2\,\pi^2.$$

Inequality (11) of Section 8.1 of course is the particular case $p=q=1$.

Remarks. (α) If a symmetry of the trilateral permutes the sides a and b, but leaves c and $\rho(s)$ invariant, then $\lambda_p^{(a)}=\lambda_p^{(b)}$ for all p, and (12) gives $\lambda_p^{(a)}\,M \leqq (2p-\tfrac{3}{2})\,\pi$. Moreover, all the inequalities in (12) follow from this one, since

$$(2p-\tfrac{3}{2})(2q-\tfrac{3}{2}) = (p+q-\tfrac{3}{2})^2 - (p-q)^2 \leqq (p+q-\tfrac{3}{2})^2.$$

(β) In the extreme case where the side b is empty, the trilateral reduces to a bilateral. Then $\lambda_q^{(b)} = \mu_q$ is a free Stekloff eigenvalue, and (12) gives

$$\lambda_p^{(a)} \mu_q M^2 \leqq (p+q-\tfrac{3}{2})^2 \pi^2. \tag{12'}$$

We shall, however, obtain a better bound than this in Section 8.5. (The restriction $f(0)=0$ is really unnecessary.)

(γ) In case both a and b are empty we have $\lambda_p^{(a)} = \mu_p$ and $\lambda_q^{(b)} = \mu_q$. By (12), therefore, $\mu_p \mu_q M^2 \leqq (p+q-\frac{3}{2})^2 \pi^2$, which is always weaker than (3) or (3'). This again is due to the unnecessary restriction $f(0)=0$.

8.3. Instead of the trilateral of Section 8.1 we now consider a simply connected domain G whose boundary is decomposed into three parts a, b, c, each part consisting of a finite number of arcs. If they consist of more than a single component, Sections 8.1 and 8.2 no longer apply. This problem is somewhat analogous to that discussed in Section 5.2. Let c have components $c_1, \ldots, c_r$ and put

$$\int_{c_i} \rho_i \, ds = M_i, \qquad m_i(s_i) = \int_{t=0}^{s_i} \rho_i(t)\, dt, \qquad m_i(L_i) = M_i.$$

We consider functions $f^{(i)}(m_i)$ defined on each c_i and vanishing at the common end points of a and c. We define u by

$$\Delta u = 0 \text{ in } G,$$

$$u=0 \text{ on } a, \qquad u = f^{(i)}(m_i) \text{ on } c_i, \qquad \partial u/\partial n = 0 \text{ on } b.$$

Every conjugate harmonic function $\tilde{u}$ is constant on each arc of b. Since only one additive constant for $\tilde{u}$ can be disposed of, we cannot in general normalize $\tilde{u}$ by the condition $\tilde{u}=0$ on b unless we impose further restrictions on u, or, what amounts to the same thing, on the boundary values f. In the case of multiply connected domains (Section 5.2) these extra conditions were required in order to make $\tilde{u}$ single-valued. In the present case $\tilde{u}$ is always single valued, since the domain is simply connected.

8.4. A bilateral is a simply connected domain with two prescribed boundary points. Let B be a bilateral, let a and c be its sides, let $\rho(s)$ be a prescribed non-negative function on the side c, and let $\rho=0$ along a.

Again let $\lambda_1^{(a)} < \lambda_2^{(a)} \leqq \cdots$ be the eigenvalues of the mixed Stekloff problem with fixed side a. We prove now the sharp inequality

$$\lambda_1^{(a)} \mu_2 M^2 \leqq \pi^2. \tag{13}$$

Some cases of equality with $\rho=$const on c are known, namely circular sectors with c the circular part of the boundary and halfstrips with c the side of finite length.

Proof of (13). We wirte λ_a for $\lambda_1^{(a)}$. Let u be the harmonic function in B with boundary values $u=0$ on a and $u(s)=\sin[\pi m(s)/M]$ on c. Obviously $\lambda_a \leqq R[u]$ by Rayleigh's principle. Let $\tilde{u}$ be the conjugate harmonic function satisfying $\int_c \tilde{u}\, dm = 0$. Then $\mu_2 \leqq R[\tilde{u}]$. By the method of Section 2.2 we obtain

$$\lambda_a \mu_2 \leqq R[u]\, R[\tilde{u}] \leqq (\pi/M)^2.$$

Remarks. (α) One could have started with the harmonic function u which equals $\cos[\pi m(s)/M]$ on c and which has vanishing normal derivative on a. Then $\mu_2 \leqq R[u]$. By considering the conjugate function $\tilde{u}$ we then obtain $\lambda_a \leqq R[\tilde{u}]$ and so forth.

(β) Another proof of (13) can be given by means of a conformal mapping of B onto a half ellipse. Because of its separate interest we include this proof here.

Let the given bilateral lie in the (x, y) plane. Let it be mapped conformally by $\tilde{x}=X(x,y)$, $\tilde{y}=Y(x,y)$ onto the upper half ellipse $\{(\tilde{x}/R_1)^2+(\tilde{y}/R_2)^2<1, \tilde{y}>0\}$ in the $(\tilde{x},\tilde{y})$ plane in such a way that the image $\tilde{a}$ of the side a is the real segment $-R_1 \leqq \tilde{x} \leqq R_1$ and the image $\tilde{b}$ of b is the upper curved boundary. The conformal mapping is not yet completely determined, since we can prescribe the image of a third boundary point. We use this freedom to impose the additional condition $\int_c \rho X\, ds=0$. (The existence of such a mapping follows from a continuity argument. Indeed, it is easily seen hat the mapping can be chosen so that the value of $\int_c \rho X\, ds$ is arbitrarily near $-MR_1$ or $+MR_1$. By continuity the mapping can be chosen in such a way that $\int_c \rho X\, ds$ assumes any value between these two extremes.) We now use the property of invariance of the Dirichlet integral under conformal transplantation:

$$D(Y)=D(\tilde{y})=\tfrac{1}{2}\pi R_1 R_2=D(\tilde{x})=D(X)$$

and

$$\lambda_a \leqq R[Y]=\frac{D(Y)}{\int_c \rho Y^2 ds}=\frac{\frac{1}{2}\pi R_1 R_2}{\int_c \rho Y^2 ds}; \qquad \mu_2 \leqq R[X]=\frac{D(X)}{\int_c \rho X^2 ds}=\frac{\frac{1}{2}\pi R_1 R_2}{\int_c \rho X^2 ds}.$$

Thus for any choice of the positive constants α and β we have

$$\frac{\alpha^2}{\lambda_a}+\frac{\beta^2}{\mu_2} \geqq \frac{2}{\pi R_1 R_2}\int_c \rho[\alpha^2 Y^2+\beta^2 X^2]\, ds.$$

In particular, let us take $\alpha=R_2^{-1}$ and $\beta=R_1^{-1}$. Then $\alpha^2 Y^2+\beta^2 X^2=1$ on c, and $\lambda_a^{-1}\alpha^2+\mu_2^{-1}\beta^2 \geqq 2M/\pi R_1 R_2=2M\pi^{-1}\alpha\beta$. Since this inequality holds for any positive values R_1 and R_2 it follows that the quadratic form $\lambda_a^{-1}\alpha^2-2M\pi^{-1}\alpha\beta+\mu_2^{-1}\beta^2$ is non-negative for all positive α and β and therefore also for all negative α and β. Since the quadratic form is positive semi-definite it follows that the discriminant $\lambda_a^{-1}\mu_2^{-1}-(M\pi^{-1})^2$ is non-negative. Thus $\lambda_a\mu_2 \leqq (\pi/M)^2$, establishing (13).

8.5. We now look for bounds for higher eigenvalues of the bilateral B. Choose a function $f(m)$ $(0 \leqq m \leqq M)$ on c with $f(0)=f(M)=0$. Let u be the harmonic function in B with boundary values zero on a and $f(m)$ on c, and let $\tilde{u}$ be the conjugate harmonic function satisfying $\int_c \tilde{u}\, dm=0$. Then we have (cf. Section 2.2)

$$R[u]\cdot R[\tilde{u}] \leqq R_c[f]. \tag{2'''}$$

The stationary values of $R_c[f]$ under the imposed conditions $f(0)=f(M)=0$ are those of the homogeneous string with fixed end-points, namely $v_n=(n\pi/M)^2$ and $f_n=\sin(n\pi m/M)$, $n=1,2,3,\ldots$. To each $f_k(m)$ corresponds a harmonic function u_k and a conjugate $\tilde{u}_k$ with $\int \tilde{u}_k\, dm=0$. Since the u_k are independent, then so are the

$\tilde{u}_k$. Let p and q be integers, $p \geqq 1$, $q \geqq 2$. We put $N = p+q-2$; $f(m) = \sum_1^N c_k f_k(m)$; $u = \sum_1^N c_k u_k$; $\tilde{u} = \sum_1^N c_k \tilde{u}_k$. There exist N constants $c_1, \ldots, c_N$ (not all vanishing) satisfying the following $N-1$ linear homogeneous conditions of orthogonality in the ρ-metric: $u \perp \varphi_1^{(a)}, \ldots, \varphi_{p-1}^{(a)}$ and $\tilde{u} \perp \varphi_2, \ldots, \varphi_{q-1}$. For these constants $c_1, \ldots, c_N$ we have $\lambda_p^{(a)} \leqq R[u]$ and $\mu_q \leqq R[\tilde{u}]$. We now apply inequality (2‴) to the present functions u and $\tilde{u}$. Then

$$\lambda_p^{(a)} \mu_q \leqq R[u]\, R[\tilde{u}] \leqq R_c[f] \leqq R_c[f_N] = \nu_N = (N\pi/M)^2,$$

and thus

$$\lambda_p^{(a)} \mu_q M^2 \leqq (p+q-2)^2 \pi^2. \tag{14}$$

Inequality (13) of Section 8.4 is the particular case $p=1$, $q=2$.

Remarks. (α) Inequality (14) always gives a better bound than the limit case (12′) of (12).

(β) In the extreme case when a is empty, we have $\lambda_p^{(a)} = \mu_p$, which implies $\mu_p \mu_q M^2 \leqq (p+q-2)^2 \pi^2$. (Compare with (3) or (3′).) Note that here we have imposed the unnecessary conditions $f(0) = f(M) = 0$.

The authors wish to thank the Swiss National Science Foundation for supporting this research. The work of L.E.P. was also supported in part by NSF Grant GP-33031X and the work of M.M.S. by NSF Grant GP-35543 and AF Contract F44620-72-C-0031.

Bibliography

1. Bandle, C., Über das Stekloffsche Eigenwertproblem: Isoperimetrische Ungleichungen für symmetrische Gebiete. Zeitschr. Angew. Math. Phys. **19**, 627–637 (1968).
2. Bergman, S., & M. Schiffer, Kernel Functions and Elliptic Differential Equations in Mathematical Physics. New York: Academic Press 1953.
3. Courant, R., & D. Hilbert, Methoden der mathematischen Physik, Vol. I. Berlin: Springer 1931.
4. Hersch, J., Caractérisation variationnelle d'une somme de valeurs propres consécutives; généralisation d'inégalités de Pólya-Schiffer et de Weyl. C. R. Acad. Sci. Paris **252**, 1714 (1961).
5. Hersch, J., On symmetric membranes and conformal radius: Some complements to Pólya's and Szegö's inequalities. Arch. Rational Mech. Analysis **20**, 378–390 (1965).
6. Hersch, J., & L. E. Payne, Extremal principles and isoperimetric inequalities for some mixed problems of Stekloff's type. Zeitschr. Angew. Math. Phys. **19**, 802–817 (1968).
7. Hilbert, D., Grundzüge einer allgemeinen Theorie der linearen Integralgleichungen, pp. 77–81. Leipzig & Berlin 1912.
8. Kuttler, J. R., & V. G. Sigillito, Inequalities for membrane and Stekloff eigenvalues. J. Math. Anal. Appl. **23**, 148–160 (1968).
9. Payne, L. E., New isoperimetric inequalities for eigenvalues and other physical quantities. Communications Pure Appl. Math. **9**, 531–542 (1956).
10. Payne, L. E., Some isoperimetric inequalities for harmonic functions. SIAM J. Math. Anal. **1**, 354–359 (1970).
11. Pólya, G., On the characteristic frequencies of a symmetric membrane. Math. Zeitschr. **63**, 331–337 (1955).
12. Pólya, G., & M. Schiffer, Convexity of functionals by transplantation. J. d'Anal. Math. **3** (2e partie), 245–345 (1953/54).
13. Schiffer, M., Various types of orthogonalization. Duke Math. J. **17**, 329–366 (1950).

14. Shamma, S. E., Asymptotic behavior of Stekloff eigenvalues and eigenfunctions. SIAM J. Appl. Math. **20**, 482–490 (1971).
15. Stekloff, M. W., Sur les problèmes fondamentaux de la physique mathématique. Ann. Sci. Ecole Norm. Sup. **19**, 455–490 (1902).
16. Szegö, G., Inequalities for certain eigenvalues of a membrane of given area. J. Rational Mech. Anal. **3**, 343–356 (1954).
17. Troesch, B. A., An isoperimetric sloshing problem. Communications Pure Appl. Math. **18**, 319–338 (1965).
18. Weinberger, H. F., An isoperimetric inequality for the N-dimensional free membrane problem. J. Rational Mech. Anal. **5**, 633–636 (1956).
19. Weinstock, R., Inequalities for a classical eigenvalue problem. J. Rational Mech. Anal. **3**, 745–753 (1954).

E. T. H.
Zurich, Switzerland

Cornell University
Ithaca, New York

Stanford University
Stanford, California

(Received September 26, 1973)

Commentary on

[109] J. Hersch, L.E. Payne and M.M. Schiffer, *Some inequalities for Stekloff eigenvalues*, Arch. Rational Mech. Anal. **57** (1975), 99–104.

The eigenvalue problem of this paper was introduced by Stekloff [St] (see also [H] and [CH, p. 401]) and has been considered by many authors. A more complete up-to-date treatment of this eigenvalue problem is given in [B, D2]. The eigenfunctions are the eigenfunctions of the Neumann function on the boundary and play an important role in the theory of orthogonalization [31, p. 361], [51, p. 397].

Following Szegő's paper [Sz], Weinstock [W] proved by conformal transplantation and a fixed point argument that

$$\mu_2 \leq \frac{2\pi}{M}, \tag{1}$$

where $M = \int_{\partial D} p(s)ds$. Equality occurs for the disk with a constant function p.

It was noticed by Hersch and Payne [HP] that Weinstock's proof contains the sharper inequality

$$\left(\frac{1}{\mu_2}+\frac{1}{\mu_3}\right)\frac{1}{M} \geq \frac{1}{\pi}, \tag{2}$$

and the starting point of the paper is to sharpen this result to

$$\mu_2\mu_3 \leq \frac{4\pi^2}{M^2}, \tag{3}$$

and to extend it to higher eigenvalues in simply and doubly connected domains. For multiply connected domains, there is another upper bound for $\mu_2\mu_3$ given using function-theoretic quantities [D2]. For simply connected domains, inequality (2) culminates in the sharp result

$$\frac{1}{\mu_2}+\frac{1}{\mu_3}+\cdots+\frac{1}{\mu_{2n+1}} \geq \frac{M}{\pi}\left(1+\frac{1}{2}+\cdots+\frac{1}{n}\right), \tag{4}$$

which corresponds to an inequality [53, p. 306, Theorem 6] for the fixed membrane problem. It should be mentioned that this result also implies, via a general result of Schur, Hardy, Littlewood, and Pólya [MO, p. 64], the inequality

$$\phi\left(\frac{1}{\mu_2}\right)+\phi\left(\frac{1}{\mu_3}\right)+\cdots+\phi\left(\frac{1}{\mu_{2n+1}}\right) \geq \phi\left(\frac{M}{\pi}\right)+\phi\left(\frac{M}{2\pi}\right)+\cdots+\phi\left(\frac{M}{n\pi}\right) \tag{5}$$

for any convex increasing function ϕ.

On the basis of [51, p. 395], it is also possible [D1] to give a formula for $\sum_2^\infty \mu_j^{-2}$, which makes it possible to calculate this sum if the conformal mapping is known and to obtain isoperimetric results like (4). Most of the results of the paper have to do with domains containing the point at infinity, but (2)–(4) are still valid for bounded domains. In a simply connected domain, for example, we have to distinguish between the Stekloff eigenvalues of the interior and the exterior domain [D2]. There are also sharp inequalities between these eigenvalues; the first one was given by Schiffer during a discussion with Bandle (see [D2], inequality (31)).

References

[B] Catherine Bandle, *Isoperimetric Inequalities and Applications*, Pitman, 1980.

[CH] R. Courant and D. Hilbert, *Methoden der mathematischen Physik*, 4. Auflage, Springer-Verlag, 1993.

[D1] Bodo Dittmar, *Sums of reciprocal Stekloff eigenvalues*, Math. Nachr. **268** (2004), 44–49.

[D2] Bodo Dittmar, *Eigenvalue problems and conformal mapping*, Handbook of Complex Analysis: Geometric Function Theory, Vol. 2, ed. R. Kühnau, Elsevier, 2005, pp. 669–686.

[HP] Joseph Hersch and Lawrence E. Payne, *Extremal principles and isoperimetric inequalities for some mixed problems of Stekloff's type*, Z. Angew. Math. Phys. **19** (1968), 802–819.

[H] D. Hilbert, *Grundzüge einer allgemeinen Theorie der linearen Integralgleichungen*, Teubner, 1912.

[MO] Albert W. Marshall and Ingram Olkin, *Inequalities: Theory of Majorization and its Applications*, Academic Press, 1979.

[St] M. W. Stekloff, *Sur les problèmes fondamentaux de la physique mathématique*, Ann. Sci. Ecole Normale Sup. **19** (1902), 455–490.

[Sz] G. Szegő, *Inequalities for certain eigenvalues of a membrane of given area*, J. Rational Mech. Anal. **3** (1954), 343–356.

[W] Robert Weinstock, *Inequalities for a classical eigenvalue problem*, J. Rational Mech. Anal. **3** (1954), 745–753.

BODO DITTMAR

[116] (with J. A. Hummel) Variational methods for Bieberbach-Eilenberg functions and for pairs

---

[116] (with J. A. Hummel) Variational methods for Bieberbach-Eilenberg functions and for pairs. *Ann. Acad. Sci. Fenn. Ser.* A I **3** (1977), 3–42.

Annales Academiæ Scientiarum Fennicæ
Series A. I. Mathematica
Volumen 3, 1977, 3—42

VARIATIONAL METHODS FOR BIEBERBACH—EILENBERG FUNCTIONS AND FOR PAIRS

J. A. HUMMEL and M. M. SCHIFFER*

1. Introduction

1.1. We let U denote the unit disc, $|z|<1$, and $\mathcal{S}$ the usual class of univalent functions, i.e., those functions $f(z)$ which are analytic and univalent in U with $f(0)=0, f'(0)=1$. A function $F(z)$ is called a *Bieberbach—Eilenberg function* if it is analytic in U, zero at $z=0$, so that it has a series development of the form

$$F(z) = b_1 z + b_2 z^2 + \ldots, \tag{1.1}$$

and is such that

$$F(z_1) F(z_2) \neq 1 \tag{1.2}$$

for any $z_1, z_2 \in U$. In this paper we will be concerned only with the subclass of these functions which are univalent in U and we will denote this class of *univalent Bieberbach—Eilenberg* functions by $\mathcal{E}$.

Two functions, F and G, are called *a pair* if they are analytic in U, $F(0)= =G(0)=0$, and such that

$$F(z_1) G(z_2) \neq 1 \tag{1.3}$$

for any $z_1, z_2 \in U$. We will use the term *pair* only in this sense in this paper. Note that if $F \in \mathcal{E}$, then $\{F, F\}$ is a univalent pair.

1.2. The class $\mathcal{E}$ was introduced by Bieberbach [7] as an aid in solving the problem of maximizing the diameter of the boundary of the image of the complement of a disc (or equivalently any simply connected domain containing ∞) under all conformal maps which carry ∞ to ∞ and have derivative 1 at ∞. He showed

* This work was supported in part by NSF grant GP 28970 az University of Maryland and by NSF grant MPS72-04967A02 at Stanford University.

that the extremal function maps onto the exterior of a line segment by proving that if $F\in\mathcal{E}$, then

$$|b_1| \leqq 1, \tag{1.4}$$

with equality if and only if $F(z)=e^{i\alpha}z$, α real. The crucial part of his argument was the observation that the function $w+1/w$ has a single valued inverse in the exterior of a continuum passing through ± 2. It therefore defines a univalent mapping from any domain in the w-plane having the property (1.2) onto the exterior of such a continuum.

Eilenberg [9] introduced the full class of (not necessarily univalent) functions satisfying (1.1) and (1.2). He showed that (1.4) held in this class with the help of a topological theorem and the subordination principle. Our use of $\mathcal{E}$ to denote the *univalent* class is therefore somewhat misleading, but we do not want to cause confusion with the standard class of bounded functions.

Rogosinski [28] simplified Eilenberg's results considerably and extended the subordination principle. Using his methods as found in [29], many results obtained in $\mathcal{E}$ can be extended to the full class. Some care is required however. See for example the discussion in § 9 of [20].

In [28], Rogosinski conjectured that all $|b_n|\leqq 1$. This was proved by Lebedev and Milin [23], later Aharonov [3] and Nehari [25] independently proved that $|b_n|\leqq$ $\leqq e^{-\gamma/2}/\sqrt{n-1}$ for all n, where γ is the Euler constant.

1.3. Many major results concerning Bieberbach—Eilenberg functions were proved by Jenkins in a series of three papers [18, 19, 20]. He studied several extremal problems by considering appropriate module problems and using symmetrization. This allowed him to obtain an explicit upper bound for $|F(r)|$ for $F\in\mathcal{E}$ and r fixed, and implicit solutions of several other problems. He also used area methods in [20] to prove some inequalities of Grunsky type.

1.4. The class $\mathcal{E}$ appeared in an unexpected fashion when Garabedian and Schiffer made a systematic search for extremal problems in $\mathcal{S}$ which would have a simple solution by variational methods. This search is described in [33]. This led to a class of inequalities of Grunsky type which in turn led to a proof of the local Bieberbach conjecture [11]. Some of the inequalities they found could be interpreted as inequalities for functions in $\mathcal{E}$, because of a close connection between $\mathcal{S}$ and $\mathcal{E}$.

If $f(z)\in\mathcal{S}$, with $1/u\notin f(U)$, then

$$F(z)=\frac{1-[1-uf(z)]^{1/2}}{1+[1-uf(z)]^{1/2}} \tag{1.5}$$

is in $\mathcal{E}$. This can be inverted, giving $f(z)=F(z)/\{F'(0)[1+F(z)]^2\}$, $F'(0)=u/4$, showing how F near the identity in $\mathcal{E}$ corresponds to f near the Koebe function. It was this fact which was important in the proof of the local Bieberbach conjecture in [11] and which indicates the potential importance of studying the class $\mathcal{E}$.

1.5. The concept of pairs was introduced by Aharonov in 1969 in [1]. In some respects, this concept is equivalent to considering functions f and g which have disjoint ranges with $f\,(=F)$ analytic and $g\;(=1/G)$ meromorphic in U. Functions with disjoint ranges had been studied by several authors earlier, for example [5, 6, 14, 20, 22]. In particular, see the supplements to [12]. However, Aharonov's class seems to be the correct one to consider as a generalization of Bieberbach—Eilenberg functions, as is evident from some of the results of the present paper.

The main previously known results about pairs are found in [1, 2, 16, 21]. Some of these are discussed in the following sections.

1.6. Variational methods are powerful tools in the study of univalent functions. They offer one of the most systematic and widely applicable methods of attacking extremal problems. The variational method in the class $\mathcal{S}$ introduced in [31] has been generalized considerably, and systematic methods of using this and similar variations have been found. See for example the discussion in [32].

The development of variational methods for special classes depends on finding a method of varying the image domain so as to preserve the desired property. In particular, it does not seem immediately obvious how to do this in the case of the class $\mathcal{E}$.

D. J. Nelson, in his thesis [26], obtained a variation in the class $\mathcal{E}$ by using (1.5) to map $F\in\mathcal{E}$ to $f\in\mathcal{S}$. The well-known variation [31] of $\mathcal{S}$ was used on f and the resulting varied function was mapped back to $\mathcal{E}$ by (1.5) again. The extra parameter u which had to be introduced could be eliminated so that he was able to obtain the differential equation satisfied by the extremal function for some specific problems. Some of the observations he made for these problems are generalized and appear in Theorem 4.2 below. In particular, Nelson obtained the differential equation (4.3) for the $F\in\mathcal{E}$ maximizing $\operatorname{Re} b_n$, and studied the specific problem of maximizing $\operatorname{Re} b_2$, obtaining some bounds on the maximum value.

1.7. In Sections 2 and 3 of this paper, we obtain variational methods for the class $\mathcal{E}$ and for univalent pairs. In Sections 4 and 5 we show how to apply these variations to quite general types of extremal problems. The remainder of the paper is devoted to studying a representative set of extremal problems.

An extremal function for such a problem is found to satisfy a differential equation of the form $Q(w)\,dw^2=R(z)\,dz^2$ where Q and R are typically rational functions which unfortunately involve unknown parameters. Each side of this equation is a quadratic differential, and we obtain information about the solution by considering the trajectories of these quadratic differentials. A simple discussion of the structure of such trajectories may be found, for example, in [4]. Here we will review some elementary facts about such trajectories.

A *trajectory* of the quadratic differential $d\Omega^2=Q(w)\,dw^2$ where $Q(w)$ is a rational function is a path $w=w(t)$, $a<t<b$ along which $d\Omega^2>0$. An *orthogonal trajectory* is a path on which $d\Omega^2<0$. A *critical point* is a point at which $Q(w)$

has a zero or pole. If w_0 is not a critical point, then $\Omega=\int\sqrt{Q(w)}\,dw$ is analytic and univalent in a neighborhood of w_0 and hence there exists a unique trajectory through w_0 (and similarly a unique orthogonal trajectory). Two trajectories can meet only at a critical point. A trajectory can be continued indefinitely unless it closes or reaches a critical point.

The structure of trajectories near a critical point can be complicated, but we will need only two special cases in this paper. If $Q(w)$ has a simple pole at w_0, then exactly one trajectory of $d\Omega^2$ leaves w_0. If $Q(w)$ has a simple zero at w_1, then three trajectories leave w_1. These make equal angles with one another.

2. A method of variation for Bieberbach—Eilenberg functions

2.1. To use the known formulas for the variation of the Green's function to obtain a variation of the mapping function, we must have a method of varying a domain so as to stay within the desired class. Define a *Bieberbach—Eilenberg domain* to be the image of a function in $\mathcal{E}$, that is, a simply connected domain containing the origin which is such that if w is in the domain, then $1/w$ is not.

Theorem 2.1. *Let D be a Bieberbach—Eilenberg domain. Let Δ be a domain whose closure does not contain 0 or ∞, which contains the boundary of D, and which is symmetric with respect to the mapping $w\to 1/w$ (i.e., $w\in\Delta$ iff $1/w\in\Delta$). Let $\Phi(w)$ be analytic in $\bar{\Delta}$ and satisfy*

$$\Phi(w) = -\Phi(1/w) \tag{2.1}$$

for all $w\in\Delta$.

Then for all ε sufficiently near 0, the function

$$w^*(w) = we^{\varepsilon\Phi(w)} \tag{2.2}$$

is univalent in Δ and maps the boundary of D onto the boundary of a Bieberbach—Eilenberg domain.

Proof. Introduce the function

$$\begin{aligned}\Psi(w,\omega) &= \frac{\Phi(w)-\Phi(\omega)}{w-\omega} \qquad w\neq\omega\\ &= \Phi'(w) \qquad w=\omega.\end{aligned}$$

This Ψ is defined, analytic, and uniformly bounded in the compact set $\bar{\Delta}\times\bar{\Delta}$.

Suppose $w^*(w)$ were not univalent in Δ. Then there would exist $w_1\neq w_2$ in Δ such that $w^*(w_1)=w^*(w_2)$. That is, $w_1-w_2=w_1[1-\exp\{\varepsilon(w_1-w_2)\Psi(w_1,w_2)\}]$. Now for any s, $|1-e^s|\leqq|s|e^{|s|}$, so this implies

$$|w_1-w_2|\leqq|\varepsilon|\cdot|w_1(w_1-w_2)\Psi(w_1,w_2)|\exp\{|\varepsilon(w_1-w_2)\Psi(w_1,w_2)|\}.$$

Divide through by $|w_1-w_2|$. The stated conditions on Δ then give a contradiction for all sufficiently small $|\varepsilon|$.

Since $w^*(w)$ is univalent in Δ, it maps the boundary of D onto the boundary of a simply connected domain D^*. It remains to show that D^* is a Bieberbach—Eilenberg domain for all sufficiently small $|\varepsilon|$.

This property is equivalent to asserting that D^* and $D_1^* = \{w\colon 1/w \in D^*\}$ do not intersect. Suppose to the contrary that they do. Then they must in fact intersect at points arbitrarily close to the boundary of D^* and we can therefore assume that there exist w_1 and w_2 in $\Delta \cap D$ such that $w^*(w_1)w^*(w_2)=1$, i.e., that $w_1 w_2 \exp\{\varepsilon[\Phi(w_1)+\Phi(w_2)]\}=1$. Using (2.1) this is equivalent to

$$(w_1 - w_2^{-1}) = w_1[1-\exp\{\varepsilon(w_1-w_2^{-1})\Psi(w_1, w_2^{-1})\}].$$

Since $w_1, w_2 \in D$, $w_1 \neq w_2^{-1}$, and we may proceed exactly as before to obtain a contradiction as $|\varepsilon| \to 0$. This proves the theorem.

2.2. We use this theorem to find a variational formula for functions in $\mathscr{E}$ using the variational formula for the Green's function [31]. The form we use is described in [32]. Let D be a domain and D^* a varied domain obtained from D by the mapping $w^* = w + \varepsilon v(w) + o(\varepsilon)$ defined and univalent near the boundary in D. If $p(w, \omega)$ and $p^*(w, \omega)$ are the analytic completions of the Green's functions of D and D^*, then

$$p^*(w, \omega) = p(w, \omega) + \varepsilon q_1(w, \omega) + o(\varepsilon)$$

where $q_1(w, \omega)$ is an analytic function of w such that

$$\text{(2.3)} \qquad \operatorname{Re}\{q_1(w, \omega)\} = \operatorname{Re}\left\{\frac{1}{2\pi i}\int_\Gamma p'(t, w)p'(t, \omega)v(t)\,dt\right\}$$

and Γ is a curve system in D bounding a subdomain of D containing w and ω.

Let F map U onto D and F^* map U onto D^* with $F(0)=F^*(0)=0$. If φ and φ^* are the inverse functions of F and F^*, then $p(w,0) = -\log\varphi(w)$, $p^*(w,0) = -\log\varphi^*(w)$ and hence $\varphi^*(w) = \varphi(w) - \varepsilon\varphi(w)q_1(w,0) + o(\varepsilon)$. Then, as in [32], it follows that

$$\text{(2.4)} \qquad F^*(z) = F(z) + \varepsilon z F'(z) q_1(F(z), 0) + o(\varepsilon).$$

Notice that the derivative at 0 is not normalized in this class.

Let $w_0 \in D$ and let α be any real number. Set

$$\text{(2.5)} \qquad \Phi(w) = e^{i\alpha}\left[\frac{1}{w-w_0} - \frac{w}{1-w_0 w}\right].$$

This Φ satisfies (2.1). The remaining hypotheses of Theorem 2.1 hold and hence the variation $w^* = we^{\varepsilon\Phi(w)} = w + \varepsilon w\Phi(w) + o(\varepsilon)$ preserves the class. We need only compute $q_1(w, 0)$ using $v(w) = w\Phi(w)$.

2.3. Let φ be the inverse function of F. Then

$$p(t, w) = -\log([\varphi(t)-\varphi(w)]/[1-\overline{\varphi(w)}\varphi(t)])$$

and hence from (2.3) we have

$$\begin{aligned}
&\operatorname{Re}\{q_1(w,0)\}\\
&= -\operatorname{Re}\left\{\frac{e^{i\alpha}}{2\pi i}\int_\Gamma \left[\frac{\varphi'(t)}{\varphi(w)-\varphi(t)} - \frac{\overline{\varphi(w)}\varphi'(t)}{1-\overline{\varphi(w)}\varphi(t)}\right]\cdot\left[\frac{\varphi'(t)}{\varphi(t)}\right]\cdot\left[\frac{t}{t-w_0}-\frac{t^2}{1-w_0 t}\right]dt\right\}\\
&= \operatorname{Re}\left\{e^{i\alpha}\frac{\varphi'(w)}{\varphi(w)}\left[\frac{w}{w-w_0}-\frac{w^2}{1-w_0 w}\right] - e^{i\alpha}\frac{w_0\varphi'(w_0)^2}{\varphi(w_0)[\varphi(w)-\varphi(w_0)]}+\right.\\
&\left.+e^{i\alpha}\frac{w_0\varphi'(w_0)^2\overline{\varphi(w)}}{\varphi(w_0)[1-\overline{\varphi(w)}\varphi(w_0)]}\right\}.
\end{aligned}$$

Here we calculated the residues at $t=w$ and $t=w_0$. There is no singularity at $t=0$.

We may conjugate the final term inside the real part. The resulting function is analytic in w and may be identified with $q_1(w,0)$. Replacing w by $F(z)$, w_0 by $F(z_0)$ and so on, (2.4) gives

Theorem 2.2. *Let $F(z)\in\mathscr{E}$, let α be any real number, and let $z_0\in U$. Then for all sufficiently small $\varepsilon>0$ there exists a function $F^*(z)\in\mathscr{E}$ such that*

$$(2.6)\quad \begin{aligned}
F^*(z) = F(z)&+\varepsilon e^{i\alpha}\left[\frac{F(z)}{F(z)-F(z_0)}-\frac{F(z)^2}{1-F(z_0)F(z)}\right]-\varepsilon e^{i\alpha}\left(\frac{F(z_0)}{z_0F'(z_0)^2}\right)\frac{zF'(z)}{z-z_0}\\
&+\varepsilon e^{-i\alpha}\overline{\left(\frac{F(z_0)}{z_0F'(z_0)^2}\right)}\frac{z^2F'(z)}{1-\bar z_0 z}+o(\varepsilon).
\end{aligned}$$

In the calculation of $q_1(w,0)$ we found a residue at w_0 since w_0 was assumed in D. There was none at $1/w_0$ since D was a Bieberbach—Eilenberg domain. Suppose neither w_0 nor $1/w_0$ were in D. Then the only residue would be at w. Thus we have

Theorem 2.3. *Let $F(z)\in\mathscr{E}$. Let α be any real number. Suppose w_0 is such that neither w_0 nor $1/w_0$ is in the closure of $F(U)$. Then for all sufficiently small $\varepsilon>0$ there exists a function $F^*(z)\in\mathscr{E}$ such that*

$$(2.7)\quad F^*(z) = F(z)+\varepsilon e^{i\alpha}\left[\frac{F(z)}{F(z)-w_0}-\frac{F(z)^2}{1-w_0F(z)}\right]+o(\varepsilon).$$

3. A method of variation for pairs

3.1. Let $\{F_1(z), F_2(z)\}$ be a univalent pair. Set $D_1=F_1(U)$, $D_2=F_2(U)$, $D_2'= \{w\colon 1/w\in D_2\}$. Suppose Δ_0 is an open set which contains the boundaries of D_1 and D_2', and that $\Phi(w)$ is analytic in the closure of Δ_0. Then one easily verifies that

$$(3.1)\quad w^*(w) = w+\varepsilon\Phi(w)$$

is univalent in Δ_0 for all sufficiently small ε. This will induce a variation of D_1 to D_1^* and hence of F_1 to F_1^*. At the same time (3.1) induces a variation of D_2' to $D_2^{*\prime}$ and hence of D_2 to D_2^* and of F_2 to F_2^* so that $\{F_1^*, F_2^*\}$ will be a new univalent pair.

As in § 2.2, we have

$$F_1^*(z) = F_1(z)+\varepsilon z F_1'(z)\, q_1(F_1(z), 0)+o(\varepsilon) \tag{3.2}$$

where $q_1(w, 0)$ is an analytic function of w satisfying

$$\operatorname{Re}\{q_1(w, 0)\} = \operatorname{Re}\left\{\frac{1}{2\pi i}\int_{\Gamma} p_1'(t, w)\, p_1'(t, 0)\, \Phi(t)\, dt\right\} \tag{3.3}$$

the quantities being defined as in § 2.2.

To find the variation of D_2 induced by (3.1), let $\omega=1/w$. Then $\omega^*=\omega+\varepsilon\Phi(\omega)$ and hence $w^*=1/\omega^*=\omega^{-1}[1+\varepsilon\Phi(\omega)/\omega]^{-1}=\omega^{-1}[1-\varepsilon\Phi(\omega)/\omega]+o(\varepsilon)=w-\varepsilon w^2\Phi(1/w)+ +o(\varepsilon)$. Hence

$$F_2^*(z) = F_2(z)+\varepsilon z F_2'(z)\, q_2(F_2(z), 0)+o(\varepsilon) \tag{3.4}$$

where q_2 is determined as before by

$$\operatorname{Re}\{q_2(w, 0)\} = -\operatorname{Re}\left\{\frac{1}{2\pi i}\int_{\Gamma_2} p_2'(t, w)\, p_2'(t, 0)\, t^2\Phi(1/t)\, dt\right\}. \tag{3.5}$$

3.2. Let w_0 be any point which is neither on the boundary of D_1 nor in the closure of D_2'. Let α be real and set

$$\Phi(w) = \frac{e^{i\alpha} w}{w-w_0}. \tag{3.6}$$

Then we may let $\Delta_0=\{w\colon |w-w_0|>\delta\}$ for some sufficiently small δ. Note that

$$-w^2\Phi(1/w) = -\frac{e^{i\alpha} w^2}{1-w_0 w}. \tag{3.7}$$

Together, (3.6) and (3.7) make up the variation used in 2.3. Thus we have split the variation for the class $\mathscr{E}$ into separate variations for each member of the pair.

There are two possibilities for w_0 in order to satisfy the stated requirements. We may have $w_0=F_1(z_0)$ for some $z_0\in U$, or w_0 may lie in the interior of the complement of $D_1\cup D_2'$ (if this set has an interior). In the latter case $\Phi(w)$ is analytic in D_1. In either case $-w^2\Phi(1/w)$ is analytic in D_2.

Proceeding as in Section 2, one easily calculates the integrals in (3.3) and (3.5), and proves

Theorem 3.1. *Let $\{F_1(z), F_2(z)\}$ be any univalent pair. Let α be any real number and let $z_0\in U$. Then for all sufficiently small $\varepsilon>0$ there exists a univalent pair*

$\{F_1^*(z), F_2^*(z)\}$ *such that*

$$F_1^*(z) = F_1(z)+\varepsilon e^{i\alpha}\frac{F_1(z)}{F_1(z)-F_1(z_0)} - \varepsilon e^{i\alpha}\left(\frac{F_1(z_0)}{z_0 F_1'(z_0)^2}\right)\frac{zF_1'(z)}{z-z_0} \tag{3.8}$$
$$+\varepsilon e^{-i\alpha}\overline{\left(\frac{F_1(z_0)}{z_0 F_1'(z_0)^2}\right)}\frac{z^2F_1'(z)}{1-\bar{z}_0 z}+o(\varepsilon),$$

$$F_2^*(z) = F_2(z)-\varepsilon e^{i\alpha}\frac{F_2(z)^2}{1-F_1(z_0)F_2(z)}+o(\varepsilon). \tag{3.9}$$

If there exists a w_0 *such that* w_0 *is not in the closure of* $F_1(U)$ *and* w_0^{-1} *is not in the closure of* $F_2(U)$, *then for all real* α *and all sufficiently small* $\varepsilon>0$ *there exists a univalent pair* $\{F_1^*(z), F_2^*(z)\}$ *such that*

$$F_1^*(z) = F_1(z)+\varepsilon e^{i\alpha}\frac{F_1(z)}{F_1(z)-w_0}+o(\varepsilon) \tag{3.10}$$

$$F_2^*(z) = F_2(z)-\varepsilon e^{i\alpha}\frac{F_2(z)^2}{1-w_0F_2(z)}+o(\varepsilon). \tag{3.11}$$

We remark that $\{F_1, F_2\}$ is a univalent pair if and only if $\{F_2, F_1\}$ is also. Hence the roles of F_1 and F_2 may be reversed in the above theorem.

4. General extremal problems in the class $\mathscr{E}$

4.1. The usual class $\mathscr{S}$ of univalent functions is normalized by the requirement that $f(0)=0$ and $f'(0)=1$ for any $f\in\mathscr{S}$. Similar normalizations are impossible in $\mathscr{E}$ since the addition of a constant or multiplication by a constant may spoil the Bieberbach—Eilenberg property of a domain. Indeed, there are only two elementary transformations of the image domain which are available in the class. If $F(z)\in\mathscr{E}$, then so are $-F(z)$ and $\overline{F(\bar{z})}$. These correspond to symmetries of the Bieberbach—Eilenberg property and are often useful.

Transformations can however be made freely in the z-plane. If $g(z)$ is univalent in U, $g(0)=0$, and $g(U)\subset U$, then $F(z)\in\mathscr{E}$ implies $F(g(z))\in\mathscr{E}$. In particular, $F(e^{i\alpha}z)\in\mathscr{E}$ for any real α. Thus we can assume $F'(0)>0$ if this seems like a useful normalization in a particular problem.

By letting α be near zero, this last transformation can be viewed as a variation in the class $\mathscr{E}$. That is, if $F(z)\in\mathscr{E}$, then so is $F^*(z)$ where

$$F^*(z) = F(e^{i\varepsilon}z) = F(z)+i\varepsilon zF'(z)+o(\varepsilon) \tag{4.1}$$

for real ε.

Another useful variation is the *slit variation* obtained by letting $g(z)$ map U onto U less a short slit. Let $k_\alpha(z)=z/(1+e^{-i\alpha}z)^2$ be the Koebe function mapping U onto the exterior of the radial slit from $e^{i\alpha}/4$ to ∞. Let $\varepsilon>0$, and define $g(z)$

by $k_\alpha(g(z))=k_\alpha(z)/(1+\varepsilon)$. This is equivalent to $(1+\varepsilon)(z^{-1}+2e^{-i\alpha}+e^{-2i\alpha}z)=(1/g+2e^{-i\alpha}+e^{-2i\alpha}g)$ and hence $g(z)=z+\varepsilon g_\alpha(z)+o(\varepsilon)$ where

$$g_\alpha(z) = -\frac{z(1+e^{-i\alpha}z)}{1-e^{-i\alpha}z}.$$

Thus, if $F\in\mathscr{E}$ and α is any real number, $F^*(z)\in\mathscr{E}$ also where

$$F^*(z) = F(g(z)) = F(z)-\varepsilon zF'(z)\frac{e^{i\alpha}+z}{e^{i\alpha}-z}+o(\varepsilon) \tag{4.2}$$

and $\varepsilon>0$. These two well-known variations will be usefull in what follows.

4.2. In considering extremal problems in $\mathscr{E}$, one must be careful since the class $\mathscr{E}$ is not compact. One easily sees that any sequence of functions in $\mathscr{E}$ has a subsequence which converges either to a function in $\mathscr{E}$ or to the constant zero. The functions $\{z/n\}_{n=1}^{\infty}$ show that the last possibility can occur. Often, the fact that no extremal function exists in $\mathscr{E}$ is not obvious. For example, if we wish to maximize $|b_2/b_1|$ where $F(z)=b_1z+b_2z^2+\dots$, we find that no maximum exists. The functions mapping U onto U slit from $-1/n$ to -1 will give $|b_2/b_1|\to 2$, but the uniqueness of the Koebe function maximizing a_2 in $\mathscr{S}$ shows there is no $F\in\mathscr{E}$ with $|b_2/b_1|=2$.

In most of the problems we consider, the existence of an extremal function will be obvious. For example, consider the maximum absolute value of b_n, the n-th coefficient of $F\in\mathscr{E}$. We can always assume $b_n>0$ and so look for F maximizing $\operatorname{Re} b_n$. A sequence of functions whose n-th coefficient converged to the supremum of such values could not tend to zero, and hence the extremal F must exist. It is instructive to see what the variation of Theorem 2.2 implies in this case.

For every α and z_0 we must have $\operatorname{Re} b_n^*\leqq \operatorname{Re} b_n$ for the extremal F. Since the real part is unchanged if we conjugate the term containing $e^{-i\alpha}$ in (2.6), the terms of order ε must have an n-th coefficient which vanishes. One easily sees that this implies that any extremal F must satisfy the differential equation

(4.3)
$$\left(\frac{zF'(z)}{F(z)}\right)^2 \sum_{k=1}^{n-1} b_n^{(k+1)}\left(\frac{1}{F(z)^k}+F(z)^k\right) = \sum_{k=1}^{n-1}\frac{(n-k)b_{n-k}}{z^k}+nb_n+\sum_{k=1}^{n-1}(n-k)\bar{b}_{n-k}z^k$$

where the $b_n^{(k)}$ are defined by $F(z)^k=\sum_{n=k}^{\infty} b_n^{(k)}z^n$.

This is exactly the differential equation which Nelson obtained in his thesis [26] for the same problem.

4.3. We now look at more general extremal problems, following the methods of [32]. Let Ψ be a continuous functional defined over $\mathscr{E}$. We assume that Ψ has a continuous Frêchet (or more generally Gâteaux) derivative L. That is, for any $F\in\mathscr{E}$ and any analytic G,

$$\Psi(F+\varepsilon G) = \Psi(F)+\varepsilon L(F;\, G)+o(\varepsilon). \tag{4.4}$$

Here $L(F; G)$ will be a continuous linear functional of G for each fixed F. Since F will remain fixed, we will usually suppress the dependence on F and write merely $L(G)$.

Define

$$(4.5)\qquad \begin{aligned} D(w) &= L\left(\frac{wF}{F-w}\right); & E(\zeta) &= L\left(\frac{\zeta z F'(z)}{z-\zeta}\right); \\ A(w) &= D(w)+L(F)+D(1/\bar{w}); & B(\zeta) &= E(\zeta)+\overline{L(zF'(z))}+\overline{E(1/\bar{\zeta})}. \end{aligned}$$

Observe that

$$(4.6)\qquad \begin{aligned} -\frac{\bar{w}F^2}{1-\bar{w}F} &= F+\frac{F}{\bar{w}(F-1/\bar{w})} \\ -\frac{\bar{\zeta}z^2F'(z)}{1-\bar{\zeta}z} &= zF'(z)+\frac{zF'(z)}{\bar{\zeta}(z-1/\bar{\zeta})} \end{aligned}$$

and hence the variation of Theorem 2.2 implies

$$\Psi(F^*) = \Psi(F)+\varepsilon e^{i\alpha}\frac{1}{F(z_0)}A(F(z_0))-\varepsilon e^{i\alpha}\frac{1}{F(z_0)}\left(\frac{F(z_0)}{z_0F'(z_0)}\right)^2 E(z_0)$$

$$-\varepsilon e^{-i\alpha}\frac{1}{\overline{F(z_0)}}\overline{\left(\frac{F(z_0)}{z_0F'(z_0)}\right)^2}[L(zF'(z))+E(1/\bar{z}_0)]+o(\varepsilon)$$

or

$$(4.7)\qquad \operatorname{Re}\Psi(F^*) = \operatorname{Re}\Psi(F)+\varepsilon \operatorname{Re}\frac{e^{i\alpha}}{F(z_0)}\left\{A(F(z_0))-\left(\frac{F(z_0)}{z_0F'(z_0)}\right)^2 B(z_0)\right\}+o(\varepsilon).$$

We are interested in problems of the type: maximize $\operatorname{Re}\Psi(F)$ among all $F\in\mathscr{E}$. Such a problem may or may not have a solution, but if it does, we can use (4.7) to characterize this solution. We will say that $F\in\mathscr{E}$ is *(locally) extremal for* $\operatorname{Re}\Psi(F)$ if $\operatorname{Re}\Psi(F^*)\leqq\operatorname{Re}\Psi(F)$ for all (nearby) $F^*\in\mathscr{E}$. Here, "nearby" is in the sense of convergence on compact subsets of U. Using (4.7) we therefore can prove:

Theorem 4.1. *Let the functional Ψ, defined over $\mathscr{E}$, have a continuous linear Frêchet derivative $L(F; G)$ as defined in* (4.4). *Let $A(w)$ and $B(\zeta)$ be defined as in* (4.5). *Suppose $F\in\mathscr{E}$ is locally extremal for* $\operatorname{Re}\Psi(F)$. *Then F satisfies the differential equation*

$$(4.8)\qquad \left(\frac{zF'(z)}{F(z)}\right)^2 A(F(z)) = B(z),\quad z\in U$$

or equivalently

$$(4.9)\qquad A(w)\frac{dw^2}{w^2} = B(z)\frac{dz^2}{z^2},\quad w = F(z),\quad z\in U.$$

Further, $L(F; zF'(z))$ is real, $B(z)$ is real and non-positive for $|z|=1$, and if $A(F(z))$ is analytic in some annulus $\varrho<|z|<1$, then F maps U onto a domain whose boundary is made up of analytic arcs (or is an analytic curve) which are trajectories of $A(w)\,dw^2/w^2$.

Proof. The fact that F satisfies the differential equation (4.8) or (4.9) follows immediately from (4.7) since α is arbitrary, and F makes $\operatorname{Re}\Psi$ a local maximum. Using the rotational variation (4.1) gives $\operatorname{Re}\Psi(F^*)=\operatorname{Re}\Psi(F)+\varepsilon\operatorname{Re} iL(zF'(z))+ +o(\varepsilon)$. Since ε can be positive or negative, $L(zF'(z))$ must be real. If $|\zeta|=1$, then (4.5) shows $B(\zeta)$ is real. Let $\zeta=e^{i\alpha}$. Then the slit variation (4.2) gives

$$\operatorname{Re}\Psi(F^*)=\operatorname{Re}\Psi(F)-\varepsilon\operatorname{Re}L\left(zF'(z)\frac{\zeta+z}{\zeta-z}\right)+o(\varepsilon).$$

Since $\varepsilon>0$, this implies $\operatorname{Re}L(zF'(z)(\zeta+z)/(\zeta-z))\geqq 0$. However, $|\zeta|=1$ and since $(\zeta+z)/(\zeta-z)=-1-2\zeta/(z-\zeta)$,

$$B(\zeta)=\operatorname{Re}B(\zeta)=L(zF'(z))+2\operatorname{Re}E(\zeta)$$
$$=-\operatorname{Re}L\left(zF'(z)\frac{\zeta+z}{\zeta-z}\right)\leqq 0.$$

If $A(F(z))$ is analytic in $\varrho<|z|<1$, then the reflection principle shows $[zF'(z)/F(z)]^2A(F(z))$ is analytic up to isolated poles on $|z|=1$. For $z=e^{it}$, $dz/z=i\,dt$ and from (4.8) the boundary of U must be a trajectory of $A(w)dw^2/w^2$.

Theorem 4.2. *Let Ψ and F be as in the previous theorem. Then $A(w)dw^2/w^2$ is invariant under the mapping $w\mapsto 1/w$. Suppose $A(w)$ is analytic up to isolated poles and is not identically zero. If w_0 and w_0^{-1} are not in $D=F(U)$, then at least one is on the boundary of D. The points 1 and -1 are on the boundary of D. The function $G(z)=F(z)+1/F(z)$ is univalent in U and maps U onto a domain D_1 (containing ∞) whose closure is the entire complex sphere.*

Proof. We remark that these observations were made by Nelson [26] for the case of the particular extremal problems he considered.

The first result is immediate from the definition of $A(w)$ in (4.5). The second follows from the fact that if w_0 and w_0^{-1} were exterior to D, Theorem 2.3 would apply and

(4.10) $$\Psi(F^*)=\Psi(F)+\varepsilon e^{i\alpha}A(w_0)+o(\varepsilon).$$

Taking the real part, this implies $A(w_0)=0$. However, this would then have to hold in an entire neighborhood of w_0, giving a contradiction.

Since 1 and -1 are such that $w_0=w_0^{-1}$, both points must be on ∂D.

Finally $w+1/w$ has a two-valued inverse, the values being w and $1/w$. Hence the Bieberbach—Eilenberg property of D makes $w+1/w$ univalent on D. Any point exterior to D_1 would have to arise from a w_0 exterior to D such that w_0^{-1} was also exterior to D. This completes the proof.

4.4. The following theorem is useful in many applications.

Theorem 4.3. *Let Ψ be a complex valued functional over $\mathscr{E}$ having a continuous Frêchet derivative L as in the hypotheses of Theorem 4.1. Let $T=\{\Psi(F)\colon F\in\mathscr{E}\}$*

and suppose $F \in \mathscr{E}$ *is such that* $\Psi(F)$ *is on the boundary of* T. *If* $E(\zeta)$ *is not constant, then there exists a complex* λ *with* $|\lambda|=1$ *such that* F *satisfies Theorems* 4.1 *and* 4.2 *as if it were locally extremal for* $\operatorname{Re}\lambda\Psi(F)$.

Proof. The proof follows the argument found on pages 495, 496 of [30], but we sketch it here since there are some differences in this case. The variation of Theorem 2.2 gives

$$\Psi(F^*) = \Psi(F)+\varepsilon e^{i\alpha}U-\varepsilon e^{-i\alpha}V+o(\varepsilon),$$

where

$$U = A\big(F(z_0)\big)/F(z_0)-\big(F(z_0)/z_0F'(z_0)\big)^2E(z_0)/F(z_0),$$

$$V = \overline{\big(F(z_0)/z_0F'(z_0)\big)^2}\cdot[L\big(zF'(z)\big)+\overline{E(1/\bar{z}_0)}]/\overline{F(z_0)},$$

these quantities being computed from (4.5). Now $s=\varepsilon e^{i\alpha}U-\varepsilon e^{-i\alpha}V$ is a linear transformation of the real and imaginary parts of $\varepsilon e^{i\alpha}$ to the real and imaginary parts of s. Hence as ε and α vary, a full neighborhood of $s=0$ will be covered unless the rank of the transformation is less than 2. Since $\Psi(F)$ is assumed to be a boundary point of T, this rank must be less than 2 and $e^{i\alpha}U-e^{-i\alpha}V$ has constant argument for all real α. That is, there exists a λ with $|\lambda|=1$ so that $\operatorname{Re}\{e^{i\alpha}\lambda U-e^{-i\alpha}\lambda V\}= =\operatorname{Re} e^{i\alpha}\{\lambda U-\bar{\lambda}\bar{V}\}=0$. Hence $\lambda U=\bar{\lambda}\bar{V}$. Then F satisfies the differential equation

$$\lambda A(w)\frac{dw^2}{w^2} = [\lambda E(z)+\bar{\lambda}\overline{L\big(zF'(z)\big)}+\bar{\lambda}\overline{E(1/\bar{z})}]\frac{dz^2}{z^2}.$$

Next, we show that $\lambda L\big(zF'(z)\big)$ is real and $B_\lambda(\zeta)=\lambda E(\zeta)+\lambda L\big(zF'(z)\big)+\bar{\lambda}\overline{E(1/\bar{\zeta})}\leqq 0$ for $|\zeta|=1$. First, consider a combination of variations (4.1) and (2.6). Let $-1\leqq u\leqq 1$, $-1\leqq v\leqq 1$ and set $F_1(z)=F(e^{i\varepsilon u}z)$. Apply variation (2.6) to F_1 with ε replaced by εv. Since $F_1=F+O(\varepsilon)$, we find

$$\begin{aligned}\Psi(F^*) &= \Psi(F_1)+\varepsilon v[e^{i\alpha}U-e^{-i\alpha}V]+o(\varepsilon)\\ &= \Psi(F)+i\varepsilon uL\big(zF'(z)\big)+\varepsilon v[e^{i\alpha}U-e^{-i\alpha}V]+o(\varepsilon)\\ &= \Psi(F)+\varepsilon W(u,v)+o(\varepsilon).\end{aligned}$$

Here, $\lambda W(u,v)=ui\lambda L\big(zF'(z)\big)+v[e^{i\alpha}\lambda U-e^{-i\alpha}\bar{\lambda}\bar{U}]+o(\varepsilon)=ui\lambda L\big(zF'(z)\big)+2vi\operatorname{Im}\{e^{i\alpha}\lambda U\}$. Again, since a neighborhood of 0 cannot be covered, we must have $\lambda L\big(zF'(z)\big)$ real. The exception would be if $U\equiv 0$. But then $V\equiv 0$ and hence $E(1/\bar{z}_0)$ would be constant. Hence $\lambda L\big(zF'(z)\big)$ is real and so is $B_\lambda(\zeta)$ if $|\zeta|=1$.

In an exactly similar way we may combine the variation (4.2) with (2.6). Then for ζ with $|\zeta|=1$, $\Psi(F^*)=\Psi(F)+\varepsilon W_1(u,v)+o(\varepsilon)$, where $\varepsilon>0$, $0\leqq u\leqq 1$, $-1\leqq v\leqq 1$, and $\lambda W_1(u,v)=-u\lambda L\big(zF'(z)(\zeta+z)/(\zeta-z)\big)+2vi\operatorname{Im}\{e^{i\alpha}\lambda U\}$. Again, $\operatorname{Im}\{e^{i\alpha}\lambda U\}$ is not identically zero. Hence, if the real part of $\lambda L\big(zF'(z)(\zeta+z)/(\zeta-z)\big)$ took on different signs for two ζ, then this combination of variations would cover a neighborhood of $\Psi(F)$. Hence $B_\lambda(\zeta)=-\operatorname{Re}\big[\lambda L\big(zF'(z)(\zeta+z)/(\zeta-z)\big)\big]$ cannot change sign.

When we chose λ above, we could have equally well chosen $-\lambda$. Hence by proper choice we can assure that $B_\lambda(\zeta)\leqq 0$ for all ζ with $|\zeta|=1$. Finally, the proof of Theorem 4.2 holds with only minor changes. If (4.10) holds, then $\Psi(F^*)$ will again cover a neighborhood of $\Psi(F)$, leading to a contradiction.

4.5. Letting $z\to 0$ in (4.8) will give a Marty relation for the given extremal problem. This can also be obtained quite simply directly from (2.6).

Theorem 4.4. *Let Ψ and L be as in Theorem* 4.1. *Suppose* $F(z)=b_1z+b_2z^2+\dots$ $\dots\in\mathscr{E}$ *is locally extremal for* $\max \operatorname{Re}\Psi(F)$. *Then F satisfies the Marty relation*

$$b_1L(F;\,1-F(z)^2) = L(F;\,F'(z))-\overline{L(F;\,z^2F'(z))}.$$

Proof. Put $z_0=0$ in (2.6). This relation follows easily.

5. General extremal problems for pairs

5.1. The pair property has one useful symmetry. If $\{F, G\}$ is a pair, then so is $\{\overline{F(\bar z)}, \overline{G(\bar z)}\}$. Although the pair property is quite restrictive, there is a wide class of transformations available for special normalization. In particular, if $\{F, G\}$ is a pair and $a\neq 0$, then $\{aF, a^{-1}G\}$ is also a pair. If $g_1(z)$ and $g_2(z)$ are any univalent functions mapping U into U with $g_1(0)=g_2(0)=0$, and $\{F, G\}$ is a univalent pair, then $\{F(g_1(z)), G(g_2(z))\}$ will be a univalent pair.

Specifically, if $\{F, G\}$ is a univalent pair and $\varrho>0$, $\alpha_0, \alpha_1, \alpha_2$ are any real numbers, then

$$\{\varrho e^{i\alpha_0}F(e^{i\alpha_1}z),\ \varrho^{-1}e^{-i\alpha_0}G(e^{i\alpha_2}z)\} \tag{5.1}$$

is also a univalent pair. Observe that by proper choice of α_0, α_1, and α_2, we can use (5.1) to make any coefficient of F and any two coefficients of G real and positive.

We also observe that the rotational and slit variations of (4.1) and (4.2) can be applied independently to each member of a univalent pair.

5.2. The class of univalent pairs is not compact, but it does have some compactness properties. Suppose $\{F_n, G_n\}$ is a sequence of univalent pairs. Then by extracting subsequences as necessary we can assume $F_n(z)/F_n'(0)\to f(z)\in\mathscr{S}$, $G_n(z)/G_n'(0)\to g(z)\in\mathscr{S}$, $F_n'(0)\to b$, and $G_n'(0)\to c$. If b and c are both finite non-zero complex numbers, then the pairs $\{F_n, G_n\}$ converge uniformly on compact subsets of U to the univalent pair $\{bf(z), cg(z)\}$.

If $b=\infty$, then quarter theorem and the pair property show that $G_n(U)\subset\subset\{w\colon |w|<4/|F_n'(0)|\}$ and hence $c=0$. In this case $F_n(z)\to\infty$ (in the sense that it converges to ∞ uniformly in compact subsets of $U-\{0\}$) and $G_n(z)\to 0$ (uniformly in compact subsets of U).

If $b=0$, then $F_n(z)\to 0$. However, $G_n(z)$ may converge to ∞, 0, or some univalent $G(z)$. This is easily seen by considering examples of $F_n(z)$ which map U onto $V_n=\{w\colon |w|<1/n\}$, or onto $V_n\cup W(1/n^2, n)$, or onto $V_n\cup W(1/n^2, 4)$ where $W(\delta, R)=\{w\colon 0<\operatorname{Re} w<R,\ |\operatorname{Im} w|<\delta\}$.

If b is finite and non-zero, then the $F_n(z)$ converge to some univalent F. The G_n may converge to a univalent G, in which case $\{F, G\}$ is a pair, or the G_n may converge to 0.

Individual extremal problems must be considered carefully to see whether an extremal univalent pair does indeed exist. For example, $(|b_1|+|c_1|)$ is easily seen to have no maximum in the set of all pairs.

5.3. Let $\Psi(F_1, F_2)$ be a continuous complex valued functional defined over all univalent pairs $\{F_1, F_2\}$. We assume that Ψ has a Frêchet derivative at $\{F_1, F_2\}$. That is,

$$\Psi(F_1+\varepsilon H_1, F_2+\varepsilon H_2) = \Psi(F_1, F_2)+\varepsilon L_1(F_1, F_2; H_1)+\varepsilon L_2(F_1, F_2; H_2)+o(\varepsilon) \tag{5.2}$$

where L_1 and L_2 are continuous linear functionals of H_1 and H_2, respectively. We will usually suppress the dependence on F_1 and F_2 and write these simply as $L_1(H_1)$ and $L_2(H_2)$. We will say that a univalent pair $\{F_1, F_2\}$ is *(locally) extremal for* $\operatorname{Re}\Psi(F_1, F_2)$ if $\operatorname{Re}\Psi(F_1^*, F_2^*)\leqq\operatorname{Re}\Psi(F_1, F_2)$ for all (nearby) univalent pairs $\{F_1^*, F_2^*\}$. Here "nearby" is in the sense of convergence on compact subsets of U.

Suppose that the univalent pair $\{F_1, F_2\}$ is locally extremal for $\operatorname{Re}\Psi(F_1, F_2)$. From Theorem 3.1 and (5.2) we must therefore have

$$\operatorname{Re} e^{i\alpha}\left\{L_1\left(\frac{F_1}{F_1-w_0}\right)-L_2\left(\frac{F_2^2}{1-w_0F_2}\right)-\left(\frac{F_1(z_0)}{z_0F_1'(z_0)^2}\right)\left[L_1\left(\frac{zF_1'(z)}{z-z_0}\right)-\overline{L_1\left(\frac{z^2F_1'(z)}{1-\bar{z}_0z}\right)}\right]\right\}\leqq 0 \tag{5.3}$$

for any $z_0\in U$ and any real α. Here $w_0=F_1(z_0)$ and we conjugated the term containing $e^{-i\alpha}$ in (3.10).

Define

$$\left\{\begin{aligned}
&D_1(w) = L_1\left(\frac{wF_1}{F_1-w}\right) && D_2(w) = L_2\left(\frac{wF_2}{F_2-w}\right)\\
&E_1(\zeta) = L_1\left(\frac{\zeta zF_1'(z)}{z-\zeta}\right) && E_2(\zeta) = L_2\left(\frac{\zeta zF_2'(z)}{z-\zeta}\right)\\
&A_1(w) = D_1(w)+L_2(F_2)+D_2(1/w) && A_2(w) = D_2(w)+L_1(F_1)+D_1(1/w)\\
&B_1(\zeta) = E_1(\zeta)+\overline{L_1(zF_1'(z))}+\overline{E_1(1/\bar{\zeta})} && B_2(\zeta) = E_2(\zeta)+\overline{L_2(zF_2'(z))}+\overline{E_2(1/\bar{\zeta})}.
\end{aligned}\right. \tag{5.4}$$

Proceeding almost exactly as in the proof Theorems 4.1 and 4.2, noting that the roles of F_1 and F_2 can be interchanged in Theorem 3.1, one easily verifies:

Theorem 5.1. *Let $\Psi(F_1, F_2)$ be a continuous complex valued functional defined over all univalent pairs, having a continuous linear Frêchet derivative defined as in* (5.2). *Let A_1, A_2, B_1, and B_2 be defined as in* (5.4). *Suppose the univalent pair $\{F_1, F_2\}$ is locally extremal for* $\operatorname{Re}\Psi(F_1, F_2)$. *Then $L_1(zF_1'(z))$ and $L_2(zF_2'(z))$ are real; $B_1(z)$*

and $B_2(z)$ are real and non-positive for $|z|=1$; and F_1 and F_2 satisfy the differential equations

$$(5.5)\quad \begin{cases} \left(\dfrac{zF_1'(z)}{F_1(z)}\right)^2 A_1(F_1(z)) = B_1(z), & z\in U \\ \left(\dfrac{zF_2'(z)}{F_2(z)}\right)^2 A_2(F_2(z)) = B_2(z), & z\in U. \end{cases}$$

If either $A_1(w)$ or $A_2(w)$ is analytic up to isolated singularities and not identically zero, then there is no w_0 exterior to $F_1(U)$ with w_0^{-1} simultaneously exterior to $F_2(U)$.

We remark that the last conclusion of theorem is equivalent to saying that the closure of $[F_1(U)]\cup[F_2(U)]'$ is the entire complex sphere, where we denote $B'=\{w\colon 1/w\in B\}$.

5.4. In most extremal problems, A_1, A_2, B_1, and B_2 are analytic up to isolated poles (and not identically zero). Since $dz^2/z^2<0$ on $|z|=1$, it follows that the boundaries of $F_1(U)$ and $F_2(U)$ lie on the trajectories

$$A_1(w)\frac{dw^2}{w^2}\geqq 0,\quad A_2(w)\frac{dw^2}{w^2}\geqq 0,$$

respectively. These boundaries must consist of a simple closed curve plus some possible slits and must be made up of analytic curves and arcs. Any branching can occur only at the critical points of A_1 and A_2. Since $w\mapsto 1/w$ carries the outer boundary of $F_1(U)$ to the outer boundary of $F_2(U)$, we must have $A_1(w)/A_2(1/w)$ real and positive for w on the analytic arcs making up the outer boundary of $F_1(U)$. In practice, we find for most simple problems that after suitable normalization $A_1(w)=A_2(1/w)$.

5.5. The obvious analog of Theorem 4.3 will hold for pairs. The proof is essentially the same.

Similarly, Marty relations will hold for an extremal pair. Putting $z_0=0$ in Theorem 3.1 gives

$$F_1'(0)[L_1(1)-L_2(F_2^2)] = L_1(F_1'(z))-\overline{L_1(z^2F_1'(z))},$$
$$F_2'(0)[L_2(1)-L_1(F_1^2)] = L_2(F_2'(z))-\overline{L_2(z^2F_2'(z))}.$$

Again these Marty relations are interesting but do not seem to be generally useful.

5.6. It is useful to demonstrate the use of Theorem 5.1 in a very simple case. Let $F_1(z)=b_1z+b_2z^2+\dots$, $F_2(z)=c_1z+c_2z^2+\dots$ and consider the problem of maximizing $|b_1c_1|$ among all univalent pairs. Here the solution $|b_1c_1|\leqq 1$ is well known [1] and is sharp, for example for the pair $F_1(z)=F_2(z)=z$. However, let us analyze the problem using the methods of this section.

From (5.1) we may assume $b_1>0$ and $c_1>0$, and hence set $\Psi(F_1, F_2)=$ $=\chi_1(F_1)\chi_2(F_2)$, where $\chi_\nu(H(z))$ represents the coefficient of z^ν in the series expansion of $H(z)$. Let $\alpha=\sup \operatorname{Re}\{\chi_1(F_1)\chi_1(F_2)\colon \{F_1, F_2\}$ is a univalent pair$\}$. There exists a sequence of pairs $\{F_{1,n}, F_{2,n}\}$ with $\chi_1(F_{1,n})=1$, $\chi_1(F_{2,n})\to\alpha$. The quarter theorem shows that the $F_{2,n}$ are bounded, hence α is finite. Since $\alpha>0$, the $F_{2,n}$ do not converge to 0. Hence an extremal pair exists for this Ψ. Observe that the important fact here is that Ψ is invariant under $\{F_1, F_2\}\mapsto\{aF_1, a^{-1}F_2\}$.

For $\Psi(F_1, F_2)=\chi_1(F_1)\chi_1(F_2)$, one easily verifies that $L_1(H_1)=c_1\chi_1(H_1)$, $L_2(H_2)=b_1\chi_1(H_2)$ where $F_1(z)=b_1z+\dots$ and $F_2(z)=c_1z+\dots$. Then from (5.4) $A_1(w)=A_2(w)=B_1(z)=B_2(z)=-b_1c_1$ and hence the extremal pair satisfy

$$\left(\frac{zF_1'(z)}{F_1(z)}\right)^2=1 \qquad \left(\frac{zF_2'(z)}{F_2(z)}\right)^2=1.$$

Thus $F_1(z)=b_1z$, $F_2(z)=c_1z$, and the relationship of the boundaries of $F_1(U)$ and $F_2(U)$ requires $|c_1|=1/|b_1|$, i.e., $\alpha=|b_1c_1|=b_1c_1=1$ for the extremal pair.

6. The value sets for $F\in\mathscr{E}$

6.1. In 1954 Jenkins showed [18] that if $F\in\mathscr{E}$ and $|z_0|=r$, then $|F(z_0)|\leqq r/(1-r^2)^{1/2}$, with the maximum being achieved at $z_0=ie^{-i\varphi}r$ by the function

$$F_{r,\varphi}(z)=\frac{(1-r^2)^{1/2}e^{i\varphi}z}{1+ire^{i\varphi}z}. \tag{6.1}$$

This function maps U onto the interior of a circle centered on the imaginary axis and passing through the points ± 1. Such a circle is invariant under the mapping $w\mapsto 1/w$ and hence $F_{r,\varphi}\in\mathscr{E}$.

In this section we study the set of possible values of a Bieberbach—Eilenberg functions at a fixed point in U. Define, for any r with $0<r<1$

$$V(r)=\{F(r)\colon F\in\mathscr{E}\}. \tag{6.2}$$

If $F(z)\in\mathscr{E}$, then so is $F(\varrho e^{i\alpha}z)$ for $0<\varrho\leqq 1$ and any real α. Hence $V(r)=\{F(re^{i\alpha})\colon F\in\mathscr{E}\}=\{F(z)\colon F\in\mathscr{E}$ and $0<|z|\leqq r\}$. If $F\in\mathscr{E}$, then so are $-F(z)$ and $\overline{F(\bar z)}$. Hence $V(r)$ is symmetric with respect to reflection in the real and imaginary axes. It thus suffices to study $V(r)$ in the first quadrant. The points $0, +1, -1$ are never in any $V(r)$, but each $V(r)$, $0<r<1$, contains a punctured neighborhood of 0. Each $V(r)$ is bounded (from the Jenkins result or using the fact that $F(z)/F'(0)$ is in $\mathscr{S}$ and $|F'(0)|\leqq 1$ so $|F(r)|\leqq r/(1-r^2)$). Thus $V(r)\cup\{0\}$ is compact. It is easy to construct examples of Bieberbach—Eilenberg domains with boundaries made up of circular arcs containing any preassigned $b\neq 0, +1, -1$. Thus, each such b is in some $V(r)$. If $F(r)=b$, then F maps a neighborhood of r onto a neighborhood of b. Hence b is an interior point of $V(r_1)$ for any $r_1>r$. That is, the domains $V(r)\cup\{0\}$ are strictly monotone increasing and $r_0=\inf\{r\colon b\in V(r)\}$ is such that $b\in\partial V(r_0)$.

6.2. Let $b \neq 0, 1$ be a boundary point of $V(r)$ in the first quadrant (i.e., $0 \leqq \arg b \leqq \leqq \pi/2$). Then $\log b$ lies on the boundary of the set of $\log F(r)$ and from Theorem 4.3, if $F \in \mathscr{E}$ with $F(r)=b$, then F will be locally extremal for $\operatorname{Re} \lambda \log F(r)$ for some λ with $|\lambda|=1$. (It is not essential to take the logarithm here, but it offers a minor convenience to do so.) Thus we consider $\Psi(F)=\lambda \log F(r)$. Then $L(H)=\lambda H(r)/F(r)$. If F is locally extremal for $\operatorname{Re} \lambda \log F(r)$, then from Theorem 4.1, F satisfies

$$\frac{\lambda(1-b^2)\,dw^2}{w(b-w)(1-bw)} = \frac{\lambda c(1-r^2)\,dz^2}{z(r-z)(1-rz)}, \qquad w = F(z), \quad z \in U, \tag{6.3}$$

where $b=F(r)$ and $c=rF'(r)/F(r)$. Theorem 4.1 shows λc is real, and since $B(1)=-\lambda c(1-r^2)/(1-r)^2$, $\lambda c>0$. Observe that $\lambda=1$ corresponds to maximizing $|F(r)|$ for $F \in \mathscr{E}$.

6.3. Set

$$\Omega = \Omega(w) = \int_0^w [w(b-w)(1-bw)]^{-1/2}\,dw. \tag{6.4}$$

This an elliptic integral and hence the inverse of a doubly periodic function. The critical points are $0, b, b^{-1}$, and ∞. One period, $2\hat{\omega}_1$, will be twice the integral from 0 to b along the line joining these points. Setting $w=bt^2$, we find

$$\begin{aligned} \hat{\omega}_1 &= 2 \int_0^1 [(1-t^2)(1-b^2t^2)]^{-1/2}\,dt \\ &= 2K(b) \end{aligned} \tag{6.5}$$

where $K(b)$ is the normal complete elliptic integral of the first kind (in Jacobi's form). This is uniquely defined for any $b \neq 0, 1$ in the first quadrant if we specify that its real part is positive to fix the sign of the root and define it by continuity for b real and greater than 1. We remark that in the last case there is no canonical path of integration. Passing above or below b^{-1} results in complex conjugate determinations of $K(b)$. We choose the determination resulting from passing above.

A second period of (6.4) is $2\hat{\omega}_2$, equal to twice the integral along the circular arc from b to b^{-1} through 1. This integral is twice the integral from b to 1 along this path since $d\Omega$ is invariant under the mapping $w \mapsto 1/w$. Set $b_1=(1-b)/(1+b)$ and $w=(1-b_1t)/(1+b_1t)$. Then

$$\begin{aligned} \hat{\omega}_2 &= \frac{4i}{(1+b)} \int_0^1 [(1-t^2)(1-b_1^2t^2)]^{-1/2}\,dt \\ &= \frac{4i}{1+b} K(b_1). \end{aligned}$$

Landen's (or Gauss') transformation gives $2K((1-b)/(1+b))=(1+b)K(\sqrt{1-b^2})=$

$=(1+b)K'(b)$ where $K'(b)$ is the associated complete elliptic integral. See [8] for example. Thus

$$\hat{\omega}_2 = 2iK'(b). \tag{6.6}$$

Other pairs of periods of (6.4) will be related to $\hat{\omega}_1$ and $\hat{\omega}_2$ by unimodular transformations and will correspond to integration over homotopically different paths in the plane less the critical points 0, b, b^{-1}.

The line segment joining 0 to $\hat{\omega}_1$ in the Ω-plane corresponds to a path $\hat{\gamma}_1$ joining 0 to b in the w-plane, along which $d\Omega$ has a constant argument, and which is homotopic to the line segment joining 0 to b. (All homotopies are relative to $\mathbf{C}-\{0, b, b^{-1}\}$.) Similarly the line joining $\hat{\omega}_1$ to $\hat{\omega}_1+(1/2)\hat{\omega}_2$ in the Ω-plane corresponds to a path $\hat{\gamma}_2$ joining b to 1 in the w-plane, along which $d\Omega$ has a constant argument, and which is homotopic to the line segment joining b to 1. Let $\hat{\gamma}_2'$ and $\hat{\gamma}_1'$ be the images of $\hat{\gamma}_2$ and $\hat{\gamma}_1$ under the mapping $w\mapsto 1/w$. Then $\hat{\gamma}_1+\hat{\gamma}_2+\hat{\gamma}_2'+\hat{\gamma}_1'$ is a path from 0 to ∞ through b, 1, and b^{-1} along each separate arc of which $d\Omega$ is constant. (6.4) defines a univalent mapping of the complement of this path onto the parallelogram with vertices $\pm\hat{\omega}_1$, $\pm\hat{\omega}_1+\hat{\omega}_2$, in which b corresponds to $\pm\hat{\omega}_1$, 1 to $\pm\hat{\omega}_1+(1/2)\hat{\omega}_2$, b^{-1} to $\pm\omega_1+\omega_2$, and ∞ to $\hat{\omega}_2$.

This parallelogram is half of the entire period parallelogram. Starting on the "other side" of $\hat{\gamma}_1$, we see that (6.4) also maps the same split plane univalently onto the other half of the period parallelogram. Finally, observe that the mapping $w\mapsto 1/w$ corresponds to symmetry with respect to $\hat{\omega}_1+(1/2)\hat{\omega}_2$ or $\hat{\omega}_1+(3/2)\hat{\omega}_2$, modulo the two periods.

6.4. In a similar way we can consider the mapping

$$Z = \int_0^z [z(r-z)(1-rz)]^{-1/2}\,dz. \tag{6.7}$$

Here $0<r<1$ and this is a much simpler integral. Much as in § 6.3 we find that (6.7) maps the interior of the unit circle slit from 0 to 1 along the real axis univalently onto the interior of the rectangle with vertices $\pm 2K(r)$ and $\pm 2K(r)+iK'(r)$. This is one fourth of the entire period parallelogram. Here $z=r$ corresponds to $\pm 2K(r)$ and $z=1$ corresponds to $\pm 2K(r)+iK'(r)$. $z=-1$ corresponds to $iK'(r)$.

6.5. Suppose $w=F(z)$ is analytic and univalent in U and satisfies (6.3) for some r, λ, and c with $\lambda c>0$, $F(r)=b$, and with 1 on the boundary of $F(U)$. Since b is interior to $F(U)$, the boundary of $F(U)$ is a trajectory of the left-hand side of (6.3) and hence a simple closed analytic curve which we denote by γ_3. We know 1 lies on γ_3 and since (6.3) is invariant under $w\mapsto 1/w$, the same is true of γ_3. Therefore $F\in\mathcal{E}$.

Let γ_1 be the image of the segment $[0, r]$ under the mapping $w=F(z)$ and let γ_2 be the image of $[r, 1]$. Then γ_1 is a trajectory of the left-hand side of (6.3) joining 0 to b while γ_2 is an orthogonal trajectory joining b to some point $w_0\in\gamma_3$.

The mapping (6.4) carries γ_1 to a line segment l_1 joining 0 to some point ω_1 which will be the same as $\hat{\omega}_1$ modulo the period parallelogram. That is, $\omega_1=n_1\hat{\omega}_1+ +n_2\hat{\omega}_2$ where n_1 is the same odd integer and n_2 is some even integer. γ_2 maps to a line segment l_2, orthogonal to l_1, joining ω_1 to some point $\omega_1+\omega_0$. γ_3 maps to a line segment l_3, parallel to l_1, from $\omega_1+\omega_0$ to $-\omega_1+\omega_0$. Thus (6.3) maps the interior of γ_3 less the slit $\gamma_1+\gamma_2$ to the interior of the rectangle R with vertices $\pm\omega_1$, $\pm\omega_1+\omega_0$. The line segment l_3 must contain a point ω_0' which is the image of 1 under (6.4). That is, $\omega_0'=\hat{\omega}_1\pm(1/2)\hat{\omega}_2+2k\hat{\omega}_1+2j\hat{\omega}_2$. It follows that there exists an ω_2 such that $\omega_0'=\omega_1+(1/2)\omega_2$ and $\omega_2=n_3\hat{\omega}_1+n_4\hat{\omega}_2$ where n_3 is some even integer and n_4 is some odd integer. The pair $2\omega_1$, $2\omega_2$ is a pair of periods for the inverse of (6.4).

The above mapping can be reflected in l_3. The exterior of γ_3 slit from w_0^{-1} to ∞ along the image of $\gamma_1+\gamma_2$ under $w\mapsto 1/w$ will map to a rectangle congruent to R and having a subsegment of l_3 in common on the boundary. It follows that the area of R is one quarter of the area of the primitive period parallelogram and hence $2\omega_1$, $2\omega_2$ is a primitive pair of periods. That is

$$(6.8)\qquad \begin{cases} \omega_1 = n_1\hat{\omega}_1+n_2\hat{\omega}_2 & n_1,\, n_4 \text{ odd integers} \\ \omega_2 = n_3\hat{\omega}_1+n_4\hat{\omega}_2 & n_2,\, n_3 \text{ even integers} \\ & n_1n_4-n_2n_3 = 1. \end{cases}$$

Here $n_1n_4-n_2n_3\neq -1$ since the mappings involved are all conformal. The transformation (6.8) thus belongs to the congruence subgroup modulo 2 of the full set of unimodular transformations.

Since $w=F(z)$ satisfies (6.3), we must have $[\lambda(1-b^2)]^{1/2}\Omega=[\lambda c(1-r^2)]^{1/2}Z$. Set $B=[\lambda(1-b^2)]^{1/2}$, $C=[\lambda c(1-r^2)]^{1/2}$. Then $B\omega_1=2CK(r)$ and $B(\omega_1+\omega_0)= =C(2K(r)+iK'(r))$. Combining these, $B\omega_0=iCK'(r)$. Since $\lambda c>0$, $C>0$ (choosing the proper root). Multiplication of Ω by B therefore rotates the rectangle R so that the image of l_1 lies on the real axis, and the image of l_3 is parallel to the real axis. Hence $\operatorname{Im}\{B\omega_0\}=\operatorname{Im}\{B\omega_0'\}=\operatorname{Im}\{B\cdot(1/2)\omega_2\}=\operatorname{Im}\{CiK'(r)\}=CK'(r)$. Hence we have shown that the following two relations hold:

$$(6.9)\qquad \begin{aligned} &[\lambda(1-b^2)]^{1/2}\omega_1 = [\lambda c(1-r^2)]^{1/2}2K(r) > 0, \\ &\operatorname{Im}\left\{\frac{\omega_2}{\omega_1}\right\} = \frac{K'(r)}{K(r)}. \end{aligned}$$

6.6. We now state the main theorem.

Theorem 6.1. *Let $b\neq 0, 1$ with $0\leqq\arg b\leqq\pi/2$ be given. Let $K(b)$ and $K'(b)$ be the normal and associated complete elliptic integrals of the first kind, with sign determinations made so that* $\operatorname{Re} K(b)\geqq 0$ *and* $\operatorname{Re}\{K'(b)/K(b)\}>0$, *and with $K(b)$ defined by continuity for b real and greater than* 1. *Then there exists a unique r_b with*

$0<r_b<1$, *a unique* λ_b *with* $|\lambda_b|=1$, *a unique* $c_b\neq 0$, *and a function* $F_b(z)\in\mathscr{E}$ *such that* $F_b(r_b)=b$ *and*

$$\text{(6.10)} \qquad \operatorname{Re}\left\{\frac{K'(b)}{K(b)}\right\} = \frac{K'(r_b)}{K(r_b)},$$

$$\text{(6.11)} \qquad [\lambda_b(1-b^2)]^{1/2}K(b) > 0,$$

$$\text{(6.12)} \qquad \lambda_b c_b > 0,$$

$$\text{(6.13)} \qquad [\lambda_b(1-b^2)]^{1/2}K(b) = [\lambda_b c_b(1-r^2)]^{1/2}K(r_b).$$

Furthermore, $w=F_b(z)$ *satisfies the differential equation* (6.3) *with* $\lambda=\lambda_b$, $r=r_b$, $c=c_b=r_bF_b'(r_b)/F_b(r_b)$. *The given* b *is on the boundary of* $V(r_b)$ *and* F_b *is the unique function in* $\mathscr{E}$ *with* $F(r_b)=b$ *unless* b *is real and greater than* 1, *in which case there is exactly one more, the function* $\overline{F_b(\bar{z})}$.

Proof. Since $K'(r)/K(r)$ is strictly monotone decreasing from ∞ to 0 as r increases from 0 to 1, there is a unique r_b, $0<r_b<1$, satisfying (6.10). Then there is a unique λ_b with $|\lambda_b|=1$ satisfying (6.11), and a unique c_b satisfying (6.12) and (6.13).

Let γ_3 be the unique trajectory of the left-hand side of (6.3), with $\lambda=\lambda_b$, through $w=1$. This is a simple closed analytic curve, invariant under the mapping $w\mapsto 1/w$, whose image by (6.4) is the line segment joining $\pm\hat{\omega}_1+\hat{\omega}_2/2$. The values which have been chosen are such that (6.4), (6.7), and $[\lambda_b(1-b^2)]^{1/2}\Omega=[\lambda_b c_b(1-r_b^2)]^{1/2}Z$ together define a univalent mapping $w=F_b(z)$ of U slit from 0 to 1 along the real axis to the interior of γ_3 slit by $\gamma_1+\gamma_2'$ where γ_2' is an orthogonal trajectory of the left-hand side of (6.3) joining b to some point of γ_3. This function is easily seen to be analytic at each point of the segment [0, 1) and hence can be continued to be univalent in all of U. Since γ_3 is invariant under $w\mapsto 1/w$, $F_b\in\mathscr{E}$. The method of construction insures that $F_b(r_b)=b$. Letting $z\to r$ in (6.3) gives $c_b=r_bF_b'(r_b)/F_b(r_b)$.

Next we show that $b\in\partial V(r_b)$. We know $b\in\partial V(r)$ for some r and since $F_b(r_b)=b$,

$$\text{(6.14)} \qquad r_b \geqq r.$$

If $b\in\partial V(r)$, then there exists an $F\in\mathscr{E}$ with $F(r)=b$ satisfying (6.3) and with the boundary of $F(U)$ passing through 1. From the discussion of § 6.5, there must exist a pair of primitive periods $2\omega_1$, $2\omega_2$, such that (6.8) and (6.9) hold. Set $\tau=\omega_2/\omega_1$, $\tau_b=\hat{\omega}_2/\hat{\omega}_1=iK'(b)/K(b)$. Then

$$\text{(6.15)} \qquad \tau = \frac{n_4\tau_b+n_3}{n_2\tau_b+n_1},$$

with n_1, n_2, n_3, and n_4 as in (6.8).

Now $\tau_b=iK'(b)/K(b)$ lies in the region $\operatorname{Im}\tau>0$, $0\leqq\operatorname{Re}\tau\leqq 1$, $|\tau-1/2|\geqq 1/2$, as is shown, for example, in [10]. It is known (see [15] for example) that this is part of the fundamental domain of the congruence subgroup, but we can verify this easily and obtain information on uniqueness at the same time.

From (6.15), $\operatorname{Im}\tau=\operatorname{Im}\tau_b/|n_2\tau_b+n_1|^2$. Set $\tau_b=x+iy$. Then $|\tau_b-1/2|\geqq 1/2$ is equivalent to $x^2-x+y^2\geqq 0$. But then $|n_2\tau_b+n_1|^2=n_2^2(x^2-x+y^2)+(n_1+n_2)^2x+ +n_1^2(1-x)\geqq x+(1-x)=1$ since n_1, being odd, is non-zero and n_1+n_2 is similarly non-zero. Therefore, $\operatorname{Im}\tau\leqq\operatorname{Im}\tau_b$ with strict inequality holding except when $n_1=\pm 1$, $n_2=0$, or when $n_1=\pm 1$, $n_2=\mp 2$, $|\tau_b-1/2|=1/2$.

In the first case, $\omega_1=\pm\hat{\omega}_1$ and from (6.9), $\lambda=\lambda_b$. It follows that $F(U)$ is bounded by $\hat{\gamma}_3$, the unique trajectory of the left-hand side of (6.3) through $w=1$, and that $\gamma_1=\hat{\gamma}_1$. Hence $F(z)=F_b(z)$.

In the second case, $b\in(1,\infty)$, $\omega_1=\pm\hat{\omega}_1\mp 2\hat{\omega}_2$. The two sign choices give the same γ_1 and γ_3, hence only one $F(z)$. It is easily seen that this transformation corresponds to taking the path of integration in computing $K(b)$ "below" $1/b$, and $F(z)=\overline{F_b(\bar{z})}$ is the unique function giving the mapping.

In every other case $K'(r)/K(r)=\operatorname{Im}\tau<\operatorname{Im}\tau_b=K'(r_b)/K(r_b)$. This implies $r>r_b$ which contradicts (6.14). This completes the proof of the theorem.

6.7. If b is given, (6.10) determines the r so that $b\in\partial V(r)$. Similarly, if r is given, (6.10) determines the set of b which lie on $\partial V(r)$. In both cases tables such as [10] are useful. It is only if λ and r are given, that there is some difficulty. The conditions of Theorem 6.1 in theory will determine b but in practice this may be impossible. However, in a few special cases something can be done.

If $\lambda=1$, corresponding to the problem of maximizing $|F(r)|$, then (6.11) requires

$$\int_0^1\left[\frac{1-b^2}{1-b^2t^2}\right]^{1/2}\frac{dt}{[1-t^2]^{1/2}}>0. \tag{6.16}$$

Here we are computing the weighted average of a set of complex numbers along a curve. The values of $(1-b^2)/(1-b^2t^2)$ lie on the circular arc from $(1-b^2)$ to 1 which, if continued, would pass through 0. Thus if b^2 is not real, then the values of $[(1-b^2)/(1-b^2t^2)]^{1/2}$ lie entirely in one half plane and (6.16) can hold only if b^2 is real. That is, only if b is positive or pure imaginary. Further, b positive and greater than 1 would also make (6.16) impossible.

Suppose b is real with $0<b<1$. Then (6.10) requires $r_b=b$, (6.11) shows $\lambda_b>0$, and the unique $F_b(z)=z$.

Suppose $b=i\beta$, $\beta>0$. Then using the imaginary modulus transformation and the reciprocal modulus transformation (see [8], for example), $K(b)=K(i\beta)=\beta_1'K(\beta_1)$ and $K'(b)=K(\sqrt{1+\beta^2})=K(1/\beta_1')=\beta_1'[K(\beta_1')+iK'(\beta_1')]=\beta_1'[K'(\beta_1)+iK(\beta_1)]$ where $\beta_1=\beta/\sqrt{1+\beta^2}$ and $\beta_1'=1/\sqrt{1+\beta^2}$. Hence (6.10) requires $K'(r)/K(r)=K'(\beta_1)/K(\beta_1)$ or $r=r_b=\beta_1$. This is equivalent to $\beta=r/\sqrt{1-r^2}$. Since $\beta>r$ in this case, this gives the maximum for $F(r)$ and we have Jenkin's result.

6.8. It is interesting to look at what can be said about r_b for b real and greater than 1. Here the transformations give $K(b)=(1/b)[K(1/b)+iK'(1/b)]$ and $K'(b)==K(i\sqrt{b^2-1})=(1/b)K(\sqrt{b^2-1}/b)=(1/b)K'(1/b)$. Hence (6.10) requires $r=r_b$ where

$$\frac{K'(r)}{K(r)}=\frac{\sigma}{1+\sigma^2},\quad \sigma=\frac{K'(1/b)}{K(1/b)}.$$

As b increases from 1 to ∞, σ increases from 0 to ∞, and $\sigma/(1+\sigma^2)$ increases from 0 to a maximum of 1/2 when $\sigma=1$ and then decreases to 0 again. If r_0 is such that $K'(r_0)/K(r_0)=1/2$, then for $r<r_0$, $V(r)$ does not intersect $(1,\infty)$. For $r=r_0$, $V(r)$ will have the single point $b=\sqrt{2}$ in common with $(1,\infty)$, and for $r>r_0$, $\partial V(r)$ will intersect $(1,\infty)$ in exactly two points. Since $K'(k)=2K(k)$ when $k=3-2\sqrt{2}$, $r_0=2(3\sqrt{2}-4)^{1/2}=0.98517\ldots$.

Finally, if $b=e^{i\theta}$, then $K(b)=(1/2)e^{-i\theta/2}[K(\cos(\theta/2))+iK'(\cos(\theta/2))]$ and $K'(b)=e^{-i\theta/2}K'(\cos(\theta/2))$. Hence $r_b=r$ where

$$\frac{K'(r)}{K(r)}=\frac{2\varrho}{1+\varrho^2},\quad \varrho=\frac{K'(\cos\theta/2)}{K(\cos\theta/2)}.$$

From this one easily verifies that $V(r)$ is contained inside the unit disc if $r<r_1=\sqrt{2}/2$. If $r=r_1$, $V(r)$ is tangent to the unit circle at $\pm i$. If $r_1<r<1$, $\partial V(r)$ intersects the unit circle at precisely one point in each quadrant.

6.9. Figure 1 shows the boundaries of $V(r)$ for the two critical r mentioned above. Only the portion in the first quadrant is shown. The full set is symmetric with respect to reflection in both axes.

The inner curve defines $V(r_1)$, $r_1=\sqrt{2}/2$. The boundary appears nearly vertical near the real axis, but its real part reaches a maximum near $b=0.7077+i0.13$.

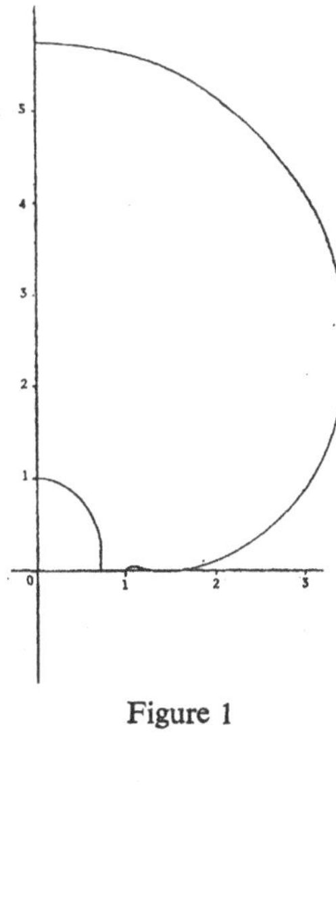

Figure 1

The outer curve defines $V(r_0)$, $r_0=2(3\sqrt{2}-4)^{1/2}$. This boundary is tangent to the real axis at $b=\sqrt{2}$. The very small loop that the boundary makes around the poiut 1 could not be determined very accurately because of the coarseness of the tables [10] used to find these boundary points.

7. The value sets for pairs

7.1. In 1969, Aharonov [1] proved that if $\{F_1, F_2\}$ is a pair and $|z_1|=r_1$, $|z_2|=r_2$, $r_1, r_2<1$, then $|F_1(z_1)F_2(z_2)|\leqq r_1 r_2(1-r_1^2)^{-1/2}(1-r_2^2)^{-1/2}$, this being sharp if $r_1=r_2$. In 1972, Jenkins [21] proved that $|F_1(z_1)F_2(z_2)|\leqq[\mu^{-1}(\nu(r_1)+\nu(r_2))]^{-1}$ where $\nu(r)$ and $\mu(R)$ are the modules of suitably defined doubly-connected domains. This result is sharp.

In this section we will study the more general problem of finding the boundary of

$$V(r_1, r_2) = \{F_1(r_1)F_2(r_2): \{F_1, F_2\} \text{ is a univalent pair}\}, \tag{7.1}$$

for any $r_1, r_2<1$.

From (5.1) we see that if $|z_1|=r_1$, $|z_2|=r_2$, then $V(r_1, r_2)=\{F_1(z_1)F_2(z_2)\colon \{F_1, F_2\}$ is a univalent pair$\}$. Much as in § 6.1 one easily sees that 0, 1 are never in any $V(r_1, r_2)$ but every other complex number is in one; each $V(r_1, r_2)\cup\{0\}$ is compact; $V(r_1, r_2)$ is symmetric with respect to reflection in the real axis; and the sets $V(r_1, r_2)\cup\{0\}$ are monotone strictly increasing with respect to either variable.

7.2. Again, it is a slight convenience to look at $\log F_1(r_1)F_2(r_2)$, so we introduce the functional

$$\Psi(F_1, F_2) = \log F_1(r_1)+\log F_2(r_2). \tag{7.2}$$

Suppose v is any complex number other than 0 or 1 and $v\in\partial V(r_1, r_2)$ for some r_1, r_2. It suffices to assume $\operatorname{Im} v\geqq 0$. Let $v=b^2$ with b in the first quadrant. Then there must exist a λ with $|\lambda|=1$ and a univalent pair $\{F_1, F_2\}$ extremal for $\operatorname{Re}\lambda\Psi(F_1, F_2)$ such that $F_1(r_1)F_2(r_2)=b^2$. One easily computes $\lambda A_1(w)=$ $=\lambda(1-b_1b_2)w/(b_1-w)(1-b_2w)$ and $\lambda A_2(w)=\lambda(1-b_1b_2)w/(b_2-w)(1-b_1w)$, where $F_1(r_1)=b_1$, $F_2(r_2)=b_2$, $b_1b_2=b^2$. The pair $\{(b_2/b_1)^{1/2}F_1, (b_1/b_2)^{1/2}F_2\}$ is extremal for the same problem, hence we may assume without loss of generality that $b_1=$ $=b_2=b$. That is, F_1 and F_2 must satisfy

$$\begin{cases} \dfrac{\lambda(1-b^2)\,dw^2}{w(b-w)(1-bw)} = \dfrac{\lambda c_1(1-r_1^2)\,dz^2}{z(r_1-z)(1-r_1z)} & w = F_1(z) \\[2ex] \dfrac{\lambda(1-b^2)\,dw^2}{w(b-w)(1-bw)} = \dfrac{\lambda c_2(1-r_2^2)\,dz^2}{z(r_2-z)(1-r_2z)} & w = F_2(z) \end{cases} \tag{7.3}$$

where $c_1=r_1F_1'(r_1)/F(r_1)$, $c_2=r_2F_2'(r_2)/F(r_2)$, and $\lambda c_1>0$, $\lambda c_2>0$.

7.3. The mappings (6.4) and (6.7) were studied in § 6.3 of the last section. The arguments of those paragraphs apply equally well here, and we have the same fundamental periods, $2\hat{\omega}_1$, $2\hat{\omega}_2$ of (6.4), $\hat{\omega}_1=2K(b)$, $\hat{\omega}_2=2iK'(b)$. Let $B^2=\lambda(1-b^2)$, so $B^2 d\Omega^2$ is the quadratic differential on the left-hand side of (7.3).

Suppose $\{F_1, F_2\}$ is a univalent pair which satisfies (7.3) for some r_1, r_2, λ, c_1, c_2 with $\lambda c_1>0$, $\lambda c_2>0$, $F_1(r_1)=b$, $F_2(r_2)=b$. Then F_1 maps the line segment $(0, r_1)$ onto the unique trajectory of $B^2 d\Omega^2$ which joins 0 to b. Since λ is the same, F_2 maps $(0, r_2)$ onto the same trajectory. Therefore, just as in § 6.5, we must have some ω_1 satisfying (6.8) such that

$$(7.4)\qquad [\lambda(1-b^2)]^{1/2}\omega_1 = [\lambda c_1(1-r_1^2)]^{1/2}2K(r_1)$$

$$= [\lambda c_2(1-r_2^2)]^{1/2}2K(r_2) > 0.$$

The boundaries of $F_1(U)$ and $F_2(U)$ are analytic curves which are also trajectories of $B^2 d\Omega^2$. These need not pass through 1, but if we assume (as in the conclusion of Theorem 5.1) that these trajectories are mapped into one another by $w\mapsto 1/w$, then they are mapped by (6.4) onto lines which are parallel to and equidistant from the line through $\pm\omega_1+(1/2)\omega_2$, where ω_2 is such that $2\omega_1, 2\omega_2$ is a pair of primitive periods satisfying (6.8). Looking at the composed mappings defined by (7.4), we must thus have $[\lambda c_1(1-r_1^2)]^{1/2}K'(r_1)+[\lambda c_2(1-r_2^2)]^{1/2}K'(r_2)=\mathrm{Im}\,[\lambda(1-b^2)]^{1/2}\omega_2$. Using (7.4), this implies

$$(7.5)\qquad \mathrm{Im}\left\{\frac{\omega_2}{\omega_1}\right\} = \frac{1}{2}\frac{K'(r_1)}{K(r_1)}+\frac{1}{2}\frac{K'(r_2)}{K(r_2)}.$$

7.4. We now state

Theorem 7.1. *Let $b\neq 0, 1$ with $0\leqq \arg b\leqq \pi/2$ be given. Let $K(b)$ and $K'(b)$ be as in the statement of Theorem 6.1. Let r_1, $0<r_1<1$ be such that $\mathrm{Re}\,\{K'(b)/K(b)\}>(1/2)\,K'(r_1)/K(r_1)$. Then there exists a unique r_2 with $0<r_2<1$, λ with $|\lambda|=1$, c_1 and $c_2\neq 0$, and a univalent pair $\{F_1, F_2\}$ with $F_1(r_1)=F_2(r_2)=b$ such that*

$$(7.6)\qquad \mathrm{Re}\left\{\frac{K'(b)}{K(b)}\right\} = \frac{1}{2}\frac{K'(r_1)}{K(r_1)}+\frac{1}{2}\frac{K'(r_2)}{K(r_2)},$$

$$(7.7)\qquad [\lambda(1-b^2)]^{1/2}K(b) > 0,$$

$$\lambda c_1 > 0,\quad \lambda c_2 > 0,$$

and

$$[\lambda(1-b^2)]^{1/2}K(b) = [\lambda c_1(1-r_1^2)]^{1/2}K(r_1) = [\lambda c_2(1-r_2^2)]^{1/2}K(r_2).$$

Further, F_1 and F_2 satisfy the differential equations (7.3) with $c_1=r_1F_1'(r_1)/F_1(r_1)$, $c_2=r_2F_2'(r_2)/F_2(r_2)$. The complex number b^2 is on the boundary of $V(r_1, r_2)$ and $\{F_1, F_2\}$ is the unique univalent pair with $F_1(r_1)=F_2(r_2)=b$ unless b is real and greater than 1, in which case there is exactly one more pair, $\{\overline{F_1(\bar{z})}, \overline{F_2(\bar{z})}\}$.

The proof of this theorem is omitted since it follows the proof of Theorem 6.1 quite closely.

7.5. Comparing Theorems 6.1 and 7.1, we easily prove:

Theorem 7.2. *Let $\{F_1, F_2\}$ be a univalent pair and let $z_1, z_2 \in U$. Then there exists a real r between $r_1 = |z_1|$ and $r_2 = |z_2|$ and an $F \in \mathcal{E}$ such that* $K'(r)/K(r) = (1/2)K'(r_1)/K(r_1) + (1/2)K'(r_2)/K(r_2)$ *and* $F(r)^2 = F_1(z_1)F_2(z_2)$.

Another theorem whose proof is immediate is

Theorem 7.3. *Let $0 < r < 1$, $|z_1| \leqq r$, $|z_2| \leqq r$. If $\{F_1, F_2\}$ is any univalent pair, then $F_1(z_1)F_2(z_2) \neq r^2$ except when $|z_1| = |z_2| = r$, $F_1(z) = c(r/z_1)z$, $F_2(z) = c^{-1}(r/z_2)z$, and $c \neq 0$ is arbitrary.*

Since $\{F(e^{i\alpha_1}z), F(e^{i\alpha_2}z)\}$ is a univalent pair for any $F \in \mathcal{E}$, Theorem 7.3 generalizes the result of Grunsky [13], that $F(z_1)F(z_2) \neq r^2$ for any $F \in \mathcal{E}$ and z_1, z_2 in U with $|z_1| < r$, $|z_2| < r$.

Maximizing $|F_1(r_1)F_2(r_2)|$ is the same as maximizing $\mathrm{Re}\log F_1(r_1)F_2(r_2)$. The extremal pair must satisfy (7.3) with $\lambda = 1$, and hence (6.16) must hold. That is, b is real, $0 < b < 1$, or is pure imaginary.

In the first case, Theorem 7.1 shows that $|F_1(r_1)F_2(r_2)| = r_1 r_2$. However, putting $F_1(z) = F_2(z) = F_{r_1, \pi/2}(z)$, the Jenkins function of (6.1) gives $F_1(r_1)F_2(r_2) = r_1 r_2/(1 - r_1 r_2) > r_1 r_2$. Hence b must be imaginary.

If b is imaginary, say $b = i\beta$, then just as in § 6.7, $|F_1(r_1)F_2(r_2)| = \beta^2 = r^2/(1 - r^2)$ where r satisfies $K'(r)/K(r) = (1/2)K'(r_1)/K(r_1) + (1/2)K'(r_2)/K(r_2)$. This is equivalent to the result of Jenkins mentioned in § 7.1.

8. The maximum of $|b_2|$ in $\mathcal{E}(|b_1|)$

8.1. For any β, $0 < \beta < 1$, let $\mathcal{E}(\beta)$ denote the set of all $F \in \mathcal{E}$ such that $|b_1| = \beta$, where as in (1.1), $F(z) = b_1 z + b_2 z^2 + \dots$. $\mathcal{E}(\beta)$ is compact for each β. We wish to find the maximum of $|b_2|$ for $F \in \mathcal{E}(\beta)$ for each fixed β.

Jenkins has solved this problem [20] in the sense that he has implicitly defined (in terms of their mappings) a one parameter family of functions which achieve the maxima. Here we shall duplicate most of his results using variational techniques and carry the analysis further to obtain the actual bounds in a more explicit form. For $\beta < 0.827\dots$ these bounds are extremely simple.

We may assume $b_1 > 0$ and look for the maximum $|b_2|$ with b_1 fixed. The extremal $F \in \mathcal{E}(b_1)$ exists and is locally extremal in $\mathcal{E}$ for $\mathrm{Re}\,\Psi(F)$ where

$$\Psi(F) = \lambda \log b_1 + \log b_2. \tag{8.1}$$

Here λ is a Lagrange multiplier. For this Ψ, $L(F; G) = \lambda\chi_1(G)/b_1 + \chi_2(G)/b_2$ where $\chi_\nu(G)$ is the ν-th coefficient of G. Then $L(zF') = \lambda + 2$. Hence from Theorem 4.1,

λ is real. Computing $A(w)$ and $B(w)$ we find that the extremal F will satisfy

$$(8.2) \qquad \left[(\lambda+1)+\frac{b_1^2}{b_2}\left(w+\frac{1}{w}\right)\right]\frac{dw^2}{w^2} = \left[(\lambda+2)+\frac{b_1}{b_2}\frac{1}{z}+\frac{b_1}{\bar{b}_2}z\right]\frac{dz^2}{z^2}$$

for $|z|<1$. Here we have changed the sign and the right-hand side of (8.2) will therefore be negative for $|z|=1$.

Let $\zeta=e^{-i\alpha}$ and consider the slit variation of (4.2). Since $\varepsilon>0$, we find $\mathrm{Re}\,[(\lambda+2)+(2b_1/b_2)\zeta^{-1}]\geqq 0$ for each ζ with $|\zeta|=1$. This implies $\lambda+2\geqq 2|b_1/b_2|>1$ since $|b_2|<2|b_1|$ for any $F\in\mathscr{E}$. That is, $\lambda>-1$.

Make the simple change of variable $\zeta=(|b_2|/\bar{b}_2)z$ and put

$$(8.3) \qquad 2W = w+\frac{1}{w}, \quad 2Z = \zeta+\frac{1}{\zeta}.$$

Then $w=F(z)$ defines W as a univalent function of Z, mapping the exterior of the line segment $[-1, 1]$ onto the exterior of a continuum Γ containing the points $W=\pm 1$, and having no interior (Theorem 4.2). Since $W^2-1=(w-1/w)^2/4$, $dw^2/w^2= =dW^2/(W^2-1)$ and hence from (8.2.), W and Z satisfy

$$(8.4) \qquad \frac{(1+\sigma W)\,dW^2}{1-W^2} = \varrho\,\frac{(1+\tau Z)\,dZ^2}{1-Z^2}$$

where

$$\sigma = \frac{2b_1^2}{b_2(\lambda+1)}, \quad \tau = \frac{2b_1}{|b_2|(\lambda+2)}, \quad \varrho = \frac{\lambda+2}{\lambda+1}.$$

We see ϱ and τ are real, $\varrho>1$, $0<\tau\leqq 1$. The last since the right-hand side of (8.4) is real and non-negative for $Z\in[-1, 1]$. If $\tau<1$, then the right-hand side of (8.4) has a zero at $-1/\tau$ and hence the continuum Γ will not contain the critical point $s=-1/\sigma$. If $\tau=1$, then Γ must contain s.

8.2. We now analyze the continuum Γ more carefully. We see that Γ contains ± 1, and may or may not contain s. Except for these points it must consist of analytic arcs which are trajectories of $d\Omega^2=(1+\sigma W)dW^2/(1-W^2)$.

If $\sigma=\pm 1$, $d\Omega^2$ simplifies. Taking $\sigma=1$, for example, $d\Omega^2=dW^2/(1-W)$. The unique trajectory of this through -1 is the infinite interval $(-\infty, +1)$ of the real axis. Hence Γ consists of a segment $[x, 1]$, $-\infty<x\leqq -1$, and F maps U onto U less a slit $[-1, \alpha]$, $-1\leqq\alpha<0$. Then, just as in the derivation of (4.2), $F/(1-F)^2=b_1z/(1-z)^2$. Comparing coefficients we find $b_2=2b_1(1-b_1)$. This is exactly the familiar Pick bound for the second coefficient of a bounded function [27]. As we shall see, this is not the extreme value. The case $\sigma=-1$ is similar and gives the same bound for $|b_2|$. Notice that in these cases $|\sigma|=1$, $\tau=1$ and hence $b_1=1/\varrho$.

Hence we assume $\sigma\neq\pm 1$. Then $d\Omega^2$ has three finite critical points. At ± 1, exactly one trajectory leaves. At $s=-1/\sigma$, exactly three leave at equal angles.

Thus Γ consists of either: ($\tau<1$) the points ± 1 and a single analytic arc not passing through s, or: ($\tau=1$) the points ± 1, s, an arc joining -1 to s, an arc joining $+1$ to s, and possibly a segment of a third arc leaving s.

Suppose γ_1 and γ_2 are two distinct arcs meeting at s and lying on the trajectories of $d\Omega^2$. Then $\int d\Omega$ on $\gamma_1+\gamma_2$ is real and is equal to the integral on $\gamma_1'+\gamma_a+\gamma_2'$ where γ_1' and γ_2' are subarcs of γ_1 and γ_2, respectively, lying outside a small circle of radius a centered at s, and γ_a is the smaller of the two arcs of this circle joining γ_1' and γ_2'. This, and indeed any path, can be altered homotopically in $\mathbf{C}-\{s, 1, -1\}$ without changing the integral of $d\Omega$.

Next we show that if J is the line segment $[-1, 1]$, then either $\int d\Omega$ over J is real, or there are two disjoint subintervals of J for which $\int d\Omega$ is real.

First, suppose $\tau<1$. Then either Γ is homotopic (in $\mathbf{C}-\{s, 1, -1\}$) to J, or the three trajectories leaving s must cross J. The above assertion therefore holds. On the other hand, if $\tau=1$, then the pair of trajectories from ± 1 to s make up a single path homotopic to J or else these, together with an arc of the third trajectory from s to some point of J, make up two paths homotopic to disjoint segments of J. Again the conclusion follows.

If $x_1, x_2 \in J$ and $\sin\theta_1=x_1$, $\sin\theta_2=x_2$, then $\int_{x_1}^{x_2} d\Omega=\int_{\theta_1}^{\theta_2}[1+\sigma\sin\theta]^{1/2}\,d\theta$. This is the weighted mean value of complex numbers on a segment of an hyperbola passing through 1. Two such disjoint integrals could be real only if σ were real.

If $\int d\Omega$ is real over J, then so is

$$\int_{-\pi/2}^{\pi/2}[1+\sigma\sin\theta]^{1/2}\,d\theta=\int_{0}^{\pi/2}[(1+\sigma\sin\theta)^{1/2}+(1-\sigma\sin\theta)^{1/2}]\,d\theta.$$

The square of the last integrand is $2+2(1-\sigma^2\sin^2\theta)^{1/2}$. If σ^2 is not real, this integral is the weighted mean of values on an arc lying entirely in one half plane except for the end point ($\theta=0$) which lies on the real axis. Since the integral is real, we conclude that σ^2 must be real. We have therefore shown that in any case σ is either pure real or pure imaginary.

8.3. Suppose σ is real. If $|\sigma|<1$, then the segment $J=[-1, 1]$ is the only trajectory joining ± 1. Hence $\Gamma=J$ and $F(U)=U$. Since $b_1>0$, $F(z)=z$ and $b_1=1$, $b_2=0$. This function however does not belong to the class $\mathcal{E}(\beta)$ with $\beta<1$.

If σ is real and $|\sigma|>1$, say $\sigma>0$, then $s=-1/\sigma$ lies on J. The only trajectory from 1 is the segment $(s, 1)$ and the only trajectory from -1 is the infinite segment $(-\infty, -1)$. Hence no bounded Γ can satisfy the requirements, and this case cannot occur. The case of $\sigma<0$ is similar.

If $\sigma=\pm 1$, we have already seen that F maps U onto U less a radial slit and $|b_2|=2b_1(1-b_1)$.

If $\sigma=i\mu$ is imaginary, the situation is more complicated. If $F(z)$ is extremal, then so is $\overline{F(\bar{z})}$ and hence we may assume $\mu>0$. One trajectory of $d\Omega^2= =(1+i\mu W)\,dW^2/(1-W^2)$ is the ray from $s=i/\mu$ to ∞ along the imaginary axis.

Thus Γ consists of an analytic arc joining ± 1 which is homotopic to J, or arcs from 1 to s, from -1 to s, and possibly a segment of the imaginary axis extending upward from s. The second case occurs only when $\tau=1$.

8.4. Suppose $\tau<1$. Then from (8.4) the integral of $d\Omega$ around Γ in the W-plane will equal the integral around $[-1, 1]$ in the Z-plane. Therefore we must have

$$q(\mu) = \sqrt{\varrho}\, p(\tau) \tag{8.5}$$

where, since Γ is homotopic to J,

$$\begin{aligned} q(\mu) &= \int_{-1}^{1} [(1+i\mu W)/(1-W^2)]^{1/2}\, dW = \int_{-\pi/2}^{\pi/2} [1+i\mu \sin\theta]^{1/2}\, d\theta \\ &= \sqrt{2} \int_{0}^{\pi/2} [(1+\mu^2 \sin^2\theta)^{1/2}+1]^{1/2}\, d\theta. \end{aligned} \tag{8.6}$$

Here we took the real part of the integrand, since the integral $q(\mu)$ is real. Similarly,

$$p(\tau) = \int_{-1}^{1} [(1+\tau Z)/(1-Z^2)]^{1/2}\, dZ = \int_{0}^{\pi} (1+\tau\cos\theta)^{1/2}\, d\theta. \tag{8.7}$$

This is only one relation among three unknowns. However, from the definition, $|\sigma/\tau| = \mu/\tau = \varrho b_1$. Hence

$$\mu = \varrho\tau b_1. \tag{8.8}$$

A third relation is obtained from (8.4) with the help of the observation that $Z=-1/\tau$ corresponds to $W=i/\mu$. Thus, the integral of $d\Omega$ in the Z-plane along the line segment $l=[-1/\tau, -1]$ must equal the integral of $d\Omega$ in the W-plane along L, the image of l. Since l lies along an orthogonal trajectory, L must be the line segment from i/μ to the point iY at which Γ crosses the imaginary axis. This integral is pure imaginary, while the integral along any part of Γ is real. Hence

$$r(\mu) = \sqrt{\varrho}\, s(\tau) \tag{8.9}$$

where

$$\begin{aligned} s(\tau) &= \operatorname{Im} \int_{-1/\tau}^{-1} [(1+\tau Z)/(1-Z^2)]^{1/2}\, dZ \\ &= \int_{1}^{1/\tau} \left[\frac{1-\tau t}{t^2-1}\right]^{1/2} dt \end{aligned} \tag{8.10}$$

and

(8.11) $$r(\mu) = \operatorname{Im} \int_{-1}^{i/\mu} d\Omega = \operatorname{Im} \int_{-1}^{0} d\Omega + \operatorname{Im} \int_{0}^{i/\mu} d\Omega$$

$$= \operatorname{Im} \int_{0}^{\pi/2} [1 - i\mu \sin\theta]^{1/2} d\theta + \int_{0}^{1} \left[\frac{1-t}{\mu^2+t^2}\right]^{1/2} dt$$

$$= \int_{0}^{1} \left[\frac{1-t}{\mu^2+t^2}\right]^{1/2} dt - \frac{1}{\sqrt{2}} \int_{0}^{\pi/2} [(1+\mu^2 \sin^2\theta)^{1/2} - 1]^{1/2} d\theta.$$

Possible ambiguity of the sign of the last integral can be resolved easily since $r(\mu)$ must be zero when i/μ is on Γ.

8.5. One easily verifies that $p'(\tau)<0$ and hence $p(\tau)$ is decreasing. We find $p(0)=\pi, p(1)=2\sqrt{2}$. Actually, $p(\tau)$ is a simple elliptic integral which can be calculated with the help of [8], for example, to be $p(\tau)=2\sqrt{1+\tau}\, E'(k)$, $k^2=(1-\tau)/(1+\tau)$ where E' is the associated complete elliptic integral of the second kind.

Similarly, $s'(\tau)<0$ and hence $s(\tau)$ is a descreasing function of τ with $s(0)=\infty$, $s(1)=0$. Again one can compute $s(\tau)=2\sqrt{1+\tau}\,[K(k)-E(k)]$, $k^2=(1-\tau)/(1+\tau)$.

The first integral of the last line of (8.11) is a decreasing function of μ which is infinite at 0 and tends toward zero as $\mu\to\infty$. The second integral is an increasing function, zero at 0 and tending toward ∞ as $\mu\to\infty$. Hence $r(\mu)$ is a decreasing function of μ which is infinite at $\mu=0$ and which decreases to 0 at some unique μ_0. The μ_0 for which $r(\mu_0)=0$ was found, with the help of numerical calculations, to be

(8.12) $$\mu_0 = 1.1622005\ldots.$$

The function $q(\mu)$ is clearly increasing. We have $q(0)=\pi$, $q(\mu_0)=q_0=3.3519319\ldots$, the last value being found with the help of numerical calculations.

8.6. Let τ, $0<\tau<1$, be given. From (8.5) and (8.9), $s(\tau)/p(\tau)=r(\mu)/q(\mu)$. The right-hand expression is a decreasing function of μ and hence this equation determines a unique μ between 0 and μ_0. Then (8.5) or (8.9) determines ϱ. (8.8) determines b_1. Finally, from the definitions of ϱ, σ, and τ we find $|b_2|=2b_1(1/\tau)(1-1/\varrho)$ and hence $|b_2|$ is determined. That is, each τ with $0<\tau<1$ determines a unique b_1 and $|b_2|$.

8.7. Suppose $\tau=1$. Then Γ passes through i/μ and hence we must have $\mu=\mu_0$. Then $b_1=\mu_0/\varrho$, $|b_2|=2b_1(1-b_1/\mu_0)$. When $\tau=1$, $d\Omega=\varrho(1-Z)^{-1/2}dZ$ in the Z-plane. This is regular at $Z=-1$, so the boundary slit, $[-1, 1]$ in the Z-plane has an "open end" at $Z=-1$, i.e., $d\Omega$ changes sign if we change directions at that point. The integral of $d\Omega$ around this slit, starting at this end, will be the same as the integral of $d\Omega$ around Γ in the w-plane, starting at the "open end" which will be i/μ or the tip of the slit extending upward along the imaginary axis. That is, $2q_0+2v=4\sqrt{2\varrho}$, where $v\geqq 0$ is the integral of $d\Omega$ from the tip of the slit to i/μ.

This implies $\varrho \geqq q_0^2/8$ with equality only if the extra slit is of zero length. We have therefore proved most of

Theorem 8.1. *Let* $F(z)=b_1z+b_2z^2+\ldots\in\mathscr{E}$ *be locally extremal for* $\mathrm{Re}\,\{\lambda \log b_1+\log b_2\}$ *with* b_1 *and* λ *real,* $0<b_1<1$, $-1<\lambda$. *Let* $\varrho=(\lambda+2)/(\lambda+1)$. *Then one of the following three holds:*

$$|b_2| = 2b_1(1-b_1), \qquad b_1 = \frac{1}{\varrho}; \tag{8.13}$$

$$|b_2| = 2b_1\left(1-\frac{b_1}{\mu_0}\right), \qquad b_1 = \frac{\mu_0}{\varrho}, \qquad \varrho \geqq \frac{q_0^2}{8}; \tag{8.14}$$

$$|b_2| = 2b_1\frac{1}{\tau}\left(1-\frac{b_1}{\mu_1}\right), \quad b_1 = \frac{\mu_1}{\varrho}, \quad \mu_1 = \frac{\mu}{\tau}, \quad \varrho < \frac{q_0^2}{8}; \tag{8.15}$$

where in (8.15) μ, τ, *and* ϱ *satisfy* (8.5) *and* (8.9).

Proof. It only remains to show that $\varrho<q_0^2/8$ in the case (8.15) when $\tau<1$. Using the formulas of § 8.5 and the well-known formulas for the derivatives of the complete elliptic integrals (see, for example, 710.00—710.05 in [8]), we find

$$\frac{d}{d\tau}\left(\frac{s(\tau)}{p(\tau)}\right) = \frac{dk}{d\tau}(k/k'^2)(KE'+K'E-KK')/E'^2.$$

The derivative of $KE'+K'E-KK'$ is easily found to be zero and since $K'(E-K)\to 0$ as $k\to 0$, this factor is the constant $\pi/2$. Hence

$$\frac{d}{d\tau}\left(\frac{s(\tau)}{p(\tau)}\right) = \frac{\pi k}{2E'^2(1-k^2)}\left(\frac{dk}{d\tau}\right), \quad k^2 = \frac{1-\tau}{1+\tau}. \tag{8.16}$$

Since $dk/d\tau<0$, we conclude that $s(\tau)/p(\tau)$ is a strictly decreasing function of τ. Hence as τ increases from 0 to 1, the μ satisfying $s(\tau)/p(\tau)=r(\mu)/q(\mu)$ increases from 0 to μ_0, and $\varrho=q(\mu)^2/p(\tau)^2$ increases from 1 to $q_0^2/8$.

8.8. How does the theorem apply to the problem of finding the maximum of $|b_2|$ in $\mathscr{E}(|b_1|)$ for a fixed b_1? Figure 2 shows the results of some numerical calculations. The dashed curve is a plot of the parabola of (8.13). If $\varrho\geqq q_0^2/8$ in (8.14), then $b_1\leqq 8\mu_0/q_0^2=0.82752416\ldots$. The arrow points to this value of b_1 and the portion of the upper curve in Figure 2 to the left of this arrow is the parabola of (8.14). Numerical computations of (8.15) gave b_1 as a decreasing function of τ and resulted in the portion of the curve to the right of $8\mu_0/q_0^2$.

Analytic proofs of the facts that (8.15) determines b_1 as a decreasing function of τ and that (8.13) gives $|b_2|$ lying below the values determined by (8.14) and (8.15) can be avoided by using the result of Jenkins in [20]. There he shows the existence of a one-parameter family of functions $F_\beta\in\mathscr{E}$ which achieve the maximum $|b_2|$ in the corresponding $\mathscr{E}(\beta)$. Theorem 8.1 must hold for each such F_β. His description of the mappings shows that they belong to our class with σ imaginary, and hence (8.13) does not hold. They are therefore exactly the mappings giving rise to (8.14) and (8.15). We conclude that for $F\in\mathscr{E}$, $|b_2|\leqq\Phi(|b_1|)$ where $\Phi(|b_1|)$ is the function determined by (8.14) if $|b_1|\leqq 8\mu_0/q_0^2$ and by (8.15) if $|b_1|>8\mu_0/q_0^2$.

We remark that the extremal functions determined in this way are unique up to the transformation $\overline{F(\bar z)}$, because of the uniqueness of the solution of the differential equations. This one ambiguity arises because we had to choose μ positive or negative. Since σ is imaginary, the trajectories of $d\Omega$ are symmetric with respect to reflection in the imaginary axis and $-F(-z)=\overline{F(\bar z)}$.

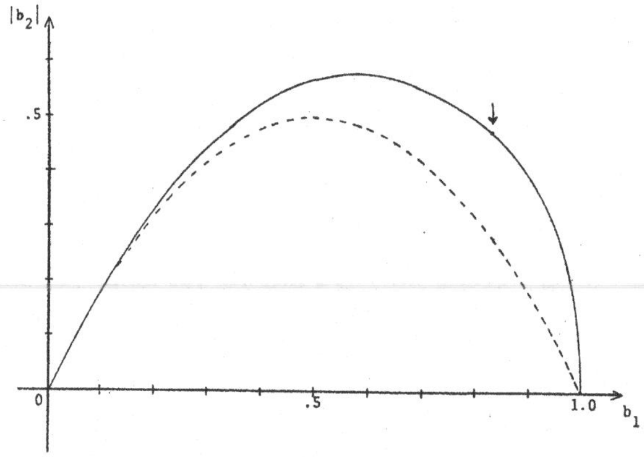

Figure 2

8.9. Let us consider the two special cases when $\lambda=0$ and $\lambda=1$.

Theorem 8.2. *Let* $F(z)=b_1z+b_2z^2+\ldots\in\mathscr{E}$. *Then* $|b_2|\leqq\mu_0/2=0.5811002\ldots$. *This is sharp.*

Proof. The extremal F must satisfy Theorem 8.1 with $\lambda=0$ and hence $\varrho=2$. If (8.15) were to hold, $\varrho\leqq q_0^2/8=1.40443\ldots$. Hence (8.14) holds with $b_1=|b_2|=$ $=\mu_0/2$.

This bound may be compared with the best previously published bound, $|b_2|\leqq$ $\leqq e^{-\gamma/2}=0.7493\ldots$, due to Nehari [24] and Aharonov [3].

Theorem 8.3. *Let* $F(z)=b_1z+b_2z^2+\ldots\in\mathscr{E}$. *Then* $|b_1b_2|\leqq 8\mu_0^2=0.4002103\ldots$. *This is sharp.*

Proof. Here $\lambda=1$ in (8.1) and $\varrho=3/2$. Again (8.15) cannot hold and hence $b_1=2\mu_0/3$. Then $|b_2|=4\mu_0/9$ and the theorem follows.

Figure 3 shows the image domain $F(U)$ for one of the two extremal functions of Theorem 8.2 (with $b_1>0$). This was obtained by setting the left-hand side of (8.2) positive and numerically integrating to find the trajectory through 1. The length of the slit is found as discussed in § 8.7 by computing the integral v required to make $q_0+v=\sqrt{8\varrho}$. The function of Theorem 8.3, or for that matter, any of the functions of (8.14), map to the same domain with different slit lengths.

8.10. As a final remark in this section, we observe that we could easily treat the problem of maximizing $\operatorname{Re} b_2$ among all $F\in\mathscr{E}$ with a fixed real b_1. This corresponds to $\Psi(F)=b_2+\lambda b_1$. Methods similar to those of this section lead to

$$\frac{2b_1^2(W-W_0)}{1-W^2}\,dW^2=\frac{2b_1(Z-Z_0)}{1-Z^2}\,dZ^2$$

where $W_0=-(b_2+\lambda b_1)/2b_1^2$, $Z_0=-(2b_2+\lambda b_1)/2b_1$ and Z_0 is real, $Z_0\leqq-1$.

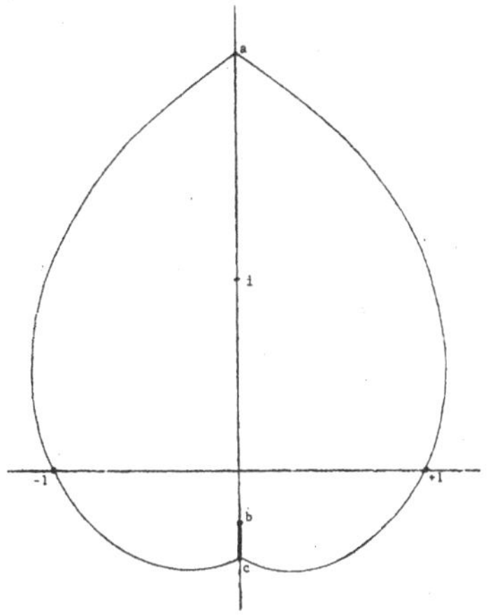

Figure 3

An analysis of the trajectories in this case leads to the conclusion that $W_0=-1$ and the extremal functions map U onto U less a radial slit from -1 to $-b_1/4$. That is, if $F\in\mathscr{E}$ and $0<b_1<1$, then $\operatorname{Re} b_2\leqq 2b_1(1-b_1)$.

9. The coefficient set of (b_1, c_1, c_2) for pairs

9.1. Let $\{F, G\}$ be a univalent pair with $F(z)=b_1z+\dots$, $G(z)=c_1z+c_2z^2+\dots$. We wish to determine the set $\mathscr{C}$ of all possible triples (b_1, c_1, c_2). We can reduce the study of this six dimensional set to that of a two dimensional set.

Let $\mathscr{C}_1$ be the set of all (c_1, c_2) such that $\{F, G\}$ is a univalent pair with $b_1=1$, $c_1>0$, and $c_2>0$. Given $(b_1, c_1, c_2)\in\mathscr{C}$, there exists a new pair obtained by the transformations (5.1) so that $(1, |b_1c_1|, |b_1c_2|)\in\mathscr{C}$, i.e., so that $(|b_1c_1|, |b_1c_2|)\in\mathscr{C}_1$. Similarly, if $\mathscr{C}_1$ is known, then from (5.1) $\mathscr{C}$ is the set of all $(xe^{i\theta_1}, x^{-1}ye^{i\theta_2}, x^{-1}ze^{i\theta_3})$ such that $x>0$, θ_1, θ_2, θ_3 are real and $(y, z)\in\mathscr{C}_1$.

For any univalent pair $|b_1c_1|\leqq 1$, [1], and $|b_1c_2|<2e^{-\gamma}/\sqrt{3}$, [2], where γ is the Euler constant. Further, since G is univalent, $|c_2/c_1|<2$. $\mathscr{C}_1$ does not contain the origin 0, but we therefore see that $\mathscr{C}_1 \cup \{0\}$ is compact. We study $\mathscr{C}_1$ by fixing c_1 and finding the maximum c_2, or, what is equivalent, finding the maximum $\operatorname{Re} b_1 c_2$ for all univalent pairs $\{F, G\}$ with fixed $b_1 c_1 > 0$ and $b_1 > 0$. If $\{F, G\}$ is extremal for this problem, it will be locally extremal for $\operatorname{Re} \Psi(F, G)$ where

$$\Psi(F, G) = \lambda b_1 c_1 + b_1 c_2 \tag{9.1}$$

and λ is the Lagrange multiplier.

9.2. We apply Theorem 5.1 to (9.1) with $F=F_1$, $G=F_2$. We find $L_1(H) = (\lambda c_1 + c_2)\chi_1(H)$, $L_2(H) = \lambda b_1 \chi_1(H) + b_1 \chi_2(H)$. Then $A_1(w) = -[\lambda b_1 c_1 + b_1 c_2 + b_1 c_1^2 w]$, $A_2(w) = -[\lambda b_1 c_1 + b_1 c_2 + b_1 c_1^2/w]$, $B_1(z) = -[\lambda b_1 c_1 + b_1 c_2]$, and $B_2(z) = -[\lambda b_1 c_1 + 2b_1 c_2 + b_1 c_1 z + b_1 c_1/z]$, where we use the fact that $b_1 c_1$, $L_1(zF_1') = \lambda b_1 c_1 + b_1 c_2$, and $L_2(zF_2') = \lambda b_1 c_1 + 2b_1 c_2$ are all real.

We can assume that the extremal pair have $b_1 = 1$, $c_1 > 0$, $c_2 \geqq 0$. Since $B_1(z) \leqq 0$ and $B_2(z) \leqq 0$ for $|z| = 1$, $\lambda c_1 + c_2 \geqq 0$ and $\lambda c_1 + 2c_2 \geqq 2c_1$. Since $c_2/c_1 < 2$, this means that λ is real and greater than -2. If $\lambda c_1 + c_2 = 0$, then $w = F_1(z)$ would satisfy $A(w)\,dw^2/w^2 = 0$, which is impossible. Hence $\lambda c_1 + c_2 > 0$.

We can therefore write the differential equations of Theorem 5.1 as

$$(1+\alpha w)\frac{dw^2}{w^2} = \frac{dz^2}{z^2}, \qquad w = F_1(z), \tag{9.2}$$

$$\left(1+\frac{\alpha}{w}\right)\frac{dw^2}{w^2} = \varrho\left[1+\frac{\tau}{2}\left(z+\frac{1}{z}\right)\right]\frac{dz^2}{z^2}, \qquad w = F_2(z), \tag{9.3}$$

where

$$\begin{cases} \alpha = \dfrac{c_1^2}{\lambda c_1 + c_2}, & \varrho = \dfrac{\lambda c_1 + 2c_2}{\lambda c_1 + c_2}, & \tau = \dfrac{2c_1}{\lambda c_1 + c_2}, \\ \alpha > 0, & \varrho > 1, & 0 < \tau \leqq 1, \\ c_1 = \dfrac{2\alpha}{\varrho\tau}, & c_2 = 2c_1 \dfrac{1}{\tau}\left(1 - \dfrac{1}{\varrho}\right), & \lambda = \dfrac{2(2-\varrho)}{\varrho\tau}. \end{cases} \tag{9.4}$$

The last three relations in (9.4) are obtained from the first three by solving for c_1, c_2, and λ.

9.3. We now study the solutions of (9.2) and (9.3). First, (9.2) can be integrated immediately. Using the boundary conditions $w(0) = 0$, $w'(0) = b_1 = 1$, we find that $w = F_1(z)$ satisfies

$$2(1+\alpha w)^{1/2} + \log\frac{(1+\alpha w)^{1/2} - 1}{(1+\alpha w)^{1/2} + 1} = 2 + \log\frac{\alpha z}{4}, \qquad w = F_1(z). \tag{9.5}$$

Since the right-hand side of (9.2) has no zeros or poles in U except at 0, $-1/\alpha \notin F_1(U)$. It may be on the boundary or be exterior; $-\alpha$ will correspondingly be on the boundary or interior to $F_2(U)$, and $\tau = 1$ or $\tau < 1$, respectively.

Suppose $0 < \tau < 1$. Then Γ_2, the boundary of $F_2(U)$ will be an analytic curve, an orthogonal trajectory of the left-hand side of (9.3). Dividing by $2\pi i$ and integrating around the unit circle, we find from (9.3) that

$$\frac{\sqrt{\varrho}}{2\pi}\int_{-\pi}^{\pi}[1+\tau\cos\theta]^{1/2}\,d\theta = \frac{1}{2\pi i}\int_{\Gamma_2}[1+\alpha/w]^{1/2}\,dw = 1.$$

Hence

(9.6) $$\sqrt{\varrho}\,p(\tau) = \pi$$

where $p(\tau)$ is defined in (8.7). Given τ, this determines a unique ϱ with $1<\varrho<<\pi^2/8$.

Let $0<\sigma<1$ be such that $2-\tau(\sigma+\sigma^{-1})=0$. Then the right-hand side of (9.3) has a zero at $-\sigma$. The segment $(-1, -\sigma)$ is a trajectory of this quadratic differential and corresponds to a segment $(-\beta, -\alpha)$ in the w-plane. Here $-\beta\in\Gamma_2$ and $\beta>0$. Integrating along these trajectories,

$$\sqrt{\varrho}\int_{-1}^{-\sigma}[1+(\tau/2)(z+1/z)]^{1/2}\,dz/z = \int_{-\beta}^{-\alpha}[1+\alpha/w]^{1/2}\,dw/w.$$

Using the substitution $z+1/z=-2t$, the left-hand integral is found to be the negative of $s(\tau)$ of (8.10). The right-hand integral is $2B+\log(1-B)/(1+B)$ where $B=(1-\alpha/\beta)^{1/2}$. This also is negative. Since $-\beta\in\partial F_2(U)$, $-1/\beta\in\partial F_1(U)$ and hence from (9.5), $-\sqrt{\varrho}\,s(\tau)=2+\log(-\alpha z_1/4)$ where z_1 is some point with $|z_1|=1$. Since this is real, $z_1=-1$ and

(9.7) $$\alpha = 4e^{-2}\exp\{-\sqrt{\varrho}\,s(\tau)\}.$$

From this c_1 and c_2 can be determined.

9.4. If $\tau=1$ so that $-1/\alpha\in\partial F_2(U)$, then the situation is much simpler. In this case $-\alpha\in\partial F_1(U)$, and from (9.5),

(9.8) $$\alpha = 4e^{-2},$$

and then from (9.4), c_1 and c_2 will be determined as a function of ϱ.

In this case (9.6) need not hold. Three orthogonal trajectories of the left-hand side of (9.3) meet at equal angles at $-1/\alpha$. One lies along the line from this point toward 0. The boundary of $F_2(U)$ will consist of a simple closed curve Γ_2, analytic except at $-1/\alpha$ plus a slit Γ_1 from $-1/\alpha$ toward 0. Γ_1 may be of zero length. Integrating around $|z|=1$ we find

$$\frac{\sqrt{\varrho}}{2\pi i}\int_{|z|=1}(1+z)\,dz/z^{1/2} = \frac{1}{2\pi i}\int_{\Gamma_2}(1+\alpha w)^{1/2}\,dw/w+\frac{2}{2\pi i}\int_{\Gamma_1}(1+\alpha w)^{1/2}\,dw/w$$

or $\sqrt{\varrho}\,p(1)/\pi=1+s$ where $s\geqq 0$. Here all integrals are positive. We start at $z=-1$ in the z-plane and at the tip of the slit in the w-plane. Thus, if $\tau=1$, we find $\varrho\geqq\pi^2/8$; and hence $c_1=8e^{-2}/\varrho\leqq 64/\pi^2e^2$. It follows that if $c_1>64/\pi^2e^2=0.8775891\ldots$, then only the $\tau<1$ case occurs.

If $\tau<1$, then $c_1=8\pi^{-2}e^{-2}p(\tau)^2\tau^{-1}\exp\{-\pi s(\tau)/p(\tau)\}$. With the help of (8.16), the formula for $p(\tau)$ of § 8.5, and the formulas for the derivatives of the elliptic integrals, one easily shows that $(d/d\tau)\log c_1 = 4^{-1}\tau^{-1}(1+\tau)^{-1}E'^{-2}(\pi^2-4E'K')$, $k^2=(1-\tau)/(1+\tau)$. However, $(d/dk)EK=(E+k'K)(E-k'K)/kk'^2$ and $(d/dk)(E-k'K)= = -(1-k')(E-K)/kk'>0$. Hence it follows that EK is an increasing function of k whose minimum at $k=0$ (and hence the minimum of $E'K'$) is $\pi^2/4$. Therefore c_1 is a strictly decreasing function of τ.

As $\tau\to 0$, $k\to 1$, $k'\to 0$, $E\to 1$, $E'\to\pi/2$, $p(\tau)\to\pi$. Also $\tau=k'^2/(2-k'^2)$. Hence

$$\lim_{\tau\to 0} c_1 = \frac{8}{\pi^2 e^2}\lim_{\tau\to 0} p(\tau)^2(2-k'^2)\exp\{-(\pi/E')(K-E+(2/\pi)E'\log k')\}$$

$$= \frac{16}{e^2}\lim \exp\{-(\pi/E')(K-\log(4/k')-E+\log 4+\log k'(2E'/\pi)-1)\}.$$

However, $K-\log(4/k')\to 0$ and $\log k'(2E'/\pi-1)\to 0$ (see 112.01 and 900.07 in [8]). Hence $c_1\to 1$ as $\tau\to 0$.

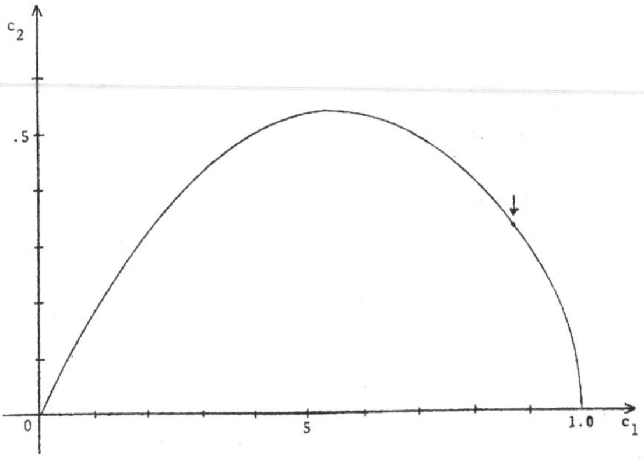

Figure 4

As $\tau\to 1$, $s(\tau)/p(\tau)\to 0$, and $p(\tau)\to 2\sqrt{2}$. Therefore $c_1\to 64/\pi^2e^2$. Putting these facts together, we have proved:

Theorem 9.1. *Let $\{F_1, F_2\}$ be a univalent pair with $F_1(z)=z+b_2z^2+\dots$, $F_2(z)=c_1z+c_2z^2+\dots$, $c_1>0$, $c_2>0$. Then for any c_1, $0<c_1\leqq 1$, we have*

$$(9.9)\qquad \begin{aligned} c_2 &\leqq 2c_1(1-e^2c_1/8) && \text{if } c_1\leqq 64/\pi^2e^2\\ &\leqq (2c_1/\tau)(1-1/\varrho) && \text{if } c_1> 64/\pi^2e^2\end{aligned}$$

where τ is the unique real number $0<\tau<1$ such that

$$c_1 = \frac{8}{\pi^2e^2}\frac{p(\tau)^2}{\tau}\exp\left\{-\pi\frac{s(\tau)}{p(\tau)}\right\},$$

and $\varrho=\pi^2/p(\tau)^2$. The functions $p(\tau)$ and $s(\tau)$ are defined in (8.7) *and* (8.10).. *These inequalities are sharp for each c_1.*

We remark that with the help of the expansion of E' near $k'=1$ (900.07 in [8]) one can easily show that $c_2 \to 0$ as $c_1 \to 1$ in (9.9). Some numerical computations were made and Figure 4 shows the bound of (9.9). The arrow points to the "joint" between the regions, i.e., at $c_1 = 64/\pi^2 e^2 = 0.8775891\ldots$. At this point $c_2 = 128(\pi^2 - 8)/\pi^4 e^2 = 0.33248428\ldots$.

9.5. The pair which maximizes $|b_1 c_2|$ must be extremal for Re $\{b_1 c_2\}$. That is, it will be one of the functions found above with $\lambda = 0$ or $\varrho = 2$. This belongs to the $\tau = 1$ case and the extremal value will be the highest point of the parabolic part of (9.9). That is,

Theorem 9.2. *If* $\{F_1, F_2\}$ *is any univalent pair with* $F_1(z) = b_1 z + \ldots$ *and* $F_2(z) = c_1 z + c_2 z^2 + \ldots$, *then*

$$|b_1 c_2| \leqq 4e^{-2}$$

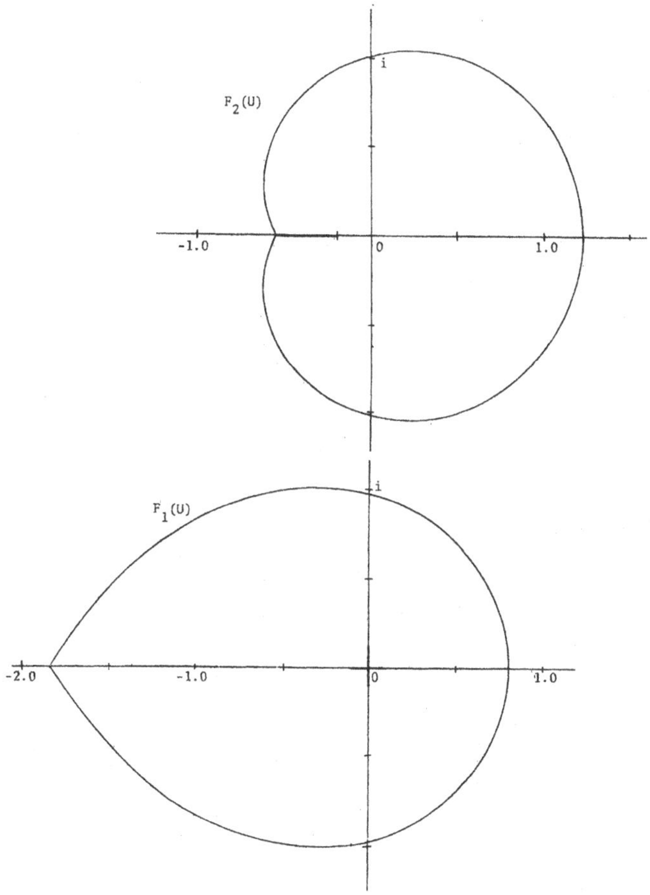

Figure 5

This result is sharp with the maximum occuring for a pair satisfying (9.2), (9.3) *and* (9.4) *with* $\varrho=2$, $\tau=1$, *and* $\alpha=4e^{-2}$. *In this case* $b_1c_1=b_1c_2=4e^{-2}$.

This bound, $4e^{-2}=0.5413411...$, may be compared with Aharonov's bound [2] $2e^{-\gamma}/\sqrt{3}=0.6483176...$, or with the bound for Bieberbach—Eilenberg functions $|b_1b_2|\leqq 8\mu_0^2/27=0.4002103...$ (Theorem 8.3).

9.6. Figure 5 shows the typical images of U by the mappings $F_1(z)$ and $F_2(z)$ which are extremal for Theorem 9.2 with $\tau=1$. The particular pair shown has $F_1'(0)=b_1=1$, hence $-\alpha=-0.5413411...$ lies on the boundary of $F_1(U)$. The boundaries and the length of the slit were obtained by numerical integration.

All extremal pairs of Theorem 9.1 in the case $\tau=1$ $(c_1\leqq 64/\pi^2e^2)$ will map onto similar domains but with slits of different lengths.

10. An inequality of Golusin type

10.1. The methods of this paper can be used to give a simple proof of an inequality of Golusin type for Bieberbach—Eilenberg functions. This is closely related to inequalities of Grunsky type proved by Jenkins [20] and Garabedian and Schiffer [11], and to the inequalities proved by Nehari in [24]. In fact, it is shown in [17] that inequalities of Golusin and Grunsky type are essentially equivalent. We include the following theorem, however, since it illustrates the reduction which occurs when the $A(w)$ in Theorem 4.1 is a perfect square

Theorem 10.1. *Let* $F(z)\in\mathscr{E}$. *Let* $z_1, z_2, \dots, z_N$ *be distinct points of* U *and let* $x_1, x_2, \dots, x_N$ *be complex constants such that*

$$\text{Im}\sum_{\nu=0}^{\infty} x_\nu = 0. \tag{10.1}$$

Then

$$\text{Re}\sum_{\nu=1}^{N}\sum_{\mu=1}^{N} x_\nu x_\mu \log\frac{F(z_\nu)-F(z_\mu)}{(z_\nu-z_\mu)[1-F(z_\nu)F(z_\mu)]} \leqq -\sum_{\nu=1}^{N}\sum_{\mu=1}^{N}\bar{x}_\nu x_\mu \log(1-\bar{z}_\nu z_\mu). \tag{10.2}$$

Remarks. *Ambiguities in the choice of branches of the logarithms disappear because of* (10.1). *The requirement that the* z_ν *be distinct can be relaxed by taking appropriate limits. When* $\mu=\nu$ *in the left-hand sum of* (10.2), *we let* $[F(z_\nu)-F(z_\mu)]/(z_\nu-z_\mu)=F'(z_\nu)$. *The right-hand side of* (10.2) *is always real.*

Proof. We apply Theorem 4.1 with

$$\Psi(F)=\sum_{\nu=1}^{N}\sum_{\mu=1}^{N} x_\nu x_\mu \log\frac{F_\nu-F_\mu}{(z_\nu-z_\mu)[1-F_\nu F_\mu]}$$

where we denote $F(z_\nu)=F_\nu$. We find with some computation that

$$A(w) = -w^2 \sum_{\nu=1}^{N} \sum_{\mu=1}^{N} \frac{x_\nu x_\mu (1-F_\nu^2)(1-F_\mu^2)}{(F_\nu - w)(F_\mu - w)(1-F_\nu w)(1-F_\mu w)}$$

$$= -w^2 \left[\sum_{\nu=1}^{N} x_\nu \frac{(1-F_\nu)^2}{(F_\nu - w)(1-F_\nu w)} \right]^2.$$

The extremal F maps U onto the interior of an analytic curve Γ through ± 1 satisfying $A(w)dw^2/w^2 > 0$. In this case we can take the square root and integrate. We find that for $w \in \Gamma$

$$\operatorname{Re} \sum_{\nu=1}^{N} x_\nu \log \frac{1-F_\nu w}{F_\nu - w} = \text{constant}.$$

Since $w=-1$ is on Γ, this constant is zero and

$$\operatorname{Re} \sum_{\nu=1}^{N} x_\nu \log \frac{F(z_\nu)-F(z)}{1-F(z_\nu)F(z)} = 0 \quad \text{if } |z|=1.$$

Then, for $|z|=1$ we have

$$\operatorname{Re} \sum_{\nu=1}^{N} x_\nu \log \frac{F(z_\nu)-F(z)}{(z_\nu - z)[1-F(z_\nu)F(z)]} = -\operatorname{Re} \sum_{\nu=1}^{N} x_\nu \log (z_\nu - z)$$

$$(10.3) \qquad = -\operatorname{Re} \sum_{\nu=1}^{N} \bar{x}_\nu \log \left(\bar{z}_\nu - \frac{1}{z} \right)$$

$$= -\operatorname{Re} \sum_{\nu=1}^{N} \bar{x}_\nu \log (1-\bar{z}_\nu z)$$

since $\sum \bar{x}_\nu \log(-1/z)$ is pure imaginary for $|z|=1$ because of (10.1).

Both sides of (10.3) are the real parts of functions which are analytic in the closure of U. Hence equality holds in all of U and

$$(10.4) \qquad \sum_{\nu=1}^{N} x_\nu \log \frac{F(z_\nu)-F(z)}{(z_\nu - z)[1-F(z_\nu)F(z)]} = -\sum_{\nu=1}^{N} \bar{x}_\nu \log (1-\bar{z}_\nu z) + ik$$

where k is some real constant.

Set $z=z_\mu$ in (10.4), multiply by x_μ, and add the resulting relations for all μ. Then $i \sum x_\mu k$ is pure imaginary and hence

$$\operatorname{Re} \sum_{\nu=1}^{N} \sum_{\mu=1}^{N} x_\nu x_\mu \log \frac{F(z_\nu)-F(z_\mu)}{(z_\nu - z_\mu)[1-F(z_\nu)F(z_\mu)]} = -\sum_{\nu=1}^{N} \sum_{\mu=1}^{N} \bar{x}_\nu x_\mu \log (1-\bar{z}_\nu z_\mu).$$

Thus, the extremal function gives equality in (10.2) and the inequality must hold for all $F \in \mathscr{E}$.

References

[1] AHARONOV, D.: A generalization of a theorem of J. A. Jenkins.-Math. Z. 110, 1969, 218—222.

[2] AHARONOV, D.: On pairs of functions and related classes.-Duke Math. J. 40, 1973, 669—676.

[3] AHARONOV, D.: On Bieberbach—Eilenberg functions.-Bull. Amer. Math. Soc. 76, 1970, 101—104.

[4] AHLFORS, L. V.: Conformal invariants. - McGraw-Hill, New York, 1973.

[5] ALENICYN, YU. E.: A contribution to the theory of univalent and Bieberbach—Eilenberg functions.-Dokl. Akad. Nauk SSSR (N.S) 109, 1956, 247—249. (Russian).

[6] ALENICYN, YU. E.: On functions without common values and the outer boundary of the domain of values of a function.- Ibid. 115, 1957, 1055—1057 (Russian).

[7] BIEBERBACH, L.: Über einige Extremalprobleme im Gebiet der konformen Abbildung. - Math. Ann. 77, 1916, 153—172.

[8] BYRD, P. F., and FRIEDMAN, M. D.: Handbook of elliptic integrals for engineers and scientists, 2nd ed.-Springer-Verlag, New York—Heidelberg, 1971.

[9] EILENBERG, S.: Sur quelques proprietés topologiques de la surface de sphère. - Fund. Math. 25, 1935, 267—272.

[10] FETTIS, H. E., and J. C. CASLIN: A table of complete elliptic integrals of the first kind for complex values of the modulus, Part I. - Applied Math. Research Lab., Aerospace Research Laboratories, USAF, Wright—Patterson Air Force Base, Ohio (Project No. 7071, Nov. 1969).

[11] GARABEDIAN, P. R., and M. SCHIFFER: The local maximum theorem for the coefficients of univalent functions. - Arch. Rational Mech. Anal. 26, 1967, 1—32.

[12] GOLUSIN, G. M.: Geometric theory of functions of a complex variable. - Transl. Math. Monographs, vol. 26, Amer. Math. Soc., Providence, R. I., 1969.

[13] GRUNSKY, H.: Einige Analoge zum Schwarzschen Lemma. - Math. Ann. 108, 1933, 190—196.

[14] GUELFER, S.: On the class of regular functions which do not take on any pair of values w and $-w$. - Mat. Sb. (N. S.) 19 (61), 1946, 33—46 (Russian).

[15] GUNNING, R. C.: Lectures on modular forms. - Ann. of Math. Studies, no. 48, Princeton University Press, Princeton, N. J., 1962.

[16] HUMMEL, J. A.: Inequalities of Grunsky type for Aharonov pairs. - J. Analyse Math. 25, 1972, 217—257.

[17] HUMMEL, J. A., and M. SCHIFFER: Coefficient inequalities for Bieberbach—Eilenberg functions. - Arch. Rational Mech. Anal. 32, 1969, 87—99.

[18] JENKINS, J. A.: On Bieberbach—Eilenberg functions. - Trans. Amer. Math. Soc. 76, 1954, 389—396.

[19] JENKINS, J. A.: On Bieberbach—Eilenberg functions. II. - Ibid. 78, 1955, 510—515.

[20] JENKINS, J. A.: On Bieberbach—Eilenberg functions. III. - Ibid. 119, 1965, 195—215.

[21] JENKINS, J. A.: A remark on "pairs" of functions. - Proc. Amer. Math. Soc. 31, 1972, 119—121.

[22] LEBEDEV, N. A.: An Application of the area principle to non-overlapping domains. - Trudy Mat. Inst. Steklov. 60, 1961, 211—231 (Russian).

[23] LEBEDEV, N. A., and I. M. MILIN: On the coefficients of certain classes of analytic functions. - Mat. Sb. (N. S.) 28(70), 1951, 359—400 (Russian).

[24] NEHARI, Z.: Some inequalities in the theory of functions. - Trans. Amer. Math. Soc. 75, 1953, 256—286.

[25] NEHARI, Z.: On the coefficients of Bieberbach—Eilenberg functions. - J. Analyse Math. 23, 1970, 297—303.

[26] NELSON, D. J.: Extremal problems in the class of Bieberbach—Eilenberg functions. - Dissertation, Stanford University, September 1972.

[27] PICK, G.: Über die konforme Abbildung eines Kreises auf ein schlichtes und zugleich beschränktes Gebiet. - Sitzungsber. Math. - Natur. Kaiserlichen Akad. der Wissensch. Wien, 126, 1917, 247—263.

[28] ROGOSINSKI, W.: On a theorem of Bieberbach—Eilenberg. - J. London Math. Soc. 14, 1939, 4—11.

[29] ROGOSINSKI, W.: On the coefficients of subordinate functions. - Proc. London Math. Soc. (2) 48, 1943, 48—82.

[30] SCHAEFFER, A. C., M. SCHIFFER, and D. C. SPENCER: The coefficient regions of schlicht functions. - Duke Math. J. 16, 1949, 493—526.

[31] SCHIFFER, M.: A method of variations within the family of simple functions. - Proc. London Math. Soc. (2) 44, 1938, 432—449.

[32] SCHIFFER, M.: Extremum problems and variational methods in conformal mapping. - Proc. International Congress Math., Edinburgh, 1958, 213—231.

[33] SCHIFFER, M.: Inequalities in the theory of univalent functions. - Inequalities-III, Academic Press, New York—London, 1972, 311—319.

University of Maryland
Department of Mathematics
College Park, Maryland 20742
USA

Stanford University
Department of Mathematics
Stanford, California 94305
USA

Received 11 September 1975

Commentary on

[116] J.A. Hummel and M.M. Schiffer, *Variational methods for Bieberbach-Eilenberg functions and for pairs*, Ann. Acad. Sci. Fenn. Ser. A I Math. **3** (1977), 3–42.

The concept of pairs [A, H1] provides the link between Bieberbach–Eilenberg (BE) functions and such closely related classes as Gel'fer functions, Grunsky–Shah functions, and disjoint functions. Recall that if F and G are functions holomorphic in the unit disc U such that $F(0) = G(0) = 0$, (F,G) is said to be a *pair* if $F(z)G(\zeta) \neq 1$ for all $z, \zeta \in U$. Clearly, then, F is a BE function precisely when (F,F) is a pair; and F is a *Grunsky–Shah function* if and only if $(F,-G)$ is a pair, where $G(z) = \overline{F(\overline{z})}$. The class of *Gel'fer functions* is defined as the collection of all functions of the form $(1+F)/(1-F)$, where F ranges over the BE functions. A.Z. Grinshpan's survey [Gr], with its extensive bibliography, discusses these subjects in considerable detail, so we focus in our comments below on certain developments not treated in [Gr], reflecting our personal point of view.

There is also a close connection between pairs and *disjoint functions*, i.e., functions whose ranges on U are disjoint, which leads naturally to the problem of finding differential operators (involving two disjoint functions) invariant under Möbius transformations. Differential operators of this type were first exhibited by Nehari [N1,N2]. Aharonov and Lavie [AL] constructed a generating function which yields an infinite sequence of such operators and used them to find necessary conditions for disconjugacy and related properties for ordinary linear differential equations.

Mention should also be made of the class of *almost bounded functions* (ABF), introduced by A.W. Goodman in 1955. W.T. Lai (1962) and later R. Miniowitz (1976) generalized basic results on BE functions to the case of specialized ABF. For a detailed survey and references, see [Go, Chap. 12].

Significant related work was done by K. Włodarczyk in a series of papers published in *Annales Polonici Mathematici* from 1980 to 1985. Consider two vectors $F = (F_1, F_2, \dots, F_m)$ and $G = (G_1, G_2, \dots, G_n)$ such that F_i and F_j are disjoint for $i \neq j$ and similarly for G and such that each (F_i, G_j) forms a pair, $1 \leq i \leq m$, $1 \leq j \leq n$. Włodarczyk generalizes many of the results of [116] to this more general situation and finds several nice additional results for related classes.

A method of symmetrization suitable for univalent pairs was found by Marcus [M], generalizing earlier results of Aharonov and Kirwan [AK] for BE functions. For further details, see Dubinin's important survey [D].

The variational methods developed in [116] were applied by Hummel [H2, H3] to study the coefficient region of univalent BE functions and have also been adapted and applied to the investigation of related classes in [J, FZ, RS, S, K, Kuh]. New directions in research in this area include the study of quasiconformal extensions of univalent pairs beyond the unit disc (cf. [Gr, Sect. 10]) and the work of Takhtajan and Teo on the Weil–Petersson metric and related topics, especially [TT], where results on pairs are used as a basic tool.

References

[A] Dov Aharonov, *A generalization of a theorem of J.A. Jenkins*, Math. Z. **110** (1969), 218–222.

[AK] Dov Aharonov and W.E. Kirwan, *A method of symmetrization and applications*, Trans. Amer. Math. Soc. **163** (1972), 369–377.

[AL] D. Aharonov and M. Lavie, *Disjoint meromorphic functions and nonoscillatory differential systems*, Trans. Amer. Math. Soc. **188** (1974), 1–14.

[D] V. N. Dubinin, *Symmetrization in the geometric theory of functions of a complex variable*, Russian Math. Surveys **49** (1994), 1–79.

[FZ] Maria Fait and Eligiusz Złotkiewicz, *A variational method for Grunsky functions*, Ann. Univ. Mariae Curie-Skłodowska Sect. A **34** (1980), 9–18.

[Go] A. W. Goodman, *Univalent Functions*, Vol. 2, Mariner Publishing Co., 1983.

[Gr] A. Z. Grinshpan, *Logarithmic geometry, exponentiation, and coefficient bounds in the theory of univalent functions and nonoverlapping domains*, Handbook of Complex Analysis: Geometric Function Theory, Vol. 1, ed. R. Kühnau, Elsevier, 2002, pp. 273–332.

[H1] J. A. Hummel, *Inequalities of Grunsky type for Aharonov pairs*, J. Analyse Math. **25** (1972), 217–257.

[H2] J. A. Hummel, *The b_1, b_2 coefficient body for Bieberbach-Eilenberg functions*, J. Analyse Math. **33** (1978), 168–190.

[H3] J. A. Hummel, *The second coefficient of univalent Bieberbach-Eilenberg functions near the identity*, Proc. Amer. Math. Soc. **90** (1980), 237–243.

[J] Halina Jondro, *Sur une méthode variationelle dans la famille des fonctions de Grunsky-Shah*, Bull. Acad. Polon. Sci. Sér. Sci. Math. **27** (1979), 541–547.

[K] S. Kirsch, *Univalent functions with range restrictions*, Z. Anal. Anwendungen **19** (2000), 1057–1073.

[Kuh] R. Kühnau, *Variation of diametrically symmetric or elliptically schlicht conformal mappings*, J. Analyse Math. **89** (2003), 303–316.

[M] Moshe Marcus, *Radial averaging of domains for Dirichlet integrals and applications*, J. Analyse Math. **27** (1974), 47–78.

[N1] Zeev Nehari, *Some inequalities in the theory of functions*, Trans. Amer. Math. Soc. **75** (1953), 256–286.

[N2] Zeev Nehari, *Some function-theoretic aspects of linear second-order differential equations*, J. Analyse Math. **18** (1967), 259–276.

[RS] A. Rost and J. Śladkowska, *Sur le fonction de Bieberbach-Eilenberg satifaisant à deux au moins équations du type de Schiffer*, Demonstratio Math. **27** (1994), 253–270.

[S] J. Śladkowska, *A variational method for generalized Gel'fer functions*, Mathematica (Cluj) **36 (59)** (1994), 229–238.

[TT] Leon A. Takhtajan and Lee-Peng Teo, *Weil-Petersson Metric on the Universal Teichmüller Space*, Mem. Amer. Math. Soc. **183** (2006), No. 861.

DOV AHARONOV

[121] (with J. A. Hummel and B. Pinchuk) Bounded univalent functions which cover a fixed disk

[121] (with J. A. Hummel and B. Pinchuk) Bounded univalent functions which cover a fixed disk. *J. Analyse Math.* **36** (1979), 118–138.

BOUNDED UNIVALENT FUNCTIONS WHICH COVER A FIXED DISC

By

JAMES A HUMMEL, BERNARD PINCHUK AND MENAHEM M. SCHIFFER*

1. Introduction

Let $U_r = \{z : |z| < r\}$ and denote the unit disc by U.

For a given d, $0 < d < 1$, let $S(d)$ denote the class of functions

$$f(z) = a_1 z + a_2 z^2 + \cdots$$

which are analytic and univalent in U and such that

$$U_d \subset D \subset U \tag{1.1}$$

where D denotes the image of U under $f(z)$ and where d and 1 are the largest and smallest radii, respectively, for which the inclusions (1.1) hold. This function class is very close to an analogous class which has been studied extensively by Netanyahu [3].

In this paper we study extremum problems for $S(d)$. In the first section we use elementary methods to solve problems such as to determine

$$\max_{f \in S(d)} |a_1| \tag{1.2}$$

and

$$\max_{f \in S(d)} |f(z)| \quad \text{where } z \in U \text{ is fixed.} \tag{1.3}$$

We also indicate how these elementary methods provide a solution to

$$\max_{f \in S(d)} |a_2| \tag{1.4}$$

* Supported in part by NSF grant MCS78–19886 at Stanford University and NSF grant MCS77–01277 at the University of Maryland.

when $d \leqq 3-2\sqrt{2}$. However, for $d > 3-2\sqrt{2}$ the elementary methods no longer suffice and we turn to other methods.

In the second section we develop a method of variations for $S(d)$. The difficulty is, of course, to construct a variation of $f \in S(d)$ in such a way as to preserve the conditions (1.1) as well as the univalence and the normalization $f(0)=0$. The variation we develop is of particular interest since it can be modified to handle side conditions more complicated than (1.1).

We then present applications of the variational method to the problem of maximizing quite general functionals over $S(d)$ and conclude with a more detailed analysis of the problem (1.4) and the question of maximizing $\mathrm{Re}\{\log f(z_0)\}$.

Let

$$K(z)=\frac{z}{(1-z)^2}$$

and

$$K_\gamma(z)=e^{i\gamma}K(e^{-i\gamma}z) \tag{1.5}$$

where $-\pi<\gamma\leqq\pi$.

These are the Koebe functions and its rotations which map U onto the w plane slit along the ray $re^{i(\pi+\gamma)}$, $r \geqq 1/4$.

Let

$$F_\gamma(z)=K_\gamma^{-1}\left[\frac{4d}{(1+d)^2}K_\gamma(z)\right]. \tag{1.6}$$

These functions, which map U onto U slit along the ray $re^{i(\pi+\gamma)}$, $d\leqq r\leqq 1$, are clearly in $S(d)$. They are, as we shall show, extremal for the problems (1.2) and (1.3). They are also extremal for (1.4) when $d\leqq 3-2\sqrt{2}$. For $d>3-2\sqrt{2}$ they cease to be extremal, and in this case the extremal domain is also slit along an arc of $|w|=d$.

Note that in view of the boundedness conditions (1.1), $S(d)$ is a normal family. While it is not a compact faimly, it can be made into one by adjoining the functions $e^{i\alpha}dz$. This now assures the existence of extremal functions, in $S(d)$, for the problems under consideration.

We also mention that we could have replaced (1.1) by

$$U_d\subset D\subset U_M, \qquad M>1.$$

However, one easily goes from $S(d)$ to these functions by division.

2. Elementary methods

Theorem 2.1. *Let* $f(z) = a_1 z + a_2 z^2 + \cdots \in S(d)$. *Then*

$$d < |a_1| \leq \frac{4d}{(1+d)^2}.$$

Equality in the upper bound only holds for the functions (1.6). *The lower bound is sharp but is not attained in* $S(d)$.

Proof. The upper bound is contained in [2], page 88. This proof for $S(d)$ is more direct.

Let D be the image of U under $f(z)$. Set

$$g_\gamma(z) = \frac{1}{a_1} K_\gamma[f(z)]. \tag{2.1}$$

Clearly, $g_\gamma(z) \in S$, the class of normalized univalent functions in U. Let z_0, $|z_0| = 1$, correspond to a boundary point w_0 of D with $|w_0| = d$. By Koebe's 1/4-Theorem for S,

$$|g_\gamma(z_0)| = \frac{1}{|a_1|} \frac{d}{|1 - e^{-i\gamma} w_0|^2} \geq \frac{1}{4}. \tag{2.2}$$

We now choose γ so that $|1 - e^{-i\gamma} w_0|^2 = (1+d)^2$ and conclude

$$|a_1| \leq \frac{4d}{(1+d)^2}.$$

Equality here can only hold if there is equality in (2.2) which is only possible if $g_\gamma(z)$ is a Koebe function (1.5). Now, if $g_\gamma(z) = K_\gamma(z)$, we find from (2.1) that

$$f(z) = K_\gamma^{-1}\left[\frac{4d}{(1+d)^2} K_\gamma(z)\right] = F_\gamma(z).$$

The lower bound for $|a_1|$ follows from Schwarz's lemma applied to the inverse of $d^{-1}f(z)$ restricted to U. Equality would imply $f(z) = de^{i\alpha}z$ which is not in $S(d)$.

To see that this bound is sharp, observe that the sequence of functions in $S(d)$ which map U onto U slit along $(-1, -d)$ and along the arc $de^{i\theta}$, $\theta \in [1/n, 2\pi]$ converges to dz as n approaches infinity and thus have first coefficients as close to d as we wish.

Theorem 2.2. *Let* $f(z) \in S(d)$. *Then*

$$d|z| < |f(z)| \leqq F_0(|z|).$$

Equality in the upper bound only holds for the functions (1.6). *The lower bound is sharp but not attained in* $S(d)$.

Proof. As in the proof of Theorem 2.1, we consider $(1/a_1)K_\gamma[f(z)]$. By the distortion theorem for S, we have for a fixed $z_0 \in U$,

$$\frac{1}{|a_1|}|K_\gamma[f(z_0)]| \leqq K(|z_0|). \tag{2.3}$$

Choose γ so that $|K_\gamma[f(z_0)]| = K(|f(z_0)|)$. Then

$$\frac{1}{|a_1|} K(|f(z_0)|) \leqq K(|z_0|)$$

or

$$|f(z_0)| \leqq K^{-1}[|a_1| K(|z_0|)]. \tag{2.4}$$

In view of Theorem 2.1 and the monotonicity of the Koebe function for positive argument, this implies

$$|f(z_0)| \leqq K^{-1}\left(\frac{4d}{(1+d)^2} K(|z_0|)\right) = F_0(|z_0|).$$

Equality here implies equality in (2.3) which is only possible if $(1/a_1)K_\gamma(f(z))$ is a Koebe function. In this case necessarily $f(z) = F_\gamma(z)$.

The lower bound follows from the maximum principle. Equality would imply $f(z) = de^{i\alpha}z$ which is not in $S(d)$. Finally, since we have seen a sequence of functions in $S(d)$ converging to dz the lower bound is sharp.

Theorem 2.3. *Let* $f(z) = a_1 z + a_2 z^2 + \cdots \in S(d)$. *Then*

$$|a_2| \leqq 2|a_1|(1 - |a_1|). \tag{2.5}$$

Equality only holds for the functions (1.6).

Proof. This is a well-known result for bounded univalent functions, see [4]. We present this simple proof for $S(d)$ for completeness.

Let

$$g(z)=\frac{1}{a_1}K[f(z)]=z+\left(\frac{a_2}{a_1}+2a_1\right)z^2+\cdots.$$

$g(z)\in S$ hence

(2.6)
$$\left|\frac{a_2}{a_1}+2a_1\right|\leqq 2.$$

Now, we can always assume that both a_1 and a_2 are real and positive. Indeed, if $\alpha=\arg a_1$ and $\beta=\arg a_2$, the function

(2.7)
$$e^{i(\beta-2\alpha)}f(e^{i(\alpha-\beta)}z)=\sum_{n=1}^{\infty}c_nz^n\in S(d)$$

and $c_1=|a_1|$ and $c_2=|a_2|$. Thus we can drop the absolute value in (2.6) and the result follows.

Equality in (2.6) implies that $g(z)$ is a Koebe function, and this proves the second statement of the theorem.

Theorem 2.4. *Let $f(z)=a_1z+a_2z^2+\cdots\in S(d)$. If $d\leqq 3-2\sqrt{2}$, then*

$$|a_2|\leqq|A_2|$$

where $F_0(z)=\sum_{n=0}^{\infty}A_nz^n$. ($F_0(z)$ is defined by (1.6).)

Proof. The function $2x(1-x)$ is increasing for $0\leqq x\leqq\frac{1}{2}$. Furthermore, for $d\leqq 3-2\sqrt{2}$ we have that $|a_1|\leqq 4d/(1+d)^2\leqq\frac{1}{2}$. Thus, as long as $d\leqq 3-2\sqrt{2}$, it follows from (2.5) that $|a_2|$ is largest when $|a_1|$ is largest and that the extremal functions are given by (1.6).

This argument breaks down for $d>3-2\sqrt{2}$ and, in fact, the functions (1.6) are no longer extremal in this case.

Indeed, let

(2.8)
$$s(z)=K_\gamma^{-1}((1-\varepsilon)K_\gamma(z))=z-\varepsilon z\frac{e^{i\alpha}+z}{e^{i\alpha}-z}+O(\varepsilon^2),\qquad \gamma=\alpha+\pi,$$

where $\varepsilon>0$. This maps U into itself less a short radial slit at $e^{i\alpha}$.

Let $f(z)=\sum_{n=1}^{\infty}a_nz^n\in S(d)$. If $z_0=e^{i\alpha}$ is any point such that $|f(z_0)|\neq d$, then

(2.9)
$$f^*(z)=f(s(z))=f(z)-\varepsilon f'(z)z\frac{e^{i\alpha}+z}{e^{i\alpha}-z}+O(\varepsilon^2),$$

where $\varepsilon > 0$, is in $S(d)$. This is the Löwner slit variation.

We find

$$\operatorname{Re}\{a_2^*\} = \operatorname{Re}\{a_2\} - \varepsilon \operatorname{Re}\{a_1 e^{i\alpha} + 2a_2 + \bar{a}_1 e^{-i\alpha}\}.$$

ε is positive, and therefore, unless we have

$$\operatorname{Re}\{a_1 e^{i\alpha} + 2a_2 + \bar{a}_1 e^{-i\alpha}\} \geqq 0, \tag{2.10}$$

we will have constructed a function $f^*(z) \in S(d)$ such that

$$\operatorname{Re}\{a_2^*\} > \operatorname{Re}\{a_2\}.$$

A simple calculation now shows that for $d > 3 - 2\sqrt{2}$ the functions (1.6) satisfy

$$\operatorname{Re}\{a_1 e^{i\theta} + 2a_2 + \bar{a}_1 e^{-i\theta}\} < 0,$$

for θ belonging to some interval, and hence these functions are not extremal for the second coefficient.

3. Variations in $S(d)$

In order to use the known formulas for the variation of the Green's function to obtain a variation of the mapping function, we must have a method of varying a domain so as to preserve conditions (1.1).

Let D be a simply connected domain bounded by the curve C in the w-plane. Let $\Delta \subset D$ be a set of subdomains bounded by analytic arcs, and let $v(w)$ be analytic in the closure of $D - \Delta$. We choose ε real and small and consider the variation

$$w^*(w) = w + \varepsilon v(w) + O(\varepsilon^2). \tag{3.1}$$

For all ε sufficiently small, (3.1) will be univalent on C and will carry C to a nearby curve C^* which bounds the varied domain D^*. The Green's function of D^* is given by the asymptotic formula (see [1], p. 202, where this is formulated for domains of finite connectivity)

$$g^*(w,\omega) = g(w,\omega) + \operatorname{Re}\left\{\frac{\varepsilon}{2\pi i}\int_{\partial\Delta} p'(t,w)p'(t,\omega)v(t)\,dt\right\} + O(\varepsilon^2) \tag{3.2}$$

where $p(t,\omega)$ is the analytic completion in t of $g(t,\omega)$, and differentiation is with respect to t. This formula holds for w and ω in Δ.

In the case that D is the image of the unit disk U in the z-plane by means of the univalent function $w = f(z) = \sum_{n=1}^{\infty} a_n z^n$, we can set

$$\Phi(w) = z, \quad g(w) = g(w,0), \quad p(t) = p(t,0), \quad p(w) = -\log \Phi(w)$$

and

$$p(w,\omega) = \log \frac{1 - \Phi(w)\overline{\Phi(\omega)}}{\Phi(w) - \Phi(\omega)}.$$

Now, set $\omega = 0$ in (3.2) and obtain

$$g^*(w) = g(w) + \operatorname{Re}\left\{\frac{\varepsilon}{2\pi i} \int_{\partial\Delta} \frac{\Phi'(t)p'(t)v(t)}{\Phi(w) - \Phi(t)}\,dt + \frac{\Phi(w)\overline{\Phi'(t)p'(t)v(t)}}{1 - \bar{\Phi}(t)\Phi(w)}\,\overline{dt}\right\} + O(\varepsilon^2).$$

We complete $g(w)$ and $g^*(w)$ to analytic functions in Δ:

$$p^*(w) = p(w) - \frac{\varepsilon}{2\pi i} \int_{\partial\Delta} \left\{\frac{1}{\Phi(w) - \Phi(t)} \frac{\Phi'(t)^2 v(t)}{\Phi(t)}\,dt \right.$$
$$\left. + \frac{\Phi(w)}{1 - \overline{\Phi(t)}\Phi(w)} \left(\frac{\Phi'(t)^2 v(t)}{\Phi(t)}\,dt\right)^{-}\right\} + O(\varepsilon^2).$$

We exponentiate and since $\Phi(w) = \exp\{-p(w)\}$, we find

$$\Phi^*(w) = \Phi(w)\left[1 + \frac{\varepsilon}{2\pi i} \int_{\partial\Delta} \left\{\frac{1}{\Phi(w) - \Phi(t)} \frac{\Phi'(t)^2 v(t)}{\Phi(t)}\,dt \right.\right.$$
$$\left.\left. + \frac{\Phi(w)}{1 - \overline{\Phi(t)}\Phi(w)} \left(\frac{\Phi'(t)^2 v(t)}{\Phi(t)}\,dt\right)^{-}\right\}\right] + O(\varepsilon^2).$$

We refer back to the z-plane. Let $z = \Phi(w)$, $\zeta = \Phi(t)$ and Σ be the preimage of Δ, i.e., $\Phi(\Sigma) = \Delta$,

$$z^* = z + \frac{\varepsilon}{2\pi i} \int_{\partial\Sigma} \left\{\frac{z}{z - \zeta} \frac{v[f(\zeta)]\,d\zeta}{\zeta f'(\zeta)} + \frac{z^2}{1 - \bar{\zeta} z} \left(\frac{v[f(\zeta)]\,d\zeta}{\zeta f'(\zeta)}\right)^{-}\right\} + O(\varepsilon^2).$$

Since z^* is the preimage of w under the varied function $f^*(z)$, we have $f^*(z^*) = f(z)$ and thus for $z \in \Sigma$

$$f^*(z)=f(z)+zf'(z)\frac{\varepsilon}{2\pi i}\int_{\partial\Sigma}\left\{\frac{v[f(\zeta)]d\zeta}{(\zeta-z)\zeta f'(\zeta)}-\frac{z}{1-\bar{\zeta}z}\overline{\left(\frac{v[f(\zeta)]d\zeta}{\zeta f'(\zeta)}\right)}\right\}+O(\varepsilon^2). \tag{3.3}$$

It is important to observe the restriction for z to lie in the preimage of the domain set Δ. Suppose that we cross from Σ through $\partial\Sigma$ into the preimage of $D-\Delta$, i.e., to $U-\Sigma$. By the Plemelj saltus condition we find

$$\begin{aligned}f^*(z)=f(z)+\varepsilon v[f(z)]+\frac{\varepsilon}{2\pi i}\int_{\partial\Sigma}\left\{\frac{v[f(\zeta)]d\zeta}{(\zeta-z)\zeta f'(\zeta)}-\frac{z}{1-\bar{\zeta}z}\overline{\left(\frac{v[f(\zeta)]d\zeta}{\zeta f'(\zeta)}\right)}\right\}\\ +O(\varepsilon^2)\end{aligned} \tag{3.3'}$$

valid for $z\in U-\Sigma$. Observe that by our assumption on $v(w)$ the term $v[f(z)]$ is analytic in the domain considered.

The function (3.3) will provide a variation for $S(d)$ if we choose $v(w)$ so that the varied domain D^* satisfies (1.1), i.e., $v(w)$ leaves the circles $|w|=d$ and $|w|=1$ unchanged and is analytic outside of a subdomain Δ of D.

If we begin with

$$w^*(w)=w\left(1+\frac{\varepsilon ae^{i\alpha}}{w-a}-\frac{\varepsilon be^{i\alpha}}{w-b}\right)\psi(w),$$

where a and b are two arbitrary but fixed points of D which are in the ring $R=\{w: d<|w|<1\}$, and $\psi(w)$ is analytic outside of Δ, we must choose $\psi(w)$ so that

$$|w^*(w)|=|w| \quad \text{on } \partial R. \tag{3.4}$$

This leads to the condition

$$\log|\psi(w)|=-\log\left|1+\frac{\varepsilon ae^{i\alpha}}{w-a}-\frac{\varepsilon be^{i\alpha}}{w-b}\right| \tag{3.5}$$

on ∂R. This is a boundary value problem for harmonic functions in R and can be solved by a generalization of the Poisson integral formula which reproduces an analytic function from the boundary values of its real part. We skip the calculations and state the result and verify that it satisfies the requirements.

Let $G_R(w,\omega)$ be the Green's function of the ring R and $P_R(w,\omega)$ its analytic completion in w. $P_R(w,\omega)$ is only determined up to an arbitrary imaginary function of ω. We must now choose this function to assure that $P_R(w,\omega)$ is harmonic in ω. Pick an arbitrary but fixed point w_0 in R and demand

$$P_R(w_0, \omega) = G_R(w_0, \omega).$$

This implies that for any other point w in the ring

$$P_R(w, \omega) = G_R(w_0, \omega) + \int_{w_0}^{w} \left(\frac{\partial G_R(w, \omega)}{\partial u} - i \frac{\partial G_R(w, \omega)}{\partial v} \right) \partial w$$

where $w = u + iv$. This guarantees the harmonicity of $P_R(w, \omega)$ as a function of ω.

Once we have shown the correct determination of $P_R(w, \omega)$, we note that

$$P_R(w, \omega) = -\log(w - \omega) + K_R(w, \omega) \tag{3.6}$$

where $K_R(w, \omega)$ is analytic for all $w \in R$. We now form

$$\begin{aligned} T_R(w, \omega) &= e^{i\alpha}\omega \frac{\partial}{\partial \omega} P_R(w, \omega) + e^{-i\alpha}\bar{\omega} \frac{\partial}{\partial \bar{\omega}} P_R(w, \omega) \\ &= e^{i\alpha} \frac{\omega}{w - \omega} + S_R(w, \omega) \end{aligned} \tag{3.7}$$

where $S_R(w, \omega)$ is analytic for $w \in R$.

Observe that the operator

$$\Omega = e^{i\alpha}\omega \frac{\partial}{\partial \omega} + e^{-i\alpha}\bar{\omega} \frac{\partial}{\partial \bar{\omega}}$$

satisfies

$$\Omega u = 2\mathrm{Re}\left\{ e^{i\alpha}\omega \frac{\partial u}{\partial \omega} \right\} \tag{3.8}$$

for real valued functions u, and is therefore a real operator. In particular,

$$\mathrm{Re}\{T_R(w, \omega)\} = \Omega G_R(w, \omega) = 0 \tag{3.9}$$

for $w \in \partial R$. Hence

$$w^*(w) = w + \varepsilon w[T_R(w, a) - T_R(w, b)] + O(\varepsilon^2) \tag{3.10}$$

is a variation of the type desired, as we now verify.

$$v(w) = w[T_R(w, a) - T_R(w, b)] \tag{3.11}$$

is analytic in R except at a and b where

$$v(w)=\frac{e^{i\alpha}a^2}{w-a}+\text{regular},\qquad v(w)=\frac{e^{i\alpha}b^2}{w-b}+\text{regular},$$

respectively. $v(w)$ is even analytic a little beyond R as follows from the Schwarz reflection principle.

$v(w)$ is single valued in R. Observe that $P_R(w,\omega)$ is not single valued in R. We have the imaginary periods

$$i\int_{|w|=d}\frac{\partial}{\partial n_w}G_R(w,\omega)dS_w=2\pi iH(\omega)$$

where $H(\omega)$ is the harmonic measure of the boundary component $|w|=d$ of R. In fact,

$$H(\omega)=\log\frac{|\omega|}{d}.$$

Hence $T_R(w,\omega)$ has the period $2\pi i\Omega H(\omega)=2\pi i/\log d$, which is independent of ω, and $v(w)$ is single valued.

From (3.9) we see that $T_R(w,a)$ and $T_R(w,b)$ are pure imaginary on ∂R. In view of (3.10) we find

$$\begin{aligned}|w^*(w)|&=|w|\,|1+\varepsilon\{T_R(w,a)\}|+O(\varepsilon^2)\\&=|w|+O(\varepsilon^2)\qquad\text{on }\partial R.\end{aligned}$$

Thus, (3.10) is an expression up to order ε^2 of the variation, whose existence is guaranteed by the argument around (3.5).

If we now choose Δ as the interior of the circles $|w-a|=\delta$, $|w-b|=\delta$ and $|w|=d-\delta$ where $\delta>0$ is chosen so small so that $d^2/\bar{a}$ and d^2/b lie inside the circle $|w|=d-\delta$, and if we choose $v(w)$ as in (3.10), then $f^*(z)$ as expressed by (3.3) will be a variation for functions in $S(d)$.

4. Applications for general functionals

Let ψ be a continuous complex valued functional defined over $S(d)$ and assume that ψ has a Gâteaux derivative, i.e.,

$$\psi(f+\varepsilon g)=\psi(f)+\varepsilon L(f;g)+o(\varepsilon)\tag{4.1}$$

where $L(f; g)$ is a continuous linear functional of $g(z)$, and this holds for all $g(z)$ which are analytic where needed in order for $\psi(f+\varepsilon g)$ to be evaluated.

We consider the problem of maximizing $\text{Re}\{\psi(f)\}$ over $S(d)$. The existence of an extremal function in $S(d)$ is assured by the compactness of $S(d)\cup\{e^{i\alpha}dz\}$. We assume first that $\psi[f]$ depends only on values of $f(z)$ in the preimage of U_d. This is, for example, the case of the various coefficient problems for the class S_d. We can then apply (3.3) as follows. Let

$$u(\zeta)=\frac{v[f(\zeta)]}{\zeta f'(\zeta)}$$

and write (3.3) as

$$f^*(z)=f(z)+\varepsilon z f'(z)\left\{\frac{1}{2\pi i}\int_{\partial\Sigma} u(\zeta)\frac{d\zeta}{\zeta-z}-\frac{1}{2\pi i}\int_{\partial\Sigma}\overline{u(\zeta)}\frac{z}{1-z\bar{\zeta}}\overline{d\zeta}\right\}+O(\varepsilon^2).$$

We obtain a useful form for $\psi(f^*)$ by letting

$$S_1(\zeta)=L\left(f;\frac{zf'(z)}{\zeta-z}\right),$$

$$S_2(\bar{\zeta})=L\left(f;\frac{z^2f'(z)}{1-z\bar{\zeta}}\right)$$

and by defining

$$A(\zeta)=\zeta S_1(\zeta)+\zeta\overline{S_2(\bar{\zeta})}. \tag{4.2}$$

Observe that in view of the linear character of $L(f; g)$ in its second argument

$$\frac{\partial}{\partial\bar{\zeta}}S_1(\zeta)=\frac{\partial}{\partial\zeta}S_2(\bar{\zeta})=0$$

and that $A(\zeta)$ is therefore analytic in ζ.

It follows that

$$\text{Re}\{\psi(f^*)\}=\text{Re}\{\psi(f)\}+\varepsilon\,\text{Re}\left\{\frac{1}{2\pi i}\int_{\partial\Sigma}u(\zeta)A(\zeta)\frac{d\zeta}{\zeta}\right\}+O(\varepsilon^2). \tag{4.3}$$

Now set $w=f(\zeta)$ and define

(4.4) $$m(w) = \left(\frac{f(\zeta)}{\zeta f'(\zeta)}\right)^2$$

and

(4.5) $$M(w) = m(w)A(\zeta).$$

Refer the integral in (4.3) back to the w plane and obtain

$$\mathrm{Re}\{\psi(f^*)\} = \mathrm{Re}\{\psi(f)\} + \varepsilon\, \mathrm{Re}\left\{\frac{1}{2\pi i}\int_{\partial\Delta} v(w)\frac{M(w)}{w^2}\,dw\right\} + O(\varepsilon^2).$$

In particular, upon choosing $v(w)$ as in (3.10) and Δ as described above, we find

(4.6) $$\mathrm{Re}\{\psi(f^*)\} = \mathrm{Re}\{\psi(f)\} + \varepsilon\, \mathrm{Re}\left\{\frac{1}{2\pi i}\int_{\partial\Delta} \frac{M(w)}{w}[T_R(w,a) - T_R(w,b)]dw\right\} + O(\varepsilon^2).$$

Suppose now that $f(z)$ maximizes $\mathrm{Re}\{\psi(f)\}$ over $S(d)$. Then

$$\mathrm{Re}\{\psi(f^*)\} \leqq \mathrm{Re}\{\psi(f)\}$$

and since ε can be positive or negative, from (4.6) we conclude that

(4.7) $$\mathrm{Re}\left\{\frac{1}{2\pi i}\int_{\partial\Delta}\frac{M(w)}{w}T_R(w,a)dw\right\} = \mathrm{Re}\left\{\frac{1}{2\pi i}\int_{\partial\Delta}\frac{M(w)}{w}T_R(w,b)dw\right\} = \text{const.}$$

for all $a \in R$. In view of the definition (3.7) of $T_R(w,a)$ we find that (4.7) yields

(4.8) $$\mathrm{Re}\left\{\frac{1}{2\pi i}\int_{\partial\Delta}\frac{M(w)}{w}\left(e^{i\alpha}a\frac{\partial}{\partial a}P_R(w,a) + e^{-i\alpha}\bar{a}\frac{\partial}{\partial \bar{a}}P_R(w,a)\right)dw\right\} = \text{constant.}$$

The path of integration $\partial\Delta$ consists of the three circles $|w-a| = \delta$, $|w-b| = \delta$ and $|w| = d - \delta$.

From the form of $P_R(w,a)$ as expressed in (3.6) we find that the value of the integral along $|w-b| = \delta$ is zero, and upon applying the residue theorem we find that the value of the integral along $|w-a| = \delta$ is $e^{i\alpha}M(a)$. Therefore, (4.8) can be written as

$$(4.9)\quad \operatorname{Re}\left\{e^{i\alpha}M(a)+\frac{1}{2\pi i}\int_{|w|=d-\delta}\frac{M(w)}{w}T_R(w,a)dw\right\}=\text{constant}\qquad \text{for } a\in R.$$

Now, in view of (3.7) and (3.8), (4.9) implies

$$\operatorname{Re}\left\{e^{i\alpha}M(a)+e^{i\alpha}2a\frac{\partial}{\partial a}\operatorname{Re}\left\{\frac{1}{2\pi i}\int_{|w|=d-\delta}\frac{M(w)}{w}P_R(w,a)dw\right\}\right\}=\text{const.}$$

for $a \in R$.

Since α is arbitrary, we conclude

$$(4.10)\qquad M(a)+2a\frac{\partial}{\partial a}\operatorname{Re}\left\{\frac{1}{2\pi i}\int_{|w|=d-\delta}\frac{M(w)}{w}P_R(w,a)dw\right\}=\text{constant}.$$

Observe that $P_R(w, a)$ depends only on the annulus R and not on the image domain D. It is harmonic as a function of a and hence the left side of (4.10) is analytic in R. Therefore, *$M(a)$ is regular and single valued in R as a function of* a.

We now assert that the image domain of a function which maximizes $\operatorname{Re}\{\psi(f)\}$ is a slit disc.

Theorem 4.1. *Let ψ be a continuous complex-valued function in $S(d)$ which depends upon the values of $f(z)$ which lie in U_d and which possesses a Gâteaux derivative. If $f(z)$ maximizes $\operatorname{Re}\{\psi[f]\}$ over $S(d)$ and if the corresponding analytic function $M(a)$ does not reduce to a constant, then the unit disc U contains no exterior points of the image domain D.*

Proof. Let s and t be exterior points of D which lie in U. In the construction of the variation above, replace the points a and b by s and t, respectively. We can now choose Δ to consist only of the disc $|w| \leq d-\delta$ since s and t are not in D. Now we find, instead of (4.10), that for all exterior points s in R,

$$s\frac{\partial}{\partial s}\operatorname{Re}\left\{\frac{1}{2\pi i}\int_{|w|=d-\delta}\frac{M(w)}{w}P_R(w,s)dw\right\}=\text{constant}.$$

By the permanence principle for analytic functions this holds then for all $s \in R$, i.e., for all a. But then (4.10) implies that $M(a)$ is constant for $a \in R$, which contradicts the hypothesis. Thus, there are no exterior points in R and since D covers all of $|w| < d$, there are no exterior points in $|w| < 1$.

The hypothesis that $M(a)$ not reduce to a constant in R is not at all restrictive, as we shall exhibit in the special cases that we consider in the next section.

In order to further study the slits which together with $|w|=1$ and part of $|w|=d$ form the boundary of the extremal domain D, we consider some elementary variations and the Löwner slit variations. We first simplify $A(\zeta)$. Notice that

$$\zeta S_1(\zeta)=L\left(f;\frac{zf'(z)}{1-z/\zeta}\right)$$

and

$$\bar{\zeta}S_2(\bar{\zeta})=L\left(f;\frac{zf'(z)}{1-z\bar{\zeta}}\right)-L(f;zf'(z)).$$

Define

$$B(\zeta)=L\left(f;\frac{zf'(z)}{1-z/\zeta}\right) \tag{4.11}$$

and

$$C=L(f;zf'(z)) \tag{4.12}$$

and obtain

$$A(\zeta)=B(\zeta)-\bar{C}+\overline{B(1/\bar{\zeta})}. \tag{4.13}$$

If $f(z)\in S(d)$ and α and β are real (and small), then

$$\begin{aligned}e^{i\alpha}f(ze^{i\beta})&=(1+i\alpha)f(z(1+i\beta))+O(\alpha^2+\beta^2)\\&=(1+i\alpha)[f(z)+i\beta zf'(z)]+O(\alpha^2+\beta^2)\in S(d).\end{aligned}$$

Therefore,

$$\mathrm{Re}\{\psi(e^{i\alpha}f(ze^{i\beta}))\}=\mathrm{Re}\{\psi(f)\}+\mathrm{Re}\{i\alpha L(f,f)+i\beta L(f,zf')\}+O(\alpha^2+\beta^2).$$

We thus conclude that if $f\in S(d)$ maximizes $\mathrm{Re}\{\psi(f)\}$, then

$$\mathrm{Im}\{L(f,f)\}=0$$

and

$$\mathrm{Im}\{L(f,zf')\}=0 \tag{4.14}$$

and in particular, C defined in (4.12) is real.

In view of (4.13) and (4.14), we conclude that $A(\zeta)$ *is real for* $|\zeta|=1$.

We now consider the Löwner variations.

The variation (2.9) constructed above,

$$f^*(z) = f(z) - \varepsilon f'(z) z \frac{e^{i\alpha} + z}{e^{i\alpha} - z} + O(\varepsilon^2), \tag{2.9}$$

$\varepsilon > 0$, corresponds to inserting a short slit at the boundary point of D which corresponds to $f(e^{i\alpha})$. The variation (2.9) is also valid for $\varepsilon < 0$ when $e^{i\alpha}$ corresponds to a tip of a slit. In this case the variation corresponds to shortening the slit. These are usable only if the variation does not take f out of $S(d)$. We find from (2.9) and (4.1),

$$\psi(f^*) = \psi(f) - \varepsilon L\left(f, zf'(z)\frac{e^{i\alpha} + z}{e^{i\alpha} - z}\right) + O(\varepsilon^2).$$

Set $e^{i\alpha} = \zeta$ and find

$$L\left(f, zf'(z)\frac{\zeta + z}{\zeta - z}\right) = L\left(f, zf'(z)\left[\frac{2\zeta}{\zeta - z} - 1\right]\right) = 2B(\zeta) - C.$$

Hence,

$$\psi(f^*) = \psi(f) - \varepsilon\{2B(\zeta) - C\} + O(\varepsilon^2). \tag{4.15}$$

Suppose now that $f(z)$ maximizes $\operatorname{Re}\{\psi(f)\}$. Then

$$\operatorname{Re}\{\psi(f^*)\} \leqq \operatorname{Re}\{\psi(f)\}. \tag{4.16}$$

The conclusions which follow from (4.15) and (4.16) differ according to whether the variation f^* corresponds to ε positive or negative.

If ζ is a point which corresponds to a boundary point of D from which a Löwner slit can be constructed without leaving $S(d)$, then ε may be chosen positive and

$$\operatorname{Re}\{2B(\zeta) - C\} = \operatorname{Re}\{A(\zeta)\} = A(\zeta) \geqq 0. \tag{4.17}$$

If ζ is a point corresponding to a tip of a slit which may be shortened without leaving $S(d)$, then ε may be chosen negative and

$$\operatorname{Re}\{A(\zeta)\} = A(\zeta) \leqq 0. \tag{4.18}$$

In particular, for a point ζ_0 corresponding to a tip of a slit w_0, $d < |w_0| < 1$, or the tip of an arc on $|w| = d$, then

$$A(\zeta_0) = 0.$$

The behavior of $A(\zeta)$ can determine the number of slits which form the boundary of D. We have, for example, the following

Theorem 4.2. *Let ψ and $f(z)$ be as in Theorem 4.1. If $A(\zeta)>0$ on $|\zeta|=1$, then f maps U onto U slit along a single arc from $|w|=1$ to $|w|=d$. The boundary contains no arc of $|w|=d$.*

We shall see how this is applicable to coefficient problems in the next section.

We can finally assert that the extremal domain is bounded by piecewise analytic slits.

Theorem 4.3. *Let ψ and $f(z)$ be as in Theorem 4.1. Then the boundary of D is piecewise analytic.*

Proof. Rewrite (4.4) and (4.5) in differential notation and find, after setting $w=f(\zeta)$,

$$M(w)\frac{dw^2}{w^2}=A(\zeta)\frac{d\zeta^2}{\zeta^2}\,. \tag{4.19}$$

According to the discussion preceding (4.17), $A(\zeta)$ is real and nonnegative for all ζ on $|\zeta|=1$ which correspond under $f(z)$ to points of R. Furthermore, we have $d\zeta^2/\zeta^2<0$ on $|\zeta|=1$. Since $w=f(\zeta)$ is a solution to the differential equation (4.19), it follows that the boundary arcs of the image domain D which lie in R must be orthogonal trajectories of the quadratic differential $M(w)dw^2/w^2$. Since $M(w)$ is analytic in R, these are piecewise analytic arcs. All other boundary points lie on $|w|=d$ or $|w|=1$ and the theorem is proven.

We close this section by considering the possibility that $\psi[f]$ depends only on values of $f(z)$ in the ring domain $d<|w|<1$. We have now to use (3.3′) and obtain

$$(4.3')\ \operatorname{Re}\{\psi[f^*]\}=\operatorname{Re}\{\psi[f]\}+\varepsilon\operatorname{Re}\left\{\frac{1}{2\pi i}\int_{\partial\Sigma}u(\zeta)A(\zeta)\frac{d\zeta}{\zeta}+L(f;v[f(z)])\right\}+O(\varepsilon^2).$$

Using the definition (3.11) of $v(w)$ and the same calculations as before, we find

$$(4.10')\quad M(a)+2a\frac{\partial}{\partial a}\operatorname{Re}\left\{\frac{1}{2\pi i}\int_{|w|=d-\delta}\frac{M(w)}{w}P_R(w,a)dw+L[f;P_r(f,a)]\right\}=\text{const.}$$

We can make similar conclusions, but it is now possible that $M(a)$ has poles in the ring domain.

Finally it is clear how we can handle functionals which depend on values of $f(z)$ in D in general.

5. Applications

Consider now the problem of maximizing $\operatorname{Re}\{a_n\}$ over the class $S(d)$. This corresponds to choosing $\psi(f) = f^{(n)}(0)/n!$. In this case we have for the extremal function $f(z)$,

$$L(f; g) = \frac{g^{(n)}(0)}{n!},$$

$$C = na_n,$$

$$B(\zeta) = \sum_{\nu=1}^{n} \nu a_\nu \zeta^{\nu-n},$$

and

$$(5.1) \quad A(\zeta) = \frac{a_1}{\zeta^{n-1}} + \frac{2a_2}{\zeta^{n-2}} + \cdots + \frac{(n-1)a_{n-1}}{\zeta} + na_n + (n-1)\bar{a}_{n-1}\zeta + \cdots + a_1\zeta^{n-1}.$$

According to (4.14) $C = na_n$ is real, and hence $A(\zeta)$ is real on $|\zeta| = 1$. Notice that $A(\zeta)$ is continuous on $|\zeta| = 1$ and can have at most $2n-2$ zeros there. Furthermore, for all ζ on $|\zeta| = 1$ not mapping to the interior of an arc of $|w| = d$, we have $A(\zeta) \geqq 0$.

Consider now the slits which form the boundary of the image domain D. Unless there is just a single slit whose intersection with $|w| = d$ is its end point, $A(\zeta)$ must vanish at each tip. Furthermore, each slit accounts for at least two zeros of $A(\zeta)$ because $A(\zeta)$ is positive at the ζ corresponding to both sides of the slit in $d < |w| < 1$ and hence must have an even number of zeros between two such points. Therefore, the functions which maximize $\operatorname{Re}\{a_n\}$ over $S(d)$ map U onto U less at most $n-1$ piecewise analytic slits.

Notice that the hypothesis of Theorem 4.1, that $M(a)$ not reduce to a constant, is trivially satisfied.

Consider now the problem of maximizing $\operatorname{Re}\{a_1\}$ over $S(d)$. In this case we have

$$A(\zeta) = a_1 > 0.$$

Hence the extemal domain is U less a single analytic slit containing no arc of $|w| = d$. Furthermore, with the use of arguments concerning the structure of the trajectories of

$$M(w)\frac{dw^2}{w^2}$$

one can obtain another proof of Theorem 2.1. We omit this proof.

Consider finally the problem of maximizing $\mathrm{Re}\{\log f(a)\}$ where $a \in U$ is arbitrary but fixed. We note that we may assume, because of the rotational invariance of $S(d)$, that $a > 0$. For this functional we find

$$L(f; g) = \frac{g(a)}{f(a)},$$

$$B(\zeta) = L\left(f, \frac{zf'(z)}{1 - \frac{z}{\zeta}}\right) = \frac{af'(a)}{f(a)}\left[\frac{1}{1 - \frac{a}{\zeta}}\right],$$

$$C = L(f, zf'(z)) = \frac{af'(a)}{f(a)},$$

and

$$A(\zeta) = \frac{af'(a)}{f(a)} \frac{(1 - a^2)\zeta}{(\zeta - a)(1 - a\zeta)}. \tag{5.2}$$

Note that in view of (4.14) we have $af'(a)/f(a)$ real and this allows the calculation leading to (5.2). For $|\zeta| = 1$, we can write (5.2) as

$$A(\zeta) = \frac{af'(a)}{f(a)} \frac{1 - a^2}{|\zeta - a|^2}. \tag{5.3}$$

Clearly $A(\zeta)$ never vanishes for $|\zeta| = 1$, hence it is positive, and the extremal function maps U onto itself less a single analytic slit, containing no arc of $|w| = d$.

We are now able to draw far-reaching conclusions with respect to the analytic function $M(w)$. By definition (4.5) and by (5.3),

$$M(w) = K\left(\frac{f(\zeta)}{\zeta f'(\zeta)}\right)^2 \frac{\zeta}{(\zeta - a)(1 - a\zeta)}, \qquad K = \frac{af'(a)}{f(a)}(1 - a^2) = \text{real}. \tag{5.4}$$

We know that if $|f(a)| < d$, $M(w)$ is regular analytic in the ring $d < |w| < 1$. If $|f(a)| > d$, we see easily that it has a simple pole at $w = f(a)$ but else is regular analytic in the ring. It is regular analytic in U_d exept for a possible simple pole $f(a)$ if $|f(a)| < d$. It is also regular analytic on the circles $|w| = d$ and $|w| = 1$ except at the points w_1 on the unit periphery where the slit enters into the ring and at the point w_0 on the circle $|w| = d$ where it ends. Moreover, it is clear that

$$M(w) = \text{real} \qquad \text{on } |w| = 1. \tag{5.5}$$

Hence, by the Schwarz reflection principle, we find easily

$$(5.6)\qquad M(w)=L\frac{(w-1)^2w}{(w-A)(1-\bar{A}w)(w-w_0)(1-\bar{w}_0w)},\qquad L=\text{real},$$

where we have used the freedom in the sign of $f(z)$ to choose $w_1=1$ and where $A=f(a)$, $w_0=de^{i\tau}$.

We end thus with the differential equation

$$(5.7)\quad\frac{(w-1)^2dw^2}{w(w-A)(1-\bar{A}w)(w-w_0)(1-\bar{w}_0w)}=r\frac{d\zeta^2}{\zeta(\zeta-a)(1-a\zeta)},\qquad r=\text{real}.$$

Along the arc of the unit periphery which corresponds to $|w|=1$, we have

$$\frac{dw^2}{w^2}(w+\bar{w}-2)\frac{1}{|w-A|^2|w-w_0|^2}=r\frac{d\zeta^2}{\zeta^2}\frac{1}{|\zeta-a|^2}.$$

This shows that $r<0$.

Next, we multiply both sides of (5.7) by $(\zeta-a)$ and let $\zeta\to a$. We obtain

$$\frac{(A-1)^2}{(1-|A|^2)}\frac{f'(a)a}{f(a)}\frac{1}{(A-w_0)(1-A\bar{w}_0)}=\frac{r}{1-a^2}$$

and conclude that

$$(5.8)\qquad \frac{(A-1)^2}{(A-w_0)(1-A\bar{w}_0)}=\text{real}.$$

This establishes one relation between the important points $A=f(a)$ and $w_0=de^{i\tau}$. Another one is given by the obvious condition

$$(5.9)\qquad \int_0^A\frac{(w-1)dw}{[w(w-A)(1-\bar{A}w)(w-w_0)(1-\bar{w}_0w)]^{\frac{1}{2}}}=\text{real}.$$

To prove that the conditions (5.8) and (5.9) imply that w_0 and A are real, we introduce the notation

$$(5.10)\qquad A=ve^{i\tau},\quad w=ve^{i\tau}t,\quad w_0=de^{i\sigma},\quad \lambda=\frac{d}{1+d^2}$$

and transform (5.9) into

$$\int_0^1 \frac{dt}{[t(t-1))(1-v^2t)]^{\frac{1}{2}}} \frac{(vte^{i\tau}-1)}{[vte^{i\tau}-\lambda(e^{i\sigma}+e^{-i\sigma}v^2t^2e^{2i\tau})]^{\frac{1}{2}}} = \text{real}.$$

Let

$$vt = e^{-s}, \quad \log\frac{1}{v} \leqq s < \infty, \quad \rho = \sigma - \tau$$

and write the square of the second factor in the integrand as

$$T = \frac{e^{i\tau - s} + e^{-(i\tau - s)} - 2}{1 - \lambda[e^{i\rho + s} + e^{-(i\rho + s)}]}. \tag{5.11}$$

The argument of T is the same as of

$$T^* = T[1 - \lambda(e^{-i\rho+s} + e^{i\rho - s})]$$
$$= e^{i\tau - s} + e^{-i\tau + s} - 2 - \lambda(e^{i(\tau-\rho)} + e^{-i(\tau+\rho)+2s} - 2e^{-i\rho+s} + e^{i(\rho+\tau)-2s} + e^{i(\rho-\tau)} - 2e^{i\rho - s}).$$

We find thus

$$\operatorname{Im}\{T^*\} = (e^{-s} - e^{s})[\sin\tau + 2\lambda(\sin\rho + \sin\sigma\cosh s)]. \tag{5.12}$$

Now observe that (5.8) states that $\operatorname{Im}\{T^*\} = 0$ for $s = \log 1/v$. Hence $\operatorname{Im}\{T^*\}$ does not change sign as s increases from $\log 1/v$ to ∞. Thus, either $\sin\tau = \sin\sigma = 0$ and $\operatorname{Im}\{T^*\} \equiv 0$ for all values of s, or T lies all the time in a half-space with fixed sign for the imaginary part of T, and the same will hold *a fortiori* for $\sqrt{T}$. Hence (5.9) is only possible if $\tau = \sigma = 0$ and A and w_0 are real.

In this case, it is easy to integrate the differential equation (5.7). We substitute

$$W = w + \frac{1}{w}, \quad \kappa = \frac{A}{1 + A^2}, \quad \lambda = \frac{w_0}{1 + w_0^2} \tag{5.13}$$

and find

$$\frac{dW^2}{(W+2)(1-\kappa W)(1-\lambda W)} = r_1 \frac{d\zeta^2}{\zeta(\zeta - a)(1 - a\zeta)}, \qquad r_1 = \text{real}. \tag{5.14}$$

The image of the unit periphery becomes now a rectilinear segment from $W = 2$ to $W = 1/\lambda = w_0 + 1/w_0$. Thus $w_0 = d > 0$. We then find

$$W[f(\zeta)] + 2 = \frac{(1+d)^2}{4d}\left(e^{i\tau}\zeta + \frac{1}{e^{i\tau}\zeta} + 2\right), \tag{5.15}$$

that is

$$K_\pi[f(\zeta)] = \frac{4d}{(1+d)^2} K_\pi(e^{i\tau}\zeta) \tag{5.16}$$

and

$$f(\zeta) = F_\pi(e^{i\tau}\zeta). \tag{5.17}$$

The condition $af'(a)/f(a)$ real yields then $e^{i\tau} = \pm 1$ and we find

$$f(\zeta) = F_\pi(\pm\zeta). \tag{5.18}$$

It is not surprising that we have two choices for the extremum function. For, had we asked for the minimum of $\log f(a)$ instead of the maximum, we would have obtained the same conditions.

We have thus completely solved the extremum problem posed. However, our main objective was to illustrate the general method by a typical example.

References

1. P L. Duren and M. M. Schiffer, *The theory of the second variation in extremum problems for univalent functions*, J. Analyse Math. **10** (1962), 193–252.

2. W. K. Hayman, *Multivalent Functions*, Cambridge University Press, Cambridge, England, 1958.

3. E. Netanyahu, *On univalent functions in the unit disk whose image contains a given disk*, J. Analyse Math. **23** (1970), 305–322.

4. G. Pick, *Über die konforme Abbildung eines Kreises auf ein schlichtes und zugleich beschränktes Gebiet*, Sitzungsber. Math.-Naturwiss, Kaiserlichen Akad. Wiss. Wien **126** (1917), 247–263.

University of Maryland
College Park, MD

Bar-Ilan University
Ramat-Gan, Israel

Stanford Univeristy
Stanford, CA

(Received July 9, 1979)

Commentary on

[121] James A. Hummel, Bernard Pinchuk and Menahem M. Schiffer, *Bounded univalent functions which cover a fixed disc*, J. Analyse Math. **36** (1979), 118–138.

This paper was written during the Spring of 1979, when Schiffer and Pinchuk, as well as Elisha Netanyahu, were all visiting Hummel at the University of Maryland.

Netanyahu [N] had studied univalent functions in the unit disc whose images cover a disc about the origin of radius $d > 0$. In [121], attention is focused on extremal problems for the class of *bounded* univalent functions with this property. Specifically, for fixed $0 < d \leq 1$, denote by $S(d)$ the class of functions

$$f(z) = a_1 z + a_2 z^2 + \ldots$$

analytic and univalent in the unit disc U such that

$$\{w : |w| < d\} \subset f(U) \subset \{w : |w| < 1\}, \qquad (1)$$

where d and 1 are the largest and smallest radii, respectively, for which the inclusions (1) hold.

Elementary methods suffice to solve the problem of maximizing $|a_1|$ over $S(d)$; but when applied to maximizing $|a_2|$, they yield the solution only for $d \leq 3 - 2\sqrt{2}$. Accordingly, the authors adopt a different approach and develop a method of variation for $S(d)$. This, the central contribution of the paper, is done in considerable detail; and [121] remains an authoritative exposition of the Schiffer variational method, especially for variations with constraints. The variational method is then applied to maximizing rather general functionals over $S(d)$, whose extremals are shown to map U onto a slit disc. It is also applied to show that the extremal functions in $S(d)$ for the functional $\mathrm{Re}\{a_n\}$ map U onto U less at most $(n-1)$ piecewise analytic slits.

In [NP], which appears immediately following [121] in *Journal d'Analyse*, Netanyahu and Pinchuk apply Baernstein's method of symmetrization [B] to characterize the mapping properties of the function maximizing $|a_2|$ over $S(d)$ and thus to identify (if somewhat implicitly) the extremal function for all $0 < d \leq 1$. Subsequently, Hummel and Pinchuk [HP] showed how the characterization of the mapping properties for the extremal function, together with the differential equation obtained for it in [NP], allows for the numerical calculation of $\max|a_2|$ over $S(d)$ to high accuracy.

References

[B] Albert Baernstein II, *Integral means, univalent functions and circular symmetrization*, Acta Math. **133** (1974), 139–169.

[HP] J.A. Hummel and B. Pinchuk, *The second coefficient of bounded univalent functions which cover a fixed disc*, J. Analyse Math. **46** (1986), 167–175.

[N] E. Netanyahu, *On univalent functions in the unit disc whose image contains a given disk*, J. Analyse Math. **23** (1970), 305–322.

[NP] E. Netanyahu and B. Pinchuk, *Symmetrization and extremal bounded univalent functions*, J. Analyse Math. **36** (1979), 139–144.

BERNARD PINCHUK

[122] (with G. Schober) The dielectric Green's function and quasiconformal mapping

[122] (with G. Schober) The dielectric Green's function and quasiconformal mapping. *J. Analyse Math.* **36** (1979), 233–243.

THE DIELECTRIC GREEN'S FUNCTION AND QUASICONFORMAL MAPPING†

By

M. SCHIFFER AND G. SCHOBER

1. Introduction

The calculus of variations is an effective technique for attacking many problems in the theory of conformal and quasiconformal (q.c.) mapping. Typically, one asks for the maximum of a continuous functional defined on a compact family of functions. By subjecting an extremal function for the problem to variations within the family, one is led to a differential relation that restricts the extremal function. With this information one attempts to characterize the extremal function and to determine the maximum value of the functional. In this way one can obtain elegant and useful inequalities for various families of conformal and q.c. mappings.

The procedure is particularly effective when the differential relation contains a factor arising from the functional derivative as the square of some analytic function. This is characteristic of functionals inherent in the classical Grunsky–Golusin inequalities. In this article we shall study closely related functionals involving the dielectric Green's function, that possess a similar character. We shall obtain inequalities which contain as special cases the Grunsky–Golusin inequalities for conformal mappings and for conformal mappings with q.c. extensions, both for simply connected and for multiply connected domains.

Sections 2 and 3 of this article review some properties of the dielectric Green's function and introduce related kernels. The extremal problem is posed in Section 4 and solved in Section 5. The resulting inequalities are contained in Section 6.

2. The dielectric Green's function

Let $\tilde{D}$ be a plane domain containing a neighborhood of ∞ and bounded by a curve system C consisting of a finite number of Jordan curves $C_1, \cdots, C_M$. Denote the interior of each C_j by D_j and the union of the M domains D_j by D.

In physical terms we consider each domain D_j to be filled with a homogeneous

† This work was supported in part by grants MCS–75–23332–A03 and MCS–77–01831–A01 from the National Science Foundation.

isotropic dielectric medium with dielectric constant ε. Then the dielectric Green's function $g_\varepsilon(z,\zeta)$ represents the electrostatic potential at z induced by a point charge at ζ. Specifically, $g_\varepsilon(z,\zeta)$ is characterized by the following properties.

(a) $g_\varepsilon(z,\zeta)$ is a harmonic function of z in $D \cup \tilde{D}$, except for $z=\zeta$ and $z=\infty$.

(b) If $\zeta \in \tilde{D}$, the function $g_\varepsilon(z,\zeta)+\log|z-\zeta|$ is harmonic at ζ, and if $\zeta \in D$, the function $g_\varepsilon(z,\zeta)+\varepsilon\log|z-\zeta|$ is harmonic at ζ.

(c) $g_\varepsilon(z,\zeta)$ is continuous across C.

(d) $g_\varepsilon(z,\zeta)$ has a (multiple-valued) harmonic conjugate $h_\varepsilon(z,\zeta)$ in $(D\cup\tilde{D})\backslash\{\zeta\}$ with jump discontinuity at C such that $\varepsilon\, h_\varepsilon(z,\zeta)$ in $\tilde{D}$ and $h_\varepsilon(z,\zeta)$ in D are continuous across C.

(e) $g_\varepsilon(z,\zeta)+\log|z|\to 0$ as $z\to\infty$ for fixed ζ.

If C_i is a sufficiently smooth curve, then the discontinuity behavior in (d) can be expressed in terms of normal derivatives as

$$\frac{\partial}{\partial n_i} g_\varepsilon(z,\zeta) = -\varepsilon \frac{\partial}{\partial \tilde{n}} g_\varepsilon(z,\zeta),$$

where n_i is directed into D_i and $\tilde{n}$ into $\tilde{D}$.

The dielectric Green's function is unique and is symmetric in its two arguments. It is known to exist classically (see, e.g., [2, 3, 4, 6]), and many of its properties were explored in [3]. In particular, if

$$p_\varepsilon(z,\zeta) = g_\varepsilon(z,\zeta) + ih_\varepsilon(z,\zeta)$$

is an analytic completion of the dielectric Green's function, then

$$F_\varepsilon(z,\zeta) = \exp[-p_\varepsilon(z,\zeta)]$$

is a univalent mapping of each of the domains $\tilde{D}, D_1, \cdots, D_M$. Furthermore, its restriction to $\tilde{D}$ interpolates between the canonical circular slit mapping as $\varepsilon\to 0$ and the canonical radial slit mapping as $\varepsilon\to\infty$.

The function $p_\varepsilon(z,\zeta)$ has logarithmic singularities at $z=\zeta$ and $z=\infty$. Because of (d) it has no periods around the curves C_i. However, it is determined only to an additive imaginary function of ζ. Therefore we normalize it further by the condition, compatible with (e), that

$$p_\varepsilon(z,\zeta)+\log(z-\zeta)\to 0 \quad \text{as } z\to\infty \quad \text{for fixed } \zeta\in D\cup\tilde{D}.$$

With this normalization it is evident that for $\varepsilon=1$

$$p_1(z,\zeta) = \log\frac{1}{z-\zeta}.$$

3. Related kernels

In general, the function $p_\varepsilon(z,\zeta)=g_\varepsilon(z,\zeta)+ih_\varepsilon(z,\zeta)$ is analytic in z and harmonic in ζ, and $g_\varepsilon(z,\zeta)$ is symmetric. Every such function has the decomposition

$$p_\varepsilon(z,\zeta)=q_\varepsilon(z,\zeta)+r_\varepsilon(z,\bar{\zeta})$$

where $q_\varepsilon(z,\zeta)$ is analytic in z and ζ and symmetric, $q_\varepsilon(z,\zeta)=q_\varepsilon(\zeta,z)$, and where $r_\varepsilon(z,\bar{\zeta})$ is analytic in z, anti-analytic in ζ, and hermitian, $r_\varepsilon(z,\bar{\zeta})=\overline{r_\varepsilon(\zeta,\bar{z})}$. The function $q_\varepsilon(z,\zeta)$ has a logarithmic singularity at $z=\zeta$, and the decomposition is unique once we specify that

$$q_\varepsilon(z,\zeta)+\log(z-\zeta)\to 0 \quad \text{as } z\to\infty \quad \text{for fixed } \zeta\in\tilde{D},$$

$$q_\varepsilon(z,\zeta)+\frac{\varepsilon+1}{2}\log(z-\zeta)\to 0 \quad \text{as } z\to\infty \quad \text{for fixed } \zeta\in D.$$

In the special case $\varepsilon=1$ it is clear that

$$q_1(z,\zeta)=\log\frac{1}{z-\zeta} \quad \text{and} \quad r_1(z,\bar{\zeta})=0.$$

For the function $p_\varepsilon(z,\zeta)$ the boundary relations (c) and (d) can be written as

$$\tilde{p}_\varepsilon(z,\zeta)=\frac{\varepsilon+1}{2\varepsilon}p_\varepsilon(z,\zeta)+\frac{\varepsilon-1}{2\varepsilon}\overline{p_\varepsilon(z,\zeta)} \quad \text{for } z\in C, \qquad \zeta\in D\cup\tilde{D}.$$

Here the superscript ~ denotes the limit as z approaches C from $\tilde{D}$; otherwise, the limit is from within D. By separating the analytic and anti-analytic functions of ζ, we find that

$$\tilde{q}_\varepsilon(z,\zeta)=\frac{\varepsilon+1}{2\varepsilon}q_\varepsilon(z,\zeta)+\frac{\varepsilon-1}{2\varepsilon}\overline{r_\varepsilon(z,\bar{\zeta})},$$

$$\tilde{r}_\varepsilon(z,\bar{\zeta})=\frac{\varepsilon+1}{2\varepsilon}r_\varepsilon(z,\bar{\zeta})+\frac{\varepsilon-1}{2\varepsilon}\overline{q_\varepsilon(z,\zeta)}$$

for $z\in C$ and $\zeta\in D\cup\tilde{D}$. At first, these equations are determined only up to an additive function of z, but this function is eliminated by the normalizations at ∞.

Define an auxiliary function

$$\pi_\varepsilon(z,\zeta)=q_\varepsilon(z,\zeta)-r_\varepsilon(z,\bar{\zeta}) \quad \text{for } z,\zeta\in D\cup\tilde{D}, \qquad z\neq\zeta.$$

This function satisfies the boundary relation

$$\tilde{\pi}_\varepsilon(z,\zeta)=\frac{\varepsilon+1}{2\varepsilon}\pi_\varepsilon(z,\zeta)-\frac{\varepsilon-1}{2\varepsilon}\overline{\pi_\varepsilon(z,\zeta)}\quad \text{for } z\in C, \qquad \zeta\in D\cup\tilde{D}.$$

That is, the real part of $\pi_\varepsilon(z,\zeta)$ has an ε-jump across C while its harmonic conjugate is continuous. From the correct logarithmic singularities, the boundary relation, and the normalization at ∞, we are able to identify

$$p_{1/\varepsilon}(z,\zeta)=\begin{cases}\pi_3(z,\zeta) & \text{for} \quad z,\zeta\in\tilde{D}\\[2ex] \dfrac{1}{\varepsilon}\pi_\varepsilon(z,\zeta) & \text{for} \quad z\in D,\zeta\in\tilde{D} \quad \text{or} \quad z\in\tilde{D},\zeta\in D\\[2ex] \dfrac{1}{\varepsilon^2}\pi_\varepsilon(z,\zeta) & \text{for} \quad z,\zeta\in D.\end{cases}$$

This interesting symmetry relation implies also that

$$q_\varepsilon(z,\zeta)=q_{1/\varepsilon}(z,\zeta) \quad \text{and} \quad r_\varepsilon(z,\bar{\zeta})=-r_{1/\varepsilon}(z,\bar{\zeta}) \quad \text{if } z,\zeta\in\tilde{D},$$

$$q_\varepsilon(z,\zeta)=\varepsilon q_{1/\varepsilon}(z,\zeta) \quad \text{and} \quad r_\varepsilon(z,\bar{\zeta})=-\varepsilon r_{1/\varepsilon}(z,\bar{\zeta})$$

$$\text{if } z\in D, \quad \zeta\in\tilde{D} \quad \text{or} \quad z\in\tilde{D}, \quad \zeta\in D$$

$$q_\varepsilon(z,\zeta)=\varepsilon^2 q_{1/\varepsilon}(z,\zeta) \quad \text{and} \quad r_\varepsilon(z,\bar{\zeta})=-\varepsilon^2 r_{1/\varepsilon}(z,\bar{\zeta}) \quad \text{if } z,\zeta\in D.$$

It is useful to have an example. For the special case that C is just the unit circle, the function $p_\varepsilon(z,\zeta)$ is given by

$$p_\varepsilon(z,\zeta)=\begin{cases}\log\dfrac{1}{z-\zeta}+\dfrac{\varepsilon-1}{\varepsilon+1}\log\dfrac{z\bar{\zeta}}{z\bar{\zeta}-1} & \text{for} \quad |z|>1,|\zeta|>1;\\[2ex] \left[\dfrac{2\varepsilon}{\varepsilon+1}\log\dfrac{1}{z-\zeta}-\dfrac{(\varepsilon-1)^2}{2(\varepsilon+1)}\log(-\zeta)\right]+\dfrac{\varepsilon-1}{2}\log(-\bar{\zeta}) & \text{for}\\ & |z|<1,|\zeta|>1;\\[2ex] \left[\dfrac{2\varepsilon}{\varepsilon+1}\log\dfrac{1}{z-\zeta}-\dfrac{(\varepsilon-1)^2}{2(\varepsilon+1)}\log(-z)\right]+\dfrac{\varepsilon-1}{2}\log(-z) & \text{for}\\ & |z|>1,|\zeta|<1;\\[2ex] \varepsilon\log\dfrac{1}{z-\zeta}+\varepsilon\dfrac{1-\varepsilon}{1+\varepsilon}\log\dfrac{1}{z\bar{\zeta}-1} & \text{for} \quad |z|<1,|\zeta|<1.\end{cases}$$

Brackets have been used to help identify the kernels $q_\varepsilon(z,\zeta)$ and $r_\varepsilon(z,\bar{\zeta})$. Of course, the dielectric Green's function is given be the real parts of this formula.

4. An extremal problem

For $0 \leqq k < 1$, let Σ_k denote the family of all q.c. homeomorphisms of the plane that are conformal in $\tilde{D}$ and k-quasiconformal in D. That is, each function f in Σ_k is conformal in $\tilde{D}$ and has complex dilatation $\mu = f_{\bar{z}}/f_z$ that satisfies

$$|\mu| \leqq k \qquad \text{a.e. in } D.$$

In addition, assume that each function f in Σ_k is normalized so that

$$f(z) = z + o(1) \qquad \text{as } z \to \infty.$$

Then Σ_k is compact in the topology of uniform convergence on compact sets.

Fix points $z_1, \cdots, z_N$ in $\tilde{D}$, complex constants $\lambda_1, \cdots, \lambda_N$, and dielectric constant ε, $0 < \varepsilon < \infty$. Corresponding to each f in Σ_k, let G_ε, P_ε, Q_ε, R_ε, and Π_ε denote the kernels of Section 3 for the domains determined by $f(C)$, that is, with dielectric constant ε in $f(D)$. Consider the functional

$$\phi[f] = \operatorname{Re} \sum_{m,n=1}^{N} \{\lambda_m \lambda_n [Q_\varepsilon(f(z_m), f(z_n)) + \log(z_m - z_n)] + \lambda_m \bar{\lambda}_n R_\varepsilon(f(z_m), \overline{f(z_n)})\}$$

where for $m = n$ we mean the obvious limit.

Since the functional ϕ is continuous on the family Σ_k, we may for the next section denote by f a function in Σ_k for which ϕ is a maximum.

5. Calculus of variations

The extremal function f will be subjected to variations generated by the Beltrami equation

$$\Phi_{\bar{w}} = ta\Phi_w, \qquad 0 < t < 1,$$

where $|a(w)| = \chi_E(w)$ is the characteristic function for a compact set E in $f(D)$. This equation has a q.c. solution of the form (cf. [8])

$$\Phi(w) = w - \frac{t}{\pi} \iint_{E} \frac{a(\zeta)}{\zeta - w}\, dm_\zeta + O(t^2),$$

which preserves the normalization of Σ_k. For the varied function $f^* = \Phi \circ f$, the dielectric Green's function has the asymptotic development (see [4])

$$G_\varepsilon^*(w^*,\omega^*)=G_\varepsilon(w,\omega)+\frac{t}{\pi\varepsilon}\operatorname{Re}\iint_E a(\zeta)P'_\varepsilon(\zeta,w)P'_\varepsilon(\zeta,\omega)\,dm_\zeta+O(t^2)$$

for $w,\omega\in f(\tilde{D})$. Since

$$\operatorname{Re}\{aP'_\varepsilon((\zeta,w)P'_\varepsilon(\zeta,\omega)\}=\operatorname{Re}\{aQ'_\varepsilon(\zeta,w)Q'(\zeta,\omega)+\overline{aR'_\varepsilon(\zeta,\bar{w})R'_\varepsilon(\zeta,\bar{\omega})}$$
$$+aQ'_\varepsilon(\zeta,w)R'_\varepsilon(\zeta,\bar{\omega})+\overline{aR'_\varepsilon(\zeta,\bar{w})Q'_\varepsilon(\zeta,\omega)}\}$$

is the real part of an analytic function of w, it follows that

$$P_\varepsilon^*(w^*,\omega^*)=P_\varepsilon(w,\omega)+\frac{t}{\pi\varepsilon}\iint_E\{a(\zeta)Q'_\varepsilon(\zeta,w)Q'_\varepsilon(\zeta,\omega)+\overline{a(\zeta)R'_\varepsilon(\zeta,\bar{w})R'_\varepsilon(\zeta,\bar{\omega})}$$
$$+a(\zeta)Q'_\varepsilon(\zeta,w)R'_\varepsilon(\zeta,\bar{\omega})+\overline{a(\zeta)R'_\varepsilon(\zeta,\bar{w})Q'_\varepsilon(\zeta,\omega)}\}dm_\zeta+O(t^2).$$

The possibility of an additional imaginary function of ω is eliminated by the normalizations at $w=\infty$. By separating the analytic and anti-analytic functions of ω, we conclude that

$$Q_\varepsilon^*(w^*,\omega^*)=Q_\varepsilon(w,\omega)+\frac{t}{\pi\varepsilon}\iint_E\{a(\zeta)Q'_\varepsilon(\zeta,w)Q'_\varepsilon(\zeta,\omega)$$
$$+\overline{a(\zeta)R'_\varepsilon(\zeta,\bar{w})R'_\varepsilon(\zeta,\bar{\omega})}\}\,dm_\zeta+O(t^2)$$

and

$$R_\varepsilon^*(w^*,\bar{\omega}^*)=R_\varepsilon(w,\bar{\omega})+\frac{t}{\pi\varepsilon}\iint_E\{a(\zeta)Q'_\varepsilon(\zeta,w)R'_\varepsilon(\zeta,\bar{\omega})$$
$$+\overline{a(\zeta)R'_\varepsilon(\zeta,\bar{w})Q'_\varepsilon(\zeta,\omega)}\}dm_\zeta+O(t^2)$$

for $w,\omega\in f(\tilde{D})$. Again, possible additive functions of w are eliminated by the normalizations at $\omega=\infty$.

We now apply these variational formulas to the functional ϕ. A straightforward calculation leads ultimately to the development

$$\phi[f^*]=\phi[f]+\frac{t}{\pi\varepsilon}\operatorname{Re}\iint_E a(\zeta)\left\{\sum_{n=1}^{N}[\lambda_nQ'_\varepsilon(\zeta,f(z_n))+\bar{\lambda}_nR'_\varepsilon(\zeta,\overline{f(z_n)})]\right\}^2dm_\zeta+O(t^2).$$

Define the function $\mathcal{J}(z)=\mathcal{U}(z)+i\mathcal{V}(z)$ by the formula

$$\mathcal{J}(z)=\int^{f(z)}\sqrt{\left\{\sum_{n=1}^{N}\left[\lambda_n Q'_\epsilon(\zeta,f(z_n))+\overline{\lambda_n}R'_\epsilon(\zeta,\overline{f(z_n)})\right]\right\}^2}\,d\zeta$$

$$=\sum_{n=1}^{N}\left[\lambda_n Q_\epsilon(f(z),f(z_n))+\overline{\lambda_n}R_\epsilon(f(z),\overline{f(z_n)})\right].$$

In this step we see the benefit of having a functional derivative that is a perfect square. Now the fundamental lemma of the calculus of variations for q.c. mappings (see [5, 7, 8]) asserts that

$$\mathcal{J}(z)-k\overline{\mathcal{J}(z)}=(1-k)\mathcal{U}(z)+i(1+k)\mathcal{V}(z)$$

is analytic in D. Since $\mathcal{J}(z)$ itself is analytic in $\tilde{D}$, except for logarithmic singularities at the points z_n and at ∞, we are led to an interesting boundary value problem for analytic functions.

In order to identify the function $\mathcal{J}(z)=\mathcal{U}(z)+i\mathcal{V}(z)$, rewrite it in the form

$$\mathcal{J}(z)=\sum_{n=1}^{N}\left[x_n P_\epsilon(f(z),f(z_n))+iy_n\Pi_\epsilon(f(z),f(z_n))\right]$$

where $\lambda_n=x_n+iy_n$. For any σ the function

$$\sum_{n=1}^{N}\left[x_n p_\sigma(z,z_n)+iy_n\pi_\sigma(z,z_n)\right]$$

has the same singularities as $\mathcal{J}(z)$ in $\tilde{D}$ and the same behavior at ∞. Since $[(1+k)/(1-k)]\mathcal{V}(z)$ is the harmonic conjugate of $\mathcal{U}(z)$ in D, the choice $\sigma=\epsilon[(1+k)/(1-k)]$ matches the discontinuity properties across C. We conclude that

$$\mathcal{J}(z)=\begin{cases}\displaystyle\sum_{n=1}^{N}\left[x_n p_\sigma(z,z_n)+iy_n\pi_\sigma(z,z_n)\right] & \text{for} \quad z\in\tilde{D}\\ \displaystyle\sum_{n=1}^{N}\left(x_n\left[p_\sigma(z,z_n)+k\overline{p_\sigma(z,z_n)}\right]+\right. & \\ \left.+\,iy_n\left[\pi_\sigma(z,z_n)-k\overline{\pi_\sigma(z,z_n)}\right]\right)/(1+k) & \text{for} \quad z\in D.\end{cases}$$

This, then, provides a relation for the extremal function f.

We rewrite the relation in the form

$$\sum_{n=1}^{N} [\lambda_n Q_\varepsilon(f(z), f(z_n)) + \overline{\lambda_n} R_\varepsilon(f(z), \overline{f(z_n)})] = \sum_{n=1}^{N} [\lambda_n q_\sigma(z, z_n) + \overline{\lambda_n} r_\sigma(z, \overline{z_n})]$$

for z in $\tilde{D}$. From this it follows that the maximum value of the functional is

$$\phi[f] = \operatorname{Re} \sum_{m,n=1}^{N} \{\lambda_m\lambda_n[q_\sigma(z_m, z_n) + \log(z_m - z_n)] + \lambda_m\overline{\lambda_n} r_\sigma(z_m, \overline{z_n})\}.$$

Due to the arbitrary nature of the complex constants $\lambda_1, \cdots, \lambda_N$, we conclude that

$$\left| \sum_{m,n=1}^{N} \lambda_m\lambda_n[Q_\varepsilon(f(z_m), f(z_n)) - q_\sigma(z_m, z_n)] \right| \leqq \sum_{m,n=1}^{N} \lambda_m\overline{\lambda_n}[r_\sigma(z_m, \overline{z_n}) - R_\varepsilon(f(z_m), \overline{f(z_n)})]$$

for every function f in Σ_k.

6. Inequalities

The following theorem provides a summary of our conclusions.

Theorem 1. *Suppose $0 \leqq k < 1$ and $0 < \varepsilon < \infty$. Corresponding to each function f in Σ_k, let $Q_\varepsilon(w, \omega)$ and $R_\varepsilon(w, \bar{\omega})$ be the dielectric kernels which have in $f(D)$ the dielectric constant ε, and let $q_\sigma(z, \zeta)$ and $r_\sigma(z, \bar{\zeta})$ be the dielectric kernels which have in D the dielectric constant $\sigma = \varepsilon[(1+k)/(1-k)]$. Then*

$$\left| \sum_{m,n=1}^{N} \lambda_m\lambda_n[Q_\varepsilon(f(z_m), f(z_n)) - q_\sigma(z_m, z_n)] \right| \leqq \sum_{m,n=1}^{N} \lambda_m\overline{\lambda_n}[r_\sigma(z_m, \overline{z_n}) - R_\varepsilon(f(z_m), \overline{f(z_n)})]$$

for all points $z_1, \cdots, z_N$ in $\tilde{D}$ and all complex constants $\lambda_1, \cdots, \lambda_N$.

One immediate conclusion is that the hermitian form on the right is nonnegative.

For the special case $\varepsilon = 1$, the kernels $Q_1(w, \omega) = \log 1/(w - \omega)$ and $R_1(w, \bar{\omega}) = 0$. The resulting Grunsky–Golusin type inequalities for conformal mappings of multiply connected domains, with q.c. extensions, were obtained first by R. Kühnau [1]:

Corollary 1. *If $f \in \Sigma_k$ and $K = (1+k)/(1-k)$, then*

$$\left| \sum_{m,n=1}^{N} \lambda_m \lambda_n \left\{ \log \frac{f(z_m) - f(z_n)}{z_m - z_n} - [q_K(z_m, z_n) - \log(z_m - z_n)] \right\} \right| \leq \sum_{m,n=1}^{N} \lambda_m \bar{\lambda}_n r_K(z_m, \bar{z}_n)$$

for all points $z_1, \cdots, z_N$ in $\tilde{D}$ and all complex constants $\lambda_1, \cdots, \lambda_N$.

By choosing the λ_n to be real and using the symmetries $q_\varepsilon(z, \zeta) = q_{1/\varepsilon}(z, \zeta)$ and $r_\varepsilon(z, \bar{\zeta}) = -r_{1/\varepsilon}(z, \bar{\zeta})$ for $z, \zeta \in \tilde{D}$, we obtain from Theorem 1 the following inequalities for the dielectric Green's function.

Corollary 2. *If $f \in \Sigma_k$ and $K = (1+k)/(1-k)$, then*

$$0 \leq \sum_{m,n=1}^{N} x_m x_n [g_{\varepsilon K}(z_m, z_n) - G_\varepsilon(f(z_m), f(z_n))]$$

and

$$0 \leq \sum_{m,n=1}^{N} x_m x_n [G_\varepsilon(f(z_m), f(z_n)) - g_{\varepsilon/K}(z_m, z_n)]$$

for all points $z_1, \cdots, z_N$ in $\tilde{D}$ and all real constants $x_1, \cdots, x_N$.

Let Σ be the class of univalent conformal mappings f of the domain $\tilde{D}$, normalized so that $f(z) = z + o(1)$ as $z \to \infty$. Since the families Σ_k are dense in Σ as $k \to 1$, all of the inequalities above apply to Σ, using $k = 1$. In this case $q_0(z, \zeta) + r_0(z, \bar{\zeta}) = p_0(z, \zeta)$ and $q_\infty(z, \zeta) + r_\infty(z, \bar{\zeta}) = p_\infty(z, \zeta)$ are obtained from the logarithms of the canonical circular and radial slit mappings as mentioned in Section 2.

Our entire development can be carried out using different dielectric constants ε_j for each of the domains $f(D_j)$ and prescribing different quasiconformality constants k_j for each of the domains D_j. The only modification of the formulas is that $\sigma_j = \varepsilon_j[(1+k_j)/(1-k_j)]$.

Since the identity mapping $f(z) = z$ belongs to the families Σ_k for all k, the inequalities of Theorem 1 imply the following monotonicity property for the kernels $q_\varepsilon(z, \zeta)$ and $r_\varepsilon(z, \bar{\zeta})$.

Corollary 3. *If $0 \leq \varepsilon \leq \sigma \leq \infty$, then*

$$\left| \sum_{m,n=1}^{N} \lambda_m \lambda_n [q_\sigma(z_m, z_n) - q_\varepsilon(z_m, z_n)] \right| \leq \sum_{m,n=1}^{N} \lambda_m \bar{\lambda}_n [r_\sigma(z_m, \bar{z}_n) - r_\varepsilon(z_m, \bar{z}_n)]$$

and

$$\left|\sum_{m,n=1}^{N} \lambda_m \lambda_n \frac{\partial}{\partial \varepsilon} q_\varepsilon(z_m, z_n)\right| \leqq \sum_{m,n=1}^{N} \lambda_m \bar{\lambda}_n \frac{\partial}{\partial \varepsilon} r_\varepsilon(z_m, \bar{z}_n)$$

for all points $z_1, \cdots, z_N$ *in* $\tilde{D}$ *and complex constants* $\lambda_1, \cdots, \lambda_N$.

The second inequality follows from the first upon dividing by $\sigma - \varepsilon$ and letting $\sigma \to \varepsilon$. Observe that the hermitian form $\sum_{m,n=1}^{N} \lambda_m \bar{\lambda}_n r_\varepsilon(z_m, \bar{z}_n)$ is a nondecreasing function of ε and that it dominates the symmetric form

$$\sum_{m,n=1}^{N} \lambda_m \lambda_n [q_\varepsilon(z_m, z_n) + \log(z_m - z_n)] \quad \text{for} \quad \varepsilon \geqq 1.$$

For $0 \leqq \varepsilon < 1$ one replaces ε by $1/\varepsilon$. These estimates could also be obtained from the development of $q_\varepsilon(z, \zeta)$ and $r_\varepsilon(z, \bar{\zeta})$ in terms of Fredholm eigenfunctions [4].

The inequalities of Theorem 1 could be interpreted in many additional ways (e.g., $N = 1$) to obtain interesting relations. However, it may be illuminating to conclude with a restatement of the inequalities for the case that C is the unit circle, since these dielectric kernels were given explicitly in Section 3.

Corollary 4. *Suppose that* $f(z) = z + \sum_{n=1}^{\infty} b_n z^{-n}$ *is univalent in* $|z| > 1$ *and has a* k*-quasiconformal extension to* $|z| \leqq 1$. *If* $K = (1+k)/(1-k)$, *then*

$$\left|\sum_{m,n=1}^{N} \lambda_m \lambda_n [Q_\varepsilon(f(z_m), f(z_n)) + \log(z_m - z_n)]\right|$$

$$\leqq \sum_{m,n=1}^{N} \lambda_m \bar{\lambda}_n \left[\frac{\varepsilon K - 1}{\varepsilon K + 1} \log \frac{z_m \bar{z}_n}{z_m \bar{z}_n - 1} - R_\varepsilon(f(z_m), \overline{f(z_n)})\right]$$

for all points $z_1, \cdots, z_N$ *in* $|z| > 1$ *and all complex constants* $\lambda_1, \cdots, \lambda_N$.

For $\varepsilon = 1$ these inequalities reduce to the more familiar Grunsky–Golusin type inequalities (see [1])

$$\left|\sum_{m,n=1}^{N} \lambda_m \lambda_n \log \frac{f(z_m) - f(z_n)}{z_m - z_n}\right| \leqq k \sum_{m,n=1}^{N} \lambda_m \bar{\lambda}_n \log \frac{z_m \bar{z}_n}{z_m \bar{z}_n - 1}.$$

REFERENCES

1. R Kuhnau, *Verzerrungssätze und Koeffizientenbedingungen vom Grunskyschen Typ für quasikonforme Abbildungen*, Math Nachr **48** (1971), 77–105

2. R Kuhnau, *Identitaten bei quasikonformen Normalabbildungen und eine hiermit zusammenhängende Kernfunktion*, Math. Nachr. **73** (1976), 73–106

3. M. Schiffer, *The Fredholm eigenvalues of plane domains*, Pacific J. Math. **7** (1957), 1187–1225.

4 M. Schiffer, *Fredholm eigenvalues of multiply-connected domains*, Pacific J. Math. **9** (1959), 211–269

5. M Schiffer, *A variational method for univalent quasiconformal mappings*, Duke Math. J. **33** (1966), 395–412

6 M. Schiffer and G Schober, *Representation of fundamental solutions for generalized Cauchy–Riemann equations by quasiconformal mappings*, Ann. Acad Sci. Fenn, Ser. A I **2** (1976), 501-531

7. M. Schiffer and G. Schober, *An application of the calculus of variations for general families of quasiconformal mappings*, Proc Colloq. Complex Analysis, Joensuu 1978, Springer–Verlag Lecture Notes in Math., to appear.

8 G Schober, *Univalent Functions—Selected Topics*, Springer–Verlag Lecture Notes in Math. #**478** (1975)

STANFORD UNIVERSITY
STANFORD, CALIFORNIA 94305 USA

INDIANA UNIVERSITY
BLOOMINGTON, INDIANA 47405 USA

(Received June 15, 1979)

Commentary on

[122] M. Schiffer and G. Schober, *The dielectric Green's function and quasiconformal mapping*, J. Analyse Math. **36** (1979), 233–243.

This is the eighth of the nine papers written by Schiffer in collaboration with Glenn Schober. Over the course of their work, the authors developed the variational method for quasiconformal mappings introduced earlier by Schiffer in [90] and applied it to a variety of extremal problems for Fredholm eigenvalues and for conformal mappings having a quasiconformal extension. Here they pose an extremal problem for the latter, obtaining a significant generalization of Goluzin-type inequalities. Thus [122] is a relatively short paper with a long lineage, extending back to [35,62,68].

The dielectric Green function $g_\varepsilon(z,\zeta)$ was introduced by Schiffer in [62] for simply connected domains and studied further in [68] for finitely connected domains, which provide the context of the present paper. As the terminology and notation suggest, a parameter $\varepsilon > 0$ enters the definition, carrying the physical interpretation that the complement of a domain D is filled with material of dielectric constant ε. Here one assumes that D contains ∞ in its interior. Unlike the usual Green function, the points z and ζ are free to range over D and its complement. Briefly, $g_\varepsilon(z,\zeta)$ is harmonic in z except at $z=\zeta$ (and at ∞) where it has a logarithmic singularity depending on ε; and it is continuous across the boundary curves, where it has a jump discontinuity in the normal derivatives, also depending on ε. The symmetry relation $g_\varepsilon(z,\zeta) = g_\varepsilon(\zeta,z)$ continues to hold. One special case is easy to identify: when $\varepsilon = 1$, one has $g_1(z,\zeta) = -\log|z-\zeta|$. There is no simple limiting relationship in ε between the dielectric Green function and the usual Green function, but there are several identities relating the two.

The connection to conformal mapping is quite striking, though not really needed for [122]. If one completes $g_\varepsilon(z,\zeta)$ to an analytic function $p_\varepsilon(z,\zeta)$, then $f_\varepsilon(z,\zeta) = \exp(-p_\varepsilon(z,\zeta))$ is univalent in D and the separate functions $f_\varepsilon(z,\zeta)$ in the domains complementary to D are also univalent. Furthermore, in the limiting cases $\varepsilon \to 0$ and $\varepsilon \to \infty$, the functions $f_0(z,0)$ and $f_\infty(z,0)$ map D onto the radial slit domain and the circular slit domain, respectively, while $f_1(z,0) = z$. See [68] for these results. Further developments of the dielectric Green function in potential theory independent of the connections with conformal mappings appear in the work of Schiffer's student John Troutman [T1, T2, T3].

The analysis in [122] is, first of all, an artful use of variational methods. Consider conformal mappings f of a finitely connected domain D (containing ∞) having a quasiconformal extension to the plane. We suppose the dilatation satisfies $|\mu| = |f_{\bar z}/f_z| \le k < 1$ and that f is normalized by $f(z) = z + o(1)$ near ∞; thus the family is compact. The functionals to be maximized over this family are assembled from functions derived from the dielectric Green function for the image $f(D)$ and its complement, principally functions derived from the analytic completion p_ε of g_ε. While one can see traces of the classical landscape in the form of the functionals—the authors are aiming for a generalization of the Goluzin inequalities, which have a variational proof—this is certainly new territory with new phenomena.

The variations of f are of the form $f^* = \Phi \circ f$, where Φ is the solution in the plane, normalized to satisfy $\Phi(w) = w + o(1)$ near ∞, of a Beltrami equation of the form

$$\Phi_{\bar w} = ta\Phi_w, \quad 0 < t < 1;$$

here $|a|$ is the characteristic function of a compact set in $f(D)$. Thus Φ is conformal near ∞ and has the representation

$$\Phi(w) = w - \frac{t}{\pi}\iint_{\mathbb{C}} \frac{a(\zeta)}{\zeta - w}\, d\xi\, d\eta + O(t^2),$$

$$\zeta = \xi + i\eta. \tag{1}$$

This formula is the basis for an expression for the variation of the dielectric Green function on $f(D)$ and hence an expression for the variation of the functional. For a concise treatment of the variational method, including a derivation of (1), see [S].

Why the dielectric Green functions for D and for $f(D)$ in the first place? The parameter ε is available to adjust. It is possible to obtain a relation for the extremal function from the variational formulas through matching the boundary discontinuities associated with the dielectric Green functions (actually a boundary value problem for associated analytic functions) by scaling ε to $\varepsilon(1+k)/(1-k)$. This relation then determines the maximum of the functional and hence the desired generalization of the Goluzin inequalities.

One corollary is worthy of special note. In the final line of the paper, the authors arrive at a sharpened form of the classical Goluzin inequalities, a beautiful result due originally to Kühnau [K], who published a series of important and influential papers on conformal mappings with a quasiconformal extension. See also his commentary on papers [62, 68, 78, 125] in these *Selecta*.

References

[K] Reiner Kühnau, *Verzerrungssätze und Koeffizientenbedingungen vom Grunskyschen Typ für quasikonforme Abbildungen*, Math. Nachr. **48** (1971), 77–105.

[S] Glenn Schober, *Univalent Functions – Selected Topics*, Lecture Notes in Math. **478**, Springer-Verlag, 1975.

[T1] John L. Troutman, *Planar dielectric potential functions*, J. Analyse Math. **25** (1972), 1–48.

[T2] John L. Troutman, *The dielectric potential operator*, Arch. Rational Mech. Anal. **42** (1971), 213–222.

[T3] John L. Troutman, *The Green function for planar dielectrics of variable strength*, J. London Math. Soc. (2) **30** (1984), 468–480.

BRAD OSGOOD

[123] (with A. Chang and G. Schober) On the second variation for univalent functions

[123] (with A. Chang and G. Schober) On the second variation for univalent functions. *J. Analyse Math.* **40** (1981), 203–238.

ON THE SECOND VARIATION FOR UNIVALENT FUNCTIONS

By

A. CHANG, M. M. SCHIFFER AND G. SCHOBER†

1. Introduction

The aim of the present paper is to derive a form of the second variation for univalent functions which allows the determination of necessary conditions in various extremum problems and which is not too complicated to be handled. In a paper by Duren–Schiffer [3] another expression for the second variation was given which was useful in finding sufficient conditions for a local extremum. Indeed, it proved quite useful in establishing the local Bieberbach theorem in the hands of E. Bombieri [1]. Here we lay greater stress on the necessary aspect in coefficient extremum problems. The formulas obtained are therefore more useful to rule out certain functions which are suspected extremum functions, as is shown in several applications.

Our discussion makes also clear the distinguished role of the Koebe function in the Bieberbach conjecture. The inequalities necessary for a proper extremum function become equalities in a wide class of variations.

The total amount of definite results in this paper is relatively small in comparison to the extensive formulas which are derived. However, these formulas lay the groundwork to a deeper study of extremum problems by variational methods.

2. The second variation in S and Σ

Let $w = f(z)$ be univalent in the unit disk $|z| < 1$ and have the normalization $f(0) = 0, f'(0) = 1$. The class of all such functions is usually denoted by S. We assume at first that $f(z)$ is still analytic on $|z| = 1$ and denote the analytic boundary of the image domain D by Γ. We vary Γ according to the formula

$$w^* = w + \varepsilon v_1(w) + \varepsilon^2 v_2(w) + \cdots, \tag{2.1}$$

where the functions $v_\nu(w)$ are defined and are analytic in a neighborhood of Γ. If ε is small enough, we will obtain a new analytic Jordan curve Γ^* which bounds a new domain D^* which also contains the origin. Let $f^*(z)$ be the univalent function in

† This work was supported in part by grants from the National Science Foundation.

$|z|<1$ which maps the disk onto D^* such that $f^*(0)=0$, $f^{*\prime}(0)>0$. In general, $f^{*\prime}(0)$ need not be 1.

Near $|z|=1$, the function

$$z^*(z)=f^{*-1}[f(z)+\varepsilon v_1[f(z)]+\varepsilon^2 v_2[f(z)]+\cdots] \tag{2.2}$$

is well defined, is analytic and takes the unit periphery into itself. Thus we can write

$$z^*(z)=z\exp[i(\varepsilon\psi_1(z)+\varepsilon^2\psi_2(z)+\cdots)]$$

where the $\psi_\nu(z)$ are analytic near $|z|=1$ and real on the circle. On the other hand, we may set up the series development for $f^*(z)$

$$f^*(z)=f(z)+\varepsilon f_1(z)+\varepsilon^2 f_2(z)+\cdots \tag{2.3}$$

and insert this into the Taylor development of $f^*(z^*)$ in ε

$$f^*(z^*(z))=f^*(z)+\varepsilon f^{*\prime}(z)\frac{\partial z^*}{\partial\varepsilon}\bigg|_{\varepsilon=0}$$

$$+\frac{1}{2}\varepsilon^2\left[f^{*\prime}(z)\frac{\partial^2 z^*}{\partial\varepsilon^2}\bigg|_{\varepsilon=0}+f^{*\prime\prime}(z)\left(\frac{\partial z^*}{\partial\varepsilon}\right)^2\bigg|_{\varepsilon=0}\right]+o(\varepsilon^2).$$

We rearrange to find

$$f^*(z^*(z))=f(z)+\varepsilon[f_1(z)+i\psi_1(z)\varphi(z)]$$

$$+\varepsilon^2\left[f_2(z)+i\psi_1(z)zf_1'(z)+\frac{1}{2}z\varphi'(z)(i\psi_1(z))^2+i\psi_2(z)\varphi(z)\right]+o(\varepsilon^2)$$

where we define $\varphi(z)=zf'(z)$. We utilize now equation (2.2) to express $f^*(z^*)$ in terms of $f(z)$. Comparing the coefficients of equal powers of ε, we obtain

$$\begin{aligned} v_1[f(z)]&=f_1(z)+i\psi_1(z)\varphi(z),\\ v_2[f(z)]&=f_2(z)+i\psi_1(z)zf_1'(z)+\frac{1}{2}(i\psi_1(z))^2 z\varphi'(z)+i\psi_2(z)\varphi(z). \end{aligned} \tag{2.4}$$

We could express all $f_\nu(z)$ correspondingly. There occur two types of functions to be determined: the $f_\nu(z)$ and the $\psi_\nu(z)$. We know from (2.3) that all $f_\nu(z)$ are regular analytic in $|z|<1$, and about the $\psi_\nu(z)$ we have the information that they are real and analytic on $|z|=1$. This is enough to determine both. For example, on $|z|=1$ we have

$$\mathrm{Re}\left\{\frac{f_1(z)}{\varphi(z)}\right\}=\mathrm{Re}\left\{\frac{v_1[f(z)]}{\varphi(z)}\right\}.$$

The right-hand side is known and by the normalization of $f^*(z)$ we have $f_1(0)=0$, $\mathrm{Im}\{f_1'(0)\}=0$; this determines $f_1(z)$ uniquely. We may either calculate it by using Poisson's integral or, as is often possible, find it directly by inspection. Once $f_1(z)$ is known, we can find $\psi_1(z)$ by (2.4). The second formula (2.4) leads then to $f_2(z)$ and $\psi_2(z)$ in the same way.

A very useful choice for $v_1[f]$ is the following:

$$v_1[f(z)] = \sum_{\nu=1}^{N} \frac{x_\nu f(z)}{[f(z)-f(\zeta_\nu)]f(\zeta_\nu)}, \tag{2.5}$$

with arbitrary complex coefficients x_ν and points ζ_ν in the unit disk. We find easily

$$\begin{aligned} f_1(z) &= \sum_{\nu=1}^{N} \frac{x_\nu f(z)}{[f(z)-f(\zeta_\nu)]f(\zeta_\nu)} \\ &\quad - \varphi(z)\left[\sum_{\nu=1}^{N} \left[\frac{x_\nu \zeta_\nu}{\varphi(\zeta_\nu)^2(z-\zeta_\nu)} - \frac{\bar{x}_\nu \bar{\zeta}_\nu z}{\overline{\varphi(\zeta_\nu)}^2(1-\bar{\zeta}_\nu z)} \right] + i\alpha \right]. \end{aligned} \tag{2.6}$$

Here α is a real constant to ensure $\operatorname{Im}\{f_1'(0)\} = 0$. By relaxing the normalisation $f^{*\prime}(0) > 0$, we may assume $\alpha = 0$. Now using (2.4), we find

$$i\psi_1(z) = \sum_{\nu=1}^{N} \left[\frac{x_\nu \zeta_\nu}{\varphi(\zeta_\nu)^2(z-\zeta_\nu)} - \frac{\bar{x}_\nu \bar{\zeta}_\nu z}{\overline{\varphi(\zeta_\nu)}^2(1-\bar{\zeta}_\nu z)} \right].$$

Next, we proceed to determine $f_2(z)$ and $\psi_2(z)$. To illustrate the general method but to avoid formulas of excessive length, we choose $N = 1$ and keep only one point ζ and one complex parameter x at our disposal. By proper choice of $v_2[f]$, we can achieve $\psi_2(z) \equiv 0$. Indeed, if $v_2[f]$ is so chosen that its singularities in $|z| < 1$ coincide with those of the known terms on the right, $\psi_2(z)$ being regular in $|z| < 1$ and real on $|z| = 1$ must be a constant, which can be chosen as zero.

Let us set up

$$v_2[f(z)] = f_2(z) + x^2 A(z) + |x|^2 B(z) + \bar{x}^2 C(z), \tag{2.7}$$

where

$$\begin{aligned} A(z) &= -\frac{\zeta\varphi(z)}{\varphi(\zeta)^2(z-\zeta)[f(z)-f(\zeta)]^2} + \frac{z\varphi(z)\zeta^2}{\varphi(\zeta)^4(z-\zeta)^3} - \frac{1}{2}\frac{z\varphi'(z)\zeta^2}{\varphi(\zeta)^4(z-\zeta)^2}, \\ B(z) &= \frac{z\bar{\zeta}\varphi(z)}{\overline{\varphi(\zeta)}^2(1-\bar{\zeta}z)[f(z)-f(\zeta)]^2} + \frac{z^2\varphi'(z)|\zeta|^2}{|\varphi(\zeta)|^4(z-\zeta)(1-\bar{\zeta}z)} \\ &\quad + \frac{z\varphi(z)|\zeta|^2(\bar{\zeta}z^2-\zeta)}{|\varphi(\zeta)|^4(z-\zeta)^2(1-\bar{\zeta}z)^2}, \\ C(z) &= -\frac{z^2\varphi(z)\bar{\zeta}^2}{\overline{\varphi(\zeta)}^4(1-\bar{\zeta}z)^3} - \frac{1}{2}\frac{z^3\varphi'(z)\bar{\zeta}^2}{\overline{\varphi(\zeta)}^4(1-\bar{\zeta}z)^2}. \end{aligned} \tag{2.8}$$

Obviously, all three functions (2.8) are analytic in $|z| < 1$ except at $z = \zeta$ where $A(z)$ and $B(z)$ have poles. We calculate their principal parts

$$A(z) = \frac{1}{2}\frac{\zeta^3\varphi'(\zeta)}{\varphi(\zeta)^4}\frac{1}{(z-\zeta)^2} + \frac{\zeta}{\varphi(\zeta)^3}\left[\frac{1}{2}\frac{\zeta\varphi'(\zeta)}{\varphi(\zeta)} - \frac{1}{6}\zeta^2\{f,\zeta\}\right]\frac{1}{z-\zeta} + \cdots,$$

where $\{f,\zeta\}$ denotes the usual Schwarzian differential parameter, and

$$B(z) = \frac{\zeta\varphi(\zeta)|\zeta|^2}{|\varphi(\zeta)|^4(1-|\zeta|^2)^2}\frac{1}{z-\zeta} + \cdots.$$

We have now to choose

$$\begin{aligned} v_2[f] = {} & \left[\frac{1}{2}\frac{\zeta\varphi'(\zeta)}{\varphi(\zeta)^2}\frac{f(z)^2}{f(\zeta)^2[f(z)-f(\zeta)]^2} + \frac{[M(\zeta)f(\zeta)-\zeta\varphi'(\zeta)]f(z)}{\varphi(\zeta)^2 f(\zeta)^2[f(z)-f(\zeta)]}\right]x^2 \\ & + \frac{|\zeta|^2 f(z)|x|^2}{(1-|\zeta|^2)^2\overline{\varphi(\zeta)}^2 f(\zeta)[f(z)-f(\zeta)]} \end{aligned} \tag{2.9}$$

with

$$M(\zeta) = \frac{1}{2}\zeta^2\left[\left(\frac{\varphi'(\zeta)}{\varphi(\zeta)}\right)^2 - \frac{1}{3}\{f,\zeta\}\right] \tag{2.10}$$

to achieve the matching of the singularities.

We have achieved our aim to determine the second variation of the given univalent function $f(z)$, namely $f_2(z)$,

$$\begin{aligned} f_2(z) = {} & \frac{x^2}{\varphi(\zeta)^4}\left[\frac{1}{2}\frac{\zeta\varphi'(\zeta)\varphi(\zeta)^2 f(z)^2}{f(\zeta)^2[f(z)-f(\zeta)]^2} + \frac{\zeta\varphi(\zeta)^2\varphi(z)}{(z-\zeta)[f(z)-f(\zeta)]^2}\right. \\ & \left. + \frac{[M(\zeta)f(\zeta)-\zeta\varphi'(\zeta)]\varphi(\zeta)^2 f(z)}{f(\zeta)^2[f(z)-f(\zeta)]} + \frac{1}{2}\zeta^2 z\frac{d}{dz}\left[\frac{\varphi(z)}{(z-\zeta)^2}\right]\right] \\ & + \frac{|x|^2}{|\varphi(\zeta)|^4}\left[\frac{|\zeta|^2\varphi(\zeta)^2 f(z)}{(1-|\zeta|^2)^2 f(\zeta)[f(z)-f(\zeta)]} - \frac{\bar{\zeta}z\varphi(\zeta)^2\varphi(z)}{(1-\bar{\zeta}z)[f(z)-f(\zeta)]^2}\right. \\ & \left. - |\zeta|^2 z\frac{d}{dz}\left[\frac{z\varphi(z)}{(z-\zeta)(1-\bar{\zeta}z)}\right]\right] + \frac{\bar{x}^2}{\overline{\varphi(\zeta)}^4}\frac{1}{2}\bar{\zeta}^2 z\frac{d}{dz}\left[\frac{z^2\varphi(z)}{(1-\bar{\zeta}z)^2}\right]. \end{aligned} \tag{2.11}$$

We have derived this result under the hypothesis that the original function $f(z)$ is analytic in the closure of the unit disk. However, since the end-result (2.11) depends only upon the functional behavior at the interior point ζ, we can extend the result by uniform convergence to the most general functions in S.

It is quite remarkable that the same formulas hold in the case that $f(z)$ belongs to the class Σ; that is, if $f(z)$ is univalent analytic in $|z|>1$ and normalized at infinity by $f(\infty)=\infty$, $f'(\infty)=1$. If we understand by the variation (2.5) the same terms but with ζ_ν in $|z|>1$ and demand that the varied function $f^*(z)$ satisfy $f^*(\infty)=\infty$, we obtain precisely the same expressions for the second variation in Σ. The difference in the theory will appear now when we will complete the normalization to hold the varied univalent functions in their respective classes.

Let us first consider the case of the class S. We have the Taylor development

$$f^*(z) = a_1^* z + a_2^* z^2 + \cdots$$

and find the value

$$a_1^* = 1 + \varepsilon x\left(\frac{1}{\varphi(\zeta)^2} - \frac{1}{f(\zeta)^2}\right) + \varepsilon^2\left\{\frac{x^2}{\varphi(\zeta)^2 f(\zeta)^2}\frac{1}{2}\left[\frac{f(\zeta)^2}{\varphi(\zeta)^2} + \frac{\zeta\varphi'(\zeta)}{f(\zeta)} - M(\zeta) - 1\right]\right.$$
$$\left. - |x|^2 \frac{|\zeta|^2}{\overline{\varphi(\zeta)}^2 f(\zeta)^2 (1-|\zeta|^2)^2}\right\} + o(\varepsilon^2).$$

We then define

$$\hat{f}(z) = \frac{1}{a_1^*} f^*(z)$$

and obtain a function of class S which is arbitrarily close to the original $f(z)$. A lengthy but elementary calculation gives

$$\hat{f}(z) = f(z) + \varepsilon\left\{x\left[\frac{f(z)^2}{[f(z)-f(\zeta)]f(\zeta)^2} - \frac{1}{\varphi(\zeta)^2}\left(f(z) + \frac{\zeta\varphi(z)}{z-\zeta}\right)\right] + \bar{x}\frac{z\bar{\zeta}\varphi(z)}{\overline{\varphi(\zeta)}^2(1-\bar{\zeta}z)}\right\}$$

$$+ \varepsilon^2\left\{x^2\left[\frac{\zeta}{\varphi(\zeta)^2 f(\zeta)^2}\frac{\varphi(z)}{z-\zeta}\frac{f(z)^2[3f(\zeta)-2f(z)]}{[f(z)-f(\zeta)^2]}\right.\right.$$

$$+\frac{1}{2}\frac{\zeta\varphi'(\zeta)}{\varphi(\zeta)^2 f(\zeta)^3}\frac{f(z)^2[3f(\zeta)-2f(z)]}{[f(z)-f(\zeta)^2]}$$

$$+\frac{f(z)^2}{[f(z)-f(\zeta)]f(\zeta)^2}\left(\frac{M(\zeta)}{\varphi(\zeta)^2} + \frac{1}{f(\zeta)^2} - \frac{1}{\varphi(\zeta)^2}\right) + \frac{\zeta\varphi(z)}{\varphi(\zeta)^4(z-\zeta)}$$

(2.12)

$$\left.+\frac{1}{2}\frac{\zeta^2 z}{\varphi(\zeta)^4}\frac{d}{dz}\left(\frac{\varphi(z)}{(z-\zeta)^2}\right) + \frac{1}{2}\frac{f(z)}{\varphi(\zeta)^4}\right]$$

$$+|x|^2\left[\frac{|\zeta|^2}{(1-|\zeta|^2)^2\overline{\varphi(\zeta)}^2}\frac{f(z)^2}{[f(z)-f(\zeta)]f(\zeta)^2}\right.$$
$$+\frac{z\bar{\zeta}\varphi(z)f(z)(f(z)-2f(\zeta))}{\overline{\varphi(\zeta)}^2 f(\zeta)^2(1-\bar{\zeta}z)[f(z)-f(\zeta)]^2} - \frac{z\bar{\zeta}\varphi(z)}{|\varphi(\zeta)|^4(1-\bar{\zeta}z)}$$

$$\left.\left.-\frac{|\zeta|^2 z}{|\varphi(\zeta)|^4}\frac{d}{dz}\left[\frac{z\varphi(z)}{(z-\zeta)(1-\bar{\zeta}z)}\right]\right] + \bar{x}^2\frac{1}{2}\frac{\bar{\zeta}^2}{\overline{\varphi(\zeta)}^4} z\frac{d}{dz}\left[\frac{z^2\varphi(z)}{(1-\bar{\zeta}z)^2}\right]\right\} + o(\varepsilon^2).$$

This is a very complicated expression, but it has one remarkable property. In the case of real ζ and real x one has $\hat{f}(z) \equiv f(z)$ for the Koebe function

$$k(z) = \frac{z}{(1-z)^2}. \tag{2.13}$$

Indeed, in this case both functions $v_1(w)$ and $v_2(w)$ are real for real values w. Since the Koebe function maps onto the whole w-plane slit along the segment $(-\infty, -\frac{1}{4}]$

of the negative real axis, the function $\hat{f}(z)$ must map onto the same region. This fact shows the distinguished role of the Koebe function in this type of second variation.

We can proceed analogously in the case of the class Σ. Now we have to look at the development near $z=\infty$. We have by (2.6) and (2.11)

$$f_1(z) = -\frac{\bar{x}}{\overline{\varphi(\zeta)}^2} z + \sigma_1 + O\left(\frac{1}{z}\right),$$

$$f_2(z) = \frac{1}{2}\frac{\bar{x}^2}{\overline{\varphi(\zeta)}^4} z + \sigma_2 + O\left(\frac{1}{z}\right),$$

with

$$\sigma_1 = \frac{x}{f(\zeta)} - \frac{x\zeta}{\varphi(\zeta)^2} - \frac{\bar{x}}{\overline{\zeta\varphi(\zeta)}^2},$$

$$\sigma_2 = \frac{x^2[M(\zeta)f(\zeta) - \frac{1}{2}\zeta\varphi'(\zeta)]}{\varphi(\zeta)^2 f(\zeta)^2} + \frac{|x|^2|\zeta|^2}{\overline{\varphi(\zeta)}^2 f(\zeta)(1-|\zeta|^2)^2}.$$

Therefore

$$\begin{aligned}\hat{f}(z) &= \frac{f^*(z) - \varepsilon\sigma_1 - \varepsilon^2\sigma_2}{1 - \varepsilon\dfrac{\bar{x}}{\overline{\varphi(\zeta)}^2} + \varepsilon^2\dfrac{1}{2}\dfrac{\bar{x}^2}{\overline{\varphi(\zeta)}^4}} + o(\varepsilon^2)\\
&= f(z) + \varepsilon\left[f_1(z) - \sigma_1 + \frac{\bar{x}}{\overline{\varphi(\zeta)}^2} f(z)\right]\\
&\quad + \varepsilon^2\left[f_2(z) - \sigma_2 + \frac{\bar{x}}{\overline{\varphi(\zeta)}^2}[f_1(z) - \sigma_1] + \frac{1}{2}\frac{\bar{x}^2}{\overline{\varphi(\zeta)}^4} f(z)\right] + o(\varepsilon^2)\\
&= f(z) + \varepsilon\hat{f}_1(z) + \varepsilon^2\hat{f}_2(z) + o(\varepsilon^2)\end{aligned} \tag{2.14}$$

lies in the class Σ, where

$$\hat{f}_1(z) = \frac{x}{\varphi(\zeta)^2}\left[\frac{\varphi(\zeta)^2}{f(z)-f(\zeta)} - \frac{\zeta\varphi(z)}{z-\zeta} + \zeta\right] + \frac{\bar{x}}{\overline{\varphi(\zeta)}^2}\left[\frac{\bar{\zeta}z\varphi(z)}{1-\bar{\zeta}z} + f(z) + \frac{1}{\bar{\zeta}}\right] \tag{2.15}$$

and

$$\begin{aligned}\hat{f}_2(z) = \frac{x^2}{\varphi(\zeta)^4}&\left[\frac{1}{2}\frac{\zeta\varphi'(\zeta)\varphi(\zeta)^2}{[f(z)-f(\zeta)]^2} + \frac{\zeta\varphi(\zeta)^2\varphi(z)}{(z-\zeta)[f(z)-f(\zeta)]^2} + \frac{\varphi(\zeta)^2 M(\zeta)}{f(z)-f(\zeta)}\right.\\
&\left.+\frac{1}{2}\zeta^2 z\frac{d}{dz}\left[\frac{\varphi(z)}{(z-\zeta)^2}\right]\right] + \frac{|x|^2}{|\varphi(\zeta)|^4}\left[\frac{|\zeta|^2\varphi(\zeta)^2}{(1-|\zeta|^2)^2[f(z)-f(\zeta)]}\right.\\
&- \frac{\bar{\zeta}z\varphi(\zeta)^2\varphi(z)}{(1-\bar{\zeta}z)[f(z)-f(\zeta)]^2} - |\zeta|^2 z\frac{d}{dz}\left[\frac{z\varphi(z)}{(z-\zeta)(1-\bar{\zeta}z)}\right]\\
&\left.+\frac{\varphi(\zeta)^2}{f(z)-f(\zeta)} - \frac{\zeta\varphi(z)}{z-\zeta} + \zeta\right]\\
&+\frac{\bar{x}^2}{\overline{\varphi(\zeta)}^4}\left[\frac{1}{2}\bar{\zeta}^2 z\frac{d}{dz}\left[\frac{z^2\varphi(z)}{(1-\bar{\zeta}z)^2}\right] + \frac{\bar{\zeta}z\varphi(z)}{1-\bar{\zeta}z} + \frac{1}{\bar{\zeta}} + \frac{1}{2}f(z)\right].\end{aligned} \tag{2.16}$$

The functions (2.15) and (2.16) are normalized first and second variations, respectively, of a function $f(z)$ in Σ. If $f(z)$ lies in the class Σ_0, which is the subclass of Σ for which $f(z)-z=o(1)$, then the varied function $\hat{f}(z)$ will lie in the same subclass.

3. Application to extremum problems for linear functionals on Σ_0

The formulas for the first and second variation allow one, in principle, to discuss extremum problems for functionals defined on S and Σ. For sake of simplicity, we shall consider now the case of linear functionals and start with the class Σ_0.

Let L denote a continuous linear functional defined on the space of all analytic functions in $|z|>1$ with the topology of locally uniform convergence. It is known [9] that such a functional may be represented by integration with respect to a compactly supported complex Borel measure in $|z|>1$. As an example, L might give the coefficient b_n in the Laurent development of f, or the value of some derivative of f at a specified point.

We shall assume that L is not constant on Σ_0 and consider the extremum problem of finding

$$\max_{f\in\Sigma_0} \operatorname{Re}\{L(f)\} = m_L .$$

Since Σ_0 is compact, there will be at least one function $f\in\Sigma_0$ for which $\operatorname{Re}\{L(f)\}=m_L$. If we compare $\operatorname{Re}\{L(\hat{f})\}$ to $\operatorname{Re}\{L(f)\}$, we can assert that

$$\varepsilon \operatorname{Re}\{L(\hat{f}_1)\} + O(\varepsilon^2) \leqq 0$$

for all $\varepsilon>0$ sufficiently small. Divide by ε and let $\varepsilon\to 0$. Then

$$\operatorname{Re}\left\{\frac{x}{\varphi(\zeta)^2}\left[L\left(\frac{\varphi(\zeta)^2}{f-f(\zeta)}-\frac{\zeta\varphi}{(z-\zeta)}+\zeta\right)+\overline{L\left(\frac{\bar{\zeta}z\varphi}{1-\bar{\zeta}z}+f+\frac{1}{\bar{\zeta}}\right)}\right]\right\}\leqq 0 .$$

Since this holds for arbitrary choice of x, we conclude that

$$L\left(\frac{\varphi(\zeta)^2}{f-f(\zeta)}-\frac{\zeta\varphi}{z-\zeta}+\zeta\right)+\overline{L\left(\frac{\bar{\zeta}z\varphi}{1-\bar{\zeta}z}+f+\frac{1}{\bar{\zeta}}\right)}=0 \quad \text{for } |\zeta|>1. \tag{3.1}$$

The linear functional applies to the functions of the variable z, which we have suppressed in our notation where possible. The variable ζ appears as an analytic parameter. Since $\varphi(\zeta)=\zeta f'(\zeta)$, we have obtained the familiar differential equation for the extremum function by use of the first variation [5, 7, 9].

The differential equation contains much information. A trivial observation is that

$$\operatorname{Im}\{L(f-zf')\}=0, \tag{3.2}$$

obtained by letting $\zeta\to\infty$. The coefficient of $1/\zeta$ in the development (3.1) leads to the Marty relation

$$L(f^2 - 3b_1 - z^2 f') + \overline{L(f' - 1)} = 0. \tag{3.3}$$

Since the functional L may be represented by an integral with a compactly supported measure in $|z| > 1$, we may consider $L(1/(f - f(\zeta)))$ and

$$\begin{aligned} q(\zeta) &\equiv L\left(\frac{\zeta\varphi}{z - \zeta} - \zeta\right) - \overline{L\left(\frac{\bar{\zeta} z\varphi}{1 - \bar{\zeta} z} + \frac{1}{\bar{\zeta}}\right)} - \overline{L(f)} \\ &= L\left(\frac{\zeta\varphi}{z - \zeta} - \zeta\right) - \overline{L\left(\frac{\varphi}{1 - \bar{\zeta} z} + \frac{1}{\bar{\zeta}}\right)} - L(f - zf') \end{aligned} \tag{3.4}$$

to be analytic functions off that compact set; in particular, in a neighborhood of $\zeta = \infty$ and for $1 < |\zeta| < \rho$ for some $\rho > 1$. Furthermore, $q(\zeta)$ is analytic in a neighborhood of $|\zeta| = 1$ and is real for $|\zeta| = 1$. Finally, we may write the identity (3.1) in the explicit differential equation form

$$(\zeta f'(\zeta))^2 L\left(\frac{1}{f(z) - f(\zeta)}\right) = q(\zeta). \tag{3.5}$$

The right side provides an analytic continuation of the left side across $|\zeta| = 1$ and exhibits the analytic character of the extremum function on that circumference.

Since we have shown that the real part of L applied to the first variation is zero, it follows that

$$\varepsilon^2 \operatorname{Re}\{L(\hat{f}_2)\} + o(\varepsilon^2) \leqq 0.$$

Divide by ε^2 and let $\varepsilon \to 0$. The resulting inequality is of the form $\operatorname{Re}\{x^2 a + |x|^2 b + \bar{x}^2 c\} \leqq 0$ and is valid for all complex x. Choose $x^2 = |a + \bar{c}|/(a + \bar{c})$ and find $|a + \bar{c}| \leqq -\operatorname{Re}\{b\}$. Thus, the second variation yields the inequality

$$\begin{aligned} &\left| L\left(\frac{1}{2}\frac{\zeta\varphi'(\zeta)\varphi(\zeta)^2}{[f - f(\zeta)]^2} + \frac{\zeta\varphi(\zeta)^2\varphi}{(z - \zeta)[f - f(\zeta)]^2} + \frac{\varphi(\zeta)^2 M(\zeta)}{f - f(\zeta)} + \frac{1}{2}\zeta^2 z \frac{d}{dz}\left[\frac{\varphi}{(z - \zeta)^2}\right]\right)\right. \\ &\left. \qquad + \overline{L\left(\frac{1}{2} z\bar{\zeta}^2 \frac{d}{dz}\left[\frac{z^2\varphi}{(1 - \bar{\zeta} z)^2}\right] + \frac{\bar{\zeta} z\varphi}{1 - \bar{\zeta} z} + \frac{1}{\bar{\zeta}} + \frac{1}{2} f\right)}\right| \\ &\leqq -\operatorname{Re}\left\{L\left(\frac{|\zeta|^2 \varphi(\zeta)^2}{(1 - |\zeta|^2)^2 [f - f(\zeta)]} - \frac{\bar{\zeta}\varphi(\zeta)^2 z\varphi}{(1 - \bar{\zeta} z)[f - f(\zeta)]^2} - |\zeta|^2 z \frac{d}{dz}\left[\frac{z\varphi}{(z - \zeta)(1 - \bar{\zeta} z)}\right]\right.\right. \\ &\qquad \left.\left. + \frac{\varphi(\zeta)^2}{f - f(\zeta)} - \frac{\zeta\varphi}{z - \zeta} + \zeta\right)\right\} \end{aligned} \tag{3.6}$$

valid for all $|\zeta| > 1$.

This inequality carries a considerable amount of additional information. For example, in the limit $\zeta = \infty$ it becomes

$$\left| L\left(\frac{1}{2} f - \varphi + \frac{1}{2} z\varphi'\right) + \overline{L\left(\frac{1}{2} f - \varphi + \frac{1}{2} z\varphi'\right)} \right| \leqq \operatorname{Re}\{L(f - 2\varphi + z\varphi')\}.$$

This implies the condition for the extremum function

$$\operatorname{Re}\{L(f - zf' + z^2 f'')\} \geqq 0. \tag{3.7}$$

Numerous additional inequalities may be obtained by developing both sides of the inequality (3.6) in series of $1/\zeta = (1/r)e^{-i\theta}$ and $1/\bar{\zeta} = (1/r)e^{i\theta}$ and using standard methods of Fourier analysis.

If $|\zeta|$ converges to 1, the leading term in (3.6) is

$$-\operatorname{Re}\left\{\frac{|\zeta|^2 \varphi(\zeta)^2}{(1-|\zeta|^2)^2} L\left(\frac{1}{f - f(\zeta)}\right)\right\} = -\frac{|\zeta|^2}{(1-|\zeta|^2)^2}\operatorname{Re}\{q(\zeta)\}.$$

We see therefore that on $|\zeta| = 1$ we have necessarily

$$q(\zeta) \leqq 0. \tag{3.8}$$

This fact can be derived by use of Löwner type variations without use of the second variation; it is noteworthy how the second variation complements the results of the first variation.

4. Forking of slits in the extremal regions for Σ_0

Let η be a specified point on $|\zeta| = 1$ and let $\omega = f(\eta)$. If the coefficient $L(1/(f-\omega))$ in the differential equation (3.5) is nonzero, then $f(\zeta)$ can be continued across an arc of the unit circle containing the point η as an analytic function of ζ. Observe that there is no danger from zeros of $q(\zeta)$ since by (3.8) such zeros are of even order and do not affect the regularity of f. Thus, in order that a fork or a singularity occur at η, it is necessary that $L(1/(f-\omega)) = 0$.

The image of $|\zeta| = 1$ under f is a compact set Γ. Since $L(1/(f-\omega))$ is an analytic function of ω on Γ, it has at most finitely many zeros on Γ. Hence, Γ can have only finitely many singularities. For now, let us assume that $L(1/(f-\omega)) = 0$ but $q(\eta) \neq 0$. We have then the asymptotic formula

$$f(\zeta) = \omega + \tau(\zeta - \eta)^{\alpha} + o((\zeta - \eta)^{\alpha})$$

near the point η. Since

$$\varphi(\zeta) = \zeta f'(\zeta) = \tau\alpha\zeta(\zeta - \eta)^{\alpha - 1} + \cdots,$$

we see that necessarily $\alpha < 1$ in order that $q(\eta) \neq 0$. We easily compute

$$\begin{gathered}\left(\frac{\varphi'(\zeta)}{\varphi(\zeta)}\right)^2 = \frac{(1-\alpha)^2}{(\zeta-\eta)^2} + o\left(\frac{1}{(\zeta-\eta)^2}\right), \quad \{f, \zeta\} = \frac{1-\alpha^2}{2(\zeta-\eta)^2} + o\left(\frac{1}{(\zeta-\eta)^2}\right), \\ M(\zeta) = \frac{5 - 12\alpha + 7\alpha^2}{12(\zeta-\eta)^2}\eta^2 + o\left(\frac{1}{(\zeta-\eta)^2}\right).\end{gathered} \tag{4.1}$$

Let $\zeta = r\eta$ and multiply (3.6) by $(1-r)^2$ and let $r \to 1$. The only terms which do not vanish in the limit are the following:

$$\lim_{r\to 1}(1-r)^2 L\left(\frac{\frac{1}{2}\zeta\varphi(\zeta)^2\varphi'(\zeta)}{[f(z)-f(\zeta)]^2}\right) = \lim_{r\to 1}(1-r)^2\left\{\frac{1}{2}\zeta^2\frac{\varphi'(\zeta)}{\varphi(\zeta)}q'(\zeta) - \zeta^2\left(\frac{\varphi'(\zeta)}{\varphi(\zeta)}\right)^2 q(\zeta)\right\}$$

$$= -(1-\alpha)^2 q(\eta),$$

$$\lim_{r\to 1}(1-r)^2 L\left(\frac{\varphi(\zeta)^2 M(\zeta)}{f(z)-f(\zeta)}\right) = \lim_{r\to 1}(1-r)^2 M(\zeta)q(\zeta) = \frac{1}{12}(5-12\alpha+7\alpha^2)q(\eta),$$

$$\lim_{r\to 1}\frac{(1-r)^2 r^2}{(1-r^2)^2} L\left(\frac{\varphi(\zeta)^2}{f(z)-f(\zeta)}\right) = \frac{1}{4}q(\eta).$$

Thus, we obtain in the limit the inequality

$$\left|q(\eta)\left(\frac{7}{12}-\alpha+\frac{5}{12}\alpha^2\right)\right| \leqq -\frac{1}{4}q(\eta).$$

Divide through by the positive quantity $-\frac{1}{12}q(\eta)$ to find

$$|7-12\alpha+5\alpha^2| \leqq 3. \tag{4.2}$$

Since $(7-12\alpha+5\alpha^2) = (7-5\alpha)(1-\alpha)$, this implies

$$\alpha \geqq \frac{2}{5}. \tag{4.3}$$

Consequently, at a forking point ω of the boundary slit Γ the joining arcs make an angle of at least $\frac{2}{5}\pi$.

In the following section we shall give an important application of our $\frac{2}{5}\pi$-result.

Similar estimates can be carried out at points η where both $q(\eta)$ and $L(1/(f-\omega))$ vanish. To deal with the case $\alpha < 1$ and $q(\eta) = 0$, we have to use an additional variation. We start with the varied function

$$\hat{f}(z) = f(z) + \varepsilon\hat{f}_1(z) + \varepsilon^2\hat{f}_2(z) + o(\varepsilon^2)$$

and subject it to the additional variation

$$\tilde{f}(z) = e^{-i\alpha}\hat{f}(e^{i\alpha}z)$$

which preserves the characteristic properties of Σ_0. We choose

$$\alpha = \rho\varepsilon$$

for fixed real ρ and develop in powers of ε. We find

$$\begin{aligned}\tilde{f}(z) &= f(z) + \varepsilon[\hat{f}_1(z) + i\rho z f'(z) - i\rho f(z)] \\ &+ \varepsilon^2\left[\hat{f}_2(z) + i\rho z\hat{f}_1'(z) - i\rho\hat{f}_1(z) - \frac{\rho^2}{2}(f(z) - zf'(z) + z^2 f''(z))\right] + o(\varepsilon^2).\end{aligned} \tag{4.4}$$

Suppose now we have to deal with a linear functional over Σ_0 and with the extremum problem

$$\max \operatorname{Re}\{L(f)\} \quad \text{in } \Sigma_0.$$

We find first the conditions

$$\operatorname{Re}\{L(\hat{f}_1)\}=0, \qquad \operatorname{Im}\{L(zf'-f)\}=0. \tag{4.5}$$

Next, we obtain

$$\operatorname{Re}\{L(\hat{f}_2)\}+\rho \operatorname{Im}\{L(\hat{f}_1-z\hat{f}_1')\}-\frac{\rho^2}{2}\operatorname{Re}\{L(f-zf'+z^2f'')\}\leqq 0.$$

This leads to the inequalities

$$\operatorname{Re}\{L(\hat{f}_2)\}\leqq 0, \qquad \operatorname{Re}\{L(f-zf'+z^2f'')\}\geqq 0, \tag{4.6}$$

and

$$[\operatorname{Im}\{L(\hat{f}_1-z\hat{f}_1')\}]^2\leqq -2\operatorname{Re}\{L(\hat{f}_2)\}\operatorname{Re}\{L(f-zf'+z^2f'')\}. \tag{4.7}$$

In particular, if $\hat{f}_1(z)$ has the form (2.15), we find

$$\begin{aligned} z\hat{f}_1'(z)-\hat{f}_1(z)=\frac{-x}{\varphi(\zeta)^2}\Big[&\frac{\varphi(\zeta)^2\varphi(z)}{[f(z)-f(\zeta)]^2}+\frac{\varphi(\zeta)^2}{f(z)-f(\zeta)}+\frac{\zeta z\varphi'(z)}{(z-\zeta)} \\ &+\frac{\zeta(\zeta-2z)\varphi(z)}{(z-\zeta)^2}+\zeta\Big] \\ &+\frac{\bar{x}}{\overline{\varphi(\zeta)^2}}\Big[\frac{\bar{\zeta}z^2\varphi'(z)}{1-\bar{\zeta}z}+\frac{(1-2\bar{\zeta}z+2\bar{\zeta}^2z^2)\varphi(z)}{(1-\bar{\zeta}z)^2}-f(z)-\frac{1}{\bar{\zeta}}\Big]. \end{aligned}$$

Hence, for $|\zeta|$ near 1 we have

$$\begin{aligned} \operatorname{Im}\{L(z\hat{f}_1'-f_1)\}= -\operatorname{Im}\Big\{\frac{x}{\varphi(\zeta)^2}\Big[&\varphi(\zeta)^2L\Big(\frac{\varphi}{[f-f(\zeta)]^2}\Big)+\varphi(\zeta)^2L\Big(\frac{1}{f-f(\zeta)}\Big) \\ &+L\Big(\frac{\zeta z\varphi'}{z-\zeta}+\frac{\zeta(\zeta-2z)\varphi}{(z-\zeta)^2}+\zeta\Big)+\overline{L\Big(\frac{\bar{\zeta}z^2\varphi'}{1-\bar{\zeta}z}+\frac{(1-2\bar{\zeta}z+2\bar{\zeta}^2z^2)\varphi}{(1-\bar{\zeta}z)^2}-f-\frac{1}{\bar{\zeta}}\Big)}\Big]\Big\}. \end{aligned} \tag{4.8}$$

Suppose now that we have the local development

$$f(z)=f(\eta)+\tau(z-\eta)^{\alpha}+o((z-\eta)^{\alpha}), \qquad |\eta|=1$$

and $q(\eta)=0$. Since $q(\zeta)\leqq 0$ on $|\zeta|=1$, this must be at least a zero of order 2. We have

$$\varphi(\zeta)=\eta\cdot\alpha\cdot\tau(\zeta-\eta)^{\alpha-1}+o((\zeta-\eta)^{\alpha-1}).$$

Observe that by (3.5)

$$\varphi(\zeta)^2L\Big(\frac{1}{f-f(\zeta)}\Big)=q(\zeta)$$

so that the leading term in $\operatorname{Im}\{L(zf_1' - \hat{f}_1)\}$ is

$$-\operatorname{Im}\left\{\frac{x}{\varphi(\zeta)^2}\left[\varphi(\zeta)^2 L\left(\frac{\varphi}{[f-f(\zeta)]^2}\right) + \text{finite terms}\right]\right\}.$$

As $\zeta \to \eta$, we have then

$$[\operatorname{Im}\{L(\hat{f}_1 - z\hat{f}_1')\}]^2 \sim \left[\operatorname{Im}\left\{\frac{x}{\varphi(\zeta)^2}\eta^2\alpha^2\tau^2(\zeta-\eta)^{2\alpha-2}L\left(\frac{\varphi}{[f-f(\eta)]^2}\right)\right\}\right]^2.$$

On the other hand, we verify easily that under the same circumstances

$$\operatorname{Re}\{L(\hat{f}_2)\} \sim \operatorname{Re}\left\{\eta^2\alpha^2\tau^2(\zeta-\eta)^{2\alpha-2}\left[\frac{x^2}{\varphi(\zeta)^4}L\left(\frac{\eta\varphi}{(z-\eta)[f-f(\eta)]^2}\right)\right.\right.$$
$$\left.\left.+\frac{|x|^2}{|\varphi(\zeta)|^4}L\left(\frac{z\varphi}{(z-\eta)[f-f(\eta)]^2}\right)\right]\right\}.$$

Thus, the left side of the inequality (4.7) goes faster to ∞ than the right side, which should dominate it, except if

$$L\left(\frac{\varphi}{[f-f(\eta)]^2}\right) = 0. \tag{4.9}$$

But even then we could not preserve the condition $\operatorname{Re}\{L(\hat{f}_2)\} \leqq 0$ except when also

$$L\left(\frac{\varphi}{(z-\eta)[f-f(\eta)]^2}\right) = 0. \tag{4.10}$$

Suppose next that

$$F(t) = L\left(\frac{\varphi}{[f-t]^2}\right)$$

has a root of order $k-1$ at $t = f(\eta)$. Then

$$\varphi(\zeta)^2 L\left(\frac{\varphi}{[f-f(\zeta)]^2}\right) = \eta^2\alpha^2\tau^2(\zeta-\eta)^{2\alpha-2}\frac{F^{(k)}(f(\eta))}{k!}\tau^k(\zeta-\eta)^{k\alpha} + \cdots$$
$$= C(\zeta-\eta)^{(2+k)\alpha-2} + \cdots.$$

If $\alpha < 2/(k+2)$, we get additional conditions.

5. The coefficient problem in Σ_0

Let $f(z) \in \Sigma_0$ and have the series development

$$f(z) = z + \sum_{n=1}^{\infty}\frac{b_n}{z^n}.$$

It is well known [2] that for $n = 1$ and 2 we have $\operatorname{Re}\{b_n\} \leqq 2/(n+1)$ and that equality occurs only for

(5.1) $$k_n(z) = z\left(1+\frac{1}{z^{n+1}}\right)^{2/(n+1)} = z + \frac{2}{n+1}\frac{1}{z^n} + \cdots.$$

However, it is also known that the maximum of $\mathrm{Re}\{b_n\}$ is larger than $2/(n+1)$ for $n = 3,4$ and all larger odd integers. In particular, one has the sharp estimate $\mathrm{Re}\{b_3\} \leqq \frac{1}{2} + e^{-6}$.

We shall now apply the result of the preceding section to show that the function $k_n(z)$ is not extremal for any $n \geqq 5$. Combined with what is known for $n = 3$ and 4, it follows that the maximum of $\mathrm{Re}\{b_n\}$ is strictly larger than $2/(n+1)$ for any $n \geqq 3$. This makes the coefficient problem for Σ very intriguing, but in the absence of even a conjecture, very difficult.

Let us formulate the results of the first and second variation for the general coefficient problem. That is, we choose the linear functional L to specify the coefficient of $1/z^n$ in the Laurent expansion for fixed n. It will be useful to define the polynomials $\Pi_\nu(t)$ and $p_\nu(t)$ through the generating functions

(5.2) $$\frac{1}{f(z)-t} = \sum_{\nu=1}^{\infty} \Pi_\nu(t) z^{-\nu}, \qquad \frac{\varphi(z)}{z-t} = \sum_{\nu=0}^{\infty} p_\nu(t) z^{-\nu}.$$

Observe the useful recursion formula

(5.3) $$p_{\nu+1}(t) = t p_\nu(t) - \nu b_\nu \quad \text{for } \nu \geqq 0.$$

The differential equation (3.5) then becomes

(5.4) $$[\zeta f'(\zeta)]^2 \Pi_n(f(\zeta)) = q_n(\zeta)$$

where

(5.5) $$q_n(\zeta) = p_{n+1}(\zeta) + \overline{p_{n+1}(1/\bar{\zeta})} + (n-1)b_n.$$

In this special case equation (3.2) becomes $L(f - zf') = (n+1)b_n =$ real, while (3.7) yields $(n+1)^2 b_n \geqq 0$. Inequality (3.6) becomes

(5.6) $$\Bigg|\frac{1}{2}\zeta\varphi'(\zeta)\varphi(\zeta)^2\Pi_n'(f(\zeta)) + \zeta\varphi(\zeta)^2\sum_{\nu=0}^{n-1} p_\nu(\zeta)\Pi_{n-\nu}'(f(\zeta)) + M(\zeta)q_n(\zeta)$$
$$- \frac{n}{2}\zeta^2 p_n'(\zeta) - \frac{n-2}{2}\overline{p_n'(1/\bar{\zeta})} - \overline{p_{n+1}(1/\bar{\zeta})} + \frac{1}{2}b_n\Bigg|$$
$$\leqq -\mathrm{Re}\Bigg\{\frac{|\zeta|^2}{(1-|\zeta|^2)^2}q_n(\zeta) + \varphi(\zeta)^2\sum_{\nu=0}^{n} p_\nu(1/\bar{\zeta})\Pi_{n-\nu+1}'(f(\zeta))$$
$$+ \frac{n|\zeta|^2}{1-|\zeta|^2}[p_{n+1}(\zeta) - p_{n+1}(1/\bar{\zeta})] + q_n(\zeta) - \zeta p_n(\zeta)\Bigg\}.$$

For the function $k_n(z) = f(z)$ defined in (5.1) we have

$$\Pi_1(t) = 1, \quad \Pi_2(t) = t, \cdots, \Pi_n(t) = t^{n-1}, \quad \Pi_{n+1}(t) = t^n - \frac{2}{n+1},$$
$$p_0(t) = 1, \quad p_1(t) = t, \cdots, p_n(t) = t^n, \quad p_{n+1}(t) = t^{n+1} - \frac{2n}{n+1}.$$

One easily verifies that

$$\Pi_n(k_n(\zeta))[\zeta k_n'(\zeta)]^2 = k_n(\zeta)^{n+1}\left[\frac{\zeta k_n'(\zeta)}{k_n(\zeta)}\right]^2 = \zeta^{n+1} + \frac{1}{\zeta^{n+1}} - 2 = q_n(\zeta)$$

so that k_n satisfies the differential equation (5.4) demanded by the first variation. The function k_n maps the circular region $|z|>1$ onto the complement of $n+1$ equal segments emanating from the origin at equal angles $2\pi/(n+1)$. All points η which satisfy $\eta^{n+1} = -1$ are carried onto the singular point at the origin. Since $q_n(\eta) = -4 \neq 0$ for all these points, the previous section implies that $2\pi/(n+1) \geqq 2\pi/5$. Consequently, k_n can satisfy the condition of the second variation only for $n \leqq 4$. That is, $k_n(z)$ is not the extremum function for the functional $\text{Re}\{b_n\}$ if $n \geqq 5$. Since we know explicitly that $\text{Re}\{b_n\} = \max$ is not achieved by $k_n(z)$ for $n = 3$ and 4, we have thus proved that only for $n = 1$ and 2 the functions $k_n(z)$ yield the maximum value. That is,

(5.7) $$\max_{\Sigma} \text{Re}\{b_n\} > \frac{2}{n+1} \qquad \text{for } n \geqq 3.$$

To show the significance and usefulness of this result consider the following problem. Let $L(f)$ be a linear functional on Σ which is invariant under a shift $f \to f + \text{const.}$, i.e., $L(1) = 0$. Let $f(z) \in \Sigma$ yield the maximum value of $\text{Re}\{L(f)\}$ in Σ. Then, as was shown in Section 3, $f(z) = w$ maps its domain $|z|>1$ onto the w-plane slit along analytic arcs with the differential equation

(5.8) $$L\left(\frac{1}{f-w}\right)(w')^2 > 0.$$

This differential equation will lead to critical points on the boundary only if $L(1/(f-w))$ vanishes there. Suppose that at some boundary point, say $w = 0$, $L(1/(f-w))$ has a zero of order k and that at that point $k+2$ different arcs of the boundary join. We will show that this fact leads to the estimate

$$|b_{k+1}| \leqq \frac{2}{k+2}$$

for the $(k+1)$-coefficient of each function $g(z) \in \Sigma_0$. Since we have just shown that this is not the case if $k>1$, we have proved that such occurrence is impossible for $k>1$. This is a very general statement concerning arbitrary linear functionals which are translation invariant.

To prove the assertion, we describe a circle of radius δ around the forking point $w = 0$. If δ is small enough, it will contain a subcontinuum Γ_δ of the boundary Γ of the extremum domain which consists of $(k+2)$ arcs joining at $w = 0$ under equal angles. We denote by ε the exterior radius of Γ_δ; that is, we can map the exterior of Γ_δ onto the circular region $|\zeta| > \varepsilon$ by the normalized function

$$\zeta(w) = w + \sum_{\nu=0}^{\infty} \frac{a_\nu(\delta)\varepsilon^{\nu+1}}{w^\nu}.$$

This expansion is valid for $|w| > \delta$. As $\delta \to 0$, it is easily seen that $\varepsilon \to 0$ and

$$\lim_{\delta\to 0} a_\nu(\delta) = \begin{cases} 0 & \text{for } \nu \leqq k \\ \dfrac{2}{k+2} e^{i\theta} & \text{for } \nu = k+1 \end{cases}$$

for some θ, in view of the asymptotic character of Γ_δ as $\delta \to 0$.

Next, let $g(z) = z + \sum_{\nu=1}^{\infty} b_\nu / z^\nu$ be an arbitrary function in Σ_0. Then

$$g_\varepsilon(\zeta) = \zeta + \sum_{\nu=1}^{\infty} \frac{b_\nu \varepsilon^{\nu+1}}{\zeta^\nu} = \varepsilon g\left(\frac{\zeta}{\varepsilon}\right)$$

is univalent and normalized in $|\zeta| > \varepsilon$. We combine the mappings to obtain a function $f^*(z) \in \Sigma$ of the form

$$f^*(z) = g_\varepsilon[\zeta(f(z))] = f(z) + \sum_{\nu=0}^{\infty} \frac{a_\nu(\delta)\varepsilon^{\nu+1}}{f(z)^\nu} + \sum_{\mu=1}^{\infty} \frac{b_\mu \varepsilon^{\mu+1}}{f(z)^\mu}\left[1 + \sum_{\nu=0}^{\infty} \frac{a_\nu(\delta)\varepsilon^{\nu+1}}{f(z)^\nu}\right]^{-\mu}$$

which competes with $f(z)$ in the maximum problem with respect to $L(f)$. We rearrange in powers of $f(z)$ to find

$$f^*(z) = f(z) + \sum_{\nu=0}^{\infty} \frac{C_\nu(\delta)\varepsilon^{\nu+1}}{f(z)^\nu}$$

for $|f(z)| > \delta$, where

$$C_\nu(\delta) = a_\nu(\delta) + b_\nu + \sum_{\mu<\nu} \pi_\mu(a_1, \cdots, a_{\nu-\mu}) b_\mu$$

and the π_μ are polynomials of their arguments which vanish if all their arguments vanish.

If δ is sufficiently small, we may calculate

$$L(f^*) = L(f) + \sum_{\nu=0}^{\infty} C_\nu(\delta)\varepsilon^{\nu+1} L(f^{-\nu}). \tag{5.9}$$

Observe that the very nature of the linear functional leads to the expansion

$$L\left(\frac{1}{f-w}\right) = \sum_{\nu=1}^{\infty} L(f^{-\nu}) w^{\nu-1}$$

near $w = 0$. We have assumed that $L(1/(f-w))$ has a zero of order k at the origin and that $L(1) = 0$. Therefore

$$L(f^{-\nu}) = 0 \qquad \text{for } \nu = 0, \cdots, k$$

and $L(f^{-k-1}) \neq 0$. Thus, we can simplify (5.9) to find

$$L(f^*) = L(f) + C_{k+1}(\delta)\varepsilon^{k+2} L(f^{-k-1}) + o(\varepsilon^{k+2}).$$

Now, combining our assumptions regarding the $a_\nu(\delta)$ and the π_μ,

$$C_{k+1}(\delta) = \frac{2}{k+2} e^{i\theta} + b_{k+1} + o(1) \qquad \text{as } \delta \to 0.$$

Since $\mathrm{Re}\{L(f^*)\} \leqq \mathrm{Re}\{L(f)\}$ by the maximality of $\mathrm{Re}\{L(f)\}$, we find in the usual way that

$$\mathrm{Re}\left\{ \left[\frac{2}{k+2} e^{i\theta} + b_{k+1} \right] L(f^{-k-1}) \right\} \leqq 0.$$

Therefore

$$\mathrm{Re}\{b_{k+1} L(f^{-k-1})\} \leqq \frac{2}{k+2} |L(f^{-k-1})|$$

for every choice of b_{k+1}. With given b_{k+1} also $e^{i\sigma} b_{k+1}$ is admissible for all real σ. This leads to the asserted estimate

$$|b_{k+1}| \leqq \frac{2}{k+2}$$

and the corresponding consequences.

6. More general extremum problems

In this section we shall consider more general extremum problems to which the variational calculus can be applied.

Let L be a continuous linear functional, and let $F(w, z)$ be an analytic function of two complex variables such that $F(f(z), z)$ is defined and analytic for each $f \in \Sigma_0$ on the support of some representing measure for L. We have in mind expressions such as $f(z)^N$ or $\log(f(z)/z)$. Consider then the extremum problem of finding

$$\max_{f \in \Sigma_0} \mathrm{Re}\{L(F(f(z), z))\}.$$

Since Σ_0 is compact, there will be at least one extremum function f for the problem. If we subject f to variations of the type $\hat{f} = f + \varepsilon \hat{f}_1 + \varepsilon^2 \hat{f}_2 + \cdots$, then

$$F(\hat{f}(z), z) = F(f(z), z) + \varepsilon F_w(f(z), z)\hat{f}_1(z)$$
$$+ \varepsilon^2 \left[\frac{1}{2} F_{ww}(f(z), z)\hat{f}_1(z)^2 + F_w(f(z), z)\hat{f}_2(z) \right] + o(\varepsilon^2).$$

Since the maximum character of $f(z)$ implies

$$\mathrm{Re}\{L(F(\hat{f}(z), z))\} \leqq \mathrm{Re}\{L(F(f(z), z))\},$$

we can derive just as before a differential equation for the extremum function. We use the variational formulas of Section 2 to find

$$
\begin{aligned}
&L\left(F_w(f(z),z)\left[\frac{\varphi(\zeta)^2}{f(z)-f(\zeta)}-\frac{\zeta\varphi(z)}{z-\zeta}+\zeta\right]\right)\\
&\qquad+\overline{L\left(F_w(f(z),z)\left[\frac{\bar\zeta z\varphi(z)}{1-\bar\zeta z}+f(z)+\frac{1}{\bar\zeta}\right]\right)}=0
\end{aligned}
\tag{6.1}
$$

for all $|\zeta|>1$. Define now

$$
q(\zeta)\equiv L\left(F_w(f,z)\left[\frac{\zeta\varphi}{z-\zeta}-\zeta\right]\right)-\overline{L\left(F_w(f,z)\left[\frac{\bar\zeta z\varphi}{1-\bar\zeta z}+\frac{1}{\bar\zeta}+f\right]\right)}
\tag{6.2}
$$

and find the differential equation

$$
[\zeta f'(\zeta)]^2 L\left(\frac{F_w(f,z)}{f-f(\zeta)}\right)=q(\zeta)
\tag{6.3}
$$

which is valid near $|\zeta|=1$.

The second variation is now still more complicated. The same techniques as before lead to the inequality

$$
\begin{aligned}
&\Bigg|L\Bigg(F_w(f,z)\left\{\frac{\frac{1}{2}\zeta\varphi(\zeta)^2\varphi'(\zeta)}{[f-f(\zeta)]^2}+\frac{\zeta\varphi(\zeta)^2\varphi}{(z-\zeta)[f-f(\zeta)]^2}+\frac{\varphi(\zeta)^2M(\zeta)}{[f-f(\zeta)]}\right.\\
&\qquad\left.+\frac{1}{2}\zeta^2 z\frac{d}{dz}\left[\frac{\varphi}{(z-\zeta)^2}\right]\right\}\\
&\qquad+\frac{1}{2}F_{ww}(f,z)\left\{\frac{\varphi(\zeta)^2}{[f-f(\zeta)]}-\frac{\zeta\varphi}{z-\zeta}+\zeta\right\}^2\Bigg)\\
&\qquad+\overline{L\left(F_w(f,z)\left\{\frac{1}{2}z\bar\zeta^2\frac{d}{dz}\left(\frac{z^2\varphi}{(1-\bar\zeta z)^2}\right)+\frac{\bar\zeta z\varphi}{1-\bar\zeta z}+\frac{1}{\bar\zeta}+\frac{1}{2}f\right\}\right.}\\
&\qquad\overline{\left.+\frac{1}{2}F_{ww}(f,z)\left\{f+\frac{\bar\zeta z\varphi}{1-\bar\zeta z}+\frac{1}{\bar\zeta}\right\}^2\right)}\Bigg|\\
&\leqq-\operatorname{Re}\Bigg\{L\Bigg(F_w(f,z)\left\{\frac{|\zeta|^2}{(1-|\zeta|^2)^2}\frac{\varphi(\zeta)^2}{[f-f(\zeta)]}-\frac{\bar\zeta\varphi(\zeta)^2 z\varphi}{(1-\bar\zeta z)[f-f(\zeta)]^2}\right.\\
&\qquad\left.-|\zeta|^2 z\frac{d}{dz}\left[\frac{z\varphi}{(z-\zeta)(1-\bar\zeta z)}\right]+\frac{\varphi(\zeta)^2}{[f-f(\zeta)]}-\frac{\zeta\varphi}{z-\zeta}+\zeta\right\}\\
&\qquad+F_{ww}(f,z)\left\{\frac{\varphi(\zeta)^2}{[f-f(\zeta)]}-\frac{\zeta\varphi}{z-\zeta}+\zeta\right\}\left\{f+\frac{\bar\zeta z\varphi}{1-\bar\zeta z}+\frac{1}{\bar\zeta}\right\}\Bigg)\Bigg\}.
\end{aligned}
\tag{6.4}
$$

Observe that the leading term as ζ converges to the unit circumference is the first term on the right-hand side, which by (6.3) is $|\zeta|^2 q(\zeta)/(1-|\zeta|^2)^2$, so that we can again conclude that $q(\zeta)\leqq 0$ for $|\zeta|=1$. We can also study the singular points of $f(\zeta)$ in view of the differential equation (6.3). These can occur only where $L(F_w(f,z)/(f-f(\zeta)))$ vanishes on $|\zeta|=1$. Let us then set up the asymptotic expansion

$$f(\zeta)=\omega+\tau(\zeta-\eta)^{\alpha}+o((\zeta-\eta)^{\alpha})\qquad \text{as } \zeta\to\eta,\quad |\eta|=1.$$

We can easily compute the asymptotic behavior of all terms in (6.4) as $\zeta\to\eta$. If $\alpha<\frac{1}{2}$, it appears that the coefficient of

$$r(\zeta)=\frac{1}{2}L\left(\frac{F_{ww}(f,z)}{[f-f(\zeta)]^2}\right), \tag{6.5}$$

which is of the order $(\zeta-\eta)^{4\alpha-4}$, dominates all other terms and tends to infinity as $\zeta\to\eta$, provided that $r(\eta)\neq 0$. This is impossible, and we can conclude that $\alpha\geqq\frac{1}{2}$ if $r(\eta)\neq 0$.

Even if $r(\zeta)$ vanishes at η, the exponent α is restricted by the requirement that $(\zeta-\eta)^{4\alpha-2}r(\zeta)$ is bounded as $\zeta\to\eta$.

Next, suppose that $(\zeta-\eta)^{4\alpha-2}r(\zeta)$ tends to zero as $\zeta\to\eta$ and that $q(\eta)\neq 0$. The significant terms in the inequality are now the same as in the case considered in Section 4, and we may conclude again that in this case $\alpha\geqq\frac{2}{5}$.

A finer analysis of the inequality (6.4) is possible if we specify the function $F(w,z)$.

7. A special application

Let us choose, for example,

$$F(f(z),z;N)=\begin{cases} f(z)^N & \text{if } N=\pm 1,\pm 2,\cdots,\\ \log\dfrac{f(z)}{z} & \text{if } N=0.\end{cases} \tag{7.1}$$

For $f(z)\in\Sigma_0$, we have the development near ∞

$$F(f(z),z;N)=\begin{cases} z^N+\displaystyle\sum_{\nu=-N+2}^{\infty}B_\nu^{(N)}z^{-\nu} & \text{if } N=\pm 1,\pm 2,\cdots,\\ \displaystyle\sum_{\nu=2}^{\infty}B_\nu^{(0)}z^{-\nu} & \text{if } N=0.\end{cases} \tag{7.2}$$

We may consider the problem of finding the maximum of $\operatorname{Re}\{B_\nu^{(N)}\}$ in the class Σ_0 for fixed ν and N, $\nu\geqq -N+2$.

We may consider the question whether one of the functions k_n defined in (5.1) can be an extremum function for this problem. The first variation leads to the necessary condition (see (6.3))

$$L\left(\frac{k_n^{N-1}}{k_n-k_n(\zeta)}\right)(\zeta k_n'(\zeta))^2=\hat{q}(\zeta) \tag{7.3}$$

where

$$\hat{q}(\zeta)=L\left(k_n^{N-1}\left[\frac{\zeta z k_n'}{z-\zeta}-\zeta\right]\right)-\overline{L\left(k_n^{N-1}\left[\frac{\bar{\zeta} z k_n'}{1-\bar{\zeta}z}+\frac{1}{\zeta}+k_n\right]\right)} \tag{7.4}$$

and the functional singles out the ν-th Laurent coefficient. As before, we conclude that $\hat{q}(\zeta) \leqq 0$ for $|\zeta| = 1$. We find by direct calculation that

$$\left(\frac{\zeta k_n'(\zeta)}{k_n(\zeta)}\right)^2 = \left(\frac{\zeta^{n+1}-1}{\zeta^{n+1}+1}\right)^2 \leqq 0 \qquad \text{for } |\zeta| = 1, \tag{7.5}$$

which yields the necessary condition

$$L\left(\frac{k_n^{N-1} k_n(\zeta)^2}{k_n - k_n(\zeta)}\right) \geqq 0 \qquad \text{for } |\zeta| = 1.$$

For $|\zeta| = 1$ we may expand

$$\frac{k_n(z)^{N-1} k_n(\zeta)^2}{k_n(z) - k_n(\zeta)} = \sum_{j=2}^{\infty} k_n(z)^{N-j} k_n(\zeta)^j.$$

Since the coefficient of $z^{-\nu}$ for $k_n(z)^{N-j}$ is zero unless $\nu + N - j$ is a nonnegative integer multiple of $n+1$, we may write

$$L\left(\frac{k_n^{N-1} k_n(\zeta)^2}{k_n - k_n(\zeta)}\right) = k_n(\zeta)^{\nu+N} + \sum_l L(k_n^{-\nu+l(n+1)}) k_n(\zeta)^{\nu+N-l(n+1)}. \tag{7.6}$$

This polynomial in $k_n(\zeta)$ must be real and nonnegative for $|\zeta| = 1$. Therefore the lowest order term that appears must be of one sign for $|\zeta| = 1$. This is the case only if the exponent of $k_n(\zeta)$ is an integer multiple of $n+1$. We conclude that $k_n(\zeta)$ can be a candidate for an extremum function only if $n+1$ divides $\nu + N$.

For later use we rewrite (7.6) as

$$L\left(\frac{k_n^{N-1} k_n(\zeta)^2}{k_n - k_n(\zeta)}\right) = k_n(\zeta)^{\nu+N} + \cdots + L(k_n^{N-n-1}) k_n(\zeta)^{n+1}. \tag{7.7}$$

Further conditions can now be obtained by studying the critical points for $k_n(\zeta)$. The points η satisfying $\eta^{n+1} = -1$ correspond to the origin, which is a forking point of the boundary continuum. To determine the value $\hat{q}(\eta)$, observe that by (7.5)

$$\lim_{\zeta \to \eta} \left(\frac{\zeta k_n'(\zeta)}{k_n(\zeta)}\right)^2 k_n(\zeta)^{n+1} = \lim_{\zeta \to \eta} \frac{(\zeta^{n+1}-1)^2}{\zeta^{n+1}} = -4. \tag{7.8}$$

Hence by (7.3), (7.8), and (7.7)

$$\hat{q}(\eta) = -4 \lim_{\zeta \to \eta} L\left(\frac{k_n^{N-1} k_n(\zeta)^{1-n}}{k_n - k_n(\zeta)}\right) = -4L(k_n^{N-n-1}). \tag{7.9}$$

If we compute the coefficient of $z^{-\nu}$ for $k_n(z)^{N-n-1}$, we obtain

$$\hat{q}(\eta) = \frac{-4\left(\frac{2N}{n+1} - 2\right)\left(\frac{2N}{n+1} - 3\right)\cdots\left(\frac{N-\nu}{n+1}\right)}{\left(\frac{\nu+N}{n+1} - 1\right)!}. \tag{7.10}$$

Thus $\hat{q}(\eta)$ is surely nonzero unless $\nu \geqq N$ and $2N/(n+1)$ is an integer larger than or equal to 2.

If $N=1$, we have already discussed the problem in Section 5. Thus, we may assume that $N \neq 1$. We shall now utilize the estimates (6.4) in this special case. If $\eta^{n+1}=-1$, let $\zeta = r\eta$ and write k_n instead of f. We need the limit value of $(1-r^2)\varphi(\zeta)^4 r(\zeta)$ in this case as $r \to 1$. Define $\hat{r}(\zeta)=(1/N)r(\zeta)$ if $N \neq 0$ and $\hat{r}(\zeta)=r(\zeta)$ if $N=0$. Then

$$\lim_{r\to 1}(1-r)^2\varphi(\zeta)^4\hat{r}(\zeta)=\lim_{r\to 1}\left(\frac{\zeta k_n'(\zeta)}{k_n(\zeta)}\right)^4 k_n(\zeta)^{2n+2}\frac{N-1}{2}(1-r)^2 L\left(\frac{k_n(\zeta)^{2-2n}k_n^{N-2}}{[k_n-k_n(\zeta)]^2}\right)$$

$$=8(N-1)\lim_{r\to 1}(1-r)^2\sum_{j=4}^{\infty}(j-3)L(k_n^{N-j})k_n(\zeta)^{j-2n-2}.$$

Since $\lim_{r\to 1}(1-r)^2 k_n(\zeta)^p=0$ for $p \geqq -n$ and $L(k_n^{N-j})=0$ for $4 \leqq j \leqq n$, we have by use of (7.9)

$$\lim_{r\to 1}(1-r)^2\varphi(\zeta)^4\hat{r}(\zeta)=8(N-1)\lim_{r\to 1}(1-r)^2(n-2)L(k_n^{N-n-1})k_n(\zeta)^{-n-1}$$

$$=8(N-1)(n-2)L(k_n^{N-n-1})\frac{-1}{(n+1)^2}$$

$$=2(N-1)\frac{n-2}{(n+1)^2}\hat{q}(\eta).$$

This is valid at least for $n \geqq 3$. Observe that $k_n(\zeta)=\tau(\zeta-\eta)^{2/(n+1)}+\cdots$ near η. Put $\alpha = 2/(n+1)$ and find finally

$$\lim_{r\to 1}(1-r)^2\varphi(\zeta)^4\hat{r}(\zeta)=(N-1)\alpha\left(1-\frac{3}{2}\alpha\right)\hat{q}(\eta). \tag{7.11}$$

We can now compare our new inequality (6.4) with the special inequality (3.6) and repeat the study of a forking point which we did in Section 4. Because of the additional term (7.11) the inequality (4.2) becomes now

$$|7-12\alpha+5\alpha^2-(N-1)(12\alpha-18\alpha^2)| \leqq 3.$$

Thus the only admissible values of N must satisfy

$$\frac{n^2+2n-12}{6(n-2)} \leqq N \leqq \frac{5n^2+10n-21}{12(n-2)}. \tag{7.12}$$

The restriction (7.12) was derived under the assumptions that $n \geqq 3$ and that it is not the case that $\nu \geqq N$ and $2N/(n+1)$ is an integer larger than or equal to 2.

The main purpose of our study in this section was to show that the inequalities from the second variation complement the information of the first variation and lead to significant information even in quite complicated problems. We can also make an interesting application in the next section.

8. The coefficient problem for the inverse functions of Σ_0

If $f \in \Sigma_0$, then the inverse function f^{-1} has an expansion

$$f^{-1}(w) = w + \sum_{N=1}^{\infty} B_N w^{-N}$$

in some neighborhood of ∞. For the odd coefficients it is known [12, 4, 10, 11] that

$$|B_{2N+1}| \leqq \frac{(2N)!}{N!(N+1)!} \qquad \text{for } 1 \leqq N \leqq 7,$$

and that equality holds only for $k_1(z) = z + 1/z$ and its rotations. G. Springer [12] conjectured that the odd function $k_1(z)$ yields the extremum value for all odd coefficients. The problem for even coefficients is apparently more difficult. It is evident that $|B_2| = |-b_2| \leqq \frac{2}{3}$, and Springer conjectured that the maximum for $|B_{2N}|$ is achieved for $k_n(z)$ where $n+1$ is the least prime divisor of $2N+1$. We shall see that this conjecture is not true.

Since

$$-NB_N = \frac{1}{2\pi i} \int_{|w|=R} wF'(w)w^{N-1}\,dw = \frac{1}{2\pi i} \int_{|z|=R} f(z)^N dz = B_1^{(N)},$$

we see that the problem of the inverse coefficients is equivalent to the first coefficient problem for the powers of f. Let us test Springer's conjecture with the results of the previous section.

We have to deal with the case $\nu = 1$. From the differential equation it follows as a first necessary condition that $n+1$ divides $N+1$. If $N+1 = \sigma(n+1)$ and $n \geqq 3$, then (7.12) leads to

$$n \leqq (n+1)\sigma - 1 = N \leqq \frac{5n^2 + 10n - 21}{12(n-2)}.$$

This is possible only for $n = 3$, $\sigma = 1$ and $n = 4$, $\sigma = 1$. However, if $n = 3$ and $\sigma = 1$, then $N = 3$ is odd, and it is known that only $k_1(z)$ provides the maximum. Thus, with one possible exception, we see that $k_n(z)$ is never an extremum function for the inverse coefficient problem in Σ_0 if $n \geqq 3$. The one exception is $k_4(z)$ for the B_4-problem; that is, it is still an open question whether $|B_4| \leqq \frac{2}{5}$.

9. The coefficient problem in S

We can establish analogous results in the case of the class S. We specialize now L as the n-th coefficient in the Taylor development of $f(z) \in S$ around the origin. If

$$f(z) = z + \sum_{n=2}^{\infty} a_n z^n,$$

we find the coefficient $\hat{a}_n$ of the varied function (2.12) as follows. We define polynomials by means of the generating functions

$$\frac{zf'(z)}{1-tz}=\frac{\varphi(z)}{1-tz}=\sum_{n=1}^{\infty}K_n(t)z^n \tag{9.1}$$

and

$$\frac{f(z)^2}{1-tf(z)}=\sum_{n=2}^{\infty}P_n(t)z^n. \tag{9.2}$$

Here $P_n(t)$ is of degree $(n-2)$ in t and $K_n(t)$ is of degree $(n-1)$. We have

$$K_n(t)=\sum_{k=1}^{n}ka_kt^{n-k} \tag{9.3}$$

and the recursion formula

$$K_{n+1}(t)=tK_n(t)+(n+1)a_{n+1}. \tag{9.4}$$

By means of these definitions, we derive from (2.12) the following asymptotic development for $\hat{a}_n$:

$$\begin{aligned}\hat{a}_n&=a_n+\varepsilon\left[\frac{x}{\varphi(\zeta)^2}\Omega_1+\frac{\bar{x}}{\overline{\varphi(\zeta)^2}}\Omega_2\right]\\&\quad+\varepsilon^2\left[\frac{x^2}{\varphi(\zeta)^4}\Omega_{11}+\frac{|x|^2}{|\varphi(\zeta)|^4}\Omega_{12}+\frac{\bar{x}^2}{\overline{\varphi(\zeta)^4}}\Omega_{22}\right]+o(\varepsilon^2)\end{aligned} \tag{9.5}$$

where

$$\Omega_1=-\frac{\varphi(\zeta)^2}{f(\zeta)^3}P_n\left(\frac{1}{f(\zeta)}\right)-a_n+K_n\left(\frac{1}{\zeta}\right),\qquad \Omega_2=K_n(\bar{\zeta})-na_n, \tag{9.6}$$

$$\begin{aligned}\Omega_{11}&=\frac{1}{2}\frac{\zeta\varphi'(\zeta)\varphi(\zeta)^2}{f(\zeta)^4}\left(3P_n\left(\frac{1}{f(\zeta)}\right)+\frac{1}{f(\zeta)}P_n'\left(\frac{1}{f(\zeta)}\right)\right)\\&\quad+\frac{\varphi(\zeta)^2}{f(\zeta)^3}P_n\left(\frac{1}{f(\zeta)}\right)\left[1-\frac{\varphi(\zeta)^2}{f(\zeta)^2}-M(\zeta)\right]\\&\quad-\frac{\varphi(\zeta)^2}{f(\zeta)^3}\sum_{k=2}^{n}kP_k\left(\frac{1}{f(\zeta)}\right)\frac{1}{\zeta^{n-k}}-K_n\left(\frac{1}{\zeta}\right)+\frac{n}{2}K_{n+1}'\left(\frac{1}{\zeta}\right)+\frac{1}{2}a_n,\end{aligned}$$

$$\begin{aligned}\Omega_{12}&=-\frac{|\zeta|^2}{(1-|\zeta|^2)^2}\frac{\varphi(\zeta)^2}{f(\zeta)^3}P_n\left(\frac{1}{f(\zeta)}\right)-\frac{\varphi(\zeta)^2}{f(\zeta)^3}\sum_{k=2}^{n-1}kP_k\left(\frac{1}{f(\zeta)}\right)\bar{\zeta}^{n-k}\\&\quad+\frac{n|\zeta|^2}{1-|\zeta|^2}\left[K_n\left(\frac{1}{\zeta}\right)-K_n(\bar{\zeta})\right]-\bar{\zeta}K_{n-1}(\bar{\zeta}),\end{aligned}$$

$$\Omega_{22}=\frac{n}{2}\bar{\zeta}^2K_{n-1}'(\bar{\zeta}).$$

Let us now deal with the problem of $\mathrm{Re}\{a_n\} = \max$. If $f(z)$ is an extremum function, we have necessarily $\mathrm{Re}\{\hat{a}_n\} \leqq \mathrm{Re}\{a_n\}$. Thus, we are led in the usual way from the first variation to the identity

$$\Omega_1 + \bar{\Omega}_2 = K_n\left(\frac{1}{\zeta}\right) + \overline{K_n(\bar{\zeta})} - (n+1)a_n - \frac{\varphi(\zeta)^2}{f(\zeta)^3} P_n\left(\frac{1}{f(\zeta)}\right) = 0. \tag{9.7}$$

Since the extremum a_n is real, we may define

$$\begin{aligned} q_n(\zeta) &= K_n\left(\frac{1}{\zeta}\right) - (n+1)a_n + \overline{K_n(\bar{\zeta})} \\ &= \frac{1}{\zeta^{n-1}} + \frac{2a_2}{\zeta^{n-2}} + \frac{3a_3}{\zeta^{n-3}} + \cdots + \frac{(n-1)a_{n-1}}{\zeta} + (n-1)a_n \\ &\quad + (n-1)\bar{a}_{n-1}\zeta + \cdots + 3\bar{a}_3\zeta^{n-3} + 2\bar{a}_2\zeta^{n-2} + \zeta^{n-1} \end{aligned} \tag{9.8}$$

and arrive at the well-known differential equation for the n-th coefficient problem

$$\frac{\varphi(\zeta)^2}{f(\zeta)^3} P_n\left(\frac{1}{f(\zeta)}\right) = \frac{\zeta^2 f'(\zeta)^2}{f(\zeta)^3} P_n\left(\frac{1}{f(\zeta)}\right) = q_n(\zeta). \tag{9.9}$$

Observe that by differentiation we obtain

$$\zeta q_n'(\zeta) = 2\frac{\zeta\varphi'(\zeta)}{\varphi(\zeta)} q_n(\zeta) - \varphi(\zeta)^3\left(\frac{3}{f(\zeta)^4} P_n\left(\frac{1}{f(\zeta)}\right) + \frac{1}{f(\zeta)^5} P_n'\left(\frac{1}{f(\zeta)}\right)\right). \tag{9.10}$$

Thus, for the extremum function we may simplify Ω_{11} and Ω_{12} as follows:

$$\begin{aligned} \Omega_{11} &= q_n(\zeta)\left[1 - \frac{\varphi(\zeta)^2}{f(\zeta)^2} + \frac{1}{2}\left(\frac{\zeta\varphi'(\zeta)}{\varphi(\zeta)}\right)^2 + \frac{1}{6}\zeta^2\{f,\zeta\}\right] - \frac{1}{2}\frac{\zeta\varphi'(\zeta)}{\varphi(\zeta)}\zeta q_n'(\zeta) \\ &\quad - \varphi(\zeta)^2 S_n\left(\frac{1}{f(\zeta)}, \frac{1}{\zeta}\right) - K_n\left(\frac{1}{\zeta}\right) + \frac{n}{2}K_{n+1}'\left(\frac{1}{\zeta}\right) + \frac{1}{2}a_n \end{aligned} \tag{9.11}$$

where

$$S_n(\omega, t) = \omega^3 \sum_{k=2}^{n} kP_k(\omega)t^{n-k}, \tag{9.12}$$

and

$$\begin{aligned} \Omega_{12} &= \frac{-|\zeta|^2}{(1-|\zeta|^2)^2} q_n(\zeta) - \varphi(\zeta)^2 S_n\left(\frac{1}{f(\zeta)}, \bar{\zeta}\right) \\ &\quad + nq_n(\zeta) + \frac{n|\zeta|^2}{1-|\zeta|^2}\left[K_n\left(\frac{1}{\zeta}\right) - K_n(\bar{\zeta})\right] - \bar{\zeta}K_{n-1}(\bar{\zeta}). \end{aligned} \tag{9.13}$$

The extremality condition in $f(\zeta)$ due to the second variation is

$$\mathrm{Re}\left\{\frac{x}{\varphi(\zeta)^4}\Omega_{11} + \frac{|x|^2}{|\varphi(\zeta)|^4}\Omega_{12} + \frac{\bar{x}}{\varphi(\zeta)^4}\Omega_{22}\right\} \leqq 0$$

for all choices of x. This implies

$$|\Omega_{11}+\bar{\Omega}_{22}|\leqq -\mathrm{Re}\{\Omega_{12}\}. \tag{9.14}$$

Hence, using also (9.4) and the definition

$$\bar{K}_n(\zeta)=\overline{K_n(\bar{\zeta})}, \tag{9.15}$$

we arrive at the condition

$$\begin{aligned}\Big| q_n(\zeta)\Big[1-\frac{\varphi(\zeta)^2}{f(\zeta)^2}+\frac{1}{2}\Big(\frac{\zeta\varphi'(\zeta)}{\varphi(\zeta)}\Big)^2+\frac{1}{6}\zeta^2\{f,\zeta\}\Big]-\frac{1}{2}\frac{\zeta\varphi'(\zeta)}{\varphi(\zeta)}\zeta q_n'(\zeta)-\varphi(\zeta)^2S_n\Big(\frac{1}{f(\zeta)},\frac{1}{\zeta}\Big)\\ -K_n\Big(\frac{1}{\zeta}\Big)+\frac{n}{2}\Big[\frac{1}{\zeta}K_n'\Big(\frac{1}{\zeta}\Big)+\zeta\bar{K}_n'(\zeta)+K_n\Big(\frac{1}{\zeta}\Big)-\bar{K}_n(\zeta)\Big]+\frac{n^2+1}{2}a_n\Big|\\ \leqq \mathrm{Re}\Big\{\frac{|\zeta|^2}{(1-|\zeta|^2)^2}q_n(\zeta)+\varphi(\zeta)^2S_n\Big(\frac{1}{f(\zeta)},\bar{\zeta}\Big)-\frac{n|\zeta|^2}{1-|\zeta|^2}\Big[K_n\Big(\frac{1}{\zeta}\Big)-\bar{K}_n(\zeta)\Big]\\ -nq_n(\zeta)+\bar{K}_n(\zeta)-na_n\Big\}.\end{aligned} \tag{9.16}$$

As a first application of the necessary maximum condition, we conclude

$$q_n(\zeta)\geqq 0 \qquad \text{for all } |\zeta|=1. \tag{9.17}$$

Next, we repeat the argument of Section 4 to study critical points on the boundary of the image domain. It is well known [6,8] that the extremum function in this problem maps the unit disk onto the entire w-plane, slit along one analytic arc which ends at ∞. However, this result is obtained by special additional arguments, among them a finite variation, comparing $f(z)$ with $f(z)(1-f(z)/w)^{-1}$, which is also in S if w is a boundary point. It is therefore interesting to see how much information can be found by infinitesimal variations alone. Let then η be a preimage on the unit circumference of a critical point w on the boundary continuum in the w-plane. Let us assume that near η

$$f(\zeta)=\omega+\tau(\zeta-\eta)^\alpha+\cdots \qquad \text{if } \alpha>0 \tag{9.18}$$

or

$$f(\zeta)=\tau(\zeta-\eta)^\alpha+o((\zeta-\eta)^\alpha) \qquad \text{if } \alpha<0.$$

We can use the same estimates (4.1) as in Section 4, and if $\alpha<0$, we must also observe

$$\Big(\frac{\varphi(\zeta)}{f(\zeta)}\Big)^2=\frac{\alpha^2\eta^2}{(\zeta-\eta)^2}+o\Big(\frac{1}{(\zeta-\eta)^2}\Big).$$

We put $\zeta=r\eta$, multiply (9.16) by $(1-r)^2$ and let $r\to 1$. In the limit, we find

$$q_n(\eta)\left|\frac{1}{2}(1-\alpha)^2+\frac{1}{12}(1-\alpha^2)\right|\leqq\frac{1}{4}q_n(\eta)\qquad\text{if }\alpha>0,$$

(9.19)

$$q_n(\eta)\left|-\alpha^2+\frac{1}{2}(1-\alpha)^2+\frac{1}{12}(1-\alpha^2)\right|\leqq\frac{1}{4}q_n(\eta)\qquad\text{if }\alpha<0.$$

Thus, supposing $q_n(\eta)\neq 0$, we conclude

$$|(5\alpha-7)(\alpha-1)|\leqq 3\qquad\text{if }\alpha>0,$$

$$|7\alpha^2+12\alpha-7|\leqq 3\qquad\text{if }\alpha<0.$$

Hence, we find the restriction

(9.20) $$\tfrac{2}{5}\leqq\alpha\leqq 2\qquad\text{if }\alpha>0$$

and

$$-\frac{6+\sqrt{106}}{7}\leqq\alpha\leqq -2\qquad\text{if }\alpha<0.$$

The last estimate shows that if $q_n(\eta)\neq 0$, the function $f(z)$ can have only a double pole at η if it becomes infinite there.

If $q_n(\eta)\neq 0$, then the differential equation (9.9) implies that $\alpha\leqq 1$. Let us assume now that $0<\alpha<1$ and $q_n(\eta)\neq 0$. Then we infer from the differential equation (9.9) that the polynomial $P_n(t)$ has a zero at $t=1/f(\eta)$. Hence, let us set up

$$P_n\left(\frac{1}{f(\zeta)}\right)=A[f(\zeta)-f(\eta)]^p+o([f(\zeta)-f(\eta)]^p).$$

From (9.18) follows

$$P_n\left(\frac{1}{f(\zeta)}\right)=B(\zeta-\eta)^{\alpha p}+\cdots,$$

and since the factor $\varphi(\zeta)^2\sim(\zeta-\eta)^{2\alpha-2}$ cancels the zero of $P_n(1/f(\zeta))$ at η, we conclude that

$$\alpha(2+p)=2,\qquad \alpha=\frac{2}{2+p},$$

where p is a positive integer. From (9.20) follows then that p can take only the values 1, 2 and 3 and hence α can only be $\tfrac{2}{3}$, $\tfrac{1}{2}$ and $\tfrac{2}{5}$. This is an additional restriction on the singularities of $f(z)$.

Consider the inequality (9.16) in the case that ζ approaches a point η which corresponds to an endpoint of the boundary slit. Here $q_n(\eta)=q_n'(\eta)=0$. Hence, near η we have the development

$$q_n(\zeta)=\frac{1}{2}(\zeta-\eta)^2q_n''(\eta)+o((\zeta-\eta)^2).$$

On the other hand, using the same calculations that lead to (9.19) and remembering that now $\alpha = 2$, we find on radial approach $\zeta = r\eta$

$$\left| -\frac{3}{8}\eta^2 q_n''(\eta) - K_n(\bar{\eta}) + n\,\mathrm{Re}\{\bar{\eta}K_n'(\bar{\eta})\} + ni\,\mathrm{Im}\{K_n(\bar{\eta})\} + \frac{n^2+1}{2}a_n \right|$$

$$\leqq \mathrm{Re}\left\{\frac{1}{8}\eta^2 q_n''(\eta) - n\bar{\eta}K_n'(\bar{\eta}) + K_n(\bar{\eta}) - na_n\right\}.$$

In particular, we find the consequence

$$0 \leqq \mathrm{Re}\left\{\frac{1}{2}\eta^2 q_n''(\eta) - 2n\bar{\eta}K_n'(\bar{\eta}) + 2K_n(\bar{\eta}) - \frac{(n+1)^2}{2}a_n\right\}.$$

Since $q_n(\eta) = K_n(\bar{\eta}) + \bar{K}_n(\eta) - (n+1)a_n = 0$, we have

$$2\mathrm{Re}\{K_n(\bar{\eta})\} = (n+1)a_n.$$

Thus finally

$$\frac{n^2-1}{2}a_n \leqq \mathrm{Re}\left\{\frac{1}{2}\eta^2 q_n''(\eta) - 2n\eta\bar{K}_n'(\eta)\right\}. \tag{9.21}$$

This estimate is an equality for the Koebe function.

Additional information is obtained in the case that $f(z)$ becomes infinite at η. It is well known that the function

$$T_n(z) = \frac{z}{(1-z^{n-1})^{2/(n-1)}} = z + \frac{2}{n-1}z^n + \cdots \tag{9.22}$$

satisfies the functional-differential equation (9.9) for the extremum function for $\mathrm{Re}\{a_n\}$. Consider then the corresponding function (9.8):

$$q_n(\zeta) = \frac{1}{\zeta^{n-1}} + 2 + \zeta^{n-1}.$$

The poles of $T_n(z)$ lie all at the roots η of the equation $\eta^{n-1} = 1$ and there $q_n(\eta) = 4$. Since $T_n(z)$ has thus infinities of order $2/(n-1)$ at points η where $q_n(\eta) \neq 0$, it cannot be an extremum function for $n > 2$. Since $2/(n-1) < n$ in this case, the result is rather obvious; but it is interesting to see that the second variation complements the first variation to eliminate spurious solutions.

10. The analytic character of the terms in the second variation

We shall now discuss the two sides of the inequality (9.16). Let

$$L(z) = 1 - \frac{\varphi(z)^2}{f(z)^2} + \frac{1}{2}\left(\frac{z\varphi'(z)}{\varphi(z)}\right)^2 + \frac{1}{6}z^2\{f, z\}. \tag{10.1}$$

We will show that the left side in (9.16)

$$V_n(\zeta)=q_n(\zeta)L(\zeta)-\frac{1}{2}\frac{\zeta\varphi'(\zeta)}{\varphi(\zeta)}\zeta q_n'(\zeta)-\varphi(\zeta)^2 S_n\left(\frac{1}{f(\zeta)},\frac{1}{\zeta}\right)+\frac{n^2+1}{2}a_n$$

(10.2)

$$-K_n\left(\frac{1}{\zeta}\right)+\frac{n}{2}\left[\frac{1}{\zeta}K_n'\left(\frac{1}{\zeta}\right)+\zeta\bar{K}_n'(\zeta)+K_n\left(\frac{1}{\zeta}\right)-\bar{K}_n(\zeta)\right]$$

is a regular analytic function of ζ for all $|\zeta|<1$; that is, that all its singularities at $\zeta=0$ cancel out. To do so, we need to recall a few important identities. We start with the identity

$$\log\frac{f(z)-f(\zeta)}{z-\zeta}=\sum_{\mu,\nu=0}^{\infty}c_{\mu\nu}z^{\mu}\zeta^{\nu} \tag{10.3}$$

which defines the Grunsky matrix $((c_{\mu\nu}))$, which occurs so often in the coefficient problem for univalent functions. Next, we define the Faber polynomials $F_n(t)$ by means of the generating function

$$\log(1-tf(z))=-\sum_{n=1}^{\infty}\frac{1}{n}F_n(t)z^n. \tag{10.4}$$

Comparing with (10.3), we find easily

$$F_n\left(\frac{1}{f(\zeta)}\right)=\frac{1}{\zeta^n}-n\sum_{\nu=0}^{\infty}c_{n\nu}\zeta^{\nu},\qquad n\geqq 1. \tag{10.5}$$

On the other hand, differentiation of (10.4) with respect to t yields

$$\frac{f(z)}{1-tf(z)}=\sum_{n=1}^{\infty}\frac{1}{n}F_n'(t)z^n, \tag{10.6}$$

whence, in view of (9.2)

$$\frac{1}{n}F_n'(t)=a_n+tP_n(t). \tag{10.7}$$

Finally, we differentiate (10.5) in ζ to obtain

$$\frac{1}{n}F_n'\left(\frac{1}{f(\zeta)}\right)\frac{\zeta f'(\zeta)}{f(\zeta)^2}=\frac{1}{\zeta^n}+\sum_{\nu=1}^{\infty}\nu c_{n\nu}\zeta^{\nu}. \tag{10.8}$$

Combining (10.7) and (10.8), we obtain the identity

$$\frac{\varphi(\zeta)}{f(\zeta)^2}\left(a_n+\frac{1}{f(\zeta)}P_n\left(\frac{1}{f(\zeta)}\right)\right)=\frac{1}{\zeta^n}+\sum_{\nu=1}^{\infty}\nu c_{n\nu}\zeta^{\nu}, \tag{10.9}$$

which is valid for all functions $f(\zeta)$ which are analytic in a neighborhood of the origin and are normalized there.

We can now calculate the term $\varphi(\zeta)^2S_n(1/f(\zeta),1/\zeta)$ by use of the definition (9.12) and the identity (10.9):

$$\varphi(\zeta)^2S_n\left(\frac{1}{f(\zeta)},\frac{1}{\zeta}\right)=\frac{\varphi(\zeta)}{\zeta^n}\frac{(n+1)n}{2}-\frac{\varphi(\zeta)^2}{f(\zeta)^2}K_n\left(\frac{1}{\zeta}\right)+\frac{\varphi(\zeta)}{\zeta^n}\sum_{k=1}^{n}\sum_{\nu=1}^{\infty}\nu k c_{\nu k}\zeta^{\nu+k}. \tag{10.10}$$

On the other hand, we have the identity

$$\lim_{z\to\zeta}\left[\frac{f'(z)f'(\zeta)}{[f(z)-f(\zeta)]^2}-\frac{1}{(z-\zeta)^2}\right]=\frac{1}{6}\{f,\zeta\} \tag{10.11}$$

and in view of (10.3),

$$\sum_{\mu,\nu=0}^{\infty} c_{\mu\nu}\mu\nu\zeta^{\mu+\nu}=\frac{1}{6}\zeta^2\{f,\zeta\}. \tag{10.12}$$

Thus, we arrive at the remarkable asymptotic relation

$$\varphi(\zeta)^2 S_n\left(\frac{1}{f(\zeta)},\frac{1}{\zeta}\right)=\frac{\varphi(\zeta)}{\zeta^n}\left[\frac{n(n+1)}{2}+\frac{1}{6}\zeta^2\{f,\zeta\}\right]-\frac{\varphi(\zeta)^2}{f(\zeta)^2}K_n\left(\frac{1}{\zeta}\right)+O(\zeta^3). \tag{10.13}$$

Observe from (9.8) and (9.3) also the asymptotic relations

$$q_n(\zeta)=\frac{\varphi(\zeta)}{\zeta^n}+O(1),\qquad K_n\left(\frac{1}{\zeta}\right)=\frac{\varphi(\zeta)}{\zeta^n}+O(\zeta). \tag{10.14}$$

Combining all these terms in $V_n(\zeta)$, one finds that all infinities at $\zeta=0$ cancel as asserted. We find thus a series development

$$V_n(\zeta)=\sum_{l=0}^{\infty}A_l^{(n)}\zeta^l \tag{10.15}$$

where the $A_l^{(n)}$ depend upon the coefficients a_n of the extremum function $f(z)$.
Consider next the right-hand term in (9.16):

$$\begin{aligned}W_n(\zeta,\bar\zeta)=&\frac{|\zeta|^2}{(1-|\zeta|^2)^2}q_n(\zeta)+\varphi(\zeta)^2S_n\left(\frac{1}{f(\zeta)},\bar\zeta\right)-n\frac{|\zeta|^2}{1-|\zeta|^2}\left[K_n\left(\frac{1}{\zeta}\right)-\bar K_n(\zeta)\right]\\&-nq_n(\zeta)+\bar K_n(\zeta)-na_n.\end{aligned} \tag{10.16}$$

By virtue of (10.9), we find

$$\begin{aligned}\varphi(\zeta)^2S_n\left(\frac{1}{f(\zeta)},\bar\zeta\right)&=\frac{\varphi(\zeta)}{\zeta^n}\sum_{k=1}^{n}k|\zeta|^{2(n-k)}+O(1)\\&=\frac{\varphi(\zeta)}{\zeta^n}\sum_{\nu=0}^{n-1}(n-\nu)|\zeta|^{2\nu}+O(1).\end{aligned} \tag{10.17}$$

Using again the asymptotic formula (10.14), we find that

$$W_n(\zeta,\bar\zeta)=\frac{\varphi(\zeta)}{\zeta^n}\left[\sum_{\nu=1}^{\infty}(\nu-n)|\zeta|^{2\nu}+\sum_{\nu=0}^{n-1}(n-\nu)|\zeta|^{2\nu}-n\right]+O(1), \tag{10.18}$$

which proves that $W_n(\zeta,\bar\zeta)$ remains finite at $\zeta=0$.
We may develop $\mathrm{Re}\{W_n(\zeta,\bar\zeta)\}$ near the origin as

$$\mathrm{Re}\{W_n(\zeta,\bar\zeta)\}=\mathrm{Re}\left\{\sum_{l=0}^{\infty}B_l^{(n)}(|\zeta|)\zeta^l\right\} \tag{10.19}$$

where the $B_l^{(n)}(|\zeta|)$ depend again on the coefficients of the extremum function $f(z)$. If $f(z)$ is real for z real and if x and ζ are real, observe that the variation (2.12) carries the real axis into itself and preserves the point at ∞. Hence, the Koebe function $k(z)=z/(1-z)^2$ goes over into itself, and by (9.5) we have in this case

$$V_n(\zeta)=W_n(\zeta,\zeta) \qquad \text{for real } \zeta \quad \text{and} \quad f(z)=k(z). \tag{10.20}$$

This is an interesting test for the dependence of the $A_l^{(n)}$ and $B_l^{(n)}(|\zeta|)$ upon the coefficients a_n.

The inequality

$$\left|\sum_{l=0}^{\infty} A_l^{(n)}\zeta^l\right| \leqq \operatorname{Re}\left\{\sum_{l=0}^{\infty} B_l^{(n)}(|\zeta|)\zeta^l\right\} \qquad \text{for } |\zeta|<1 \tag{10.21}$$

implies many inequalities that add conditions on the coefficients of the extremum function. Observe that in view of the differential equation (9.9) all coefficients a_ν with $\nu>n$ can be recursively expressed in terms of $a_2, a_3, \cdots, a_n$. These are the so-called "Marty relations". Hence, all estimates in (10.21) lead to inequalities involving the first n coefficients of the extremum function. Thus, a large set of necessary conditions is added by considering the second variation.

We give the first few terms in the developments of $V_n(\zeta)$ and $\operatorname{Re}\{W_n(\zeta,\bar\zeta)\}$:

$$\begin{gathered} A_0^{(n)}=A_1^{(n)}=0, \\ A_2^{(n)}=\frac{(n+1)(n+2)}{2}a_{n+2}-3a_3a_n-2a_2(n-1)\bar a_{n-1}+\frac{(n-1)(n-2)}{2}\bar a_{n-2}, \end{gathered} \tag{10.22}$$

and

$$\begin{gathered} B_0^{(n)}(|\zeta|)=(n^2-1)a_n|\zeta|^2-2(n-1)a_2a_{n-1}|\zeta|^2+O(|\zeta|^4), \\ B_1^{(n)}(|\zeta|)=O(|\zeta|^2), \qquad B_2^{(n)}(|\zeta|)=O(|\zeta|^2). \end{gathered} \tag{10.23}$$

It may also be observed that one can easily derive analogous necessary extremum conditions from the second variation in the case of an arbitrary continuous linear extremum problem. These conditions will be of particular interest when the Koebe function is not the extremum function.

11. Further necessary maximum conditions for the coefficient problem in S

Had we chosen the variation (2.7) of Section 2 with N arbitrary points ζ_ν in the unit disk and N complex parameters x_ν, a lengthy but straightforward calculation would lead to the following inequality:

$$\operatorname{Re}\left\{\sum_{\mu,\nu=1}^{N} y_\mu \bar y_\nu L_n(\zeta_\mu,\bar\zeta_\nu)\right\} \geqq \left|\sum_{\mu,\nu=1}^{N} y_\mu y_\nu M_n(\zeta_\mu,\zeta_\nu)\right| \tag{11.1}$$

where

$$L_n(\zeta_\mu,\bar\zeta_\nu)=q_n(\zeta_\mu)\frac{\zeta_\mu\bar\zeta_\nu}{(1-\zeta_\mu\bar\zeta_\nu)^2}+\varphi(\zeta_\mu)^2S_n\left(\frac{1}{f(\zeta_\mu)},\bar\zeta_\nu\right)-nq_n(\zeta_\mu)$$

(11.2)

$$-n\frac{\zeta_\mu\bar\zeta_\nu}{1-\zeta_\mu\bar\zeta_\nu}\left[K_n\left(\frac{1}{\zeta_\mu}\right)-K_n(\bar\zeta_\nu)\right]+K_n(\bar\zeta_\nu)-na_n$$

and

$$2M_n(\zeta_\mu,\zeta_\nu)=n\frac{\zeta_\mu\zeta_\nu}{\zeta_\mu-\zeta_\nu}\left[K_{n+1}\left(\frac{1}{\zeta_\mu}\right)-K_{n+1}\left(\frac{1}{\zeta_\nu}\right)\right]+K_n\left(\frac{1}{\zeta_\mu}\right)+K_n\left(\frac{1}{\zeta_\nu}\right)$$

$$+\varphi(\zeta_\mu)^2S_n\left(\frac{1}{f(\zeta_\mu)},\frac{1}{\zeta_\nu}\right)+\varphi(\zeta_\nu)^2S_n\left(\frac{1}{f(\zeta_\nu)},\frac{1}{\zeta_\mu}\right)$$

(11.3)

$$-q_n(\zeta_\nu)\left[\frac{\varphi(\zeta_\mu)^2}{[f(\zeta_\mu)-f(\zeta_\nu)]^2}-\frac{\zeta_\mu\zeta_\nu}{(\zeta_\mu-\zeta_\nu)^2}+\left(1-\frac{\varphi(\zeta_\mu)^2}{f(\zeta_\mu)^2}\right)\right]$$

$$-q_n(\zeta_\mu)\left[\frac{\varphi(\zeta_\nu)^2}{[f(\zeta_\mu)-f(\zeta_\nu)]^2}-\frac{\zeta_\mu\zeta_\nu}{(\zeta_\mu-\zeta_\nu)^2}+\left(1-\frac{\varphi(\zeta_\nu)^2}{f(\zeta_\nu)^2}\right)\right]$$

$$-a_n-n\frac{\zeta_\mu\zeta_\nu}{\zeta_\mu-\zeta_\nu}[\bar K_{n-1}(\zeta_\mu)-\bar K_{n-1}(\zeta_\nu)].$$

In the case $\mu=\nu$ one has to interpret $M_n(\zeta_\mu,\zeta_\nu)$ as the L'Hôpital limit. It coincides, of course, with $-V_n(\zeta_\nu)$.

We may assert again that the inequality (11.1) becomes an equality for the Koebe function if we specialize the variation (2.5) to the symmetric expression

$$v_1[f(z)]=\sum_{\nu=1}^{N}\left(\frac{x_\nu f(z)}{[f(z)-f(\zeta_\nu)]f(\zeta_\nu)}+\frac{\bar x_\nu f(z)}{[f(z)-f(\bar\zeta_\nu)]f(\bar\zeta_\nu)}\right). \tag{11.4}$$

In this case it is geometrically evident that $\hat f(z)\equiv f(z)$ if f is the Koebe function.

We have obtained an inequality between a quadratic form and a hermitian form as a necessary extremum condition. Such inequalities occur frequently in the theory of univalent functions.

12. Some applications

We know that the extremum function maps the circle $|z|<1$ onto a slit domain bounded by an analytic arc. We may then find points η_1 and η_2 in the unit periphery $|z|=1$ where $f(z)$ is analytic and such that

$$f(\eta_1)=f(\eta_2). \tag{12.1}$$

By differentiation in η_1 and changing η_2 such that (12.1) remains true, we find

$$f'(\eta_1)=f'(\eta_2)\frac{d\eta_2}{d\eta_1}.$$

Observe that if we put $\eta_\nu = e^{i\theta_\nu}$, we have

$$\frac{d}{d\theta_\nu} f(\eta_\nu) = i\eta_\nu f'(\eta_\nu)$$

so that $i\eta_\nu f'(\eta_\nu)$ are in tangential directions, but opposite to each other. Define

$$\rho = \left|\frac{f'(\eta_1)}{f'(\eta_2)}\right| \quad \text{and} \quad \dot{\rho} = \frac{d}{d\theta_1}\rho(e^{i\theta_1}). \tag{12.2}$$

Then we have

$$\varphi(\eta_1) = \eta_1 f'(\eta_1) = -\rho\eta_2 f'(\eta_2) = -\rho\varphi(\eta_2). \tag{12.3}$$

It follows from this identity that

$$\frac{d\eta_2}{d\eta_1} = -\rho\frac{\eta_2}{\eta_1} \tag{12.4}$$

and

$$\frac{\dot{\rho}}{i\rho} = \frac{\eta_1\varphi'(\eta_1)}{\varphi(\eta_1)} + \rho\frac{\eta_2\varphi'(\eta_2)}{\varphi(\eta_2)}. \tag{12.5}$$

Next choose $\zeta_1 = r_1\eta_1$, $\zeta_2 = r_2\eta_2$, and let $r_\nu \to 1$. We find that the coefficients in the inequality (11.1) tend to infinity. The worst terms in (11.2) are

$$q_n(\zeta_\nu)\frac{r_\nu^2}{(1-r_\nu^2)^2} = Q_\nu \tag{12.6}$$

and in (11.3) the terms

$$P = -\left[q_n(\zeta_2)\frac{\varphi(\zeta_1)^2}{[f(\zeta_1)-f(\zeta_2)]^2} + q_n(\zeta_1)\frac{\varphi(\zeta_2)^2}{[f(\zeta_1)-f(\zeta_2)]^2}\right]. \tag{12.7}$$

If we replace (11.1) by the weaker inequality

$$\operatorname{Re}\left\{\sum_{\mu,\nu=1}^{2} y_\mu \bar{y}_\nu L_n(\zeta_\mu, \bar{\zeta}_\nu) + y_\mu y_\nu M_n(\zeta_\mu, \zeta_\nu)\right\} \geqq 0 \tag{12.8}$$

and choose the change of the r_ν properly, we can achieve that the various terms tending to infinity compensate each other to yield finite limits. Indeed, if we choose

$$y_1 = \frac{1}{\rho}e^{i\alpha}, \quad y_2 = \rho e^{-i\alpha}, \quad (1-r_2) = \rho(1-r_1) \tag{12.9}$$

where ρ is defined in (12.2), one verifies easily that

$$\lim_{r_1\to 1} \operatorname{Re}\{|y_1|^2 Q_1 + |y_2|^2 Q_2 + (y_1 y_2)P\} = A \tag{12.10}$$

exists and is finite. One has here to use the facts that (12.5) is purely imaginary and that because of (9.9), (12.1) and (12.3) one has

$$\frac{q_n(\eta_1)}{q_n(\eta_2)} = \frac{\varphi(\eta_1)^2}{\varphi(\eta_2)^2} = \rho^2. \tag{12.11}$$

Just as in (12.5), it follows from this identity that

$$\frac{2\dot{\rho}}{i\rho} = \frac{\eta_1 q_n'(\eta_1)}{q_n(\eta_1)} + \rho\frac{\eta_2 q_n'(\eta_2)}{q_n(\eta_2)}. \tag{12.12}$$

We will write for short

$$\begin{gathered} q_n(\eta_1) = q_1, \quad q_n(\eta_2) = q_2, \quad \varphi(\eta_1) = \varphi_1, \quad \varphi(\eta_2) = \varphi_2, \\ S_n\left(\frac{1}{f}, \frac{1}{\eta_1}\right) = S_1, \qquad S_n\left(\frac{1}{f}, \frac{1}{\eta_2}\right) = S_2. \end{gathered} \tag{12.13}$$

After laborious calculations one finds

$$\begin{aligned} A = {} & \frac{3}{32} q_2 \left[(1-\rho)^2 + \left(\frac{\dot{\rho}}{\rho}\right)^2\right] - \frac{1}{2} q_2 \rho \operatorname{Re}\left[\frac{\eta_1 q_1'}{q_1}\frac{\eta_2 \varphi_2'}{\varphi_2} + \frac{\eta_2 q_2'}{q_2}\frac{\eta_1 \varphi_1'}{\varphi_1}\right. \\ & \left. - 2\frac{\eta_1 \varphi_1'}{\varphi_1}\frac{\eta_2 \varphi_2'}{\varphi_2}\right] - \frac{1}{6}\operatorname{Re}[q_2 \eta_1^2\{f, \eta_1\} + q_1 \eta_2^2\{f, \eta_2\}]. \end{aligned} \tag{12.14}$$

The remaining terms in the inequality have finite limit values. A long rearrangement of terms leads to the final estimate

$$\operatorname{Re}\{T_0 + e^{2i\alpha} T_1 + e^{-2i\alpha} T_2\} \geqq 0 \tag{12.15}$$

where

$$\begin{aligned} T_1 = {} & \varphi_1^2 S_2 + \varphi_2^2 S_1 + q_1 \frac{\eta_1 \eta_2}{(\eta_1 - \eta_2)^2} - n q_1 + n\frac{\eta_1}{\eta_1 - \eta_2}\left(K_n\left(\frac{1}{\eta_1}\right) - K_n\left(\frac{1}{\eta_2}\right)\right) \\ & + K_n\left(\frac{1}{\eta_2}\right) + \frac{1}{\rho^2} K_n\left(\frac{1}{\eta_1}\right) - \frac{1}{6} q_2 \eta_1^2\{f, \eta_1\} - \frac{1}{2} q_2 \left(\frac{\eta_1 \varphi_1'}{\varphi_1}\right)^2 + \frac{1}{2} q_2 \frac{\eta_1 q_1'}{q_1}\frac{\eta_1 \varphi_1'}{\varphi_1} \\ & - \frac{n}{2}\left[\frac{1}{\rho^2}\left(\frac{1}{\eta_1} K_n'\left(\frac{1}{\eta_1}\right) + K_n\left(\frac{1}{\eta_1}\right)\right) + \rho^2\left(\frac{1}{\eta_2} K_n'\left(\frac{1}{\eta_2}\right) - K_n\left(\frac{1}{\eta_2}\right)\right)\right] \\ & - q_2\left(1 - \frac{\varphi_1^2}{f^2}\right) - \left(n + \frac{1}{2}\frac{1}{\rho^2} + \frac{n^2}{2}\rho^2\right) a_n \end{aligned} \tag{12.16}$$

and

$$\begin{aligned} T_2 = {} & \varphi_1^2 S_2 + \varphi_2^2 S_1 + q_2 \frac{\eta_1 \eta_2}{(\eta_1 - \eta_2)^2} - n q_2 + n\frac{\eta_2}{\eta_1 - \eta_2}\left(K_n\left(\frac{1}{\eta_1}\right) - K_n\left(\frac{1}{\eta_2}\right)\right) \\ & + K_n\left(\frac{1}{\eta_1}\right) + \rho^2 K_n\left(\frac{1}{\eta_2}\right) - \frac{1}{6} q_1 \eta_2^2\{f, \eta_2\} - \frac{1}{2} q_1 \left(\frac{\eta_2 \varphi_2'}{\varphi_2}\right)^2 + \frac{1}{2} q_1 \frac{\eta_2 q_2'}{q_2}\frac{\eta_2 \varphi_2'}{\varphi_2} \\ & - \frac{n}{2}\left[\rho^2\left(\frac{1}{\eta_2} K_n'\left(\frac{1}{\eta_2}\right) + K_n\left(\frac{1}{\eta_2}\right)\right) + \frac{1}{\rho^2}\left(\frac{1}{\eta_1} K_n'\left(\frac{1}{\eta_1}\right) - K_n\left(\frac{1}{\eta_1}\right)\right)\right] \\ & - q_1\left(1 - \frac{\varphi_2^2}{f^2}\right) - \left(n + \frac{1}{2}\rho^2 + \frac{n^2}{2\rho^2}\right) a_n. \end{aligned} \tag{12.17}$$

Finally,

$$T_0 = T_1 + T_2 + \frac{1}{2}(n-1)^2\left(\rho - \frac{1}{\rho}\right)^2 a_n + \frac{3}{32}(1-\rho)^2 q_2 + \frac{19}{32}\left(\frac{\dot\rho}{\rho}\right)^2 q_2. \tag{12.18}$$

Thus, (12.15) implies

$$\begin{aligned}\operatorname{Re}\{(1+e^{2i\alpha})((T_1+\bar T_2)\} + \frac{1}{2}(n-1)^2\left(\rho-\frac{1}{\rho}\right)^2 a_n \\ + \frac{3}{32}(1-\rho)^2 q_2 + \frac{19}{32}\left(\frac{\dot\rho}{\rho}\right)^2 q_2 \geqq 0,\end{aligned} \tag{12.19}$$

for arbitrary α. This is a complicated but unexpected relation between the points η_1 and η_2 which map into the same points of the boundary slit.

As another application, let us choose $(2n)$ arbitrary points ζ_ν in the unit disk and restrict the coefficients y_ν by the conditions

$$\sum_{\nu=1}^{2n} y_\nu \zeta_\nu^{-k} = \sum_{\nu=1}^{2n} y_\nu \zeta_\nu^{k} = 0, \qquad k = 0, 1, \cdots, n-1. \tag{12.20}$$

Then we have

$$\sum_{\nu=1}^{2n} y_\nu S_n\left(\frac{1}{f(\zeta)}, \frac{1}{\zeta_\nu}\right) = \sum_{\nu=1}^{2n} \bar y_\nu S_n\left(\frac{1}{f(\zeta)}, \bar\zeta_\nu\right) = 0$$

and

$$\sum_{\nu=1}^{2n} y_\nu q_n(\zeta_\nu) = \sum_{\nu=1}^{2n} y_\nu K_n\left(\frac{1}{\zeta_\nu}\right) = \sum_{\nu=1}^{2n} \bar y_\nu K_n(\bar\zeta_\nu) = 0.$$

This simplifies the inequality (11.1) considerably. We find

$$\begin{aligned}&2\operatorname{Re}\left\{\sum_{\mu,\nu=1}^{2n} y_\mu \bar y_\nu \left[\frac{\zeta_\mu\bar\zeta_\nu}{(1-\zeta_\mu\bar\zeta_\nu)^2} q_n(\zeta_\mu) - n\frac{\zeta_\mu\bar\zeta_\nu}{1-\zeta_\mu\bar\zeta_\nu}\left(K_n\left(\frac{1}{\zeta_\mu}\right) - K_n(\bar\zeta_\nu)\right)\right]\right\} \\ &\geqq \left|\sum_{\mu,\nu=1}^{2n} y_\mu y_\nu\left[q_n(\zeta_\mu)\left(\frac{\zeta_\mu\zeta_\nu}{(\zeta_\mu-\zeta_\nu)^2} - \frac{\varphi(\zeta_\nu)^2}{[f(\zeta_\mu)-f(\zeta_\nu)]^2}\right)\right.\right. \\ &\quad + q_n(\zeta_\nu)\left(\frac{\zeta_\mu\zeta_\nu}{(\zeta_\mu-\zeta_\nu)^2} - \frac{\varphi(\zeta_\mu)^2}{[f(\zeta_\mu)-f(\zeta_\nu)]^2}\right) \\ &\quad\left.\left. + \frac{n}{\zeta_\mu-\zeta_\nu}\left(\zeta_\nu K_n\left(\frac{1}{\zeta_\mu}\right) - \zeta_\mu K_n\left(\frac{1}{\zeta_\nu}\right) - \zeta_\nu\bar K_n(\zeta_\mu) + \zeta_\mu\bar K_n(\zeta_\nu)\right)\right]\right|.\end{aligned} \tag{12.21}$$

Since

$$\frac{1}{\zeta_\mu-\zeta_\nu}\left(\zeta_\nu K_n\left(\frac{1}{\zeta_\mu}\right) - \zeta_\mu K_n\left(\frac{1}{\zeta_\nu}\right)\right) = -\sum_{k=1}^{n} k a_k \sum_{\alpha=0}^{n-k} \zeta_\mu^{-\alpha}\zeta_\nu^{\alpha-(n-k)},$$

$$\frac{1}{\zeta_\mu-\zeta_\nu}\left(\zeta_\nu\bar K_n(\zeta_\mu) - \zeta_\mu\bar K_n(\zeta_\nu)\right) = \sum_{k=1}^{n} k\bar a_k \sum_{\alpha=1}^{n-k-1} \zeta_\mu^{\alpha}\zeta_\nu^{n-k-\alpha},$$

$$\frac{\zeta_\mu\bar\zeta_\nu}{1-\zeta_\mu\bar\zeta_\nu}\left(K_n\left(\frac{1}{\zeta_\mu}\right) - K_n(\bar\zeta_\nu)\right) = \sum_{k=1}^{n} k a_k \sum_{\alpha=1}^{n-k} \zeta_\mu^{\alpha-(n-k)}\bar\zeta_\nu^{\alpha},$$

we can use (12.20) to simplify (12.21) to

$$\sum_{\mu,\nu=1}^{2n} y_\mu \bar{y}_\nu \frac{\zeta_\mu \bar{\zeta}_\nu}{(1-\zeta_\mu \bar{\zeta}_\nu)^2}[q_n(\zeta_\mu)+\overline{q_n(\zeta_\nu)}]$$

$$\geqq \left| \sum_{\mu,\nu=1}^{2n} y_\mu y_\nu \left[q_n(\zeta_\mu)\left(\frac{\varphi(\zeta_\nu)^2}{[f(\zeta_\mu)-f(\zeta_\nu)]^2}-\frac{\zeta_\mu\zeta_\nu}{(\zeta_\mu-\zeta_\nu)^2}\right)\right.\right. \tag{12.22}$$

$$\left.\left. + q_n(\zeta_\nu)\left(\frac{\varphi(\zeta_\mu)^2}{[f(\zeta_\mu)-f(\zeta_\nu)]^2}-\frac{\zeta_\mu\zeta_\nu}{(\zeta_\mu-\zeta_\nu)^2}\right)\right]\right|.$$

This estimate is very similar to the well-known Grunsky type inequalities

$$\sum_{\mu,\nu=1}^{N} y_\mu \bar{y}_\nu \frac{\zeta_\mu \bar{\zeta}_\nu}{(1-\zeta_\mu \bar{\zeta}_\nu)^2} \geqq \left| \sum_{\mu,\nu=1}^{N} y_\mu y_\nu \left(\frac{\varphi(\zeta_\mu)\varphi(\zeta_\nu)}{[f(\zeta_\mu)-f(\zeta_\nu)]^2}-\frac{\zeta_\mu\zeta_\nu}{(\zeta_\mu-\zeta_\nu)^2}\right)\right|, \tag{12.23}$$

valid for all univalent functions and all $y_1, \cdots, y_N$. However, it seems to be not quite of the same kind.

Let $p(x)=\sum_{\nu=0}^{2n} \alpha_\nu x^\nu$ be a polynomial of degree $2n$ whose roots ζ_ν $(\nu=1,\cdots,2n)$ lie all inside the unit disk. We shall use the reduction formula

$$\zeta_\nu^m = \sum_{k=0}^{2n-1} A_{mk}\zeta_\nu^k \tag{12.24}$$

where the A_{mk} are well-known symmetric expressions in terms of the α_ν. First determine $2n$ numbers y_ν which satisfy the conditions (12.20) and let $x_\nu = y_\nu \zeta_\nu^{-n}$. Then

$$\sum_{\nu=1}^{2n} x_\nu \zeta_\nu^k = 0 \qquad \text{for } k=1,\cdots,2n-1$$

and we may impose the normalization

$$\sum_{\nu=1}^{2n} x_\nu = 1.$$

Now compute

$$\sum_{\mu,\nu=1}^{2n} y_\mu \bar{y}_\nu \frac{\zeta_\mu \bar{\zeta}_\nu}{(1-\zeta_\mu\bar{\zeta}_\nu)^2}\zeta_\mu^k = \sum_{l=1}^{\infty} l \sum_{\mu=1}^{2n} x_\mu \zeta_\mu^{l+n+k}\overline{\left(\sum_{\nu=1}^{2n} x_\nu \zeta_\nu^{l+n}\right)}$$

$$= \sum_{l=1}^{\infty} l A_{l+n+k,0}\bar{A}_{l+n,0}.$$

If we define

$$\pi_k(\alpha) = \sum_{l=1}^{\infty} l A_{l+n+k,0}\bar{A}_{l+n,0}, \tag{12.25}$$

then $\pi_k(\alpha)$ is a symmetric function of the coefficients α_ν of the polynomial $p(x)$, and we find

$$\sum_{\mu,\nu=1}^{2n} y_\mu \bar{y}_\nu q_n(\zeta_\mu)\frac{\zeta_\mu \bar{\zeta}_\nu}{(1-\zeta_\mu\bar{\zeta}_\nu)^2} = \sum_{k=1}^{n} k a_k \pi_{k-n}(\alpha) + (n-1)a_n \pi_0(\alpha) \tag{12.26}$$
$$+ \sum_{k=1}^{n} k \bar{a}_k \pi_{n-k}(\alpha).$$

By (12.22), this expression must have nonnegative real part for every choice of the $n+1$ parameters α_ν, subjected only to the condition that all roots of $p(x)$ lie inside the unit disk. This is an interesting restriction on the coefficients a_k of the extremum function.

13. A general formula for the second variation of a_n

The estimates of Section 11 suggest the consideration of the two kernels

$$L_n(t,\bar{\tau}) = q_n(t)\frac{t\bar{\tau}}{(1-t\bar{\tau})^2} + \varphi(t)^2 S_n\left(\frac{1}{f(t)}, \bar{\tau}\right) - nq_n(t) \tag{13.1}$$
$$- n\frac{t\bar{\tau}}{1-t\bar{\tau}}\left[K_n\left(\frac{1}{t}\right) - K_n(\bar{\tau})\right] + K_n(\bar{\tau}) - na_n$$

and

$$2M_n(t,\tau) = n\frac{t\tau}{t-\tau}\left[K_{n+1}\left(\frac{1}{t}\right) - K_{n+1}\left(\frac{1}{\tau}\right)\right] + K_n\left(\frac{1}{t}\right) + K_n\left(\frac{1}{\tau}\right)$$
$$+ \varphi(t)^2 S_n\left(\frac{1}{f(t)}, \frac{1}{\tau}\right) + \varphi(\tau)^2 S_n\left(\frac{1}{f(\tau)}, \frac{1}{t}\right)$$
$$- q_n(t)\left[\frac{\varphi(\tau)^2}{[f(t)-f(\tau)]^2} - \frac{t\tau}{(t-\tau)^2} + \left(1 - \frac{\varphi(\tau)^2}{f(\tau)^2}\right)\right] \tag{13.2}$$
$$- q_n(\tau)\left[\frac{\varphi(t)^2}{[f(t)-f(\tau)]^2} - \frac{t\tau}{(t-\tau)^2} + \left(1 - \frac{\varphi(t)^2}{f(t)^2}\right)\right]$$
$$- a_n - n\frac{t\tau}{t-\tau}(\bar{K}_{n-1}(t) - \bar{K}_{n-1}(\tau)).$$

It is easily verified by the same methods used in the case $L_n(t,\bar{t})$ and $M_n(t,t)$ that both kernels are analytic in their variables in the interior of the unit disk, even at the origin.

We can now express the inequality (11.1) as follows:

$$\mathrm{Re}\left\{\int_\Gamma\int_\Gamma L_n(t,\bar{\tau})a(t)\overline{a(\tau)}\,dt\,d\bar{\tau}\right\} \geqq \left|\int_\Gamma\int_\Gamma M_n(t,\tau)a(t)a(\tau)\,dt\,d\tau\right| \tag{13.3}$$

with

$$a(t) = \frac{1}{2\pi i}\sum_{\nu=1}^{N}\frac{y_\nu}{t-\zeta_\nu}. \tag{13.4}$$

The contour Γ may be taken as a circle slightly inside the unit circle. Since the ζ_ν in the unit disk and the parameters y_ν are quite arbitrary, we can approximate every function $a(t)$ which is analytic in $|t| \geqq 1$ and hence obtain the condition (13.3) for each such function. Since for every function $a(t)$ which is analytic inside the unit disk the above integrals vanish because of the Cauchy integral theorem, we obtain as a necessary maximum condition that the inequality (13.3) must hold for every function $a(t)$ which is analytic on $|t| = 1$.

REFERENCES

1. E. Bombieri, *On the local maximum property of the Koebe function*, Invent. Math. **4** (1967), 26–67.
2. P. L. Duren, *Coefficients of univalent functions*, Bull. Amer. Math. Soc. **83** (1977), 891–911.
3. P. L. Duren and M. Schiffer, *The theory of the second variation in extremum problems for univalent functions*, J. Analyse Math. **10** (1962/63), 193–252.
4. Y. Kubota, *Coefficients of meromorphic univalent functions*, Kōdai Math. Sem. Rep. **28** (1977), 253–261.
5. M. Schiffer, *A method of variation within the family of simple functions*, Proc. London Math. Soc. **44** (1938), 432–449.
6. M. Schiffer, *On the coefficients of simple functions*, Proc. London Math. Soc. **44** (1938), 450–452.
7. M. Schiffer, *Variation of the Green function and theory of the p-valued functions*, Amer. J. Math. **65** (1943), 341-360.
8. M. Schiffer, *On the coefficient problem for univalent functions*, Trans. Amer. Math. Soc. **134** (1968), 95-101.
9. G. Schober, *Univalent Functions — Selected Topics*, Springer-Verlag Lecture Notes in Math. # 478, (1975).
10. G. Schober, *Coefficients of inverses of meromorphic univalent functions*, Proc. Amer. Math. Soc. **67** (1977), 111-116.
11. G. Schober, *Coefficients of inverses of univalent functions with quasi-conformal extensions*, Kodai Math. J. **2** (1979), 411–419.
12. G. Springer, *The coefficient problem for schlicht mappings of the exterior of the unit circle*, Trans. Amer. Math. Soc. **70** (1951), 421–450.

UNIVERSITY OF CALIFORNIA
LOS ANGELES, CA 90024 USA

STANFORD UNIVERSITY
STANFORD, CA 94305 USA

INDIANA UNIVERSITY
BLOOMINGTON, IN 47405 USA

(Received June 28, 1981)

Commentary on

[123] A. Chang, M. M. Schiffer and G. Schober, *On the second variation for univalent functions*, J. Analyse Math. **40** (1981), 203–238.

Following Schiffer's pioneering work [5,6] of 1938, other methods of variation were developed and applied to extremal problems in geometric function theory. In 1943, Schiffer himself introduced the method of interior variation [13]. In a series of papers beginning in 1946, Goluzin developed his equally powerful method of annular variation (see [G] or [P]). Around the same time, Schaeffer and Spencer devised special methods for application to the coefficient problem (see [SS]). All of these early investigations were focused upon the vanishing of the first variation, making no use of the second variation to obtain additional information.

The first systematic attempt to exploit the second variation came with the paper [77], where Schiffer and I worked with interior variation and laboriously calculated an expression for the second variation of functions in the class S. Bombieri [Bo] later used these results to prove a local form of the Bieberbach conjecture. Independently of [77], Babenko (see [B]) developed and applied a more general form of the second variation. In their paper [123], the authors present a new and remarkably simple variational method that allows a very easy derivation (for instance) of Schiffer's differential equation for a function maximizing $\text{Re}\{a_n\}$. Not surprisingly, such an approach is more amenable than that of [77] to calculation of the second variation, and new results follow. The method makes essential use of the Riemann mapping theorem and so is restricted to simply connected domains, but it applies equally well to the classes S and Σ_0. I recall that the authors were quite excited over their discovery, and for good reason!

The new formula for the second variation is applied in [123] to uncover previously unknown properties of solutions to general classes of extremal problems in the families S and Σ_0. Perhaps the most striking result concerns the set Γ of arcs omitted by an extremal function in Σ_0. The authors use the second variation to show that wherever Γ has a fork, the arcs must meet at an angle of opening at least $\frac{2\pi}{5}$. It follows that for $n \geq 5$, the function

$$g(z) = z\left(1+z^{-(n+1)}\right)^{2/(n+1)} = z + \frac{2}{n+1}z^{-n} + \dots$$

does not maximize $\text{Re}\{b_n\}$ in Σ_0, since its omitted set consists of $n+1$ radial line segments meeting at the origin with equal angles $\frac{2\pi}{n+1} < \frac{2\pi}{5}$. Clearly, this disproves the conjecture $|b_n| \leq \frac{2}{n+1}$ for every $n \geq 5$. (See the commentary on [59].) Although the proof in [123] is new and elegant, this last result had been obtained independently around the same time by Tsao [T1, T2], also by application of a second variation. Tsao developed an expression for the second variation in Goluzin's annular variation method and applied it in a different way to show that for $n \geq 5$, the conjectured extremal function g does not provide even a local maximum for $\text{Re}\{b_n\}$, but is a saddle point. Thus the conjecture $|b_n| \leq \frac{2}{n+1}$ fails for every $n \geq 3$, since it had been proved false for $n = 3$ by Bazilevich [Ba] (and, much later, by Garabedian and Schiffer [59]) and for $n = 4$ by Kubota [K]. The commentaries on [4] and [59] contain further discussion of this problem.

References

[B] K. I. Babenko, *The Theory of Extremal Problems for Univalent Functions of Class S*, Proc. Steklov. Inst. Math., No. 101, Izdat "Nauka", Moscow 1972; English transl., American Mathematical Society, 1975.

[Ba] I. E. Bazilevich, *Supplement to the papers "Zum Koeffizientenproblem der schlichten Funktionen" and "Sur les théorèmes de Koebe–Bieberbach"*, Mat. Sb. **2** (44) (1937), 689–698 (in Russian).

[Bo] Enrico Bombieri, *On the local maximum property of the Koebe function*, Invent. Math. **4** (1967), 26–67.

[G] G. M. Goluzin, *Geometric Theory of Functions of a Complex Variable*, second edition, Izdat. "Nauka", Moscow 1966; English transl., American Mathematical Society, 1969.

[K] Yoshihisa Kubota, *On the fourth coefficient of meromorphic univalent functions*, Kōdai Math. Sem. Rep. **26** (1974/75), 267–288.

[P] Ch. Pommerenke, *Univalent Functions*, Vandenhoeck & Ruprecht, 1975.
[SS] A. C. Schaeffer and D. C. Spencer, *Coefficient Regions for Schlicht Functions*, American Mathematical Society, 1950.
[T1] Anna Tsao, *Coefficients of meromorphic univalent functions*, Doctoral dissertation, University of Michigan, 1981.
[T2] Anna Tsao, *Disproof of a coefficient conjecture for meromorphic univalent functions*, Trans. Amer. Math. Soc. **274** (1982), 783–796.

PETER DUREN

[126] (with D. Aharonov and L. Zalcman) Potato kugel

[126] (with D. Aharonov and L. Zalcman) Potato kugel. *Israel J. Math.* **40** (1981), 331–339.

ISRAEL JOURNAL OF MATHEMATICS, Vol. 40, Nos. 3–4, 1981

POTATO KUGEL

BY

DOV AHARONOV, M. M. SCHIFFER AND LAWRENCE ZALCMAN

ABSTRACT

Let P be a solid, homogeneous, compact, connected "potato" in space which attracts each point outside it (according to Newton's law) as if all its mass were concentrated at a single point. Answering a question of Lee Rubel, we show that P is a ball. The same conclusion is also obtained under substantially weakened hypotheses.

1. Some time ago, Lee Rubel posed the following problem. Let P be a solid, homogeneous, compact, connected "potato" in space which gravitationally attracts each point outside it as if all its mass were concentrated at a single point, say 0. Must P be spherical, i.e. a ball? In this note we offer an affirmative answer and obtain some extensions.

Surprising connections between potatoes and Peano arithmetic have already been noted in [13]; see also [10, p. 4]. Other points of contact with previous papers, notably [8] and [11], should be fairly evident. It is therefore appropriate to emphasize that our results are altogether unrelated to [7].

2. By assumption, the uniform mass distribution dx on P and the point mass $c\delta_0$ induce identical gravitational fields outside P. Thus [12, p. 4]

$$\int_P \frac{x-y}{|x-y|^3}\,dx = -c\frac{y}{|y|^3}, \qquad y \notin P. \tag{1}$$

The corresponding gravitational potentials differ by at most a constant on each component of the complement of P. Since that complement is, by assumption, connected we have

$$\int_P \frac{dx}{|x-y|} = \frac{c}{|y|} + d, \qquad y \notin P$$

Received June 5, 1981

for an appropriate constant d. Making $|y| \to \infty$ shows that $d = 0$, so that

$$\int_P \frac{dx}{|x-y|} = \frac{c}{|y|}, \qquad y \notin P. \tag{2}$$

The left-hand side of (2), being the convolution of the locally integrable function $1/|x|$ with a bounded measurable function of compact support (*viz.* the characteristic function of P), defines a bounded continuous function *on all of* $\mathbf{R}^3$. It follows that 0 must lie in the interior of P, for otherwise the right-hand side of (2) would be unbounded.

Suppose u is harmonic on (a neighbourhood of) P. We claim that

$$\int_P u(x)dx = cu(0). \tag{3}$$

Indeed, given such a function, we can modify it to be smooth and of compact support, preserving the harmonicity near P. Then [12, p. 13]

$$\begin{aligned}\int_P u(x)dx &= \int_P \left(-\frac{1}{4\pi}\int \frac{\Delta u(y)}{|x-y|}\,dy\right) dx \\ &= -\frac{1}{4\pi}\int \Delta u(y)\left(\int_P \frac{dx}{|x-y|}\right) dy \\ &= -\frac{1}{4\pi}\int \Delta u(y)\frac{c}{|y|}\,dy \\ &= cu(0)\end{aligned}$$

as claimed.

Now choose a point $x_0 \in \partial P$ which lies closest to 0 and take x_n exterior to P, $x_n \to x_0$. The functions

$$v_n(x) = (|x|^2 - |x_n|^2)/|x - x_n|^3$$

are each harmonic on P and tend pointwise to

$$v(x) = (|x|^2 - |x_0|^2)/|x - x_0|^3$$

on $P \setminus \{x_0\}$. Since the $\{v_n\}$ form a uniformly integrable family [6, p. 11], we have

$$\int_P v\,dx = \lim \int_P v_n dx = \lim cv_n(0) = cv(0)$$

so v satisfies (3).

The rest of the argument follows [4]. Set

$$U(x) = 1 + |x_0|\,v(x)$$

and let B be the ball centered at 0 having radius $|x_0|$. Then $B \subset P$ and

$$0 = cU(0) = \int_P U\,dx = \int_B U\,dx + \int_{P \setminus B} U\,dx$$

$$= 0 + \int_{P \setminus B} U\,dx.$$

Since $U(x) > 0$ for $|x| > |x_0|$ and P is assumed to be solid (so that $\overline{P^0} = P$), we obtain a contradiction unless $P \subset B$. Thus $P = B$ as required.

As usual, an analogous result obtains in $\mathbf{R}^n$, with the kernel $|x - y|^{-1}$ replaced by $|x - y|^{2-n}$ when $n \geqq 4$ and by $-\log|x - y|$ for $n = 2$. In the sequel we shall continue to state our results for potatoes in $\mathbf{R}^3$ with the understanding that similar results hold, *mutatis mutandis*, in spaces of arbitrary dimension.

It is worth noting that the condition (2) can be relaxed very considerably. It is enough, for instance, to assume that (2) holds only for y sufficiently large (or, indeed, for y belonging to any open set in the complement of P). For both sides of (2) are harmonic off P, so the full force of (2) follows by harmonic continuation. Observe also that the constant c in (2) is the volume of P. This is physically obvious and, in any case, follows by multiplying both sides of (2) by $|y|$ and making $y \to \infty$.

To what extent does the result of this section mirror a more general phenomenon? In other words, is the map

$$\mathrm{pot} : \mathrm{POT} \to \mathrm{Pot},$$

which associates to each object in the (discrete) category POT of potatoes its external potential or, better, the corresponding object in the category Pot of (germs of) potentials at infinity, a monofunctor? Obviously, an affirmative solution would constitute a notable extension of the result obtained above. Unfortunately, however, the answer is negative. When $n = 2$, this follows from an example due to Sakai [9]. As pointed out to us by B. Weiss, Sakai's example can be easily modified to yield examples in higher dimensions as well.

3. In the remaining sections we shall be concerned with extending the result of the preceding paragraph by relaxing the various hypotheses (connectedness, compactness, homogeneity, and solidity) initially placed on our potato. On physical grounds, we shall continue to assume that P is both connected and solid (i.e., $P = \overline{P^0}$, the closure of its interior, and $\mathbf{R}^3 \setminus P$ is connected).

Actually, an amusing aspect of the previous proof is that the assumption of connectedness played no role whatsoever and could accordingly be dropped.

This suggests consideration of the more physically meaningful configuration of a "sack of potatoes", i.e. a finite collection of disjoint (connected) potatoes.

Accordingly, let $S = \bigcup_{j=1}^{l} P_j$, where the P_j are pairwise disjoint potatoes. Suppose there exist points $x_1, x_2, \cdots, x_l$ in $\mathbf{R}^3$ and scalars $a_1, a_2, \cdots, a_l$ such that

$$\int_S \frac{dx}{|x-y|} = \sum_{j=1}^{l} \frac{a_j}{|y-x_j|}, \qquad y \notin S. \tag{4}$$

Arguing as before, we see that if $a_j \neq 0$, x_j must lie in the interior of S and that

$$\int_S u(x)dx = \sum_{j=1}^{l} a_j u(x_j) \tag{5}$$

for any function u harmonic in a neighbourhood of S. Suppose some component P_j of S fails to meet the set $X = \{x_1, \cdots, x_l\}$. Choosing u to be 1 on a neighbourhood of P_j and 0 elsewhere, we obtain a contradiction to (5). Thus each P_j contains at least, and hence exactly, one point of X. Renumbering if necessary, we have $x_j \in P_j$. We may now take u in (5) to be an arbitrary function harmonic in a neighbourhood of P_j and vanishing on a neighbourhood of the rest of S. Then (5) becomes

$$\int_{P_j} u(x)dx = a_j u(x_j)$$

which is, essentially, (3). The situation is thus reduced to that of a single potato, and the previous argument applies. It follows that S is a collection of disjoint balls

$$P_j = \{|x - x_j| \leqq r_j\}, \qquad \text{where } r_j = \sqrt[3]{3a_j/4\pi}.$$

Conclusions are possible also when the number of points in X exceeds the number of components of S. In this case, the points of X are partitioned among the potatoes in S, and each potato satisfies an appropriate quadrature identity [1] with respect to the points of X it contains. Detailed discussion of the geometric consequences of this fact is deferred to a later occasion.

4. Only slight variations are required to adapt the argument of §2 to the case of an unbounded potato. Suppose, for instance, that the integral $\int_P |x|^{-1}dx$ is finite, so that the potential in (2) exists. Fixing y exterior to P, we have by (2)

$$\int_P \left[\frac{1}{|x-\rho y|} - \frac{1}{|\rho x - y|}\right] dx = \frac{c}{\rho |y|} - \frac{c}{|y|} \tag{6}$$

for $\rho > 0$ sufficiently close to 1. Differentiating (6) with respect to ρ and setting $\rho = 1$, we obtain

$$\int_P \frac{|x|^2 - |y|^2}{|x-y|^3}\, dx = -\frac{c}{|y|}.$$

We may now let y tend to x_0 and complete the argument as before. This reasoning also provides an attractive alternate route to the result of §2.

Of course, Rubel's question makes sense even if the potential fails to converge, so long as the field of (1) remains finite. In that case, different techniques seem required to settle the issue.

5. One cannot hope to dispense entirely with the hypothesis that P is homogeneous. Indeed, any measure for which the analogue of (3) holds will obviously satisfy the analogue of (2); and, on *any* potato, such measures exist in abundance. On the other hand, consideration of rotation-invariant measures allows us to weaken this hypothesis very considerably. This is evident already from the reasoning in §2, which applies *verbatim* if the uniform mass distribution is replaced by $f(|x|)dx$, where f is a bounded positive function. Elaboration of the argument in [3] enables us to go further and consider mass distributions of variable sign. We turn directly to the details.

Accordingly, let $P \subset \mathbf{R}^3$ be a compact, connected potato containing 0 and let $f = f(|x|)$ be a smooth radial function, defined in a neighbourhood of P, *which does not vanish identically on any open set which intersects* ∂P. We call a function with this property *admissible* for P. Set

$$\psi(y) = \int_P \frac{f(|x|)}{|x-y|}\, dx.$$

Then $\psi \in C^1(\mathbf{R}^3)$ since

$$\nabla\psi(y) = -\int_P f(|x|)\frac{x-y}{|x-y|^3}\, dx,$$

the (vector-valued) convolution of a bounded measurable function of compact support with the locally integrable function $x/|x|^3$. Suppose that the force field induced by f on the exterior of P is Newtonian, i.e.

$$\nabla\psi(y) = -c\frac{y}{|y|^3}, \qquad y \notin P. \tag{7}$$

We claim that P is a ball centered at 0.

It suffices to show that ψ is a radial function on P, i.e. $\psi(x)=\psi(r)$ where $r=|x|$, $x\in P$. Indeed by (7) and the continuity of $\nabla\psi$, we have

$$\nabla\psi(x)=\psi'(r)\frac{x}{r}=-c\frac{x}{r^3},\qquad x\in\partial P \tag{8}$$

so

$$\psi'(r)=-\frac{c}{r^2} \tag{9}$$

on ∂P. If P is not a ball, we can find $x_1, x_2\in\partial P$ with $r_1=|x_1|<|x_2|=r_2$. The open shell $S=\{r_1<|x|<r_2\}$ then intersects both ∂P and P^0. Since ψ is radial on P we have, by (9),

$$\nabla^2\psi(x)=\psi''(r)+\frac{2}{r}\psi'(r)=0,\qquad x\in S\cap P^0.$$

But ([12, p. 16])

$$\nabla^2\psi(x)=-4\pi f(x),\qquad x\in P^0. \tag{10}$$

Thus $f(x)=0$ for $x\in S\cap P^0$, and this contradicts the assumption that f is admissible for P.

The proof that ψ is radial on P hinges on the following lemma, which may be of some interest in itself.

LEMMA. *Let $D\subset\mathbf{R}^n$ be a bounded domain containing 0 and let $u\in C^1(\bar{D})$ be harmonic on D. Assume that the gradient vector ∇u is parallel to the radius vector at each point of ∂D, i.e.*

$$\nabla u(x)=\lambda(x)x,\qquad x\in\partial D. \tag{11}$$

Then u is constant on D.

PROOF. Set

$$g(x)=g_{ij}(x)=x_i\frac{\partial u}{\partial x_j}-x_j\frac{\partial u}{\partial x_i},\qquad 1\leqq i<j\leqq n.$$

A simple calculation shows that $\nabla^2 g=0$ on D, and by (11) $g=0$ on ∂D. Thus, $g=g_{ij}$ vanishes identically on D for each pair of indices i,j. It follows that $\nabla u(x)=\lambda(x)x$ throughout D, i.e. ∇u is parallel to the radius vector at each point of D. Let S be a sphere contained in D, centered at 0. At each point $x\in S$, S is orthogonal to the radius vector at x and hence to $\nabla u(x)$. Thus, at each point of S the directional derivatives of u vanish in all directions tangent to S. It follows that u is constant on S, so u is a function of $|x|$ near the origin. Thus

$$u(x) = \begin{cases} a|x|^{2-n} + c, & n \geqq 3 \\ a \log|x| + c, & n = 2 \end{cases}$$

for x sufficiently small. Since u is not singular at 0, $a = 0$ and $u(x) = c$ for small x. By continuation, $u(x) = c$ throughout D.

Returning to the proof that ψ is radial, we set

$$\varphi(y) = \int_B \frac{f(|x|)}{|x-y|}\, dx,$$

where B is a ball centered at the origin which contains P. This function is obviously radial and continuously differentiable on $\mathbf{R}^3$; thus

$$\text{(12)} \qquad \nabla\varphi(x) = \frac{\varphi'(r)}{r}\, x, \qquad r = |x|$$

for *all* $x \neq 0$ and hence, in particular, for $x \in \partial P$. Moreover,

$$\text{(13)} \qquad \nabla^2\varphi(x) = -4\pi f(x), \qquad x \in B^0.$$

It follows from (10), (13), (8), and (12) that $\psi - \varphi$ satisfies the hypotheses of the lemma on $D = P^0$. Hence $\psi(x) = \varphi(x) + c$ throughout P, and ψ is radial on P as claimed.

The regularity assumptions on f can be relaxed considerably. It suffices, for instance, to assume that f is a bounded measurable function. Equations (10) and (13) then hold in the sense of distributions [12, p. 18], and the harmonicity of $\psi - \varphi$ follows from Weyl's lemma.

The argument above also applies to the case of a sack of potatoes, considered in §3. Indeed, suppose we are given disjoint potatoes $P_1, P_2, \cdots, P_l$ each endowed with a (not necessarily positive) mass distribution f_j which is radial with respect to some fixed point $x_j \in P_j$ and which is admissible for P_j. Let ψ be the corresponding potential and suppose that

$$\psi(y) = \sum_{j=1}^{l} \frac{c_j}{|y - x_j|}, \qquad y \notin \bigcup_{j=1}^{l} P_j.$$

Then for fixed k, the function

$$\psi_k(y) = \psi(y) - \sum_{j \neq k} \frac{c_j}{|y - x_j|}$$

satisfies

$$\nabla^2 \psi_k(x) = -4\pi f_k(\; x - x_k \;), \qquad x \in P_k^0,$$

$$\nabla \psi_k(x) = -\frac{c_k}{x - x_k{}^{3}}(x - x_k), \qquad x \in \partial P_k,$$

and we may argue exactly as before to conclude that P_k is a ball centered at x_k.

Finally, let us note that the argument of this section is easily adapted to handle the case of potatoes "with holes", i.e. for which $\tilde{P} = \mathbf{R}^3 \setminus P$ is not connected. Indeed, suppose f is admissible for $\tilde{P}$ and that (7) holds; we allow different values of c on different components of P and require, for physical reasons, that $c = 0$ on that component (if any) which contains 0. The reasoning used in the lemma shows that ψ is constant on each component of $S \cap P$ for any sphere S about the origin. It then follows as before that each component of ∂P is a sphere about the origin. Since P is itself connected, it can be only a ball or a spherical shell. Details are left to the reader.

6. *Dessert.* See [2] and [5].

Acknowledgements

This paper was written while the second author was visiting professor at the Technion and the third author visiting professor at Bar-Ilan University and the Hebrew University of Jerusalem. Research of the third author was supported by the National Science Foundation, the General Research Board of the University of Maryland, and the Lady Davis Fellowship Trust.

References

1. Dov Aharonov and Harold S. Shapiro, *Domains on which analytic functions satisfy quadrature identities*, J. Analyse Math. **30** (1976), 39–73.

2. L. E. Dubins and E. H. Spanier, *How to cut a cake fairly*, Amer. Math. Monthly **68** (1961), 1–17.

3. Bernard Epstein and M. M. Schiffer, *On the mean-value property of harmonic functions*, J. Analyse Math. **14** (1965), 109–111.

4. U. Kuran, *On the mean-value property of harmonic functions*, Bull. London Math. Soc. **4** (1972), 311–312.

5. Raphael Loewy and Hans Schneider, *Positive operators on the n-dimensional ice cream cone*, J. Math. Anal. Appl. **49** (1975), 375–392.

6. G. Lumer, *Algèbres de fonctions et espaces de Hardy*, Lecture Notes in Mathematics #75, Springer-Verlag, 1968.

7. Allan Ollis and Sam Griffiths, *A final slice of pork*, Math. Gaz. **64** (1980), 283–286.

8. J. F. Ramaley, *Buffon's noodle problem*, Amer. Math. Monthly **76** (1969), 916–918.

9. Makoto Sakai, *A problem on Jordan domains*, Proc. Amer. Math. Soc. **70** (1978), 35–38.

10. A. Seidenberg, *km, a widespread root for ten*, Arch. History Exact Sci. **16** (1976/7), 1–16.

11. A. H. Stone and J. W. Tukey, *Generalized sandwich theorems*, Duke Math. J. **9** (1942), 356–359.

12. John Wermer, *Potential Theory*, Lecture Notes in Mathematics #408, Springer-Verlag, 1974.

13. Lawrence Zalcman, *Counting: It's child's play*, J. Irr. Res. **23** (1978), 5–7.

TECHNION — ISRAEL INSTITUTE OF TECHNOLOGY
HAIFA, ISRAEL

STANFORD UNIVERSITY
STANFORD, CA 94305 USA

UNIVERSITY OF MARYLAND
COLLEGE PARK, MD 20742 USA

Commentary on

[126] Dov Aharonov, M.M. Schiffer and Lawrence Zalcman, *Potato kugel*, Israel J. Math. **40** (1981), 331–339.

According to a famous theorem of Newton, the gravitational attraction exerted by a solid homogeneous ball on any point outside it is identical to that exerted by a point mass (of mass equal to that of the ball) placed at the center of the ball. Lee Rubel asked if the converse obtains: must a solid, homogeneous, compact, connected "potato" in space which gravitationally attracts each point outside it as if all its mass were concentrated at a single point be a solid ball? I mentioned the problem to Dov Aharonov, and he passed it on to Schiffer, who was visiting the Technion at that time. Thinking about the problem individually, we came up with a variety of approaches to the (positive) solution, with [126] as the result. These various approaches lead to generalizations in different directions. For instance, there is a corresponding result for a "sack of potatoes," i.e., a collection of disjoint homogeneous solids; and even nonhomogeneous potatoes can be handled if the density is radial. The result also holds for an unbounded potato, as long as the potential (or even just the gravitational field [Z, p. 338]) it defines is finite.

The term "potato" was part of Rubel's original formulation. Since the result asserts that the potato must be a ball ("Kugel" in German), the title "Potato Kugel" suggested itself.[1] Of course, this is a bit of a joke, as potato kugel is also the name of one of the signature dishes of eastern European Jewish cooking, a savory pudding of potatoes and eggs, often served as part of traditional Sabbath meals. Accordingly, when I spoke on the results of our paper at the Mathematics Colloquium of the Hebrew University, I arranged (with appropriate advance publicity) for generous servings of potato kugel to be available at the tea preceding the lecture. That must have been the best attended pre-Colloquium tea in the history of the Hebrew University Mathematics Department!

The pun in the title led to further tomfoolery, mainly in the list of references, many of which are cited simply because they mention food in their titles. Max was not especially enthusiastic about such attempts at humor, but he was a good sport and did not object. In any case, such silliness did not prevent the paper from being listed as one of "seven small pearls" from convexity in Peter Gruber's survey [G] or being singled out for special mention by Bob Osserman in his review [O] of Heinz Hopf's *Differential Geometry in the Large*.

Interesting questions in this area remain. For instance, let P_1 and P_2 be homogeneous solids (of the same density) whose Newtonian potentials agree near ∞ (and hence, by harmonic continuation, agree on the unbounded component of $\mathbb{R}^3 \setminus (P_1 \cup P_2)$). In [Z], examples are given to show that, without additional assumptions, there may be uncountably (!) many homogeneous solids having the same potential near ∞. Since the basic result of [126] may be formulated as saying that if P_1 is a ball, then $P_2 = P_1$, it is natural to conjecture that if P_1 is convex, then $P_2 = P_1$. If P_1 and P_2 are *both* convex, this is a classical result of Novikov [N]. Recently, Gardiner and Sjödin [GS] proved that if the potentials agree throughout $\Omega = \mathbb{R}^3 \setminus (P_1 \cup P_2)$ (and not merely on the unbounded component of Ω) and P_1 is convex, then P_1 and P_2 must coincide; see also [S]. Further historical results, as well as a simple elementary proof of the basic result, can be found in [Z].

[1] Shortly after the paper appeared, we received a reprint request from the Central Potato Research Institute in India. As pointed out to us by Barry Simon, this was entirely appropriate, as the main result of the paper is that the potato is central!

References

[GS] Stephen J. Gardiner and Thomas Sjödin, *Convexity and the exterior inverse problem of potential theory*, Proc. Amer. Math. Soc. **136** (2008), 1699–1703.

[G] Peter M. Gruber, *Seven small pearls from convexity*, Math. Intelligencer **5** (1983), 16–19.

[N] P.S. Novikoff, *On uniqueness for the inverse problem of potential theory*, C.R. (Dokl.) Acad. Sci. URSS (N.S.) **18** (1938), 165–168.

[O] Robert Osserman, Review of *Differential Geometry in the Large*, by Heinz Hopf, Amer. Math. Monthly **93** (1986), 71–74.

[S] Henrik Shahgholian, *Convexity and uniqueness in an inverse problem of potential theory*, Proc. Amer. Math. Soc. **116** (1992), 1097–1100.

[Z] Lawrence Zalcman, *Some inverse problems of potential theory*, Integral Geometry, Contemp. Math. **63**, Amer. Math. Soc., 1987, pp. 337–350.

LAWRENCE ZALCMAN

[127] (with P. L. Duren and Y. J. Leung) Support points with maximum radial angle

[127] (with P. L. Duren and Y. J. Leung) Support points with maximum radial angle. *Complex Variables Theory Appl.* **1** (1982/83), 263–277.

Complex Variables, 1983, Vol. 1, pp. 263–277
0278-1077/83/0102-0263 $18.50/0

Support Points with Maximum Radial Angle

P. L. DUREN

Department of Mathematics, University of Michigan, Ann Arbor, MI 48109 U.S.A.

Y. J. LEUNG

Department of Mathematics, State University of New York at Albany, Albany, NY 12222 U.S.A.

and

M. M. SCHIFFER

Department of Mathematics, Stanford University, Stanford, CA 94305 U.S.A.

(*Received July 16, 1982*)

It has been known for some time that every support point of class S of univalent functions must map the disk onto the complement of an analytic arc whose radial angle is less than $\pi/4$ in magnitude except perhaps at the finite endpoint. It was discovered only recently that the bound $\pi/4$ is best possible and can actually be attained at the tip of the slit. The problem therefore arises to describe the functionals which generate support points whose omitted arcs have maximum radial angle, and to describe the corresponding omitted arcs. The present paper contains several results which shed light on this problem and suggest that the phenomenon is rather unusual.

AMS (MOS): 30C70

1. INTRODUCTION

Let $\mathscr{H}(\mathrm{D})$ be the linear space of all functions analytic in the unit disk D, with the usual topology of uniform convergence on compact subsets. Let S be the class of all functions $f \in \mathscr{H}(\mathrm{D})$ which are univalent in D and have the form

$$f(z) = z + a_2 z^2 + a_3 z^3 + \dots, \qquad |z| < 1.$$

263

Then S is a compact subset of $\mathscr{H}(\mathsf{D})$, and every real-valued continuous functional attains a maximum value on S.

A *support point* of S is a function which maximizes $\operatorname{Re}\{L\}$ over S for some complex-valued functional L which is linear and continuous on $\mathscr{H}(\mathsf{D})$ and is not constant on S. It is well known [8, 1, 9, 5] that if f is a support point of S associated with L, then f maps D onto the complement of an analytic arc Γ extending with increasing modulus from a finite point w_0 to infinity and satisfying the differential equation

$$\Phi(w)\frac{dw^2}{w^2} > 0, \qquad \Phi(w) = L\left(\frac{f^2}{f-w}\right). \tag{1}$$

This function Φ is analytic on Γ, and $\operatorname{Re}\{\Phi(w)\} > 0$ at all interior points of Γ. The *radial angle* at a point $w \in \Gamma$ is defined as the angle $\arg\{dw/w\}$ between the radius and tangent vectors. In view of (1), the property $\operatorname{Re}\{\Phi(w)\} > 0$ implies that Γ has a radial angle less than $\pi/4$ in magnitude everywhere except perhaps at its finite tip w_0.

The Koebe function $k(z) = z(1-z)^{-2}$ and its rotations are support points of S, but there are many others. Schober [9] observed, for example, that no rotation of the Koebe function can maximize $\operatorname{Re}\{L\}$ if either $L(f) = a_3 + ia_2$ or $L(f) = f(\zeta)$ for fixed $\zeta \in \mathsf{D}$ off a certain segment of the real axis. (See also Schaeffer and Spencer [7], p. 152.) Brown [2, 3] studied these two extremal problems and established various geometric properties of the arcs Γ omitted by the support points. Among other things, Brown found numerically that for a certain choice of $\zeta \in \mathsf{D}$, each support point associated with the point-evaluation functional $L(f) = f(\zeta)$ omits an arc Γ having a radial angle of magnitude approximately equal to $\pi/4$ (to five decimal places) at its tip. This was a strong indication that $\pi/4$ in the general description of support points is best possible. Pearce [6] confirmed this by showing that a certain functional of the form $L(f) = e^{-i\sigma}f'(\zeta)$ generates a support point whose omitted arc Γ is a half-line with radial angle equal to $\pi/4$ at its tip. Duren [4] made a more complete study of the support points arising from linear functionals of this form.

In the present paper we establish some general properties of support points whose omitted arc Γ has a radial angle of $\pm\pi/4$ at its tip. We show that in this case Γ satisfies a second differential equation independent of (1). If Φ is a rational function, as it is for most functionals of interest, the two differential equations combine to show that Γ is an algebraic curve. Under the additional hypothesis that Φ has no double zeros, we are able to show that Γ must be a half-line. We also find that a radial angle of $\pm\pi/4$ cannot occur for any point-evaluation functional $L(f) = f(\zeta)$ or for any coefficient functional $L(f) = \sum_{\nu=2}^{n} \lambda_\nu a_\nu$. For a derivative functional $L(f) = e^{-i\sigma} f'(\zeta)$, it can occur only in the case discovered by Pearce, where the extremal problem is equivalent to the (nonlinear) problem of maximizing or minimizing $\arg f'(\zeta)$. Finally, if for general Φ the arc Γ has a radial angle of $\pm\pi/4$ at its tip, then Γ is confined to a certain quarter-plane with vertex at its tip.

2. THE SECOND DIFFERENTIAL EQUATION

Let f be a support point of S which maximizes $\text{Re}\{L\}$. Suppose that the radial angle of its omitted arc Γ has magnitude $\pi/4$ at the finite tip w_0. According to the differential equation (1), this is equivalent to the assumption that $\text{Re}\{\Phi(w_0)\} = 0$. (We shall see that $\Phi(w_0) \neq 0$.)

Now define $\tilde{f} = w_0 f/(w_0 - f)$. Then $\tilde{f} \in S$ and

$$f - \tilde{f} = \frac{f^2}{f - w_0}, \tag{2}$$

so the hypothesis $\text{Re}\{\Phi(w_0)\} = 0$ implies $\text{Re}\{L(\tilde{f})\} = \text{Re}\{L(f)\}$. This shows that $\tilde{f}$ also maximizes $\text{Re}\{L\}$, so $\tilde{f}$ maps D onto the complement of an arc $\tilde{\Gamma}$ which satisfies

$$\tilde{\Phi}(\tilde{w}) \frac{d\tilde{w}^2}{\tilde{w}^2} > 0, \qquad \tilde{\Phi}(\tilde{w}) = L\left(\frac{\tilde{f}^2}{\tilde{f} - \tilde{w}}\right). \tag{3}$$

It is clear from the definition of $\tilde{f}$ that $\tilde{\Gamma}$ is related to Γ by the mapping

$$\tilde{w} = \frac{w_0 w}{w_0 - w}, \qquad w \in \Gamma. \tag{4}$$

The identity

$$\frac{\tilde{w}\tilde{f}}{\tilde{w} - \tilde{f}} = \frac{wf}{w - f} \tag{5}$$

is easily verified. Note also the identities

$$f - \frac{wf}{w-f} = \frac{f^2}{f-w}\,; \tag{6}$$

$$\tilde{f} - \frac{\tilde{w}\tilde{f}}{\tilde{w}-\tilde{f}} = \frac{\tilde{f}^2}{\tilde{f}-\tilde{w}}\,. \tag{7}$$

Now subtract (7) from (6), use the identities (2) and (5), and apply the linear functional L to obtain

$$\Phi(w_0) = L(f) - L(\tilde{f}\,) = \Phi(w) - \tilde{\Phi}(\tilde{w}), \tag{8}$$

where w and $\tilde{w}$ are related by (4). On the other hand, differentiation of (4) gives

$$\frac{d\tilde{w}^2}{\tilde{w}^2} = \left(1 - \frac{w}{w_0}\right)^{-2}\frac{dw^2}{w^2}\,. \tag{9}$$

The relations (8) and (9) transform the differential equation (3) to

$$\left[\Phi(w) - \Phi(w_0)\right]\left(1 - \frac{w}{w_0}\right)^{-2}\frac{dw^2}{w^2} > 0, \qquad w \in \Gamma. \tag{10}$$

In this form it is a second differential equation for the omitted arc Γ, derived under the assumption that $\operatorname{Re}\{\Phi(w_0)\} = 0$.

Because $\operatorname{Re}\{\Phi(w)\} > 0$, it is clear that $\Phi(w) \neq 0$ at all interior points of Γ. It may now be observed that $\Phi(w_0) \neq 0$ unless f is a rotation of the Koebe function. Indeed, if $\Phi(w_0) = 0$, then Γ satisfies both (1) and (10). Division of (1) by (10) produces the simple relation

$$\left(1 - \frac{w}{w_0}\right)^2 > 0, \qquad w \in \Gamma, \quad w \neq w_0\,,$$

which implies that Γ lies entirely on the line through 0 and w_0.

In general, the division of (1) by (10) gives

$$\frac{\Phi(w)}{\Phi(w) - \Phi(w_0)}\left(1 - \frac{w}{w_0}\right)^2 > 0, \qquad w \in \Gamma, \quad w \neq w_0\,. \tag{11}$$

If Φ is a rational function, it follows that Γ is an algebraic curve.

The results may be summarized as follows.

THEOREM 1 *Let f be a support point of S which maximizes* $\mathrm{Re}\{L\}$, *and suppose that its omitted arc* Γ *has a radial angle of* $\pm\pi/4$ *at its finite tip* w_0. *Let* $\Phi(w) = L(f^2/(f-w))$. *Then* $\Phi(w_0) \neq 0$, $\mathrm{Re}\{\Phi(w_0)\} = 0$, *and* Γ *satisfies the two differential equations* (1) *and* (10). *In addition,* Γ *satisfies the relation* (11).

3. POINT-EVALUATION FUNCTIONALS

Theorem 1 will now be specialized to point-evaluation functionals. Fix $\zeta \in \mathrm{D}$ with $\zeta \neq 0$ and let $f \in S$ be a function for which $\mathrm{Re}\{g(\zeta)\}$ is largest. Suppose that the arc Γ omitted by f has a radial angle of $\pm\pi/4$ at its tip w_0. Then Γ satisfies (11), where $\Phi(w) = B^2/(B-w)$ and $B = f(\zeta)$. This relation simplifies to

$$(B - w_0)w_0^{-2}(w - w_0) > 0, \qquad w \in \Gamma, \quad w \neq w_0,$$

which implies that Γ is a half-line.

On the other hand, Γ is a solution of the differential equation (1), which in the case of a point-evaluation functional reduces to

$$\frac{B^2}{B-w}\,\frac{dw^2}{w^2} > 0.$$

We claim that this equation has no half-line solution unless B is real and the half-line is radial. (This was previously observed by Johnny Brown.) Indeed, any half-line solution would coincide near infinity with the asymptotic half-line (see [1], [2], or [4])

$$w = \frac{B}{3} - B^2 t, \qquad t \geqslant 0.$$

Substituting this into the differential equation, we find after slight manipulation

$$B^{-3}(2 + 3Bt)(1 - 3Bt)^2 > 0,$$

or

$$2B^{-3} - 9B^{-2}t + 27t^3 > 0$$

for all $t > t_0 \geqslant 0$. It follows that B^2 and B^3 are real, so B is real. This implies that Γ is a radial half-line.

Thus the assumption that Γ has a radial angle of $\pm\pi/4$ at its tip has led to the conclusion that Γ is a radial half-line, which has a radial angle of 0 at every point. This contradiction establishes the following theorem.

THEOREM 2 *Given an arbitrary nonzero point $\zeta \in \mathbf{D}$, let f maximize* $\operatorname{Re}\{g(\zeta)\}$ *for all $g \in S$, and let Γ be the arc omitted by f. Then the radial angle of Γ is less than $\pi/4$ in magnitude at the tip.*

In reconciling Theorem 2 with Brown's numerical evidence cited in §1, we are forced to the remarkable conclusion that the support points generated by point-evaluation functionals have omitted arcs whose radial angles can be within 10^{-5} of $\pi/4$, but can never be equal to $\pi/4$. This demonstrates once more that numerical evidence may be misleading.

4. COEFFICIENT FUNCTIONALS

The Bieberbach conjecture asserts that all support points associated with coefficient functionals of the form $L(f) = a_n$ have omitted arcs which are radial half-lines. We now show that no such omitted arc can have a radial angle of $\pm\pi/4$ at its tip. More generally, we shall prove the following theorem.

THEOREM 3 *Let L be a functional of the form $L(f) = \sum_{\nu=2}^{n}\lambda_\nu a_\nu$, where λ_ν are complex constants with $\lambda_n \neq 0$. Let f maximize* $\operatorname{Re}\{L\}$ *over S, and let Γ be the arc omitted by f. Then the radial angle of Γ is less than $\pi/4$ in magnitude at the tip.*

Proof For $\nu = 2, 3, \ldots,$ let P_ν be the monic polynomial of degree $\nu - 1$ defined by the expansion

$$\frac{[f(z)]^2}{f(z) - w} = -\sum_{\nu=2}^{\infty} P_\nu\left(\frac{1}{w}\right) z^\nu.$$

Let

$$\mathscr{P}_n(\zeta) = \sum_{\nu=2}^{n} \lambda_\nu P_\nu(\zeta).$$

Then $\Phi(w) = -\mathscr{P}_n(1/w)$, and Γ satisfies the differential equation

$$\mathscr{P}_n\left(\frac{1}{w}\right)\frac{dw^2}{w^2} < 0. \tag{12}$$

If Γ has a radial angle of $\pm\pi/4$ at its endpoint w_0, then by Theorem 1 it also satisfies

$$\left[\mathscr{P}_n\left(\frac{1}{w}\right) - \mathscr{P}_n\left(\frac{1}{w_0}\right)\right]\left(1 - \frac{w}{w_0}\right)^{-2}\frac{dw^2}{w^2} < 0. \tag{13}$$

Let z_0 and z_1 be the points on the unit circle $\mathbb{T}$ where $f(z_0) = w_0$ and $f(z_1) = \infty$.

Now introduce the parametrization $w = f(e^{it})$ and conclude in a standard manner from (12) and (13) that $w = f(z)$ satisfies the differential equations

$$\mathscr{P}_n\left(\frac{1}{w}\right)\frac{z^2}{w^2}\left(\frac{dw}{dz}\right)^2 = \mathscr{Q}_n(z) \tag{14}$$

and

$$\left[\mathscr{P}_n\left(\frac{1}{w}\right) - \mathscr{P}_n\left(\frac{1}{w_0}\right)\right]\left(1 - \frac{w}{w_0}\right)^{-2}\frac{z^2}{w^2}\left(\frac{dw}{dz}\right)^2 = \mathscr{R}_n(z), \tag{15}$$

where $\mathscr{Q}_n$ and $\mathscr{R}_n$ are rational functions which are real and nonnegative on $\mathbb{T}$. The function $\mathscr{Q}_n$ has a double zero at z_0 but is otherwise positive on $\mathbb{T}$. Similarly, $\mathscr{R}_n$ has a double zero at z_1 but is otherwise positive on $\mathbb{T}$, as may be seen by tracing the equation (15) back to its primitive form (3). Indeed, if the curve $\tilde{\Gamma}$ is parametrized by $\tilde{w} = \tilde{f}(e^{it})$, the equation (15) is found to have the form

$$\tilde{\mathscr{P}}_n\left(\frac{1}{\tilde{w}}\right)\frac{z^2}{\tilde{w}^2}\left(\frac{d\tilde{w}}{dz}\right)^2 = \tilde{\mathscr{Q}}_n(z), \qquad \tilde{w} = \tilde{f}(z),$$

where $\tilde{\mathscr{Q}}_n$ vanishes at the point $\tilde{z}_0$ for which $\tilde{f}(\tilde{z}_0) = \tilde{w}_0$, the finite tip of $\tilde{\Gamma}$. However, it is easily seen from (4) and the definition of $\tilde{f}$ that $\tilde{z}_0 = z_1$.

Aside from poles of order $n-1$ at the origin and at infinity, the functions $\mathscr{Q}_n$ and $\mathscr{R}_n$ are analytic everywhere in the complex plane.

The differential equations (14) and (15) generate a common analytic continuation of f to a (possibly multiple-valued) function F analytic in the extended complex plane aside from a finite number of singular points which are either poles or algebraic branch points. Indeed, the division of (14) by (15) shows that F is an algebraic function.

We now show that $F(0)=0$ in every branch of the analytic continuation. Otherwise, F would have one of the local structures

$$w = F(z) = \omega + cz^{\alpha} + \dots, \qquad \omega \neq 0, \quad c \neq 0, \quad \alpha > 0, \tag{16}$$

or

$$w = F(z) = cz^{-\alpha} + \dots, \qquad c \neq 0, \quad \alpha > 0. \tag{17}$$

Putting (16) into the equation (14), we see that the left-hand side is $O(z^{2\alpha})$ as $z \to 0$, while $\mathcal{Q}_n(z) \sim \lambda_n z^{-n+1}$. This is impossible. Similarly, the substitution of (17) into (14) produces a left-hand side which is $O(z^{\alpha})$, since $\mathcal{P}_n(0)=0$. This again is impossible, so $F(0)=0$. Similar considerations show that each branch of F is analytic and locally univalent at the origin.

The next step is to show that all branches of F coincide with the original function f in some neighborhood of the origin. For this purpose we express (14) in the form

$$\mathcal{P}_n\left(\frac{1}{w}\right)\left(1 - \frac{w}{w_0}\right)^{-2}\left(1 - \frac{2w}{w_0} + \frac{w^2}{w_0^2}\right)\left(\frac{w'}{w}\right)^2 = \frac{1}{z^2}\mathcal{Q}_n(z)$$

and subtract it from (15) to get

$$\left\{\left(\frac{2w}{w_0} - \frac{w^2}{w_0^2}\right)\mathcal{P}_n\left(\frac{1}{w}\right) - \mathcal{P}_n\left(\frac{1}{w_0}\right)\right\}\left(1 - \frac{w}{w_0}\right)^{-2}\left(\frac{w'}{w}\right)^2$$
$$= z^{-2}\left[\mathcal{R}_n(z) - \mathcal{Q}_n(z)\right]. \tag{18}$$

Now divide (18) by (14) to obtain an algebraic equation of the form

$$\frac{2w}{w_0} + O(w^2) = \mathcal{R}_n(z)/\mathcal{Q}_n(z) - 1 = O(z),$$

or $\varphi(w) = \psi(z)$, where φ and ψ are analytic functions with $\varphi(0) = \psi(0) = 0$ and $\varphi'(0) \neq 0$. Thus $w = \varphi^{-1}(\psi(z))$ in some neighborhood of the origin. This uniquely determines the values of F near the origin.

It now follows that F is single-valued. Suppose, on the contrary, that some pair of continuations from the origin to a point ζ along paths C_1 and C_2 lead to different values $F(\zeta) = w_1$ and $F(\zeta) = w_2$. We have just shown that the continuation along the closed path $C_1 - C_2$ must return F to its original function element at the origin. A reversal of the part of this continuation which extends over $-C_2$ then shows that the continuation along C_2 from 0 to ζ produces the value $F(z_1) = w_1 \neq w_2$. This contradiction proves that F is single-valued. Thus F is rational function.

However, it is known (Schaeffer and Spencer [7], p. 157) that the only rational functions $f \in S$ which satisfy a differential equation of the form (14) are given by

$$f(z) = \frac{z}{(1 - \alpha z)(1 - \beta z)}, \qquad |\alpha| = |\beta| = 1,$$

and map D onto the complement of one or two collinear rays. Because f is a support point, it can omit only one ray. Thus f must be a rotation of the Koebe function. This contradicts our assumption that the arc Γ omitted by f has a radial angle of $\pm\pi/4$ at its tip. In other words, the magnitude of the radial angle must be less than $\pi/4$, as the theorem asserts.

The essential idea of this proof is due to Schaeffer and Spencer ([7], p. 156), who applied it to a similar situation where a function satisfies two independent differential equations.

5. RATIONAL SUPPORT POINTS

The method of analytic continuation just applied to coefficient functionals shows ultimately that a support point which satisfies a second differential equation must be a rational function. In applying the method to other functionals, it will be helpful to appeal to the following general theorem, which gives an elegant description of the rational support points of S.

THEOREM 4 *Let $f \in S$ be a rational function which maps the unit disk onto the complement of an arc Γ extending to infinity. Then Γ is a half-line and f is a Marty transform of the Koebe function.*

This theorem was conjectured several years ago by Donald Wilken and was proved quite recently by Uri Srebro [10]. By a Marty transform of f we mean the composition of f with a self-mapping of the disk, renormalized to belong to S. Only the following corollary will be needed here.

COROLLARY *Every rational support point of S maps the unit disk onto the complement of a half-line.*

In fact, with the additional requirement that the omitted half-line have a radial angle no larger than $\pi/4$ in magnitude, this corollary gives a complete description of the rational support points of S. The work of Pearce [6] on certain support points associated with derivative functionals actually leads to a converse. In other words, if Γ is a half-line with the $\pi/4$-property, whose complement $\mathbb{C} - \Gamma$ has unit conformal radius, then some (rational) support point of S maps the disk onto the complement of Γ. This was pointed out to us by Louis Brickman.

6. FUNCTIONALS OF RATIONAL TYPE

A continuous linear functional L is said to be of *rational type* if

$$\Phi(w) = L\left(\frac{f^2}{f-w}\right)$$

is a rational function for each $f \in S$. For instance, every coefficient functional $L(f) = a_n$ is of rational type, and more generally so is every functional $L(f) = f^{(n)}(\zeta)$ given by point-evaluation of a derivative at a point $\zeta \in \mathbb{D}$. Every finite linear combination of functionals of rational type is again of rational type.

We shall prove the following theorem.

THEOREM 5 *Let f be a support point of S which maximizes* $\operatorname{Re}\{L\}$, *where L is a linear functional of rational type whose corresponding rational function Φ has no zeros of order two. If the arc Γ omitted by f has a radial angle of $\pm\pi/4$ at its tip, then Γ is a half-line.*

Proof As before, let z_0 and z_1 be the points on the unit circle at which $f(z_0) = w_0$ and $f(z_1) = \infty$, where w_0 is the tip of Γ. Under the hypothesis that Γ has a radial angle of $\pm\pi/4$ at w_0, it follows from Theorem 1 (as in the proof of Theorem 3) that $w = f(z)$ satisfies both of the differential equations

$$\Phi(w)\left(\frac{w'}{w}\right)^2 = \frac{1}{z^2}\,Q(z) \tag{19}$$

and

$$\left[\Phi(w) - \Phi(w_0)\right]\left(1 - \frac{w}{w_0}\right)^{-2}\left(\frac{w'}{w}\right)^2 = \frac{1}{z^2}\,R(z), \tag{20}$$

where Q and R are rational functions. Aside from double zeros of Q at z_0 and of R at z_1, both functions are real, positive, and finite on $\mathbb{T}$. (It is known in general that $L(f^2) \neq 0$, so Φ has a simple zero at infinity.)

Observe first that $\Phi'(w_0) \neq 0$. This is seen by substituting $w = f(z)$ into (20) and letting z approach z_0. If $\Phi'(w_0) = 0$, then the left-hand side of (20) must tend to zero (since $f'(z_0) = 0$), while the right-hand side goes to $z_0^{-2}R(z_0) \neq 0$.

The rest of the proof is similar to that of Theorem 3, with the point z_0 now playing the role of the origin. Dividing (19) by (20), we obtain the algebraic equation

$$\frac{(w - w_0)^2\Phi(w)}{w_0^2\left[\Phi(w) - \Phi(w_0)\right]} = \frac{Q(z)}{R(z)}\,. \tag{21}$$

Because $\Phi'(w_0) \neq 0$, this equation has the form

$$\frac{\Phi(w_0)}{w_0^2\Phi'(w_0)}(w - w_0) + O\big((w - w_0)^2\big) = O\big((z - z_0)^2\big)$$

in some neighborhood of (z_0, w_0). It follows that under the condition $g(z_0) = w_0$, the equation (21) has a unique local solution $w = g(z)$, valid in some neighborhood of z_0. This function g is an analytic continuation of f to a full neighborhood of z_0.

Now let F be the algebraic function obtained by analytic continuation of f through the differential equations (19) and (20). The theorem will be proved if we can show that $F(z_0) = w_0$ in every continuation. Indeed, it will then follow from our discussion of the algebraic equation (21) that F returns in every continuation to the same function element f at z_0. This will imply that F is single-valued and is therefore a rational function. The desired result will then follow from the corollary to Theorem 4.

If some continuation produces a value $F(z_0) \neq w_0$, then F must have one of the three local structures

$$w = F(z) = c(z - z_0)^{\alpha} + \dots, \qquad c \neq 0, \quad \alpha > 0; \tag{22}$$

$$w = F(z) = c(z - z_0)^{-\alpha} + \dots, \qquad c \neq 0, \quad \alpha > 0; \tag{23}$$

$$w = F(z) = \omega + c(z - z_0)^{\alpha} + \dots, \qquad c \neq 0, \quad \alpha > 0, \quad \omega \neq 0, w_0. \tag{24}$$

The structures (22) and (23) are easily eliminated. If F has the form (22), then (19) implies that $\Phi(0) = 0$, in which case the left-hand side of (20) has a pole at z_0. If F has the form (23), then because Φ has a simple zero at infinity the equation (19) gives $\alpha - 2 = 2$, or $\alpha = 4$; while the equation (20) gives $2\alpha - 2 = 0$, or $\alpha = 1$.

Now suppose that F has the local form (24). If Φ has a pole of order m at ω, then (19) gives $-m\alpha + 2(\alpha - 1) = 2$, while (20) gives $-m\alpha + 2(\alpha - 1) = 0$. This is impossible. If Φ is analytic at ω and has a zero of order m there, then (19) gives $m\alpha + 2(\alpha - 1) = 2$ and (20) gives $2(\alpha - 1) = 0$. This implies $\alpha = 1$ and $m = 2$. But by hypothesis, Φ has no zeros of order two. Finally, suppose Φ is analytic at ω and $\Phi(\omega) \neq 0$. Then (19) gives $2(\alpha - 1) = 2$, or $\alpha = 2$; while (20) gives $2(\alpha - 1) \leqslant 0$, or $\alpha \leqslant 1$. Having now eliminated all three of the local structures (22), (23), and (24), we have shown that $F(z_0) = w_0$ in every sheet of the analytic continuation. This concludes the proof.

The hypothesis that Φ has no double zeros is probably not essential. If Theorem 5 can be strengthened by elimination of this hypothesis, then Theorem 3 will be a direct consequence, since the only rational support points satisfying a differential equation of the form (14) are the Koebe function and its rotations.

7. DERIVATIVE FUNCTIONALS

As an application of Theorem 5, we now consider a derivative functional of the form $L(f) = e^{-i\alpha} f'(\zeta)$. Suppose that f maximizes

$\operatorname{Re}\{L\}$ over S, and let Γ be the arc omitted by f. A calculation gives

$$\Phi(w) = L\left(\frac{f^2}{f-w}\right) = \frac{e^{-i\sigma}BC(B-2w)}{(w-B)^2},$$

where $B = f(\zeta)$ and $C = f'(\zeta)$. This is a rational function with no double zeros, so Theorem 5 tells us that Γ cannot have a radial angle of $\pm\pi/4$ at its tip unless it is a half-line. On the other hand, Duren [4] showed that Γ is a half-line only in the case noted by Pearce [6], where the problem of maximizing $\operatorname{Re}\{L\}$ is equivalent to that of maximizing or minimizing $\arg f'(\zeta)$.

8. A QUARTER-PLANE CONTAINING Γ

We now return to the general case, where Φ need not be a rational function, and prove the following theorem.

THEOREM 6 *Let f be a support point of S, and let Γ be its omitted arc. If Γ has a radial angle of $\pm\pi/4$ at its tip w_0, then Γ lies in a certain quarter-plane with vertex at w_0. Furthermore, the half-line which bisects this quarter-plane is tangent to Γ at w_0 and is parallel to the asymptotic half-line of Γ.*

Proof Let Γ be parametrized with increasing modulus by $w = w(t)$, $0 \leqslant t < \infty$, so that $w(0) = w_0$. If $\arg\{w'(0)/w_0\} = \pm\pi/4$, then $\operatorname{Re}\{\Phi(w_0)\} = 0$ and by Theorem 1 the function $w(t)$ satisfies

$$\Phi(w)\left(\frac{w'}{w}\right)^2 > 0 \tag{25}$$

and

$$[\Phi(w) - \Phi(w_0)]\left(1 - \frac{w}{w_0}\right)^{-2}\left(\frac{w'}{w}\right)^2 > 0. \tag{26}$$

Suppose first that $\arg\{w'(0)/w_0\} = \pi/4$. Then we claim that Γ lies entirely in the quarter-plane with vertex w_0 bounded by the half-lines

$$w = w_0 + w_0 t, \qquad 0 \leqslant t < \infty, \tag{27}$$

and

$$w = w_0 + i w_0 t, \qquad 0 \leqslant t < \infty. \tag{28}$$

In fact, Γ enters this quarter-plane at w_0 and has no other points on either of the half-lines (27) or (28). To see this, divide (26) by (25) to obtain

$$\left[1 - \frac{\Phi(w_0)}{\Phi(w)}\right]\left(1 - \frac{w}{w_0}\right)^{-2} > 0, \qquad w \in \Gamma, \quad w \neq w_0. \tag{29}$$

If Γ meets the half-line (27) at a point $w \neq w_0$, then $(1 - w/w_0)^2 > 0$ and (29) implies $1 - \Phi(w_0)/\Phi(w) > 0$. In particular, $\Phi(w_0)/\Phi(w)$ is real. But this is impossible, since $\Phi(w_0) \neq 0$, $\operatorname{Re}\{\Phi(w_0)\} = 0$, and $\operatorname{Re}\{\Phi(w)\} > 0$. Similarly, Γ cannot meet the half-line (28), because $(1 - w/w_0)^2 < 0$ there. For the same reason, Γ cannot meet the half-line

$$w = w_0 - iw_0 t, \qquad 0 \leqslant t < \infty. \tag{30}$$

Therefore, if $\arg\{w'(0)/w_0\} = -\pi/4$, the arc Γ must enter and remain in the quarter-plane bounded by the half-lines (27) and (30).

It is clear from the construction that in either case the bisector of the containing quarter-plane is tangent to Γ at w_0. To show that it is parallel to the asymptotic half-line of Γ, we need only show that

$$\arg\left\{\frac{w - w_0}{w_0}\right\} \to \arg\left\{\frac{w'(0)}{w_0}\right\} = \pm\frac{\pi}{4} \tag{31}$$

as $w \to \infty$ along Γ. For this purpose we consider again the second support point $\tilde{f}$ (introduced in §2) and its omitted arc $\tilde{\Gamma}$. Recall that Γ and $\tilde{\Gamma}$ are related by (4):

$$\tilde{w} = \frac{w_0 w}{w_0 - w}, \qquad w \in \Gamma. \tag{32}$$

In particular, $\tilde{w} \to \tilde{w}_0 = -w_0$ as $w \to \infty$ along Γ. Thus equation (8) shows that $\Phi(w_0) = -\tilde{\Phi}(\tilde{w}_0)$. In particular, $\operatorname{Re}\{\tilde{\Phi}(\tilde{w}_0)\} = 0$ and $\operatorname{Im}\{\tilde{\Phi}(\tilde{w}_0)\} = -\operatorname{Im}\{\Phi(w_0)\}$. This shows that as $\tilde{w} \to \tilde{w}_0$ on $\tilde{\Gamma}$,

$$\arg\left\{\frac{\tilde{w} - \tilde{w}_0}{\tilde{w}_0}\right\} \to -\arg\left\{\frac{w'(0)}{w_0}\right\}.$$

However, the relation (32) gives

$$\frac{\tilde{w} - \tilde{w}_0}{\tilde{w}_0} = \frac{w_0}{w - w_0},$$

so

$$\arg\left\{\frac{w - w_0}{w_0}\right\} = -\arg\left\{\frac{\tilde{w} - \tilde{w}_0}{\tilde{w}_0}\right\} \to \arg\left\{\frac{w'(0)}{w_0}\right\}$$

as $w \to \infty$ along Γ. This establishes (31) and completes the proof.

It may be remarked that the last assertion of this theorem is a strong suggestion that Γ is actually a half-line in every case, whenever its radial angle has magnitude $\pi/4$ at the tip.

Acknowledgements

This research was supported in part by grants from the National Science Foundation. The second-named author wishes to acknowledge helpful conversations with Louis Brickman and Donald Wilken.

References

[1] L. Brickman and D. Wilken, Support points of the set of univalent functions, *Proc. Amer. Math.* **42** (1974), 523–528.
[2] J. E. Brown, Geometric properties of a class of support points of univalent functions, *Trans. Amer. Math. Soc.* **256** (1979), 371–382.
[3] J. E. Brown, Univalent functions maximizing $\text{Re}\{a_3 + \lambda a_2\}$, *Illinois J. Math.* **25** (1981), 446–454.
[4] P. L. Duren, Arcs omitted by support points of univalent functions, *Comment. Math. Helv.* **56** (1981), 352–365.
[5] P. L. Duren, *Univalent Functions*, Springer-Verlag, Heidelberg and New York, to appear in 1983.
[6] K. Pearce, New support points of S and extreme points of HS, *Proc. Amer. Math. Soc.* **81** (1981), 425–428.
[7] A. C. Schaeffer and D. C. Spencer, *Coefficient Regions for Schlicht Functions*, Amer. Math. Soc. Colloq. Publ., vol. 35, 1950.
[8] M. Schiffer, A method of variation within the family of simple functions, *Proc. London Math. Soc.* **44** (1938), 432–449.
[9] G. Schober, *Univalent Functions—Selected Topics*, Lecture Notes in Mathematics, No. 478, Springer-Verlag, 1975.
[10] U. Srebro, Is the slit of a rational slit mapping in S straight?, *Proc. Amer. Math. Soc.*, to appear.

Commentary on

[127] P. L. Duren, Y. J. Leung, and M. M. Schiffer, *Support points with maximum radial angle*, Complex Variables Theory Appl. **1** (1982/83), 263–277.

Univalent functions do not constitute a linear space, but the class S is a compact subset of the linear space $\mathscr{H}(\mathbb{D})$ of analytic functions in the unit disk $\mathbb{D}$, endowed with the topology of local uniform convergence. A *support point* is a function $f \in S$ which maximizes $\operatorname{Re}\{L\}$ for some continuous linear functional L on $\mathscr{H}(\mathbb{D})$ that is not constant on S. A standard application of Schiffer's method of boundary variation [5] shows that each support point f maps the disk conformally onto the complement of a network Γ of analytic arcs that are trajectories of a quadratic differential:

$$\Phi(w)\frac{dw^2}{w^2} > 0, \qquad \Phi(w) = L\left(\frac{f^2}{f-w}\right).$$

By the time [127] was published, it was known that Γ is a single analytic arc extending to infinity with increasing modulus. Schiffer [95] proved $L(f^2) \neq 0$, so that the quadratic differential has a simple pole at infinity and Γ terminates there. It was also known that $\operatorname{Re}\{\Phi(w)\} > 0$ at all interior points of Γ; and so its *radial angle*, the angle $\arg\{dw/w\}$ between radius and tangent vectors, has magnitude less than $\pi/4$ everywhere on Γ except perhaps at its finite tip w_0. Brown [Bw1, Bw2] studied the support points arising from point-evaluation functionals $L(f) = f(\zeta)$ and found compelling numerical evidence that the bound $\pi/4$ is sharp. Then Pearce [P] observed that the functional $L(f) = -f'(\sin\pi/8)$ generates a support point for which Γ is a half-line with radial angle equal to $\pi/4$ at w_0. (See [Sch, D] for further discussion.)

Pearce's example suggested the problem of describing all support points whose omitted arcs have maximum radial angle $\pi/4$ at the finite tip w_0. In [127], we found that because each such support point f solves a second extremal problem, its omitted arc Γ must also be a trajectory of a second quadratic differential. As a consequence, the relation

$$\frac{\Phi(w)}{\Phi(w)-\Phi(w_0)}\left(1-\frac{w}{w_0}\right)^2 > 0$$

holds for every point $w \in \Gamma$ except w_0. In particular, if L is of *rational type*, meaning that $\Phi(w)$ is a rational function, then Γ is an algebraic curve. Under the additional hypothesis (later removed by Zhang and Ma [ZM]) that $\Phi(w)$ has no zeros of order 2, we concluded that f is a rational function, and hence Γ is a half-line whenever it has radial angle $\pm\pi/4$ at w_0. Brickman [Br] showed that functionals of rational type are finite linear combinations of functionals of the form $L(f) = f^{(n)}(\zeta)$, where $\zeta \in \mathbb{D}$ and $n = 0, 1, 2, \ldots$. Surprisingly, although Brown had found a curve Γ whose radial angle at w_0 equals $\pi/4$ to 5 decimal places, the analysis in [127] shows that for point-evaluation functionals the angle is always less than $\pi/4$.

The method of proof in [127] can be traced to Schaeffer and Spencer [SS, p. 157], who used a similar argument to show that if a function $f \in S$ maximizes $\operatorname{Re}\{L\}$ and $\operatorname{Re}\{M\}$ for two independent coefficient functionals

$$L(f) = \lambda_2 a_2 + \cdots + \lambda_n a_n, \quad \lambda_n \neq 0$$

and

$$M(f) = \mu_2 a_2 + \cdots + \mu_m a_m, \quad \mu_m \neq 0,$$

where $n-1$ and $m-1$ are relatively prime, then f is a rotation of the Koebe function. Bakhtin [B1] drew the same conclusion (before the Bieberbach conjecture was proved) when f maximizes $\operatorname{Re}\{a_n\}$ and $\operatorname{Re}\{a_m\}$ for any $n \neq m$. Those results prompted the *two-functional conjecture*: if f maximizes $\operatorname{Re}\{L\}$ and $\operatorname{Re}\{M\}$ for any pair of continuous linear functionals L and M which are independent in the sense that $L \neq cM$ for all $c > 0$, then f is a rotation of the Koebe function. Bakhtin [B2] and Goh [G1, G2, G3] have verified the conjecture for large classes of functionals of rational type, but the general case remains open.

References

[B1] A. K. Bakhtin, *Some properties of functions of the class S*, Ukrain. Mat. Zh. **33** (1981), 154–159 (in Russian); English translation, Ukrainian Math. J. **33** (1981), 122–126.

[B2] A. K. Bakhtin, *Extrema of linear functionals*, Akad. Nauk Ukrain. SSR Inst. Mat. Preprint **1986**, no. 25 (in Russian).

[Br] Louis Brickman, *Functionals of rational type over the class S*, Proc. Amer. Math. Soc. **92** (1984), 372–376.

[Bw1] Johnny E. Brown, *Linear extremal problems in the class of univalent functions*, Doctoral dissertation, University of Michigan, 1979 .

[Bw2] Johnny E. Brown, *Geometric properties of a class of support points of univalent functions*, Trans. Amer. Math. Soc. **256** (1979), 371–382.

[D] Peter L. Duren, *Univalent Functions*, Springer-Verlag, 1983.

[G1] Say Song Goh, *On the two-functional conjecture for univalent functions*, Complex Variables Theory Appl. **20** (1992), 197–206.

[G2] Say Song Goh, *Support points and double poles*, Proc. Amer. Math. Soc. **122** (1994), 463–468.

[G3] Say Song Goh, *Functionals of higher derivative type*, J. London Math. Soc. (2) **58** (1998), 111–126.

[P] Kent Pearce, *New support points of S and extreme points of HS*, Proc. Amer. Math. Soc. **81** (1981), 425–428.

[SS] A. C. Schaeffer and D. C. Spencer, *Coefficient Regions for Schlicht Functions*, American Mathematical Society, 1950.

[Sch] Glenn Schober, *Univalent Functions – Selected Topics*, Lecture Notes in Math. **478**, Springer-Verlag, 1975.

[ZM] Yulin Zhang and Jinxi Ma, *A note on support points with maximum radial angle*, J. Math. Anal. Appl. **160** (1991), 598–601.

PETER DUREN

[135] (with P. L. Duren) Univalent functions which map onto regions of given transfinite diameter

[135] (with P. L. Duren) Univalent functions which map onto regions of given transfinite diameter. *Trans. Amer. Math. Soc.* **323** (1991), 413–428.

TRANSACTIONS OF THE
AMERICAN MATHEMATICAL SOCIETY
Volume 323, Number 1, January 1991

UNIVALENT FUNCTIONS WHICH MAP ONTO REGIONS OF GIVEN TRANSFINITE DIAMETER

P. L. DUREN AND M. M. SCHIFFER

ABSTRACT. By a variational method, the sharp upper bound is obtained for the second coefficients of normalized univalent functions which map the unit disk onto regions of prescribed transfinite diameter, or logarithmic capacity.

0. INTRODUCTION

Let S be the usual class of functions $f(z) = z + a_2 z^2 + \cdots$ analytic and univalent in the unit disk $\mathbb{D}$. Many years ago, Pick [7] used Bieberbach's inequality $|a_2| \leq 2$ to obtain the sharp bound $|a_2| \leq 2(1 - 1/M)$ for functions $f \in S$ with $|f(z)| < M$ for some number $M > 1$. The extremal functions map the unit disk onto the disk $|w| < M$ minus a radial slit of suitable length.

Our purpose is to solve a related problem: to find the maximum of $|a_2|$ among all functions $f \in S$ whose range has a given transfinite diameter R, $1 < R < \infty$. This problem is considerably harder than Pick's, but we shall see that for "large" R the solution is somewhat similar. The sharp inequality is

$$|a_2| \leq 2 - e^2/4R$$

and the extremal function maps the disk onto a certain Jordan region minus a radial spire. For small R the situation is rather different. The sharp bound is then determined implicitly through elliptic integrals, and the extremal function maps the disk onto a certain Jordan region with analytic boundary. The division occurs at the transfinite diameter

$$R = e^2\pi^2/64 = 1.139\ldots.$$

This unexpected phenomenon, the dual nature of the solution, gives the problem a particular interest.

A more detailed summary of our results appears in the theorem stated at the end of the paper. The proof uses a special variational method devised to

Received by the editors October 18, 1988 and, in revised form, January 20, 1989. Presented to the American Mathematical Society at Louisville, Kentucky, on January 18, 1990.

1980 *Mathematics Subject Classification* (1985 *Revision*). Primary 30C70, 30C50, 30C85, 33A25.

Key words and phrases. Univalent functions, second coefficient, extremal problems, variational methods, quadratic differentials, transfinite diameter, logarithmic capacity, elliptic integrals.

The research of the first author was supported in part by the National Science Foundation under Grant DMS-8701751.

0002-9947/91 $1.00 + $.25 per page

preserve both the class S and the transfinite diameter of the range. It combines the standard methods of interior variation and boundary variation, and it leads ultimately to a quadratic differential for the boundary arcs in the extremal situation. The trajectory structure then allows three possible cases, each of which is eventually seen to occur for suitable values of R.

A few preliminary remarks about transfinite diameter (or logarithmic capacity) are in order. We begin by recalling the original definition. For a bounded set E in the complex plane, the nth diameter is

$$d_n(E) = \left\{ \sup_{z_1, \dots, z_n \in E} \prod_{j<k} |z_j - z_k| \right\}^{2/n(n-1)}.$$

It can be shown that $d_{n+1}(E) \le d_n(E)$. The transfinite diameter of E is defined by $d(E) = \lim_{n\to\infty} d_n(E)$.

The transfinite diameter of a circular disk is its radius. According to a result of Pólya ([8]; see [4, Chapter VII, §2]), the outer area of a set E is bounded above by πR^2, where $R = d(E)$. Thus for functions $f \in S$ with $f(\mathbb{D})$ of transfinite diameter R, Pólya's theorem gives the inequality

$$1 + \sum_{n=2}^{\infty} n|a_n|^2 \le R^2.$$

It follows that $R > 1$ unless f is the identity mapping. Incidentally, Pólya's inequality also gives the crude estimate

$$|a_2| \le 2^{-1/2}(R^2 - 1)^{1/2}.$$

It is important to recall a result of Szegö [13] which equates transfinite diameter with logarithmic capacity. Given a compact set E of positive transfinite diameter in the plane, let

$$g(w) = g(w, \infty) = \log|w| + \gamma + O(1/w)$$

be Green's function of the unbounded component of the complement of E. The number γ is known as the Robin constant of E, and $e^{-\gamma}$ is its logarithmic capacity. Szegö's theorem asserts that $R = e^{-\gamma}$, or $\gamma = -\log R$, where $R = d(E)$. Because of the conformal invariance of Green's function, this basic relation accounts for the natural occurrence of transfinite diameter in function-theoretic problems.

Szegö's theorem also allows the interpretation of the transfinite diameter as a conformal radius. Suppose for simplicity that the compact set E is simply connected. Let

$$\zeta = h(w) = w + c_0 + c_1/w + \cdots$$

map the complementary domain $\tilde{E}$ conformally onto a circular region $|\zeta| > \rho$. Then Green's function of $\tilde{E}$ is

$$g(w, \infty) = \log|h(w)| - \log\rho,$$

which shows that $\rho = R$. An equivalent statement is that a function of the form

$$\tilde{f}(z) = Rz + b_0 + b_1/z + \cdots$$

maps $|z| > 1$ conformally onto $\widetilde{E}$.

For basic facts about transfinite diameter and related topics, the reader may consult Goluzin [4, Chapter VII] or Hille [5, Chapter 16]. Further information is in Pólya and Szegö [9, Part IV, Chapter 2]. See also Tsuji [14].

1. The variational method

Let S_R be the class of functions $f \in S$ whose range has transfinite diameter $R < \infty$. Although S_R is a normal family, it is not compact. Indeed, a "pinching" construction shows that a sequence of functions $f_n \in S_R$ may converge uniformly on compact sets to a function $f \in S_Q$ for some $Q < R$. We claim, however, that $\widehat{S}_R = \bigcup_{Q \leq R} S_Q$ is a compact normal family. To prove this, let the functions $f_n \in \widehat{S}_R$ map $\mathbb{D}$ onto a sequence of regions Δ_n with transfinite diameters $R_n \leq R$. Let $E_n = \overline{\Delta}_n$, and let

$$\tilde{f}_n(z) = R_n z + b_{0n} + b_{1n}/z + \cdots$$

map $|z| > 1$ conformally onto $\widetilde{E}_n$. We may choose a subsequence $\{n_k\}$ such that $f_{n_k}(z) \to f(z)$ and $\tilde{f}_{n_k}(z) \to \tilde{f}(z)$, uniformly on compact sets. Then $f \in S$ and $\tilde{f}$ maps $|z| > 1$ onto the complement of a compact set E with transfinite diameter $Q \leq R$. Since $f(\mathbb{D}) \subset E$, it follows that $f \in S_P$ for some $P \leq Q$. This proves the compactness of $\widehat{S}_R$.

The compactness of $\widehat{S}_R$ ensures that $\operatorname{Re}\{a_2\}$ attains a maximum within the family. We may safely assume that the maximum occurs for a function $f \in S_R$, because the extremal value of $\operatorname{Re}\{a_2\}$ will be seen to increase monotonically with the transfinite diameter of $f(\mathbb{D})$.

The extremal functions will be found by applying a variation which preserves the class S_R. Such a variation is constructed as follows. Given $f \in S_R$, fix three distinct points w_1, w_2, $w_3 \notin \overline{f(\mathbb{D})}$ and consider the perturbation

$$w^* = w + \sum_{j=1}^{3} \varepsilon_j \frac{w}{w - w_j}, \tag{1}$$

where the ε_j are small complex parameters. If $\varepsilon = \max |\varepsilon_j|$ is sufficiently small, then clearly

$$f^*(z) = f(z) + \sum_{j=1}^{3} \varepsilon_j \frac{f(z)}{f(z) - w_j} = \sum_{n=1}^{\infty} a_n^* z^n \tag{2}$$

is univalent in $\mathbb{D}$. Observe that

$$a_1^* = 1 - \sum_{j=1}^{3} \varepsilon_j w_j^{-1}, \tag{3}$$

$$a_2^* = a_2 - \sum_{j=1}^{3} \varepsilon_j (a_2 w_j^{-1} + w_j^{-2}). \tag{4}$$

In view of (3), the condition

$$\sum_{j=1}^{3} \varepsilon_j w_j^{-1} = 0 \tag{5}$$

will ensure that $f^* \in S$.

The transfinite diameter R^* of the region $f^*(\mathbb{D})$ can be calculated by the method of interior variation, which gives a formula for Green's function $g^*(w, \infty)$ of the complement of $\overline{f^*(\mathbb{D})}$. Let $g(w, \infty)$ be Green's function of the complement of $\overline{f(\mathbb{D})}$ with pole at infinity, and let $p(w) = p(w, \infty)$ be its (multiple-valued) analytic completion. Then we have the variational formula (cf. [11])

$$R^* = R\left(1 - \operatorname{Re}\left\{\sum_{j=1}^{3} \varepsilon_j w_j p'(w_j)^2\right\}\right) + O(\varepsilon^2).$$

In order to have $R^* = R$, we must first require

$$\sum_{j=1}^{3} \varepsilon_j w_j p'(w_j)^2 = 0. \tag{6}$$

Then under conditions (5) and (6), the variation (2) may be modified by adding a term of order ε^2 to achieve $R^* = R$ while retaining $a_1^* = 1$. The modified variation preserves the class S_R and changes formula (4) for a_2^* only by the addition of a term of order ε^2. (This last step is based on the implicit function theorem and is a standard technique for constrained variation. See [6 or 12] for a more detailed discussion.)

Now hold the ε_j fixed at values (not all zeros) satisfying (5) and (6), and let ε_j be replaced by $\rho e^{i\theta} \varepsilon_j$, where $e^{i\theta}$ is an arbitrary constant of unit modulus and ρ is a small positive parameter. Note that conditions (5) and (6) are unaffected. After modifying the variation (2) as indicated above, we obtain $f^* \in S_R$ and

$$a_2^* = a_2 - \rho e^{i\theta} \sum_{j=1}^{3} \varepsilon_j (a_2 w_j^{-1} + w_j^{-2}) + O(\rho^2). \tag{7}$$

Suppose now that f maximizes $\operatorname{Re}\{a_2\}$ in S_R. Then $\operatorname{Re}\{a_2^*\} \le \operatorname{Re}\{a_2\}$, and (7) gives

$$\sum_{j=1}^{3} \varepsilon_j (a_2 w_j^{-1} + w_j^{-2}) = 0.$$

In view of (5), this reduces to

$$\sum_{j=1}^{3} \varepsilon_j w_j^{-2} = 0. \tag{8}$$

Conditions (5), (6), and (8) constitute a linear homogeneous system of equations with a nontrivial solution $(\varepsilon_1, \varepsilon_2, \varepsilon_3)$. Therefore, the determinant of coefficients vanishes:

$$\begin{vmatrix} w_1^{-1} & w_2^{-1} & w_3^{-1} \\ w_1^{-2} & w_2^{-2} & w_3^{-2} \\ w_1 p'(w_1)^2 & w_2 p'(w_2)^2 & w_3 p'(w_3)^2 \end{vmatrix} = 0. \tag{9}$$

This is a necessary condition for the extremal function f, valid for each choice of distinct points w_j outside the closure of the range of f.

In order to exploit condition (9), we hold w_2 and w_3 fixed ($w_2 \neq w_3$) and regard $w_1 = w$ as variable. Expansion of determinant (9) along the first column gives

$$\lambda_1 w^{-1} + \lambda_2 w^{-2} + \lambda_3 w p'(w)^2 = 0,$$

where $\lambda_j = \lambda_j(w_2, w_3)$ and $\lambda_3 \neq 0$ since $w_2 \neq w_3$. Thus we have

$$p'(w)^2 = aw^{-2} + bw^{-3} \tag{10}$$

for some constants a and b.

Recall now that $p(w)$ is the analytic completion of Green's function. It has the form

$$p(w) = \log w + c_0 + c_1 w^{-1} + c_2 w^{-2} + \cdots$$

near infinity, and so

$$p'(w)^2 = w^{-2} - 2c_1 w^{-3} + \cdots. \tag{11}$$

A comparison of (10) and (11) shows that $a = 1$. Thus

$$w^2 p'(w)^2 = 1 + b/w, \qquad b \in \mathbb{C}. \tag{12}$$

On the other hand, $p(w)$ is purely imaginary on the boundary of $\overline{f(\mathbb{D})}$, and so we may conclude informally that the boundary is composed of arcs $w = w(t)$ which satisfy the differential equation

$$[p'(w(t))w'(t)]^2 < 0.$$

In view of (12), this suggests that the outer boundary of $f(\mathbb{D})$ lies on trajectories of the quadratic differential

$$-\left(1 + \frac{b}{w}\right)\frac{dw^2}{w^2} > 0. \tag{13}$$

In fact, the entire boundary of $f(\mathbb{D})$ satisfies (13). For a proof which will apply also to internal spires (if any), we now develop a special kind of boundary variation which preserves the family S_R. This is an auxiliary variation which will make use of expression (12) for $p'(w)$.

Let Γ be the boundary of $f(\mathbb{D})$. Choosing an arbitrary point $w_0 \in \Gamma$, we begin with the boundary variation (see [10] or [3, Chapter 10])

$$w^* = w + \frac{a\rho^2 w}{w_0(w - w_0)} + O(\rho^3), \tag{14}$$

which is analytic and univalent outside a small subcontinuum of Γ containing w_0, and vanishes at the origin. Next we choose distinct points $w_1, w_2 \notin \overline{f(\mathbb{D})}$ and modify the variation (14) by adding two terms $\varepsilon_j w(w - w_j)^{-1}$, where the ε_j are of order ρ^2 and will be specified later. This produces a variation

$$w^* = V_\rho(w) = w + \frac{a\rho^2 w}{w_0(w - w_0)} + \frac{\varepsilon_1 w}{w - w_1} + \frac{\varepsilon_2 w}{w - w_2} + O(\rho^3)$$

which is analytic and univalent outside a small subcontinuum of Γ near w_0 except for small disks centered at w_1 and w_2. Then $f^* = V_\rho \circ f$ is analytic and univalent in $\mathbb{D}$, and $f^*(0) = 0$. In order to make $f^{*\prime}(0) = 1$, we must require that $V_\rho'(0) = 1$, or

$$a\rho^2 w_0^{-2} + \varepsilon_1 w_1^{-1} + \varepsilon_2 w_2^{-1} = 0. \tag{15}$$

Then $f^* \in S$.

The transfinite diameter of $f^*(\mathbb{D})$ is

$$R^* = R[1 - \mathrm{Re}\{\varepsilon_1 w_1 p'(w_1)^2 + \varepsilon_2 w_2 p'(w_2)^2\}] + O(\rho^3).$$

In order to make $R^* = R$, we put a second demand

$$\varepsilon_1 w_1 p'(w_1)^2 + \varepsilon_2 w_2 p'(w_2)^2 = 0 \tag{16}$$

on the variation. We then adjust the error term $O(\rho^3)$ to have $R^* = R$ and $a_1^* = 1$. Thus $f^* \in S_R$.

In view of (12), condition (16) takes the form

$$\frac{\varepsilon_1}{w_1}\left(1 + \frac{b}{w_1}\right) + \frac{\varepsilon_2}{w_2}\left(1 + \frac{b}{w_2}\right) = 0.$$

Combining this with (15), we obtain

$$a\rho^2 w_0^{-2} = b(\varepsilon_1 w_1^{-2} + \varepsilon_2 w_2^{-2}). \tag{17}$$

A calculation now gives

$$\begin{aligned} a_2^* &= a_2 - \frac{a\rho^2}{w_0^2}\left(a_2 + \frac{1}{w_0}\right) - \frac{\varepsilon_1}{w_1}\left(a_2 + \frac{1}{w_1}\right) \\ &\quad - \frac{\varepsilon_2}{w_2}\left(a_2 + \frac{1}{w_2}\right) + O(\rho^3) \\ &= a_2 - \frac{a\rho^2}{bw_0^2}\left(1 + \frac{b}{w_0}\right) + O(\rho^3), \end{aligned} \tag{18}$$

where (15) and (17) have been used. The assumption that $b \neq 0$ is justified by (12). Indeed, if $w^2 p'(w)^2 = 1$, then Green's function takes the form $g(w) = \log|w| - \log R$, and the circle $|w| = R$ is the outer boundary of $f(\mathbb{D})$. This returns us to Pick's problem (see §0), so the sharp bound would be $|a_2| \leq 2(1 - 1/R)$. Later, however, it will be apparent that $|a_2|$ can always be larger for functions in the class S_R.

Since $f^* \in S_R$ and f is extremal, we have $\operatorname{Re}\{a_2^*\} \leq \operatorname{Re}\{a_2\}$. Thus (18) gives

$$\operatorname{Re}\left\{\frac{a\rho^2}{bw_0^2}\left(1 + \frac{b}{w_0}\right) + O(\rho^3)\right\} \geq 0$$

for each $w_0 \in \Gamma$. It now follows from the fundamental lemma of the method of boundary variation ([10]; see also [3, p. 297]) that Γ consists of analytic arcs which satisfy

$$-\frac{1}{b}\left(1 + \frac{b}{w}\right)\frac{dw^2}{w^2} > 0. \tag{19}$$

It will be seen presently that $b > 0$, so that (19) and (13) are equivalent.

2. Analysis of the quadratic differential

The quadratic differential (19) has a simple zero at $-b$, a triple pole at the origin, and a double pole at infinity. Parametrizing Γ by $w = f(e^{it})$, we conclude from (19) that

$$F(z) = \frac{z^2 f'(z)^2}{bf(z)^3}[f(z) + b] \geq 0 \tag{20}$$

on $|z| = 1$. Since $f \in S$, it is clear that F is analytic in $\mathbb{D}$ except for a simple pole at the origin with residue 1. Taking into account all possible singularities of Γ at singular points of the quadratic differential, one shows that F is analytic and nonvanishing on the unit circle, except perhaps for one double zero. By the Schwarz reflection principle, F has a meromorphic continuation to the Riemann sphere which satisfies $F(1/\bar{z}) = \overline{F(z)}$. Thus F has the form

$$F(z) = z + 2c + 1/z,$$

where $c \geq 1$. But a short calculation gives $2c = a_2 + 1/b$. Since $0 < a_2 < 2$, we conclude that $b > 0$. In particular, conditions (19) and (13) are equivalent.

Because $b > 0$, the trajectories of (13) are symmetric with respect to the real axis, and the real segment $-b < w < 0$ is a trajectory. The structure of the trajectories is shown in Figure 1.

There are now three cases to consider.

Case I. Γ is an analytic Jordan curve surrounding the point $-b$. Then F is analytic on the unit circle and $F(z) \neq 0$ in $\overline{\mathbb{D}}$ except for a simple zero at the point $-\eta$ $(0 < \eta < 1)$ where $f(-\eta) = -b$.

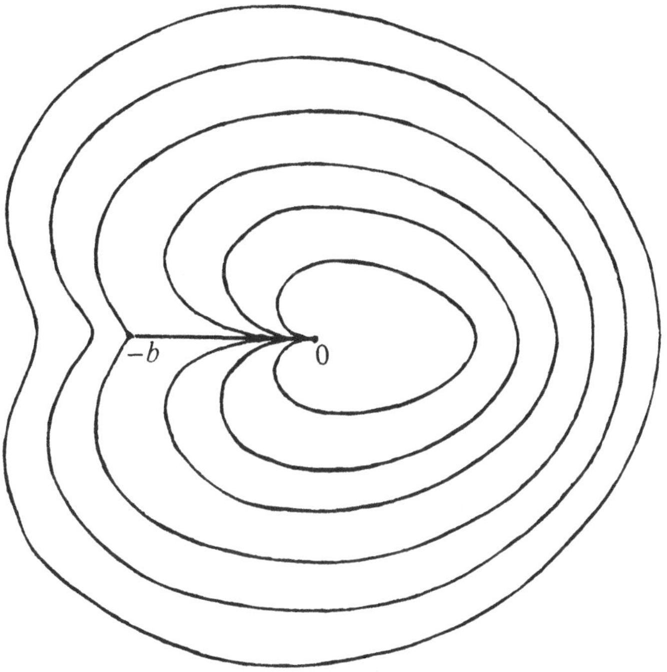

FIGURE 1. Trajectories of the quadratic differential

Case II. Γ is a Jordan curve passing through $-b$. Then F is analytic and nonvanishing on the unit circle except at the point -1 where $f(-1) = -b$. A calculation shows that F has a double zero at -1. Except for the pole at the origin, F is analytic and nonvanishing in $\mathbb{D}$.

Case III. Γ consists of a Jordan curve through $-b$ plus a segment $[-b, \tau]$ of the real axis, where $-b < \tau < -\frac{1}{4}$. Then we find that F is analytic and nonvanishing in $\overline{\mathbb{D}}\backslash\{0\}$ except for a double zero at the point -1 where $f(-1) = \tau$.

In Cases II and III, the double zero on the boundary implies that $c = 1$ and so

$$F(z) = z(1 + 1/z)^2 \tag{21}$$

and $b = (2 - a_2)^{-1} > \frac{1}{2}$. Because of the perfect square, integration of the differential equation is elementary in Cases II and III. This will be carried out in §3, while Case I will be deferred to §§4 and 5.

3. CASES II AND III: CORNERS AND INTERNAL SPIRES

Suppose now that either Case II or III prevails, so that F is given by (21) and $b = (2 - a_2)^{-1}$. The differential equation for $w = f(z)$ may then be expressed in the form

$$\left(1 + \frac{b}{w}\right)^{1/2} \frac{dw}{w} = \sqrt{b}\left(1 + \frac{1}{z}\right) \frac{dz}{\sqrt{z}}.$$

Integration of the left-hand side gives

$$
\begin{aligned}
G(w) &= \int_{-b}^{w}\left(1+\frac{b}{w}\right)^{1/2}\frac{dw}{w} \\
&= -2\left(1+\frac{b}{w}\right)^{1/2} + \log\frac{1+(1+b/w)^{1/2}}{1-(1+b/w)^{1/2}} \qquad (22)\\
&= \log w + \log(-4/b) - 2 + O(1/w).
\end{aligned}
$$

Recalling that Γ is a trajectory of (13), we see that $\operatorname{Re}\{G(w)\} = 0$ on Γ. Furthermore, G is analytic outside $\overline{f(\mathbb{D})}$ apart from the logarithmic singularity at infinity. It follows that $\operatorname{Re}\{G(w)\}$ is Green's function

$$g(w) = \log|w| - \log R + O(1/w)$$

of the complement of $\overline{f(\mathbb{D})}$, and that

$$\log R = \log b - \log 4 + 2, \qquad (23)$$

or $R = be^2/4$. But $b = (2-a_2)^{-1}$, so $a_2 = 2 - e^2/4R$. Because f maximizes $\operatorname{Re}\{a_2\}$, this gives the sharp bound

$$|a_2| \le 2 - e^2/4R \qquad (24)$$

whenever Case II or III occurs.

It remains to determine the values of R for which Case II or III can occur as an extremal configuration. These will correspond to the values of $b = 4R/e^2$ for which the Jordan curve of Case II bounds a region Ω_b of inner radius $r_b \ge 1$. If $r_b > 1$, then Ω_b can be modified by the insertion of a radial slit (as in Case III) to reduce its inner radius to 1. If $r_b < 1$, however, no admissible modification can change Ω_b to the range of a function in S, and Case I must prevail.

It is important, therefore, to calculate the inner radius r_b of the region Ω_b. Let

$$w = \varphi(z) = \alpha z + \beta z^2 + \cdots, \qquad \alpha = r_b > 0,$$

map $\mathbb{D}$ conformally onto Ω_b. Parametrizing the boundary of Ω_b by $w = \varphi(e^{it})$ and invoking (13), we obtain as in §2

$$\left(1+\frac{b}{w}\right)\frac{dw^2}{w^2} = \frac{b}{\alpha z}\left(1+\frac{1}{z}\right)^2 dz^2. \qquad (25)$$

For small $\varepsilon > 0$, integration of (25) gives

$$
\begin{aligned}
G(f(-\varepsilon)) &= \int_{-b}^{f(-\varepsilon)}\left(1+\frac{b}{w}\right)^{1/2}\frac{dw}{w} = \left(\frac{b}{\alpha}\right)^{1/2}\int_{-1}^{-\varepsilon}(1+z)z^{-3/2}\,dz \\
&= 2\left(\frac{b}{\alpha}\right)^{1/2}(\varepsilon^{1/2} - \varepsilon^{-1/2} - 2)i.
\end{aligned}
$$

But by (22),

$$G(f(-\varepsilon)) = -2i\left(\frac{b}{\alpha\varepsilon}\right)^{1/2} - \pi i + O(\sqrt{\varepsilon}). \qquad (26)$$

Comparing the two expressions and letting ε tend to 0, we obtain

$$r_b = \frac{16b}{\pi^2} = \frac{64R}{e^2\pi^2}.$$

Thus $r_b < 1$ precisely when

$$R < e^2\pi^2/64 = 1.139\ldots.$$

This proves that the extremal function cannot have the form of Case II or III if $1 < R < e^2\pi^2/64$. In other words, *each extremal function must have the form of Case* I *if* $1 < R < e^2\pi^2/64$.

So far it is unclear whether Case I or III prevails if $R > e^2\pi^2/64$, or whether Case I or II prevails if $R = e^2\pi^2/64$. In the next two sections we shall show that if an extremal function has the form of Case I, then $1 < R < e^2\pi^2/64$. Since we have just proved the converse, this will imply that Case I prevails if $R < e^2\pi^2/64$, Case II if $R = e^2\pi^2/64$, and Case III if $R > e^2\pi^2/64$.

4. Case I: analytic Jordan curves

Suppose now that Case I describes an extremal function $f \in S_R$ for some R. Then f maps $\mathbb{D}$ onto the interior of an analytic Jordan curve Γ which is a trajectory of the quadratic differential (13) for some $b > 0$. Furthermore, f satisfies the differential equation

$$F(z) = z + 2c + 1/z \tag{27}$$

for some $c = \frac{1}{2}(a_2 + 1/b) > 1$, where F is defined by (20). The point $-\eta$ where $f(-\eta) = -b$ is a zero of F, so the quadratic equation gives $\eta = c - (c^2 - 1)^{1/2}$.

Now consider the function

$$H(w) = \int_{w_0}^{w} \left\{1 + \frac{b}{w}\right\}^{1/2} \frac{dw}{w} = G(w) - G(w_0),$$

where $w_0 = f(-1)$ is the point at which Γ meets the negative real axis. Because Γ satisfies (13), it follows as before that $\operatorname{Re}\{H(z)\}$ is Green's function of the region outside Γ. Comparison with (22) gives

$$\log R = \log b - \log 4 + 2 + \operatorname{Re}\{G(w_0)\}.$$

But by (27)

$$G(w_0) = \int_{-b}^{w_0} \left\{1 + \frac{b}{w}\right\}^{1/2} \frac{dw}{w} = \sqrt{b} \int_{-\eta}^{-1} \left\{z + 2c + \frac{1}{z}\right\}^{1/2} \frac{dz}{z}.$$

Hence

$$\log R = 2 + \log(b/4) + \sqrt{b}\, I, \tag{28}$$

where

$$I = \int_{\eta}^{1} \left\{2c - \left(t + \frac{1}{t}\right)\right\}^{1/2} \frac{dt}{t}. \tag{29}$$

A further relation is obtained from the fact that Γ must enclose a region of inner radius 1. For small $\varepsilon > 0$ we have by (27)

$$b^{-1/2}G(f(-\varepsilon)) = \int_{-\eta}^{-\varepsilon}\left(t + 2c + \frac{1}{t}\right)^{1/2}\frac{dt}{t}$$
$$= -i\int_{\varepsilon}^{\eta}\{(1-2ct+t^2)^{1/2}-1\}t^{-3/2}\,dt + 2i(\eta^{-1/2}-\varepsilon^{-1/2}).$$

Referring to (26) with $\alpha = 1$, we conclude that

$$-b^{-1/2}\pi = 2\eta^{-1/2} - J, \tag{30}$$

where

$$J = \int_0^{\eta}\{(1-2ct+t^2)^{1/2}-1\}t^{-3/2}\,dt. \tag{31}$$

The integral (31) is easily simplified. Integration by parts gives

$$J = 2\eta^{-1/2} + 2\int_0^{\eta}(1-2ct+t^2)^{-1/2}(t-c)t^{-1/2}\,dt.$$

Observe now that the integrand has an analytic extension to the closed unit disk with a branch cut along the segment from 0 to η. By Cauchy's integral theorem, the contour may be deformed to the unit circle $\mathbb{T}$. Thus (30) becomes

$$\pi b^{-1/2} = \int_{\mathbb{T}}\left(z - 2c + \frac{1}{z}\right)^{-1/2}(z-c)\frac{dz}{z},$$

or

$$\int_0^{2\pi}(c+\cos\theta)^{1/2}\,d\theta = \frac{\pi\sqrt{2}}{\sqrt{b}}, \tag{32}$$

where

$$2c = a_2 + 1/b > 2. \tag{33}$$

Equations (28), (32), and (33) allow the computation of a_2 and R, with c as a parameter. The computation is facilitated by expressing the integrals in terms of standard elliptic integrals. This will be done in the next section.

5. Elliptic integrals

The integral I defined by (29) may be transformed as follows. Let $k = (1-\eta)/(1+\eta)$ and make the substitution $x = (1-t)/(1+t)$. Then

$$2I = \int_{\eta}^{1/\eta}\left\{2c - \left(t + \frac{1}{t}\right)\right\}^{1/2}\frac{dt}{t}$$
$$= 2\sqrt{2}\int_{-k}^{k}\left\{\frac{1+k^2}{1-k^2} - \frac{1+x^2}{1-x^2}\right\}^{1/2}\frac{dx}{1-x^2},$$

since $2c = \eta + 1/\eta$. Making the further substitution $x = k\sin\theta$, we then obtain

$$I = 4k^2(1-k^2)^{-1/2}\int_0^{\pi/2}\cos^2\theta(1-k^2\sin^2\theta)^{-3/2}\,d\theta,$$

where $0 < k < 1$.

The last integral can be expressed in terms of the standard complete elliptic integrals

$$K = K(k) = \int_0^{\pi/2}(1-k^2\sin^2\theta)^{-1/2}\,d\theta$$

and

$$E = E(k) = \int_0^{\pi/2}(1-k^2\sin^2\theta)^{1/2}\,d\theta$$

of the first and second kinds, respectively (see [1] or [2]). It may be verified that

$$I = (4/k')(K-E), \tag{34}$$

where $k' = (1-k^2)^{1/2}$ is the complementary modulus. Substituting (34) into (28), we have

$$\log R = 2 + \log(b/4) + 4\sqrt{b}(K-E)/k'. \tag{35}$$

The integral in (32) can also be reduced to standard form. The half-angle formula gives

$$\int_0^{2\pi}(c+\cos\theta)^{1/2}\,d\theta = \frac{4\sqrt{2}}{k'}E',$$

where

$$K' = K'(k) = K(k'); \qquad E' = E'(k) = E(k').$$

Thus (32) reduces to

$$4E' = \pi k'/\sqrt{b}, \tag{36}$$

while (33) takes the form

$$a_2 = 2\frac{1+k^2}{1-k^2} - \frac{16E'^2}{\pi^2 k'^2}. \tag{37}$$

Combining (35) and (36) to eliminate b, we obtain the basic relation

$$\log R = \log(e^2\pi^2/64) + 2\log k' - 2\log E' + \pi(K-E)/E'. \tag{38}$$

Observe first that for $k = 0$ (a limiting value for Case I), $K = E = \pi/2$ and $E' = 1$, so that (38) gives $R = e^2\pi^2/64$. For $k = 1$ we have $E = 1$ and $E' = \pi/2$, while

$$\lim_{k\to 1}\{K(k) - \log(4/k')\} = 0. \tag{39}$$

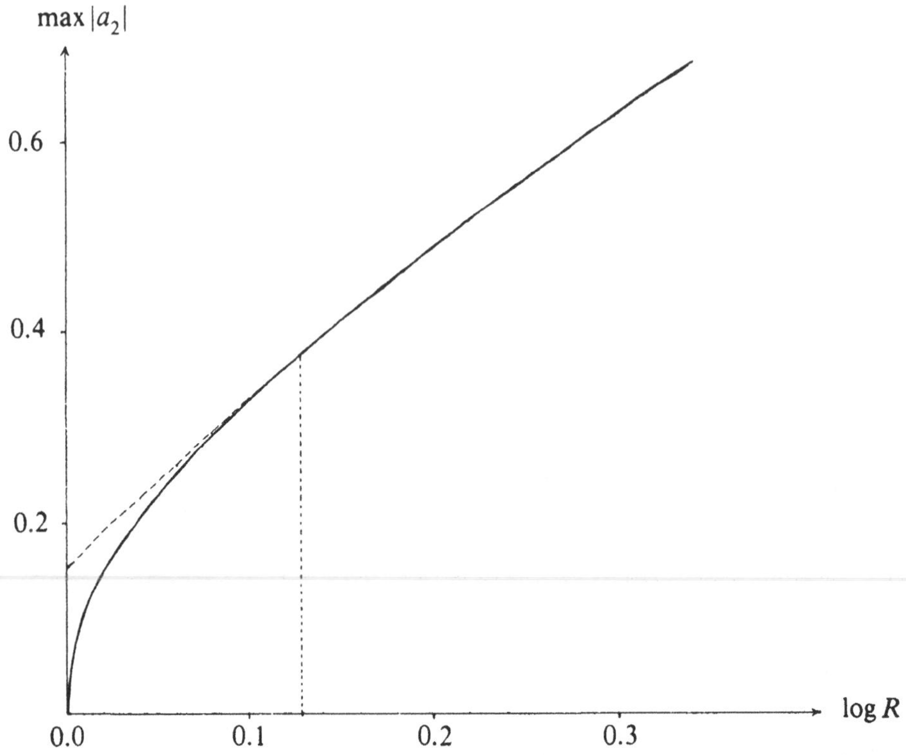

FIGURE 2. The solution for small R

(See [2, #112.01].) Thus a short calculation based on (38) shows that $R \to 1$ as $k \to 1$.

We shall now show on the basis of (38) that R is a strictly decreasing function of k, $0 < k < 1$. This will establish our assertion that $1 < R < e^2\pi^2/64$ if the extremal function has the form of Case I. It will also show that R uniquely determines k, b, c, and a_2 via equations (38), (36), and (37).

In using (38) to calculate the derivative of $\log R$ with respect to k, we must use the formulas [2, #710.00 and #710.02]

$$\frac{dK}{dk} = \frac{E - k'^2 K}{kk'^2}; \qquad \frac{dE}{dk} = \frac{E - K}{k}. \tag{40}$$

Using also the simple formula $dk'/dk = -k/k'$, we find after some manipulation

$$\begin{aligned} k'^2 E'^2 \frac{d(\log R)}{dk} &= -2kE'K' + \pi k(E'K + EK' - KK') \\ &= k(\pi^2/2 - 2E'K'), \end{aligned} \tag{41}$$

where Legendre's relation [2, #110.10] was used in the last step. But the Schwarz inequality shows immediately that $E'K' > \pi^2/4$. This completes the proof that

$$\frac{d(\log R)}{dk} < 0, \qquad 0 < k < 1.$$

Consequently, we have shown that $1 < R < e^2\pi^2/64$ *whenever Case* I *prevails.* Hence in view of the results in §3, we have proved the following theorem.

Theorem. *For all functions $f \in S$ with range of transfinite diameter R, the sharp inequality $|a_2| \le 2 - e^2/4R$ holds if $R \ge e^2\pi^2/64$. If $1 < R < e^2\pi^2/64$, the sharp inequality involves elliptic integrals and is implicitly determined by* (37) *and* (38), *with k $(0 < k < 1)$ acting as a parameter. In all cases the extremal function is unique up to a rotation, and the function maximizing* $\mathrm{Re}\{a_2\}$ *maps the disk onto a region bounded by trajectories of the quadratic differential* (13), *as shown in Figure* 1, *where $b = 4e^{-2}R$ if $R \ge e^2\pi^2/64$ and $b = \pi^2(1-k^2)E'(k)^{-2}/16$ if $1 < R < e^2\pi^2/64$. If $R > e^2\pi^2/64$, the range of the extremal function is bounded by a Jordan curve which is analytic except for a corner at the point $-b$, plus a slit along the negative real axis from $-b$ toward the origin. The length of this linear slit is determined by the requirement that the curves bound a region of inner radius* 1. *If $R = e^2\pi^2/64$, the slit is not present. If $R < e^2\pi^2/64$, the extremal range is bounded by an analytic Jordan curve surrounding the point $-b$. This trajectory of* (13) *is determined by the requirement that its interior have inner radius* 1.

The relations (37) and (38), together with a table of complete elliptic integrals, can be used to compute numerical values for the sharp bound on $|a_2|$ in S_R for $1 < R < e^2\pi^2/64$. A sample is shown in Table 1, and a graph is plotted in Figure 2. In Figure 2 the vertical dotted line marks the junction of the two curves at $R = e^2\pi^2/64$, and the dashes to the left indicate the continuation of the curve $a_2 = 2 - e^2/4R$. The graph suggests that the two curves have a common tangent line at the junction, and a calculation confirms this. One may compute da_2/dk from (37), using (40). Equation (41) gives dR/dk. Taking the ratio and letting k tend to 0, one finds from (39) that

$$\frac{da_2}{dR} = \frac{1024}{e^2\pi^4} = 1.4222\ldots$$

at $R = e^2\pi^2/64$. The curve $a_2 = 2 - e^2/4R$ is easily seen to have the same derivative there.

Note added in proof. Professor J. A. Jenkins has pointed out to us that the theorem of this paper is implicit in Theorem 7.3 of his book *Univalent functions and conformal mapping* (Springer-Verlag, 1958). His derivation relies on the General Coefficient Theorem. Our variational approach has other applications; for instance, R. Cunningham has applied it to find the maximum of $|f(\zeta)|$ for fixed $\zeta \in \mathbb{D}$ as f ranges over the class S_R.

TABLE 1. Maximum of $|a_2|$ for $1 \le R \le e^2\pi^2/64$

k^2	R	$\max\|a_2\|$
1.00	1.0000	0.0000
.90	1.0002	0.0131
.80	1.0008	0.0278
.70	1.0020	0.0444
.60	1.0040	0.0633
.50	1.0073	0.0853
.40	1.0124	0.1115
.30	1.0206	0.1437
.20	1.0343	0.1856
.15	1.0448	0.2125
.10	1.0599	0.2459
.08	1.0680	0.2622
.06	1.0779	0.2809
.04	1.0905	0.3031
.02	1.1077	0.3315
.01	1.1198	0.3501
.00	1.1395	0.3789

REFERENCES

1. Bateman Manuscript Project (A. Erdélyi, ed), *Higher transcendental functions*, Vol. II, McGraw-Hill, New York, 1953.
2. P. F. Byrd and M. D. Friedman, *Handbook of elliptic integrals for engineers and physicists*, Springer-Verlag, Berlin and New York, 1954.
3. P. L. Duren, *Univalent functions*, Springer-Verlag, Berlin and New York, 1983.
4. G. M. Goluzin, *Geometric theory of functions of a complex variable*, Moscow, 1952; German. transl., Deutscher Verlag, Berlin, 1957; 2nd ed., Moscow, 1966; English transl., Amer. Math. Soc., Providence, R. I., 1969.
5. E. Hille, *Analytic function theory*, Vol. II, Ginn, Boston, Mass., 1962; 2nd ed., Chelsea, New York, 1987.
6. J. A. Hummel, *Lagrange multipliers in variational methods for univalent functions*, J. Analyse Math. **32** (1977), 222–234.
7. G. Pick, *Uber die konforme Abbildung eines Kreises auf ein schlichtes und zugleich beschränktes Gebiet*, S.-B. Kaiserl. Akad. Wiss. Wien **126** (1917), 247–263.

8. G. Pólya, *Beitrag zur Verallgemeinerung des Verzerrungssatzes auf mehrfach zusammenhängende Gebiete*. I, II, S.-B. Preuss. Akad. Wiss. (1928), 228–232; 280–282.

9. G. Pólya and G. Szegö, *Aufgaben und Lehrsätze aus der Analysis*, Zweiter Band, Springer-Verlag, Berlin, 1925; English transl. *Problems and theorems in analysis*, vol. II, Springer-Verlag, New York, 1976.

10. M. Schiffer, *A method of variation within the family of simple functions*, Proc. London. Math. Soc. **44** (1938), 432–449.

11. ———, *Hadamard's formula and variation of domain-functions*, Amer. J. Math. **68** (1946), 417–448.

12. M. Schiffer and D. C. Spencer, *Functionals of finite Riemann surfaces*, Princeton Univ. Press, Princeton, N. J., 1954.

13. G. Szegö, *Bermerkungen zu einer Arbeit von Herrn M. Fekete: Über die Verteilung der Wurzeln bei gewissen algebraischen Gleichungen mit ganzzahligen Koeffizienten*, Math. Z. **21** (1924), 203–208.

14. M. Tsuji, *Potential theory in modern function theory*, Maruzen, Tokyo, 1959; 2nd ed., Chelsea, New York, 1975.

DEPARTMENT OF MATHEMATICS, UNIVERSITY OF MICHIGAN, ANN ARBOR, MICHIGAN 48109

DEPARTMENT OF MATHEMATICS, STANFORD UNIVERSITY, STANFORD, CALIFORNIA 94305

Commentary on

[135] P. L. Duren and M. M. Schiffer, *Univalent functions which map onto regions of given transfinite diameter*, Trans. Amer. Math. Soc. **323** (1991), 413–428.

Of the ten papers that Max Schiffer and I wrote together, [135] is my personal favorite. I take special pleasure in remembering this work. We set about to explore the relationship of the second coefficient a_2 of a function $f \in S$ with the transfinite diameter $d(f(\mathbb{D}))$ of its range. After experimenting with other formulations, we settled on asking for the maximum value of $|a_2|$ for functions $f(z) = z + a_2z^2 + \ldots$ in the class $S_R = \{f \in S : d(f(\mathbb{D})) = R\}$, where $1 < R < \infty$. In that form, the problem recalls the classical bound $|a_2| \leq 2(1 - 1/M)$ for functions $f \in S$ with $|f(z)| \leq M$, found by Pick [P], where equality occurs for mappings onto the disk $|w| < M$ with a radial slit of suitable length. But the problem for the class S_R lies deeper, and its solution demonstrates the great power of Schiffer's variational method. It turns out that for $R > e^2\pi^2/64$, the solution is similar to that of Pick's problem: $|a_2| \leq 2(1 - e^2/8R)$, with equality for mappings onto a certain Jordan region with a radial slit. For $R < e^2\pi^2/64$, on the other hand, the extremal functions map onto a Jordan region with analytic boundary, without a slit, and the sharp bound is given in terms of elliptic integrals.

Needless to say, this solution could not have been anticipated. We had essentially worked out the variation that preserves the class S_R before I traveled to Stanford for a week during which we hoped to solve the extremal problem. As usual, thanks to Schiffer's "fundamental lemma" of boundary variation, it was not difficult to conclude that the boundary of the extremal region $f(\mathbb{D})$ lies on trajectories of a certain quadratic differential. The difficulty lay in integrating the differential equation to obtain quantitative information that would solve the extremal problem. In the case of large R, the integration was elementary; but for small R it led to the complete elliptic integrals

$$K(k) = \int_0^{\pi/2} \frac{1}{\sqrt{1-k^2\sin^2\theta}}\,d\theta$$

and

$$E(k) = \int_0^{\pi/2} \sqrt{1-k^2\sin^2\theta}\,d\theta$$

of the first and second kinds, respectively, with *modulus* k, where $0 < k < 1$. As things turned out, Max was called away unexpectedly, and I was left to face the elliptic integrals alone. For me this was a daunting prospect; but after a day of concentrated effort, I was surprised to find that order had emerged and the problem was under control. Amazingly, the key to the final stages of calculation was the *Legendre relation*

$$K'E + E'K - K'K = \pi/2,$$

where $K' = K'(k) = K(k')$ and $k' = \sqrt{1-k^2}$ is the *complementary modulus*, and E' is defined in a similar way. I had marveled at Legendre's relation when coming across it in lists of formulas for elliptic integrals and now felt impelled to learn its proof and explore its historical origins. Schiffer encouraged my quest and suggested that Enneper's book [E] would be helpful. (It was.) My efforts led to the expository account in [D1]. A more elementary discussion appears in [D2].

The variation developed in [135] has been used to study other extremal problems for functions whose range has prescribed transfinite diameter. Cunningham [C1] applied it to find the maximum modulus $|f(\zeta)|$ for functions $f \in S_R$, where ζ is a prescribed point in $\mathbb{D}$. In [C2], he considered the minimum modulus problem and arrived at the analogue of the classical Koebe one-quarter theorem for the family S_R. Looking at a problem slightly different from the one studied in [135], he then found in [C3] the sharp bound $|a_2| \leq 4R/e^2$ for functions f univalent in $\mathbb{D}$ whose range has transfinite diameter $R > 0$; here there is no restriction on $a_1 = f'(0)$ other than the necessary condition $|a_1| \leq R$. Finally, Cunningham turned to a variant of the maximum modulus problem solved in [C1], and obtained the sharp upper bound for $|f(\zeta)|/d(f(\mathbb{D}))$ for functions $f \in S$. It turns out [C4] that the bound is expressed in terms of the elliptic integral K, while the extremal function f maps $\mathbb{D}$ onto the interior of an ellipse with foci at 0 and $f(\zeta)$, quite an elegant result.

After [135] went to press, Jenkins informed us that our main theorem is implicit in [J, Theorem 7.3]. There, as an application of his General Coefficient Theorem, Jenkins investigates the trajectory structure of the quadratic differential $(z-\mu)z^{-2}dz^2$, where $\mu > 0$. His calculations are somewhat similar to those in [135] (even the Legendre relation is invoked), but he makes no explicit connection with the extremal problem of maximizing $|a_2|$ for functions in S whose range has prescribed transfinite diameter.

References

[C1] Robin Cunningham, *Univalent functions of given transfinite diameter: a maximum modulus problem*, Ann. Acad. Sci. Fenn. Ser. A I Math. **18** (1993), 249–271.

[C2] Robin Cunningham, *A finite capacity analogue of the Koebe one-quarter theorem*, Proc. Amer. Math. Soc. **119** (1993), 869–875.

[C3] Robin Cunningham, *Extremal results for a class of univalent functions*, Complex Variables Theory Appl. **24** (1994), 271–279.

[C4] Robin Cunningham, *An extremal modulus problem for univalent functions*, Complex Variables Theory Appl. **29** (1996), 221–224.

[D1] Peter Duren, *The Legendre relation for elliptic integrals*, Paul Halmos: Celebrating 50 Years of Mathematics, eds. J. H. Ewing and F. W. Gehring, Springer-Verlag, 1991, pp. 305–315.

[D2] Peter Duren, *Invitation to Classical Analysis*, American Mathematical Society, 2012.

[E] Alfred Enneper, *Elliptische Functionen: Theorie und Geschichte*, zweite Auflage, Louis Nebert, 1890.

[J] J. A. Jenkins, *Univalent Functions and Conformal Mapping*, Springer-Verlag, 1958.

[P] Georg Pick, *Über die konforme Abbildung eines Kreises auf ein schlichtes und zugleich beschränktes Gebiet*, Sitzungsberichte der Kaiserlichen Akademie der Wissenschaften in Wien, Math.-Naturwiss. Kl. Abt. IIa **126** (1917), 247–263.

PETER DUREN

[137] (with P. L. Duren) Robin functions and distortion of capacity under conformal mapping

[137] (with P. L. Duren) Robin functions and distortion of capacity under conformal mapping. *Complex Variables Theory Appl.* **21** (1993), 189–196.

Complex Variables, 1993, Vol. 21, pp. 189–196
Reprints available directly from the publisher
Photocopying permitted by license only

Printed in the United States of America

Robin Functions and Distortion of Capacity Under Conformal Mapping*

PETER DUREN
Department of Mathematics, University of Michigan, Ann Arbor, MI 48109

M. M. SCHIFFER
Department of Mathematics, Stanford University, Stanford, CA 94305

Under normalized conformal mappings of a multiply connected domain Ω it is found that the sharp lower bound for the capacity of the image of a given set $A \subset \partial\Omega$ is the Robin capacity of A with respect to Ω. The result is generalized to a sharp inequality for quadratic forms involving Green's function and the Robin function associated with A and Ω.

AMS No. 30C85, 30C35, 30C70, 31A15
Communicated: F. W. Gehring
(Received January 31, 1992)

1. INTRODUCTION

Some years ago, one of the authors [10] applied a variational method to find the minimum logarithmic capacity of the image of an arc on the unit circle under normalized conformal mappings

$$f(z) = z + b_0 + b_1 z^{-1} + \cdots, \qquad |z| > 1, \tag{1}$$

of the exterior of the disk. Specifically, if an arc I on the circle subtends an angle α at the center of the disk, then the sharp inequality $d(f(I)) \geq \sin^2(\alpha/4)$ holds, where d denotes the capacity. Observing that the capacity of I is $d(I) = \sin(\alpha/4)$, Pommerenke [7, 8] later used the Goluzin inequalities to show more generally that $d(f(A)) \geq d(A)^2$ for an arbitrary closed set A on the unit circle.

In the present paper we show that Pommerenke's estimate, with suitable interpretation, can be generalized to an arbitrary multiply connected domain. A variational method leads to the sharp estimate $d(f(A)) \geq \delta(A)$, where f is any conformal mapping of the domain with the form (1) near infinity and $\delta(A)$ is the "Robin capacity" of the boundary set A, to be defined below. When the domain is the exterior of the unit disk, it turns out that $\delta(A) = d(A)^2$, which explains the elegance of Pommerenke's result.

Our definition of Robin capacity is potential-theoretic. Ahlfors and Beurling [2, 1] showed that another definition can be given in terms of "reduced extremal distance". This approach leads also to the inequality $d(f(A)) \geq \delta(A)$ but does not establish its sharpness.

*Dedicated to the memory of Glenn Schober.

The paper concludes with a sharp inequality (Theorem 4) comparing Green's function with Robin's function, in analogy with previous results in [5]. A particular case is the pointwise domination of Green's function by Robin's function, which gives the inequality $d(f(A)) \geq \delta(A)$ as a special corollary.

2. MINIMUM CAPACITY

Let Ω be a finitely connected domain in the extended complex plane $\hat{\mathbb{C}}$, containing the point at infinity and bounded by smooth Jordan curves $\Gamma_1, \Gamma_2, \ldots, \Gamma_n$. Let $A \subset \partial\Omega$ be an arbitrary closed subset of the boundary and let $B = \partial\Omega \setminus A$ be its complement. For a fixed finite point $\zeta \in \Omega$, the *Robin function* $R(z,\zeta)$ is defined by the following requirements:

(i) $R(z,\zeta)$ is harmonic in Ω except for a logarithmic singularity at ζ, so that $R(z,\zeta) + \log|z-\zeta|$ is harmonic near ζ;
(ii) $R(z,\zeta)$ is continuous together with its first partial derivatives in the closure of Ω;
(iii) $R(z,\zeta) = 0$ for all $z \in A$, while the normal derivative $\partial R/\partial n(z,\zeta) = 0$ for all $z \in B$.

For $\zeta = \infty$ the definition is modified by requiring $R(z,\infty) - \log|z|$ to be harmonic in a neighborhood of infinity.

In the case where A is the whole boundary, the Robin function is simply Green's function $g(z,\zeta)$ of Ω. The existence of Robin's function was established in [5] only for the special case where A is the union of some collection of entire boundary components Γ_j, but the proof extends to the more general situation. Note that the Robin function is of interest even for a simply connected domain. Some basic properties of Robin functions were pointed out in [5], but only its conformal invariance needs to be mentioned here. If f is a conformal mapping of Ω onto a smoothly bounded domain $\tilde{\Omega}$, and if $\tilde{R}(w,\omega)$ is Robin's function of $\tilde{\Omega}$ with respect to $\tilde{A} = f(A)$, then $R(z,\zeta) = \tilde{R}(f(z), f(\zeta))$. This identity allows the Robin function to be defined even for domains with rough boundary, where the normal direction has no meaning.

Actually, the Robin function (like Green's function) is defined only for nondegenerate sets A. It is defined directly for unions of finitely many arcs, and the definition is then extended to more general sets by a process of approximation.

The *logarithmic capacity* (or *transfinite diameter*) of a compact set $A \subset \mathbb{C}$ is $d(A) = e^{-\gamma}$, where

$$\gamma = \lim_{z\to\infty} \{g(z,\infty) - \log|z|\}$$

and $g(z,\infty)$ is Green's function of the unbounded component of $\hat{\mathbb{C}} \setminus A$. If $A \subset \partial\Omega$, the *Robin capacity* of A with respect to Ω is defined by $\delta(A) = e^{-\rho}$, where

$$\rho = \lim_{z\to\infty} \{R(z,\infty) - \log|z|\}$$

and $R(z,\infty)$ is the Robin function of Ω with respect to A. The Robin capacity is preserved under *admissible* conformal mappings; i.e., under mappings with the form (1) near infinity. Thus if f is an admissible mapping of Ω onto $\tilde{\Omega}$ which carries A

onto a set $\tilde{A} \subset \partial\tilde{\Omega}$, then $\delta(\tilde{A}) = \delta(A)$. In other words, the Robin capacity of $\tilde{A}$ with respect to $\tilde{\Omega}$ is equal to the Robin capacity of A with respect to Ω.

Strictly speaking, the set $\tilde{A} = f(A)$ requires a definition if the boundaries of Ω and $\tilde{\Omega}$ are not sufficiently smooth. It could be defined, for instance, as a cluster set. However, the Robin capacity of $f(A)$, as well as the ordinary logarithmic capacity, can be defined simply by the conformal invariance property of the Robin function.

We can now state our theorem on the distortion of capacity under conformal mapping.

THEOREM 1 *Let $\Omega \subset \hat{\mathbb{C}}$ be a finitely connected domain containing the point at infinity and bounded by smooth Jordan curves $\Gamma_1, \Gamma_2, \ldots, \Gamma_n$. Let $A \subset \partial\Omega = \Gamma_1 \cup \cdots \cup \Gamma_n$ be a closed subset of the boundary, and let $\delta(A)$ be the Robin capacity of A with respect to Ω. Then under all admissible conformal mappings f of Ω, the sharp inequality $d(f(A)) \geq \delta(A)$ holds. Thus the Robin capacity of A is the sharp lower bound for the logarithmic capacity of the image of A under all such f.*

Proof Let $\mathcal{F}(\Omega)$ be the equivalence class of domains $\tilde{\Omega} = f(\Omega)$ obtained from Ω by admissible mappings f. There is no loss of generality in taking A to be the union of a finite number of closed subarcs of the curves Γ_j, as the general inequality can then be deduced by an approximation argument (cf. [6], [7], [8]). Under the mapping f, the original decomposition $\partial\Omega = A \cup B$ induces a decomposition $\partial\tilde{\Omega} = \tilde{A} \cup \tilde{B}$, where $\tilde{A} = f(A)$ and $\tilde{B} = f(B)$.

Consider the extremal problem of minimizing $d(\tilde{A})$ among all $\tilde{\Omega} \in \mathcal{F}(\Omega)$. Since the admissible mappings with $b_0 = 0$ constitute a compact normal family, it is clear that an extremal domain $\tilde{\Omega}$ exists. Let $R = d(\tilde{A})$ be the minimum capacity.

For fixed $w_0 \in \tilde{B}$, we now introduce the boundary variation (see [9], [3])

$$w^* = V_\rho(w) = w + \frac{a\rho^2}{w - w_0} + O(\rho^3). \tag{2}$$

The function V_ρ is analytic and univalent in the complement of a small subarc of $\partial\Omega$ containing w_0. In particular, it acts as an interior variation of the larger domain $\hat{\Omega} \supset \tilde{\Omega}$ bounded only by $\tilde{A}$. Let $\tilde{A}^* = V_\rho(\tilde{A})$ and $R^* = d(\tilde{A}^*)$. By a known variational formula [10],

$$R^* = R[1 - \mathrm{Re}\{a\rho^2 \hat{p}'(w_0)^2\} + O(\rho^3)], \tag{3}$$

where $\hat{p}(w)$ is the analytic completion of Green's function $\hat{g}(w) = \hat{g}(w, \infty)$ of $\hat{\Omega}$. But $R^* \geq R$ by the extremal property of $\tilde{\Omega}$, so

$$\mathrm{Re}\{a\rho^2 \hat{p}'(w_0)^2\} + O(\rho^3) \leq 0. \tag{4}$$

Since $\hat{p}'$ is analytic and not identically zero on $\tilde{B}$, we may now invoke the basic lemma of the method of boundary variation ([9]; see also [3], Chapt. 10) to conclude from (4) that $\tilde{B}$ is the union of analytic arcs lying on trajectories of the quadratic differential

$$\hat{p}'(w)^2 \, dw^2 > 0.$$

Parameterizing each segment of $\tilde{B}$ by $w = w(s)$, we conclude that

$$\frac{d}{ds}\hat{p}(w(s)) = \hat{p}'(w(s))w'(s)$$

is real, or that $(d/ds)\text{Im}\{\hat{p}(w)\} = 0$. By the general form of the Cauchy–Riemann equations, this implies that

$$\frac{\partial}{\partial n}\hat{g}(w) = \frac{\partial}{\partial n}\text{Re}\{\hat{p}(w)\} = 0 \quad \text{on} \quad \tilde{B}.$$

On the other hand, $\hat{g}(w) = 0$ on $\tilde{A}$ by the definition of Green's function. Thus $\hat{g}$ is actually Robin's function of $\tilde{\Omega}$; that is, $\hat{g}(w,\infty) = \tilde{R}(w,\infty)$. In view of the definition of Robin capacity, this shows that $d(\tilde{A}) = \delta(\tilde{A}) = \delta(A)$, the Robin capacity of A with respect to Ω. Because $\tilde{\Omega}$ was assumed to be an extremal domain it follows that $d(f(A)) \geq \delta(A)$ for all admissible mappings f, and the bound is sharp. This completes the proof.

An immediate corollary is that $\delta(A) \leq d(A)$ for every closed set $A \subset \partial\Omega$, since the identity mapping is admissible. This also follows from the inequality $\tilde{R}(w,\omega) \geq \hat{g}(w,\omega)$, to be proved in more general form in Section 4.

For the special case where Ω is bounded by a single Jordan curve Γ and A consists of finitely many subarcs of Γ, it is seen that one extremal domain $\tilde{\Omega}$ is the complement of a disk $|w| \leq \delta$ and a system of radial spires corresponding to $B = \Gamma \backslash A$. The Robin function of $\tilde{\Omega}$ is then

$$\tilde{R}(w,\infty) = \log|w| - \log\delta,$$

so that δ is the Robin capacity of A. Intuitively speaking, the radial spires help to maintain the capacity of the full boundary while the subset A is compressed into a set of smallest possible capacity. If Ω is the domain outside the unit circle and A is a single arc of the circle with central angle α, the mapping can be constructed explicitly (cf. [10]) and it turns out that $\delta = \sin^2(\alpha/4)$. It can be shown that $d(A) = \sin(\alpha/4)$, so that $\delta(A) = d(A)^2$ in this case. In fact, this last relation remains valid for any subset of the circle.

THEOREM 2 *Let $\Omega = \{|z| > 1\}$ be the domain outside the unit circle, and let A be an arbitrary closed subset of the circle. Then $\delta(A) = d(A)^2$; that is, the Robin capacity of A with respect to Ω is the square of the ordinary logarithmic capacity of A.*

Proof Let $g(z,\zeta)$ be Green's function of $\hat{\mathbb{C}}\backslash A$. Then for $\zeta \in \Omega$,

$$G(z,\zeta) = g(z,\zeta) - g(1/\bar{z},\zeta)$$

is Green's function of Ω and

$$R(z,\zeta) = g(z,\zeta) + g(1/\bar{z},\zeta)$$

is Robin's function of Ω. In particular,

$$\log|z| = g(z,\infty) - g(1/\bar{z},\infty) \qquad \text{and} \qquad R(z,\infty) = g(z,\infty) + g(1/\bar{z},\infty).$$

Thus

$$-\log d(A) = \lim_{z\to\infty}\{g(z,\infty) - \log|z|\} = g(0,\infty)$$

and

$$-\log\delta(A) = \lim_{z\to\infty}\{R(z,\infty) - \log|z|\} = -2\log d(A),$$

or $\delta(A) = d(A)^2$.

As mentioned above, Robin capacity is closely connected with the notion of reduced extremal distance, defined in terms of extremal length (see [2], [1, pp. 78–80], [6, p 238 ff.]). Once this connection is established, the inequality $d(f(A)) \geq \delta(A)$ of Theorem 1 is a simple corollary. However, it does not seem possible to show in this way that the inequality is sharp. The connection between Robin capacity and extremal length is further exploited in [4]. Theorem 2 is implicit in Ohtsuka [6, p. 248], whose proof is similar to ours. (We thank the referee for this reference.) It is also implicit in [11].

3. SUM OF TWO CAPACITIES

Again let Ω be a finitely connected domain with smooth boundary, and let $\partial\Omega = A \cup B$. Let us ask for the minimum of the sum $d(\tilde{A}) + d(\tilde{B})$ among all domains $\tilde{\Omega}$ in the equivalence class $\mathcal{F}(\Omega)$. As a trivial consequence of Theorem 1, we see that the sum is never smaller than $\delta(A) + \delta(B)$, the sum of the Robin capacities of A and B with respect to Ω. It seems remarkable that this estimate is actually sharp for all configurations.

THEOREM 3 *Let $\Omega \subset \hat{\mathbb{C}}$ be a domain containing infinity and bounded by a finite number of smooth Jordan curves. Let $\partial\Omega = A \cup B$ be a partition of the boundary into disjoint sets. Then*

$$d(f(A)) + d(f(B)) \geq \delta(A) + \delta(B)$$

for all admissible conformal mappings f of Ω, and the inequality is sharp.

Proof Again we may take A and B to be unions of finitely many arcs on the boundary. For notational simplicity, let us suppose that Ω is already an extremal domain, so that $d(A) + d(B)$ renders the sum $d(\tilde{A}) + d(\tilde{B})$ a minimum among all $\tilde{\Omega} \in \mathcal{F}(\Omega)$. Choose $w_0 \in B$ and apply the variation $w^* = V_\rho(w)$ as given in (2). Then $d(B^*) = d(B)$ since V_ρ is an admissible mapping of the domain $\hat{\mathbb{C}}\backslash\overline{B}$. Thus $d(A^*) \geq d(A)$, and we conclude as before that $g_A(w,\infty) = R_A(w,\infty)$. In other words, Green's function of $\hat{\mathbb{C}}\backslash\overline{A}$ is the Robin function of Ω which vanishes on A. Thus $d(A) = \delta(A)$. Similarly choosing $w_0 \in A$, one sees by the same argument that $g_B(w,\infty) = R_B(w,\infty)$, so that $d(B) = \delta(B)$. Therefore, $d(A) + d(B) = \delta(A) + \delta(B)$ in the extremal case, and the theorem is proved.

As an illustration, let $\Omega = \{|z| > 1\}$ and again let A be a single arc of the circle with central angle α. Then B is an arc with angle $\beta = 2\pi - \alpha$. Suppose for simplicity that A is symmetric with respect to the real axis. Then the admissible function $f(z) = z + 1/z$ maps Ω onto $\tilde{\Omega} = \hat{\mathbb{C}}\backslash[-2,2]$, and it carries A and B onto segments of length $4\sin^2(\alpha/4)$ and $4\sin^2(\beta/4)$, respectively. Thus $d(\tilde{A}) = \delta(A)$ and $d(\tilde{B}) = \delta(B)$.

4. QUADRATIC FORMS

In our previous paper [5] it was shown that Robin's function dominates Green's function in various ways. In particular, $g(z,\zeta) < R(z,\zeta)$ for all $z \in \Omega$, $z \neq \zeta$. This inequality will now be sharpened and generalized to quadratic forms.

THEOREM 4 *Let $\Omega \subset \hat{\mathbb{C}}$ be a domain bounded by a finite number of smooth Jordan curves. Let $R(z,\zeta)$ be the Robin function of Ω with respect to a given partition $\partial\Omega = A \cup B$ of the boundary. Let $z_1, z_2, \ldots, z_n$ be distinct points of Ω, and let $x_1, x_2, \ldots, x_n$ be arbitrary real parameters. Then*

$$\sum_{j=1}^{n}\sum_{k=1}^{n} x_j x_k [g_A(z_j,z_k) - R(z_j,z_k)] \leq 0, \tag{5}$$

where $g_A(z,\zeta)$ is Green's function of the unbounded component of $\hat{\mathbb{C}}\backslash\overline{A}$. The inequality is sharp in the equivalence class of Ω with respect to admissible mappings.

Remark For $j = k$ both g_A and R are infinite, but their difference is to be interpreted as

$$g_A(\zeta,\zeta) - R(\zeta,\zeta) = \lim_{z\to\zeta}[g_A(z,\zeta) - R(z,\zeta)].$$

Proof of Theorem It may again be supposed that A is the union of finitely many closed arcs of the boundary. Fixing the points z_j and the parameters x_j, consider the functional

$$\phi(f) = \sum_{j=1}^{n}\sum_{k=1}^{n} x_j x_k [\tilde{g}_{\tilde{A}}(w_j,w_k) - \tilde{R}(w_j,w_k)],$$

where f is an admissible mapping of Ω onto $\tilde{\Omega}$. Here $\tilde{R}$ denotes the Robin function of $\tilde{\Omega}$ with respect to the induced boundary partition $\partial\tilde{\Omega} = \tilde{A} \cup \tilde{B}$, while $\tilde{g}_{\tilde{A}}$ is Green's function of the unbounded component of $\hat{\mathbb{C}}\backslash\tilde{A}$, and $w_j = f(z_j)$. Let us pose the extremal problem of maximizing $\phi(f)$ over all admissible mappings f. Suppose for notational convenience that Ω already gives the maximum value. Suppose, in other words, that $\phi(f) \leq \phi(f_0) = \phi_0$ for all admissible mappings f, where f_0 is the identity mapping. Note that ϕ_0 is the value of the quadratic form in (5).

Now select a point $z_0 \in B$ and apply the boundary variation $z^* = V_\rho(z)$ as given in (2). Because V_ρ is admissible, it follows from the extremal property of Ω that $\phi^* = \phi(V_\rho) \leq \phi_0$. By the conformal invariance of the Robin function, $R^*(z^*,\zeta^*) = R(z,\zeta)$. On the other hand, V_ρ acts as an interior variation of $\hat{\Omega}$, the unbounded component of $\hat{\mathbb{C}}\backslash A$; and the variational formula for Green's function is known [10] to be

$$\hat{g}^*(z^*,\zeta^*) = \hat{g}(z,\zeta) + \mathrm{Re}\{a\rho^2\hat{p}'(z_0,z)\hat{p}'(z_0,\zeta)\} + O(\rho^3),$$

where $\hat{g}(z,\zeta) = g_A(z,\zeta)$ and $\hat{p}(z,\zeta)$ is again the analytic completion of $\hat{g}(z,\zeta)$. A calculation gives

$$\phi^* = \phi_0 + \mathrm{Re}\left\{a\rho^2\left[\sum_{k=1}^{n} x_k\hat{p}'(z_0,z_k)\right]^2\right\} + O(\rho^3),$$

so that

$$\operatorname{Re}\left\{a\rho^2\left[\sum_{k=1}^{n} x_k \hat{p}'(z_0, z_k)\right]^2\right\} + O(\rho^3) \leq 0.$$

The fundamental lemma of the method of boundary variation now shows that B is the union of analytic arcs lying on trajectories of the quadratic differential

$$\left[\sum_{k=1}^{n} x_k \hat{p}'(z, z_k)\right]^2 dz^2 > 0. \tag{6}$$

Parametrizing each arc of B by $z = z(s)$, one concludes from (6) that

$$\operatorname{Im}\left\{\frac{d}{ds}\sum_{k=1}^{n} x_k \hat{p}(z, z_k)\right\} = 0,$$

whereupon the Cauchy–Riemann equations give

$$\frac{\partial}{\partial n}\sum_{k=1}^{n} x_k \hat{g}(z, z_k) = 0 \quad \text{on} \quad B.$$

But by the definition of Green's function,

$$\sum_{k=1}^{n} x_k \hat{g}(z, z_k) = 0 \quad \text{on} \quad A.$$

The Robin function has the same properties. Thus the function

$$u(z) = \sum_{k=1}^{n} x_k[\hat{g}(z, z_k) - R(z, z_k)]$$

is harmonic in Ω and vanishes on A, while its normal derivative vanishes on B. This implies that $u(z) \equiv 0$ in Ω, because of the identity

$$\iint_{\Omega} |\nabla u|^2 \, dx\, dy = \int_{\partial\Omega} u \frac{\partial u}{\partial n} \, ds$$

for harmonic functions. In particular, it follows that $\phi_0 = 0$. But this was assumed to be the maximum value, so $\phi(f) \leq 0$ for all admissible f. This proves the theorem.

If in (5) we choose $n = 1$, $x_1 = 1$, and $z_1 = \infty$, we obtain the inequality

$$-\log d(f(A)) + \log \delta(f(A)) \leq 0,$$

or $d(f(A)) \geq \delta(A)$. This shows that Theorem 1 is a very special case of Theorem 4.

ACKNOWLEDGMENTS

The work of the first-named author was supported in part by a grant from the National Science Foundation. The authors are grateful to the referee for helpful comments.

References

[1] L. V. Ahlfors, *Conformal Invariants: Topics in Geometric Function Theory* (McGraw-Hill, New York, 1973).

[2] L. V. Ahlfors and A. Beurling, Conformal invariants, in *Construction and Applications of Conformal Maps: Proceedings of a Symposium* (E. F. Beckenbach, ed.), National Bureau of Standards, Appl. Math. Series No. 18 (U.S. Government Printing Office, Washington, D.C., 1952), 243–245.

[3] P. L. Duren, *Univalent Functions* (Springer-Verlag, Heidelberg and New York, 1983).

[4] P. Duren and J. Pfaltzgraff, Robin capacity and extremal length, *J. Math. Anal. Appl.*, to appear.

[5] P. L. Duren and M. M. Schiffer, Robin functions and energy functionals of multiply connected domains, *Pacific J. Math.* **148** (1991), 251–273.

[6] M. Ohtsuka, *Dirichlet Principle, Extremal Length and Prime Ends* (Van Nostrand Reinhold, New York, 1970).

[7] Ch. Pommerenke, On the logarithmic capacity and conformal mapping, *Duke Math. J.* **35** (1968), 321–325.

[8] Ch. Pommerenke, *Univalent Functions* (Vandenhoeck & Ruprecht, Göttingen, 1975).

[9] M. M. Schiffer, A method of variation within the family of simple functions, *Proc. London Math. Soc.* **44** (1938), 432–449.

[10] M. M. Schiffer, Hadamard's formula and variation of domain-functions, *Amer. J. Math.* **68** (1946), 417–448.

[11] M. M. Schiffer, Some distortion theorems in the theory of conformal mapping, *Atti Accad. Naz. Lincei Mem.* **10** (1970), 1–19.

Commentary on

[137] P. L. Duren and M. M. Schiffer, *Robin functions and distortion of capacity under conformal mapping*, Complex Variables Theory Appl. **21** (1993), 189–196.

The *Robin function* $R(z,\zeta)$ of a smoothly bounded finitely connected domain Ω is harmonic in Ω except for a logarithmic singularity at ζ; it vanishes on a prescribed subset A of the boundary $\partial\Omega$, while its normal derivative vanishes elsewhere on $\partial\Omega$. Named by Bergman and Schiffer [51] for the French mathematical physicist Gustave Robin (1855–1897), Robin's function is a generalization of Green's function $G(z,\zeta)$ and plays a similar role of resolvent kernel in the solution of mixed boundary-value problems. The logarithmic capacity (= transfinite diameter) of a closed bounded set E is $d(E) = e^{-\gamma}$, where $\gamma = \lim_{z\to\infty}\{G(z,\infty) - \log|z|\}$ and $G(z,\zeta)$ is Green's function of the domain Ω complementary to E. Analogously, the *Robin capacity* of A with respect to Ω is defined in [137] by $\delta(A) = e^{-\rho}$, where $\rho = \lim_{z\to\infty}\{R(z,\infty) - \log|z|\}$ and $R(z,\zeta)$ is Robin's function of Ω for the given set $A \subset \partial\Omega$. The existence of the Robin function was proved (for a special case) in [136], where it was also shown that $G(z,\zeta) \le R(z,\zeta)$, which implies that $\delta(A) \le d(A)$.

Although the capacity of the full boundary is invariant under admissible conformal mappings of Ω, that is, by mappings having the form $f(z) = z + b_0 + b_1 z^{-1} + \ldots$ near ∞, the capacity of a subset $A \subset \partial\Omega$ may be distorted. The main result of [137] is that the minimum of $d(f(A))$ under all admissible mappings f of Ω is equal to the Robin capacity $\delta(A)$. The proof proceeds by a variational method, which shows that Green's function of the extremal domain is the Robin function. It is also proved in [137], by an argument implicit in [99], that $\delta(A) = d(A)^2$ when Ω is the exterior of the unit disk. These two results combine to "explain" a theorem of Pommerenke [P], obtained via the Goluzin inequalities, that $d(f(A)) \ge d(A)^2$ for $A \subset \partial\mathbb{D}$ and $f \in \Sigma$. The variational approach shows further that the lower bound is sharp. For the special case where A is a single arc on the unit circle of length α, the result goes back to [17], where Schiffer used a variational method to derive the sharp inequality $d(f(A)) \ge \sin^2(\alpha/4)$. Schiffer [99] later generalized the inequality $d(f(A)) \ge d(A)^2$, involving other conformal invariants, for sets A that are the union of finitely many closed arcs (see commentary on [99]).

Since [137], the theory of Robin capacity has been extended and applied in many directions. In [DP1], an equivalent formulation was found in terms of extremal length, generalizing a known expression for ordinary capacity. A physical interpretation was given in [DPT], where the classical minimum energy formulation was generalized with a Neumann potential replacing the usual logarithmic potential. The theory of capacity and Robin capacity is carried out in the settings of hyperbolic and elliptic geometry, respectively, in [DP2] and [DK]. In [DP2], the authors pointed out a gap in the proof of the main theorem in [137]; the existence of an extremal domain was in question. The gap was filled for a special case in [DP2] and, for the general case, independently in [OT] and [DS].

Meanwhile, Thurman [T1] addressed the problem of *maximum* capacity $d(f(A))$ for all admissible mappings f, expressing it in terms of harmonic measures of boundary components. In [T2], he offers an ingenious extremal length description in terms of a notion of "bridged extremal distance." In [T3], he presents a minimum energy interpretation of maximal capacity; and he reinterprets the Neumann potential, used in [DPT] to describe the Robin capacity, as a double-layer potential.

A survey of developments up to 1997 appears in [D]. More recently, the concept of Robin capacity has proved important in a variety of applications. Nasyrov [N1, N2] has applied it to the lift of airfoils. Vasil'ev [V] found it relevant to a problem of Hele-Shaw flow. Dubinin and Vuorinen have brought Robin functions and Robin capacity to bear on problems of geometric function theory (see, for instance, [Db, DV]). In [B], Betsakos studied the effect of polarization, a form of symmetrization, on Robin capacity. Stiemer [S1, S2] has developed effective methods for numerical calculation of the Robin function and Robin capacity.

References

[B] Dimitrios Betsakos, *Polarization, conformal invariants, and Brownian motion*, Ann. Acad. Sci. Fenn. Math. **23** (1998), 59–82.

[DS] B. Dittmar and A. Yu. Solynin, *Distortion of the hyperbolic Robin capacity under a conformal mapping and extremal configurations*, Zap. Nauchn. Sem. S.-Peterburg Otdel. Mat. Inst. Steklov. (POMI) **263** (2000), 49–69, 238 (in Russian); English transl., J. Math. Sci. (New York) **110** (2002), 3058–3069.

[Db] V. N. Dubinin, *On quadratic forms generated by Green and Robin functions*, Mat. Sb. **200** (2009), no. 10, 25–38 (in Russian); English transl., Sb. Math. **200** (2009), 1439–1452.

[DV] V. N. Dubinin and M. Vuorinen, *Robin functions and distortion theorems for regular mappings*, Math. Nachr. **283** (2010), 1589–1602.

[D] Peter Duren, *Robin capacity*, Computational Methods and Function Theory 1997, eds. N. Papamichael, St. Ruscheweyh, and E. B. Saff, World Scientific Publishing, 1999, pp. 177–190.

[DK] Peter Duren and Reiner Kühnau, *Elliptic capacity and its distortion under conformal mapping*, J. Analyse Math. **89** (2003), 317–335.

[DP1] Peter Duren and John Pfaltzgraff, *Robin capacity and extremal length*, J. Math. Anal. Appl. **179** (1993), 110–119.

[DP2] Peter Duren and John Pfaltzgraff, *Hyperbolic capacity and its distortion under conformal mapping*, J. Analyse Math. **78** (1999), 205–218.

[DPT] Peter Duren, John Pfaltzgraff, and Robert E. Thurman, *Physical interpretation and further properties of Robin capacity*, Algebra i Analiz **9** (1997), No. 3, 211–219 = St. Petersburg Math. J. **9** (1998), 607–614.

[N1] Samyon Nasyrov, *Robin capacity and lift of infinitely thin airfoils*, Complex Variables Theory Appl. **42** (2002), 93–107.

[N2] S. R. Nasyrov, *Variations of Robin capacities and their applicatons*, Sibirsk. Mat. Zh. **49** (2008), 1128–1146 (in Russian); English transl., Sib. Math. J. **49** (2008), 894–910.

[OT] Michael D. O'Neill and Robert E. Thurman, *Extremal domains for Robin capacity*, Complex Variables Theory Appl. **41** (2000), 91–109.

[P] Ch. Pommerenke, *On the logarithmic capacity and conformal mapping*, Duke Math. J. **35** (1968), 321–325.

[S1] Marcus Stiemer, *A representation formula for the Robin function*, Complex Variables Theory Appl. **48** (2003), 417–427.

[S2] Marcus Stiemer, *Extremal point methods for Robin capacity*, Comput. Methods Funct. Theory **4** (2004), 475–496.

[T1] Robert E. Thurman, *Upper bound for distortion of capacity under conformal mapping*, Trans. Amer. Math. Soc. **346** (1994), 605–616.

[T2] Robert E. Thurman, *Bridged extremal distance and maximal capacity*, Pacific J. Math. **176** (1996), 507–528.

[T3] Robert E. Thurman, *Maximal capacity, Robin capacity, and minimum energy*, Indiana Univ. Math. J. **46** (1997), 621–636.

[V] Alexander Vasil'ev, *Robin's modulus in a Hele-Shaw problem*, Complex Variables Theory Appl. **49** (2004), 663–672.

PETER DUREN

[138] Issai Schur: Some personal reminiscences

[138] Issai Schur: Some personal reminiscences, in *Mathematik in Berlin*, Vol. II, (H. Begehr, editor), Shaker Verlag, Aachen (1998) 177–181. [Manuscript of lecture presented at Tel Aviv University in 1986.]

Issai Schur
Some Personal Reminiscences*

by
MENAHEM MAX SCHIFFER

My talk will not deal with the mathematical achievements of Issai Schur, but with my own memories of a great man and a wonderful teacher.
I feel justified in doing so, since there are not many people left who experienced the force of his personality and his influence.
In 1930 I enrolled at the Friedrich-Wilhelms-University in Berlin with the intention of studying physics and applied science. For this reason I also had to take an intensive study course in mathematics; and thus I came into early contact with the teachers of this subject.
There were three full professors (ordinarii) in this field. Schur represented algebra, Schmidt – analysis and Bieberbach – geometry. The instruction consisted of lecture series, exercises classes and seminars. Those of Schur and Schmidt were of an outstanding level and organization.
First, of course, I took the lectures. Schur taught algebra, number theory, determinant theory, invariant theory, Galois theory, analytic number theory and ideal theory. His problem sessions were led by his assistant, at that time Alfred Brauer. Schur was always very careful in his selection of assistants, and he influenced and inspired them deeply. I remember of them Szegö, Löwner, Richard Brauer and his brother Alfred.
Schur's lectures were of extraordinary clarity, yet, of great depth. It was my luck that I started early with his course on invariant theory. You can measure the quality of this course by the fact that these lectures were published in the sixties by Grunsky as a book in the Springer series on mathematics, and that it was a great success.
My interest in the course was originally inspired by the idea that invariance is an important concept in physics, so that this course would be

*Manuscript of a lecture presented in 1986 at a symposium dedicated to the memory of Issai Schur at Tel Aviv University.

very valuable for my studies. My teacher in physics, Erwin Schrödinger, was of the same opinion and encouraged me in this choice. But soon the subject fascinated me beyond this purpose and I fell in love with pure mathematical research. I thought out a method to construct orthogonal invariants by certain integral operations and I showed my notes to Schur. I was then a very young and inexperienced student and the integrals which I used were not convergent. Schur called me into his office after a class and showed me my error; but he told me, that my idea was good, that I should not be discouraged and that I should try to fix up my method.

Now, he started something which I wish all teachers should imitate. I was sitting in his seminar where many problems in algebra came up for discussion. Whenever a question in invariant theory arose, he would say: „We have somebody here who is an expert in this field“ and pointed me out. I felt then such a pride and responsibility to live up to this rather undeserved distinction that I studied all the available literature and publications in the field.

I fixed up my faulty proof and prepared my first paper for the printers. Before I sent the manuscript for publication, Schur called me again and gave me an hour's lecture on mathematical research, technique of publishing, quoting references, style, etc.

In all this excitement I was still studying physics as my main subject, till one day Schrödinger called me in and told me to decide between physics and mathematics. Two immense subjects could not be mastered by one person. „We have enough trouble from one Hermann Weyl“, he said. So I became a mathematician, though I love physics to this day.

Under the influence of Schur, I continued my work on invariant theory and discovered an elegant method of proving the finiteness theorem in invariant theory for large classes of groups. I wrote a long paper which Schur liked very much. Unfortunately, the year was 1933, when Hitler came to power. I soon left Germany and Schur promised to have the paper published. It was refused by the editor on the grounds that it was too abstract and we both forgot about it. However, Schur always had many ways of making ideas known. He met many important algebraists and told them news of interesting research. Thus, Richard Brauer learned of my results; he told them to Hermann Weyl and in 1945 my method was finally published in an appendix to Weyl's book on the *Classical Groups and their Invariants.*

I tell this story only to show the close network of algebraists in which Schur was the leading spirit. He was very interested in modern abstract algebra, but was somewhat critical of that development. His own algebra course was completely classical and he charged Alfred Brauer with the responsibility of teaching a complementary course in abstract algebra. He was quite conservative with respect to mathematical education and training. I remember a lecture which he gave at the Hebrew University in which he lamented the lack of knowledge of young mathematicians in analysis. „Imagine“, he said, „how many young analysts have not even heard of the beautiful series development of $(\arcsin x)^2$.“ I felt very guilty and ran home to derive this simple series.

Now, the year 1933 was a decisive cut in the life of every German Jew. In April of that year all Jewish government officials were dismissed, a boycott of Jewish businesses was decreed and antisemitic legislation was begun. When Schur's lectures were cancelled there was an outcry among the students and professors, for Schur was respected and very well liked. The next day Erhard Schmidt started his lecture with a protest against this dismissal and even Bieberbach, who later made himself a shameful reputation as a Nazi, came out in Schur's defense. Schur went on quietly with his work on algebra at home. He asked me once to visit him there, to discuss my paper.

I remember that visit well. Our discussion was interrupted by the visit of a lady who had come from England on a mission for the English Jewish Emergency Council to investigate the situation of Jewish scientists in Germany. Schur asked me to stay on and to participate in the conversation. For thirty years he had been active in mathematics and knew practically everybody in the field. So we went carefully through the list of many Jewish colleagues and their problems.

In the end, the lady asked Schur whether and where he wanted to go, because for a man of his reputation all doors would be open. But Schur responded that he did not intend to go; for he did not want to enable the Nazis to say, that many Jewish professors just left for better jobs. Besides, there were many younger colleagues who needed help much more urgently, and he would not take away their chances. He would stick it out in Berlin, for the craze of the Hitlerites could not last long. I benefited from that meeting in that I was offered a scholarship to the Hebrew University in Jerusalem.

At that time I had already decided to leave Germany, but did not yet know where to go. Schur had told me that I could get a job in Groningen in Holland where Weitzenböck was an outstanding expert in invariant theory and had written the definitive textbook in the field. As assistant to him I would be able to continue my work. But he advised me to go to Jerusalem. He was very wise; Weitzenböck was later one of the worst quislings under the German occupation of Holland.

I accepted the stipendium for Jerusalem and Schur wrote me letters of recommendation to the two professors there: Fränkel and Fekete. Schur was a good Jew but not a very orthodox one. When he was reading through the letters again he suddenly realized that the letter to Fränkel was dated on a Saturday and got very anxious that Fränkel might check the date and that this would do me harm.

In my last days in Berlin, we talked a lot about Palestine. Schur was skeptical about the Zionist future. He thought Jews were too individualistic and obstinate to be able to form a democratic state.

I left Berlin in October 1933 and started my studies in Jerusalem. There was nobody there interested in invariant theory. In order to have scientific exchange I began to work in function theory and conformal mapping which was close to Fekete's interest. Later Toeplitz came to Jerusalem and I learned a lot about modern analysis from him.

I returned in 1938 from a short stay in Paris and met Toeplitz, he told me that Schur had arrived in Tel-Aviv. He said, „Go and see Schur; he needs it very much." Of course, I went to see him at once. I was really shocked at our first meeting. He was in a terrible state of depression, fear and despair. He told me of his experiences during the last few years. Many colleagues and former friends had cut off all contact with him. He had been a member of many distinguished academies and learned societies; for example, the Prussian, Bavarian, Saxonian Academies of Science and many more. He had been ejected from each of them. The height of despair came by the end of 1938 with the infamous progrom, called by the Nazis „Kristall Nacht". Synagogues were burned, Jewish stores destroyed and all Jews they could catch were sent to concentration camps. Good friends of Schur called him by telephone, not daring to give their names, but warning him to hide. So he did, for several days, till the worst was over. When he came home, some of his closest friends had the courage to visit him; von Laue, the physicist; Erhard Schmidt, the mathematician and several other.

Now, here is something which in the present day atmosphere is hard to believe. When he complained bitterly to Schmidt about the Nazi actions and Hitler, Schmidt defended the latter. He said, „Suppose we had to fight a war to rearm Germany, unite with Austria, liberate the Saar and the German part of Czechoslovakia. Such a war would have cost us half a million young men. But everybody would have admired our victorious leader. Now, Hitler has sacrificed half a million Jews and has achieved great things for Germany. I hope some day you will be recompensed but I am still grateful to Hitler."

So spoke a great scientist, a decent man, and a loyal friend. Imagine the feelings of a German Jew at that time. Schur told me that the only person at the Mathematical Institute in Berlin who was kind to him was Grunsky, then a young lecturer.

Long after the war, I talked to Grunsky about that remark and he literally started to cry: „You know what I did? I sent him a postcard to congratulate him on his sixtieth birthday. I admired him so much and was very respectful in that card. How lonely he must have been to remember such a small thing."
Worst of all, in 1938 Schur was convinced that Hitler would conquer the world and that his respite in Palestine would not last long. So, he was in complete despair.
But slowly he recovered somewhat and returned to his beloved mathematics. He even agreed to give a lecture at the Hebrew University and this I will never forget. He spoke about an interesting inequality in polynomial theory with his customary clarity and elegance. Suddenly, in the middle of his talk he sat down, bent his head and was silent. We, in the audience, did not understand what was going on; we sat quietly and respectfully. After a few minutes he got up and finished his talk in his usual manner.
I was sitting next to a physician from the Hadassah Hospital who had come to see this famous man. He was quite upset; after the lecture he told me that Schur had obviously had a heart attack and he could not understand the self-discipline which had enabled Schur to finish his talk. That was the man Schur, for you!
Eventually his mathematical creativity came back and he forgot his fears. In 1940, we received a paper by Grunsky which seemed to us important and very promising. Within a few weeks Schur wrote three papers to allow further extension of Grunsky's ideas. But each is of importance on its own: *Formal development in power series, Normal form of quadratic forms in the complex domain, On the coefficients of Faber polynomials.* At the same time, he advised Theodore Motzkin on some problems in number theory and started to interact with younger men at the Mathematics Institute.
In 1941 he celebrated his last birthday. His son Georg had volunteered for the British army and was already in uniform on that occasion. The excitement was probably too much for Schur. Soon after, he had his final heart attack and we lost him. His work in algebra, group representations, invariant theory, function theory and integral equations lives on.
My aim in this talk has been to give you some idea of the human dimensions of the man and the difficult and even tragic circumstances under which some of his work was accomplished.

Commentary on

[138] *Issai Schur: Some personal reminiscences*, Mathematik in Berlin, Vol. II, ed. Heinrich Begehr, Shaker Verlag, Aachen, 1998, pp. 177–181.

For a number of years during the last quarter of the twentieth century, Tel Aviv University sponsored a series of biennial symposia on mathematics, held (in alternate years) in memory of two distinguished mathematicians who found refuge in Palestine from Nazi persecution in the late 1930s, Issai Schur and Otto Toeplitz. This moving memoir of Schiffer's teacher Schur was originally written to be presented (under the title "Personal memories of Issai Schur") on Monday, May 26, 1986, as the introduction to the Issai Schur Memorial Lectures of that year. Other speakers scheduled that day were E. Bombieri and B. Mazur. As a result of a misunderstanding concerning the timing of the symposium, Schiffer was unable to attend; and I ended up reading his lecture to those assembled.

For an account of the difficulties faced by Jewish mathematicians in Germany during the Nazi period, see the very extensive treatment in [SS2], as well as [SS3] and now also the remarkable exhibition catalog [TT]. The incident involving Schmidt (to whom Schiffer and Bergman would eventually dedicate [35]) is quoted in [SS2, p. 86]; even today, three quarters of a century later, it makes one shudder. (For a slightly different, though still admiring, perspective on Schmidt, see [W, p. 53].) By way of contrast, Grunsky's extremely decent behavior under such dreadful circumstances, exhibited on more than one occasion, seems all the more remarkable in view of his close connection with Bieberbach; see [SS1] for further details.

References

[SS1] Reinhard Siegmund-Schultze, *Helmut Grunsky (1904–1986) in the Third Reich*, Helmut Grunsky: Collected Papers, eds. Oliver Roth and Stephan Ruscheweyh, Heldermann Verlag, 2004, pp. xxxi–l.

[SS2] Reinhard Siegmund-Schultze, *Mathematicians Fleeing Nazi Germany: Individual Fates and Global Impact*, Princeton University Press, 2009.

[SS3] Reinhard Siegmund-Schultze, *Landau and Schur – documents of a friendship until death in an age of inhumanity*, Eur. Math. Soc. Newsl. **84** (2012), 31–36.

[TT] *Transcending Tradition: Jewish Mathematicians in German-Speaking Academic Culture*, eds. Birgit Bergmann, Moritz Epple and Ruti Ungard, Springer Verlag, 2012.

[W] André Weil, *The Apprenticeship of a Mathematician*, Birkhäuser Verlag, 1992.

LAWRENCE ZALCMAN

Printed in the United States
By Bookmasters